KT-161-769

FUNDAMENTAL

NEUROSCIENCE

As seen in this unretouched photograph of a small myelinated axon, mitochondria may assume a variety of sizes, shapes, and orientations.

FUNDAMENTAL NEUROSCIENCE

Edited by Duane E. Haines, Ph.D.

Professor and Chairman, Department of Anatomy
The University of Mississippi Medical Center, Jackson, Mississippi

Contributors

M. D. ARD, PH.D.
J. R. BLOEDEL, M.D., PH.D.
P. B. BROWN, PH.D.
N. F. CAPRA, PH.D.
R. B. CHRONISTER, PH.D.
J. J. CORBETT, M.D.
J. D. DICKMAN, PH.D.
O. B. EVANS, M.D.
S. G. P. HARDY, PH.D.
C. K. HENKEL, PH.D.
J. B. HUTCHINS, PH.D.
J. C. LYNCH, PH.D.
T. P. MA, PH.D.
P. J. MAY, PH.D.
G. A. MIHAILOFF, PH.D.
J. P. NAFTEL, PH.D.
R. W. ROCKHOLD, PH.D.
R. D. SWEAZEY, PH.D.
S. WARREN, PH.D.
R. P. YEZIERSKI, PH.D.

Illustrators

M. E. KIRKMAN, B.A., M.S.M.I., AND M. P. SCHENK, B.S., M.S.M.I.

CHURCHILL LIVINGSTONE

A Division of Harcourt Brace & Company
New York, Edinburgh, London, Philadelphia, San Francisco

Churchill Livingstone
A Division of Harcourt Brace & Company
The Curtis Center
Independence Square West
Philadelphia, PA 19106

Library of Congress Cataloging-in-Publication Data

Fundamental neuroscience / edited by Duane E. Haines ; contributors,
M.D. Ard . . . [et al.] ; illustrators, M.E. Kirkman and M.P. Schenk.
p. cm.
Includes bibliographical references and index.
ISBN 0-443-08874-8
1. Neurosciences. I. Haines, Duane E. II. Ard, M. D. (March D.)
[DNLM: 1. Nervous System—anatomy & histology. 2. Nervous System–
–physiology. 3. Neurons. WL 101 F981 1996]
QP355.2.F86 1996
612.8—dc20
DNLM/DLC 96-18576

Distributed in the United Kingdom by Churchill Livingstone, Robert Stevenson House, 1–3 Baxter's Place, Leith Walk, Edinburgh EH1 3AF, and by associated companies, branches, and representatives throughout the world.

Accurate indications, adverse reactions, and dosage schedules for drugs are provided in this book, but it is possible that they may change. The reader is urged to review the package information data of the manufacturers of the medications mentioned.

The Publishers have made every effort to trace the copyright holders for borrowed material. If they have inadvertently overlooked any, they will be pleased to make the necessary arrangements at the first opportunity.

Printed in Singapore

Last digit is the print number: 7 6 5 4 3 2 1

In Appreciation of

NORMAN CROOKS NELSON, M.D.

Vice Chancellor for Health Affairs and
Dean of the School of Medicine (1973–1994)
The University of Mississippi Medical Center
for
Fostering an academic environment in which teaching, research,
and scholarly activity in the basic and clinical neurosciences can flourish

Contributors

MARCH D. ARD, PH.D.
Department of Anatomy, The University of Mississippi Medical Center, Jackson, Mississippi

JAMES R. BLOEDEL, M.D., PH.D.
Division of Neurobiology, St. Joseph's Hospital and Medical Center, Barrows Neurological Institute, Phoenix, Arizona

PAUL B. BROWN, PH.D.
Department of Physiology, West Virginia University Health Sciences Center, Morgantown, West Virginia

NORMAN F. CAPRA, PH.D.
Department of Oral and Craniofacial Biological Sciences, The University of Maryland School of Dentistry, Baltimore, Maryland

ROBERT B. CHRONISTER, PH.D.
Department of Anatomy, University of South Alabama College of Medicine, Mobile, Alabama

JAMES J. CORBETT, M.D.
Chairman, Department of Neurology, The University of Mississippi Medical Center, Jackson, Mississippi

J. DAVID DICKMAN, PH.D.
Department of Surgery, The University of Mississippi Medical Center, Jackson, Mississippi

OWEN B. EVANS, M.D.
Chairman, Department of Pediatrics, The University of Mississippi Medical Center, Jackson, Mississippi

S. G. PATRICK HARDY, PH.D.
Department of Physical Therapy, School of Health Related Professions, The University of Mississippi Medical Center, Jackson, Mississippi

DUANE E. HAINES, PH.D.
Chairman, Department of Anatomy,The University of Mississippi Medical Center, Jackson, Mississippi

CRAIG K. HENKEL, PH.D.
Department of Neurobiology and Anatomy, Bowman Gray School of Medicine of Wake Forest University, Winston-Salem, North Carolina

JAMES B. HUTCHINS, PH.D.
Department of Anatomy, The University of Mississippi Medical Center, Jackson, Mississippi

JAMES C. LYNCH, PH.D.
Department of Anatomy, The University of Mississippi Medical Center, Jackson, Mississippi

TERENCE P. MA, PH.D.
Department of Anatomy, The University of Mississippi Medical Center, Jackson, Mississippi

PAUL J. MAY, PH.D.
Department of Anatomy, The University of Mississippi Medical Center, Jackson, Mississippi

GREGORY A. MIHAILOFF, PH.D.
Department of Anatomy, The University of Mississippi Medical Center, Jackson, Mississippi

JOHN P. NAFTEL, PH.D.
Department of Anatomy, The University of Mississippi Medical Center, Jackson, Mississippi

ROBIN W. ROCKHOLD, PH.D.
Department of Pharmacology and Toxicology, The University of Mississippi Medical Center, Jackson, Mississippi

ROBERT D. SWEAZEY, PH.D.
Department of Anatomy, Indiana University School of Medicine, Fort Wayne Center for Medical Education, Fort Wayne, Indiana

SUSAN WARREN, PH.D.
Department of Anatomy, The University of Mississippi Medical Center, Jackson, Mississippi

ROBERT P. YEZIERSKI, PH.D.
The Miami Project, University of Miami, Miami, Florida

Preface

One feature common to all neuroscience courses is the need to prepare students for the next phase of their education. For example we, as instructors, prepare first-year medical students to become second-year medical students, and we give them information that is *essential* to their successful negotiation of Step 1 of the USMLE. We are not teaching specialists (neurologists, oral surgeons, hand therapists) or, for that matter, even family practitioners. Rather, we provide information to "generalist" students.

The organization of *Fundamental Neuroscience* recognizes two important points. First, many neuroscience courses begin from their own reference point. There is no universally agreed upon topic that should appear first. Some consider the electrical properties of nerve cell membranes first; others start with embryology or with the anatomy of the telencephalon. Second, students in contemporary neuroscience courses must master *regional neuroanatomy* and *systems neurobiology*, as these are the bases for understanding brain and spinal cord function. Students in professional programs will find *regional neuroanatomy* essential to reading MRI and CT scans and will find *systems neurobiology* to be the foundation for the successful diagnosis of the neurologically impaired patient. In an attempt to address these points, the three sections were designed to be used as independent, but interrelated, segments of information, in whatever sequence most closely fits a given educational program.

SECTION I: ESSENTIAL CONCEPTS

This section deals with neuron structure, function, and communication and with development. Each chapter is a free-standing block of information that may be used at the beginning of, or at any time during, a neuroscience course. The instructor who elects to start with principles of neurophysiology may do so, while another may choose to initially emphasize concepts of neurohistology or neuropharmacology. A variety of approaches may be used without compromising continuity.

SECTION II: REGIONAL NEUROANATOMY

Viewing an MRI or CT scan of a *normal* brain is simply looking at brain anatomy *in situ*. The shapes and contours of each structure, and its relationship to adjacent structures, are characteristic and unique. Viewing an MRI or CT scan of an *abnormal* brain is simply evaluating how trauma, or a pathologic process, has altered brain structure. In light of the ever-increasing levels of detail provided by contemporary clinical imaging methods, *understanding brain anatomy in its clinical context* is an essential part of the modern neuroscience educational experience.

Section II covers the ventricles, meninges, and external vasculature of the central nervous system, followed by considerations of the spinal cord and each division of the brain. Some of the unique approaches in chapters of this section are

1. The development of the spinal cord and of each division of the brain is summarized at the beginning of the appropriate chapter.
2. The internal distribution of blood vessels is presented in association with internal structures.
3. Clinical examples are correlated with the morphology of the central nervous system.
4. Photographs and color line drawings are of excellent clarity.

This book introduces some terms that, although not common in many textbooks, are standard fare in the clinical setting. For example, what is termed the *horizontal plane* by morphologists is called the *axial plane* by the clinical neuroscientists. Since many users of this book will find themselves in a clinical curriculum, it seems entirely appropriate to make these correlations early in their training.

SECTION III: SYSTEMS NEUROBIOLOGY

Successful evaluation of the neurologically impaired patient requires a thorough understanding of systems neurobiology. This book describes how systems function under normal conditions, what information they convey and how they do so, and what deficits appear when systems are damaged. The patient with occlusion of a major cerebral vessel may, with time, display increased proteins in the CSF or xanthochromia of the CSF. However, the information that is frequently most valuable for early treatment is gained by identifying neural system dysfunction as soon as possible. This thinking has guided the organization of this section.

Section III provides comprehensive coverage of all sensory, motor, and integrative systems. The innovative approaches used in these chapters include

1. Emphasis on the topographical organization of sensory and motor pathways.
2. A review of the blood supply to most pathways at key levels of the neuraxis.
3. Integration of anatomy, physiology, and pharmacology where appropriate.
4. Original artwork specifically designed to help students understand neural systems.
5. Clinical information that is integrated within the text in close conjunction with the basic science information to which it is related.

Fundamental Neuroscience was written by individuals with significant teaching and research expertise in their respective areas. Consequently, the information is contemporary, integrated across specialties (anatomy, physiology, pharmacology), and attuned to the needs of students taking a modern neuroscience course in professional and graduate programs.

This book is most appropriate for students taking human neurobiology courses in medical, graduate, and dental programs. Its emphasis on basic concepts and systems neurobiology should prove quite useful to students in physical and/or occupational therapy, to other allied health students, or to residents needing a clear, succinct review as they prepare for their specialty board examinations in the clinical neurosciences.

Recognizing that both teaching and learning are multidimensional and on-going processes, the authors welcome comments, suggestions, and corrections from our students, our colleagues, and from other users of this book.

D. E. Haines, Ph.D.

Acknowledgments

It is a great pleasure to recognize those individuals who listened to ideas, offered suggestions, reviewed preliminary artwork and/or text, or in other ways contributed to the completion of this project. We express our sincere thanks to Drs. Vinod K. Anand, Dora E. Angelaki, Ronald H. Baisden, Anthony J. Castro, Steven C. Crawford, Espen Dietrichs, Jonathan T. Ericksen, William C. Hall, Rosemary Hoffman, James S. King, W. Michael King, George R. Leichnetz, George F. Martin, Inglis J. Miller, Richard S. Nowakowski, Dudley F. Peeler, Alan Peters, John D. Porter, Jose A. Rafols, William A. Roy, Laura F. Schweitzer, Daniel L. Tolbert, and Michael L. Woodruff. Some of these individuals provided photographs and their kindness is acknowledged in the appropriate figure caption. The photographs of Golgi-stained material are from the Clement A. Fox Collection at Wayne State University and through the courtesy of Dr. Rafols. The photograph on the half-title page was provided by Drs. Ross Kosinski and Greg Mihailoff.

Several individuals went out of their way to offer help. Dr. Frank A. Raila (Neuroradiology) tracked down MRIs, checked the evaluations and diagnoses, and made sure that structures were labeled correctly. Mr. Allen C. Terrell (Chief MRI technologist) identified many good examples of normal and abnormal scans and was most responsive to finding things on our "want list." We greatly appreciate their outstanding help. Dr. Andrew D. Parent offered us access to some of his pediatric neurosurgery cases along with the enthusiastic assistance of his nurse, Ms. Teresa McMillan. Dr. R. Brent Harrison has, over the years, allowed the editor free access to the CT and MRI facilities at the University of Mississippi Medical Center (UMMC); this kindness is greatly appreciated.

All of the artwork and photography (excepting specifically acknowledged photographs) was done in the Department of Medical Illustration and the Department of Biomedical Photography, respectively, at UMMC. The authors are deeply indebted to Ms. Myriam E. Kirkman-Oh for the outstanding artwork that contributes so significantly to most of the chapters, to Mr. Michael P. Schenk (Director) for his excellent work in the early chapters and for his able oversight, and to Ms. Diane Johnson and Mr. Ricky Manning for their numerous and important contributions to the completion of many pieces of the artwork that collectively form this project. The visual impact of the artwork is largely due to the individual skills and collective teamwork of this group. The photography was completed by Mr. G. William Armstrong (Director), Mr. William H. deVeer (former Director), and Mr. Paul A. Buki. The editor is enormously appreciative of their patience and cooperation in getting the best quality photographs for this book.

Ms. Gail Rainer typed the entire manuscript and Ms. Ellen Murray created the index. Their patience, cooperation, and willingness to go the extra mile and to be attentive to the details are greatly appreciated.

Production of this finely done and visually appealing book would not have been accomplished without the enthusiastic support of Churchill Livingstone. The editor is grateful to Mr. William Schmidt, under whom the project was inititated, and Dr. Kerry Willis, under whom most of the project was completed. We also express our sincere appreciation to Ms. Margot Otway, Developmental Editor, for her many constructive suggestions; to Ms. Ann Ruzycka, Assistant Editor, for transcribing modifications to the text and her many good natured responses to many questions; to Mr. Larry Meyer, Vice President and Production Director; to Ms. Jeannette Jacobs, Art Director; to Ms. Donna Balopole, Production Editor; and to Mr. Charlie Lebeda, Design Resources Manager. We also thank Dr. William F. Marovitz, President of Churchill Livingstone International, and Ms. Jennifer Mitchell, Vice President for Professional and Reference Publishing, for the corporate support that allowed this project to come to fruition.

Figure Acknowledgments

The editor acknowledges the following publishers for permission to borrow illustrations from some of his publications:

From Haines DE: On the question of a subdural space. Anat Rec 230:3–21, 1991.

Fundamental Neuroscience Figure 7-3 (adapted from the original)

From Haines DE: *Neuroanatomy: An Atlas of Structures, Sections, and Systems.* 4th Ed., Williams & Wilkins, Baltimore, 1995.

Fundamental Neuroscience Figures 9-5 (photographs only), 15-9B, and 15-11 (MRI only)

Fundamental Neuroscience Figures 7-2, 8-6, 8-7, 8-13, 30-5, and 30-6 (adapted from the originals)

Fundamental Neuroscience Figures 13-4, 13-15, 14-5, 14-12, 15-14E&F, 15-16A&B, 19-14A, 25-3A&B, 29-2B (modified from the originals)

From Haines DE, Frederickson RG: The meninges. In Al–Mefty O (ed): *Meningiomas*. Raven Press, New York, 1991.

Fundamental Neuroscience Figures 7-4, 7-6, and 7-9 (adapted from the originals)

Additional borrowed illustrations in *Fundamental Neuroscience* are acknowledged within their figure captions.

Contents

CHAPTER ONE

Orientation to the Central Nervous System

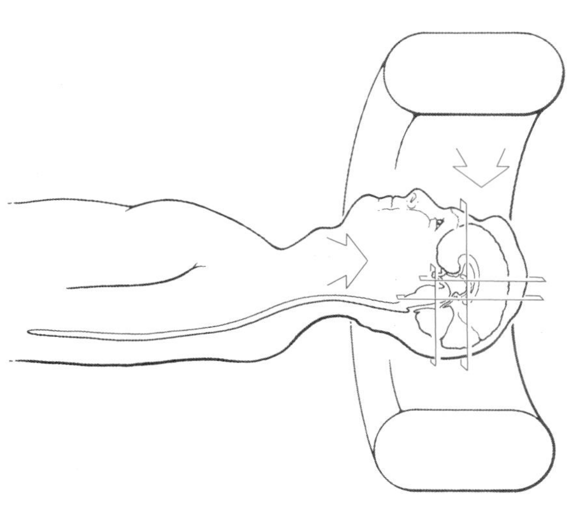

D. E. HAINES

Our nervous system makes us what we are. Personality, outlook, intellect, coordination (or lack thereof), and the *many* other characteristics that are unique to each of us are the result of complex interactions within our nervous system. Information is received from the environment by sensory receptors and transmitted into the brain and/or spinal cord. Once inside the brain or spinal cord, this sensory information is processed and integrated, and an appropriate response is initiated.

The nervous system can be viewed as a scale of structural complexity. Microscopically, the individual structural and functional unit of the nervous system is the *neuron*, or nerve cell. Interspersed among the neurons of the central nervous system are supportive elements called *glial cells*. At the macroscopic end of the scale are the large divisions (or parts) of the nervous system that can be handled and studied without magnification. These two extremes are not independent but form a continuum; functionally related neurons aggregate to form small structures, which combine to form larger structures, and so on. Communication takes place at many different levels, the end result being a wide range of productive or life-sustaining, nervous activities.

OVERVIEW

Central, Peripheral, and Autonomic Nervous Systems The human nervous system is divided into the *central nervous system* (CNS) and the *peripheral nervous system* (PNS) (Fig. 1-1A). The CNS consists of the brain and spinal cord. Because of their locations in the skull and vertebral column, these structures are the most protected in the body. The PNS is made up of nerves that connect the brain and spinal cord with peripheral structures. These nerves innervate muscle (skeletal, cardiac, smooth) and glandular epithelium and contain a variety of sensory fibers. These sensory fibers enter the spinal cord via the dorsal (posterior) root, and motor fibers exit through the ventral (anterior) root. The *spinal nerve* is formed by the joining of dorsal (sensory) and ventral (motor) roots and is, consequently, a *mixed nerve* (Fig. 1-1B). In the case of *mixed cranial nerves*, the sensory and motor fibers are combined into a single root.

The *autonomic nervous system* (ANS) is a functional division of the nervous system that has parts in both the CNS and the PNS (Fig. 1-1). It is made up of neurons that innervate smooth muscle, cardiac muscle, or glandular epithelium or combinations of these tissues. These individual *visceral tissues*, when combined, make up *visceral organs* such as the stomach. The ANS is also called the *visceral motor* or *vegetative nervous system* because it regulates motor responses outside the realm of conscious control.

Neurons At the histologic level, the nervous system is composed of *neurons* and *glial cells*. As the basic structural and functional units of the nervous system, neurons are specialized to receive information, transmit electrical impulses, and influence other neurons or effector tissues.

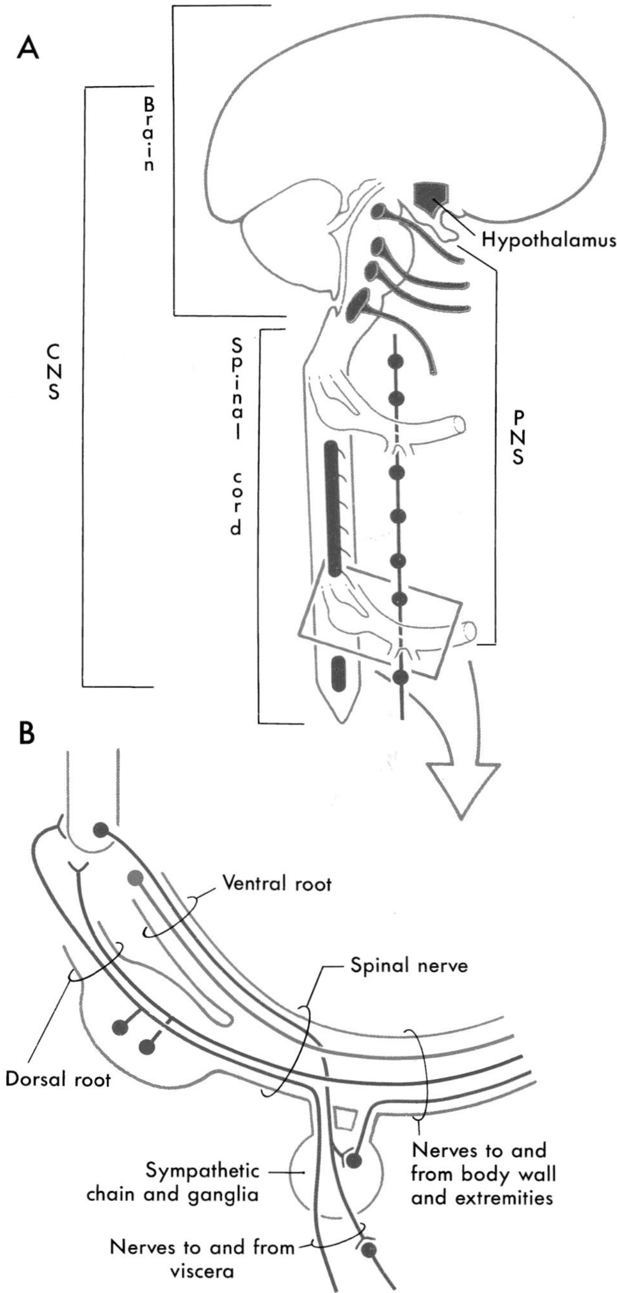

FIGURE 1-1 *(A) The general relationships of the central (CNS), peripheral (PNS), and autonomic nervous systems. The autonomic regions of the CNS and PNS are shown in red. (B) Enlargement of the boxed area in A, showing the relationship of efferent (outgoing, motor) and afferent (incoming, sensory) fibers to the spinal nerves and roots. The motor fibers are general visceral efferent (autonomic; red) and general somatic efferent (green); the sensory fibers are general somatic afferent (blue) and general visceral efferent (black).*

In many areas of the nervous system, neurons are structurally modified to serve particular functions. At this point, however, we will only consider the neuron as a general concept.

A *neuron* consists of a *cell body* (*perikaryon* or *soma*) and the processes that emanate from the cell body (Fig. 1-2). Collectively, neuronal cell bodies constitute the *gray matter* of the CNS. Named, and usually function-specific, clusters of cell bodies in the CNS are called *nuclei* (singular, *nucleus*). Typically, *dendrites* are those processes that ramify in the vicinity of the cell body, whereas a single, longer process called the *axon* carries impulses to a more remote destination. The *white matter* of the CNS consists of bundles of axons that are wrapped in a sheath of insulating lipoprotein called *myelin*. The axon terminates at specialized structures called *synapses* or, in the case of those neurons that innervate muscles, as *motor end plates* (*neuromuscular junctions*), which function much like synapses.

The generalized synapse shown in Figure 1-2 is the most common type seen in the CNS and is sometimes called an *electrochemical synapse*. It consists of a *presynaptic element*, which is usually part of an axon, a gap called the *synaptic cleft*, and the *postsynaptic region* of the innervated neuron or effector structure. Communication across this synapse is accomplished as follows. An electrical impulse (the *action potential*) causes the release of a neuroactive substance (a *neurotransmitter*, *neuromodulator*, or *neuromediator*) from the presynaptic element into the synaptic cleft. This substance is stored in *synaptic vesicles* in the presynaptic element and is released into the synaptic space by the fusion of these vesicles with the cell membrane (Fig. 1-2).

The neurotransmitter diffuses rapidly across the synaptic space and binds to receptor sites on the postsynaptic membrane. Based on the action of the neurotransmitter at receptor sites the postsynaptic neuron may be excited (lead to generation of an action potential) or inhibited (prevent generation of an action potential). Neurotransmitter residues in the synaptic cleft are rapidly inactivated by other chemicals found in this space. In this brief example we see that (1) the neuron is structurally specialized to receive and propagate electrical signals, (2) this propagation is accomplished by a combination of electrical and chemical events, and (3) the transmission of signals across the synapse is in one direction; that is, from the presynaptic neuron to the postsynaptic neuron.

Figure 1-2 shows the convention that will be used for illustrating neurons as elements of reflex arcs and pathways in this book. The dendrites and cell body (the

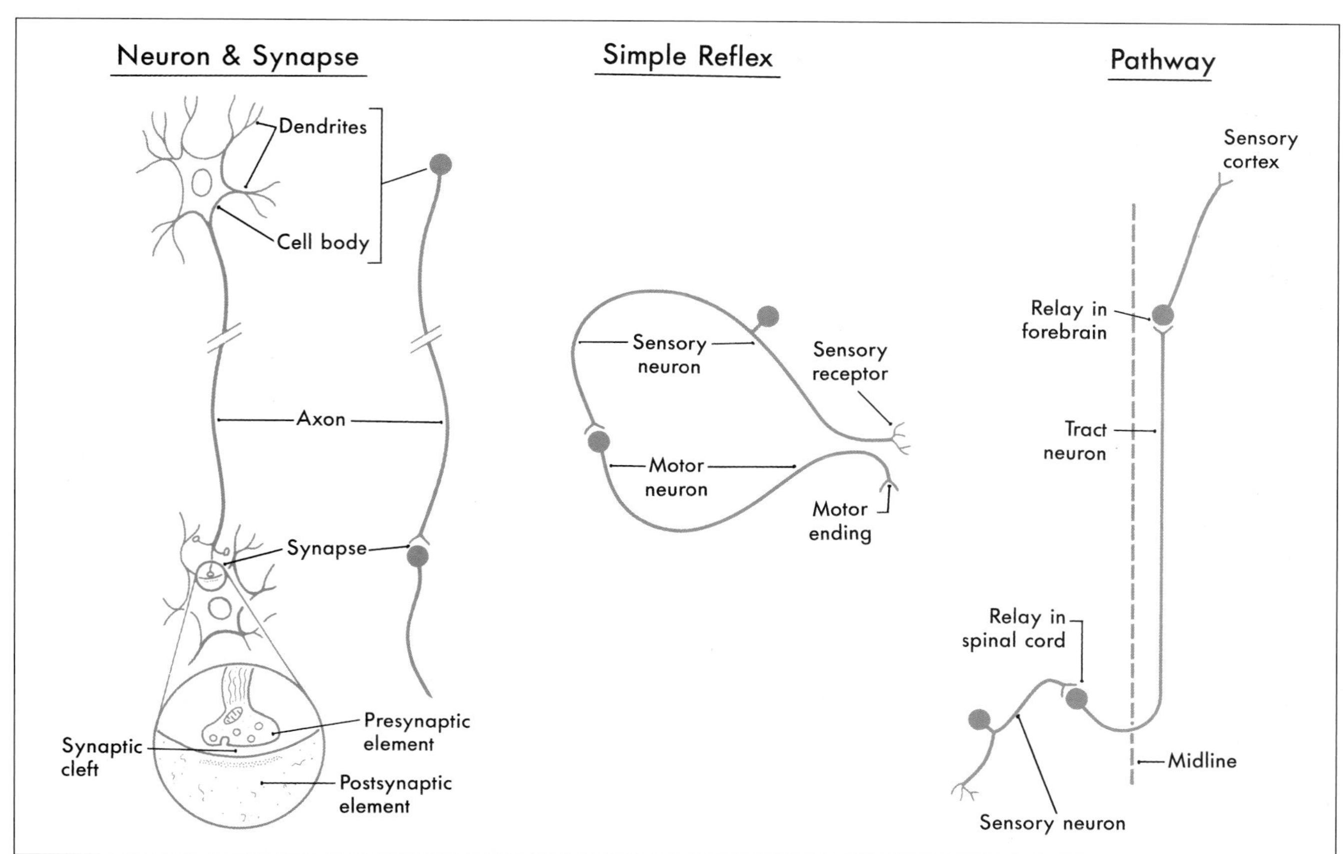

FIGURE 1-2 *A representative neuron and synapse, a simple (monosynaptic) reflex, and a pathway.*

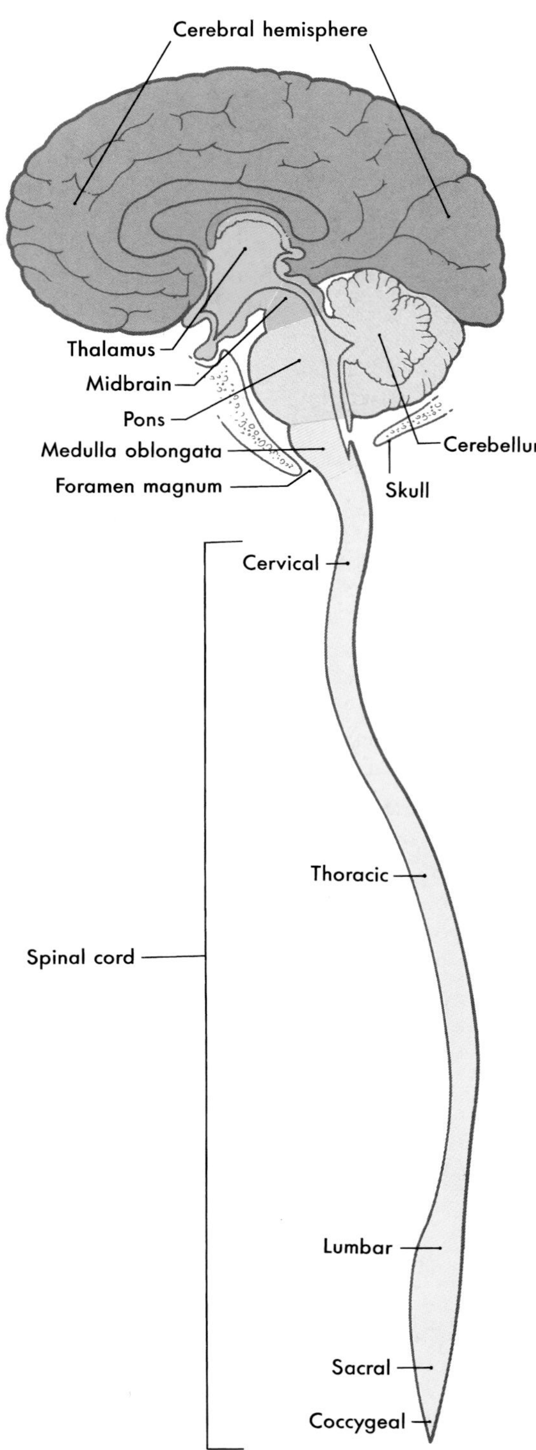

FIGURE 1-3 *The basic divisions of the central nervous system.*

"receiving" parts of the neuron) are represented by a large dot, and the axon (the "sending" part of the neuron) is represented by a line, which terminates in a fork or Y at the synapse.

Reflexes and Pathways The function of the nervous system is based on neurons interacting with each other. Figure 1-2 illustrates one of the simplest types of neuronal circuits, a reflex arc composed of only two neurons. This is called a *monosynaptic reflex arc* because only one synapse is involved. In this example, the peripheral end of a sensory fiber responds to a particular type of input. The resulting action potential is conducted by the sensory fiber into the spinal cord, where it influences a motor neuron. The axon of the motor neuron conducts a signal from the spinal cord to the appropriate skeletal muscle, which responds by contracting. *Reflexes are involuntary responses to a particular bit of sensory input.* For example, the physician taps on the patellar tendon and the leg jerks; the patient does not think about it—the motor response just happens.

Building on these summaries of the *neuron* and of the *basic reflex arc*, we will briefly consider what neuronal elements constitute a *pathway*. If the patient bumps her knee and not only hits the patellar tendon but also damages the skin over the tendon, two things happen (Fig. 1-2). First, impulses from receptors in the tendon travel through a reflex arc that causes the leg to jerk (*knee jerk* or *patellar reflex*). The synapse for this reflex arc is located in the lumbosacral spinal cord. Second, impulses from pain receptors in the damaged skin are transmitted in the lumbosacral cord to a second set of neurons, which convey them via ascending axons to the forebrain. As can be seen in Figure 1-2, these axons cross the midline of the spinal cord and form an ascending tract on the contralateral side. In the forebrain, these signals are passed to a third group of neurons, which distributes them to a region of the cerebral cortex specialized to interpret them as pain from the knee. This three-neuron chain constitutes a *pathway*, a series of neurons designed to carry a specific type of information from one site to another (Fig. 1-2). Some pathways carry information to a level of conscious perception (we not only recognize pain but know that it is coming from the knee), and others convey information that does not reach the conscious level.

REGIONS OF THE CNS

Spinal Cord The spinal cord is located inside the vertebral canal and is rostrally continuous with the medulla oblongata of the brain (Fig. 1-3). An essential link between the peripheral nervous system and the brain, it conveys sensory information originating from the body wall, extremities, and gut and distributes motor impulses to these areas. Impulses enter and leave the spinal cord through the 31 pairs of spinal nerves (see Fig. 1-1). The spinal cord contains sensory fibers and motor neurons involved in reflex activity and ascending and descending pathways (also called *tracts*) that link spinal centers with other parts of the CNS. Ascending pathways convey sensory information to higher centers, whereas descending pathways influence the activity of neurons in the spinal cord gray matter.

Medulla Oblongata At the level of the foramen magnum, the spinal cord is continuous with the most caudal part of the brain, the *medulla oblongata*, commonly called the medulla (Fig. 1-3). The medulla consists of (1) neurons that perform functions associated with the medulla and (2) ascending and descending tracts that pass through the medulla on their way from or to the spinal cord. In general, fibers that descend through the medulla are involved in motor functions, whereas fibers that ascend through it carry sensory information. Some of the neuronal cell bodies of the medulla are organized into nuclei associated with specific cranial nerves. The medulla contains the nuclei for the glossopharyngeal (cranial nerve IX), vagus (X), and hypoglossal (XII) nerves, as well as portions of the nuclei for the trigeminal (V), vestibulocochlear (VIII), and spinal accessory (XI) nerves. It also contains important relay centers and nuclei that are essential to the regulation of respiration, heart rate, and various visceral functions.

Pons and Cerebellum Embryologically, the pons and cerebellum originate from the same segment of the developing neural tube. However, in the adult the pons forms part of the *brainstem* (the other parts being the midbrain and medulla), and the cerebellum is a *suprasegmental* structure because it is located dorsal to the brainstem (Fig. 1-3).

Like the medulla, the pons contains many neuronal cell bodies, some of which are organized into cranial nerve nuclei, and it is traversed by ascending (sensory) and descending (motor) tracts. The pons contains the nuclei of the abducens (VI) and facial (VII) nerves and portions of the nuclei for the trigeminal (V) and vestibulocochlear (VIII) nerves. The ventral part of the pons contains large populations of neurons that form a relay station between the cerebral cortex and cerebellum and descending motor fibers that travel to all spinal levels.

The cerebellum is connected with diverse regions of the CNS. Functionally, the cerebellum is considered part of the motor system. It serves to coordinate the activity of individual muscle groups to produce smooth, purposeful, synergistic movements.

Midbrain Rostrally, the pons is continuous with the midbrain. This latter part of the brain is, quite literally, the link between the brainstem and the forebrain. Ascending or descending pathways to or from the forebrain must traverse the midbrain. The nuclei for the oculomotor (III) and trochlear (IV) cranial nerves, as well as part of the trigeminal (V) complex, are found in the midbrain. Other midbrain centers are concerned with visual and auditory reflex pathways, motor function, the transmission of pain, and visceral functions.

Thalamus The forebrain consists of the *cerebral hemispheres* and the large groups of neurons that comprise the *basal ganglia* and the *thalamus* (Fig. 1-3). We will see later that the thalamus actually consists of several regions, for example, the hypothalamus, subthalamus, epithalamus, and dorsal thalamus.

The thalamus is rostral to the midbrain and almost completely surrounded by elements of the cerebral hemisphere. Individual parts of the thalamus can be seen in detail only when the brain is cut in coronal or axial (horizontal) planes.

With the exception of olfaction, all sensory information that eventually reaches the cerebral cortex must pass through the thalamus. One function of the thalamus, therefore, is to receive sensory information of many sorts (temperature, pain, vision, etc.) and to distribute it to the specific regions in the cerebral cortex that are specialized to decode it. Other areas of the thalamus receive input from pathways conveying information on, for example, position sense or the tension in a tendon or muscle. This input is relayed to areas of the cerebral cortex that function to generate smooth purposeful movements.

Although quite small, the hypothalamus is extremely important. It functions in sexual behavior, feeding, hormonal output of the pituitary gland, body temperature regulation, and a wide range of autonomic functions. Through descending connections, the hypothalamus influences visceral centers in the brainstem and spinal cord.

Cerebral Hemispheres The largest and most obvious parts of the human brain are the two cerebral hemispheres. Each hemisphere is composed of three major subdivisions. First, the *cerebral cortex* is a layer of neuronal cell bodies about 0.5 cm thick that covers the entire surface of the hemisphere. This layer of cells is thrown into elevations called *gyri* (singular, *gyrus*) separated by creases called *sulci* (singular, *sulcus*). The second major part of the hemisphere is the *subcortical white matter*, which is made up of myelinated axons that carry information to or from the cerebral cortex. The largest and most organized part of the white matter is the *internal capsule*. The third major component of the hemisphere is a prominent group of neuronal cell bodies collectively called the *basal ganglia*. These prominent forebrain centers are involved in motor function. Parkinson's disease, a neurologic disorder associated with the basal ganglia, is characterized by a profound impairment of movement.

The gyri and sulci that make up the cerebral cortex are named, and many are associated with particular functions. Some gyri receive sensory input, such as vision or general sensation, whereas others give rise to motor fibers that project to the spinal cord and motor nuclei of cranial nerves. The cerebral cortex also has association areas that are essential for analysis and cognitive thought.

FUNCTIONAL SYSTEMS AND REGIONS

A *functional system* is a set of neurons linked together to convey a particular block of information or accomplish a particular task. In this respect, *systems* and *pathways*, in some cases, may be quite similar, and occasionally their meaning may overlap.

Anatomic parts of the CNS, such as the medulla and pons, are commonly called *regions*. The study of their structure and function, called *regional neuroanatomy*, is the focus of Section II of this book. *Systems* and *pathways*, however, generally traverse more than one region. The system of neurons and axons that allows you to feel the edge of this page, for example, crosses every region of the nervous system between your fingers and the somatosensory cortex. The study of functional systems, called *systems neurobiology*, is the focus of Section III. It is important to remember that the *functional characteristics of regions coexist with those of systems*.

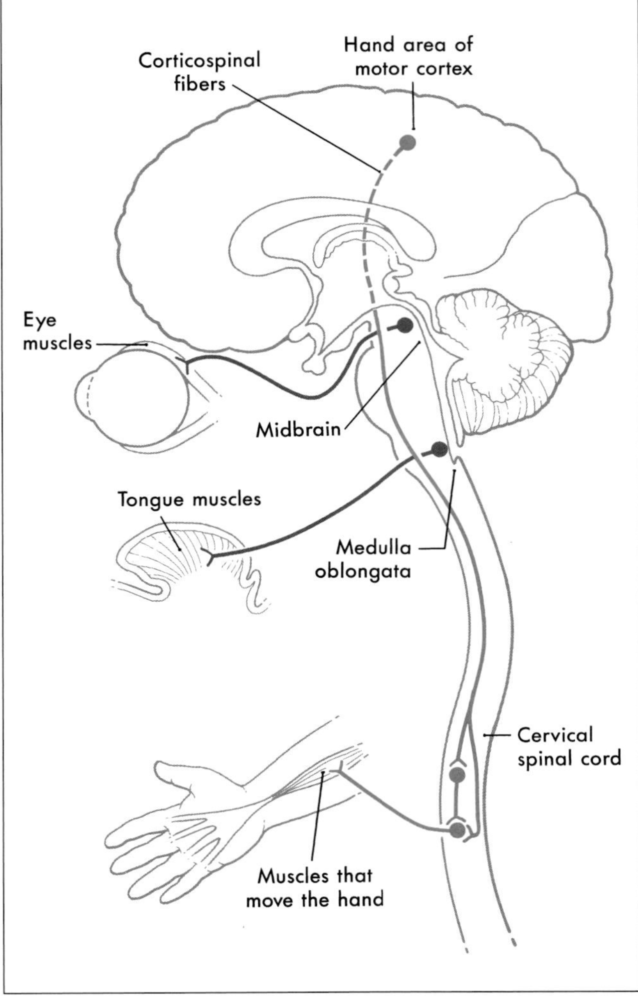

FIGURE 1-4 *An example of the relation of systems to regions. Fibers of the motor system that control hand movement descend from the motor cortex to the cervical spinal cord. In the cord, these fibers influence motor neurons that control hand and forearm muscles. Injury at any point along the way can damage fibers of the system and/or structures specific to the region. For example, injury to the midbrain could damage both fibers to the hand and fibers to the eye muscles, whereas injury to the medulla could damage both fibers to the hand and fibers to the tongue musculature.*

Let us consider an example of how the interrelation of systems and regions can be important clinically. The signals that influence movements of the hand originate in the cerebral cortex. Neurons in the hand area of the *motor cortex* send their axons to cervical levels of the spinal cord, where they influence spinal motor neurons that innervate muscles of the forearm. These are called *corticospinal fibers* because their cell bodies are in the cerebral cortex (*cortico-*), and their axons end in the spinal cord (*-spinal*). These fibers pass through the subcortical white matter, the entire brainstem, and upper levels of the cervical spinal cord. En route they pass near nuclei and fiber tracts that are specific to that particular region (Fig. 1-4). In the midbrain, for example, they pass near fibers of the oculomotor nerve, which originate in the midbrain and control certain extraocular muscles. In the medulla, they pass near fibers that originate in the medulla and innervate the musculature of the tongue. An injury to the midbrain, therefore, could cause motor problems in the hand (*systems damage*) combined with partial paralysis of eye movement (*regional damage*). In similar fashion, an injury to the medulla could cause the same hand problem but now in association with partial paralysis of the tongue. As we study the nervous system, we will see that *successful diagnosis of patients with neurologic disorders will, among other things, depend on a good understanding of both regional and systems neurobiology*.

DORSAL, VENTRAL, AND OTHER DIRECTIONS IN THE CNS

By convention, directions in the human CNS—such as *dorsal* (*posterior*) and *ventral* (*anterior*), *medial* and *lateral*, and *rostral* and *caudal*—are absolute with respect to the central axis of the brain and spinal cord. In similar manner, the anatomic orientation of the body in space is related to its central axis. For example, if the patient is lying on his stomach, the dorsal surface of the trunk is up and its ventral surface is down (Fig. 1-5). If the patient rolls over, his back remains the dorsal surface of his body even though it now faces down.

As shown in Figure 1-6, the spinal cord and the brainstem (medulla, pons, and midbrain) form a nearly straight line, which is roughly parallel with the superoinferior axis of the body. Therefore, anatomic direc-

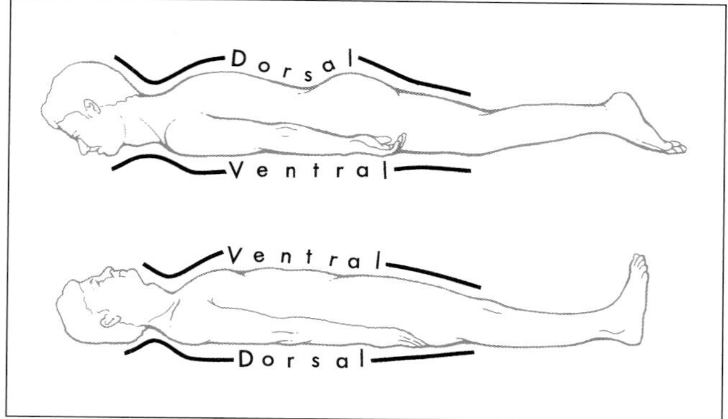

FIGURE 1-5 *An illustration showing that the anatomic directions of the body are absolute with respect to the axes of the body, not with respect to the position of the body in space.*

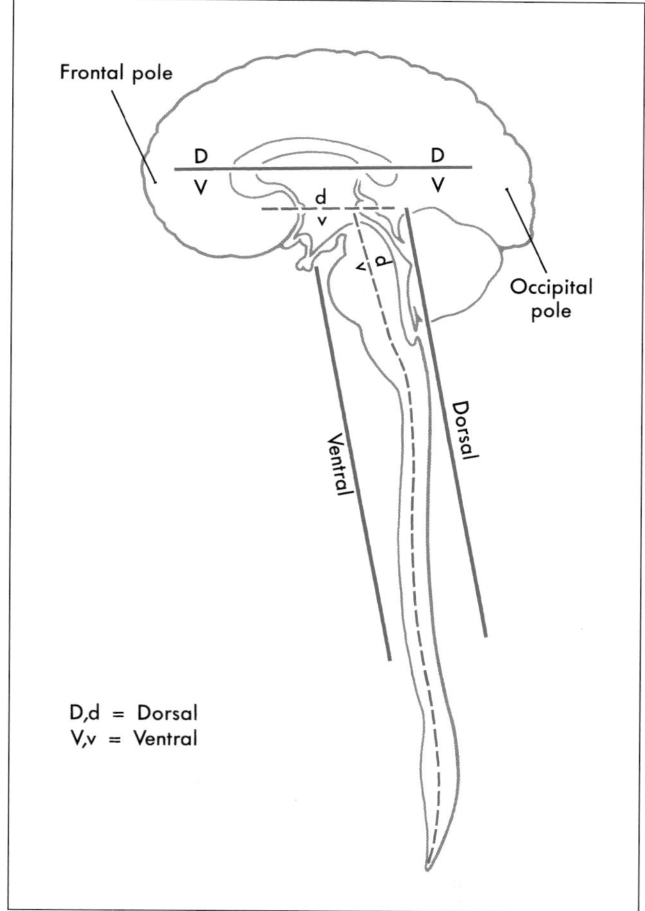

FIGURE 1-6 *The central axis and anatomic directions of the CNS. The dashed line shows the long (rostrocaudal) axis of the CNS. The long axis of the spinal cord and brainstem forms a sharp angle with the long axis of the forebrain. Dorsal and ventral orientations are also shown.*

tions in these regions of the CNS coincide roughly with those of the body as a whole. The pons is rostral to the medulla, for instance, and the cerebellum is dorsal to the pons. The situation is different in the forebrain, because during embryonic development, the forebrain rotates (at the cephalic flexure) relative to the midbrain until its rostrocaudal axis corresponds to a line drawn from the forehead to the occiput (from the frontal to the occipital poles of the cerebral hemispheres). This rotation creates a sharp angle in the long axis of the CNS at the midbrain-thalamus junction. Anatomic directions in the forebrain relate to its long axis; therefore, the dorsal side of the forebrain structures faces the vertex of the head (Fig. 1-6).

Clinical Images of the Brain Modern technology has given us the tools to view the living CNS in some detail. The most routinely used methods to image the brain are computed tomography (CT) and magnetic resonance imaging (MRI). We will use MRI images to illustrate some important anatomic relationships.

Patients usually lie on their back for imaging of the brain or spinal cord (Fig. 1-7). In this position, the dorsal surface of the brainstem and spinal cord and the caudal surface (occipital pole) of the cerebral hemispheres face down. The ventral surface of the brainstem and spinal cord and the frontal pole are face up.

Images of the brain are commonly made in *coronal*, *axial* (horizontal), and *sagittal* planes. To illustrate the basic orientation of the CNS in situ, we will look at examples of coronal and axial images, shown as they would appear in the clinical setting (Fig. 1-7). Coronal imaging planes are oriented perpendicular to the rostrocaudal axis of the forebrain but are nearly parallel to the rostrocaudal axis of the brainstem. Therefore, a coronal image taken at a relatively rostral level of the cerebral hemispheres (Fig. 1-7A) will show only forebrain structures, and these structures will appear in cross section (perpendicular to their long axis). As the plane of imaging is moved caudally, brainstem structures enter the picture (Fig. 1-7B), but the brainstem is cut nearly parallel to its rostrocaudal axis. Axial images, in contrast, are oriented parallel to the rostrocaudal axis of the cerebral hemispheres but nearly perpendicular to the long axis of the brainstem. Consequently, an axial image taken midway through the cerebral hemispheres (Fig. 1-7C) will show only forebrain structures, with the rostral end of the forebrain at the top of the image and the caudal end at the bottom. As the plane of imaging is moved further ventrally relative to the forebrain, the brainstem appears (Fig. 1-7D). The brainstem, however, is cut in near cross section and is oriented with the ventral surface "up" (toward the top of the image) and the dorsal surface "down."

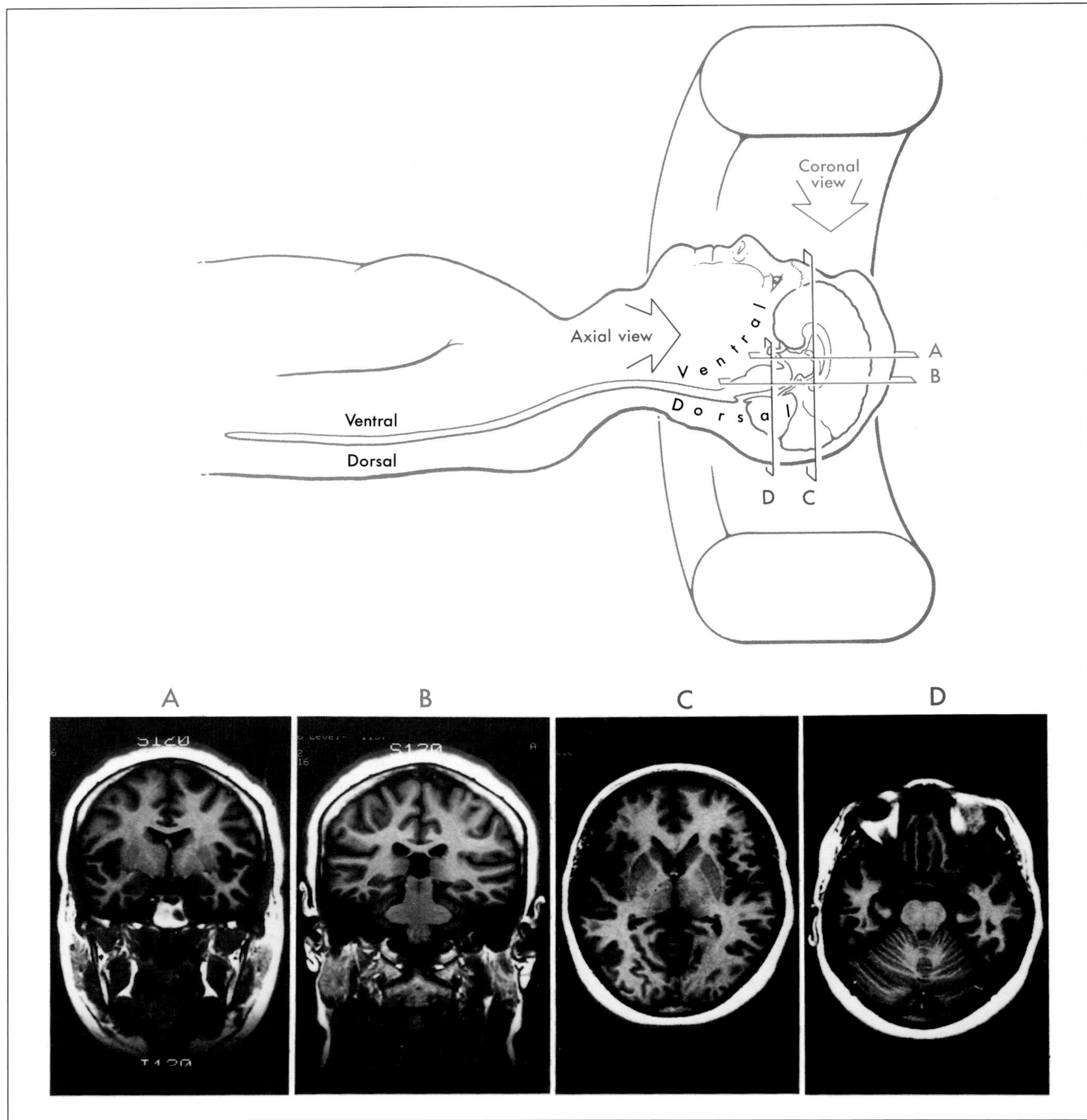

FIGURE 1-7 *The relation of imaging planes to the brain. The diagram shows the usual orientation of a patient in an MRI machine and the planes of the four scans that are shown. (A,B) Coronal scans; (C,D) axial scans.*

A point also needs to be made about how the clinician looks at scans such as those in Figure 1-7. Coronal scans are viewed as though you are looking the patient in the face, whereas axial scans are viewed as though you are standing at the patient's feet looking toward his head while he lies on his back in the machine. Axial scans, in other words, show the cerebral hemispheres from ventral to dorsal, with the patient's orbits at the top of the image and the occiput at the bottom. In both coronal and axial views, the patient's left side is to the observer's right.

CHAPTER TWO

The Cell Biology of Neurons and Glia

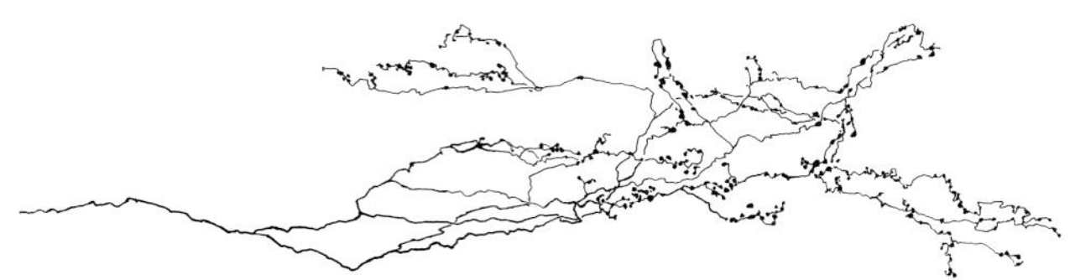

J. B. HUTCHINS • J. P. NAFTEL • M. D. ARD

The number of cells in the adult human central nervous system (CNS) has been estimated at 100 billion. All arise from a relatively small population of precursors, yet a diversity of cell types is seen in the adult. Their most basic classification is as *neurons* and *glia*.

OVERVIEW

Nerve cells (neurons) manipulate information. Doing so involves changes in the *bioelectrical* and/or *biochemical* properties of the cell, and these changes require a vast expenditure of energy for each cell. The nervous system, compared to other organs, is the greatest consumer of oxygen and glucose. These energy requirements arise directly from the metabolic demand placed on cells, which have a large surface area and concentrate biomolecules and ions against an energy gradient. Along with maintaining its metabolism, each neuron (1) *receives information* from the environment or from other nerve cells, (2) *processes information*, and (3) *sends information* to other neurons or effector tissues.

Unlike neurons, *glial cells* do not receive and transmit information with point-to-point specificity. Rather, their primary function is control of the environment within the CNS. They *shuttle nutritive molecules* from blood vessels to neurons, *remove waste* products, and *maintain the electrochemical* surroundings of neurons. Glia are also essential in the early development of the CNS for *guiding developing neurons* to their correct locations, and, in the adult, glia provide *structural support* for nerve cells.

THE STRUCTURE OF NEURONS

The archetypical neuron is bounded by a continuous plasma membrane and consists of a *cell body* or *soma* from which *dendrites* and an *axon* arise (Fig. 2-1). In most neurons, information normally flows from the dendrites to the cell body to the axon (and its terminals), and then to the next neuron or effector tissue. We will describe these components of the neuron in the same order.

Dendrites Dendrites usually branch extensively in the vicinity of the cell body, giving the appearance of a tree or bush (Figs. 2-1, 2-2A). They receive signals either from other neurons through contacts (*synapses*) made on their surfaces, or from the environment via specialized receptors. Information travels from distal to proximal along dendrites to converge at the cell body.

Small budlike extensions (*dendritic spines*) of a variety of shapes are frequently seen on the more distal branches of the dendritic tree (Fig. 2-2B,C). These are usually the sites of *synaptic contacts* (discussed below). As the dendrite progresses toward the cell body, its thickness increases as it anastomoses with other branches.

Small dendrites contain cytoskeletal elements but appear devoid of other organelles. Larger dendrites contain numerous *microtubules*, *neurofilaments* (a type of intermediate filament exclusively present in neurons), *mitochondria*, some saccules of endoplasmic reticulum, and collections of polyribosomes and free ribosomes (Fig. 2-2D,E). In many nerve cells, the distal dendrites collect into large, trunklike *primary dendrites* that contain the same organelles as the cell body. The microtubular and neurofilamentous skeletons of the dendritic tree are continuous throughout its extent and help to maintain its branched structure.

Cell Body The cell body of a neuron is also called the *soma* (pl. *somata*) or *perikaryon* (pl. *perikarya*) (Figs. 2-1, 2-3). The perikaryon is the *metabolic center* of the nerve cell. Abundant mitochondria reflect the high energy consumption of the cell. Active protein synthesis is indicated by other structural features. A large nucleus contains diffuse chromatin (euchromatin) and typically at least one prominent nucleolus. In the cytoplasm, *ribosomes* are abundant, and the *rough endoplasmic reticulum* (rER) and *Golgi complex* are extensive. The rER is basophilic (binds basic dyes) as a result of the RNA content of the ribosomes attached to the endoplasmic membrane, and the stacked layers of rER are seen as patches of basic staining (called *Nissl substance*) in histologic preparations of nerve cells.

Neurons are classified into three broad types on the basis of the shape of the cell body and the pattern of processes emerging from it. These types are the multipolar, pseudounipolar, and bipolar cells (Fig. 2-1; Table 2-1). In *multipolar* neurons, multiple dendrites emerge from the cell body, giving it a polygonal shape. A single axon, usually of small diameter, also arises from the cell body. Different kinds of multipolar cells have characteristic patterns of processes; some are listed in Table 2-1.

The cell body of a *pseudounipolar* (or *unipolar*) neuron gives rise to a single process and is round, with a centrally located nucleus (Fig. 2-1D). The single process divides close to the cell body into two branches: a *peripheral* branch, which carries sensory information from the periphery, and a *central* branch, which relays the information onward to its target in the CNS. The two processes thus function as a combined axon and dendrite. The distal end of the peripheral process is dendritelike, and one or more of its branches either constitute or contact sensory receptors. The central process terminates in either the spinal cord or the brainstem. The cell bodies of pseudounipolar cells are found primarily in the sensory ganglia of cranial and spinal nerves.

Bipolar neurons have a round or oval-shaped perikaryon with a single large process emanating from each end of the cell body (Fig. 2-1E,F). They are commonly found in sensory structures. In the retina, bipolar cells are interposed between receptor cells and output

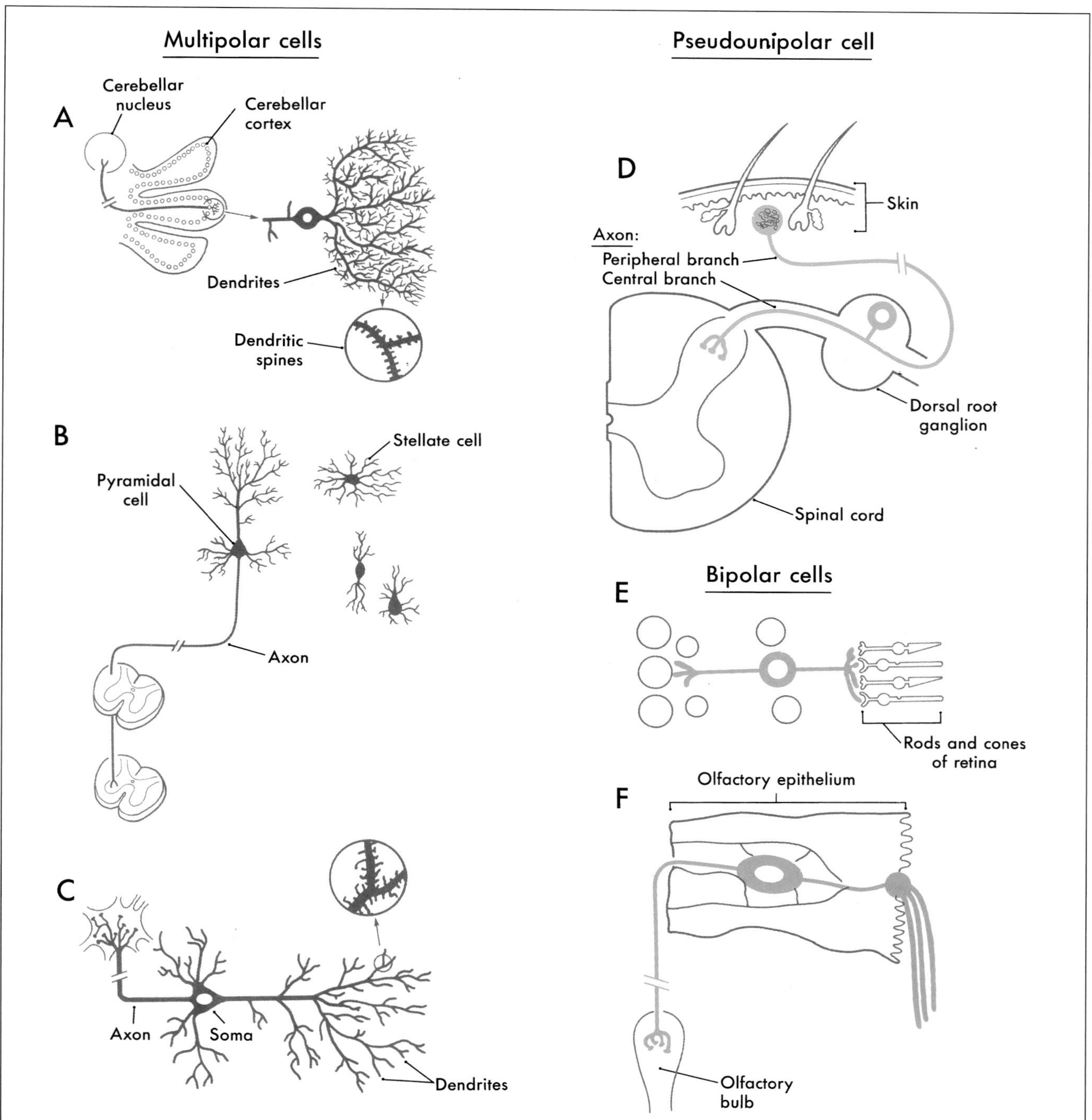

FIGURE 2-1 *Examples of various types of neurons showing the dendrites, somata, and axons of multipolar cells: from the cerebellar cortex (A), from the cerebral cortex (B), and a pyramidal cell of the cerebral cortex (C). Compare these with the a pseudounipolar cell of the dorsal root ganglion (D), and bipolar cells from the retina (E) and olfactory epithelium (F).*

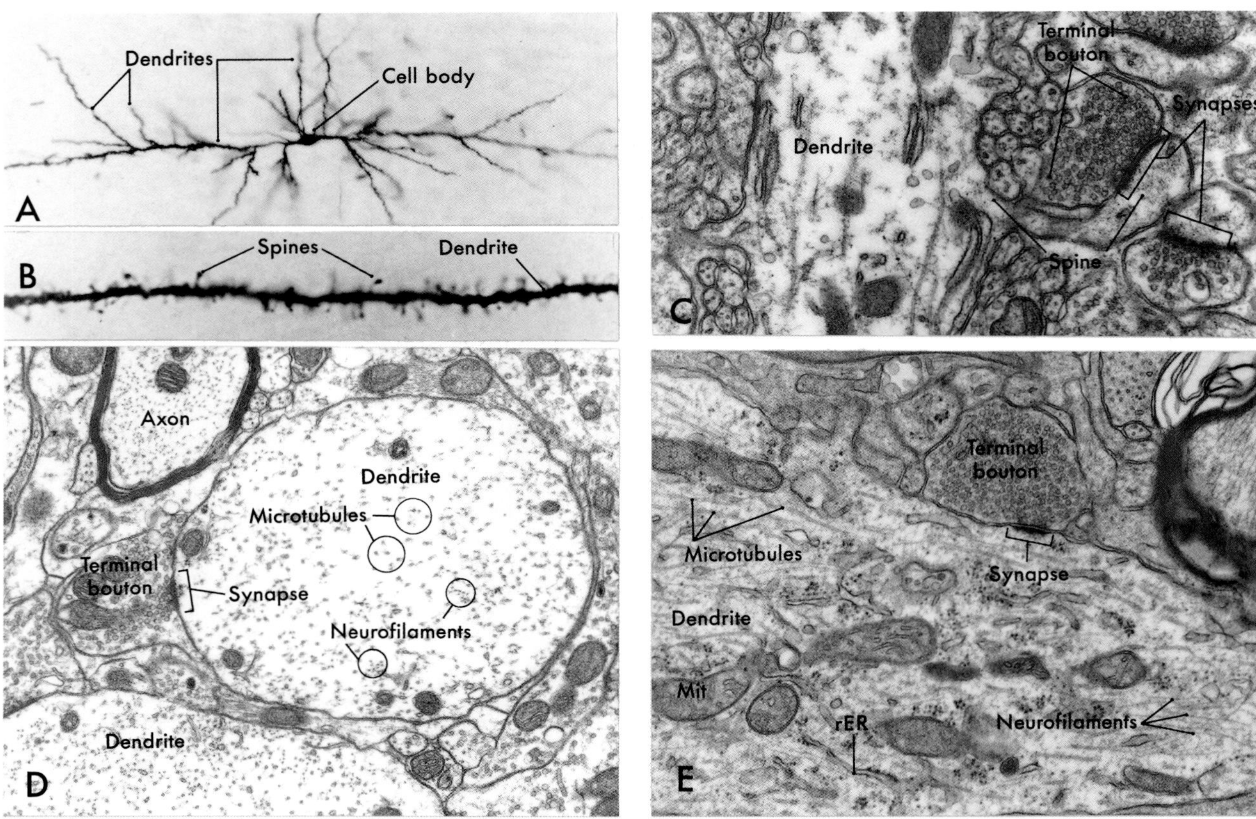

FIGURE 2-2 *Elements of dendrite structure. Dendritic tree of a multipolar neuron (A) and dendritic spines (B) both in Golgi-stained cortical tissue. Ultrastructural features of dendrites, showing an axonal terminal bouton synapsing on a dendritic spine (C), a cross section of a dendrite with characteristic cytoskeletal elements and organelles (D) and a longitudinal section of a dendrite in the anterior horn of the spinal cord (E). (A and B courtesy of Dr. Jose Rafols; D courtesy of Dr. Alan Peters.)*

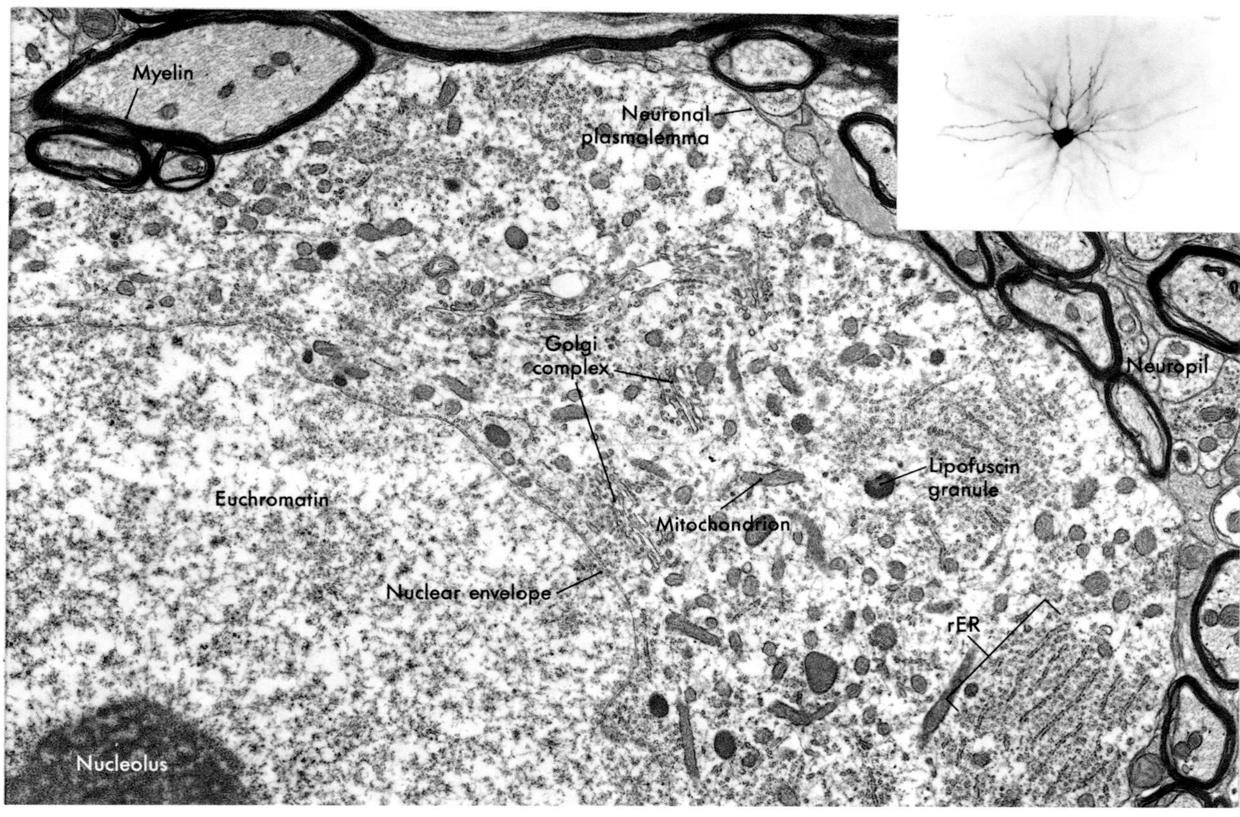

FIGURE 2-3 *The cell body of a multipolar neuron as seen in an electron micrograph and in a Golgi-stained preparation (Inset, courtesy of Dr. Jose Rafols).*

TABLE 2-1

A FEW OF THE NEURONAL TYPES FOUND IN THE NERVOUS SYSTEM

TYPE OF NEURON	TYPICAL LOCATION(S)
Pseudounipolar	Dendrite: specialized nerve ending in periphery Cell body: dorsal root or cranial nerve ganglion Axon: terminates in spinal cord or brainstem
Bipolar	Retina Olfactory epithelium Vestibular ganglion Auditory (spiral) ganglion
Multipolar	
stellate ("star-shaped")	Many areas of CNS
fusiform ("spindle-shaped")	Many areas of CNS
pyriform ("pear-shaped")	Many areas of CNS
pyramidal	Hippocampus; layers II, III, V & VI of cerebral cortex
Purkinje	Cerebellar cortex
mitral	Olfactory bulb
chandelier	Visual areas of cerebral cortex
granule	Cerebral and cerebellar cortex
amacrine ("axonless")	Retina

cells. In the olfactory system, they constitute the receptor cells themselves, and in the vestibular and auditory systems they are the output cells that send information to the brainstem.

Axons and Axon Terminals The *axon* arises from the cell body at a small elevation called the *axon hillock*. The proximal part of the axon, adjacent to the axon hillock, is the *initial segment*. The cytoplasm of the axon (axoplasm) contains dense bundles of *microtubules* and *neurofilaments* (Fig. 2-4). These function as structural elements, and they also play key roles in the transport of metabolites and organelles along the axon. Axons are typically devoid of ribosomes, a feature that distinguishes them from dendrites at the ultrastructural level.

In contrast to dendrites, axons may extend for long distances before branching and terminating. For example, the axon of a motor neuron innervating the adductor hallucis muscle in a tall individual is well over a meter in length. To place this in scale, if the cell body of this neuron were the size of a softball, the axon (with a diameter of about 6 mm) would extend about 10 km. Although the cell body is the nutritive center of the neuron, its volume is a small fraction of the collective volume of the dendrites and axon. The surface area of an axon can be several thousand times the surface area of the parent cell body.

Axons in the CNS often end in fine branches known as *terminal arbors* (Fig. 2-4C). In most neurons, each axon terminal is capped with small *terminal boutons* (*boutons termineaux*, terminal buttons; Fig. 2-2C–E). These correspond to functional points of contact (synapses) between nerve cells. In some cells, boutons are found along the length of the axon, where they are called *boutons en passant*. Other axons contain swellings, or *varicosities*, which are not buttonlike but still can represent points of cell-to-cell information transfer.

The site at which an axon terminal communicates with a second neuron, or with an effector tissue, is called a *synapse*. In general the synapse can be defined as the apposition of a process (usually an axon) of one neuron to the dendrites, cell body, or axon of a second neuron, or to an effector cell such as a skeletal muscle fiber. Synapses are considered later in this chapter in the section **Neurons as Information Transmitters**.

Axonal Transport Nerve cells have an elaborate transport system that moves organelles and macromolecules from the cell body out to the axon terminals. Transport in the axon occurs in both directions (Table 2-2; Fig. 2-5). Axonal transport from the cell body toward the terminals is called *anterograde* or *orthograde*; transport from the terminals toward the cell body is called *retrograde*.

Anterograde axonal transport is classified into *fast* and *slow* components. Fast transport, at speeds of up to 400 mm/day, is based on the action of a protein called *kinesin*. Kinesin, an ATPase, moves macromolecule-containing vesicles and mitochondria along microtubules, much like a small insect crawling along a straw. Slow transport carries important structural and metabolic components from the cell body to axon terminals; its mechanism is less well understood.

Retrograde axonal transport allows the neuron to respond to molecules, for example, growth factors, which are taken up near the axon terminal by either *pinocytosis* or *receptor-mediated endocytosis*. In addition, this form of transport functions in the continual recycling of components of the axon terminal. Retrograde transport along axonal microtubules is driven by the protein dynein rather than by kinesin.

Axonal transport is important in the pathogenesis of some human neurologic diseases. The *rabies virus* replicates in muscle tissue at the site of a bite by a rabid animal and is then *transported retrogradely* to the cell body of neurons innervating the muscle. The infected cells produce and shed copies of the rabies virus, which, in

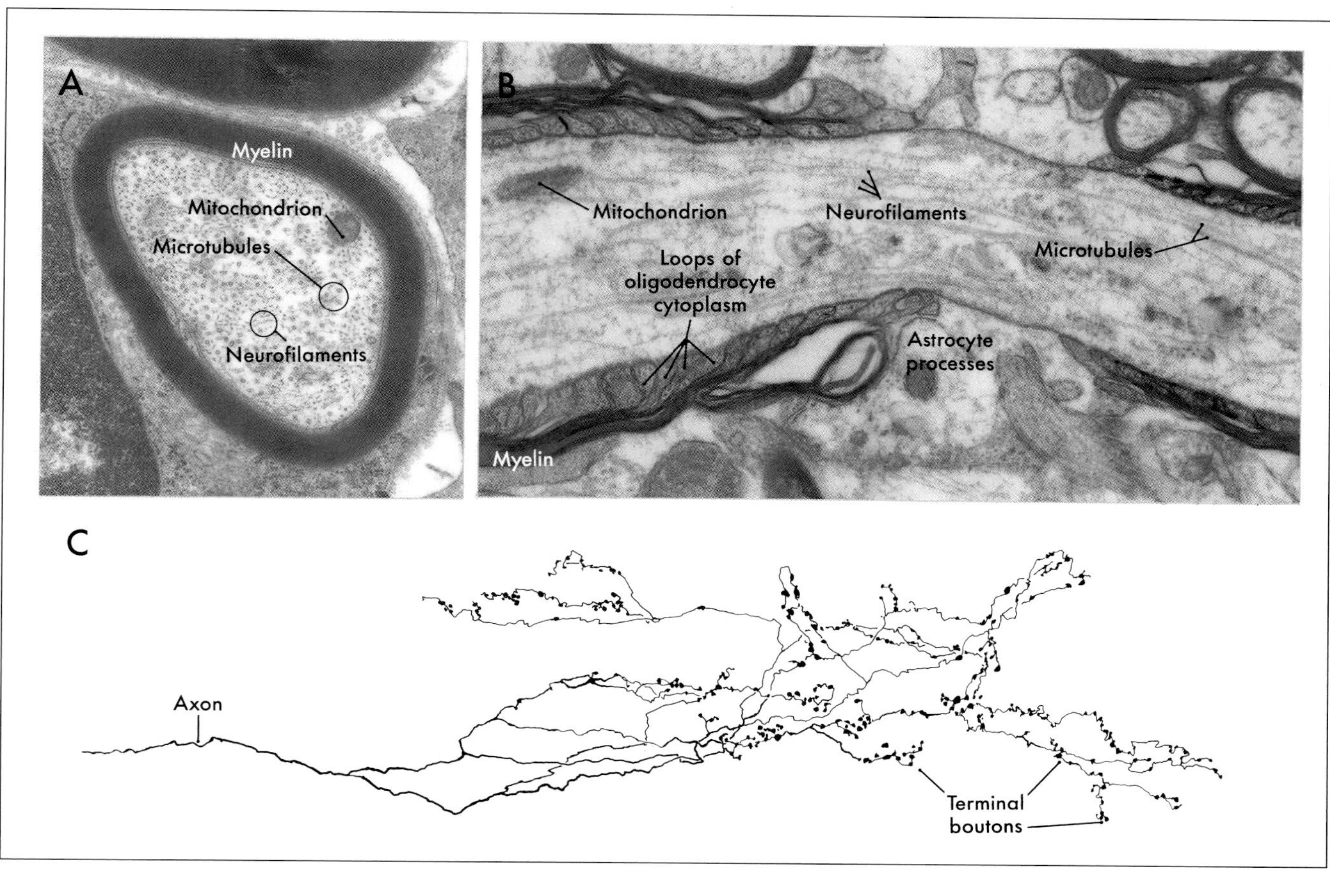

FIGURE 2-4 *Elements of axon structure. Ultrastructural features of a small myelinated axon in a cross section of a peripheral nerve (A) and a longitudinal view at a node of Ranvier of a myelinated axon in the CNS (B). Drawing of the complete terminal arbor of an axon in the thalamus, reconstructed from serial sections (C, courtesy of Dr. Ed Lachica).*

TABLE 2-2

CHARACTERISTICS OF AXONAL TRANSPORT

DIRECTION OF TRANSPORT	SPEED OF TRANSPORT	PROPOSED MECHANISM	SUBSTANCES CARRIED
Anterograde	Fast (100–400 mm/day)	Kinesin/microtubules	Proteins in vesicles
		Neurotransmitters in vesicles, mitochondria	
	Slow (~ 1 mm/day)	Unknown	Cytoskeletal protein components (actin, myosin, tubulin) Neurotransmitter-related cytosolic enzymes
Retrograde	Fast (50–250 mm/day)	Dynein/microtubules	Macromolecules in vesicles, "old" mitochondria Pinocytotic vesicles from axon terminal

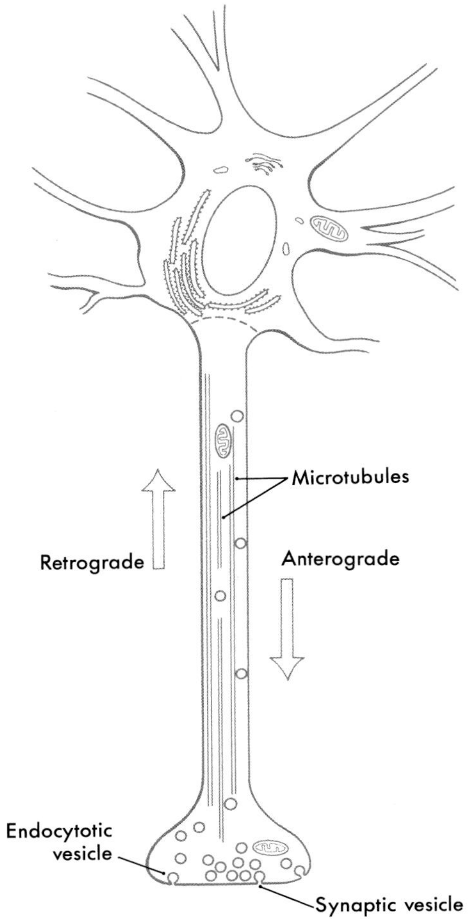

FIGURE 2-5 *Anterograde and retrograde axonal transport.*

turn, are taken up by the terminals of adjacent cells. In this way the infection becomes distributed throughout the CNS, causing the behavioral changes associated with this disease. From the CNS, the virus travels to the salivary glands by means of anterograde axonal transport in neurons innervating these glands. The infected salivary glands shed the virus in the saliva.

The toxin produced by the bacterium *Clostridium tetani* is also transported retrogradely in nerve cells whose axons terminate in the site of infection. *Tetanus toxin* is released from the nerve cell body and taken up by the terminals of neighboring neurons. However, unlike the rabies virus, which is replicated in the cell body, the tetanus toxin is diluted as it passes from cell to cell. In spite of this dilution effect, patients infected with *C. tetani* may suffer a range of neurologic deficits. About one-half of infected individuals die.

Axonal Transport as a Research Tool The ability of neurons to transport intracellular materials in anterograde and retrograde directions is exploited in investigations of neuronal connections. For example, when the enzyme *horseradish peroxidase* (HRP) or a *fluorescent substance* is injected into regions containing axon terminals, it is taken up by these processes and transported retrogradely to the cell body. After histologic preparation, the cell bodies containing these retrograde tracers are visualized. The presence of the label in a cell body suggests that the neuron has axon terminals at the site of injection.

Tracer studies can also exploit the anterograde transport system of neurons. For example, if radioactively labeled amino acids are injected into a group of neuronal cell bodies, they will be incorporated into neuronal proteins and transported in an anterograde direction. The axons containing the labeled proteins can then be detected by autoradiography. Another commonly used anterograde tracer is HRP conjugated to the glycoprotein-binding molecule (lectin) *wheat germ agglutinin* (WGA-HRP). Anterograde tracers are used to identify the distribution patterns of axons arising from a specific population of neuronal cell bodies.

The fact that the cell body is the trophic center of the neuron provides two other methods to study connections in the nervous system. If the cell body is destroyed, the axon undergoes *anterograde* (*Wallerian*) degeneration. These degenerated axons can be visualized when neural tissue is impregnated with silver nitrate. Variations on this method make it possible to conduct studies on human material obtained at autopsy. Injury to the axon, on the other hand, will result in a set of changes in the cell body that are referred to as *chromatolysis*. The cell body swells, the nucleus assumes an eccentric position, and the Nissl substance disperses. (This breakup of the dye-binding parts of the cell gives chromatolysis its name.) This technique also has been used in animal experimentation and in human autopsy material.

CLASSIFICATION OF NEURONS AND GROUPS OF NEURONS

Functionally related nerve cell bodies and axons are often aggregated to form distinct structures in the nervous system. Table 2-3 lists the main terms used for such structures. In the CNS, a cluster of functionally related nerve cell bodies is most commonly called a *nucleus* (pl. *nuclei*), although a layer of cells may be called a *layer*, *lamina*, or *stratum*; and columnar groups of cells may be called *columns*. The latter term is used for two sets of structures. In the cerebral cortex, it refers to a group of cells that are related by function and by the location of the stimulus that drives them and that form a column oriented perpendicular to the plane of the cortex. In the spinal cord, it refers to a longitudinal group of functionally related cells that extends for part or all of the length of the brainstem or spinal cord. Bundles of axons in the CNS are called *tracts*, *fasciculi*, or *lemnisci*. These are typically composed of specific populations of functionally related fibers (as in the

TABLE 2-3

TERMS USED TO DESCRIBE GROUPINGS OF NEURONAL COMPONENTS

NAME	DESCRIPTION	EXAMPLES
CNS STRUCTURES		
Nucleus (plural: *nuclei*)	A group of functionally related nerve cell bodies in the CNS	Inferior olivary nucleus, nucleus ambiguus, caudate nucleus
Column	In the cerebral cortex, a group of nerve cell bodies that are related in function and in the location of the stimulus that drives them and that form a column oriented perpendicular to the plane of the cortex.	The ocular dominance and orientation columns of the visual cortex
	In the spinal cord, a group of functionally related nerve cell bodies that form a longitudinal column extending through part or all of the length of the spinal cord.	Clarke's column
Layer, lamina (laminae), stratum (strata)	A group of functionally related cells that form a layer oriented parallel to the plane of the larger neural structure that includes it	Layer IV of cerebral cortex, the stratum opticum of superior colliculus
Tract, fasciculus (fasciculi), lemniscus (lemnisci) (*fasciculus* is Latin for "bundle")	A bundle of parallel axons in the CNS	Optic tract, corticospinal tract, medial longitudinal fasciculus, fasciculus gracilis
Funiculus (funiculi) (Latin for "cable")	A group of several parallel tracts or fasciculi	Anterior, posterior, and lateral funciculi of spinal cord
PNS STRUCTURES		
Ganglion (ganglia)	A group of nerve cell bodies located in a peripheral nerve or root; it forms a visible knot	Dorsal root ganglia, trigeminal ganglion
Nerve, ramus (rami), root	A peripheral structure consisting of parallel axons plus associated cells	Facial nerve, ventral roots of spinal nerves, gray and white rami of spinal nerve roots

corticospinal tract and medial lemniscus). A group of several tracts or fasciculi is called a *funiculus* or, in certain cases, a *system*. In the PNS, collections of cell bodies form a *ganglion* (pl. *ganglia*), which may be either *sensory* (dorsal root, cranial nerve) or *motor* (visceromotor or autonomic); and axons make up *nerves*, *rami*, or *roots*.

Neurons may also be classified on the basis of functional characteristics. A neuron that conducts signals from the periphery toward the CNS is called *afferent*; one that conducts signals in the opposite direction is called *efferent*. Neurons with long axons that convey signals to a distant target are called *projection neurons*, whereas neurons that act locally (because their dendrites and axon are limited to the vicinity of the cell body) are called *interneurons* or *local circuit cells*.

Neurotransmitter specificity also can be used to describe neurons and their axons. For example, cells that contain the neurotransmitter *dopamine* are called *dopaminergic* neurons. The neurons whose axons form the corticospinal tracts produce the neurotransmitter *glutamate*, and they and their axons are consequently *glutaminergic*.

The lines between categories based on shape, projection type, or transmitter type are not as clear as noted above. For example, most neurons will not resemble the "ideal" multipolar cell. In addition, neurons may overlap several categories of classification. In practice, reference to ganglia, nuclei, and tracts commonly uses a blend of these terms. For example, *dorsal root ganglion* cells are *pseudounipolar* (their shape), *sensory* (type of input), *afferent* (information conveyed toward CNS), and many are *peptidergic* (they contain peptides such as substance P).

ELECTRICAL PROPERTIES OF NEURONS

The electrical characteristics of neurons are fully discussed in Chapter 3. Some of these concepts are briefly introduced here in relation to features of the cell membrane.

Nerve cells are electrically charged relative to the fluids bathing them. This is not a unique feature, as virtually all cells in the body have an electrical charge. However, nerve cells are unique in their ability to *manipulate* the flow of charges across the cell membrane of other neurons or effector cells.

The plasma membrane of a nerve cell controls ion transport so that Na^+ and Cl^- are more concentrated outside the cell than inside it (Fig. 2-6). The interior of the cell ends up with a relative excess of negative charges, so a voltage exists across the cell membrane. This voltage is called the *resting membrane potential*; in a typical neuron it has a value of about –70 mV.

Ion Channels The voltage across the neural cell membrane is manipulated by the opening and closing of *ion channels* in the membrane. Each kind of channel permits the flow of one species of ion, while largely excluding others (Fig. 2-7). An ion channel is a transmembrane protein or protein complex that has a central pore that is selective for the ion in question. When the channel is in the "open" conformation, the ion can flow through the pore; when the channel is in the "closed" conformation, the pore is blocked. Channels change from one conformation to the other in response to changes in the electrical or chemical environment (see Chapter 3). In some cases the transition is rapid in both directions. In other cases, the channel becomes temporarily *refractory* after changing to a particular conformation—that is, it pauses in that state for several milliseconds before being able to change back.

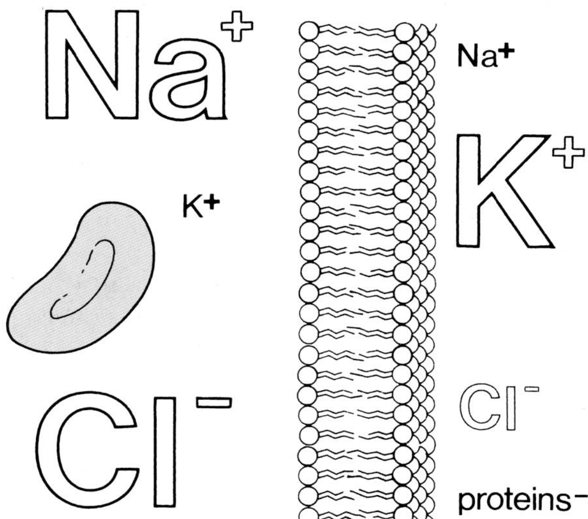

FIGURE 2-6 *Relative ion concentrations in the neuronal cytoplasm (right) versus the extracellular fluid (left). The concentration of K^+ is relatively high intracellularly, and the concentrations of Na^+ and Cl^- are relatively high extracellularly.*

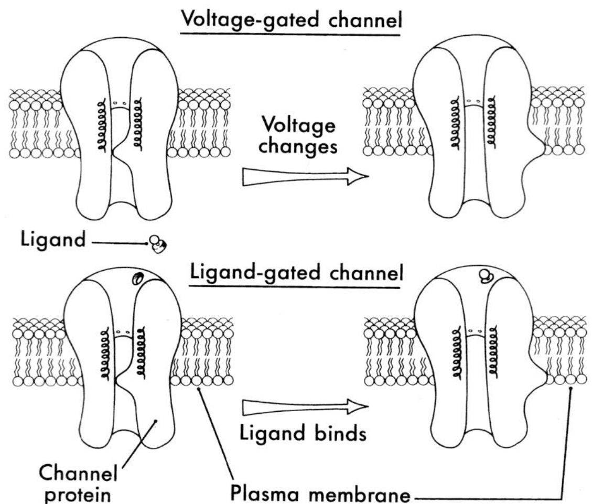

FIGURE 2-7 *Comparison of voltage-gated (upper) and ligand-gated (lower) ion channels. In a voltage-gated channel, alterations in membrane voltage result in a conformational change opening or closing the channel. In a ligand-gated channel, binding of a specific molecule such as a neurotransmitter induces a similar change.*

These channels fall into groups depending on the ions that pass through them (e.g., *sodium* or *potassium channels*). They also may be grouped according to the type of signal that triggers conformational changes (Fig. 2-7). *Voltage-gated ion channels* change their configuration in response to differences in the voltage across the cell membrane. *Ligand-gated ion channels* change their permeability in response to the presence of *neurotransmitters* (*chemical messengers*) (Table 2-4).

In some instances, a disease can selectively involve one type of ion channel. For example, in *Lambert-Eaton syndrome*, a tumor in the lung (*oat cell*, or *small cell carcinoma*) causes the production of antibodies to voltage-gated calcium channels necessary for the function of a synapse (see below). In many cases of *myasthenia gravis*, the patient's immune system produces antibodies to the nicotinic acetylcholine receptor, a ligand-gated channel found at the synapse between nerve and muscle cells. These antibodies result in pathologic destruction of these *neuromuscular junctions*, which in turn causes the muscle weakness characteristic of the disease.

Channel-Mediated Changes in Membrane Potential As mentioned, the nerve cell interior has a negative charge with respect to the extracellular fluid, so the resting membrane potential is negative. If the charge difference across the membrane is reduced, so that the membrane potential moves closer to zero, the membrane is said to have undergone *depolarization*. On the other hand, if the charge difference across the membrane is increased, making the membrane potential more negative, the membrane is said to be *hyperpolarized*.

TABLE 2-4

VOLTAGE CHANGES THAT CAN RESULT FROM CHANGES IN PERMEABILITY OF IONS

ION	DIRECTION OF CHANGE IN PERMEABILITY	EFFECT ON MEMBRANE VOLTAGE
K^+	Increase	Hyperpolarize (IPSP)
K^+	Decrease	Depolarize (EPSP)
Na^+	Increase	Depolarize (EPSP)
Na^+	Decrease	Hyperpolarize (IPSP)
Cl^-	Increase	Holds membrane near resting voltage
Cl^-	Decrease	Little change
Na^+ & K^+	Increase	Depolarize (EPSP)

All values assume typical values for the resting membrane potential and equilibrium potentials for each ion (see Chapter 3). Each cell's values will vary from these "ideals." At synapses, depolarizations are called excitatory postsynaptic potentials (EPSPs), and hyperpolarizations are called inhibitory postsynaptic potentials (IPSPs).

Changes in membrane potential are produced by the opening or closing of populations of ion channels. The change is local, affecting the vicinity of ion channels; a nerve cell membrane may thus have distinct zones of hyperpolarization and depolarization.

There are two functionally different kinds of membrane potential change. A *graded potential* can have any of a range of values, depending on the intensity and duration of the ion channel response and the preexisting membrane potential. It gradually decays back to the resting potential as the membrane restores its resting state. Graded potentials result from the activity in a single kind of ion channel or a group of different kinds of channels. An *action potential*, in contrast, is a large, abrupt spike of depolarization and repolarization that is stereotyped in magnitude and shape and is able to propagate itself along the membrane. It results from the opening and closing of voltage-gated sodium and potassium channels that are found only on the axon. An action potential that is triggered anywhere on an axon travels the length of the axon; the resting potential is restored in its wake.

Changes in ionic permeabilities of the cell membrane are the basis for the reception, propagation, and transmission of information within the nerve cell.

The Sodium/Potassium Pump The resting potential of a neuron is maintained and restored by the activity of a *sodium/potassium pump* located in the plasma membrane. This pump uses ATP energy to actively expel Na^+ while actively importing K^+, thus maintaining the resting concentrations of these ions (Na^+ concentrated on the outside of the cell; K^+ concentrated inside the cell).

NEURONS AS INFORMATION RECEIVERS

Neurons collect, transform, and transmit information. Collection of information by the nerve cell is the first step in this chain of events. Neurons can receive input from other neurons or directly from the environment. *Sensory information* enters the nervous system by the latter of the two routes.

Sensory Neural Information *Primary sensory neurons* receive information from the environment; these include *photoreceptors*, *chemoreceptors*, *mechanoreceptors*, *thermoreceptors*, and *nociceptors*. Further information on these receptor types is found in the chapters describing sensory systems.

The process of converting sensory input into a form interpretable by the nervous system is *transduction*. Each type of sensory receptor transduces a physical stimulus into electrical or chemical changes, which then can be transmitted within the nervous system.

The rod and cone photoreceptors of the retina are specialized for transducing light energy (*photons*). As few as three photons (possibly even a single photon!) can be detected by a trained human observer. As a photon strikes the photoreceptor, it sets into motion a complex chain of events culminating in the closing of a large number of normally open sodium channels. As a result, the photoreceptor cell becomes *hyperpolarized*. This makes the photoreceptor unique among sensory cells, in that the membrane potential becomes more negative upon application of the stimulus, rather than more positive.

In humans, the taste and olfactory receptor cells mediate the two primary types of chemoreception. Both receptor types respond to the presence of specific chemicals dissolved in a solution. Also included in this category are receptors in the hypothalamus, which sense low blood glucose, low oxygen tension, or changes in blood pH; similar O_2 and pH receptors are found in the aortic sinus and the carotid body.

Mechanoreceptors transduce various qualities of physical force into electrical signals that are transmitted by sensory neurons. Such receptors are found in the vestibular, auditory, and somatosensory systems.

Other types of sensory receptors include *nociceptors*, which transduce painful stimuli, and *thermoreceptors*, which sense temperature changes in the skin and viscera.

Other Neural Information Although sensory neurons transduce external stimuli, most nerve cells rely on other neurons for input. In general, the direction of information flow in a neuron is from dendrites to soma to axon, but most cells also receive information at their cell bodies, and many receive information at the axon terminal. In all these cases, the reception of information is mediated by *synapses* (from the Greek word for "to clasp"). Synapses are the point of information transfer from one neuron to another.

NEURONS AS INFORMATION TRANSMITTERS

Synapses The *synapse* is the location at which a process of one neuron (usually an axon terminal) communicates with a second neuron or an effector (gland or muscle) cell. Although the synapse is actually a physiologic entity, it has traditionally been defined by its morphologic characteristics. In general, there are two broad categories of synapses, *chemical* and *electrical* (or *electrotonic*). The vast majority of synapses in the mammalian CNS are of the chemical type.

Chemical Synapses The most common type of CNS synapse involves an axon that is apposed to a dendrite or dendritic spine of a second neuron. The prototypical chemical synapse consists of a *presynaptic element*, a *postsynaptic element*, and the intervening space (the *synaptic cleft*), which is 20 to 50 nm wide (Fig. 2-8). The presynaptic element typically takes the form of an axonal bouton. The bouton contains mitochondria, which supply energy for synaptic processes, and also a prominent collection of *vesicles*, which contain the neurotransmitter that will be released into the synapse. These vesicles are often aggregated near sites on the presynaptic membrane called *active sites* (or *zones*), which are the sites of neurotransmitter release. Directly across the synapse is the postsynaptic membrane, which often appears thick and dark on electron micrographs (Fig. 2-8). Mitochondria are typically present in its vicinity.

As explained in detail in Chapter 4, communication across the synapse is mediated by the neurotransmitter stored in the presynaptic vesicles (Fig. 2-9). Neurotransmitter release is initiated by the arrival of an action potential, which depolarizes the presynaptic terminal. In the terminal, depolarization causes *calcium channels* to open. The resulting influx of Ca^{2+} into the

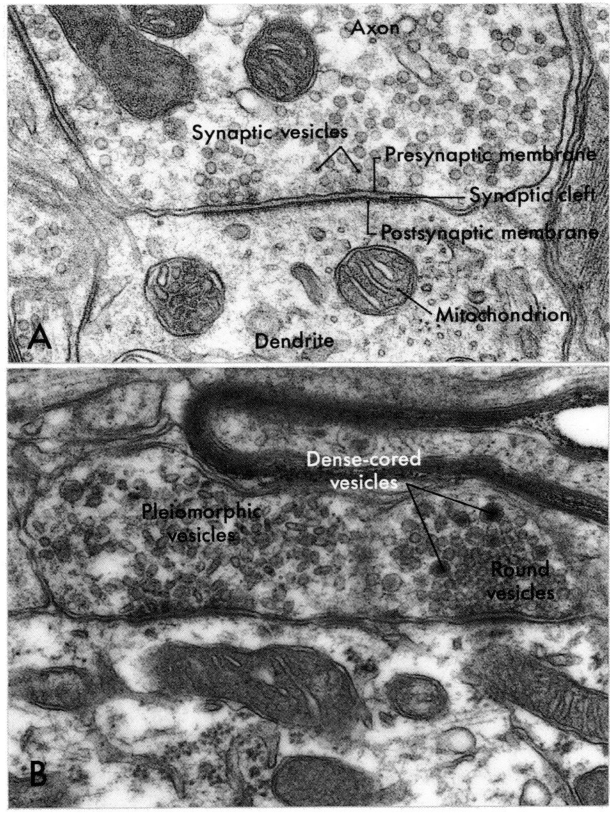

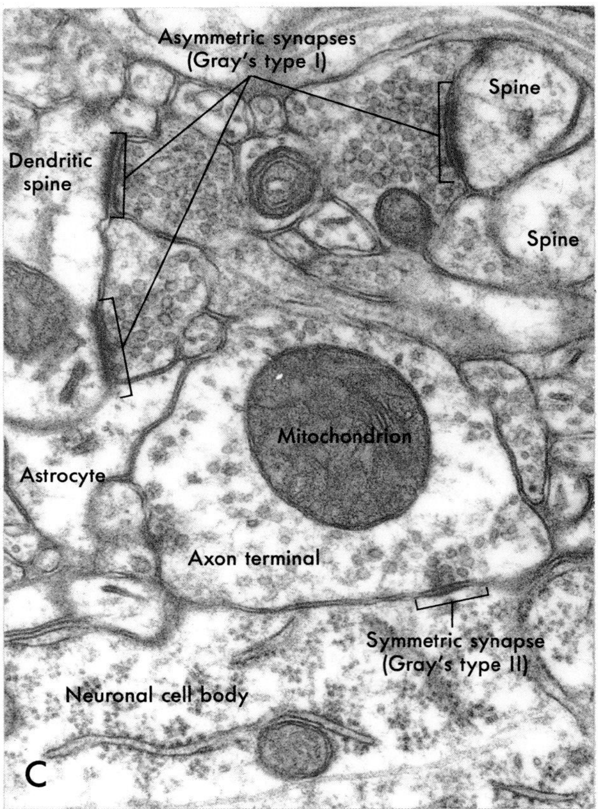

FIGURE 2-8 *Ultrastructural elements of a chemical synapse (A), and three principal forms of synaptic vesicles (B). Examples of asymmetric and symmetric synapses in visual cortex (C, courtesy Dr. Alan Peters).*

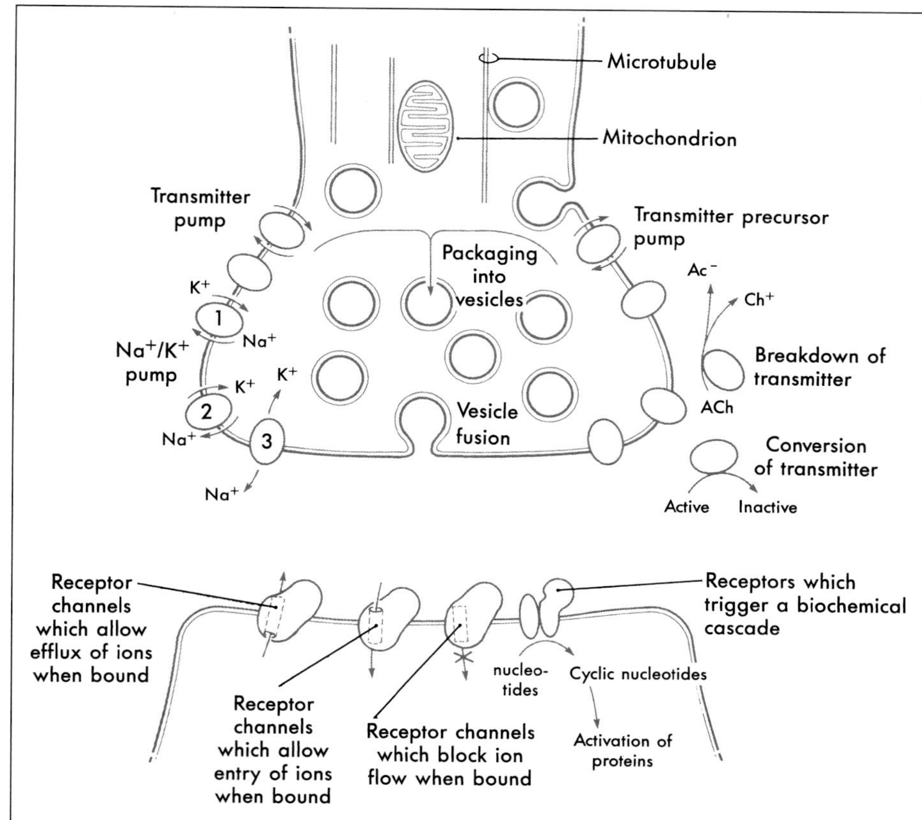

FIGURE 2-9 *Diagrammatic representation of the events occurring at a generalized chemical synapse. Numbers 1–3 indicate the sequence of events as K^+ is brought into the cell and Na^+ is expelled.*

cell causes synaptic vesicles to fuse with the presynaptic plasma membrane and release their neurotransmitter into the cleft (see Chapter 4 for more detail). The transmitter diffuses across the cleft to bind to specific *receptors* on the postsynaptic membrane, and this event triggers an electrochemical and/or biochemical change in the postsynaptic cell. This change represents the information as received by the postsynaptic cell.

Two functional properties of chemical synapses should be noted. First, they are *unidirectional*; that is, they transmit information only in the direction from the presynaptic cell to the postsynaptic cell. This directionality results because only the presynaptic cell releases the neurotransmitter, and only the postsynaptic cell expresses the receptor protein that will elicit the normal postsynaptic response to it. Second, the strength of the effect on the postsynaptic membrane is variable and depends on the amount of neurotransmitter released into the synapse. Each synaptic vesicle contains a fixed amount of neurotransmitter (called a *quantum*), so the amount of neurotransmitter released depends on the number of vesicles that fuse with the membrane in response to Ca^{2+} influx.

Chemical synapses were once visible only as the terminal boutons of nerve cell axons. In the early years of electron microscopy, two basic morphologic types of synapse became apparent. They were named *Gray's type I* and *type II* synapses. The characteristics of these synapse types are listed in Table 2-5 and illustrated in Figure 2-8C. At one time it was believed that type I synapses were excitatory in function and type II synapses were inhibitory. We now know that the excitatory or inhibitory function of a synapse depends on the receptors present on the postsynaptic membrane and cannot be reliably predicted from the ultrastructural characteristics of the presynaptic bouton. Nevertheless, the scheme is still a useful way to classify chemical synapses. For example, synapses using *acetylcholine* often have the Gray's type I morphology, whereas those using γ-aminobutyric acid (GABA) usually resemble Gray's type II.

Although the vesicles of Gray's type I and type II synapses differ in size and shape, in both cases the centers of the vesicles appear clear (electron-lucent). Vesicles with an electron-dense core are also seen in some synaptic endings; these dense-cored vesicles are generally thought to contain neuropeptides or serotonin as a neurotransmitter. This type of synapse is not included in the Gray's classification.

NEUROTRANSMITTERS

As we have seen, *neurotransmitters* are a means by which information is exchanged among nerve cells as well as between nerve cells and effector cells. Neurotransmitters are considered fully in Chapter 4 and are mentioned here briefly in relation to the structure of a typical neuron.

Neurotransmitters may be *biogenic amines* (e.g., acetylcholine, dopamine, and norepinephrine), *amino acids* (e.g., glutamate and GABA), *nucleotides* (e.g., adenosine), *neuropeptides* (e.g., substance P, cholecystokinin and/or somatostatin), or even *gases* (e.g., nitric oxide and carbon monoxide). Many of these neurotransmitters are stored in and released from synaptic

TABLE 2-5

MORPHOLOGIC CHARACTERISTICS OF GRAY'S TYPE I AND TYPE II SYNAPSES

Gray's type I synapses
Dense material present on postsynaptic membrane but not presynaptic membrane (so that the synapse is visibly asymmetric)
Synaptic cleft 30 nm wide
Synaptic vesicles round and large (30 to 60 nm), with clear centers
Synaptic region up to 1 to 2 μm long
Gray's type II synapses
Dense material present on both the presynaptic and the postsynaptic membrane (so that the synapse appears symmetric)
Synaptic cleft 20 nm wide
Synaptic vesicles oval, flattened, or pleiomorphic (variable) in shape
Synaptic region less than 1 μm long

vesicles in the axon terminal as described earlier, but in other cases, such as nitric oxide, nonvesicular release is clearly present.

Biogenic amines (acetylcholine) and amino acid neurotransmitters (GABA) are synthesized in the axon terminal, although the enzymes necessary for their synthesis are produced in the cell body and shipped to the terminal by axonal transport. It should be remembered that the axon lacks the machinery to synthesize proteins (or membrane lipids) and thus must obtain these materials from the cell body. Thus, *axonal transport* is always necessary to support synaptic function.

Disorders of Neurotransmitter Metabolism Disorders of neurotransmitter metabolism account for a large variety of neurologic and psychiatric illnesses, but in many cases the etiology is not well understood. This category of diseases is under intensive investigation, and three examples are briefly discussed here.

Parkinson's disease affects dopamine-synthesizing neurons located in an area of the brainstem known as the *substantia nigra*. For unknown reasons, these dopaminergic cells begin dying at an accelerated rate. The loss of dopamine results in a characteristic tremor and inability to properly control movement. Originally, therapy involved administering supplements of L-DOPA, a precursor for dopamine. This treatment increases dopamine synthesis by mass action but loses its effectiveness with time. Currently, therapy involves a combination of L-DOPA with *carbidopa*, which inhibits the enzyme L-aromatic amino acid decarboxylase. Because carbidopa cannot cross the *blood-brain barrier* (defined later), it decreases the metabolism of L-DOPA in peripheral tissues, making more L-DOPA available to the CNS for dopamine synthesis in the remaining neurons.

Manic-depressive disorder affects several million Americans and appears to be caused by imbalances in the phosphatidylinositol (PI)-linked neurotransmitter systems. An increase in PI turnover is a biochemical change triggered by some subcategories of acetylcholine, serotonin, norepinephrine, and histamine receptors. It is thought that a pathologic increase in PI turnover beyond normal levels may result in mood changes. The drug *lithium carbonate* decreases PI turnover and thereby stabilizes the patient's mood.

Alzheimer's disease (*presenile dementia of the Alzheimer's type*) affects more than 1 million Americans. A definitive diagnosis can be made only by postmortem microscopic examination of brain tissue. Alzheimer's disease is characterized by the degeneration of neurons in the cerebral cortex and hippocampus and the presence of pathologic structures called *neurofibrillary tangles* and *senile plaques*. Cortical cells normally receive terminals from *cholinergic* (acetylcholine-releasing) cells. In Alzheimer's disease, these terminals are lost, and the activity of choline acetyltransferase (the enzyme responsible for acetylcholine synthesis) in the cortex and hippocampus of diseased patients is extremely low.

GLIA

Unlike neurons, glia do not propagate action potentials, and their processes are not specialized to receive and transmit electrical signals. Rather, they provide neurons with structural support and maintain an appropriate microenvironment for neuronal function.

Glia account for most of the cells in the nervous system, and normal brain function depends critically on them. Glia in the CNS (Fig. 2-10; Table 2-6) are *astro-*

cytes and *oligodendrocytes*, derived from neuroectoderm, and *microglia*, derived from mesoderm. The analogous cell types in the PNS are the satellite cells, Schwann cells, and macrophages.

ASTROCYTES

Astrocytes occur throughout the CNS. They are highly branched, and many of their processes end in expansions called *end feet* (Fig. 2-10). Most of the free surface of neuronal dendrites and cell bodies, as well as some axonal surfaces, are covered with apposed astrocyte end feet. These feet join together to completely line the interfaces between the CNS and other tissues. The outer surface of the brain and spinal cord, where they meet the inner surface of the *pia mater* (the innermost of the meningeal membranes that enclose the CNS), is covered with a coating of joined end feet called the *glia limitans* (also called the glial limiting membrane). Similarly, every blood vessel in the CNS is jacketed by a layer of end feet that separate it from the neural tissue.

As shown in Figure 2-10, the astrocytes of gray matter, called *protoplasmic astrocytes*, differ in shape from the astrocytes of white matter, called *fibrous astrocytes*. Astrocytes can be distinguished immunohistochemically (for purposes of research and diagnosis) by the presence in intermediate filaments of a distinctive marker protein, *glial fibrillary acidic protein* (GFAP).

Structural Support and Response to Injury During development, astrocytes (in the form of radial glia) provide a framework for neuronal migration. In the adult brain, astrocytes frame certain clusters of neurons, for example, the barrels of the somatosensory cortex of rodents. In white matter, they also enclose bundles of unmyelinated axons.

Current research on astrocytes indicates that they secrete growth factors vital to the support of some neurons. In disease processes, astrocytes may secrete cytokines, which regulate the function of immune cells invading CNS tissue.

If injury to the CNS results in cell loss, the space created by the breakdown of debris is filled by proliferation and/or hypertrophy of astrocytes, resulting in the

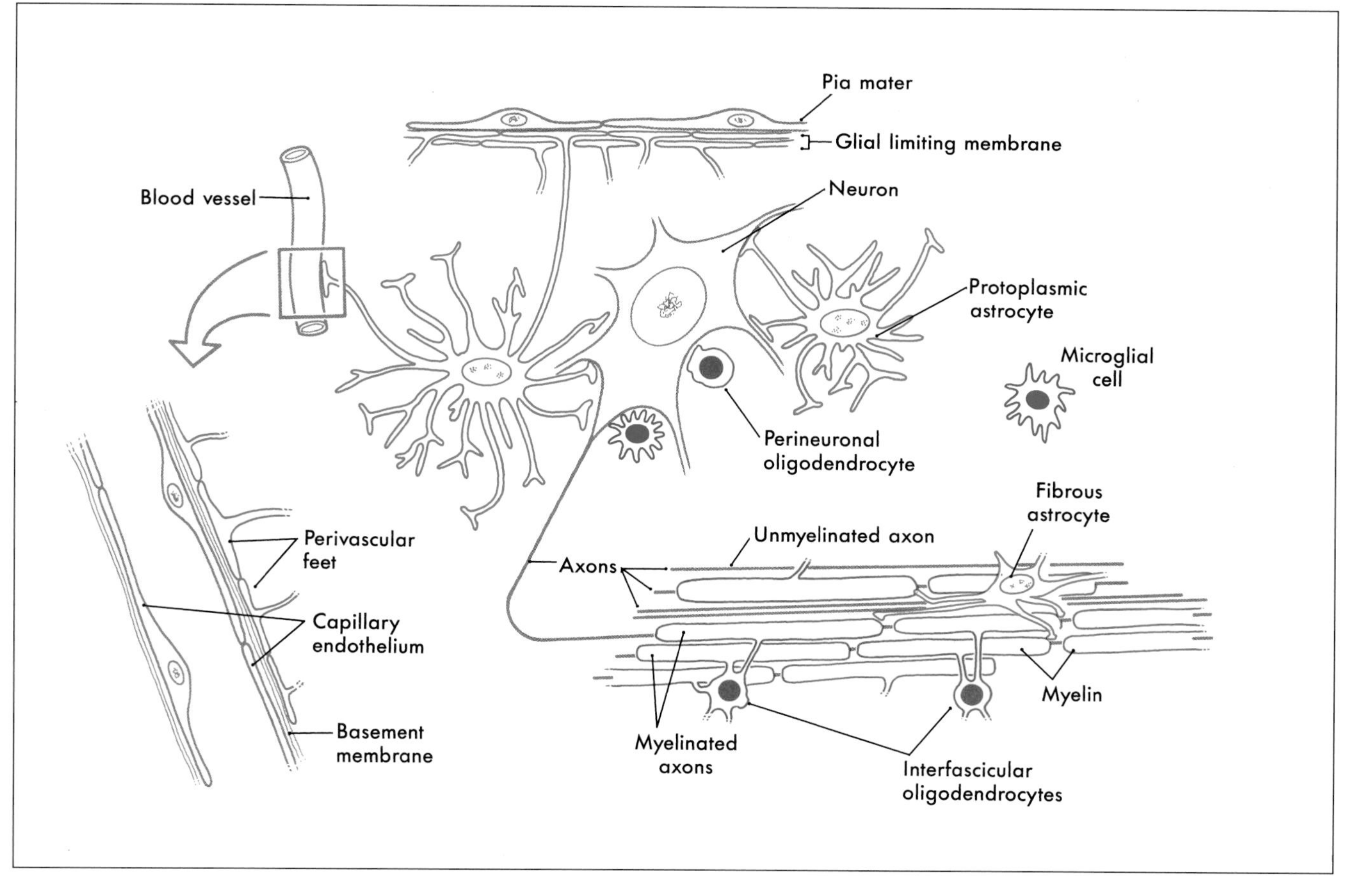

FIGURE 2-10 *Relationships of CNS glial cells to neuronal cell bodies, axons, blood vessels, and pia mater.*

TABLE 2-6

TYPES OF GLIA, THEIR LOCATIONS, AND FUNCTIONS

CELL TYPE	LOCATION	FUNCTIONS
CNS		
Astrocytes	Throughout the CNS; contact neuronal cell bodies, dendrites, and axons and form a complete lining around the external surfaces of the CNS and around CNS blood vessels. Gray matter astrocytes are called *protoplasmic* and white matter astrocytes are called *fibrous*	Mainenance of extracellular ionic environment; secretion of growth factors; structural and metabolic support of neurons
Oligodendrocytes		
Myelinating	Form myelin sheaths around CNS axons	Myelination
Satellite cells	Surround CNS neuronal cell bodies	Unknown
Microglia	Gray and white matter of CNS	Scavenging and phagocytosis of debris following cell injury and death; secretion of cytokines
PNS		
Schwann cells	Form myelin sheaths around myelinated axons and ensheath unmyelinated axons	Myelination; biochemical and structural support of myelinated axons
Satellite cells	Surround neuronal cell bodies in PNS ganglia	Unknown

formation of an *astrocytic scar*. That astrocytes retain the ability to proliferate in the mature brain (and thus are susceptible to events that disrupt the control of cell division) explains why the majority of CNS tumors are of astrocytic origin.

Environmental Control The ionic environment and pH of the extracellular space are buffered by astrocytes. These cells have ion channels in their membranes that are different from those in neurons. For example, potassium ions released from neurons during firing of an action potential are removed from the extracellular space by astrocytes via plasma membrane ion channels. Astrocytes are connected to each other by gap junctions and may act electrically as a syncytium.

Metabolism Astrocytes also participate in neurotransmitter metabolism. Their membranes have receptors for some neuroactive substances and uptake systems for others. Astrocytic uptake systems terminate the postsynaptic effect of some neurotransmitters by removing them from the synaptic cleft. For example, the amino acid neurotransmitter glutamate is taken up by astrocytes and is then inactivated by the enzymatic addition of ammonia to produce glutamine (catalyzed by the enzyme *glutamine synthetase*). Glutamine released from astrocytes can be reconverted to glutamate in neurons. This astrocytic pathway also detoxifies ammonia in the CNS.

Regional Heterogeneity Astrocytes vary biochemically between gray matter (protoplasmic astrocytes) and white matter (fibrous astrocytes), and from one region of gray matter to another. White matter astrocytes differ from gray matter astrocytes in terms of their ion channels, neurotransmitter receptors and uptake systems, and other special properties. For unknown reasons, *astrocyte tumors of particular types occur in characteristic distributions rather than with random frequency throughout the CNS*. For example, the malignant tumor glioblastoma multiforme develops most frequently in the frontal or temporal lobe of the cerebral cortex.

THE BLOOD-BRAIN BARRIER

In many tissues, solutes can pass freely between the capillary plasma and the interstitial space by diffusing through gaps between endothelial cells. In CNS vessels (with a few exceptions), the endothelial cells are sealed together by tight junctions, and solutes can reach the neural tissue only by passing through endothelial cells (Fig. 2-11). The resulting restricted exchange constitutes the *blood-brain barrier*. Water, gases, and lipid-soluble small molecules can diffuse across the endothelial cells; other substances must be carried across by transport systems, and their exchange is highly selective. The blood-brain barrier is of major clinical importance because it largely excludes many drugs from the CNS. CNS vessels are induced to form tight junctions by the surrounding jacket of astrocyte end feet.

The exchange across the endothelium of CNS vessels is also rendered selective by a reduction in pinocytotic transport. In most parts of the body, endothelial cells transport solutes nonspecifically from the plasma to the perivascular space in these vesicles. In CNS endothelial cells the vesicles are relatively few.

OLIGODENDROCYTES

Oligodendrocytes, the other major type of CNS glia, also occur in both gray matter and white matter (Fig. 2-10). The only proven function of oligodendrocytes is *myelination*, that is, the provision of an electrochemically insulating sheath around some axons in the white matter (Figs. 2-10 and 2-12). Other oligodendrocytes lie adjacent to and surround neuronal cell bodies in the gray matter, but the significance of this arrangement is not well understood.

A myelin sheath is a membranous wrapping around an axon that greatly increases the speed of conduction of action potentials along the axon. Large-diameter axons have thick myelin sheaths and high conduction velocities; smaller-diameter axons have thinner myelin sheaths and slower conduction velocities; and the smallest axons are unmyelinated and have the slowest conduction velocity.

Myelin is formed by a cell-cell interaction in which an axon destined for myelination is recognized by proteins on the oligodendrocyte surface. The oligodendrocyte responds by producing a flattened, sheet-like process that wraps repeatedly around the axon (Fig. 2-12). As the layers of membrane accumulate, all cytoplasm is excluded. The mature myelin sheath consists of layers of oligodendrocyte plasma membrane firmly pressed together. Cytoplasm remains only in the innermost and outermost turns of the oligodendrocyte process.

The myelin sheath surrounding an axon is not continuous along its entire length. Rather, the axon is covered by a series of myelin segments, each formed by an interfascicular oligodendroglial cell. The interruptions between segments are called *nodes of Ranvier* (Fig. 2-4B). Morphologic specializations at the nodes include a dense undercoating of the axonal membrane, as seen at the initial segment of the axon, and contact by an astrocyte process. The rapid ionic exchanges essential for generating the action potential and propagating it down the axon occur at these sites. The depolarization is then passively conducted along the axon (as a graded potential) to the next node. This method, *saltatory conduction*, is faster than having ionic exchanges occur continuously along the length of the axon.

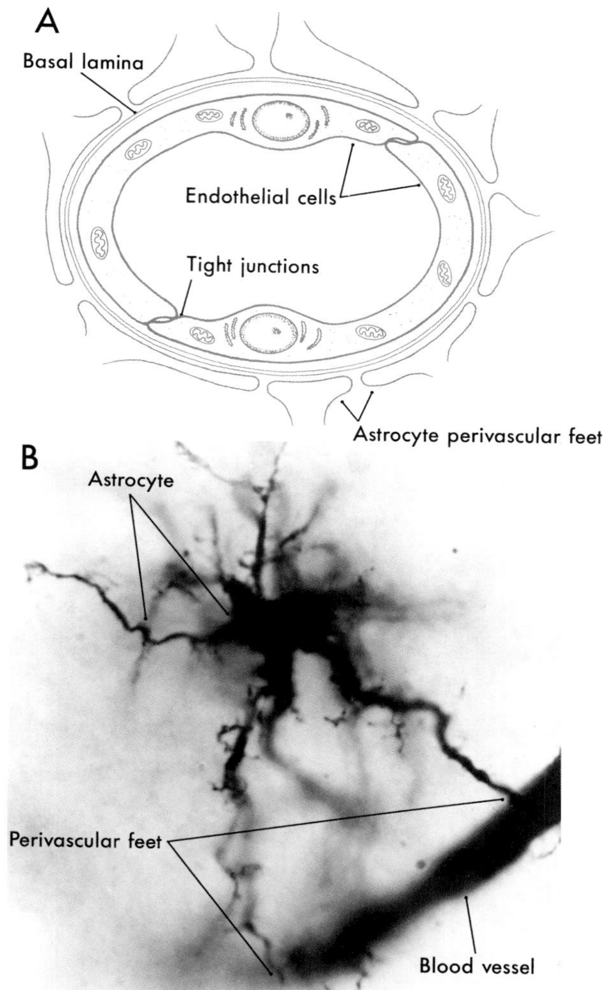

FIGURE 2-11 *The relationship of astrocytes to CNS blood vessels. Perivascular feet cover blood vessels of the CNS (A). Golgi-stained astrocyte with feet apposed to a blood vessel (B, courtesy of Dr. Jose Rafols).*

The segments of myelin between adjacent nodes of Ranvier are called *internodal segments* or *internodes*. Although the name *oligodendrocyte* means "few branches," some of these cells give rise to myelinating processes forming internodal segments on as many as 40 axons.

In *demyelinating* diseases, such as *multiple sclerosis* (*MS*), groups of oligodendroglia and their corresponding myelin segments degenerate and are replaced by

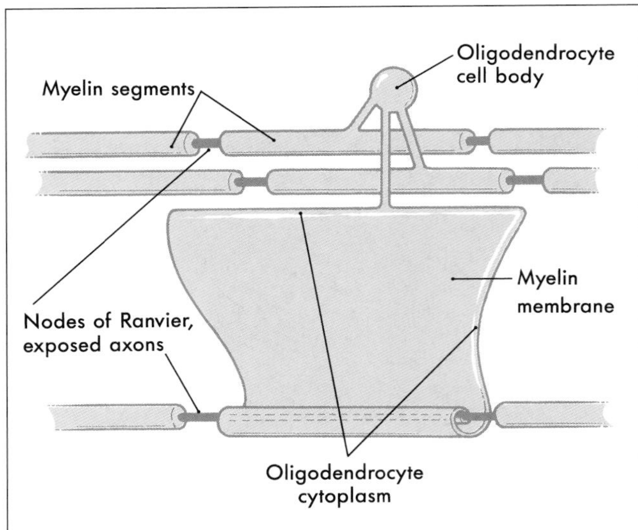

FIGURE 2-12 *Myelin sheath formation by processes of an oligodendrocyte. The cytoplasm of the oligodentrocyte is trapped in the edges of the cell membrane as it wraps around the axon. (Data from Butt and Ransom, 1989.)*

astrocytic plaques. This loss of myelin results in an interruption of the propagation of the action potential down these axons. The array of motor, visual, or general sensory losses in patients with MS reflects the locations of the demyelination.

MICROGLIA

The embryonic origin of the microglia is controversial, but the prevailing view is that they develop from blood cells of the monocyte-macrophage lineage, which migrate into the CNS during development. They make up about 1% of the CNS cell population (Fig. 2-10).

Like macrophages, microglia are able to become phagocytic scavengers when activated. When the CNS suffers injury, activated microglia migrate to the site of damage, where they proliferate and phagocytose cell debris. In tissue culture, microglia, like macrophages, produce cytokines such as interleukins. Microglia may also be able to act as antigen-presenting cells and participate in autoimmune disease processes.

SUPPORTING CELLS OF THE PNS

As mentioned earlier, the PNS contains supporting cells called *satellite cells* and *Schwann cells*, which are analogous to astrocytes and oligodendrocytes, respectively. Satellite cells surround the cell bodies of neurons in sensory and autonomic ganglia, and Schwann cells ensheath the axons in peripheral nerves (Fig. 2-13).

In the PNS, large- and intermediate-diameter axons have myelin sheaths, and the smallest-diameter axons are unmyelinated. Schwann cells produce these myelin sheaths and also envelop the unmyelinated axons. The myelin sheaths are similar to the CNS type, consisting of a tight spiral wrapping of fused plasma membrane. They are also formed similarly, by a Schwann cell that is attracted to an axon segment and wraps repeatedly around it to produce a compact sheath (Fig. 2-13). As in the CNS, cytoplasm remains only in the innermost and outermost layers of this wrapping (Fig. 2-13).

There are some differences between CNS and PNS myelin. Small pockets of cytoplasm known as *Schmidt-Lanterman clefts* are found at irregular intervals in PNS myelin. A *basal lamina* covers the external surface of the Schwann cell. This structure is formed by the Schwann cell and may help to stabilize it during the process of myelin formation. In addition, every internode on a PNS axon represents an individual Schwann cell, in contrast to the CNS, where oligodendrocytes send out numerous processes that each form a myelin internode. Unmyelinated axons in the PNS are enclosed in canals formed by Schwann cells (Fig. 2-13). These Schwann cells also are covered by a basal lamina.

External to the Schwann cell basal lamina, peripheral nerve fibers are covered by three connective tissue sheaths (Fig. 2-14). The innermost of these, the *endoneurium*, consists of thin, type III collagen fibrils and occasional fibroblasts between individual nerve fibers. At the second level, a distinctive sheath, the *perineurium*, surrounds each group (fascicle) of axons. The perineurium is composed of several concentric layers of flattened fibroblasts, which are unusual because they have a basal lamina and an abundance of pinocytotic vesicles (Fig. 2-14). Perineurial cells also are connected to each other by tight junctions. This forms a protective barrier against diffusion of substances into peripheral nerve fascicles. Lastly, the entire peripheral nerve is covered by *epineurium*, a dense connective tissue sheath of type I collagen and typical fibroblasts.

Tumors of peripheral nerve are usually of Schwann cell origin. The type called a *schwannoma* arises singly and, because it is encapsulated and does not include nerve fibers, is easily excised. The type known as a *neurofibroma* is usually multiple. Neurofibromas are generally difficult to remove because they are encapsulated and infiltrate along nerve bundles.

DEGENERATION AND REGENERATION

In the adult mammalian nervous system, neurons do not proliferate (except in the olfactory epithelium), and neuronal precursors (stem cells) are not present. Therefore, neurons lost through disease or trauma are not replaced.

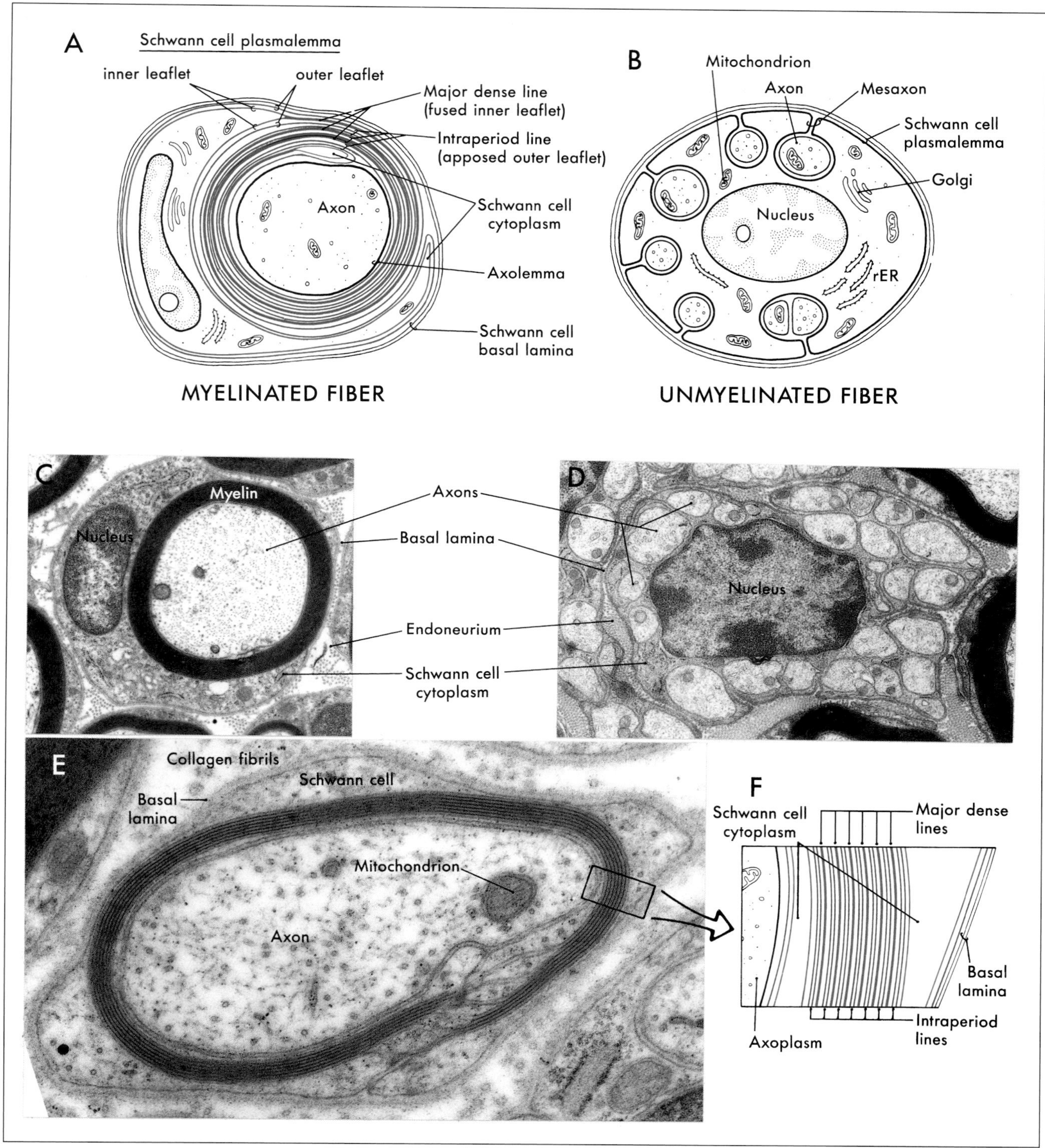

FIGURE 2-13 *Diagrammatic representation of myelinated (A) and unmyelinated (B) fibers. Schwann cells ensheath all peripheral nerve axons. Multiple wraps of the plasmalemma of a Schwann cell fuse to form compact myelin (see also Fig. 2-12). The inner leaflet of the plasmalemma (red) fuses to form major dense lines, and the outer leaflets (blue) of each adjacent wrap contact each other to form intraperiod lines (A). In an unmyelinated fiber, small axons occupy troughs formed by invaginations of the Schwann cell plasmalemma (B). Electron micrographs of a myelinated fiber (C) and an unmyelinated fiber (D) composed of a single Schwann cell supporting more than 20 axons. A small myelinated fiber sectioned through part of a Schmidt-Lantermann cleft reveals the membrane composition of myelin (E). The layers of the boxed segment of myelin in E are diagrammed in F.*

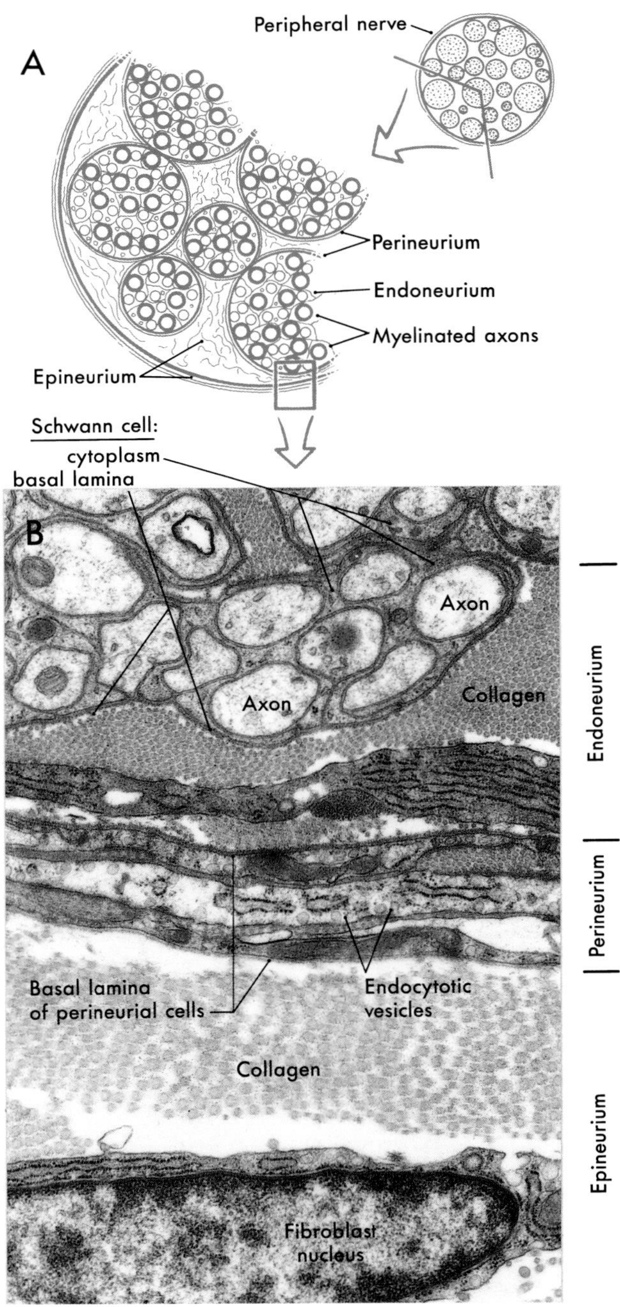

FIGURE 2-14 *The connective tissue layers of peripheral nerves (A). An electron micrograph corresponding to the area enclosed by the box in A reveals ultrastructural elements characteristic of each of the three sheaths (B).*

However, if the axon is damaged but the cell body remains intact, regeneration and return of function can occur in some circumstances.

The chance of axonal regeneration is best when a peripheral nerve is compressed or crushed but not severed. Although the crush kills the axons distal to the site of injury, the neuronal cell bodies, which are in the spinal cord or in sensory or autonomic ganglia, usually survive. These cell bodies may undergo chromatolysis in response to the trauma. Days to weeks later, *axonal sprouting* starts at the point of injury, and the axons grow distally. Meanwhile, in the distal part of the nerve, axons die and are removed by macrophages, but the Schwann cells remain. They lose their myelin but keep their basal lamina. Within these tubes of basal lamina, Schwann cells proliferate, forming cordons called *bands of Büngner*. These Schwann cells and basal lamina tubes guide the distally growing axonal sprouts. Regeneration depends on a variety of influences, including neurotrophins (neuronal growth factors) and the basal lamina. The growth rate of sprouting axons is about 1 mm/day.

In a crush injury, the proximal axon sprouts and distal bands of Schwann cells remain in their original orientation, so nerve fibers are lined up just as they were before the injury. Therefore, when the axons regenerate, they will find their original positions within the nerve and are more likely to accurately reconnect with their proper targets.

When a peripheral nerve is severed rather than crushed, regeneration is less likely to occur. Sprouting occurs at the proximal end of the axon, and the axon grows, but it may not reach its distal target. As axons grow from the proximal stump toward the distal stump, some may enter appropriate bands of Büngner and may be directed to their correct peripheral targets. These nerve fibers will become functional. Some axons may enter bands of Büngner that lead them to incorrect targets, so normal function does not return. Other axons may fail to enter the Schwann cell tubes, instead ending blindly in connective tissue to form a *neuroma*. Mechanical stimulation of these blindly ending sensory axons may be the cause of "phantom pain" in individuals with amputated limbs.

In axon tracts of the CNS, little or no regeneration can be expected, and in humans there is *no regeneration to a functional state*. Basal lamina guides, such as those found in the PNS, are not available. When a CNS axon is severed, the neuron mounts a sprouting response. Astrocytes hypertrophy and proliferate at the site of injury and fill any space left by the injury or by degeneration of the damaged nerve tissue. The responding astrocytes grow in a random orientation and form a scar rather than a pathway. Furthermore, astrocytes may not secrete adequate growth factors to sustain regrowing axons. The astrocytic scar appears to be a barrier rather than a guidance mechanism for axonal sprouts. It has been hypothesized that specific molecules present in oligodendrocyte myelin may also inhibit axonal regrowth. Eventually the axonal sprouts are retracted, and the loss of function associated with the severed pathway is permanent.

SOURCES AND ADDITIONAL READING

Butt AM, Ransom BR: Visualization of oligodendrocytes and astrocytes in the intact rat optic nerve by intracellular injection of lucifer yellow and horseradish peroxidase. Glia 2:470–475, 1989.

Casagrande VA, Hutchins JB: Methods for analyzing neuronal connections in mammals. *Methods in Neurosciences*, Vol 3: *Quantitative and Qualitative Microscopy*. Academic Press, San Diego, 1990, pp 188–207.

Dani JW, Chernjavsky, Smith SJ: Neuronal activity triggers calcium waves in hippocampal astrocyte networks. Neuron 8:429–440, 1992.

Gehrmann J, Matsumoto Y, Kreutzberg GW: Microglia: intrinsic immuneffector cell of the brain. Brain Res Rev 20:269–287, 1995.

Hertz L: Neuronal-astrocytic interactions in brain development, brain function and brain disease. Adv Exp Med Biol 296:143–159, 1991.

Kettenmann H, Ransom BR (eds): *Neuroglia*. Oxford University Press, New York, 1995.

Peters A, Palay SL, Webster H deF: *The Fine Structure of the Nervous System*, 3rd Ed. Oxford University Press, New York, 1991.

Shepherd GM (ed): *The Synaptic Organization of the Brain*, 3rd Ed. Oxford University Press, New York, 1990.

Stevens CF: The neuron. Sci Am 241:55–65, 1979.

Steward O, Banker GA: Getting the message from the gene to the synapse: sorting and intracellular transport of RNA in neurons. Trends Neurosci 15:180–186, 1992.

Unwin N: Neurotransmitter action: opening of ligand-gated ion channels. Cell 72:31–42, 1993.

CHAPTER THREE

The Electrochemical Basis of Neuronal Integration

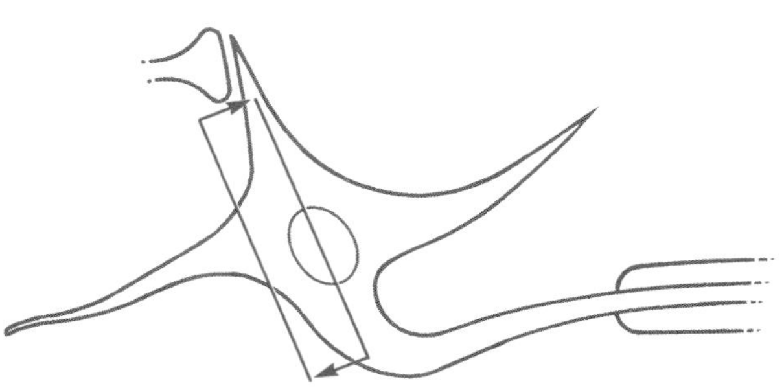

P. B. BROWN

The nervous system is the organ of behavior. Neural circuits process and store information to produce the full range of behavior, including voluntary and involuntary movement, homeostatic processes, and cognitive functions such as perception, emotion, and thinking. Sherrington, the founder of modern neurophysiology, called this information processing the *integrative action of the nervous system*. This diverse repertoire of activities is supported by cellular mechanisms that include molecular and electrochemical processes.

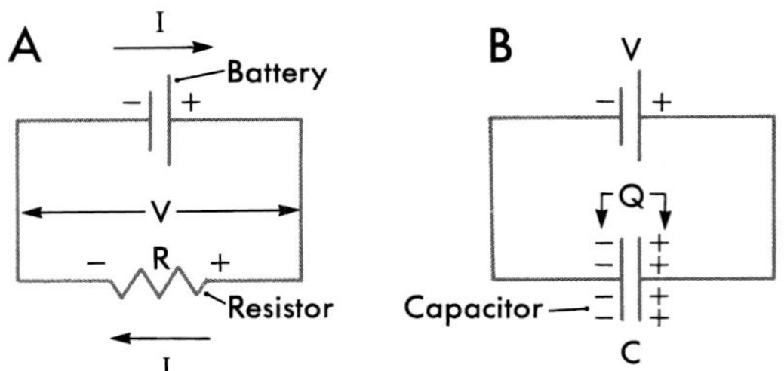

FIGURE 3-1 *Simple circuits showing a battery and resistor (A) and a battery and capacitor (B).*

OVERVIEW

Each of the billions of information-processing elements, the neurons, makes hundreds or thousands of connections with other neurons. Each neuron continuously produces output signals based on its many inputs. Because all these neural computations take place simultaneously, the information processing is highly *parallel*.

The synaptic connections among neurons are organized so that parallel streams of information are segregated in separate tracts and nuclei. This arrangement leads to a *localization of function* within certain combinations of nuclei and tracts. Information is relayed from one nucleus to another in a *serial* fashion, with each nucleus performing a specific set of manipulations on its inputs. This permits the extraction of more and more abstract information from the data provided by sensory receptors, in what is referred to as a *hierarchical* processing scheme.

The storage of information in the central nervous system (CNS) is accomplished in a *distributed* fashion, by modification of the properties of synaptic connections. Thus the storage of information involves altering the properties of many synapses, such that a particular input-output association is stored in many synapses, and an individual synapse is involved in storing many associations. Typically, many neurons perform similar computations on similar sets of inputs, so the loss of individual neurons has no observable effect on the operation of the nervous system. This *redundancy* is essential because many of the neurons we are born with die in the course of our lifetime and are not replaced.

Although different parts of the nervous system perform diverse operations, they do so by means of a limited repertoire of physiologic mechanisms, namely the generation and propagation of all-or-none action potentials, synaptic transmission, and the production of graded sensory and synaptic potentials.

ELECTRICAL CIRCUITS

The electrical activity of nerve cells is best understood in terms of electrical circuits. In a simple circuit consisting of a *battery* and a *resistor* (Fig. 3-1A), the battery is a *voltage source*, producing a voltage or *potential difference*, V, across the resistor. This voltage causes a flow of charged particles (current, I, carried by electrons in metals and by ions in solutions), such that *positive* current moves from the battery's negative pole to its positive pole through the battery, and from the positive pole to the negative pole through the resistor. This closed loop of current is called an *electrical circuit*. Positively charged cations flow in the direction of positive current; negatively charged anions and electrons flow in the opposite direction. Because the charge leaving any point in the circuit is equal to the charge entering it, there is no net transfer, accumulation, or depletion of charge.

The relationship among current, voltage (the term *voltage* is used interchangeably with *potential*), and resistance is expressed by *Ohm's law:*

$$V = IR,$$

where V is the potential difference across the resistor (*in volts*,V), I is the current through the resistor (in *amperes*, *A*), and R is the resistance of the resistor (in *Ohms*, Ω). Resistance is the tendency to impede the flow of current, similar to the resistance to flow in a blood vessel. *Conductance* (G) is the inverse of resistance:

$$G = I/R,$$

where conductance is expressed in units of *siemens* (*S*). A *conductor* is a resistor with a negligible resistance (very high conductance), and an *insulator* is a resistor with a negligible conductance (very high resistance). The lines connecting the battery to the resistor are conductors (e.g., wires), and the empty space surrounding the symbols is an insulator (e.g., the air surrounding a real circuit).

Although there is no net transfer or separation of charge in current through a resistor, charge separation does occur when a voltage is applied across a *capacitor* (Fig. 3-1B), which consists of two conductors separated by an insulator. When a battery is attached to a capacitor, positive charge leaves the positive pole of the battery and accumulates on the capacitor, repelling positive charge on the opposite side of the capacitor and causing an equal charge to enter the negative pole of the battery.

The only net charge transfer is on the two sides of the capacitor. The amount of charge, Q, stored in the capacitor is:

$$Q = CV,$$

where Q is expressed in *coulombs* (C), and capacitance, C, is expressed in *farads* (F). When a voltage V is initially attached to a capacitor that has any charge other than $Q = CV$ on it, charge must move; this is a current. In fact, current can be expressed as movement of charge per second: 1 ampere = 1 coulomb/second.

THE RESTING POTENTIAL

The plasma membrane separates two electrolyte solutions with different ionic concentrations. *Membrane channels* are formed by proteins resident in the membrane that have hydrophilic cores through which hydrated ions can move (Fig. 3-2A). Because of differences in the size and charge of this central pore, the channels are selective for the passage of certain ions and vary in their permeability to these ions. In the resting membrane, the channels are relatively permeable to K^+ and Cl^- but relatively impermeable to Na^+. In addition, the neuron contains proteins and amino acids to which the membrane is impermeable, and more of these molecules are negatively charged than positively charged.

For each ion, the concentration difference thus produces a driving force (the *diffusion potential*) on the ion. If an equal and opposite electrical potential is applied across the membrane, there is a net zero force on the ions, and the same number diffuse in each direction, for zero net ion flow (an *equilibrium*). This voltage is the *equilibrium potential*. The ionic gradient acts as a battery, with a voltage equal to the equilibrium potential and an internal resistance whose conductivity is proportional to the permeability of the membrane for that ion (Fig. 3-2A,B).

In the case of the neuron, three important ions are unequally distributed between the outside and inside of the cell membrane. These concentration differences are maintained by a membrane *pump*, which uses the energy from ATP to transport three Na^+ ions out of the cell for every two K^+ ions it transports into the cell. Because two cations are pumped in for every three pumped out, this is not an electrically neutral process, and the pump current results in a slight depolarization, which is added to the potentials produced by the ion batteries. Although Cl^- is not actively pumped, it too is unequally distributed across the membrane owing to the unequal distributions of Na^+ and K^+.

Each ion has its own "battery," with a voltage corresponding to the ion's equilibrium potential and with its own internal resistance corresponding to the membrane permeability. The equilibrium potential E_{ion} for a monovalent ion is calculated with the *Nernst equation*:

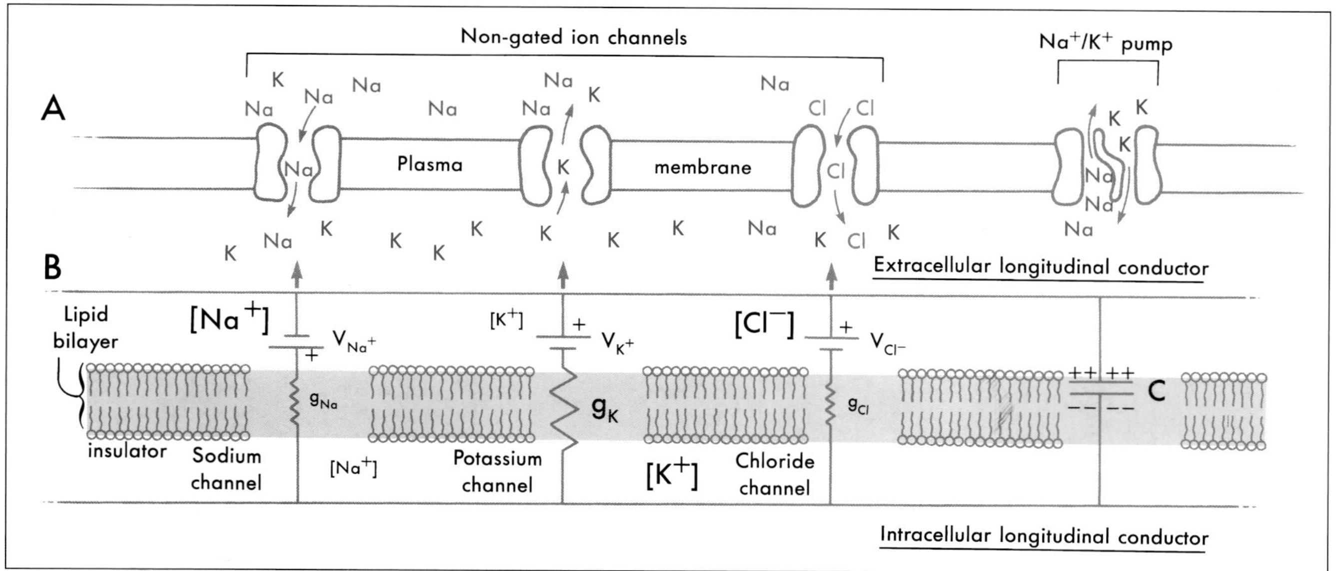

FIGURE 3-2 *Electrical equivalent circuit of nerve cell membrane. The lipid bilayer is an insulator separating the conducting extracellular and intracellular fluids. Na^+, K^+, and Cl^- channels penetrate the membrane, permitting flow of ions (A). Na^+ and Cl^- have highest concentrations outside, and K^+ has highest concentration inside the cell (B). These potential gradients determine the magnitudes and polarities of the three-ion "batteries." In the resting membrane, the K^+ channel has the highest permeability, so the K^+ potential dominates the membrane potential. The capacitance C, is composed of the insulating membrane and the conductive extracellular and intracellular electrolytes.*

$$E_{ion} = RT/zF \ln [ion]_{out}/[ion]_{in},$$

where R is the gas constant, T the absolute temperature, F the Faraday constant, and z the charge of the ion. Substituting and converting to base 10 at body temperature,

$$E_{ion} = 61 (\log_{10} [ion]_{out}/[ion]_{in}).$$

For Na^+, which has external and internal concentrations of 140 and 15 mM, respectively, the equilibrium potential comes to 60 mV, inside positive.

Neurons have lower K^+ concentrations outside than inside, with an equilibrium potential of about –70 mV. They also have a higher $[Cl^-]_{out}$ than $[Cl^-]_{in}$, with a Cl^- Nernst potential of about –60 mV.

There are thus three batteries with different potentials across the membrane. If the resistances of the Na^+ and Cl^- batteries are relatively high (insulators), their voltages do not contribute much to the membrane potential. If the resistance of the K^+ battery is relatively low (a conductor) the K^+ battery voltage dominates the membrane potential. In the resting state, the K^+ permeability is the highest of the three, so the transmembrane potential is close to the K^+ equilibrium potential. The Cl^- is distributed passively, so it is close to its equilibrium potential at the resting potential. The net resting membrane potential *RP* is calculated with the *Goldman-Hodgkin-Katz equation*:

$$RP = 61 \log_{10} \frac{p_K[K^+]_o + p_{Na}[Na^+]_o + p_{Cl}[Cl^-]_i}{p_K[K^+]_i + p_{Na}[Na^+]_i + p_{Cl}[Cl^-]_o}$$

where the *p's* are membrane permeabilities, the subscript *o* indicates "out," and the subscript *i* indicates "in."

The membrane is not in equilibrium because the ions are not at their equilibrium potentials, and in the absence of the Na^+/K^+ pump there would be a net flow of ions to equalize their equilibrium potentials. However, it is in a steady state with a zero net flow of current.

Intracellular and extracellular electrolytes are good conductors, and the lipid membrane is a good insulator, separating the extracellular and intracellular conductors. This makes the membrane a capacitor, and a charge is stored on the membrane that is proportional to the transmembrane voltage.

CABLE PROPERTIES OF MEMBRANES

If the conductance of the Na^+ channel is abruptly increased (Fig. 3-3A) so that it is greater than the conductance of the K^+ channel (Fig. 3-3B), then Na^+ begins to flow into the cell (Fig. 3-3C), discharging the membrane capacitance. The membrane potential moves toward the Na^+ equilibrium potential (Fig. 3-3D). The membrane potential shift is slowed by the membrane capacitance. The capacitor is initially charged negatively on the inside and positively on the outside, reflecting the polarity of the K^+ battery. As Na^+ flows into the cell, much of the current is used to reverse this capacitive charge. As the capacitor's charge is modified, the voltage

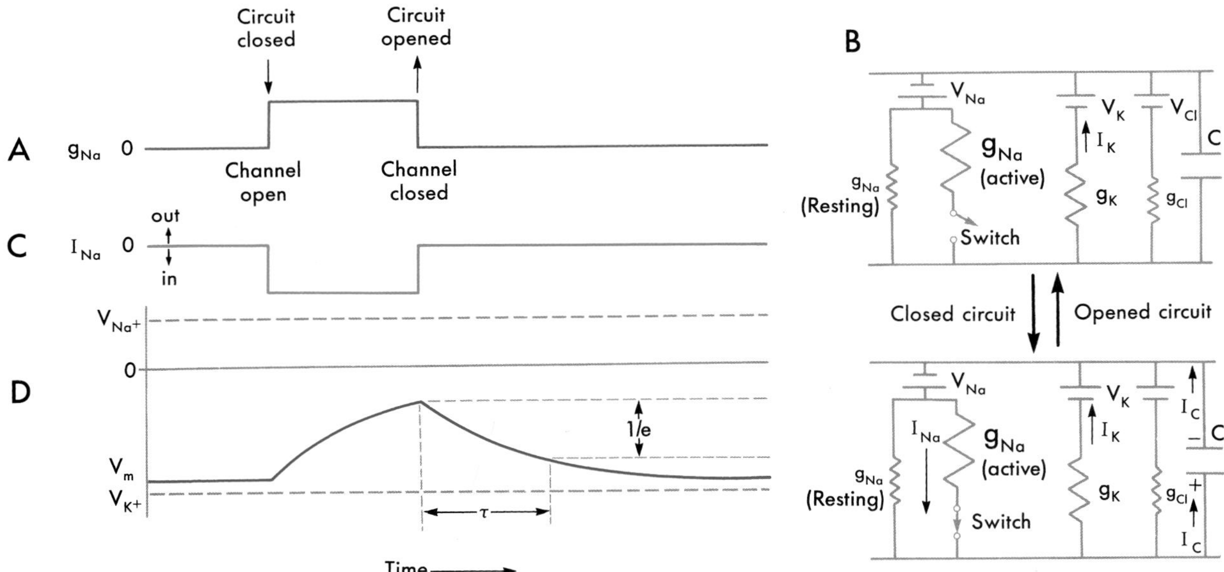

FIGURE 3-3 *The effect of a sudden increase in Na^+ permeability. Time course of permeability change (A). Electrical model of permeability change (B); switch on high-conductance path goes from open to closed (a closed switch is the same as an open channel, and an open switch is the same as a closed channel). Na^+ current (C). Membrane potential shifts from vicinity of K^+ potential toward Na^+ potential (D). Membrane capacitance slows the voltage change.*

changes exponentially with time, approaching the new steady state asymptotically. The rate of change is expressed as the time τ (tau) it takes for an increase or decrease by the factor e, the base of the natural logarithms. For any resistor-capacitor (RC) circuit,

$$\tau = RC.$$

This is called the *time constant*.

The tube-shaped axons and dendrites of nerve cells contain more than one ionic battery and associated resistor. Because the ion channels and pumps that generate the resting potential are distributed throughout the membranes of the axons and dendrites, a longitudinal strip of nerve fiber membrane can be thought of as a series of ionic batteries with associated resistors (Fig. 2-4A). The nerve fiber thus is like an undersea cable in which the internal conductor is connected to the conductive seawater by a leaky insulator. For this reason, the passive electrical properties of axons and dendrites are referred to as their *cable properties*.

A current pulse applied across the membrane produces a slower voltage pulse that spreads down the membrane, reaching a peak later and later the further it goes down the membrane (Fig. 3-4B). The peak amplitude declines exponentially as a function of distance (Fig. 3-4C). The distance in which it declines to $1/e \times$ the amplitude at the origin is referred to as the *length constant* or *space constant* λ (lambda). The length constants of dendrites and unmyelinated axons are a few tens or hundreds of micrometers. Consequently, graded potentials such as receptor and synaptic potentials cannot conduct more than a few hundred micrometers.

This *decremental* spread of current and voltage does not involve any permeability changes; therefore it is referred to as *passive*. A synonym for passive conduction is *electrotonic conduction*.

A decrease in the intracellular longitudinal resistance will cause the space constant to lengthen, and an increase in the transmembrane resistance will have the same effect. Similarly, a decrease in membrane capacitance will decrease the time constant. Any of these changes will cause the peak of the voltage pulse to propagate more rapidly and to decline in amplitude more slowly with distance. Mammalian nervous systems use all three effects to speed propagation of action potentials.

ACTION POTENTIALS

At receptor endings or synapses, a localized process produces a *generator potential*. This potential, referred to as a *graded potential* because it varies continuously in amplitude,

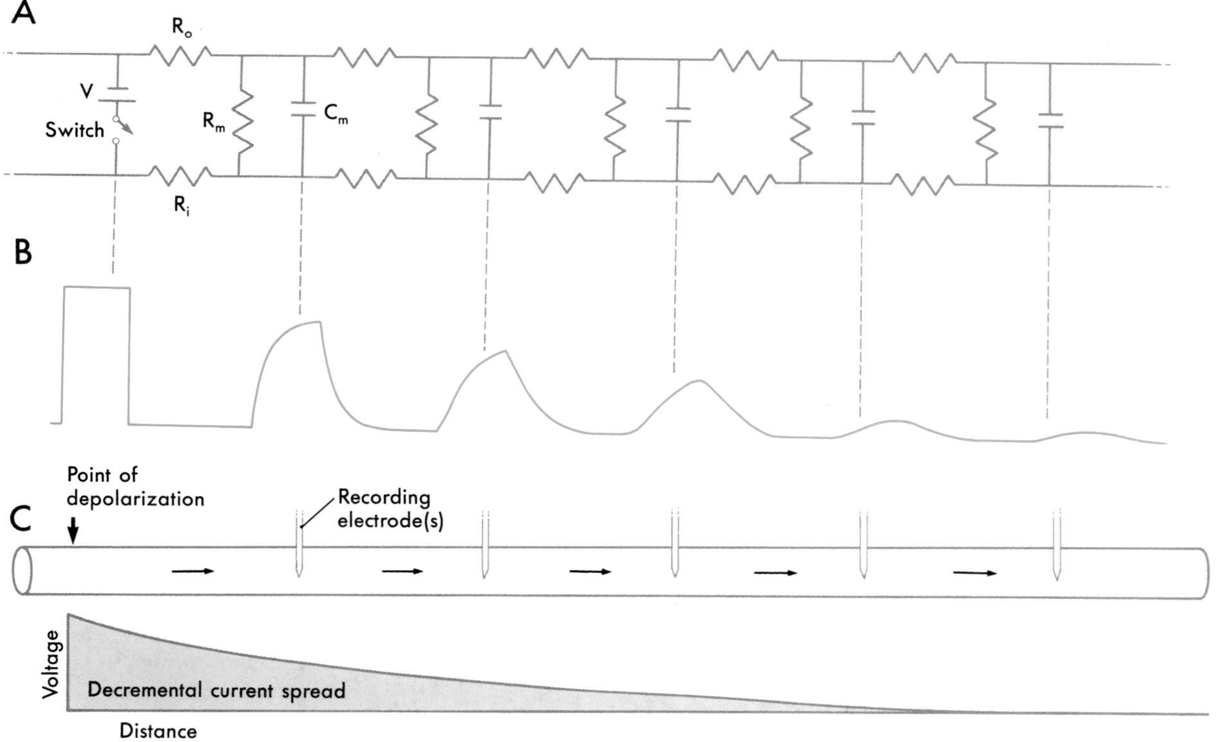

FIGURE 3-4 *Passive electrical properties of the membrane. Electrical model (A) were R_o = external longitudinal resistance, R_i = internal longitudinal resistance, C_m = membrane capacitance. Voltage pulse is applied at switch, resulting signals are recorded at capacitors (B). Pulse undergoes decremental conduction, decreasing in amplitude with distance from the switch (B,C). Pulse is also slowed progressively with distance.*

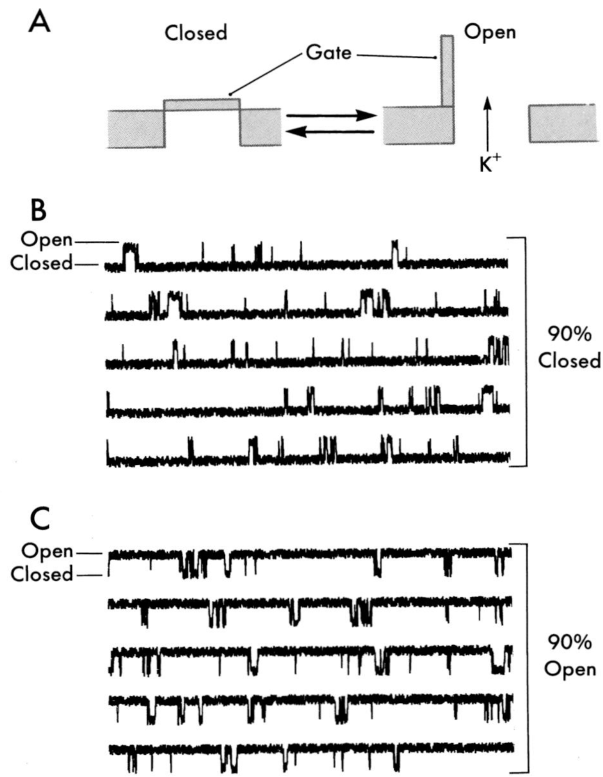

FIGURE 3-5 *A simulated voltage-modulated K+ channel. Closed and open states (A). Five channels, closed 90% of the time (B). Same five channels, open 90% of the time (C).*

passively spreads to an electrically excitable region of membrane called the *trigger zone*, where it may produce an *action potential* if it exceeds a *threshold* level. The propagation of all-or-none action potentials permits the transmission of information over much longer distances than would be possible for graded potentials, which are *local* (rapidly diminishing in amplitude with propagation distance).

Ion Channels The basis for most ionic flow across the lipid plasma membrane is the *ion channel.* Ion channels are intramembranous protein complexes, which are stabilized in the membrane by uncharged lipophilic amino acid residues. These channels have a hollow core or *pore* lined with charged, hydrophilic residues. The permeability to a given ion is determined by the charge distribution and diameter of the pore. Different kinds of channels differ in their permeabilities to specific ions and therefore show considerable ion specificity.

Gated channels open and close, switching rapidly between low ("closed") and high ("open") permeability states. *Voltage-gated* channels change configuration under the influence of depolarization or hyperpolarization, so that they spend a larger fraction of the time in the open or closed state.

The action potentials of most nerve cells arise from the behavior of voltage-gated Na^+ and K^+ channels. The K^+ channel is relatively simple: It has a single gate that responds to depolarization with a shift in equilibrium toward the open state (Fig. 3-5). This effect, called *activation* of the channel, results in a greater permeability to K^+ ions, which respond by flowing down their concentration gradient (that is, out of the cell). By contrast, the Na^+ channel has *two* voltage-sensitive gates: an activation gate and an inactivation gate (Fig. 3-6). The *activation gate* behaves like the gate of the K^+ channel, responding to depolarization with a shift in equilibrium toward the open state. The *inactivation gate*, by contrast, responds to

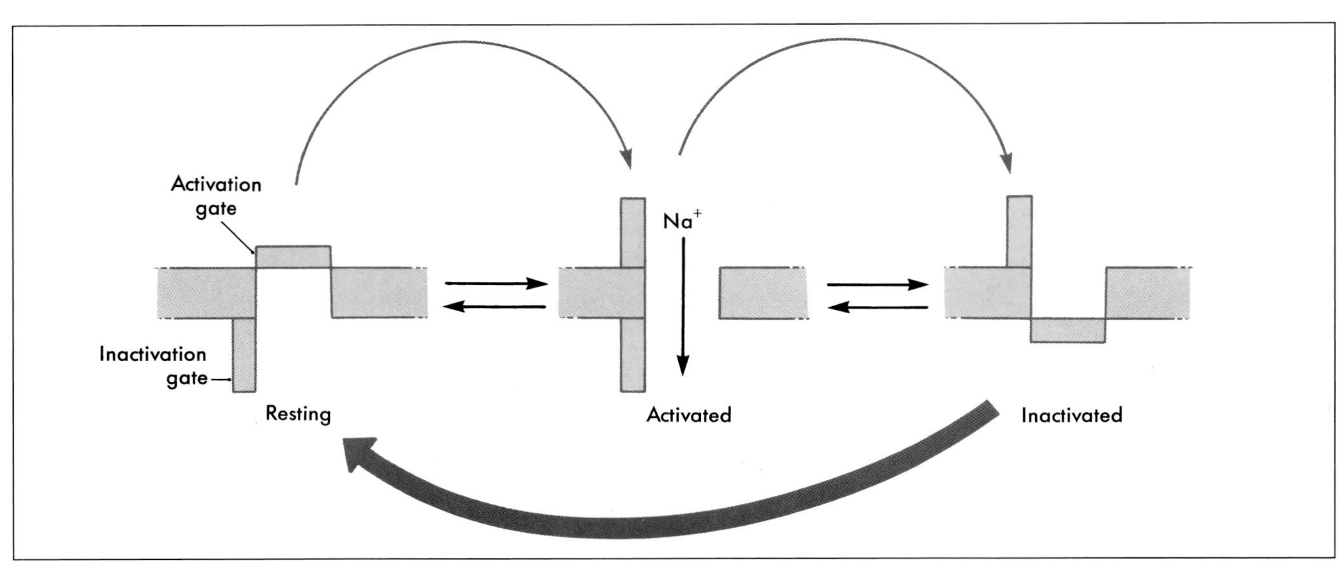

FIGURE 3-6 *A model of the voltage-modulated Na+ channel. Unlike the K+ channel, the Na+ channel has two parts and therefore three possible states.*

depolarization with a shift in equilibrium toward the *closed* state. The gates do not simply cancel each other out, however, because the inactivation gate tends to respond more slowly to depolarization than the activation gate. There is a delay between the opening of the activation gate in response to depolarization and the closing of the inactivation gate. Depolarization of the membrane thus will produce an initial brief phase in which both gates are likeliest to be open, after which the inactivation gates are likeliest to be closed while the activation gates are likeliest to be open. The three states of the Na^+ channel are called *resting* (activation gate closed, inactivation gate open), *activated* (both gates open), and *inactivated* (activation gate open, inactivation gate closed). The transition between each pair of states has its own equilibrium. When the channel is in the activated state, Na^+ is free to flow down its concentration gradient into the cell.

The sequence of events underlying an action potential is best described as two regenerative electrochemical cycles: the depolarization and repolarization cycles.

The Depolarization Cycle An action potential is a self-perpetuating all-or-none wave of depolarization that spreads along the nerve fiber membrane. An action potential is triggered by a depolarization of the resting membrane that reaches or exceeds a certain threshold amplitude; weaker depolarizations remain local and die out passively (Fig. 3-7). The existence of the threshold results from the behavior of the K^+ and Na^+ channels. A weak depolarization (curve 1 in Fig. 3-7A) causes mild activation of the Na^+ and K^+ channels. Because the Na^+ channels respond faster, there is an initial net influx of Na^+, which displaces some of the negative charge on the membrane capacitance, further depolarizing the membrane. However, the resting K^+ permeability is great enough to prevent this process from becoming self-perpetuating. In 1 msec or so, Na^+ inactivation and K^+ activation catch up with the Na^+ activation and overwhelm it, returning the membrane potential to its resting value, and hence returning voltage-dependent Na^+ and K^+ permeabilities to their resting values. This transient process is called a *local response*.

The *threshold* depolarization is defined as the level of depolarization that results in an action potential 50% of the time. A depolarization of this magnitude activates the Na^+ channels just strongly enough so that the inward flux of Na^+ is not automatically overwhelmed by the growing outward K^+ flux, but instead comes into balance with it. A slight perturbation one way or the other will determine the outcome; either the depolarization will die out (curve 2 in Fig. 3-7), or it will develop into an action potential (curve 3). A suprathreshold depolarization always causes an action potential (curve 4). In this case, the inward Na^+ flux equals the outward K^+ flux at the moment when the depolarization passes through the threshold value (open arrow in Fig. 3-7B).

Above threshold, the inward Na^+ current discharges the membrane capacitance fast enough to result in a depolarization that increases the Na^+ activation, keeping it ahead of the developing K^+ activation and Na^+ inactivation (Fig. 3-7A). Once this occurs, the process is self-regenerating, and the depolarization cycle (the *Hodgkin cycle*) develops explosively, to the point where Na^+ permeability exceeds K^+ permeability and the potential reverses, going positive inside and approaching the Na^+ equilib-

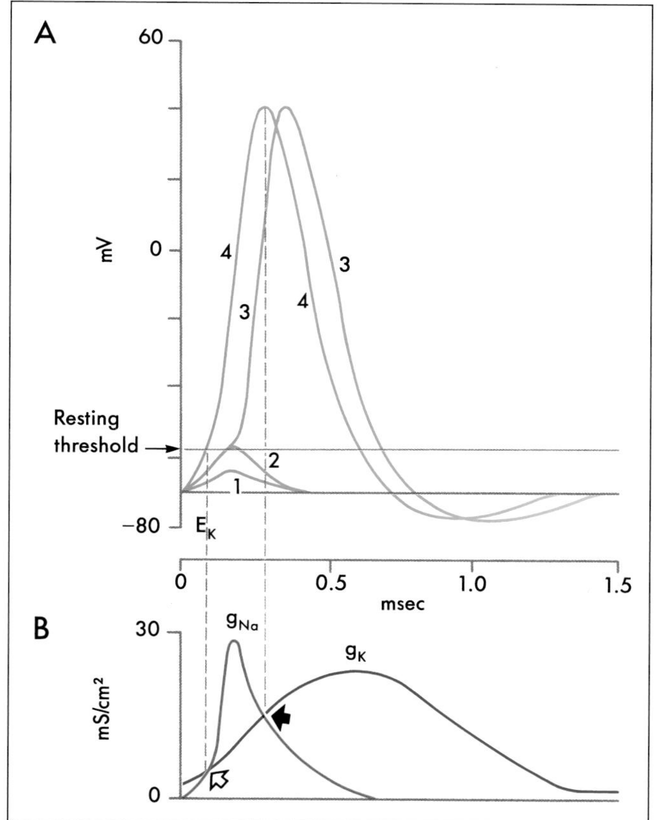

FIGURE 3-7 *Generation of the action potential. Four potentials resulting from different amplitude depolarization pulses (A). Curve 1, local response produced by subthreshold depolarization. Although an active process occurs, K^+ permeability remains greater than Na^+ permeability, and potential returns to resting level without development of an action potential. Threshold depolarization produces a local response (curve 2) half the time and an action potential (curve 3) half the time. At peak of active process, Na^+ permeability is equal to K^+ permeability. Suprathreshold depolarization (curve 4) always produces an action potential the same size as 3. Na^+ and K^+ permeabilities are equal at threshold (B, open arrow) and at the peak of the action potential (B, closed arrow). Na^+ and K^+ permeabilities during the action potential of curve 4 in A. Na^+ current is inward at all times, and K^+ current is outward at all times.*

rium potential. Much of the current is absorbed by the membrane capacitance in the active portion of the membrane, so that a positive charge develops inside the cell as the membrane potential reverses sign. Some of the current spreads down the axoplasm in both directions away from the active zone. Some serves to discharge membrane capacitance, and some crosses the membrane in the form of a K^+ current (because K^+ permeability is higher than Na^+ permeability in resting membrane). This is cable current, resulting in a passive spread of the depolarization caused by the action potential.

Repolarization At the peak of the action potential, inward Na^+ current is again equal to outward K^+ current (Fig. 3-7B, closed arrow), but only briefly, because Na^+ activation has passed its peak and K^+ activation and Na^+ inactivation are still increasing. This pulls the membrane back to resting potential. As the membrane approaches the resting potential, Na^+ and K^+ permeability begin to return to resting levels. If the resting potential is reached before permeabilities return to resting levels, the membrane potential undershoots slightly, approaching the K^+ equilibrium potential. This *after-hyperpolarization* persists until permeabilities have returned to their resting states. Once again, most of the current is absorbed by the membrane capacitance, recharging it.

The Excitability Cycle The resting threshold is the level of depolarization needed to produce action potentials with a probability of 50% when the membrane permeabilities all start off in their resting states. Once an action potential is generated, the membrane cycles through a stereotyped sequence of changes in threshold. During the rising phase of the action potential and part of the falling phase, additional inward current cannot influence the amplitude of the action potential (it is said to be *all or none*) or produce a second action potential (Fig. 3-8). The threshold is considered to be infinite, and the membrane is absolutely refractory (unresponsive). This phase of the excitability cycle is therefore called the *absolute refractory period.* At some point in the recovery cycle, it becomes possible to initiate a second action potential if a sufficiently large depolarizing current is used (several times the resting threshold current). During this period, the *relative refractory period*, the continuing Na^+ inactivation and K^+ activation represent an additional barrier that must be overcome, but a new Hodgkin cycle can be initiated. As the recovery from the first action potential proceeds, the threshold continues to come down toward the resting threshold, which is reached once the membrane potential and permeabilities have reached their resting states. Owing to residual Na^+ inactivation and K^+ activation, action potentials generated during the relative refractory period are smaller than those produced from a resting state.

Action Potential Propagation For the action potential to travel from one end of an axon to the other, the depolarization cycle at the active zone must depolarize adjacent resting membrane to threshold. Then the inactive membrane can develop an action potential, and its depolarization cycle can depolarize the next section of inactive membrane. This process, which proceeds in a wavelike fashion down the axon, is called *propagation* or *conduction* of the action potential. The depolarization of inactive membrane occurs as a result of the net inward current in the active zone (Fig. 3-9). Some of the inward current spreads longitudinally by passive spread and flows out of the portion of the membrane that is undergoing repolarization, recharging the membrane capacitance. Some also flows out of the resting membrane in the direction of propagation. This discharges the membrane capacitance, depolarizing the membrane to threshold. The ratio of the amount of current flowing into the inactive membrane to that needed to exceed threshold is referred to as the *safety factor* and must exceed 1 for propagation to succeed. In large axons, it can reach a value of 7, but in smaller ones such as those found at the ends of repeatedly branching terminal axon collaterals, the safety factor may approach 1 or even fall to less than 1 on occasion, resulting in a block of conduction.

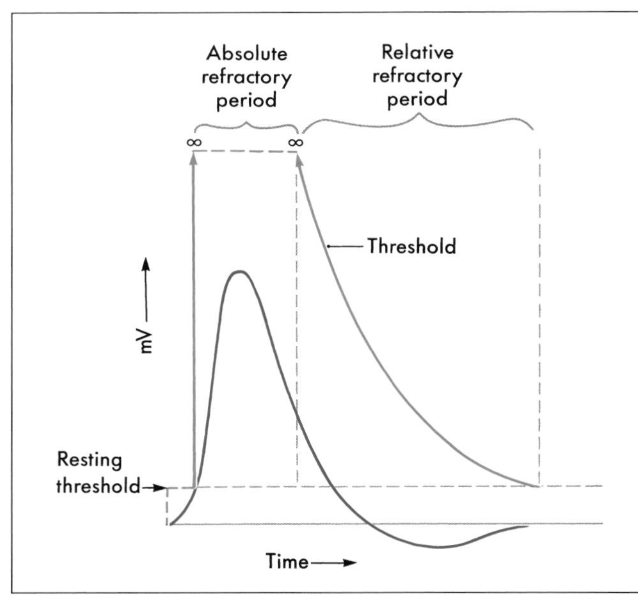

FIGURE 3-8 *Recovery cycle correlated with refractory periods. Upper curves are threshold changes during the recovery cycle and lower curve is an action potential. During the rising and part of the falling phases of the action potential, the threshold is effectively infinite, and the cell is absolutely refractory to depolarization. During the rest of the recovery cycle, the threshold is elevated, but a sufficiently strong depolarization can trigger an action potential. This is the relative refractory period.*

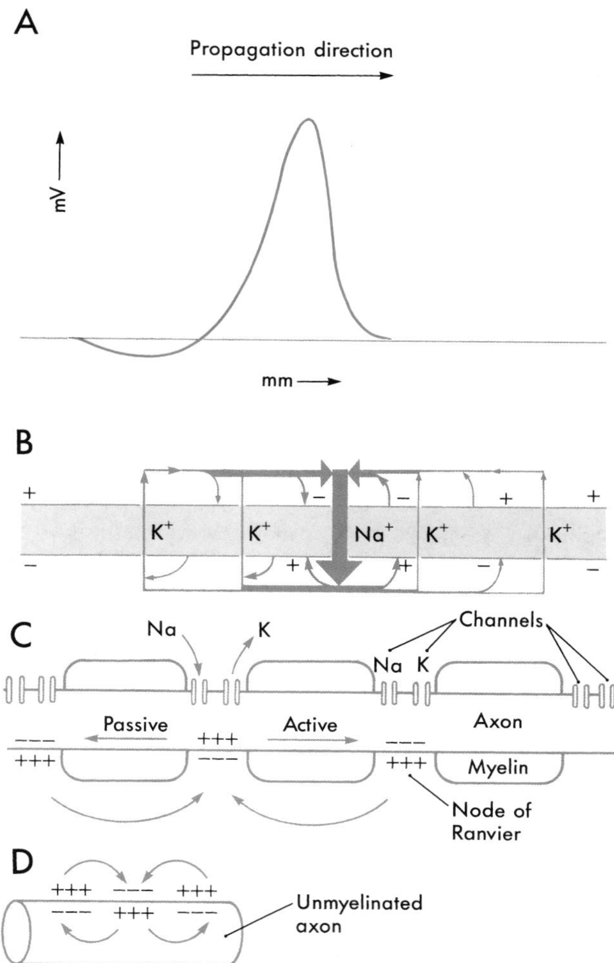

FIGURE 3-9 *Current paths during the propagation of the action potential (A) are shown in an electrical circuit of a membrane (B) and in myelinated (C) and unmyelinated (D) axons.*

The conduction velocities of unmyelinated axons increase with their longitudinal conductances, which are proportional to the squares of their diameters. Above 1 μm diameter, modest conduction velocity increases can only be achieved with large diameter increases. In vertebrates, axons more than 1 μm in diameter have myelin sheaths. Myelin increases the resistance between the intracellular and extracellular conductances, and capacitance is decreased by enlarging the gap between extracellular and intracellular conductors. These changes result in improved cable properties. Current from an active node bypasses the high-resistance myelinated portions of the membrane, flowing mostly through adjacent nodes of Ranvier (Fig. 3-9C). Action potentials therefore jump from node to node, a process called *saltatory conduction*.

Conduction velocity in myelinated axons is proportional to diameter because the internodal distances are proportional to diameter. The ratio of conduction velocity to diameter is 6 m/sec per μm outer diameter of myelin sheath.

If an action potential is artificially induced in the middle of an axon, for example by electrical stimulation or mechanical trauma (as happens when the "funny bone" is hit), action potentials course in both directions from the site of initiation. Conduction in the normal direction is referred to as *orthodromic*, and in the opposite direction as *antidromic*. The normally evoked action potential starts at one end of the axon. It does not "backfire" and send a stream of action potentials traveling backwards from its trailing edge. The membrane behind the active zone area is undergoing repolarization and is absolutely refractory or relatively refractory, with a safety factor of less than 1.

Conduction Block Substances that interfere with the operation of the voltage-gated channels can block conduction. *Tetrodotoxin*, produced by the Japanese puffer fish, blocks the voltage-gated Na^+ channels. Local anesthetics also work by blocking action potential propagation. It is also possible to block nerves by applying current. If a hyperpolarizing current is applied, it will oppose the depolarization wave produced by the propagating action potential. This is referred to as a *hyperpolarization block* or *anodal block* because it occurs at the anode of a bipolar pair of electrodes applied to a whole nerve. Interestingly, a conduction block can also be produced with a cathodal, depolarizing current. Prolonged depolarization holds the K^+ channels in an activated state and the Na^+ channels in an inactivated state.

Accommodation In some axons, a prolonged depolarizing generator potential causes K^+ activation and Na^+ inactivation to develop. The resulting increased threshold (which can also be thought of as decreased *sensitivity*, the reciprocal of threshold) is called *accommodation*. It is similar to depolarization block or the refractoriness found in the wake of an action potential. Accommodation can result in decreased rate of firing in response to a maintained stimulus, or complete inexcitability.

RECEPTOR POTENTIALS

Sensory receptors produce local, graded potentials whose amplitudes parallel the amplitudes of stimuli. Receptors produce their graded potentials by modulating the open times and hence permeabilities of ion channels, shifting the membrane toward or away from the equilibrium potentials for the ions they conduct. In some mechanoreceptors, channel permeability is modulated directly by mechanical stimuli, which deform the channel molecules by stretching the membrane. In other mechanoreceptors, the stimulus initiates a cascade of biochemical reactions that alters channel permeabilities indirectly. In the case of chemical receptors, the specific binding of a chemical substance (*agonist*) to membrane receptor molecules can modulate the permeability of an attached ion channel directly, or through a cascade of biochemical reactions.

Receptors As Transducers The role of a receptor is to convert physical energy or ligand concentration into receptor current. The conversion of one form of energy into another is called *transduction*. Receptors for different sensory systems are most sensitive to different forms of stimulus energy (or to different ligands, in the case of taste or olfaction). This specificity ensures that they respond best to the appropriate form of stimulus for the sensory system they serve (*the law of specific nerve energy*). The class of sensation to which stimulation of a receptor gives rise is known as its *modality*. Thus, light receptors in the retina are specific for light and subserve the visual modality. This modality specificity is maintained in the central connections of sensory axons, so stimulus modality is represented by the set of receptors, afferent axons, and central pathways that it activates. Modality representation (or "coding") is therefore a *place code*. The sensory system activated by a stimulus determines the nature of the sensation.

There is also specialization for different types of stimuli within a modality. For example, different cutaneous receptors respond best to different stimuli, evoking the different sensations of touch, itch, tickle, flutter, vibration, fast and slow, pain, cold, and heat. These different subclasses of sensations are referred to as *submodalities*. Submodalities are place coded in the same way as modalities; that is, the neurons and connections that serve a given submodality are kept appropriately segregated.

Ionic Mechanisms In many receptors, the depolarization that constitutes the receptor potential is produced by a generalized increase in the permeability of the membrane to small ions (e.g., Na^+, K^+, Cl^-). The steady-state potential for such a generalized increase in conductance is close to zero mV. As the stimulus is increased or decreased, the equilibrium shifts toward more or fewer open channels, producing a graded potential that reflects the stimulus amplitude.

Because the channels producing receptor potentials are not voltage-modulated, they cannot produce an action potential. For this reason, the receptor potential is *local*. Therefore, except for very small cells, whose outputs are less than 1 mm from their inputs, a receptor potential cannot be passed on unless it reaches an area of membrane containing voltage-modulated channels—a trigger zone—and initiates an action potential.

Action Potential Generation The amplitudes of graded potentials represent information, for example the magnitude of force applied to a cutaneous mechanoreceptor. The amplitudes of action potentials, however, cannot be used to represent information because they are independent of the amplitudes of generator potentials used to produce them.

The currents produced by receptor channels spread to nearby membrane that contains voltage-modulated channels (the trigger zone), and if the threshold for action potentials is exceeded, an action potential is produced. During a prolonged receptor potential depolarization, action potentials may be generated repetitively at a rate that is related to the depolarization. The action potential frequency "encodes" or represents the magnitude of the receptor potential, and hence the stimulus amplitude.

Adaptation The frequency of discharge may decrease during sustained presentation of a fixed-amplitude stimulus. Afferent axons that show a substantial drop in discharge rate during a maintained stimulus are called *phasic* or *rapidly adapting*. Those that show little or no adaptation are *tonic* or *slowly adapting*. Adaptation may result from accommodation, but it also may have other causes. Some sensory receptors have specializations that cause them to be rapidly adapting. For example, *pacinian corpuscles* have layers of fluid-filled compartments surrounding their transducer membranes, which can couple rapid movements only to the mechanically gated channels. These axons will only produce one or two action potentials at the onset of a skin displacement and another when the stimulus is removed. Some slowly adapting cutaneous receptors will continue to respond indefinitely to a maintained skin contact.

REPRESENTATION OF INFORMATION

For the CNS to process information, it must have some means of representing and manipulating it.

Place Codes Besides the place codes that represent modality and submodality, a place code is used to represent the location of a stimulus on a sheet of sensory receptors. One such sheet is the skin. Cutaneous sensory axons branch a few times in the skin, with all the branches of a particular axon terminating in the same type of receptor. The area of skin served by the receptor endings of a given axon, the area over which an appropriate stimulus will generate a response in the axon, is called the *receptive field*. In a peripheral nerve, the distribution of responding axons represents the location of a cutaneous stimulus.

The visual system has a sheet of receptors on the retina. The location of a stimulus in the visual field is encoded by the locations of the responding receptors on the retina. The auditory system has a sheet of receptors laid out along a frequency-tuned membrane, such that the location of a receptor along the membrane determines the auditory frequencies to which it will respond best. Thus the position of responding receptors along this membrane represents auditory frequency (pitch).

In all three sensory systems, the place code representing the locus of stimulation on the sensory sheet is preserved through the ascending pathway to the cortex by means of selective connections. This place code must not be confused with the place codes for modality or submodality. All submodalities for cutaneous sensation are found everywhere in the skin, for example, and they are distinguished by activity in *labeled lines*, groups of cells and axons that specialize in representing a specific type of stimulus (e.g., mechanical vs thermal). To represent location of a stimulus on the skin, connections of afferent axons are laid out in a spatially ordered fashion in each nucleus, so that their nearest-neighbor relations are the same at each end of the nerve or tract. As a result, the spatial patterns of responding cells in a nucleus resembles the spatial patterns of stimuli applied to receptors on the skin, and the nucleus can be thought of as a *map* of the skin. Electrical stimulation of sites in these maps will elicit sensations that are perceived as originating at the corresponding sites in the receptor sheet. Thus stimulation of a point on the somatosensory cortex would produce a vivid perception of a stimulus located on the skin rather than in the CNS.

Frequency Coding Although discharge frequency can represent stimulus amplitude, the relationship generally is not linear. In reality, two transformations are involved: from stimulus intensity to receptor potential amplitude and from receptor potential amplitude to frequency of discharge. These two can be combined in a diagram of frequency of discharge as a function of stimulus intensity (Fig. 3-10).

Some important generalizations can be reached from these graphs. Receptor potentials have thresholds, but they may be as little as one quantum of energy (e.g., one photon in the visual system). Action potentials also have depolarization thresholds. In some modalities, we can perceive stimuli that elicit as little as one action potential in one sensory axon. Of course, we cannot perceive stimulus intensities below the action potential threshold of the most sensitive axon being stimulated.

Both receptor potential amplitudes and action potential frequencies have upper limits; this is called *saturation*. We cannot perceive changes of stimulus above saturation for the axon with the highest saturation level.

The range of stimuli over which we can perceive variations of stimulus intensity from threshold to saturation is called the *dynamic range* for perception of that stimulus. Every neuron has a dynamic range for its inputs, from its threshold to its point of saturation. Our perceptual dynamic range falls within the limits of the dynamic ranges of all axons capable of responding to the stimulus.

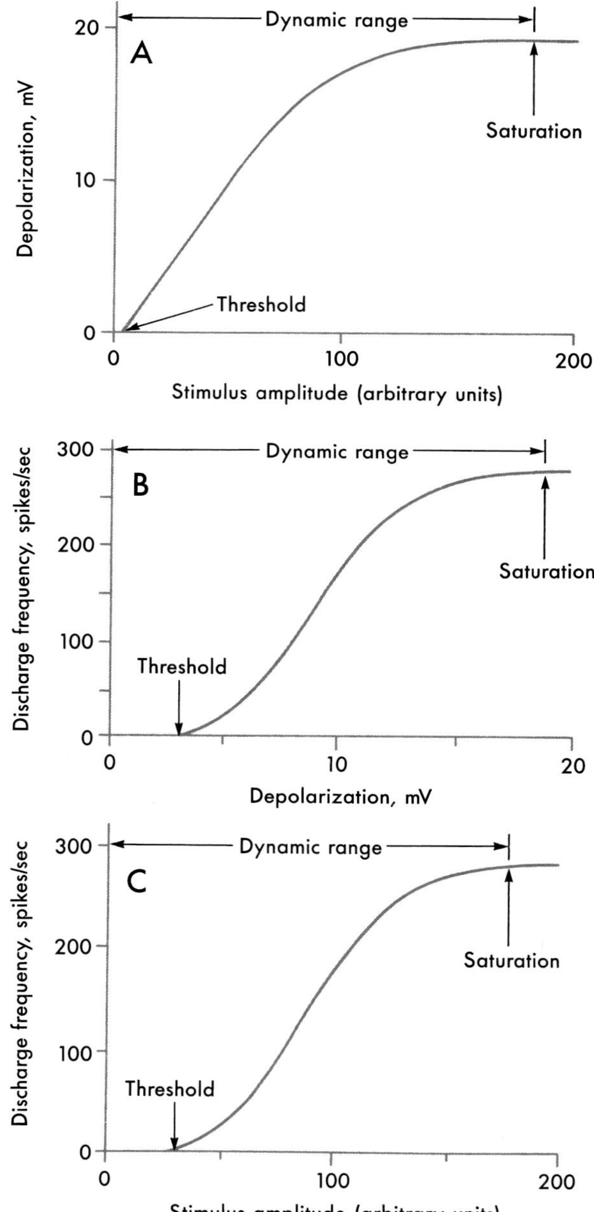

FIGURE 3-10 *Stimulus-response curves for a single axon. Receptor potential as a function of stimulus amplitude (A). Action potential frequency as a function of receptor potential (B). Action potential frequency as a function of stimulus amplitude (C).*

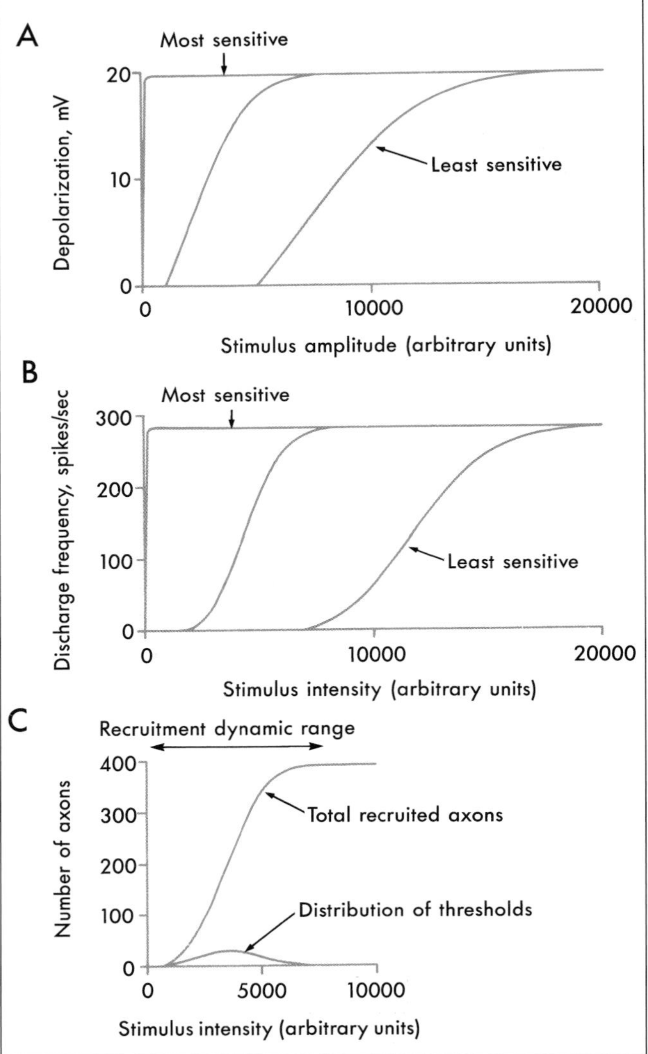

FIGURE 3-11 *Stimulus-response curves for a population of axons. Receptor potential as a function of stimulus amplitude for three axons, including least and most sensitive in population (A). Most sensitive is the same axon as Figure 3-10. Discharge frequency as a function of stimulus intensity for the same three axons (B). Distribution of thresholds for population of axons (bell-shaped curve), and cumulative number of axons recruited to discharge (S-shaped curve), both as a function of stimulus intensity (C).*

Recruitment Different afferent axons of the same modality can have different dynamic ranges, so increasing stimulus amplitude causes more afferent axons to discharge, a phenomenon called *recruitment* (Fig. 3-11). Thus recruitment can be added to discharge frequency as a way in which stimulus amplitude can be encoded. Recruitment makes it possible to encode and discriminate stimulus amplitude over a range larger than the dynamic range of a single axon.

CONDUCTION VELOCITY GROUPS

Axon Diameters In a peripheral nerve or in a tract, axon diameters can range from less than 1 μm to more than 20 μm (Fig. 3-12A). In general, axon diameters cluster into three or four groups.

Conduction Velocities Because different diameter nerve fibers have different conduction velocities (cv), synchronous electrical stimulation of all the axons in a nerve will result in their action potentials arriving some distance (x) from the stimulus after different conduction times (t): $t = x/cv$. Thus one *conduction velocity group* corresponds to each axon diameter group. This is evident when recording from an entire peripheral nerve. The extracellular wave forms of all the individual axons' action potentials add together to give a *compound action potential* (Fig. 3-12B). Because the action potentials of different-diameter axons arrive at different

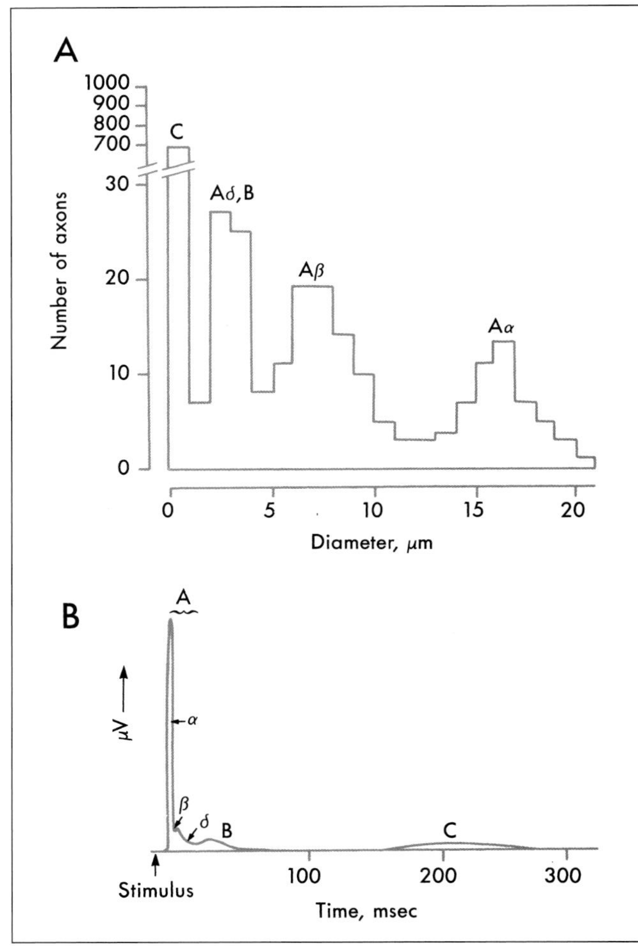

FIGURE 3-12 *The relation of axon diameter to conduction velocity for groups of axons in a nerve. Conduction velocity groups. Distribution of axon diameters for a hypothetical mixed (sensory and motor) nerve (A). Compound action potential recorded from the whole nerve, using a stimulus amplitude adequate to stimulate all the axons in the nerve (B).*

TABLE 3-1

CONDUCTION VELOCITY GROUPS

TYPE	GROUP	DIAMETER (MM)	VELOCITY (METERS/SEC)	FUNCTION
SENSORY AXONS				
Ia & Ib	Aα	12–22	72–132	motor afferent: annulospiral, Golgi tendon organ
II	Aβ	5–12	30–72	motor afferent: flower spray sensory afferent: touch, pressure, hair, joint, vibratory receptors
III	Aδ	2–5	12–30	sensory afferent: hair, free nerve endings (fast pain, temperature)
	B	1–3	3–15	sympathetic
IV	C	0.3–1.3	0.6–2.3	sympathetic, sensory afferent: slow pain, temperature, mechanoreceptors
MOTOR AXONS				
	α	3.5–8.5	50–100	α-motor neurons
	γ	2.5–6.5	10–40	γ-motor neurons

times, the compound action potential is longer than the individual action potentials, and it has multiple peaks that correspond to the different conduction velocity groups.

Although small-diameter axons far outnumber large-diameter axons in a nerve, the largest peak in the compound action potential is produced by the largest axon size class. This occurs because the contribution of each action potential to the compound action potential depends on the amount of current the action potential generates, which is proportional to the surface area of the membrane involved in the action potential. Large axons therefore contribute disproportionately to the compound action potential.

Functional Groupings Different conduction velocity groups have different biophysical properties. The larger the diameter of an axon, the lower its threshold for response to external electrical stimulation. Therefore if a whole nerve is stimulated with electrical stimuli of gradually increasing amplitude, the largest axons will be recruited first and the smallest last. The smallest axons are most sensitive to local anesthesia, which is why local anesthetics produce analgesia before complete anesthesia. The largest axons are most sensitive to hypoxia, as demonstrated by the unpleasant sequence of returning sensation after a limb has "fallen asleep."

The axons in different conduction velocity groups also subserve different functions (Table 3-1). Generally, the largest diameter sensory axons subserve submodalities with the lowest stimulus-energy thresholds. Two systems are used for classifying peripheral axons. First, these axons may be classified into *groups A, B,* and *C* on the basis of their contribution to a compound action potential in a mixed spinal nerve, where the group A axons produce the first peak and the group C axons produce the last peak (Fig. 3-12). Group A is divided into subgroups α, β, and δ, where α is fastest and δ slowest. Axons of touch receptors and muscle afferents contribute to the Aα and Aβ peaks. Axons of some hair receptors, thermoreceptors, and nociceptors contribute to the Aδ peak. Second, sensory fibers may be grouped on the basis of axon diameter and myelin thickness into *groups I, II, III,* and *IV.* Class I axons are large and heavily myelinated; classes II and III are progressively smaller and less myelinated; and class IV axons are small and unmyelinated. Class I is divided into subclasses Ia and Ib, where the fastest Ia fibers supply muscle spindles and the slowest Ib fibers supply Golgi tendon organs. The two schemes can be correlated because axon diameter and myelination determine conduction velocity, which in turn determines the position of a fiber's contribution to the compound action potential. For example, groups C and IV would be equivalent.

CHEMICAL SYNAPSES

There are two types of synapses, electrical and chemical. At electrical synapses, pairs of ion channels are precisely apposed on opposite sides of the extracellular gap to form structures called *connexons* (Fig. 3-13A). Ions can flow either way through connexons, which thus form a low-resistance pathway between the two cells. Some of the longitudinal current spreading passively in advance of the action potential in one cell flows through the connexon, producing a depolarization in the other cell. The threshold is exceeded, and an action potential is produced. In effect, the action potential can propagate between cells as though they were one. Electrical synapses are rare in mammals.

Chemical synapses share certain common characteristics. The presynaptic cell manufactures a compound called a *transmitter*, which is packaged in small membrane-bound *synaptic vesicles* in nerve terminals. When invaded by an action potential, the presynaptic terminal releases transmitter, which diffuses across the *synaptic cleft* and binds to receptor molecules in the postsynaptic membrane. The transmitter-receptor complex is short-lived, and free transmitter molecules are removed from the synaptic gap by a variety of mechanisms. As a consequence of transmitter-receptor binding, a modification of postsynaptic ion channel permeability occurs, causing a synaptic current, which produces depolarization or hyperpolarization of the postsynaptic membrane. Because of the specialization of presynaptic and postsynaptic elements, chemical synapses are unidirectional.

The release and diffusion of transmitter takes time, so there is a *synaptic delay* between the arrival of the presynaptic action potential and the onset of synaptic current in the postsynaptic neuron. This delay lasts about 0.5 to 1.2 msec. The time needed to remove synaptic transmitter sets a lower limit of about 1 to 2 msec on the duration of synaptic currents. The passive membrane properties of postsynaptic neurons slow down the postsynaptic potentials produced by this current, so the fastest potentials last about 5 msec. Thus postsynaptic potentials can be longer than the shortest intervals between action potentials in presynaptic axons (around 1 msec), which has important consequences for information processing at synapses.

Synaptic currents in a postsynaptic neuron can be evoked by activity in thousands of synapses. These currents interact, producing a net hyperpolarization or depolarization. This interaction is the basis for much of the information processing in the nervous system.

Presynaptic Mechanisms Synaptic vesicles accumulate near the release sites where they bind to the cell membrane (Fig. 3-13B, event 1). There is a large variety of transmitter molecules (e.g., see Table 4-2). A presynaptic neuron may make one or several transmitters.

The presynaptic action potential produces depolarization of the terminal membrane (Fig. 3-13B, event 2). Voltage-modulated Ca^{2+} channels open during depolarization, admitting Ca^{2+} ions into the terminal (event 3). The increase of intracellular Ca^{2+} enables *docking* of synaptic vesicles to release sites (event 4; see Chapter 4 for more detail). The vesicles fuse with the membrane and open into the extracellular space (event 5). The con-

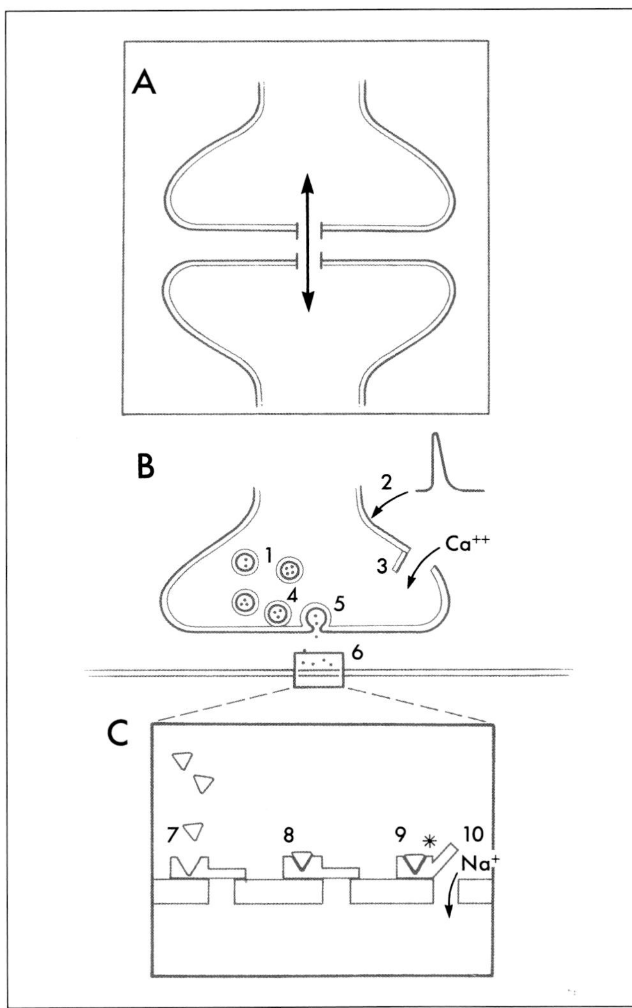

FIGURE 3-13 *General properties of synapses. An electrical synapse (A) has a connexon providing a low-resistance bridge between the two nerves . At chemical synapses presynaptic events (B) are (1) transmitter is synthesized and stored in membrane-bound vesicles, (2) action potential depolarizes the terminal, (3) depolarization causes Ca^{++} influx through voltage-modulated Ca^{++} channels, (4) Ca^{++} influx enables synaptic vesicle docking at release sites, (5) vesicles exocytose their neurotransmitter into the synaptic cleft, and (6) transmitter molecules diffuse across the cleft. Postsynaptic events at chemical synapses (C) are (7) receptor and transmitter dissociated, (8) transmitter-receptor complex, (9) activated complex causes channel to open, and (10) channel current.*

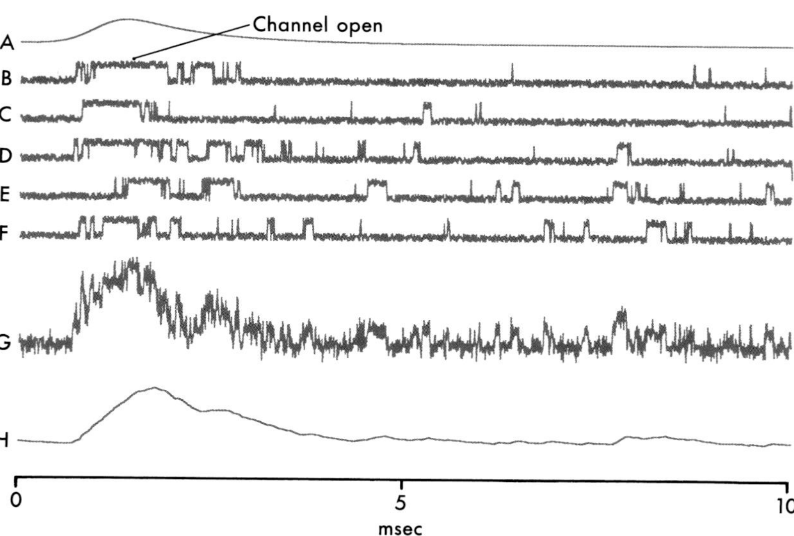

FIGURE 3-14 *Production of a synaptic potential by transmitter-modulated channels. Transmitter concentration in the vicinity of a postsynaptic membrane (A). Currents in five transmitter-modulated channels (B–F). Probability of the open state is proportional to transmitter concentration, so each channel goes from mostly closed in the resting state to mostly open during maximum transmitter concentration. Sum of channel currents for the five simulated channels (G). Membrane capacitance smooths the resulting synaptic potential, which is a slightly delayed reflection of transmitter concentration (H). With larger numbers of channels typically responding during synaptic transmission, the postsynaptic potential is even smoother.*

tents of one vesicle is the smallest unit (*quantum*) of transmitter release. The total transmitter released is equal to an integral number of quanta.

Once a vesicle has exocytosed its contents, the vesicle membrane is recycled for reuse. The raised intracellular Ca^{2+} is lowered quickly by an active pump that sequesters Ca^{2+} in the endoplasmic reticulum (ER). The Ca^{2+} is eventually removed from the ER to the extracellular space.

Typically, released transmitter that diffuses into the synaptic cleft (event 6) binds reversibly to postsynaptic membrane receptor molecules. Free transmitter is removed from the synaptic cleft by (1) *inactivation* by membrane-bound enzymes or (2) *reuptake* by endocytosis at the presynaptic membrane. The degradation product produced by enzymatic inactivation is also taken up by the presynaptic membrane, and is used for resynthesis of more transmitter. Recycled and resynthesized transmitter is then reincorporated into vesicles.

Transmitter-Receptor Interactions A three-state equilibrium is established between the receptor proteins in the postsynaptic membrane and transmitter in the synaptic cleft (Fig. 3-13C): dissociated (event 7), transmitter-receptor complex (event 8); and activated (event 9). In the simplest postsynaptic mechanisms, the activated receptor causes an associated channel to change configuration, producing altered permeability and synaptic currents (event 10).

When activated, the receptor protein may produce a change in the permeability of a directly coupled channel, or it may initiate a cascade of biochemical reactions. Such biochemical cascades can produce altered ion channel permeabilities, modulation of intracellular Ca^{++} stores, activation of enzymes, and even regulation of gene transcription.

Therefore it is convenient to categorize postsynaptic mechanisms in two classes: *single-messenger* synapses, in which the activated transmitter-receptor complex directly modulates channel permeability, and *second-messenger* synapses, in which the activated complex produces a biochemical cascade.

Single Messenger Synapses The single messenger in these synapses is the neurotransmitter. Because the activated transmitter-receptor complex directly causes the change in permeability, the synaptic current at a particular channel lasts only as long as the activated complex, usually a fraction of a millisecond (Fig. 3-14).

Second Messenger Synapses Second-messenger synapses act through a *guanosine nucleotide-binding protein (G protein)* (Fig. 3-15A). The second messenger is a substance produced as a result of activation of the G protein (see Chapter 4).

Three types of second messenger G protein systems are most common. In the first (Fig. 3-15B), an adenylyl cyclase coupled to the G protein produces cyclic AMP (cAMP), which acts as the second messenger to produce the permeability change. The second (Fig. 3-15C) couples phospholipase C to the G protein, and the enzyme is used to convert phosphatidylinositol 4,5-biphosphate (PI_2) to diacylglycerol (DAG) and inositol 1,4,5-triphosphate (IP_3), both of which act as second messengers. The third (Fig. 3-15D) couples the G protein to phospholipase A, which initiates a cascade of biochemical reactions starting with the second messenger arachidonic acid and producing a number of subsequent messengers through a variety of enzymatic reactions.

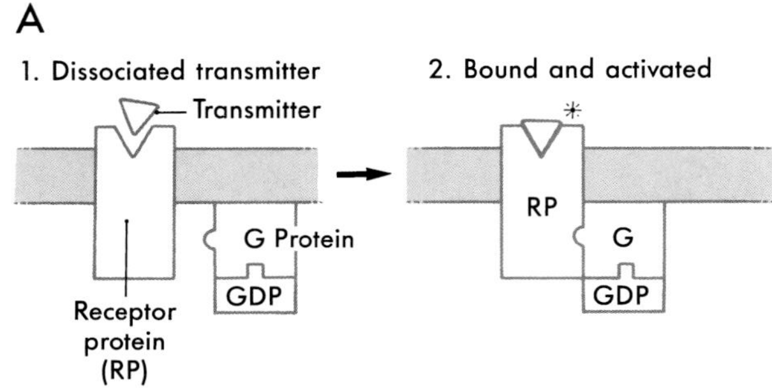

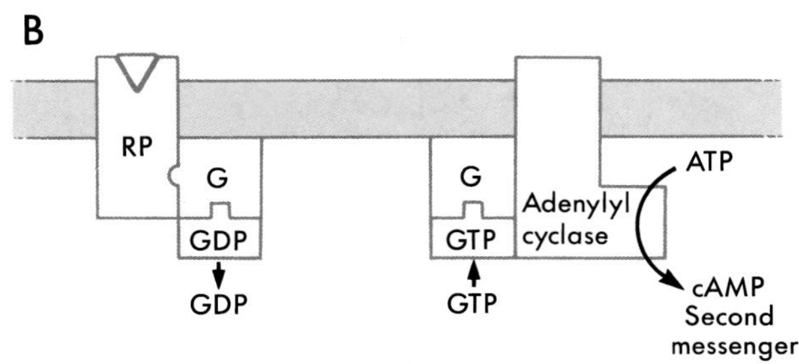

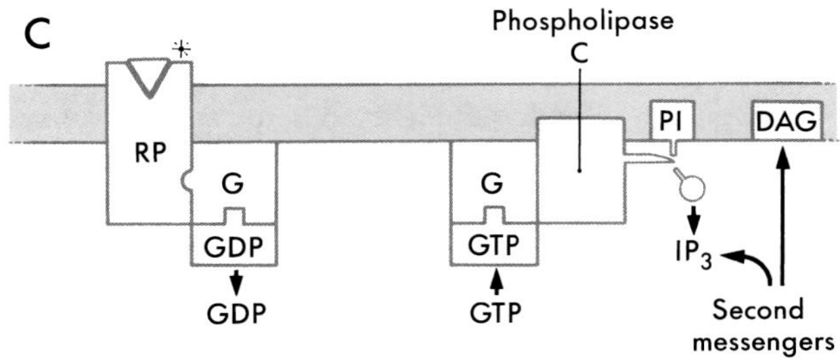

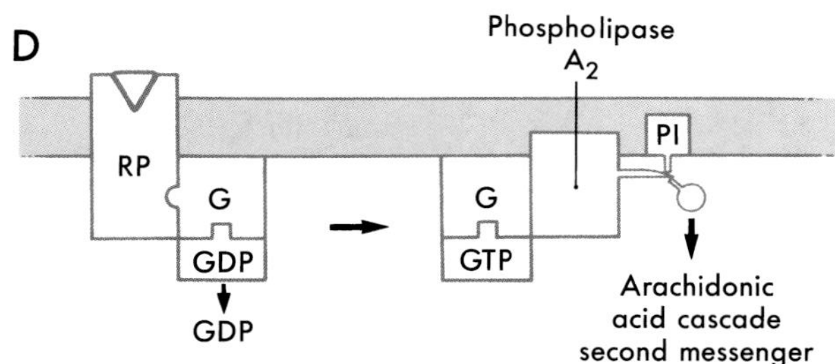

FIGURE 3-15 *Second-messenger synapses, in which transduction is mediated by a G protein. Interactions between the transmitter (first messenger), the receptor protein, and the G protein (A). Three kinds of second-messenger synapses (B–D). The activated G protein (B) is coupled with adenylate cyclase, which produces cAMP as the second messenger. The activated G protein (C) is coupled with phospholipase C, which cleaves phosphatidylinositol 4,5-bisphosphate (PI_2) to produce the second messengers inositol 1,4,5-trisphosphate (IP_3) and diacylglycerol (DAG). The activated G protein (D) is coupled with phospholipase A_2, which initiates a cascade of biochemical reactions starting with the second messenger arachidonic acid.*

Second-messenger synapses produce many molecules of second messenger per molecule of first messenger, and the second messengers persist after removal of the first messenger, so that the signal is amplified in both strength and duration.

SYNAPTIC INFORMATION PROCESSING

Currents and Potentials Synaptic potentials can be hyperpolarizing (inhibitory, stabilizing) or depolarizing (excitatory), fast (single messenger) or slow (second messenger). A permeability change causes ions to run down their concentration gradients to bring the local membrane area closer to their equilibrium potentials. To complete a circuit, the active currents produce distributed passive return currents through nearby membrane areas. The area of membrane close to the excitatory synapses that has the lowest threshold for action potential generation (e.g., because it has the most voltage-modulated Na^{++} channels) is the site where action potentials are generated, the *trigger zone*. In many neurons, it is located at the axon hillock.

Excitatory synapses generally involve increased permeability to Na^+ or to a combination of ions whose net combined channel current flow is inward, discharging membrane capacitance and producing depolarization (Fig. 3-16). This brings the membrane closer to, or farther above, the threshold for action potentials at the trigger zone.

At inhibitory synapses, different ionic permeability changes occur that result in a net combined outward current. This can produce a hyperpolarization, or if the membrane is already at the steady-state potential for this set of ions, it can result in a lowered membrane resistance with no voltage change. The most important effect is the production of a low-resistance path for the outward flow of passive return current from excitatory synapses. This *shunt* path deflects current away from the trigger zone, decreasing the depolarization produced by excitatory synaptic action (Fig. 3-17).

Potential changes at the trigger zone are thus the effect of interactions of currents generated at all the synapses within passive conduction distance of the trigger zone. For convenience, physiologists subdivide such interactions into *spatial* and *temporal*. In spatial summation (Fig. 3-18A), currents from multiple inputs add algebraically. In temporal summation, presynaptic action potentials arrive at intervals shorter than the durations of the postsynaptic potentials they produce (Fig. 3-18B). Thus there is a cumulative effect as postsynaptic potentials overlap in time, and their charges "pile up," or summate, on the membrane capacitance.

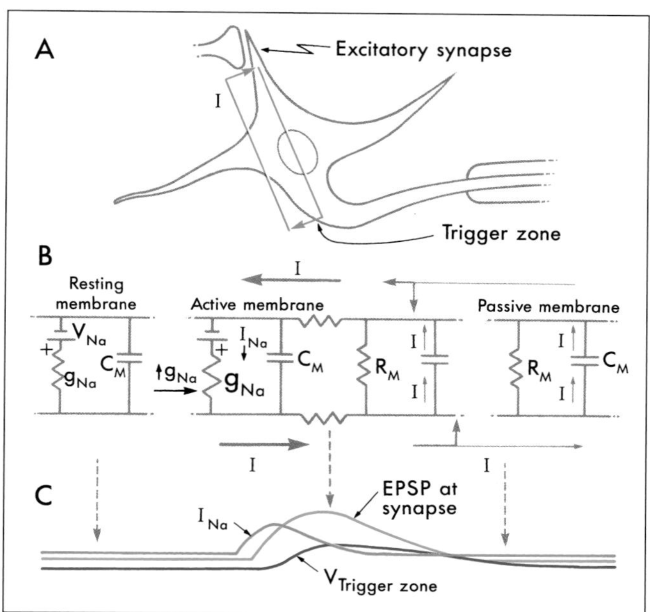

FIGURE 3-16 *The effects of excitatory postsynaptic potentials (EPSPs) on the trigger zone. Current flow between an excitatory synapse and the trigger zone (A). An equivalent circuit (B) showing the change from resting to active membrane during an EPSP (on left) and the passive effects at the trigger zone (on right). Comparison of the latencies and amplitudes of the current (I_{Na}) synaptic voltage (EPSP) and trigger zone voltage (shown in C).*

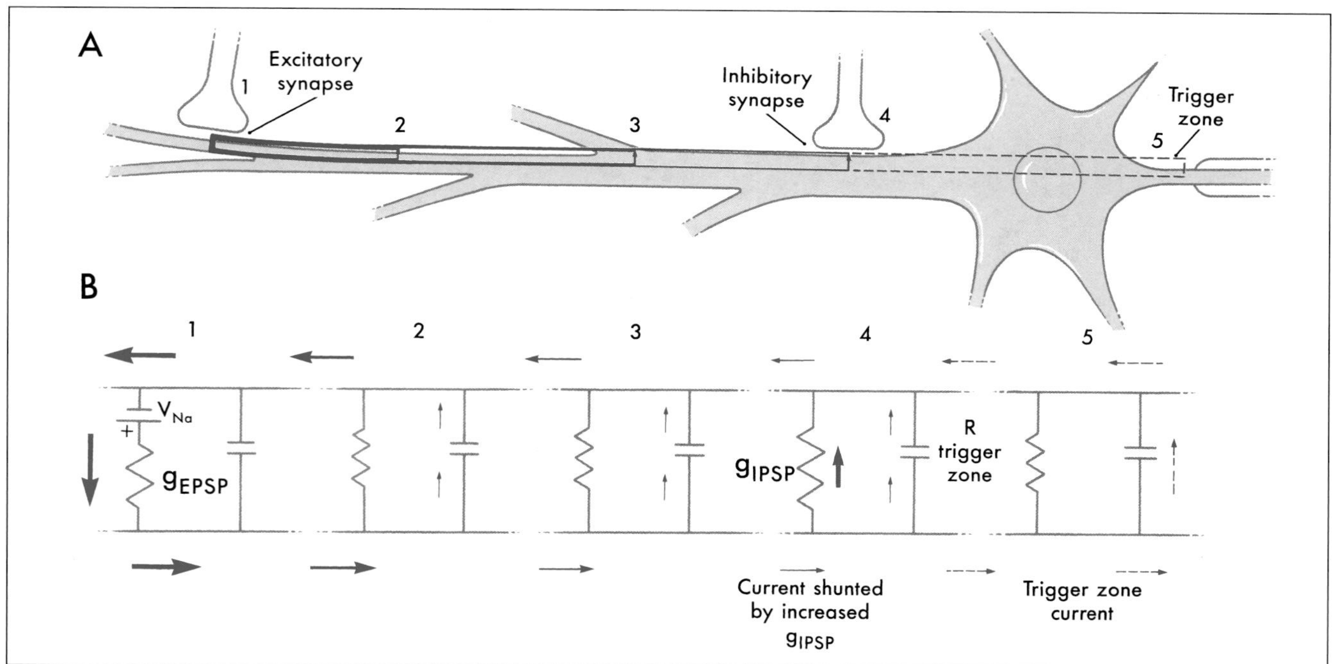

FIGURE 3-17 *Neuron (A) and equivalent circuit (B) showing a shunt path. The current produced at the EPSP (at 1) decrements as it flows passively down the dendrite (1 to 2 to 3 to 4) as represented by the decreasing line and arrow thickness toward the soma. When the inhibitory synapse (at 4) is active, current is shunted across the membrane and does not reach the trigger zone. When this synapse is inactive, the current (dashed lines and arrows) spreads to the trigger zone (at 5) and can produce an action potential.*

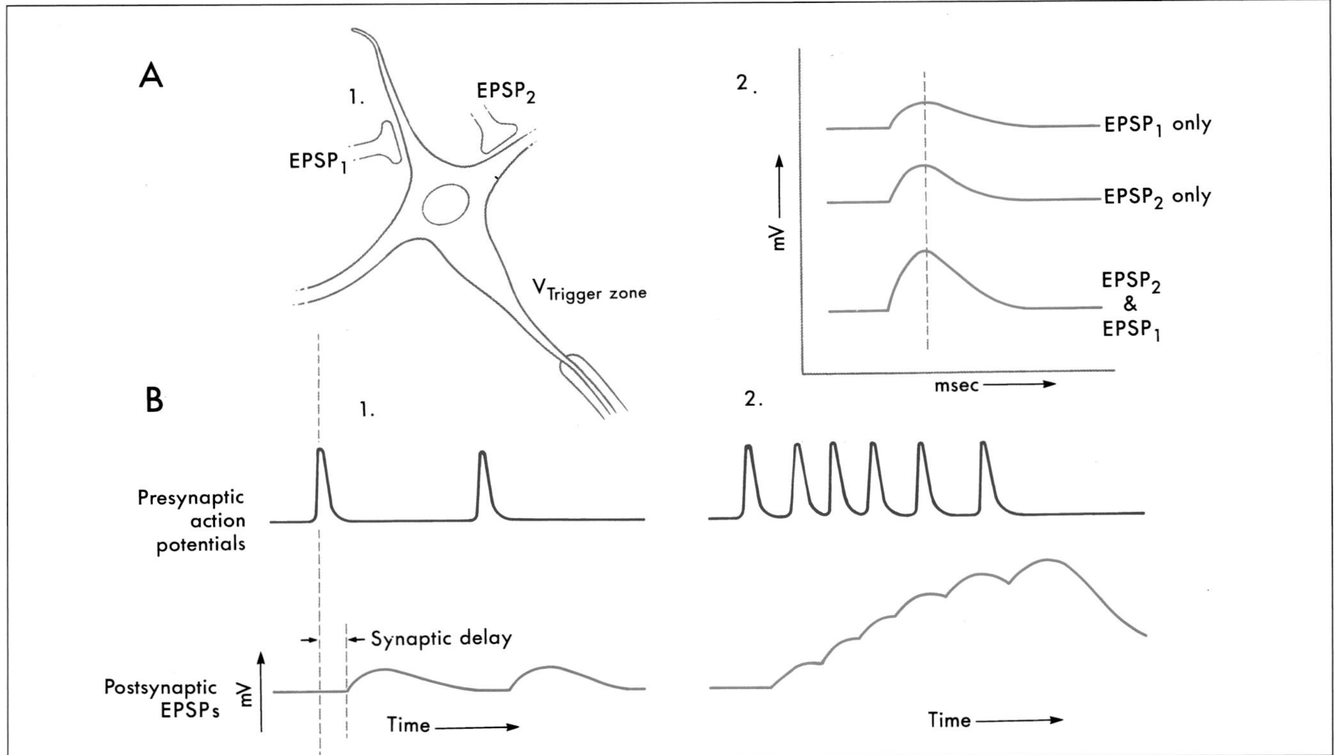

FIGURE 3-18 *Spatial (A) and temporal (B) summation. Locations of two excitatory synapses on the same postsynaptic neuron (A, 1). EPSP$_1$ only, EPSP$_2$ only, and both (A, 2). Currents sum to produce a larger EPSP (A, 2). Low presynaptic action potential frequency produces EPSPs that do not overlap in time and cannot summate (B, 1). High presynaptic action potential frequency produces EPSPs that overlap in time, causing summation (B, 2).*

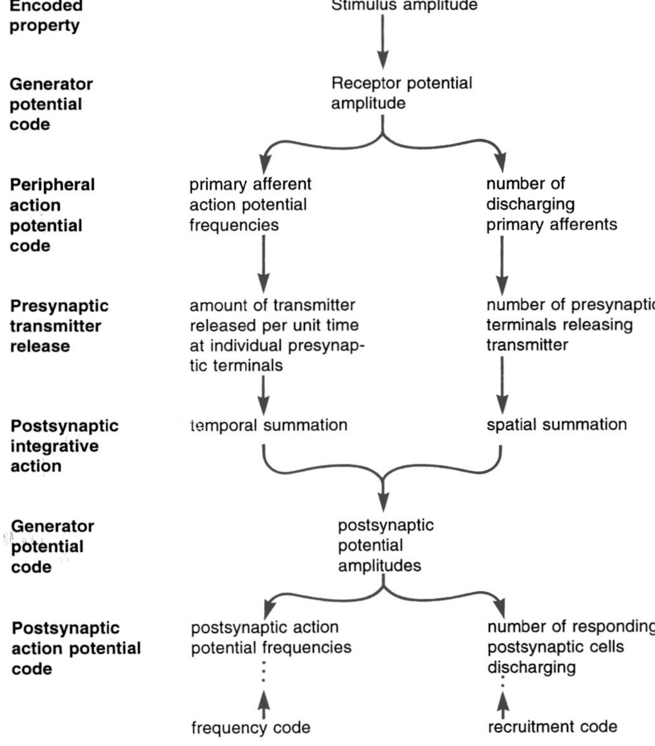

Action Potential Generation Different neurons in a postsynaptic population have different dynamic ranges with regard to the level of activity of neurons in the presynaptic population. Consider the simple case of a single excitatory presynaptic population of sensory afferent axons (Fig. 3-19). An increase in stimulus intensity within the dynamic range for the population results in increased discharge rates and recruitment of afferent axons. This causes increased temporal and spatial summation of postsynaptic potentials. The net depolarization of neurons in the postsynaptic population will be increased, resulting in increased frequency of discharge of already firing cells, and a recruitment of postsynaptic cells into the discharging population. Therefore the recruitment and frequency codes present in the presynaptic population are propagated through the synaptic relay to produce similar recruitment and frequency codes in the output population.

FIGURE 3-19 *Successive transformations of the representation of stimulus amplitude in receptor potentials, afferent axon population, synaptic transmitter release, postsynaptic integration, postsynaptic potentials, and postsynaptic population discharges.*

Information Processing In most CNS nuclei, cells receive input from several different tracts, and local circuits mediate interactions among the cells of the nucleus. Each cell typically receives both excitatory and inhibitory inputs from different sources, and the balance of inhibition and excitation determines the net output. This type of interaction allows control of the sign and magnitude of motor reflexes, detection of features in a visual stimulus such as the locations of edges of visual objects, and comparison of the timing of sounds arriving at the two ears to determine the direction of a sound source.

SOURCES AND ADDITIONAL READING

Dowling JE: *Neurons and Networks*. Harvard University Press, Cambridge, MA, 1992.

Kuffler SW, Nicholls JG, Martin AR: *From Neuron to Brain*, 2nd Ed. Sinauer Associates, Sunderland, MA, 1984.

Levitan IB, Kaczmarek LK: *The Neuron: Cell and Molecular Biology*. Oxford University Press, New York, 1991.

Shepherd GM: *Neurobiology*, 3rd Ed. Oxford University Press, New York, 1994.

CHAPTER FOUR

The Chemical Basis for Neuronal Communication

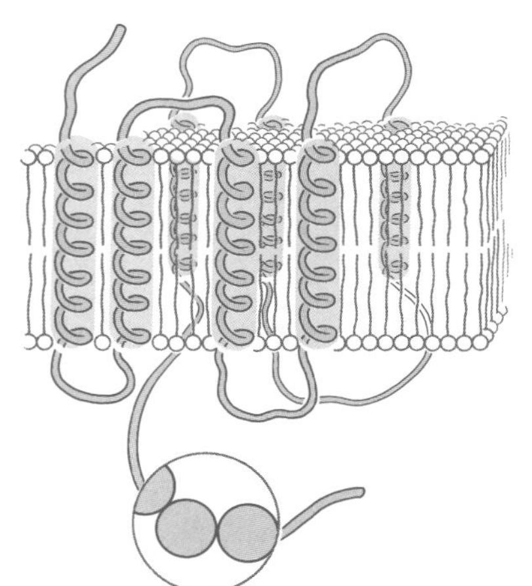

R. W. ROCKHOLD

Neurons in the human brain communicate primarily by the release of small quantities of *chemical messengers*, most of which are commonly called *neurotransmitters*. These chemicals alter the electrical activity of neurons after they interact with receptors on cell surfaces. Therapeutic alteration of brain function requires an understanding of the processes that regulate the synthesis and release of neurotransmitters, and the means by which receptors alter neuronal electrical activity and biochemical function.

OVERVIEW

The brain contains approximately 100 billion (10^{11}) neurons, each of which makes as many as 1,000 terminal contacts. Thus, it has been estimated that the human brain contains 10^{14} to 10^{15} connections between neurons. Communication at most of these connections is mediated by *chemical messengers*. The transfer of information between neurons takes place at structurally and functionally specialized locations called *synapses*. Most synapses use chemical messengers that are released in discrete units (*quanta*) from presynaptic axonic or dendritic terminals, in response to depolarization of the terminal.

Rapid diffusion of a chemical messenger across the *synaptic cleft* is followed by binding of this substance to receptors spanning the postsynaptic membrane. There is a resultant alteration in the electrical, biochemical, or genetic properties of that neuron. Less frequently, chemical messengers may also be released at sites without synaptic specializations. These messengers diffuse more widely than neurotransmitters released at synaptic sites, and they influence receptors located at distant sites and on more than one neuron. Whether synaptic or nonsynaptic, chemical communication in the nervous system depends on (1) the nature of the presynaptically released *chemical messenger*, (2) the type of postsynaptic *receptor* to which it binds, and (3) the mechanism that couples receptors to effector systems in the target cell.

FUNDAMENTALS OF CHEMICAL NEUROTRANSMISSION

Neurotransmitters Specific criteria that define whether a chemical messenger can be identified as a *neurotransmitter* are listed in Table 4-1. Although a wide variety of putative neurotransmitters have been identified, these criteria have been met for only a few chemical substances. Generally, these transmitters can be categorized as *small molecule messengers* (having fewer than 10 carbon atoms) or larger *neuropeptides* (containing 10 or more carbon atoms).

TABLE 4-1

CRITERIA NECESSARY TO DEFINE A SUBSTANCE AS A NEUROTRANSMITTER

I. Localization	A putative neurotransmitter must be localized to the presynaptic elements of an identified synapse and must be present also within the neuron from which the presynaptic terminal arises.
II. Release	The substance must be shown to be released from the presynaptic element upon activation of that terminal and simultaneously with depolarization of the parent neuron.
III. Identity	Application of the putative neurotransmitter to the target cells must be shown to produce the same effects as stimulation of the neurons in question.

Small molecular chemical messengers are classed as biogenic amines, amino acids, and nucleotides or nucleosides (Table 4-2). The *biogenic amines* include the familiar neurotransmitters acetylcholine, dopamine, norepinephrine, epinephrine, serotonin, and histamine. *Amino acid* neurotransmitters include γ-aminobutyric acid (GABA), glycine, aspartate, and glutamate. The nucleotide/nucleoside class includes adenosine and adenosine triphosphate (ATP). Recently, *nitric oxide*, which functions as an endogenous *nitrovasodilator* in the cardiovascular system, has been identified as a putative neurotransmitter.

More than 40 neuropeptides have been identified in brain tissue. These include methionine enkephalin (met-enkephalin) and leucine enkephalin (leu-enkephalin), as well as larger peptides, such as endorphins, calcitonin gene-related peptide (CGRP), arginine vasopressin, cholecystokinin, and many others (Table 4-2).

With a few exceptions, one being nitric oxide, *the chemical messengers used by neurons are stored in secretory vesicles and released from them by exocytosis*. In the case of neurotransmitters, these vesicles are found mainly in the presynaptic nerve terminals.

Fast and Slow Synaptic Transmission The diffusion of a chemical message across the synaptic cleft can be quite rapid. At the neuromuscular junction, for example, it takes only about 50 μsec for acetylcholine to reach the postsynaptic membrane. Total *synaptic delay*, the time from presynaptic release of neurotransmitter to the activation or inhibition of the postsynaptic neuron, is variable. This variability is influenced by the transduction mechanisms in the postsynaptic neuron.

TABLE 4-2

SUBSTANCES BELIEVED TO ACT AS CHEMICAL MESSENGERS IN THE CENTRAL NERVOUS SYSTEM

I. SMALL MOLECULES	II. NEUROPEPTIDES
A. **Biogenic amines**	A. **Opioid peptides**
acetylcholine	methionine-enkephalin
monoamines	leucine-enkephalin
catecholamines	β-endorphin
dopamine	dynorphin(s)
norepinephrine	neoendorphin(s)
epinephrine	
serotonin	B. **Posterior pituitary peptides**
histamine	arginine vasopressin
	oxytocin
B. **Amino acids**	
GABA	C. **Tachykinins**
glycine	substance P
glutamate	kassinin
aspartate	neurokinin A
homocysteine	neurokinin B
taurine	eledoisin
C. **Nucleotides and nucleosides**	D. **Glucagon-related peptides**
adenosine	vasoactive intestinal peptide
ATP	glucagon
	secretin
	growth hormone releasing hormone
D. **Other**	
nitric oxide	
	E. **Pancreatic polypeptide-related peptides**
	neuropeptide Y
	F. **Other**
	somatostatin
	corticotropin-releasing factor
	calcitonin gene-related peptide
	cholecystokinin
	angiotensin II

Transduction mechanisms can be divided into fast and slow types. *Fast chemical neurotransmission* operates with a total synaptic delay of only a few milliseconds, whereas *slow chemical neurotransmission* usually requires hundreds of milliseconds. In both cases, the receptors on the postsynaptic membranes are glycoproteins that span the lipid bilayer membrane and transduce an extracellular chemical signal into a functional change in the target neuron. The difference relates to the complexity of the transduction mechanism.

In *fast chemical neurotransmission*, the postsynaptic receptor is itself an ion channel. This type of transmission is associated exclusively with *small-molecule neurotransmitters*. The binding of transmitter stimulates the channel to open, permitting a flux of ions across the membrane that alters the membrane potential. The process is fast because it is direct. Ion channels in this type of neurotransmission are called *ligand-gated* or *receptor-gated ion channels*; the ions normally involved are Na^+, K^+, Ca^{++} or Cl^-. Movement of these ions causes a change in the *transmembrane electrical potential*, which, if it exceeds threshold, may lead to generation of an action potential.

In *slow chemical neurotransmission*, the signal is transduced by a mechanism involving G-protein coupled receptors. These proteins and their action are discussed later in the chapter. Briefly, the binding of the transmitter (frequently a *neuropeptide*) causes the receptor to activate a G-protein, which in turn binds to and influences an effector protein, which elicits the cellular effect. In some cases, the effector protein is an ion channel, which is induced to open or close. Transduction in these cases can be almost as rapid as in fast neurotransmission. More often, the effector is an enzyme that produces an intracellular second messenger, such as cyclic AMP (cAMP), whose cytoplasmic concentration is altered in response to the reception of a signal (binding of the transmitter) at the cell surface and which elicits intracellular responses to the signal. Second messengers can produce a plethora of cellular responses, ranging from the opening or closing of membrane ion channels to alterations in gene expression. These effects are mediated by complex sequences of chemical events, which is why they are relatively slow.

Information Flow Across Chemical Synapses Transmission of information at a chemical synapse involves the following general sequence of events (Fig. 4-1): (1) secretory vesicle synthesis and transport to the synaptic terminal; (2) for small-molecule neurotransmitters, loading of the transmitter into the vesicle; for neuropeptides, this step accompanies vesicle synthesis; (3) depolarization of the presynaptic terminal; (4) vesicle docking with the presynaptic membrane, exocytosis of its contents, and transsynaptic diffusion of the transmitter; (5) binding of transmitter to, and activation of, the postsynaptic receptor; (6) transduction of the signal resulting in a postsynaptic response and one or two terminal steps; (7) active reuptake of the transmitter by the presynaptic cell and/or (8) enzymatic degradation of the transmitter in the synaptic cleft. These final events eliminate transmitter from the synaptic cleft and thereby terminate its action.

In many synapses, the amount of transmitter that a presynaptic terminal releases in response to an action potential can be regulated from outside the cell. Two reg-

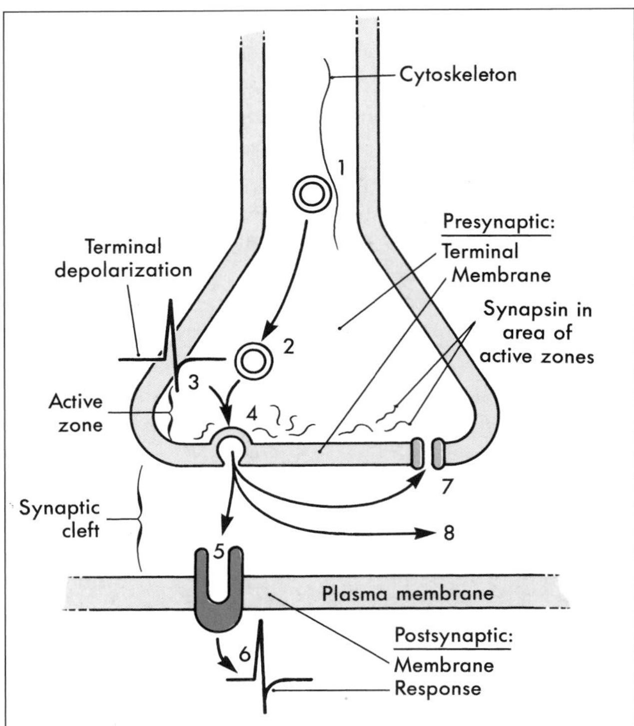

FIGURE 4-1 *A generalized scheme for chemical synaptic transmission. The main steps are as numbered: (1) Proximodistal axonal transport of a secretory vesicle; (2) synthesis and loading of small molecule messengers in synaptic vesicles (neuropeptides are synthesized and loaded into large dense-cored vesicles in the soma); (3) depolarization of the presynaptic terminal by an arriving action potential, which causes (4) fusion of vesicles with the plasma membrane and exocytosis of the vesicle contents; (5) binding of transmitter with a postsynaptic receptor to produce (6) a postsynaptic response; and finally, elimination of the transmitter from the synapse by either (7) uptake into a cell (here, the presynaptic cell) or (8) enzymatic degradation in the synaptic cleft.*

ulatory mechanisms are (1) *presynaptic receptor mediated autoregulation* and (2) *retrograde transmission*. In *presynaptic mediated autoregulation* the neuron self-regulates the subsequent quantal release of its own chemical messenger. As a neurotransmitter enters the synaptic cleft, it stimulates not only postsynaptic receptors, but also receptors located on the membranes of the terminal from which it was released. This constantly updates the presynaptic neuron concerning neurotransmitter synthesis, release, and the efficiency of information transfer. In most cases, autoregulation is inhibitory. Loss or reduction of this input is interpreted as a reduction in signaling ability and the presynaptic neuron increases the subsequent synthesis and/or release of stored neurotransmitter.

In *retrograde transmission*, the postsynaptic neuron responds to synaptic activation by releasing a second chemical messenger. This messenger diffuses back across the synapse and alters the function of the presynaptic terminal. Nitric oxide is currently the best example of a mediator of retrograde transmission.

SYNTHESIS, STORAGE, AND RELEASE OF CHEMICAL MESSENGERS

Neuronal chemical messengers are stored in two types of vesicles: *small vesicles* (also called *synaptic vesicles*) and *large dense-cored vesicles*. *Synaptic small* vesicles (~50 nm in diameter), appear clear and empty in electron micrographs and contain small molecule chemical messengers such as GABA, glutamate, and acetylcholine. A subset of these small vesicles, with electron-dense cores, are found in both central and peripheral neurons. These vesicles contain the catecholamine family of biogenic amines (dopamine, norepinephrine, and epinephrine). Synaptic vesicles cluster near the exocytotic surface of a presynaptic nerve terminal in regions called *active zones* (Fig. 4-1).

Large, dense-cored vesicles (~75 to 150 nm in diameter) are less numerous and appear in other intraneuronal locations, as well as in the axon terminal. The electron opaque, dense core is composed of soluble proteins that are mainly one or more neuropeptides. This core may also contain a small chemical messenger; often a biogenic amine, *co-stored* with a neuropeptide.

Neurons in certain hypothalamic nuclei contain a third type of vesicles called the *neurosecretory vesicles*. These vesicles are large (~150 to 200 nm in diameter), contain neurohormones, and are especially concentrated in axon terminals in the neurohypophysis (the posterior pituitary).

Composition of Vesicle Membranes All vesicles are composed of a lipid bilayer membrane, spanned by a variety of proteins. Some proteins are common to both large dense-cored vesicles and synaptic vesicles, such as those that form *calcium channels*, and the proteins *synaptotagmin* and *SV2*. Other proteins are found in high concentrations only in synaptic vesicles; these include *synaptophysin* and *synaptobrevin*. The differences in protein content reflect the different roles that large dense-cored vesicles and synaptic vesicles play in neurons.

Vesicles also contain proteins that act to accumulate small chemical messengers. These take the form of membrane pumps or transporters, most of which are coupled to the transport of protons. Synaptic vesicles contain at least four classes of *proton-coupled transporters* for chemical messengers, each specific for a different type of messenger. One class drives the accumulation of biogenic amines, including the catecholamines dopamine, norepinephrine, and epinephrine, as well as the monoamine serotonin. Others are specific for acetylcholine, glutamate, and GABA/glycine. Large, dense-cored vesicles can also accumulate small chemical messengers in addition to their neuropeptides. However, it is believed that the transporters involved are different from those used by synaptic vesicles.

Biosynthesis In terms of biosynthesis, an important *difference between synaptic vesicles and large, dense-cored vesicles is that the former can be recycled and refilled in the axon terminal, whereas the latter are both made and filled in the neuronal soma and are not recycled.* This reflects the fact that small-molecule neurotransmitters can be synthesized in axon terminals, whereas neuropeptides, because they are synthesized on ribosomes and processed through the endoplasmic reticulum and Golgi complex, can only be made in the soma (Fig. 4-2). The *cis face* of the Golgi complex (also called the *proximal* or *forming face*) is prototypically concave toward the nucleus of the cell, whereas the *trans face* (*distal* or *maturation face*) is convex (Fig. 4-2). Peptides from the endoplasmic reticulum enter the *cis* face of the Golgi complex and are sorted and packaged into vesicles that bud from its *trans* face.

Large dense-cored vesicles contain neuropeptide messengers and are filled during the process of vesicle synthesis in the Golgi complex. These vesicles are translocated, by *fast axonal transport* (range 4 to 17 mm/hr), from the cell body to axonal or dendritic release sites (Fig. 4-3). Frequently, neuropeptides are synthesized in the form of large precursor peptides that may be cleaved to yield more than one secreted bioactive neuropeptide. Maturation of neuropeptides can require covalent chemical modification of amino acid side chains, often with the addition of small chemical groups. Examples of the types of chemical modifications include the addition of methyl groups (methylation), sugar moieties (glycosylation), and sulfate groups (sulfation). This process of maturation can occur within the endoplasmic reticulum, during packaging of peptides into large dense-cored vesicles within the Golgi complex, or during axonal transport.

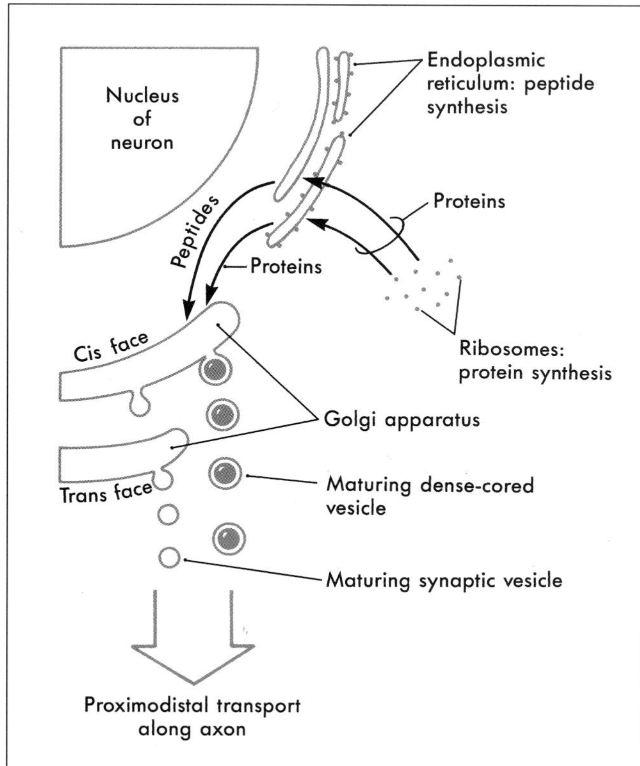

FIGURE 4-2 *The synthesis of large, dense-cored vesicles and of synaptic vesicles in the neuron cell body. Synaptic vesicles are formed without existing stores of neurotransmitter; these are synthesized as the vesicles move into the nerve terminal. Large dense-cored vesicles are formed with existing stores of neuropeptide messengers as electron-dense cores.*

In general, synaptic vesicles are formed initially by budding from the Golgi apparatus within the cell body (Figs. 4-2, 4-4). After transport to and release from the presynaptic terminal, however, the lipoprotein membrane components of the synaptic vesicles are *recycled* in a continuous process that occurs within nerve terminals (Fig. 4-4). *Synthesis of the chemical messenger in a synaptic vesicle can occur while the vesicle is in the nerve terminal, rather than in the cell body.*

Some small-molecule neurotransmitters are synthesized in the cytosol of the axon and axon terminal and then transported into synaptic vesicles, whereas others are synthesized in the vesicle itself. The synthesis of acetylcholine is an example of the first of these mechanisms. The soluble enzyme *choline acetyltransferase* (*CAT*) catalyzes the acetylation of choline from acetyl CoA to yield the neurotransmitter acetylcholine. A high affinity vesicular membrane transport protein concentrates this transmitter in cholinergic synaptic vesicles. Synthesis of the catecholamine norepinephrine is an example of the second mechanism. In the case of norepinephrine, synthesis occurs within the synaptic vesicle. The immediate precursor to norepinephrine, dopamine, is concentrated within the *noradrenergic* synaptic vesicle by a transporter specific for biogenic amines. Only then is dopamine converted to norepinephrine by the action of the enzyme, *dopamine β-hydroxylase*, which is attached to the luminal border of the vesicular membrane.

Transporters for small chemical messengers concentrate compounds inside the vesicle to levels 10 to 1,000 times higher than those found in the cytosol. The energy required for this transport is derived from an ATP-driven proton pump. The exchange of protons for the chemical messenger allows accumulation of the latter inside the vesicle.

Localization As mentioned earlier, *synaptic vesicles* are preferentially concentrated in *active zones* of the nerve terminal (Figs. 4-4, 4-5). These zones are biochemically and anatomically specialized for neurotransmitter release. Large numbers of voltage-sensitive calcium channels are clustered in the plasma membrane of active zones. Consequently, depolarization of the axon terminal (or in special cases the dendrites) results in a high local concentration of Ca^{++}. This calcium causes synaptic vesicles to bind to the plasma membrane and stimulates exo-

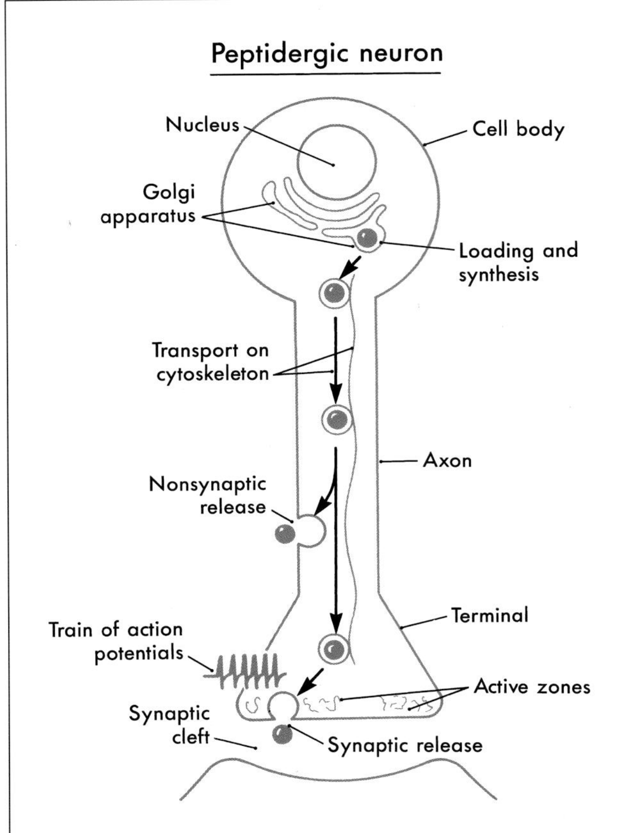

FIGURE 4-3 *The formation, transport, and use of large dense-cored vesicles (containing neuropeptides) in a representative peptidergic neuron.*

cytotic release of vesicle contents into the synaptic cleft. Active zones also contain high concentrations of the filamentous protein *synapsin*, which aids in the clustering of synaptic vesicles.

Although *large dense-cored vesicles* may accumulate in active zones, they also bind to the plasma membrane and release their contents from other sites in the terminal and axon that lack active zones (Fig. 4-3). As with synaptic vesicles, exocytosis depends on a local increase in Ca^{++} concentration. However, the release mechanisms for large dense-cored vesicles appear to be more sensitive to Ca^{++} than those for synaptic vesicles. Therefore, sites of release do not require the high density of Ca^{++} channels found in active zones. Release sites outside active zones (such as those associated with large dense-cored vesicles) also do not have anchoring proteins such as synapsins.

Release The essential structural elements critical for synaptic vesicle release are depicted in Figure 4-5. Proteins in the vesicle wall interact with cytoskeletal proteins to propel vesicles into the active zone. The surface of the synaptic vesicle contains two groups of proteins that are crucial for exocytotic release: *docking proteins* and elements of the *fusion pore*. A rise in intracellular Ca^{++} levels causes the vesicular docking proteins to interact with docking proteins on the cell membrane, creating a *docking complex* that brings the two membranes into apposition. Complimentary proteins on both membranes then interact to form the *fusion pore*, and the lipid bilayers of the two membranes fuse at this site to form a rapidly expanding hole. Stored neurotransmitter in the vesicle begins to leak out through the fusion pore and exits in bulk (complete exocytosis) as the pore expands.

It is likely that large, dense-cored vesicles use somewhat different proteins and mechanisms for docking and exocytosis than do synaptic vesicles. Also, while synaptic vesicles undergo exocytosis in response to single nerve impulses (Fig. 4-4), large, dense-cored vesicles respond preferentially to high-frequency trains of impulses (Fig. 4-3). In experiments using peripheral nerves, a stimulation frequency of 10 Hz is often required to elicit neuropeptide release. Frequencies of that magnitude occur natu-

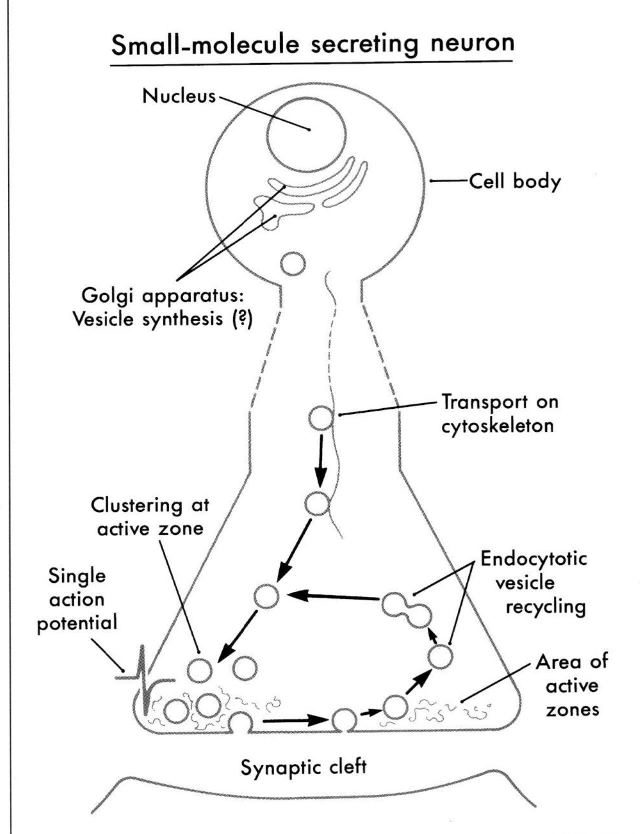

FIGURE 4-4 *The formation, transport, and cycling of synaptic vesicles (containing small-molecule neurotransmitters) in a representative neuron.*

FIGURE 4-5 *The protein components that mediate transport and docking of synaptic vesicles and the probable formation of the fusion pore. Depolarization opens voltage-dependent calcium channels, allowing ingress of Ca^{++}, which facilitates formation of the docking complex. Once docking is accomplished, additional proteins, under the influence of elevated intracellular Ca^{++} levels, associate to form a fusion pore.*

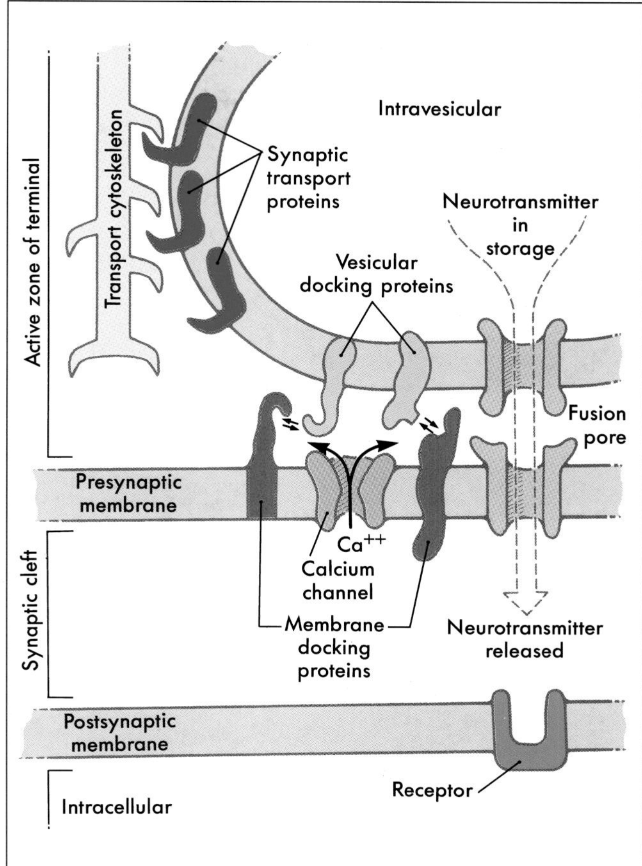

rally in the autonomic nervous system under conditions of extreme behavioral or physiologic stress. Consequently, neuropeptides may play a role in stress responses.

SIGNAL TRANSDUCTION

Chemical messengers, once released from a presynaptic site, must interact with a postsynaptic neuron to transmit information. The postsynaptic membrane contains target molecules that exhibit an affinity for individual chemical messengers; these molecules are known as *receptors*. Most receptors are transmembrane glycoprotein chains. The binding of a messenger with its receptor precipitates a change in the architecture (*conformation*) of the glycoprotein chain that begins the process of information transfer. Some exceptions do exist. For example, there are *intracellular receptors* for testosterone. To be activated, drugs such as testosterone must first traverse the plasma membrane to gain access to the receptor.

Receptors and Receptor Subtypes The *receptor* is capable of altering intracellular function in response to a change in the concentration of a specific chemical messenger in the environment. Thus a receptor transduces a chemical signal (i.e., the concentration of a chemical messenger) into an intracellular event. *Receptors are characterized by the response of a cell or tissue to a series of chemicals of different molecular structure.* Each compound in the series produces identical cell or tissue responses. However, the compounds can be ranked according to their individual potencies (i.e., to the concentration required to elicit the desired response). It is common to rank potencies in terms of the concentration of an agent that produces 50% of the maximal biologic responses, or the *effective concentration*$_{50}$ (EC_{50}).

Receptors are grouped according to the type of chemical messenger to which they respond. For example, all the receptors that respond to physiologically relevant concentrations of acetylcholine are called *cholinergic receptors*. *Adrenergic receptors* respond to the catecholamine neurotransmitters epinephrine (previously called adrenaline) and norepinephrine (noradrenaline).

The receptors that respond to a given transmitter can often be divided into *subtypes* that elicit different biologic responses. For example, cholinergic receptors are divided into *nicotinic* and *muscarinic* subtypes. A cholinergic synapse with nicotinic receptors is commonly excitatory, whereas one with muscarinic receptors is commonly inhibitory. These receptor subtypes are named after plant compounds that stimulate them selectively and helped lead to their discovery. Nicotinic receptors are named after the nicotine of tobacco, and muscarinic receptors are named after muscarine, a substance found in the toxic mushroom, *Amanita muscaria*. Both nicotinic and muscarinic receptors have subtypes of their own. For example, five different muscarinic receptor subtypes, termed M1 to M5, have been recognized. Adrenergic receptors are classified broadly into α- and β-receptor subtypes, each of which can then be classified further (e.g., α_1, α_2; β_1, β_2, β_3).

Structure and Function Most membrane-coupled receptors fall into the two classes: ligand-gated ion channels and G-protein coupled receptors. Transmembrane receptor proteins have a general structure that is based on glycoprotein chains that fold into the neural membrane in multiple loops (Fig. 4-6). The protein is held in the membrane by several hydrophobic membrane-spanning segments (usually *alpha [α] helices* but in specialized regions having a *beta [β]-sheet* conformation), which are connected by loops that project into the aqueous environment on either side of the membrane. The N-terminal and C-terminal segments also

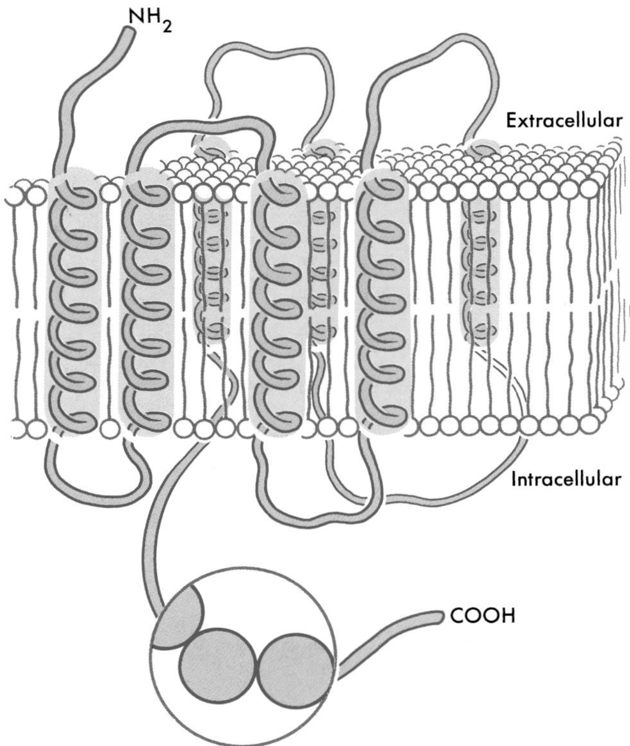

FIGURE 4-6 *A membrane-linked receptor protein (human β-adrenergic receptor), embedded in a neuronal plasma membrane. Hydrophobic transmembrane amino acid sequences are coiled in an α helical array and form a cluster of seven transmembrane columns. The G protein, although not shown here, would associate with intracellular loops of the receptor protein. The enlarged area denotes the fact that the protein is composed of linked amino acids.*

project into the aqueous environment; they are typically relatively straight. In some transmembrane receptors, such as the β-adrenergic receptor, the N-terminal segment projects extracellularly and the C-terminal segment projects intracellularly (Fig. 4-6). In contrast, in voltage-gated ion channels the N- and C-terminal segments usually both project intracellularly. G-protein coupled receptors are formed from a single polypeptide chain, whereas most ligand-gated ion pores are multisubunit structures.

In G-protein coupled receptors, the transmembrane segments of the protein (usually seven in number) form a cluster that contains the binding site or sites for chemical messengers. The site is usually in a relatively hydrophobic pocket in the cluster, although it is sometimes on the extracellular surface of the protein. The receptor binds with its G-protein transducer through multiple cationic sites on intracellular hydrophilic regions. The β-adrenergic receptors are the best characterized of the G-protein coupled receptors; their functioning is discussed later in the chapter.

Ligand-gated Ion Channels Ligand-gated ion channels are formed by several structurally distinct protein subunits called *channel subunits* (Fig. 4-7). Each channel subunit is a transmembrane glycoprotein (as described previously) with membrane-spanning segments connected by intracellular and extracellular loops. The channel subunits complex to form a roughly cylindrical structure that encloses a water-filled transmembrane channel. As exemplified by the nicotinic cholinergic receptor, the external face of the channel is enlarged and cuplike (Fig. 4-7). The channel narrows as it crosses the membrane, reducing the inner diameter such that it can selectively pass small cations (Na^+, K^+, and Ca^{++}) or anions (Cl^-). The internal face of the channel widens again as it emerges from the lipid bilayer. The inner surface of the pore is blocked at rest by amino acid residues that project into the aqueous lumen of the pore and prevent the conductance of charged ions. This part of the channel is termed the *gate*. *Binding sites*, which most commonly occur at relatively hydrophobic regions within the transmembrane region of the channel, are specific for a chemical messenger. When a binding site is filled, conformational changes occur within the channel protein to open the gate and permit selective passage of ions across the membrane.

Two gene superfamilies of ligand-gated ion channels have been identified. One contains nicotinic cholinergic, serotonin (5-hydroxytryptamine), GABA, and glycine receptors; and the other encodes the receptors for the excitatory neurotransmitter glutamate. The segregation

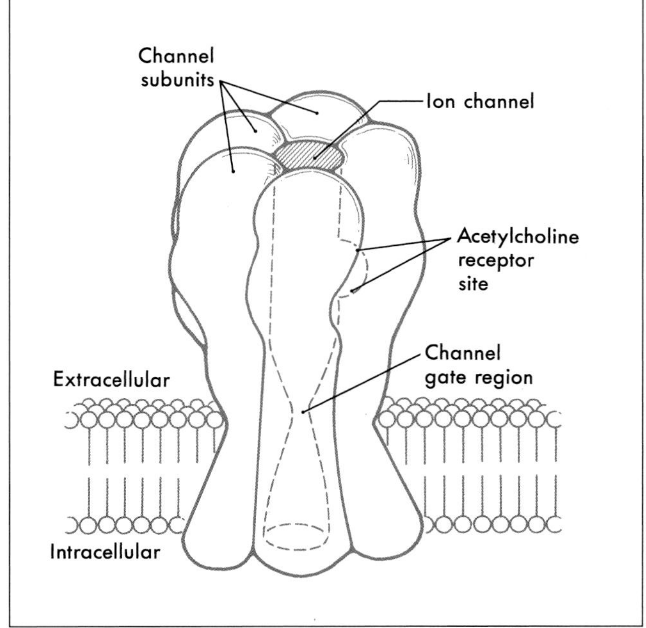

FIGURE 4-7 *A typical ligand-gated ion channel, the nicotinic cholinergic receptor. This receptor is composed of five nonhomologous channel subunit proteins, each containing four hydrophobic membrane-spanning regions. Acetylcholine binds to the nicotinic receptor in the central core.*

of receptors into different superfamilies is based on the degree of homology of amino acid sequences. The subunits of the various receptors in a superfamily have about 20 to 40% sequence homology with each other. The subunits of any given receptor generally have sequence homologies greater than 40%.

G-protein Coupled Receptors The G-protein receptors introduce a further level of complexity to chemical transmission (Fig. 4-8). *About three-fourths of all chemical messengers transmit their information through G-protein coupled receptors.* The basic elements of this system include a *receptor*, which must face the external surface of the membrane, the *guanosine triphosphate (GTP) binding protein*, which consists of α, β, and γ subunits, and an *effector protein*, which may be an enzyme that alters the concentrations of intracellular *second messengers* (such as Ca^{++}; inositol 1, 4, 5 trisphosphate; diacylglycerol; or members of the eicosanoid family), or may be an ion channel (Fig. 4-8A–D). The responses mediated by these receptors are generally slow (hundreds of milliseconds to minutes). The G-protein coupled receptor complex transduces an extremely wide range of chemical messages. About 100 different receptors have been identified that can link to a G-protein, and at least 20 distinct G-proteins have a similar number of effector proteins. The biogenic amines, bioactive peptides, the eicosanoids, light (one of the first characterized G-protein coupled receptors was rhodopsin), and odorants all interact with G-protein coupled receptors.

The G-protein functions to amplify a signal received by a transmembrane receptor, transmitting that message to effector proteins within a neuron. Each G-protein exists as a complex (a *heterotrimer*) formed by α, β, and γ subunits (Fig. 4-8A–D). The αβγ heterotrimer maintains a loose association with the receptor glycoprotein, but is not covalently bound to the receptor. Within the heterotrimer, the α subunit determines the nature of the G-protein. It has the ability to bind GTP, detach from the

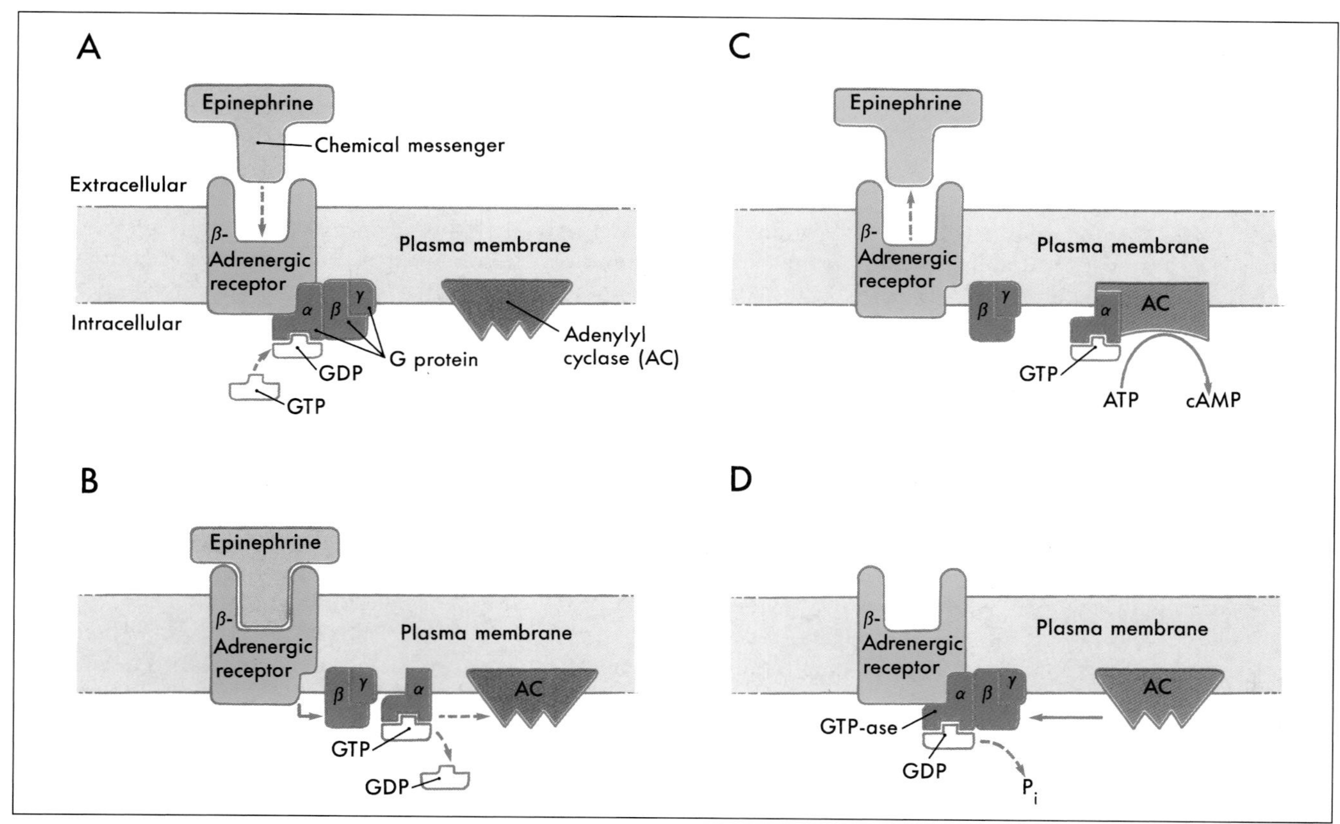

FIGURE 4-8 *A typical example of G-protein action. The receptor is the β-adrenergic receptor (activated in this case by epinephrine). It is coupled to the G_s-type of G-protein, which has a stimulatory action on the effector, adenylyl cyclase. When stimulated, adenylyl cyclase produces the second messenger cAMP from ATP. The cycle of G-protein action is as follows: (A) In the resting state, the G-protein is in the form of an αβγ heterotrimer, and the α subunit carries bound GDP. (B) Epinephrine binding to, and conformation change of, the receptor protein and the G-protein. The GDP, which has been bound to the α subunit, exchanges for GTP as the α subunit dissociates from the βγ complex and from the receptor. (C) Binding of the α subunit (with GTP attached) to adenylyl cyclase alters the conformation of the enzyme. This increases enzymatic catalysis of the substrate ATP to cAMP. (D) Slow enzymatic dephosphorylation of GTP to GDP by the GTP-ase allows the α subunit to return to its resting conformation. As this occurs, the α subunit dissociates from adenylyl cyclase and reassociates with the βγ subunit. The heterotrimeric reassociates with the receptor protein.*

coupled $\beta\gamma$ complex and alter the activity of an effector protein. The effector proteins whose activity can be modulated by α subunits are diverse and include the enzymes such as adenylyl cyclase as well as ion channels for calcium and potassium. In contrast, the $\beta\gamma$ complex anchors α subunits to membrane sites and inhibits the GTP/guanosine diphosphate (GDP) exchange that activates the α subunit.

The resting form of a G-protein exists as the heterotrimer, with GDP bound to the α subunit and the α subunit by the $\beta\gamma$ complex (Fig. 4-8A). When activated, the α subunit exchanges GDP for GTP and dissociates from the $\beta\gamma$ complex. The α subunit is then free to bind with and alter the activity of the effector protein; in this example it is adenylyl cyclase (Fig. 4-8B,C). The α subunit has an integral slow GTPase activity, which eventually hydrolyzes the bound GTP to bound GDP (usually in about 3–15 seconds). Reassociation of the α subunit/GDP complex with the $\beta\gamma$ complex then completes the cycle (Fig. 4-8C,D).

The most completely characterized G-protein coupled receptor is the β_2-adrenergic receptor. The endogenous ligand for this receptor is the catecholamine, epinephrine. The β_2-adrenergic receptor is coupled to a G-protein that stimulates activity of adenylyl cyclase, catalyzing formation of cAMP from intracellular ATP stores (Fig. 4-8A–D). The stimulatory G-protein associated with the β_2-adrenergic receptor is called G_s.

Effector Proteins As mentioned earlier, G-proteins can interact with two main kinds of effector proteins: *ion channels* (called *G-protein coupled ion channels*) and *enzymes that alter the level of intracellular second messenger compounds*. The action of the G-protein on its target may be positive or negative. That is, it may cause a channel to open or, more rarely, close, or it may stimulate or inhibit a target enzyme. Most synaptic G-protein responses are mediated through second messenger systems. Second messengers can elicit a variety of cellular responses, including the opening or closing of ion channels in the cell membrane (note that this indirect mechanism is in addition to the mechanism by which G-proteins can interact directly with ion channels), the release or reuptake of Ca^{++} from intracellular storage sites, alterations in the activity of key cellular enzymes, and alterations in the expression of specific genes. Many of these effects are mediated by *protein kinases*, enzymes that regulate the activity of other proteins by phosphorylation. A given G-protein coupled receptor may activate more than one mechanism and produce multiple coordinated effects. G-protein coupled systems therefore have the capacity to mediate complex changes in neuronal function.

The best-known second messenger systems involve the enzymes *adenylyl cyclase* (Fig. 4-8) and *guanylate cyclase*. Adenylyl cyclase produces the second messenger cAMP from cytoplasmic ATP, and guanylate cyclase produces the second messenger cyclic GMP (cGMP) from cytoplasmic GTP. Another important second messenger system involves the enzyme phospholipase C, which hydrolyzes the membrane phospholipid *phosphatidylinositol 4,5-bisphosphate* to produce the second messengers *inositol 1,4,5-trisphosphate* (*IP_3*) and *diacylglycerol* (*DAG*). G-proteins can also interact with *phospholipase A_2*, which stimulates the formation of members of the eicosanoid family, and with another enzyme called phospholipase D.

REGULATION OF NEURONAL EXCITABILITY

As we have seen, neurotransmitters cause the opening or closing of ion channels in the postsynaptic membrane. If the transmitter signal is transduced by a G-protein mechanism, it also may have other effects. The result will be a transient, local change in the polarization of the postsynaptic membrane, called a *synaptic potential*. This potential consists of either a depolarization or a hyperpolarization of the membrane relative to the resting potential. (Membrane potentials and the electrical properties of neuronal cell membranes are discussed in more detail in Chapter 3.) Synaptic potentials are graded in amplitude, reflecting the varying strengths of the incoming synaptic signals that elicit them. They generally do not exceed 20 mV. Local potentials of this type spread passively over the membrane of the postsynaptic cell, gradually losing amplitude and dying out. However, if they reach a *trigger zone*—a locus at which action potentials can be initiated—they may contribute to the production or suppression of action potentials. An action potential is triggered whenever the membrane is depolarized beyond a certain threshold potential. Therefore *depolarizing* synaptic potentials tend to promote action potentials and are called *excitatory postsynaptic potentials* (*EPSPs*). Conversely, *hyperpolarizing* synaptic potentials inhibit the production of action potentials and are called *inhibitory postsynaptic potentials* (*IPSPs*).

In the central nervous system (CNS), a neuron is constantly bombarded by neurotransmitters, each of which can generate or modify a synaptic potential. Neurotransmitters that move the membrane toward depolarization (by reducing the –70 mV resting potential), with the resultant production of an action potential, are commonly called *excitatory neurotransmitters*. Neurotransmitters that move the membrane away from depolarization (by making the resting membrane more negative, the membrane is hyperpolarized) are frequently referred to as *inhibitory neurotransmitters*. *Because the postsynaptic response is actually elicited by the receptor rather than by the transmitter, the postsynaptic receptor determines whether a given neurotransmitter will be excitatory or inhibitory*. Some neurotransmitters can have either effect, depending on the type of postsynaptic receptor present.

Excitatory neurotransmitters act by promoting the opening of channels selective for cations that flow into the cell and further depolarize the membrane (usually Na^+ but sometimes Ca^{++}). A major excitatory neurotransmitter in the vertebrate CNS is the amino acid *glutamate*. Several distinct glutamate receptor subtypes have been identified, all of which are excitatory. One of the best characterized is the *NMDA receptor*, which is named after *N*-methyl-D-aspartate, an amino acid analog that is a potent stimulant of the receptor. Inhibitory neurotransmitters act by opening channels for K^+ or Cl^-. Important examples of inhibitory neurotransmitters are the amino acids *γ-aminobutyric acid (GABA)* and *glycine*.

It is essential to recognize that *a single chemical messenger can evoke either an EPSP or an IPSP, depending on the receptor to which it binds*. A good example is the neurotransmitter norepinephrine. Like glutamate, norepinephrine binds to multiple receptor subtypes. In the CNS receptors for norepinephrine fall into two categories, α-adrenergic and β-adrenergic receptors. Both types are G-protein coupled. The G-protein to which β-adrenergic receptors are coupled is of a type called G_i, which *stimulates* the activity of adenylyl cyclase and thus produces a rise in intracellular cAMP. This rise in cAMP leads to an EPSP. In contrast, the G-protein to which the α_2-adrenergic receptor subtype is coupled, called G_i, *inhibits* the activity of adenylyl cyclase. The resulting fall in intracellular cAMP leads to an IPSP. In both cases, cAMP acts through enzymes called *cAMP-dependent protein kinases*. In the pathway under discussion, the final targets are membrane ion channels, which open or close in response to the phosphorylation of sites on their cytoplasmic domains. Consequently, norepinephrine can elicit either an excitatory or inhibitory response, depending on the receptor.

MAINTENANCE OF THE SYNAPTIC ENVIRONMENT

The concentration of a chemical messenger in the synaptic cleft is critical to information transfer. However, the time frame during which a chemical message is active must be limited if a temporally discrete signal is to be produced. This is particularly true when neurons fire at rates of more than several depolarizations per second. Simple diffusion out of the synaptic cleft is rarely adequate to effectively terminate the postsynaptic signal. Accordingly, active mechanisms exist to reduce or eliminate chemical messengers in the synaptic cleft. The principal mechanisms are *enzymatic degradation of transmitter in the cleft* and *transporter-mediated uptake* across cell membranes.

Acetylcholine and the neuropeptides are examples of transmitters that are neutralized by enzymatic degradation in the cleft. Acetylcholine is cleaved by the enzyme *acetylcholinesterase*, which is synthesized by the neuron and inserted into the postsynaptic membrane near receptor sites. Neuropeptides are degraded through hydrolysis by the action of multiple *peptidases*, which are found in extracellular fluid.

The neurotransmitters whose action is terminated by uptake from the synaptic cleft include the monoamines (such as serotonin, histamine and the catecholamines) and the amino acid neurotransmitters GABA, glycine, glutamate, and aspartate. This uptake is accomplished by the action of specific membrane-bound *transport proteins*. The monoamine class of biogenic amines (including the catecholamines, serotonin, and histamine) are avidly removed from the synaptic space by such transport proteins.

In the case of norepinephrine, *reuptake* into the cytoplasm of the presynaptic terminal (a process known as *uptake 1*) is primarily responsible for terminating the action of the transmitter (Fig. 4-9). After reuptake, some norepinephrine is enzymatically degraded by the mitochondrial enzyme *monoamine oxidase* (MAO), whereas an additional fraction is retained in a cytoplasmic pool. The norepinephrine in this pool is an important target for drug action. Norepinephrine can also be removed through the action of a transporter on the postsynaptic membrane (*uptake 2*), although this process is usually less effective (Fig. 4-9). Norepinephrine transported into the postsynaptic neuron is degraded by the enzyme, *catechol-O-methyltransferase* (*COMT*).

In the CNS, glial cells, primarily astrocytes, express transporter proteins on their membranes and can remove transmitters from the synaptic cleft. The neurotransmitter dopamine is transported by glial cells in the substantia nigra. These glia metabolize the transported dopamine to inactive products by means of a form of monoamine oxidase unique to the CNS, *MAO-B*. Metabolism by MAO-B is essential to the activation of a dopaminergic neurotoxin called MPTP (1-methyl-4-phenyl-1,2,3,6-tetrahydropyridine). This substance destroys dopamine neurons in the substantia nigra, resulting in a syndrome markedly similar to that seen in patients with Parkinson's disease. The actions of the amino acids, GABA, glycine, glutamate, and aspartate are all terminated by active transport into neurons and glia. No active uptake mechanisms have been found that terminate the action of neuropeptides.

The mechanisms of termination of some other chemical messengers, such as adenosine, ATP, and nitric oxide, are less well understood. Nitric oxide is very labile; it undergoes redox reactions with membrane and cytoplasmic sulfhydryl moieties, reducing them and becoming oxidized itself. Specific ATPases may terminate the action of ATP functioning as a neurotransmitter.

PHARMACOLOGIC MODIFICATION OF SYNAPTIC TRANSMISSION

Drugs can alter virtually every level of neuronal and synaptic function. Therapeutic effects are most commonly achieved by actions of drugs on *neurotransmitter synthesis, vesicular uptake and storage, depolarization-induced exocytosis, neurotransmitter-receptor binding,* and *termination of neurotransmitter action.* Increasingly, drugs are being developed that modify neurotransmitter action through interaction with *postsynaptic effector systems.*

The Noradrenergic Synapse The noradrenergic synapse is used to illustrate the range of pharmacologic agents that can modify synaptic transmission (Fig. 4-9). Noradrenergic synapses use the neurotransmitter norepinephrine, and their postsynaptic receptors fall into the two classes introduced earlier, the α- and β-adrenergic receptor. In the peripheral nervous system, norepinephrine is a critical neurotransmitter in the regulation of the *sympathetic* division of the *autonomic nervous system.* In the CNS, norepinephrine is synthesized in neurons concentrated in several discrete brainstem regions; these areas send noradrenergic axons throughout the brain and exert widespread effects.

Norepinephrine is synthesized from the amino acid *tyrosine* in a sequence of three enzymatic reactions (Fig. 4-9). The first two occur in the cytoplasm; the final reaction takes place within the synaptic vesicle. Tyrosine is accumulated in the terminal by a membrane-bound amino acid carrier and is converted to dopa by *tyrosine hydroxylase* (Fig. 4-9, step 1). The rate of this conversion is the rate-limiting step in norepinephrine synthesis. Regulation of tyrosine hydroxylase is accomplished by phosphorylation of the enzyme by intracellular protein kinases, which increases the rate of enzymatic catalysis. The drug *α-methyltyrosine* can limit noradrenergic function by acting as a competitive inhibitor of tyrosine hydroxylase, thereby reducing the rate of norepinephrine synthesis.

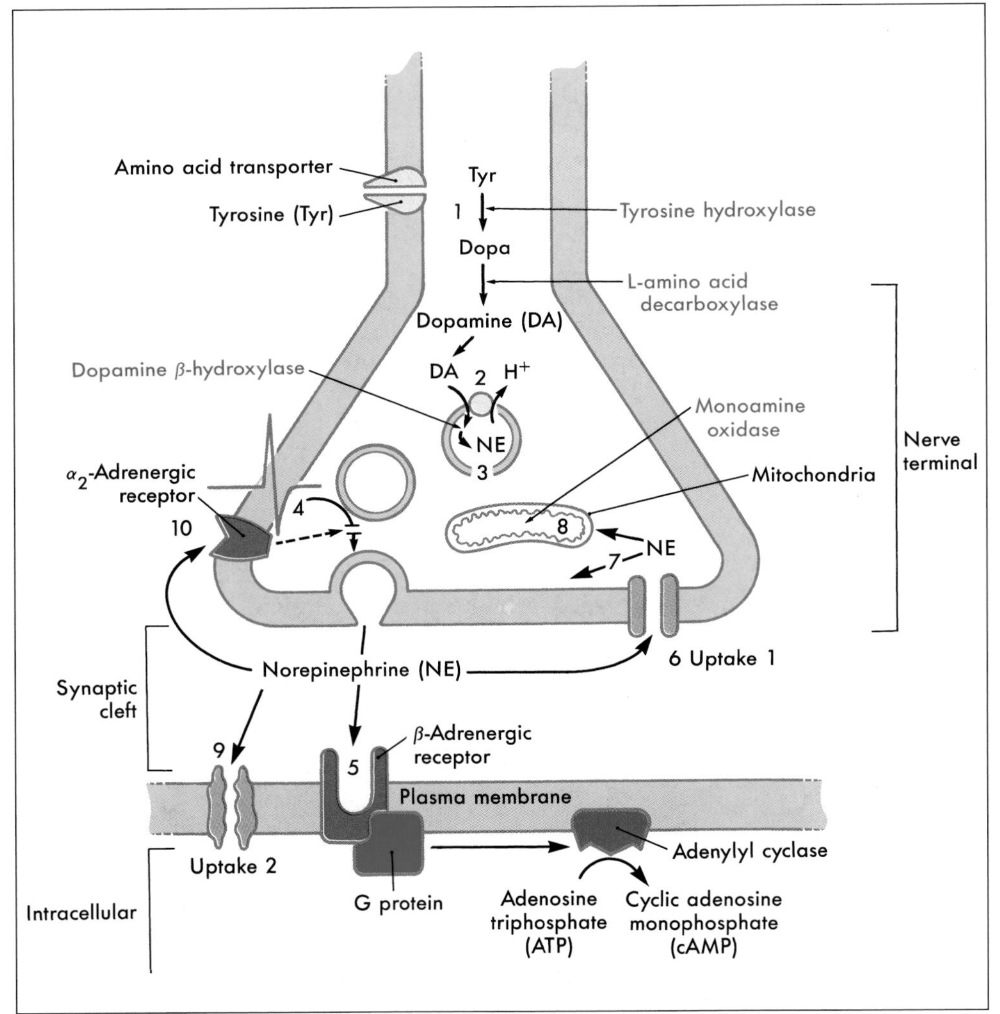

FIGURE 4-9 *The noradrenergic β_2-adrenergic receptor synapse. Pharmacologic agents are identified by the numerals. Synthetic and degradative enzymes are shown in red, membrane receptors, transporters, and ion channels in green, and the postsynaptic effector G-protein in blue:* (1) α-methyltyrosine *competitively inhibits tyrosine hydroxylase;* (2) reserpine *irreversible inhibits the monoamine-H^+ vesicular transport pump;* (3) α-methyldopa *acts as a false transmitter, displacing norepinephrine in the synaptic vesicle;* (4) guanethidine *blocks the ability of membrane depolarization to cause exocytotic release of vesicle contents;* (5) propranolol *is a competitive antagonist at β-adrenergic receptors;* (6) cocaine *blocks synaptic membrane reuptake of norepinephrine (Uptake 1);* (7) tyramine *displaces norepinephrine from a cytoplasmic storage pool back into the synaptic cleft;* (8) pargyline *blocks the degradation of norepinephrine by mitochondrial monoamine oxidase;* (9) corticosterone *prevents uptake of norepinephrine by the postsynaptic membrane (Uptake 2);* (10) yohimbine *is a competitive antagonist at presynaptic, autoinhibitory α_2-adrenergic receptors.*

Dopa is converted to *dopamine* (a neurotransmitter in its own right) by *L-amino acid decarboxylase*. Dopamine is transported into, and concentrated within, the synaptic vesicle by a monoamine-H^+ transporter (Fig. 4-9, step 2). The accumulation of dopamine (and ultimately, norepinephrine) can be prevented by *reserpine*, a plant alkaloid that irreversibly inactivates the vesicular transporter (Fig. 4-9, step 2). The inability to fill vesicles with neurotransmitter results in a progressive reduction in the level of transmitter in the axon terminal, which inhibits neurotransmission.

Within the vesicle, dopamine is converted to norepinephrine by *dopamine β-hydroxylase*. The accumulation of both dopamine and norepinephrine can also be reduced by administration of *α-methyldopa*. This dopa analog is enzymatically converted in successive steps to α-methyldopamine and α-methylnorepinephrine, which takes the place of the normal synthetic products, resulting in reduction of noradrenergic transmission (Fig. 4-9). Elevated sympathetic nerve activity contributes to hypertensive cardiovascular disease. Thus reserpine and α-methyldopa are important drugs in the management of this disease.

The drug *guanethidine* interferes with the coupling between excitation of the nerve terminal and exocytotic release of norepinephrine, thereby reducing the amount of norepinephrine released. In addition, guanethidine acts like reserpine to inactivate vesicle transport. This drug, too, is useful in the treatment of hypertension.

Once released into the synaptic cleft, norepinephrine can bind to two sets of receptors: (1) postsynaptic α- or β-adrenergic receptors, which elicit the postsynaptic response, or (2) presynaptic receptors, partially but not exclusively, of the α_2-adrenergic subtype, which are involved in autoregulation. Eventually, the norepinephrine is removed from the synapse by reuptake into the presynaptic terminal or uptake into the postsynaptic cell. *Propranolol* is an example of a drug that interferes with the binding of norepinephrine to postsynaptic β-adrenergic receptors. This drug binds competitively to the receptor and prevents its activation, thereby blocking the postsynaptic response (in this case, the rise in intracellular cAMP mediated by G_s activation of adenylyl cyclase). Propranolol is used widely in cardiovascular medicine.

The autoregulatory presynaptic receptors for norepinephrine exert an inhibitory effect over the amount of norepinephrine released in response to an action potential. They influence both the synthesis of norepinephrine and its exocytotic release. The α_2-adrenergic receptors that are responsible for these effects can be blocked by the drug *yohimbine*. The resulting loss of autoinhibition increases the amount of norepinephrine released and enhances noradrenergic function.

Drugs such as *cocaine* inhibit the presynaptic reuptake of norepinephrine, thus prolonging the synaptic activity of the transmitter and resulting in exaggerated postsynaptic responses. As explained earlier, norepinephrine that has undergone reuptake can either be degraded by the mitochondrial enzyme *monoamine oxidase* or retained in a cytoplasmic pool. Monoamine oxidase inhibitors such as *pargyline* increase the amount of norepinephrine in the cytoplasmic pool and thereby enhance noradrenergic transmission. Agents such as *tyramine* displace norepinephrine from the cytoplasmic pool back into the synapse, also enhancing noradrenergic activity.

SOURCES AND ADDITIONAL READINGS

Cooper JR, Bloom FE, Roth RH: *The Biochemical Basis of Neuropharmacology*. Oxford University Press, New York, 1991.

Hall ZW: *An Introduction to Molecular Neurobiology*. Sinauer Associates, Sunderland, MA, 1992.

Jessell TM, Kandel ER: Synaptic transmission: A bidirectional and self-modifiable form of cell-cell communication. Cell 72/Neuron 10 (Suppl):1–30, 1993.

Kelly RB: Storage and release of neurotransmitters. Cell 72/Neuron 10 (Suppl):43–53, 1993.

Kobilka B: Adrenergic receptors as models for G protein-coupled receptors. Annu Rev Neurosci 15:87–114, 1992.

Rahmann H, Rahmann M: *The Neurobiological Basis of Memory and Behavior*. Springer-Verlag, New York, 1992.

Trimble WS, Linial M, Scheller RH: Cellular and molecular biology of the presynaptic nerve terminal. Annu Rev Neurosci 14:93–122, 1991.

Zoli M, Agnati LF, Hedlund PB, Li XM, Ferre S, Fuxe K: Receptor-receptor interactions as an integrative mechanism in nerve cells. Mol Neurobiol 7:293–334, 1993.

CHAPTER FIVE

Development of the Nervous System

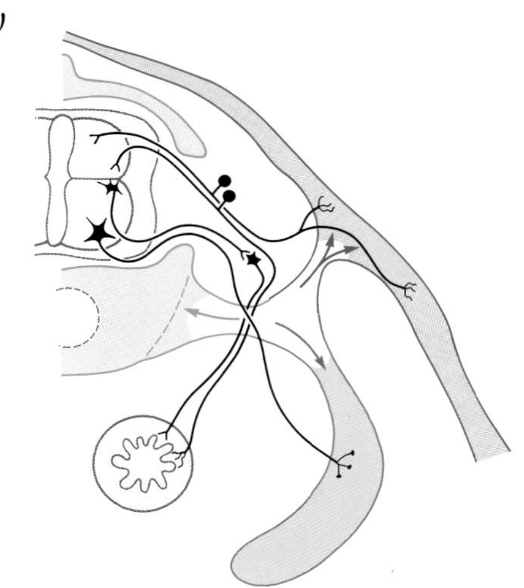

O. B. EVANS • J. B. HUTCHINS

The central nervous system (CNS) develops from primitive ectoderm, one of the three germ layers of the embryo. From a few dozen cells, the brain becomes an organ weighing about 800 g at birth, 1,200 g at 6 years of age, and about 1,400 g in the adult. Most, but not all, neurons undergo their last cell division before birth. The development of a fully functional nervous system requires division and migration of nerve cells and the formation of synaptic connections.

OVERVIEW

Considering the complex embryology of the human CNS, it is remarkable that there are so few congenital CNS defects. Although 3% of births are associated with major malformations of the CNS, most fetuses and infants in this category do not survive. About 75% of spontaneously aborted fetuses and 40% of infants who die within the first year of life have major CNS malformations.

The basic form of the human CNS is complete by about the sixth week of gestation. The next phases, which include cellular proliferation and migration, are most prominent in the second trimester of gestation but continue until term. Myelination peaks during the third trimester but continues until adulthood. The development of synaptic connections between neurons and the response of the brain to its experiences result in its functional maturity.

From a few primordial cells, about 100 billion neurons develop, each with thousands of contacts with other neurons. Through this network of interconnecting neurons, the human brain is capable not only of directing the movement of the body and sensing the environment but also of thinking, reasoning, experiencing emotions, and dreaming.

BRAIN DEVELOPMENT

The first neural tissue appears at the end of the third week of embryonic development, when the embryonic disc is composed of *ectoderm*, *mesoderm*, and *endoderm*. A specialized part of the ectoderm, the *neuroectoderm*, gives rise to the brain, spinal cord, and peripheral nervous system (Fig. 5-1).

Induction The *notochord* arises from axial mesoderm at about 16 days and is completely formed by the beginning of the fourth week. It defines the longitudinal axis of the embryo, determines the orientation of the vertebral column, and persists as the *nucleus pulposus* of the intervertebral discs. One important function of the notochord is *to induce the overlying ectoderm to form the neural plate* (Fig. 5-1A,B). Associated with this process is a production of *cell adhesion molecules* in the notochord. These molecules diffuse from the notochord into the neural plate and function to join the primitive neuroepithelial cells into a tight unit.

Within the neuroectoderm, some neuroepithelial cells elongate and become spindle-shaped. This cellular elongation, also induced by the notochord, forms the *neural plate* and is completed by the end of the third week of gestation (Fig. 5-1A). The neural plate gives rise to most of the nervous system.

Primary Neurulation The CNS develops from a hollow structure called the *neural tube*, which is produced by *neurulation*. There are two neurulation processes. Most of the neural tube forms from the neural plate by a process of infolding called *primary neurulation*. This part of the neural tube will give rise to the brain and to the spinal cord through lumbar levels. The caudalmost portion of the neural tube, which will give rise to sacral and coccygeal levels of the cord, is formed by a process called *secondary neurulation*. Secondary neurulation is described in the next section. By about the eighteenth day after fertilization, the neural plate begins to thicken at its lateral margins (Fig. 5-1B). This elevates the edges of the neural plate to form *neural folds*. At about 20 days, the neural folds first contact each other to begin the formation of the *neural tube*. This fusion initially takes place on the dorsal midline at what will become cervical levels of the spinal cord and proceeds, zipper-like, in rostral and caudal directions (Fig. 5-1C,D). During the process, the lumen of the neural tube, called the neural canal, is open to the amniotic cavity both rostrally and caudally (Fig. 5-1D). The rostral opening, the *anterior neuropore*, closes on about 24 days, and the caudal opening, the *posterior neuropore*, closes about 2 days later.

Neurulation is brought about by morphologic changes in the *neuroblasts*, the immature and dividing future neurons. As mentioned above, these cells are elongated and are oriented at right angles to the dorsal surface of the neural plate, which will be the inner wall of the neural canal. Microfilaments in each cell form a circular bundle parallel to the future luminal surface, whereas microtubules extend along the length of the cell. The contraction of the circular bundle of microfilaments causes the microtubules to splay out like the rays of a fan. This forms an elongated conical cell with its apex at the neural groove and its base at the edge of the neural fold. Neurulation does not occur in embryos exposed to colchicine, which depolymerizes microtubules, or to cytochalasin, which inhibits microfilament-based contraction.

Congenital malformations associated with defective neurulation are called *dysraphic defects*. Because of the intimate relationship of neural tissue to the surrounding bone, meninges, muscles, and skin and because of their interdependence via inductive factors, failure of neurulation also impairs the formation of these surrounding structures.

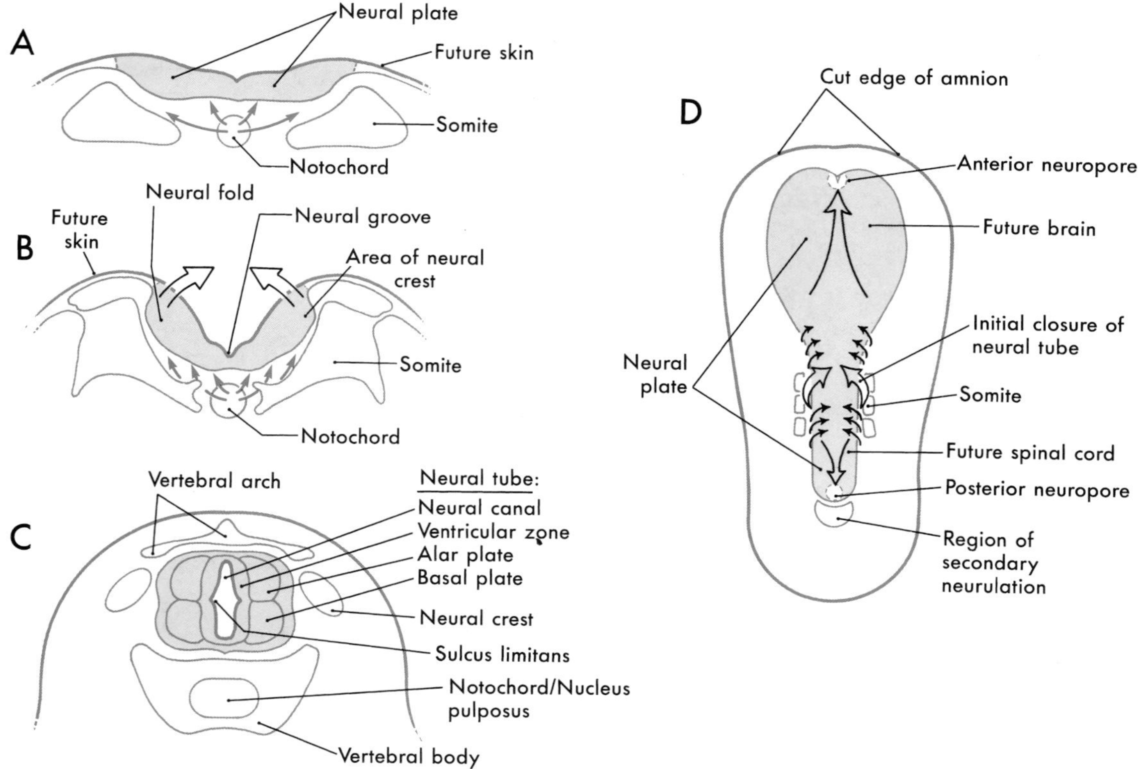

FIGURE 5-1 *Basic development of the nervous system. Cross sections (A–C) showing the transition from neural plate (A) to neural tube (C). A dorsal view (D) of the neural plate shows the point of initial closure and the direction of closure (small arrows) toward anterior and posterior neuropores. Green arrows (A, B) represent induction of neural tube formation.*

Most dysraphic disorders occur at the location of the anterior or posterior neuropore. Failure of the anterior neuropore to close causes *anencephaly* (Fig. 5-2). In this defect, the brain is not formed, the surrounding meninges and skull may be absent, and there are facial abnormalities. The defect extends from the level of the *lamina terminalis*, the site of anterior neuropore closure, to the region of the *foramen magnum*. Anencephaly occurs in about 5 of every 10,000 live births and results in death.

An *encephalocele* is a herniation of intracranial contents through a defect in the cranium (*crania bifidum*) (Fig. 5-3A). The cystic structure may contain only meninges (*meningocele*), meninges plus brain (*meningoencephalocele*), or meninges plus brain and a part of the ventricular system (*meningohydroencephalocele*) (Fig. 5-3B–D). Encephaloceles are most common in the occipital region, but they may also occur in frontal or parietal locations.

The *Arnold-Chiari malformation* is a congenital herniation of the cerebellar vermis through the foramen magnum, which usually causes pressure on the medulla oblongata and cervical spinal cord (Fig. 5-4). This defect may go unnoticed until early adulthood and is often associated with a cavitation of the spinal cord (*syringomyelia*) or of the medulla (*syringobulbia*).

Defects in the closure of the posterior neuropore cause a range of malformations known collectively as *myeloschisis*. The defect always involves a failure of the vertebral arches at the affected levels to form completely

A B

FIGURE 5-2 *Lateral (A) and frontal (B) views of anencephaly. Note the associated cranial and facial abnormalities. (A, courtesy of Dr. J. Fratkin.)*

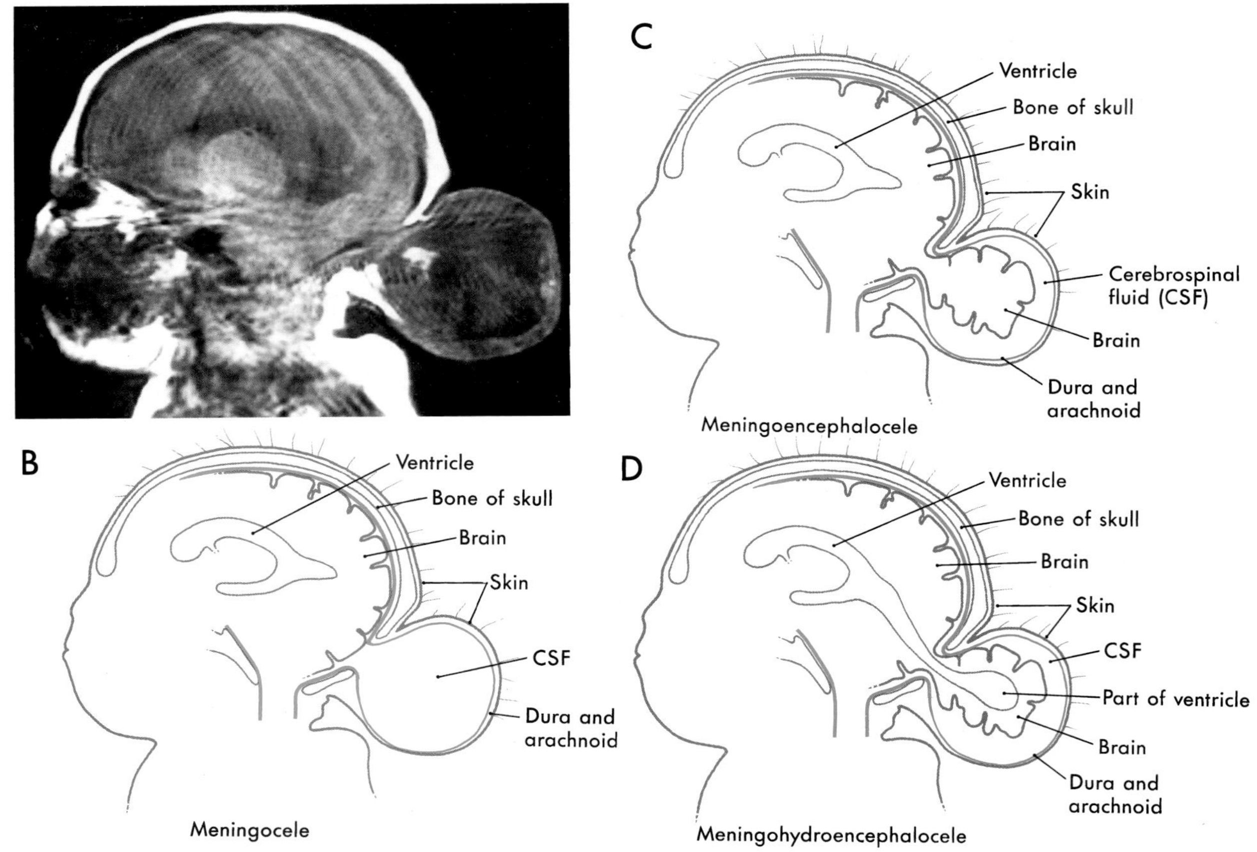

FIGURE 5-3 *Sagittal views of occipital encephaloceles. Magnetic resonance image (MRI) of meningohydroencephalocele (A) and drawings of meningocele (B), meningoencephalocele (C), and meningohydroencephalocele (D).*

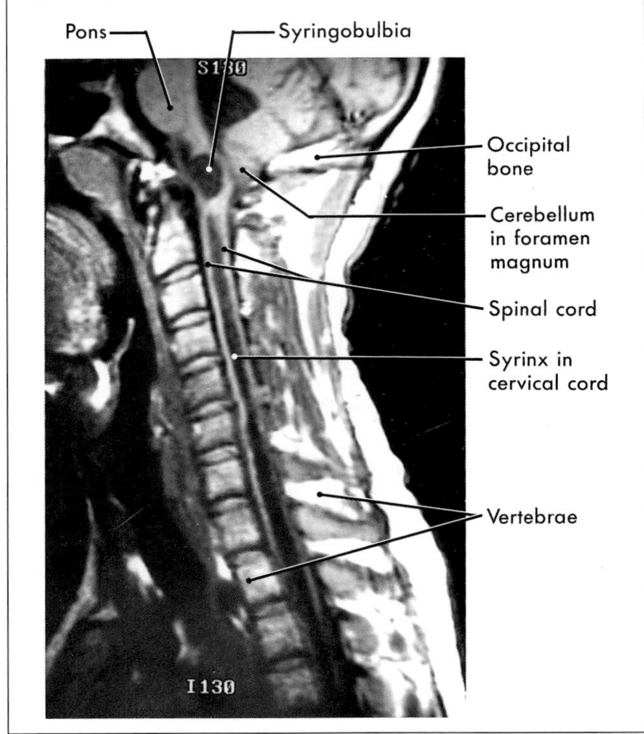

FIGURE 5-4 *Sagittal MRI of a patient with Arnold-Chiari malformation and with cavitations in the medulla (syringobulbia) and cervical spinal cord (syringomyelia).*

and fuse to cover the spinal cord (*spina bifida*). If that is the only defect, and the skin is closed over it, the condition is called *spina bifida occulta* (Fig. 5-5A,B). The site of the defect is usually marked by a patch of dark, coarse hairs. If the skin is not closed over the vertebral defect, the malformation is called *spina bifida aperta*.

As with occipital encephaloceles, a cystic mass (*spina bifida cystica*) may also accompany spina bifida (Fig. 5-5C,D). This saccular structure may contain only meninges and cerebrospinal fluid (CSF) (*meningocele*) or meninges and CSF plus spinal neural tissue (*meningomyelocele*). In the latter case, the neural tissue may be the lower part of the spinal cord or, more commonly, a portion of the cauda equina. Infants with meningomyelocele may be unable to move their lower limbs (*paraplegic*) or to perceive pain (*anesthetic*) at the level of the lesion. These infants may also have other CNS malformations, such as *hydrocephalus*. The incidence of meningomyelocele is approximately 5 per 10,000 births.

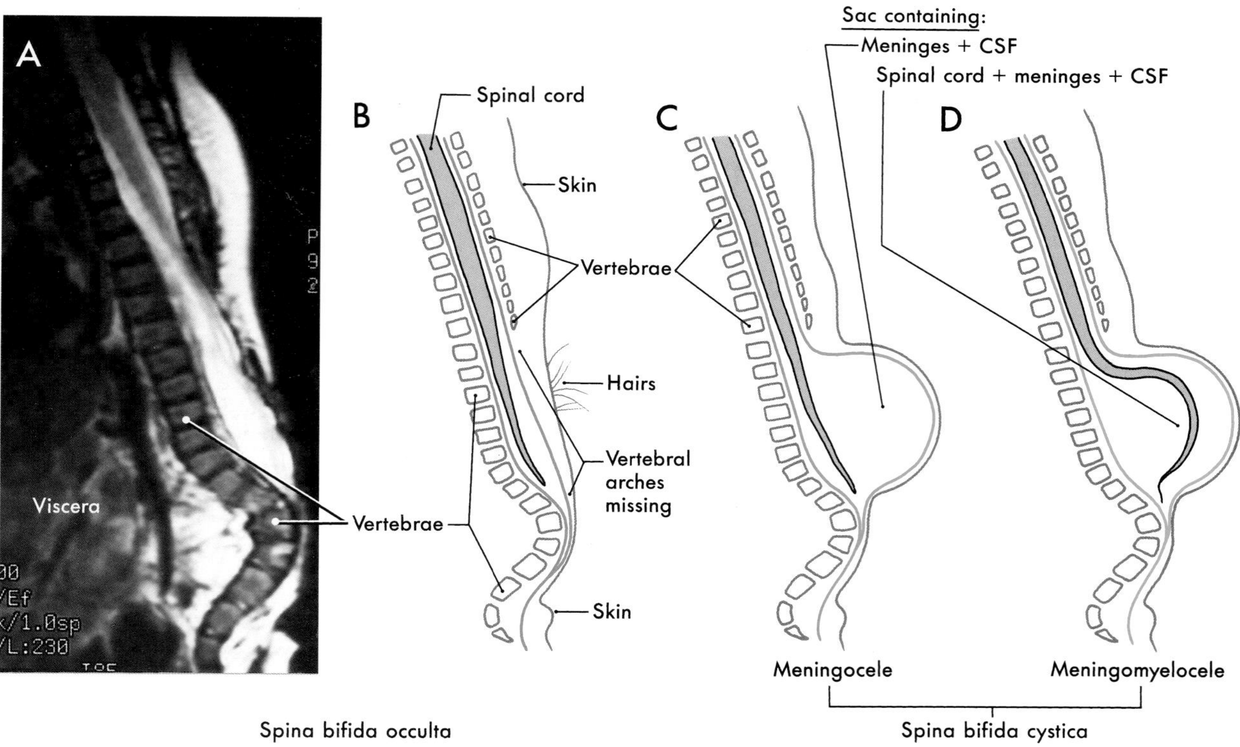

FIGURE 5-5 *Sagittal views of spina bifida malformations. MRI (A) and corresponding views showing spina bifida occulta (A, B) and spina bifida cystica (C, meningocele; D, meningomyelocele).*

Secondary Neurulation The sacral and coccygeal segments of the spinal cord and their corresponding dorsal and ventral roots are formed by *secondary neurulation* (see Fig. 5-1D). This process begins on day 20 and is complete by about day 42. A cell mass, the *caudal eminence*, appears just caudal to the neural tube and then enlarges and cavitates. The caudal eminence joins the neural tube, and its cavity becomes continuous with the neural canal.

Myelodysplasia refers to malformations of the parts of the neural tube formed by secondary neurulation. In most cases the malformation is covered with skin, but the site may be marked by unusual pigmentation, hair growth, *telangiectasias* (large superficial capillaries), or a prominent dimple. A common abnormality is *tethered cord syndrome*, in which the conus medullaris and filum terminale are abnormally fixed to the defective vertebral column. The sustained traction damages the cord, with subsequent loss of sensations from the legs and feet, and problems of bladder control.

Primary Brain Vesicles During the fourth week after fertilization, in which the anterior neuropore closes, there is rapid growth of neural tissue in the cranial region. The three *primary brain vesicles* formed are *prosencephalon* (forebrain), *mesencephalon* (midbrain), and *rhombencephalon* (hindbrain) (Fig. 5-6A,B). At the rhombencephalon-spinal cord junction there is a slight bend in the developing neural tube; this is the *cervical flexure*. A second bend in the neural tube at the level of the mesencephalon is the *cephalic* (or *mesencephalic*) *flexure*.

Secondary Brain Vesicles During the fifth week, the appearance of additional flexures divides the three primary brain vesicles into five *secondary brain vesicles* (Fig. 5-6C,D). The pontine flexure appears dorsally and divides the hindbrain into the *myelencephalon* and *metencephalon*, whereas the mesencephalon remains undivided at the level of the cephalic flexure. The forebrain is divided into the *diencephalon* and *telencephalon* by the budding of the latter vesicle from the prosencephalon (Fig. 5-6C,D). It subsequently expands caudally, dorsally, and laterally; and the bend separating it from the diencephalon is sometimes called the *telencephalic flexure*.

Diencephalon and Cerebral Hemispheres The main structures of the forebrain develop during the second month of gestation. Because the mesoderm in this region is simultaneously forming facial structures, abnormalities of forebrain development are often associated with facial defects (Fig. 5-2). The process of forebrain development is often referred to as *central induction*.

At about the end of the fifth week, the telencephalon gives rise to two lateral expansions, called the *telencephalic (cerebral) vesicles* (Fig. 5-6C,D). These are the pri-

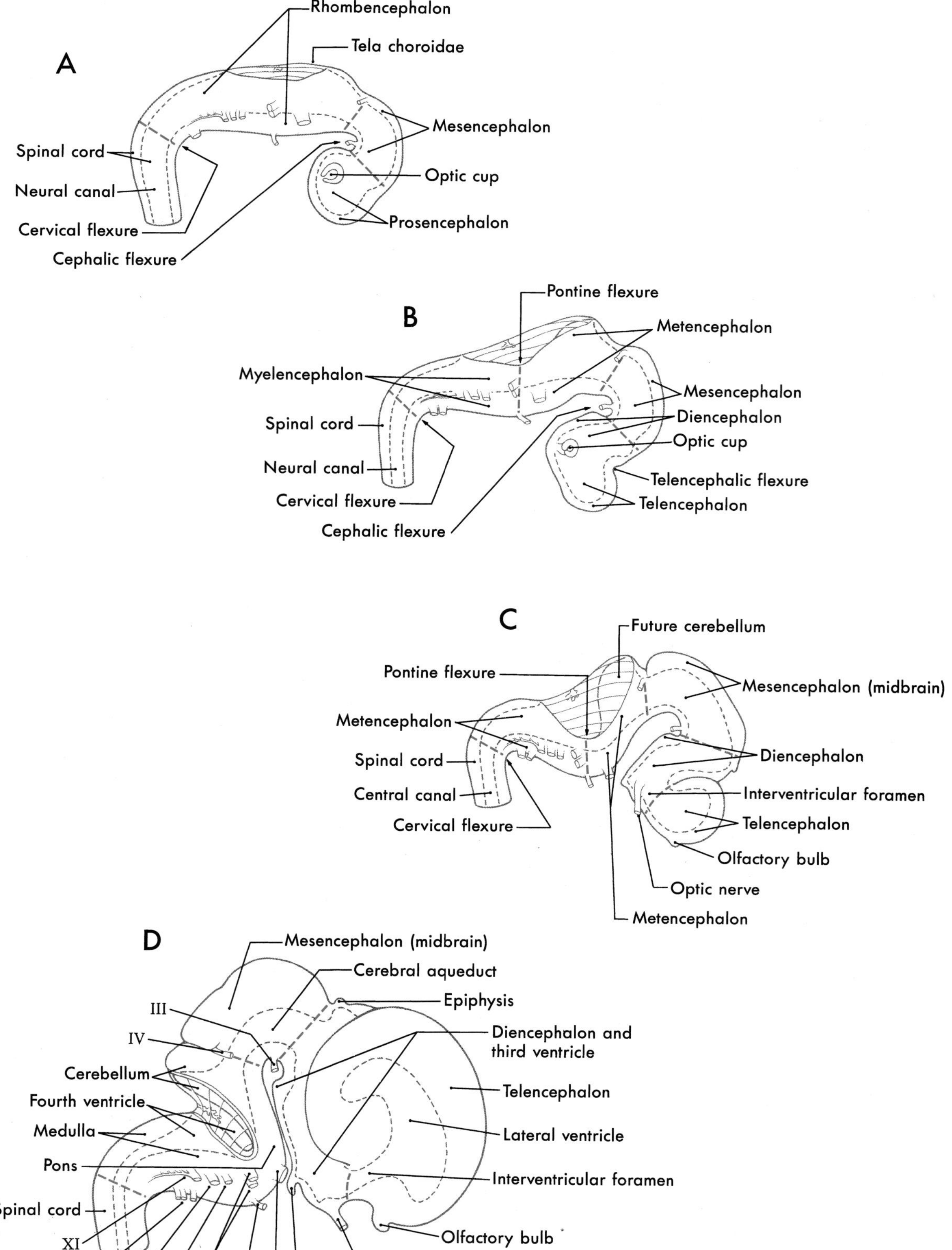

FIGURE 5-6 *Developmental sequence from three to five brain vesicles. Three brain vesicles (A, about 4.75 weeks of gestation) divide into five vesicles (B, about 6 weeks of gestation) with the appearance of additional flexures. In subsequent stages (C, about 6.5 weeks of gestation; D, about 8.5 weeks gestation) there is rapid enlargement of forebrain regions, especially the telencephalon. Note the ventricular spaces (dash lines, A–D) follow the shape changes in the brain.*

mordia of the cerebral hemispheres. Their adult derivatives include the cerebral cortex and the subcortical white matter (including the internal capsule), the olfactory bulb and tract, the basal ganglia, the amygdala, and hippocampus. The *diencephalon* develops into the thalamic nuclei and associated structures and also gives rise to the optic cup, which eventually forms the optic nerve and retina. By 10 weeks of development, the major structures of the CNS are clearly recognizable by their morphology, and all structures in the brain are present by the end of the first trimester.

The sequence of events by which the primitive prosencephalon differentiates into the diencephalic and telencephalic vesicles is called *prosencephalization*. Failure of the prosencephalon to undergo cleavage results in a malformation called *holoprosencephaly* (Fig. 5-7A). In its most severe form (*alobar holoprosencephaly*) there is a large single ventricle in this part of the brain, the thalamus is poorly developed, and many structures (corpus callosum, interhemispheric sulcus and falx cerebri, olfactory structures) are lacking. In *semilobar holoprosencephaly* (Fig. 5-7B,C), there is some separation of the hemispheres (more prominent in occipital areas) and partial development of the falx cerebri. The hemispheres are partially divided into lobes and gyri, and there are rudimentary, but enlarged, lateral and third ventricles. Most infants with holoprosencephaly also have facial malformations. These may be as subtle as mild *hypotelorism* (unusually close-set eyes) or as obvious as the presence of only a single, midline eye (*cyclops*) accompanied by a rudimentary nasal structure (*proboscis*). In general, the more severe the brain malformation, the more severe the facial defect.

Ventricular System The ventricular system is an elaboration of the lumen of cephalic portions of the neural tube, and its development parallels that of the brain (Figs. 5-6A–D; 5-8A–D). This process, also discussed in Chapter 6, is summarized here. The cavities of the telencephalic vesicles become the *lateral ventricles*; the diencephalic cavity becomes the *third ventricle*; and the rhombencephalic cavity becomes the *fourth ventricle*. The cavity of the mesencephalon becomes the narrow *cerebral aqueduct* (*of Sylvius*) connecting the third and fourth ventricles, and the openings between the lateral ventricles and the third ventricle become the *intraventricular foramina* (*of Monro*).

The ventricular system is lined with ependymal cells. Each ventricle originally has a thin roof composed of an internal layer of ependyma and an outer layer of delicate connective tissue (pia mater). In each ventricle, blood vessels invaginate this membrane to form the *choroid plexus*.

Openings that arise in the caudal roof of the fourth ventricle during development form a communication between the ventricular system and the subarachnoid space. These are the midline *medial aperture* (*foramen of Magendie*) and the paired lateral *foramina of Luschka*. Although these foramina develop slowly, they are patent by the end of the first trimester. CSF, most of which is produced by the choroid plexi of the lateral and third ventricles, passes through the ventricular system and into the subarachnoid space, from which it is absorbed into the venous system.

If the flow of CSF through the ventricles is obstructed during *in utero* development, the ventricular system can become markedly dilated, a condition called *congenital hydrocephalus* (Fig. 5-9A). The cerebral aqueduct, being only about 0.5 mm in diameter, is a prime point at which CSF flow may be blocked. Congenital *atresia* (failure to form) of the aqueduct can occur as an isolated event, be inherited, or be associated with CNS deformities (Fig. 5-8E). *Stenosis*, or total obstruction,

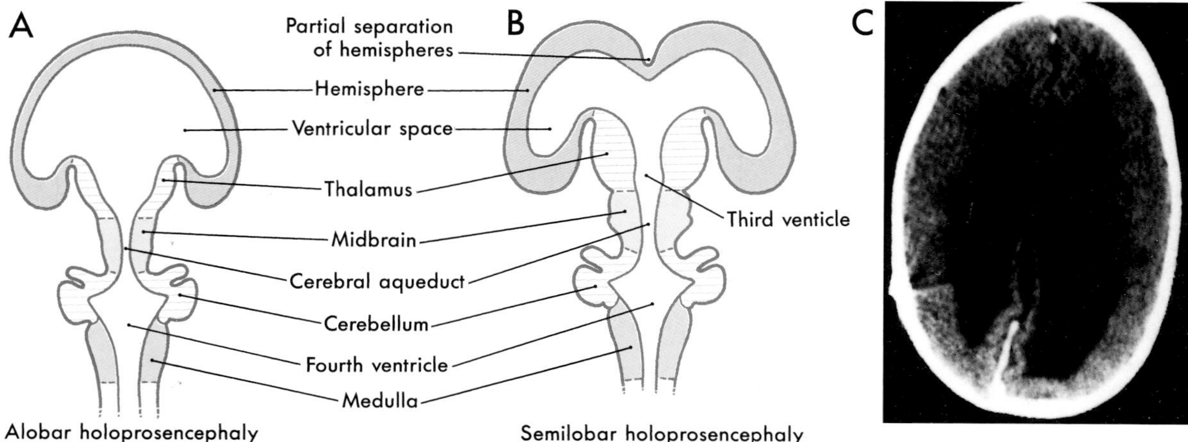

FIGURE 5-7 *Alobar (A) and semilobar (B, C) holoprosencephaly. In the alobar form (A) the single brain vesicle has a horseshoe-shaped ventricle and many major brain structures are absent. Although ventricles are present in the semilobar form (B, C-axial MRI), they are enlarged and many major structures are only partially developed.*

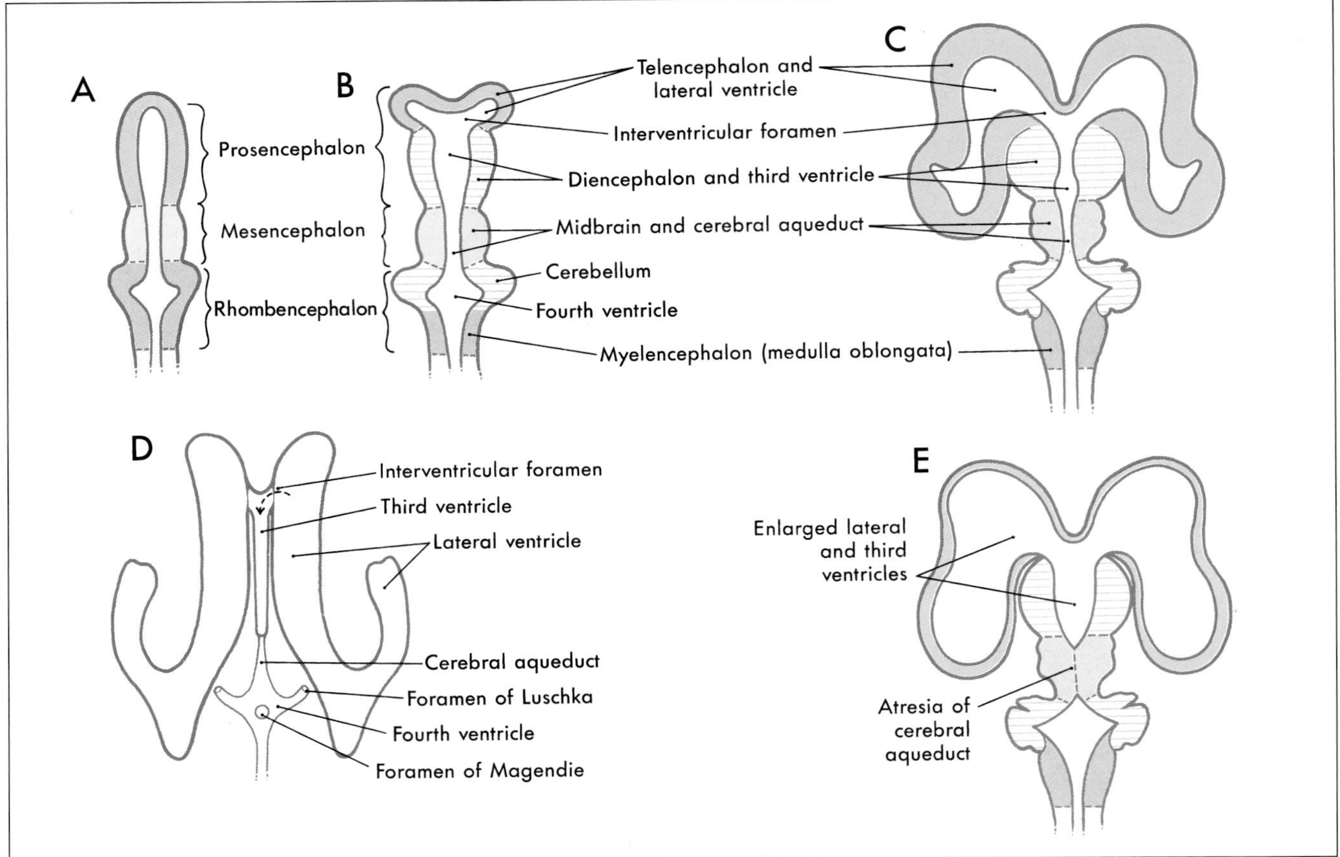

FIGURE 5-8 *Development of the ventricular system and associated brain divisions (A–C) and the general adult pattern (D) as seen from the dorsal perspective. Failure of the cerebral aqueduct to form causes the third and lateral ventricles to enlarge (E).*

from cellular debris associated with an infection or from an *intraventricular hemorrhage*, may also occlude this narrow passage.

Enlarged ventricles are also seen in the *Dandy-Walker malformation*. These patients have a cystic dilation of the fourth ventricle accompanied by a variable degree of *aplasia* (absence or defective development) of the *cerebral vermis* (Fig. 5-9B). In some cases, there is also obstruction of the foramina of the fourth ventricle.

PERIPHERAL NERVOUS SYSTEM

Neural Crest The *peripheral nervous system* develops mostly from cells of the *neural crest* (Figs. 5-1, 5-10). These cells arise from the lateral edge of the neural plate, detach, and migrate lateral to the spinal cord and to the developing brain. The ganglia of certain cranial nerves also receive a contribution from clumps of specialized epidermal cells called *placodes* in the cephalic region. Specifically, these are the ganglia of cranial nerves V, VII, VIII, IX, and X. The neural crest gives rise to essentially all the remaining cells of the peripheral nervous system, as well as to a number of other structures (Table 5-1). The two basic types of neural-crest-derived neurons considered here are the *pseudounipolar* cells of sensory ganglia and the *multipolar* cells of sympathetic and parasympathetic ganglia.

Pseudounipolar Cell Development Neural crest cells that form neurons of *sensory ganglia* are initially fusiform and apolar in shape (Fig. 5-10). These *neuroblasts* each give rise to a *peripheral (distal) process* that innervates the appropriate target tissue and to a *central (proximal) process* that enters the spinal cord or brainstem. As development proceeds, the processes near the cell body fuse to form a T-shaped structure. This gives peripheral sensory cell bodies their characteristic appearance and explains why they are called *pseudounipolar cells*. These cell bodies reside in the *dorsal (or posterior) root ganglia* adjacent to the spinal cord and in the sensory *cranial nerve ganglia*. A notable exception is the pseudounipolar cells that form the *mesencephalic nucleus of the trigeminal nerve*. These are sensory cells that failed to migrate with the neural crests.

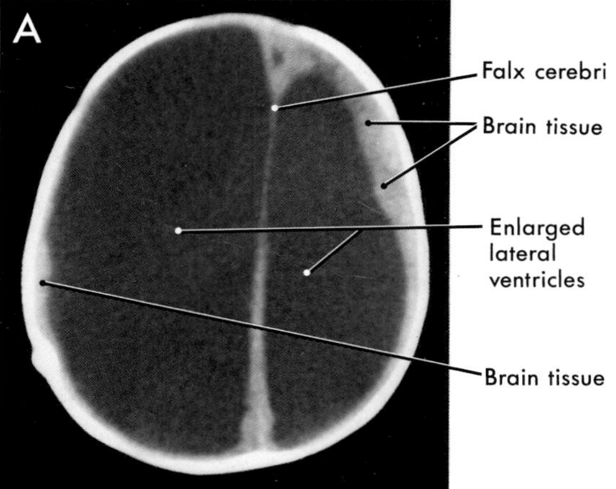

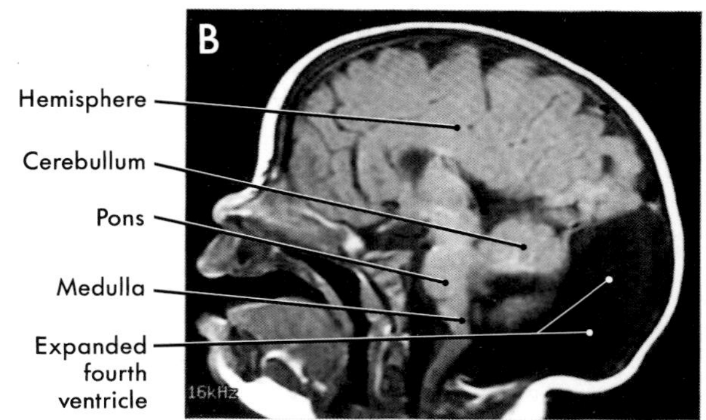

FIGURE 5-9 *Axial MRI (A) of a 9-month-old boy with congenital hydrocephalus. The brain is compressed as a thin rim on the inner surface of the skull and on the falx cerebri. Sagittal MRI (B) of a 5-month-old boy with a Dandy-Walker malformation. In addition to aplasia of the cerebellum and a pronounced enlargement of the fourth ventricle, this patient also has almost complete agenesis of the corpus callosum.*

Cranial Nerve Ganglia Cranial nerves V (trigeminal), VII (facial), IX (glossopharyngeal), and X (vagus) have sensory ganglia that originate from neural crest and placode cells and contain pseudounipolar cell bodies. These are the *trigeminal* or *semilunar* (V) ganglion, the *geniculate* ganglion (VII), the *superior* and *inferior* (IX) ganglia of the glossopharyngeal nerve, and the *jugular* and *nodose* (X) ganglia. The distal processes of these cranial nerves travel as the sensory components of the corresponding cranial nerve, whereas the proximal processes innervate the appropriate cranial nerve nuclei in the brainstem.

The cell bodies of the ganglia of cranial nerve VIII (the *vestibulocochlear nerve*) arise primarily from the *otic placode*, with a small contribution from neural crest. These ganglion cells retain a bipolar shape in the adult (Fig. 5-10).

TABLE 5-1

PRINCIPAL STRUCTURES DERIVED FROM NEURAL CREST CELLS

Neural Elements
Neurons of:
dorsal root ganglia
paravertebral (sympathetic chain) ganglia
prevertebral (preaortic) ganglia
enteric ganglia
parasympathetic ganglia of cranial nerves VII, IX, and X
sensory ganglia of cranial nerves V, VII, VIII, IX, and X*
Non-neural Elements
Schwann cells
Melanocytes
Odontoblasts
Satellite cells of peripheral ganglia
Cartilages of the pharyngeal arches
Ciliary and pupillary muscles
Chromaffin cells of the adrenal medulla
Pia and arachnoid of the meninges

*Some of the sensory cells in these ganglia arise from placodes.

Dorsal (Posterior) Root Ganglia Pseudounipolar cells of *dorsal* or *posterior root ganglia* are derived from neural crest. Each spinal nerve and its corresponding ganglion are associated with a segment (or *somite*) of the developing embryo (Fig. 5-11). As the somites grow out to form portions of the body's connective tissue and musculature, the peripheral processes of the developing pseudounipolar cells of the corresponding dorsal root ganglia grow distally, using the extracellular matrix of the underlying tissue as a guide (Fig. 5-11).

The matrix molecules *fibronectin* and *laminin* contain the amino acid sequence arginine-glycine-aspartate (called the *RGD sequence* after the one-letter abbreviations for these amino acids). This sequence is recognized by proteins known as *integrins* on the surface of neural crest cells. The selective adhesion of the peripheral process of a neural crest cell to the RGD sequence of the extracellular matrix is one mechanism probably involved in guiding the distal processes to their correct targets.

The segmental nature of the embryo is reflected in the segmental sensory innervation of the body surface (Fig. 5-11; see also Fig. 17-4). These segments, known as *dermatomes* (Latin: "skin slices"), are important in the diagnosis of many neurologic disorders.

Visceral Motor System The post-ganglionic sympathetic and parasympathetic neurons of the visceral motor system are also derived from the neural crests. These neurons, although multipolar in adults, originate from apolar neuroblasts (Fig. 5-10). Some of these cells remain near their site of origin to form the *sympathetic chain ganglia* adjacent to the vertebral column. Other cells migrate with branches of the aorta to form the sympathetic *prevertebral ganglia*.

Most of the autonomic (visceromotor) neurons of the digestive tract (*Auerbach's* and *Meissner's* plexi) are formed by neural crest cells that migrate from the area of the rhombencephalon. Consequently these cells receive vagal innervation in the adult. Visceromotor (autonomic) neurons of the descending colon and pelvic structures are derived from neural crest cells that arise from sacral cord levels during secondary neurulation.

The human syndrome *congenital megacolon* (*Hirschsprung's disease*) closely resembles an animal model that has excess extracellular matrix molecules in the colon. This results in aberrant migration of neural crest derived cells. In this animal model, and in Hirschsprung's disease, the enteric ganglia fail to migrate into the lower bowel. In the absence of these cells, no sensory signal indicating the presence of feces in the colon, is sent to the CNS. Therefore no motor signal is sent to control expulsion of feces. Another clinical entity, *familial dysautonomia*, also reflects aberration in the development of neural crest derivatives. These patients have both sensory symptoms (impaired pain and temperature perception) and autonomic symptoms (cardiovascular instability, gastrointestinal dysfunction).

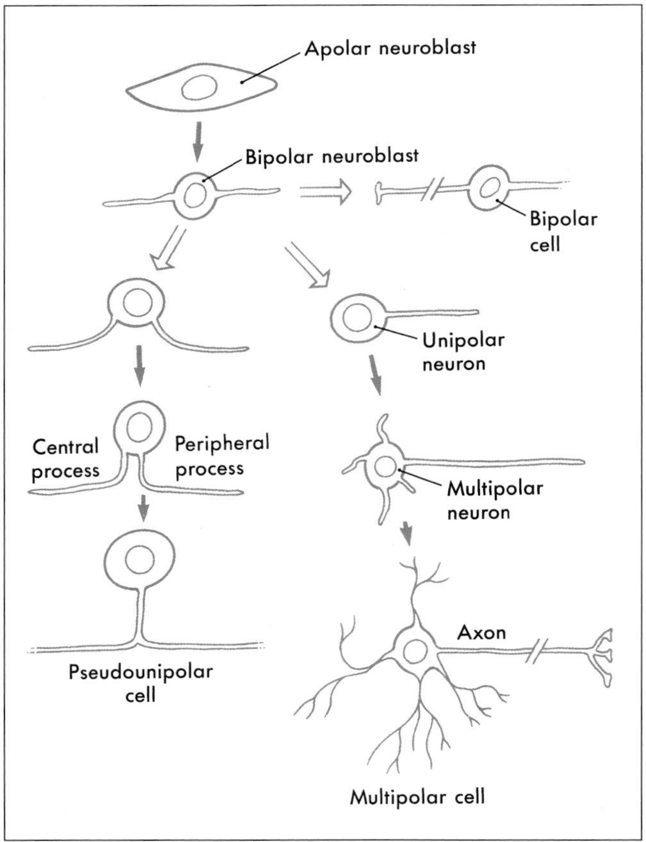

Schwann Cells The Schwann cells, which ensheathe and myelinate axons in the peripheral nervous system, are also derived from neural crest cells. Schwann cells migrate in a segmental fashion, accompanying the growing processes of the peripheral nerve fibers they will eventually ensheathe.

CENTRAL NERVOUS SYSTEM

Basic Features In general, CNS neuroblasts arise at the ventricular surface of the developing brain (i.e., the luminal surface of the neural tube). Just after the neural tube forms, there is no apparent cell differentiation in a cross section at any level. At this time, the neural tube is a pseudostratified columnar epithelium. As development proceeds and the wall of the neural tube thickens, however, dividing cells cluster at the ventricular surface, leaving a zone without cell bodies at the abluminal surface. This region is called the *marginal zone*.

As cells undergo their last division, they begin to migrate away from the luminal (ventricular) surface on transient glial cell guides called *radial glia*. As they migrate, they form a moving front of cell bodies between the marginal and ventricular zones called the *intermediate zone*. (Note that the older literature calls this moving front of cells the *mantle layer* or *mantle zone*, a term that has been discouraged since the 1970 Boulder Committee meeting.) These features are common to all parts of the developing neuraxis; other elaborations are possible and are outlined in the following sections.

After cells migrate and take up their final positions in the developing brain, they begin to extend processes and form connections with other neurons or muscle cells. Dendritic processes begin to receive information from other developing cells. Meanwhile, an axonal process tipped by a spadelike extension called the *growth cone*, begins to drive its way through intervening regions to reach distant targets.

Although the idea is controversial, both neurons and glia seem to originate from a single precursor cell population. Two main lineages arise: a *neuroblastic* lineage that generate neurons and a *glioblastic* lineage that includes precursors of radial glia, astroglia, and oligodendrocytes. The glioblastic lineage is believed to split into three main branches: (1) the *type 1 astrocyte* progenitor, (2) the

FIGURE 5-10 *Transformation of the apolar neuroblast into pseudounipolar, bipolar, and multipolar neurons.*

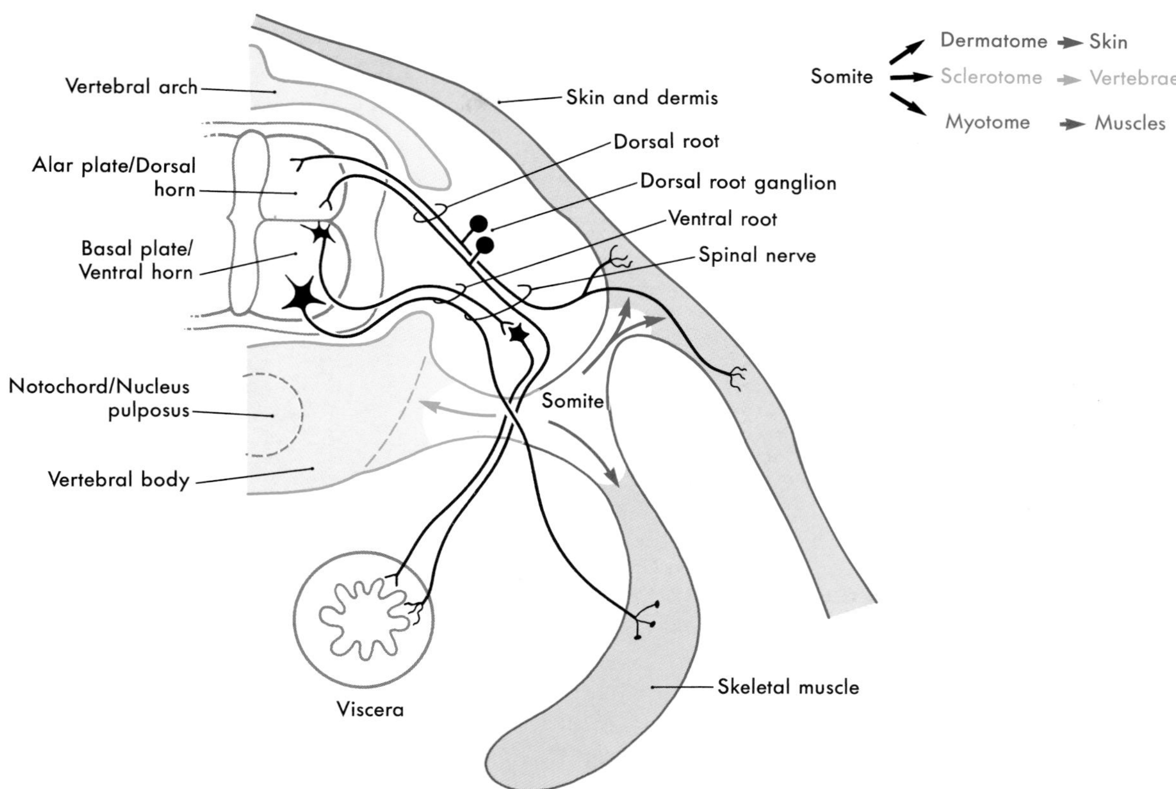

FIGURE 5-11 *Derivatives of a somite and the corresponding innervation of structures that originate from the dermatome and myotome.*

oligodendrocyte/type 2 astrocyte precursor (called the "*O2A progenitor*," and (3) the *radial glia* progenitor. Whereas all other cell types persist into adulthood, radial glial cells in most regions of the brain appear to be converted to astrocytes, *ependymal cells*, or *tanycytes*. In the cerebellum and retina (respectively), radial glia retain most of their features and persist as *Bergmann glia* or as *Müller glia*. Differentiation of glia is influenced by a variety of *growth factors*, such as platelet-derived growth factor, ciliary neurotrophic factor, and fibroblast growth factor, which are secreted by neighboring glia and neurons.

Spinal Cord The adult spinal cord gray matter is butterfly shaped and consists of ventral and dorsal horns. At some levels of the spinal cord, an *intermediate zone* and *lateral horn* of the gray matter lie halfway between dorsal and ventral horns.

The spinal cord develops from caudal portions of the neural tube (Fig. 5-1). The neural canal in this region will become the central canal of the spinal cord (Fig. 5-6). Neuroblasts that give rise to spinal cord neurons are produced between the fourth and twentieth weeks of development by a burst of proliferation in the ventricular layer lining the neural canal. These cells migrate peripherally to form four longitudinal *plates*, which will become the gray matter of the spinal cord: a pair of ventrally located *basal plates* and a pair of dorsally located *alar plates*. The basal and alar plates on each side are separated by a longitudinal groove called the *sulcus limitans* in the lateral wall of the central canal. The *basal plate* develops into the *ventral horn* of the spinal cord and the *alar plate* will become the *dorsal horn* of the spinal cord (Fig. 5-11). Development in the basal plate somewhat precedes that in the alar plate; postmitotic neurons are clearly evident in the basal plate during week 20 of development. That portion of the adult spinal cord commonly called the intermediate zone (and the lateral horn) originate from the interface of the alar and basal plates.

As the basal plate develops, axons of nascent motor neurons form the developing ventral roots that will innervate peripheral structures. Ventral horn motor neurons innervate skeletal muscle and are classified as *general somatic efferent* (*GSE*) cells. The lateral horn motor neurons project to autonomic (visceromotor) ganglia and are classified as *general visceral efferent* (*GVE*) cells. The categories GSE, GVE, etc., are referred to as *functional components*. The cord regions devoted to GSE and GVE functional components can be thought of as constituting distinct longitudinal *cell columns* in the gray matter (see Fig. 5-13). The GSE column runs the full length of the spinal cord, whereas the GVE column extends from T1

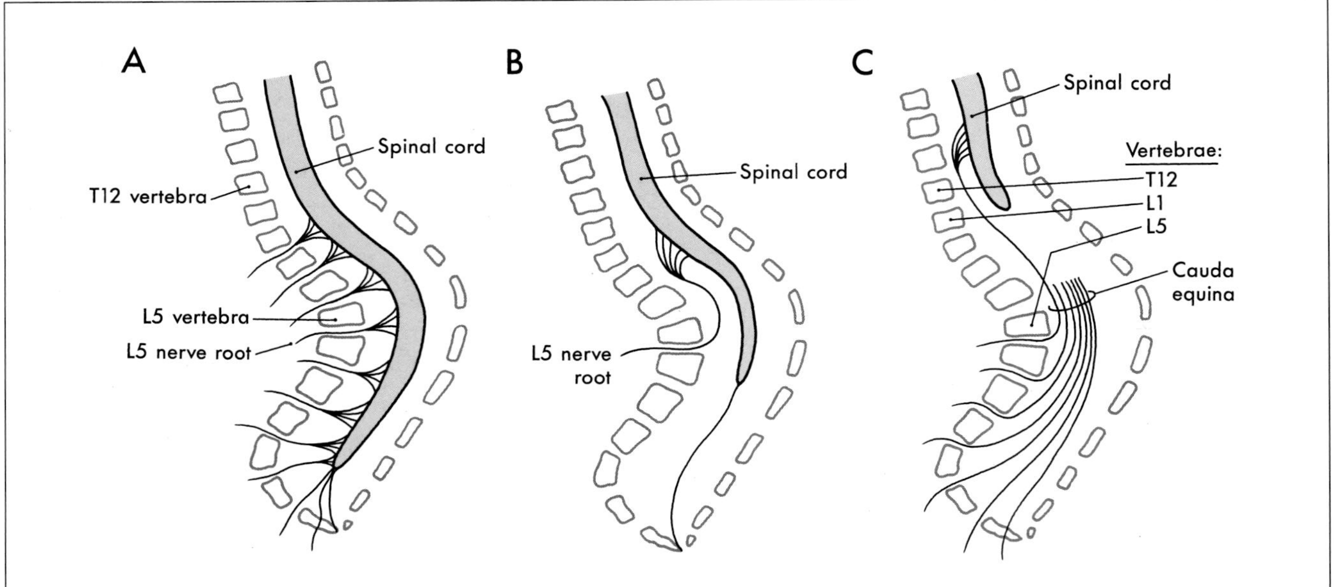

FIGURE 5-12 *The formation of the cauda equina as a result of the differential growth of the vertebral column and spinal cord. The relationship of the cord to the vertebral column is shown diagrammatically at about 12 weeks of gestation (A), 16 to 18 weeks of gestation (B), and at about 1 year of age (C).*

through L2, where it is called the *intermediolateral cell column*, and from S2 through S4, where it is called the *sacral visceromotor nucleus*.

Neurons of the alar plates receive the central processes of developing dorsal root ganglion (sensory) cells. Sensory neurons whose peripheral processes innervate the skin and receptors in joint capsules, tendons, and muscles are classified as *general somatic afferent* (*GSA*) cells. Those that innervate receptors in visceral structures, such as the stomach, are classified as *general visceral afferent* (*GVA*) cells. Like the GSE and GVE regions, the GSA and GVA cells constitute separate columns (see Fig. 5-13).

As each somite develops, it subdivides into a *sclerotome*, which forms vertebrae, a *dermatome*, which forms skin and dermis, and a *myotome*, which forms muscles (Fig. 5-11). The derivatives of the dermatomes and myotomes are innervated, respectively, by the axons of the dorsal (sensory) and ventral (motor) roots of the corresponding spinal cord levels. These roots join at about the level of the future *intervertebral foramina* to form the *spinal nerves* (Fig. 5-11). The spinal nerves thus show the same segmental pattern as the dermatomes and myotomes they innervate. By contrast, the vertebrae develop between the spinal nerves and are thus *intersegmental* in position, even though they originate from the segmental sclerotomes. This situation comes about because the sclerotomes each split into cranial and caudal halves, and the vertebral rudiments are formed by the union of the caudal half of one sclerotome with the cranial half of the next posterior sclerotome.

Relationship of Spinal Cord to Vertebral Column Although the spinal cord retains its general shape from the third trimester into adulthood, its physical relationship to the vertebral column alters dramatically (Fig. 5-12). By the end of the first trimester, the spinal cord, its meningeal coverings, and the surrounding vertebral arches are fully formed. The spinal nerves exit at about right angles to the spinal cord and pass through the intervertebral foramina. As development proceeds, the vertebral column grows faster than the spinal cord. The net result is that the cord is "pulled" rostrally by its attachment to the brain. The intervertebral foramina, containing the spinal nerves, move caudally; and the dorsal and ventral roots from lumbar, sacral, and coccygeal levels are significantly lengthened to form a bundle called the *cauda equina* (Fig. 5-12).

Brainstem The brainstem consists of the *myelencephalon* (medulla oblongata), the *pons* (a part of the metencephalon), and the *mesencephalon* (midbrain). Although developmentally the cerebellum is a part of the metencephalon, it is considered a "suprasegmental" structure and not a part of the brainstem.

As one travels from the rostral part of the spinal cord to the caudal part of the brainstem (*medulla oblongata*), two changes take place. First, the appearance of the cerebellum and the flaring open of the central canal into the fourth ventricle force the dorsal portion of the neural tube (*alar plate*) to rotate dorsolaterally (Fig. 5-13). This results in a lateral-to-medial orientation of sensory (*alar plate*) versus motor (*basal plate*) areas of the developing brainstem, in contrast to their dorsoventral relationship in the spinal cord. Second, the *sulcus limitans*, which disappears in the spinal cord during development, is retained as an important landmark in the floor of the fourth ventricle (Fig. 5-13).

The basal plates in the brainstem give rise to cranial nerve motor nuclei, and the alar plates give rise to cranial nerve sensory nuclei. As in the spinal cord, these portions of the basal and alar plates differentiate into rostrocaudally oriented cell columns, each of which is associated with a specific functional component. As shown in Figure 5-13, however, there are six cell columns and seven corresponding functional components in the brainstem, as compared to four in the spinal cord. This difference occurs because "special" (SVE, SVA, and SSA) functional components are unique to the head. As development proceeds, some of these columns fragment into distinct, separate nuclei. The nuclei derived from a given cell column have the same functional component, and they generally remain aligned along the same rostrocaudal axis but may lie in different parts of the brainstem (Fig. 5-13).

The alar plate gives rise to four cranial nerve nuclei (Fig. 5-13; see also Fig. 10-6). The *spinal trigeminal nucleus* forms a continuous cell column from the cord-medulla junction to midpontine levels, whereas the *principal sensory trigeminal nucleus* is found in the rostral pons anterior to the spinal nucleus. Both of these nuclei receive GSA input via cranial nerves V, VII, IX, and X. The *solitary nucleus* extends the entire length of the medulla and receives GVA and taste (special visceral afferent-SVA) input via cranial nerves VII, IX, and X. Although the *vestibular* and *cochlear nuclei* also arise from the alar plate, the peripheral fibers projecting to these nuclei originate primarily from the *otic placode*. These fibers are classified as special somatic afferent (SSA). The alar plate also gives rise to other brainstem cell groups, such as the inferior olivary nucleus of the medulla, the basilar pontine nuclei, and the substantia nigra of the midbrain.

Motor neurons in the brainstem originate from the *basal plate*. In contrast to cranial nerve nuclei derived from the alar plate, most of which form continuous cell columns, those cell groups that arise from the basal plate form separate nuclei (Fig. 5-13). The nuclei of the most medial column have a GSE functional component and innervate muscles that originate from occipital somites (the tongue) or from mesoderm in the vicinity of the optic cup (the eye muscles). These are the *hypoglossal nucleus* (XII, in the medulla), the *abducens nucleus* (VI, pons), and the *oculomotor* and *trochlear nuclei* (III, IV) of the midbrain. The next lateral column of nuclei have a GVE functional component and innervate visceral motor ganglia, which, in turn, innervate visceral structures. These nuclei are the *dorsal motor vagal nucleus* and the *inferior salivatory nucleus* of the medulla, the *superior salivatory nucleus* of the pons,

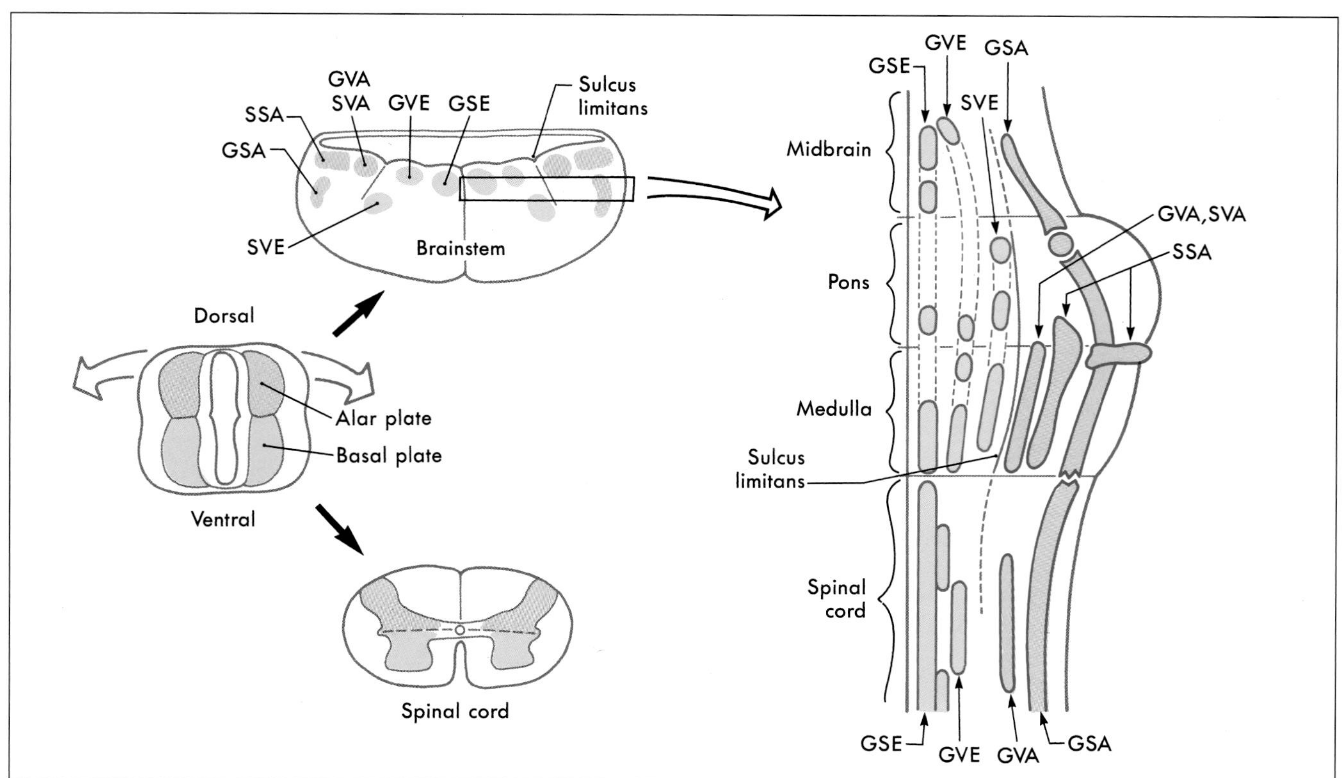

FIGURE 5-13 *The derivatives of the alar (in blue) and basal (in red) plates in the spinal cord and brainstem. Nuclei are grouped into rostrocaudally oriented cell columns that correspond to their functional components.*

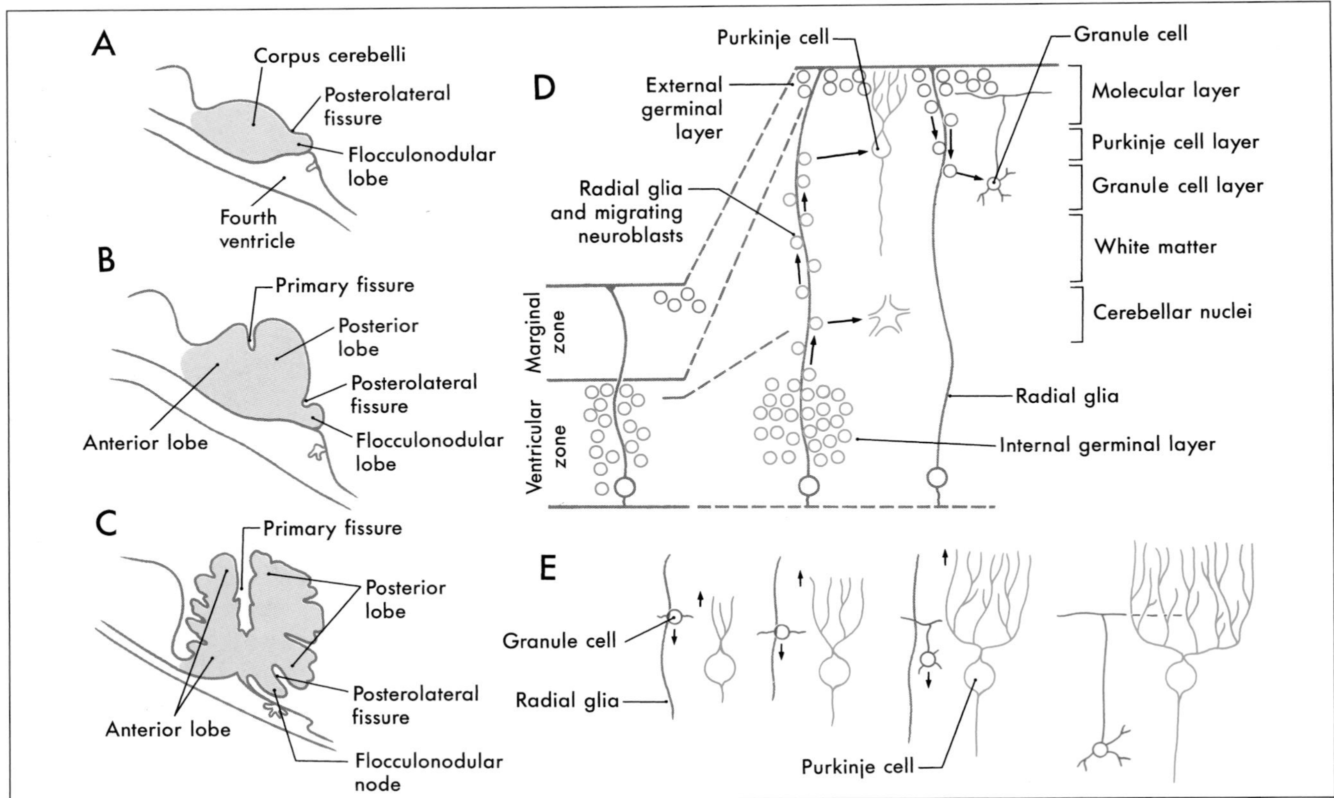

FIGURE 5-14 *Development of the cerebellum. Sagittal sections (A–C) show the growth of the cerebellum and the development of the main lobes and fissures. Cytodifferentiation of the cerebellar cortex and nuclei (D). Note the migration of neuroblasts on radial glia. Formation of connections between granule cell axons and Purkinje cell dendrites (E). As granule cells migrate inward, they trail an axon that forms contacts with the dendrites growing outward from Purkinje cells. These axons become the parallel fibers of the cerebellar cortex.*

and the *Edinger-Westphal nucleus* of the midbrain. These nuclei project via the vagal (X), glossopharyngeal (IX), facial (VII), and oculomotor (III) nerves, respectively. The most lateral and ventral column of nuclei originating from the basal plate innervates muscles that arise from the mesoderm of the *pharyngeal arches*, and, consequently, they are designated as special visceral efferent (SVE). These cell groups are the *ambiguus nucleus* (IX, X) of the medulla and the *facial nucleus* (VIII) and *trigeminal motor nucleus* (V) of the pons.

Along with the dorsal-ventral division of brainstem into alar and basal plates, there is a rostral-caudal segmentation of the developing rhombencephalon into *rhombomeres*. These are clusters of neuroblasts separated from each other by thin, transversely oriented bands of neuroepithelial cells. Cells in one rhombomere give rise to a specific motor nucleus (or nuclei) but will not migrate into adjacent rhombomeres. In general, the cell clusters forming the rhombomeres represent the rostral continuation of the alar and basal plates of the developing spinal cord. Indeed, motor nuclei originate from specific rhombomeres, and input from the corresponding sensory ganglia enters the corresponding rhombomere.

Rhombomeres are also sites of *homeobox gene* expression. These genes (abbreviated Hox) are "master switches" which control the formation of large blocks of tissue. For example, the gene *Hox 2.1* is only expressed in the rhombomeres that give rise to cranial nerves X and XII; *Hox 2.9* is only expressed in the region of the developing facial nerve. There is considerable sequence similarity between homeobox genes from widely divergent species (such as flies and humans). This implies that expression of homeobox genes is an essential element of neural development that has been preserved in evolution.

Cerebellum The cerebellum arises from the *rhombic lip*, an alar plate structure that forms part of the wall of the fourth ventricle. The rostral part of the rhombic lip forms the cerebellum, whereas the caudal part gives rise to the *inferior olivary*, *cochlear*, and *pontine* nuclei.

The rhombic lips join dorsal to the developing fourth ventricle to form the *cerebellar plate*. During the histogenesis of the cerebellar cortex, fissures appear that divide the cerebellum into its main lobes (Fig. 5-14A–C). The first, the *posterolateral fissure*, divides the cerebellar plate into the *flocculonodular lobe* and the *corpus cerebelli*.

The *primary fissure* is the second to appear, and it divides the corpus cerebelli into *anterior* and *posterior lobes*. The advent of additional fissures divides the anterior and posterior lobes into the lobules characteristic of the adult.

The process of cerebellar cortical development involves the migration of neuroblasts to form the cells characteristic of the adult (Fig. 5-14D). Initially the cerebellar primordium is composed of a *ventricular zone*, a layer of primitive neuroblasts formerly called the mantle layer and now called the *intermediate zone*, and a *marginal zone*. By the end of the first trimester, a second layer of neuroblasts has appeared in the outer part of the marginal layer. This is called the *external germinal* (or *granular*) *layer*, and the intermediate zone is now called the *internal germinal* (or *granular*) *layer*.

Radial glial cells extend from the ventricular zone to the surface of the marginal layer and are necessary for the proper migration of developing neurons (Fig. 5-14D). Neuroblasts of the internal germinal layer migrate outward along the radial glia to form the *cerebellar nuclei* and the *Purkinje cells* and *Golgi cells* of the cerebellar cortex. Neuroblasts of the external germinal layer migrate inward along radial glia to form the *granule cells*. Other cells of the external germinal layer congregate just external to the Purkinje cell layer, where they will differentiate into the *stellate cells* and *basket cells* of the *molecular layer*.

Developing cerebellar neurons participate in other important cell interactions in addition to those with the radial glia. For example, as the granule cells migrate inward, they sprout axons that form synaptic contacts with the Purkinje cell dendrites that are growing into the molecular layer (Fig. 5-14E). Continued growth and development of Purkinje cell dendrites depend on these contacts with granule cell axons. Purkinje cell dendrites are stunted in the mutant mouse *weaver*, in which the granule cells die during development. These and other experimental mutant animals with cerebellar defects show the symptoms characteristic of cerebellar disease in humans: *ataxia*, *hypotonia*, and *tremor*.

Thalamus The gray matter of the diencephalon develops from a continuation of the brainstem alar plates; there is no homolog of the basal plate in the diencephalon. This alar plate is recognizable at 4 weeks of gestation, and by 6 weeks of gestation it has differentiated into three main areas of the diencephalon: the *epithalamus*, *thalamus* (*dorsal thalamus*), and *hypothalamus* (Fig. 5-15A,B). These structures are visible as swellings in the wall of the third ventri-

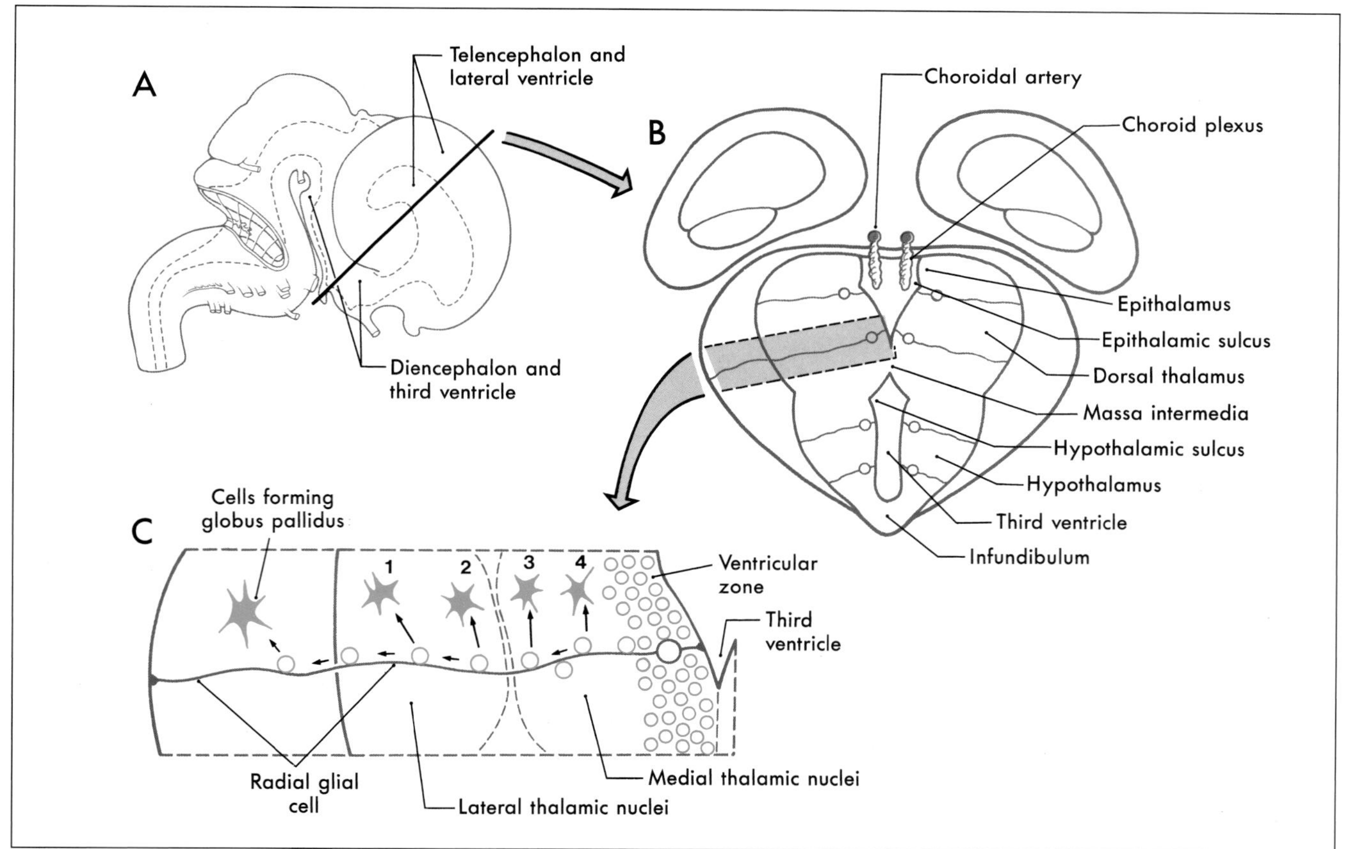

FIGURE 5-15 *Development of the diencephalon. The diencephalon is shown in a lateral view (A) and in a coronal view (B, plane from A) at about 8 weeks of gestation. Details of cell migration (C, detail from B) show that lateral cell groups (C, #1) are formed first while more medial cell groups (C, #4) are formed last.*

cle, where they are separated from each other by the *epithalamic* and *hypothalamic sulci.* As development progresses, the epithalamic area becomes quite small, whereas the hypothalamus, and especially the thalamus, enlarge. In about 80% of individuals, the two thalami fuse across the third ventricle to form the *interthalamic adhesion (massa intermedia).*

The basic concepts of the development of the thalamus are the same as for other CNS regions. *Radial glia* extend from the third ventricle to the pial surface, and developing neurons migrate along these guides (Fig. 5-15B,C). The development of the thalamus occurs in an *"outside first"* sequence. That is, the first neurons to undergo their final cell division migrate to the outermost portion of the thalamus, where they mature. This means that the most lateral of the thalamic nuclei, such as the geniculate nuclei and the lateral and ventral nuclei, are generated first. The most medial thalamic nuclei, such as the dorsomedial nucleus, are the last to develop.

An important process in the development of the thalamic relay nuclei is the establishment of orderly maps of the sensory world. For example, a retinotopic (vision) map is formed in the lateral geniculate nucleus. As retinal ganglion cells send axons to the lateral geniculate nucleus, the arrangement of axonal contacts on cells in this visual relay center must accurately reflect the positions of ganglion cells in the retina. In this way, the map of visual space on the retina is maintained in the lateral geniculate nucleus and, ultimately, in the visual cortex. Similar maps are formed in the medial geniculate nucleus (tonotopic mapping) and ventral posterolateral nucleus (somatotopic mapping).

Cerebral Cortex The cerebral cortex is generated using the basic mechanisms described above for other regions. Except during mitosis and cytokinesis, cortical neuroblasts retain connections to both the ventricular and the pial surfaces of the developing brain and thus have a fusiform shape (Fig. 5-16). The nucleus engages in a peculiar cycle of migration within the cell, however. During the G_1 phase of the cell cycle (before DNA replication), the nucleus travels from near the ventricular pole of the cell to near the pial pole. During the G_2 phase (after DNA replication), it reverses direction and migrates back to a ventricular position. At the start of mitosis, the cell loses contact with the pial surface, but after cytokinesis the daughter cells grow processes that reconnect with the pial surface (Fig. 5-16). The cycle is then ready to repeat.

Unlike other regions of the brain, in the cerebral cortex the first cells to migrate will disembark from the radial glial cell and take up positions close to the ventricular surface. Successive "waves" of neuroblasts, migrating along radial glia, force their way through the differentiated cell layers to take up positions progressively closer to the pial surface. This sequence is called an *"inside-out" pattern* of development. These ranks of cells form the *cortical plate*; the axons of previously settled cells grow toward their targets.

The cerebral cortex proper is formed from expansion of the superficial part of the intermediate zone, the *subplate* and *cortical plate* (Fig. 5-17). Developing axons, originating in regions such as the thalamus that innervate the cortex, send out and form transient synaptic contacts in the subplate. The subplate is a transient structure that does not persist into adulthood. Neuronal cell bodies vacate the area between the subplate and the ventricular surface; most of the remaining cell bodies are glia. This region forms the *white matter* of the adult nervous system. The ventricular zone is reduced to a single layer of *ependymal cells* that line the lateral ventricles in the adult.

During peak periods of cellular migration, the hemispheric fissures appear and mold the telencephalic surface into the gyri and sulci characteristic of the adult. By the end of the first trimester, the *interhemispheric, Sylvian,* and *transverse cerebral fissures* are recognizable. The secondary sulci are completed by 32 weeks of development, with tertiary sulci completed during the last month of gestation.

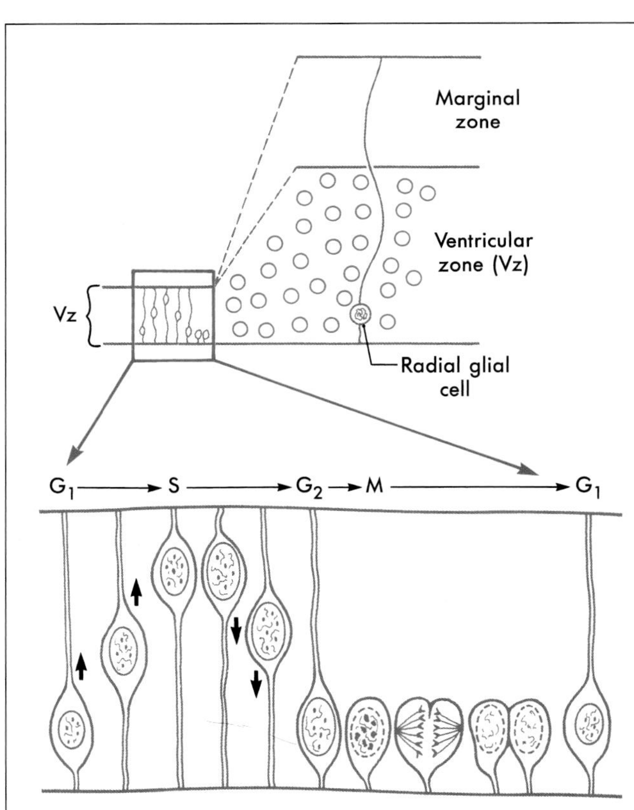

FIGURE 5-16 *Early development of the cerebral cortex. The nuclei of the neuroblasts undergo a cycle of outward and inward migratins as the neuroblasts progress through their cell cycle. Phases of the cell cycle: G_1, first gap phase; S, DNA replication; G_2, second gap phase; M, mitosis.*

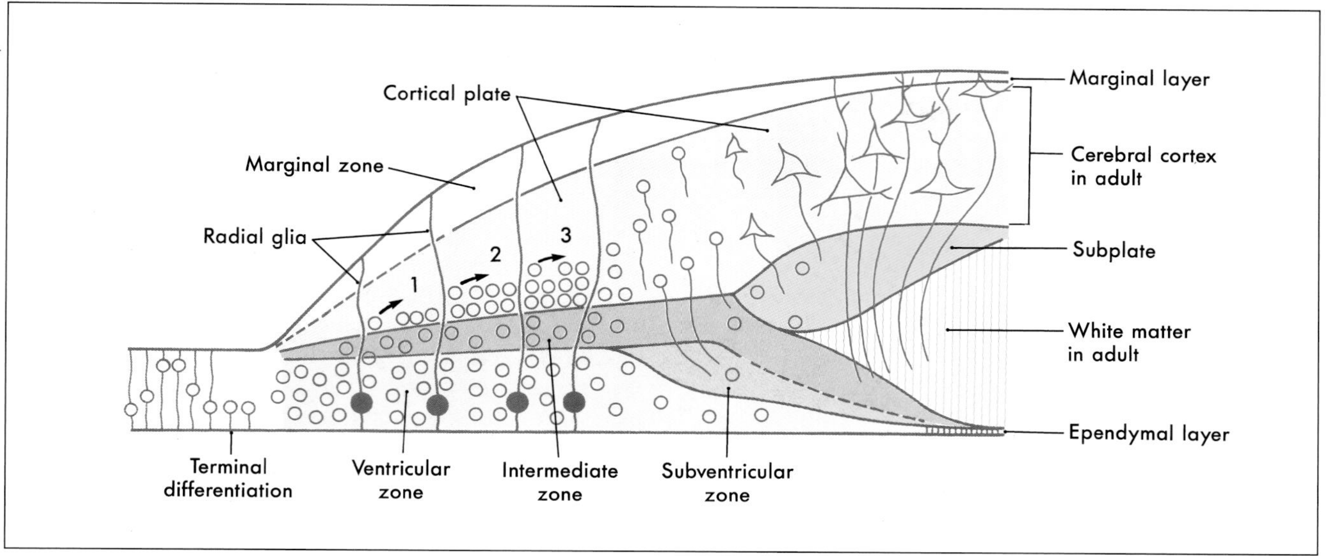

FIGURE 5-17 *Later development of the cerebral cortex. Inner cellular layers (#1) are formed first and progressively more superficial layers (#2, #3) are formed later.*

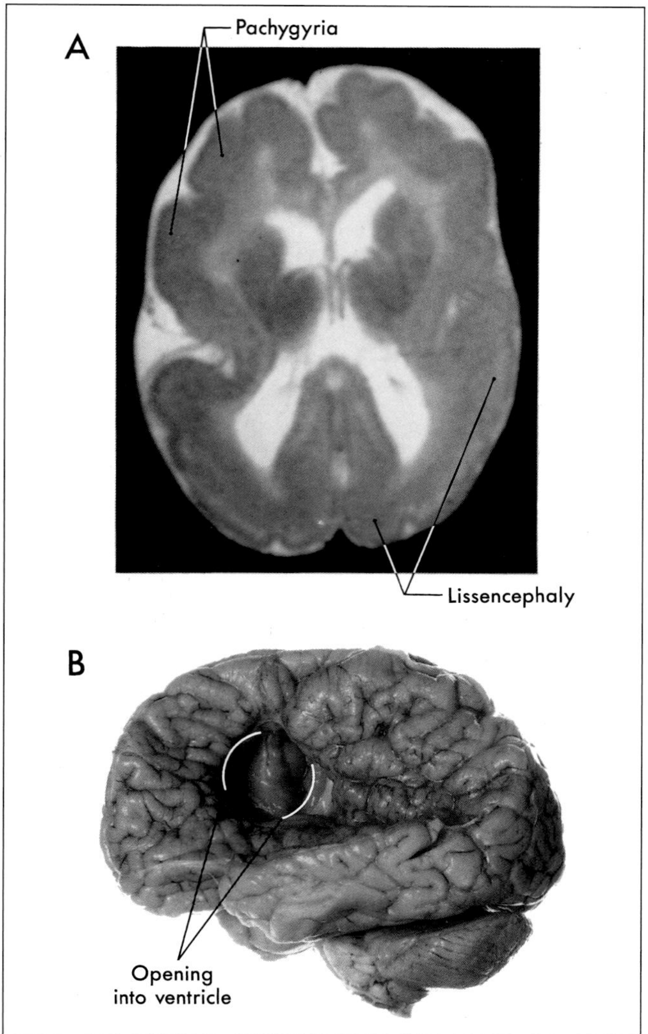

Gross Abnormalities of Cortical Development Abnormal patterns of gyri and sulci are caused by disorders of cell migration in the developing cerebral cortex (Fig. 5-18A). If gyri fail to form, the cerebral cortex will have a smooth surface, a condition called *lissencephaly*. Unusually large gyri constitute *pachygyria*, and unusually small gyri constitute *microgyria*. Any of these conditions may affect the whole cerebrum or may be localized, and they may coexist in the same patient (Fig. 5-18A).

Abnormal patterns of sulcal and gyral development are seen in *schizencephaly*, a condition in which there are unilateral or bilateral clefts in the cerebral hemispheres (Fig. 5-18B). The resulting defect may be localized and small, consisting of a thin spot in the hemisphere where the pia and ependyma come abnormally close. In severe cases, the defect is large, resulting in a substantial loss of brain tissue and producing an open channel between the ventricular cavity and the subarachnoid space. Schizencephaly may result from a profound failure of cell migration. An alternative explanation, which may apply particularly to severe cases, is that the affected region did not receive an adequate blood supply during development. The result would be a central area of necrosis, which would become a thin spot or an open channel, surrounded by a zone of abnormal neuroblast migration (Fig. 5-18B).

FIGURE 5-18 *Axial MRI of a 6-month-old female patient with cortical malformations (A). Brain of a child with schizencephaly (B) showing an open channel from the brain surface into the lateral ventricle.*

CELLULAR EVENTS IN BRAIN DEVELOPMENT

The organization of the brain ultimately determines its function. Three important parameters in brain organization are (1) the density of neurons, (2) the pattern of axon and dendrite branching, and (3) the pattern of synaptic contacts. These characteristics begin to develop toward the end of the peak period of neuronal migration at the sixth month of gestation. Although neuronal density and the basic patterns of axonal and dendritic growth are determined within the first 2 to 3 years after birth, remodeling of synaptic connections continues throughout life.

Overproduction of Neurons and Apoptosis Embryogenesis produces one and a half to two times more neurons than are present in the mature brain. By 24 weeks of gestation, almost all these neurons have been produced. Subsequent to this, there is selective death of neurons.

Genetically programmed cell death (*apoptosis*) of neurons is a feature of cellular development in many areas of the brain. In contrast to cell death resulting from injury, apoptosis requires protein synthesis and, therefore, is an active cellular process.

Some growth factors interrupt the normal process of apoptosis. *Nerve growth factor*, *brain-derived neurotrophic factor*, and *fibroblast growth factor* are known to limit cell death. This has led to the idea that the administration of growth factors may block the neuronal cell death that occurs in certain degenerative diseases. For example, the use of *ciliary neurotrophic factor* is under investigation in patients with *amyotrophic lateral sclerosis*, a devastating progressive degeneration of large motor neurons in the spinal cord.

Axonal Outgrowth After neuroblasts complete their final cell division and migrate to their final location, they begin to extend a single axon with one or more distal elaborations known as *growth cones*. This spade-shaped extension of the growing axon is capable of driving through fields of developing nervous or mesenchymal tissue to reach distant targets. The guidance of growth cones is influenced by both *tropic factors* (which guide a cell toward a particular target) and *trophic factors* (which maintain the metabolism of a cell or its processes). As an axon grows, it may send out branches, each with its own growth cone. Some branches may terminate in sites that will not ultimately be innervated by the cell. For example, cells of the motor cortex that send axons into the spinal cord to innervate motor neurons also send transient branches to structures of the brainstem; these ectopic connections are not normally maintained.

Synaptogenesis Once an axonal growth cone arrives at its site of termination, it undergoes biochemical and morphologic changes to become a *presynaptic terminal*. Similarly, the area of the target neuron contacted by the presynaptic process begins expressing the characteristic postsynaptic machinery, such as neurotransmitter receptors and second messenger molecules.

One view of synapse development is that there is competition for the synaptic space available on target neurons. A synapse that forms will persist only if it exchanges the right cues with the target cell. Many synapses that form are subsequently lost and are replaced by other synapses that may be more successful. Thus only a subset of the large number of synapses that form are ultimately retained. This is the concept of *synaptic stabilization*, and it requires (1) a signal generated by the presynaptic cell, possibly the neurotransmitter to be used at the adult synapse; (2) a means for the postsynaptic cell to respond to the presynaptic signal; and (3) a "retrograde signal" from the postsynaptic cell to the presynaptic cell to indicate which contacts are to remain in maturity.

Plasticity and Competition An important process related to the development of neuronal organization is *plasticity*. The developing brain is not as vulnerable to injury as is the mature brain. Infants who suffer significant cortical injury in the prenatal or early postnatal period may show surprising functional recovery, ending up with few or no obvious deficits. The mechanism of plasticity relates to alterations in selective neuronal death and axonal simplification and to the retention of transient axonal branches and synapses that would otherwise be lost, as discussed above.

One example of plasticity and the competition for synaptic space is the development of visual cortical connections. Fibers conveying visual input from each eye arrive in overlapping territories in the visual cortex during the fetal period. Normally the synaptic space within this region is equivalently distributed to terminals carrying input from each eye. However, if the input from one eye is lost or if it is not functionally equivalent to the input from the other eye, the terminals from the good eye will experience a competitive advantage and occupy a larger share of the available synaptic space. The period of time during which these types of plastic changes can occur is called the *critical period*. Other areas of cortex have their own critical periods; the duration and time of occurrence of the critical period vary from region to region.

The concept of a critical period has clinical implications. If the input from one eye is dysfunctional during the critical period for visual system development (for example, if one eye is severely myopic), the axons carrying information from that eye are at a disadvantage as they compete for synaptic space. Left untreated, the "good" eye has exclusive access to visual cortex and input from the "bad" eye is ignored. This condition is called *amblyopia*. If the myopia is corrected later in life, no signals can pass from the retina to the visual cortex because the appropriate synaptic connections were not formed during the critical period. As a result the eye remains functionally blind. This blindness can be avoided by clinical procedures that equalize competition for synaptic territory during the critical period.

Synaptic development occurs in parallel with cellular proliferation and migration. Dendritic spines are the site of many synaptic contacts, especially in cortical neurons. The rate of spine formation varies in different parts of the brain, but is usually maximal during the sixth month after birth. Many children with mental retardation, including those with *Down's syndrome*, have fewer and less complex axonal and dendritic ramifications and fewer dendritic spines than normal children. In some patients, there is a disturbance of the cytoskeletal structure that supports the architecture of axonal processes. Axonal and synaptic development are especially vulnerable to *perinatal hypoxia*, *malnutrition*, and *environmental toxins*.

Myelination Oligodendroglia myelinate neuronal axons in the CNS. Myelination begins at about the sixth month of development and peaks between birth and the first year of life, but it continues into adulthood. A delay in myelination can result in a delay in functional development. The best example is a congenital cortical blindness that resolves during the first year of life.

There is a definite hierarchy in the regional maturation of myelin formation. The motor and sensory tracts throughout the nervous system mature early, whereas the association tracts mature relatively late.

Several neurodegenerative diseases (*leukodystrophies*) affect the formation of myelin. Many other inborn errors of amino and organic acid metabolism impair myelination, notably phenylketonuria. Finally, inadequate nutrition also can impair myelination.

SOURCES AND ADDITIONAL READING

Barkovich AJ: *Pediatric Neuroimaging*, 2nd Ed. Raven Press, New York, 1995.

Boulder Committee: Embryonic vertebrate central nervous system: Revised terminology. Anat Rec 166:257–262, 1970.

Evans OB: *Manual of Child Neurology*. Churchill Livingstone, New York, 1987.

Jacobson M: *Developmental Neurobiology*, 3rd Ed. Plenum Press, New York, 1991.

McConnell SK: The determination of neuronal fate in the cerebral cortex. Trends Neurosci 12:342–349, 1989.

Noden DM: Vertebrate craniofacial development: the relation between ontogenetic process and morphological outcome. Brain Behav Evol 38:190–225, 1991.

Purves D, Lichtman JW: *Principles of Neural Development*. Sinauer Associates, Sunderland MA, 1985.

Rakic P: Principles of neural cell migration. Experientia 46:882–891, 1990.

Shatz C: The developing brain. Sci Am 267:61–67, 1992.

Walsh C, Cepko CL: Clonally related cortical cells show several migration patterns. Science 241:1342–1345, 1988.

CHAPTER SIX

The Ventricles, Choroid Plexus, and Cerebrospinal Fluid

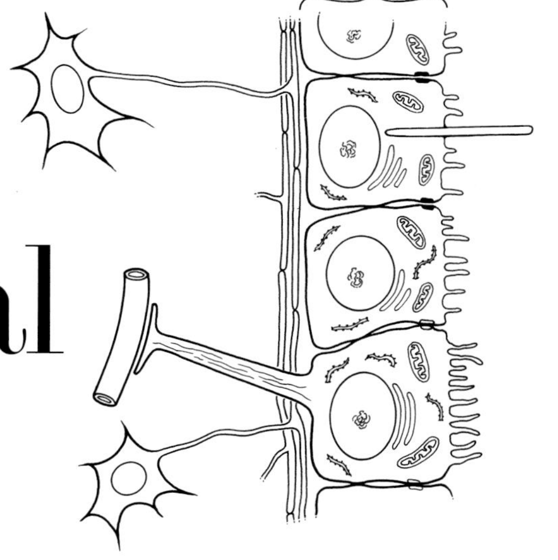

J. J. CORBETT • D. E. HAINES • M. D. ARD

The ventricular spaces of the brain are the adult elaborations of the neural canal of early developmental stages. These spaces, the choroid plexuses in them, and the cerebrospinal fluid (CSF) produced by the choroid plexus are essential elements in the normal function of the brain.

OVERVIEW

By about the third week of development, the nervous system consists of a tube closed at both ends and somewhat hook-shaped rostrally (Fig. 6-1). The cavity of this tube, the *neural canal*, eventually gives rise to the *ventricles* of the adult brain and the *central canal* of the spinal cord. The former becomes quite elaborate as the various parts of the brain differentiate, whereas the latter becomes progressively smaller.

The choroid plexus, which secretes the CSF that fills the ventricles and the subarachnoid space, arises from tufts of cells that appear in the wall of each ventricle during the first trimester. These cells are specialized for a secretory function. The production of CSF is an active process that requires an expenditure of energy by the choroidal cells.

Any condition that causes CSF to accumulate, such as overproduction or an obstruction of its movement through the ventricle system, produces serious neurologic deficits. Perhaps the most widely recognized example is *hydrocephalus* as seen in a fetus or in a newborn infant. This condition is usually caused by an obstruction of CSF flow and resultant enlargement of the ventricular spaces. The bones of the developing skull move apart, and the head may enlarge significantly. In most of these cases some type of surgical diversion of CSF flow (a shunting procedure) is necessary.

DEVELOPMENT

At about 22 to 24 days of gestational age, the *anterior* and *posterior neuropores* close. At this point, the neural tube is lined by *neuroepithelial cells* undergoing waves of cell division. Some of these precursor cells give rise to the cuboidal *ependymal cells* that line the developing (and mature) ventricular system and the central canal.

The brain is initially composed of three primary brain vesicles (*rhombencephalon*, *mesencephalon*, and *prosencephalon*), each containing a portion of the cavity of the neural tube (Fig. 6-1A,D). The appearance of the pontine flexure in the rhombencephalon and the telencephalic flexure in the prosencephalon divides these three vesicles into the five brain vesicles (*myelencephalon*, *metencephalon*,

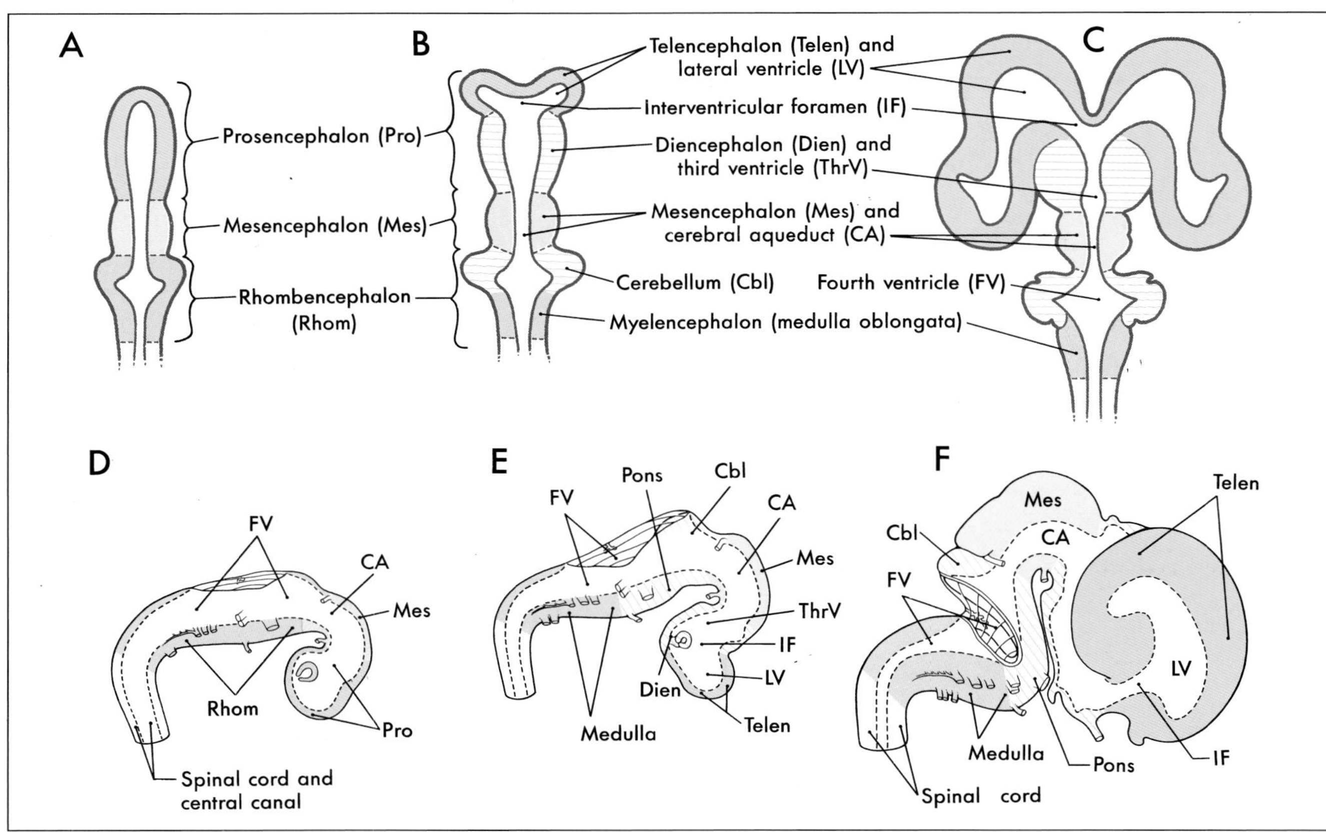

FIGURE 6-1 *The early development of the brain and ventricular system, showing how brain growth and the configuration of the ventricles interrelate. Diagrammatic dorsal veins (A–C) correlate in general with lateral views (D–F) at about 5 weeks (D), 6 weeks (E), and 8.5 weeks of gestation (F). The outlines of the ventricles are shown in D–F as dashed lines.*

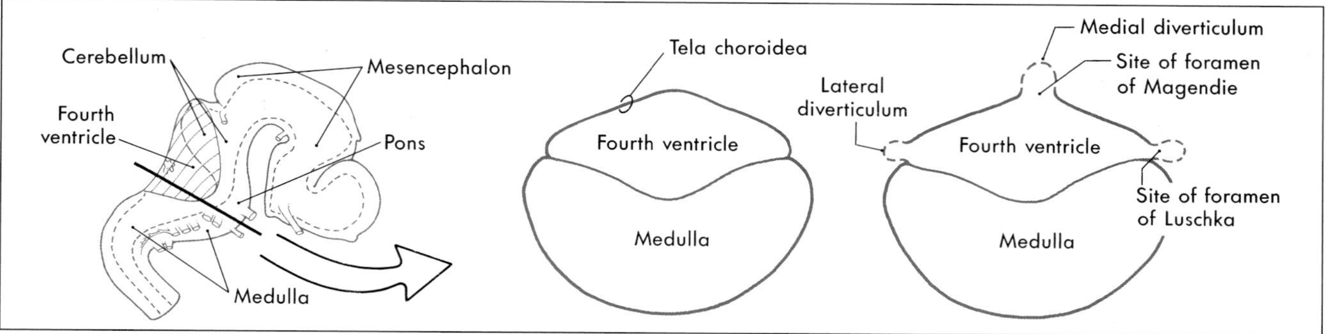

FIGURE 6-2 *Development of the foramina of Luschka and Magendie in the fourth ventricle.*

mesencephalon, diencephalon, telencephalon) characteristic of the adult (Fig. 6-1B,E). With subsequent development, the telencephalic (cerebral) vesicles enlarge significantly (Fig. 6-1C,F). As the brain enlarges from three to five vesicles, each part pulls along a portion of the cavity of the primitive neural tube. These spaces in each brain vesicle form the ventricle of that part of the brain in the adult. Consequently, the shape of the ventricular system conforms, in general, to the changes in configuration of the surrounding parts of the brain.

The lateral ventricles follow the enlarging cerebral hemispheres, and the third ventricle remains a single midline space (Fig. 6-1A–C). The communications between the lateral ventricles and the third ventricle, the *interventricular foramina* (of *Monro*), are initially large but become small as development progresses (Fig. 6-1C,F).

The neural tube in the mesencephalon narrows to form the *cerebral aqueduct* (Fig 6-1C,F). This creates a constricted region in the ventricular system and thus a point at which the flow of CSF may be easily blocked. Occlusion of the cerebral aqueduct during development may be caused by a glial scar (*gliosis*) as a result of an infection or because of failure of the aqueduct to form. Occlusion results in a lack of communication between the third and fourth ventricles. Caudally the cerebral aqueduct flares open into the fourth ventricle (Fig. 6-1B,C,E,F).

Foramina of the Fourth Ventricle The ventricles and central canal of the spinal cord form a closed system when they first arise. However, in the second and third months of development, three openings form in the roof of the fourth ventricle and render the ventricular system continuous with the subarachnoid space surrounding the brain and spinal cord. The caudal part of the roof of the fourth ventricle consists of a layer of ependymal cells internally and a delicate layer of connective tissue externally (Fig. 6-2). The future apertures first appear in the form of small bulges in the caudal roof and at the lateral extremes of the fourth ventricle. The membrane forming the roof at these points becomes thinned and breaks down. The resultant openings are the medial *foramen of Magendie* (also called the *median aperture*) and the lateral *foramina of Luschka* (Fig. 6-2).

Formation of the Choroid Plexus In the adult, the *choroid plexus* is found in both lateral ventricles and in the third and fourth ventricles (see Fig. 6-4). The development of this structure is essentially the same in all of these spaces and is illustrated here for the fourth ventricle.

The caudal roof of the fourth ventricle is composed of ependymal cells on the luminal surface and a delicate external layer of connective tissue, the pia mater. These structures collectively form the *tela choroidea* (Fig. 6-3). Developing arteries in the immediate vicinity invaginate the roof of the ventricle to form a narrow groove, the

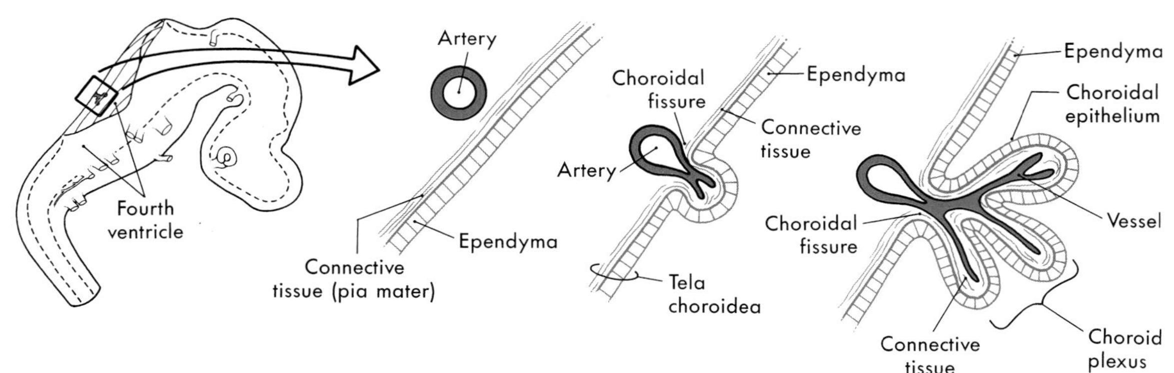

FIGURE 6-3 *Development of the choroid plexus.*

choroidal fissure, in the tela choroidea (Fig. 6-3). The involuted ependymal cells, along with vessels and a small amount of connective tissue, represent the primordial choroid plexus inside the ventricular space. As development progresses, the choroid plexus enlarges, forms many small elevations called *villi*, and begins to secrete CSF (Fig. 6-3; see also Fig. 6-12). By the end of the first trimester, the choroid plexus is functional, the openings in the fourth ventricle are patent, and there is circulation of CSF through the ventricular system and into the subarachnoid space.

The choroid plexuses of the third and lateral ventricles originate in the same manner. A choroid fissure appears in the roof of the third ventricle and in the medial wall of the lateral ventricle where the covering of each (the tela choroidea) is thin. The choroid plexus develops along these lines, bulges into the respective space, and is continuous from lateral to third ventricles through the interventricular foramen (Fig. 6-4).

VENTRICLES

Lateral Ventricles The cavities of the telencephalon are the *lateral ventricles*, of which there is one in each hemisphere (Fig. 6-4A,B). As the growth of the hemispheres creates the frontal, temporal, and occipital lobes, the lateral ventricles are pulled along and thus acquire their

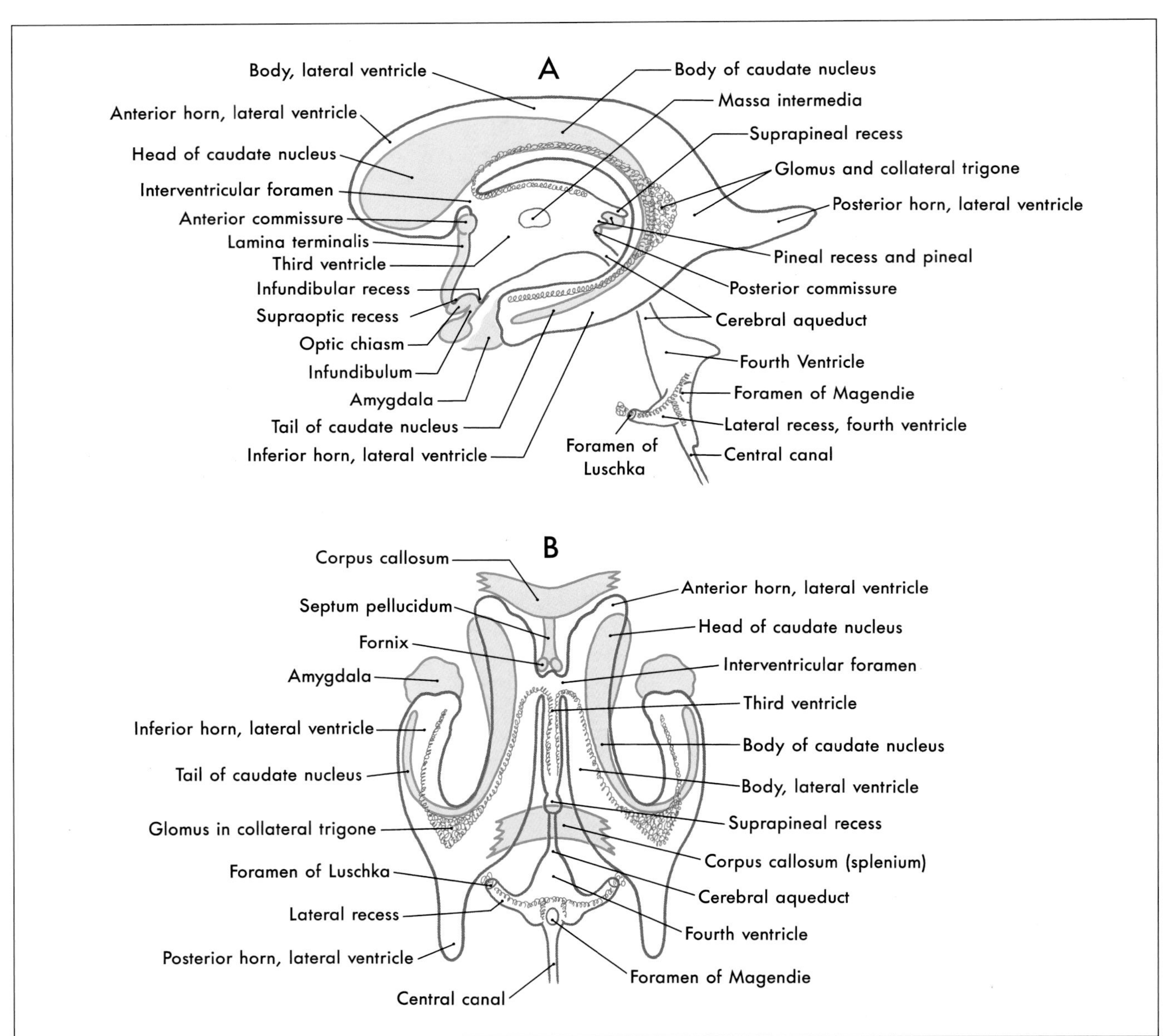

FIGURE 6-4 *Lateral (A) and dorsal (B) views of the lateral, third, and fourth ventricles and the cerebral aqueduct. Structures that border on the various parts of the ventricular system are shown in green and the choroid plexus is shown in red.*

definitive adult shape of a flattened C with a short tail (Fig. 6-4). This shape is present by birth. The lateral ventricle consists of an *anterior horn*, a *body*, and *posterior* and *inferior horns* (Fig. 6-4A,B). The junction of the body with the posterior and inferior horns constitutes the *collateral trigone* (or *atrium*). An especially large clump of choroid plexus, the *glomus* (or *glomus choroideum*) is found in the collateral trigone. In adults and especially in elderly individuals, the glomus may contain calcifications that are visible (as white spots) on radiographs or computed tomography scans. Shifts in the position of the glomus, usually accompanied by alterations in the volume of the surrounding ventricle, indicate some type of ongoing pathologic process.

The elaborate shape of the lateral ventricle means that different structures border on different parts of this space. The anterior horn and body of the lateral ventricle are bordered medially by the *septum pellucidum* (at rostral levels) and by a bundle of fibers called the *fornix* (at caudal levels), and dorsally by the *corpus callosum* (Figs. 6-4, 6-5). The floor of the body of the lateral ventricle is made up of the *thalamus*, and the *caudate nucleus* is characteristically found in the lateral wall of the lateral ventricle throughout its extent (Figs. 6-4, 6-5). In the temporal lobe, the inferior horn of the lateral ventricle contains the tail of the *caudate nucleus* in its lateral (dorsolateral) wall, the *hippocampus* in its ventral (ventromedial) wall, and a large group of cells (the *amygdaloid complex*) in its anterior end (Figs. 6-4, 6-5). The openings between lateral and third ventricles, the *interventricular foramina* (of Monro), are located between the fornix and the anterior and medial end of the thalamus.

Third Ventricle The *third ventricle*, the cavity of the diencephalon, is a narrow, vertically oriented midline space that communicates rostrally with the lateral ventricles and caudally with the *cerebral aqueduct* (Figs. 6-4, 6-7). The third ventricle has an elaborate profile in sagittal view (Fig. 6-4A), but it is quite narrow in the coronal and axial planes (Fig. 6-6A,B).

The boundaries of the third ventricle are formed by a variety of structures, most important the dorsal thalamus and hypothalamus, and by structures that form small outpocketings called recesses (Figs. 6-4A, 6-7). These are the *supraoptic recess* (dorsal to the optic chiasm, and sometimes called the *optic* recess), the *infundibular recess* (in the infundibulum, the stalk of the pituitary), the *pineal recess* (in the stalk of the pineal), and the *suprapineal recess* (above the pineal). The rostral wall of the third ventricle is formed by a short segment of the *anterior commissure* and a thin membrane, the *lamina terminalis*, which extends from the anterior commissure ventrally to the rostral edge of

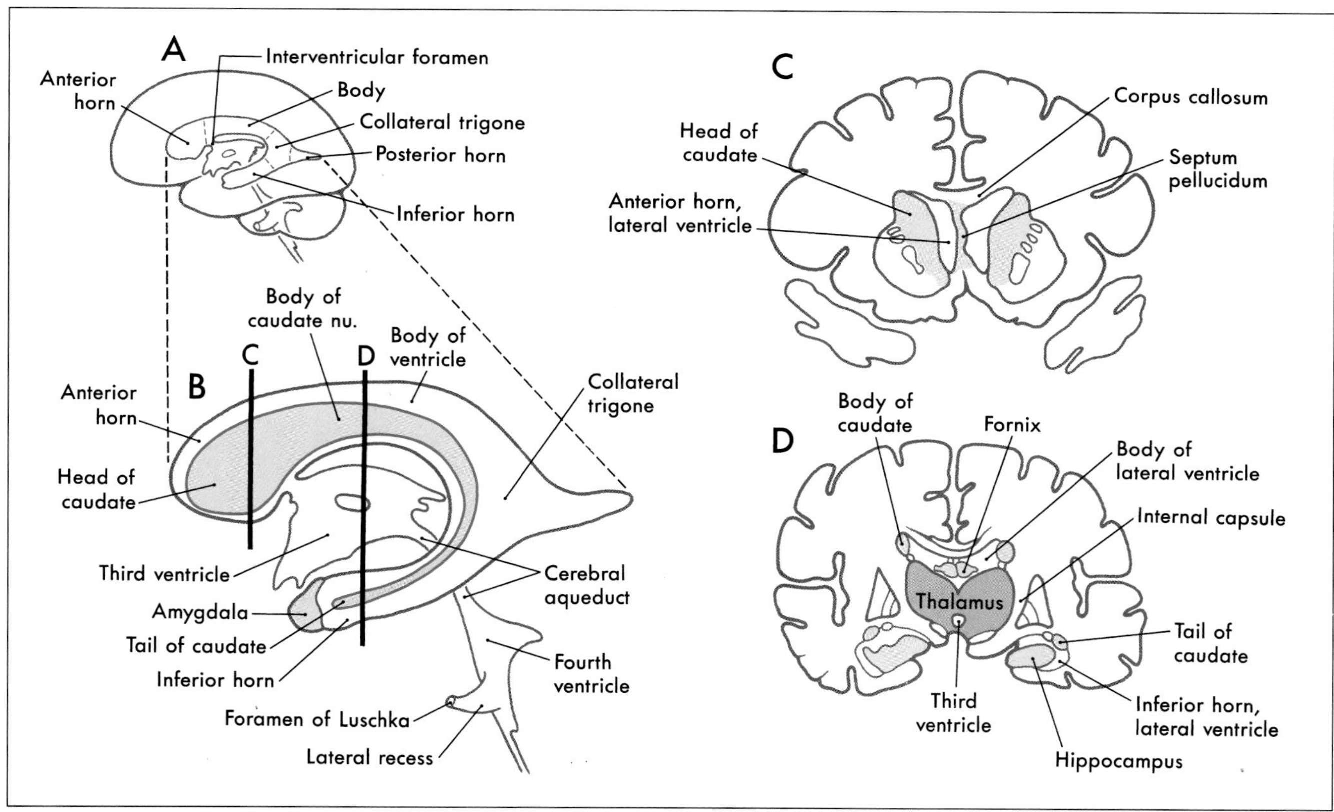

FIGURE 6-5 *Lateral view of the ventricles (A, B) and representative cross sections (C and D, details from B) showing the lateral and third ventricles, and the major structures that border on these spaces.*

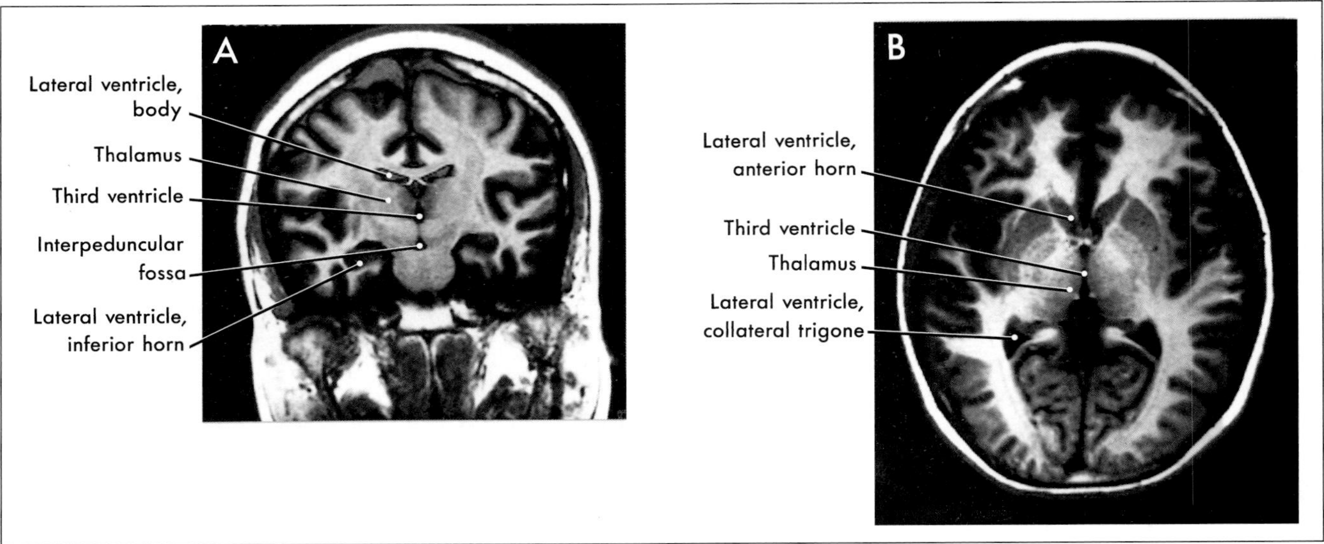

FIGURE 6-6 *Magnetic resonance images (MRIs) of the third ventricle in coronal (A) and axial (B) views.*

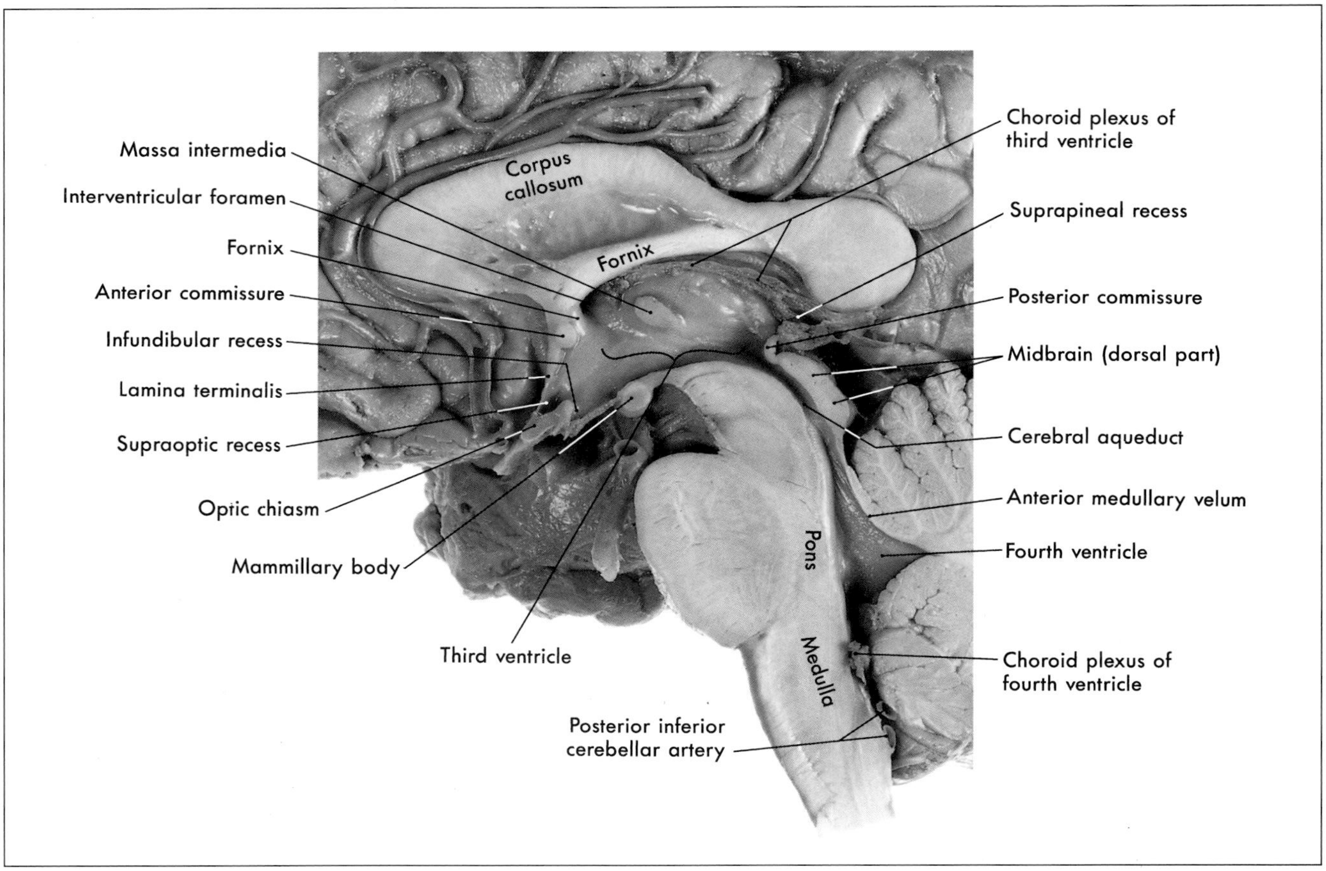

FIGURE 6-7 *Midsagittal view of the brain showing the third ventricle, cerebral aqueduct, and fourth ventricle, and structures closely related to these spaces.*

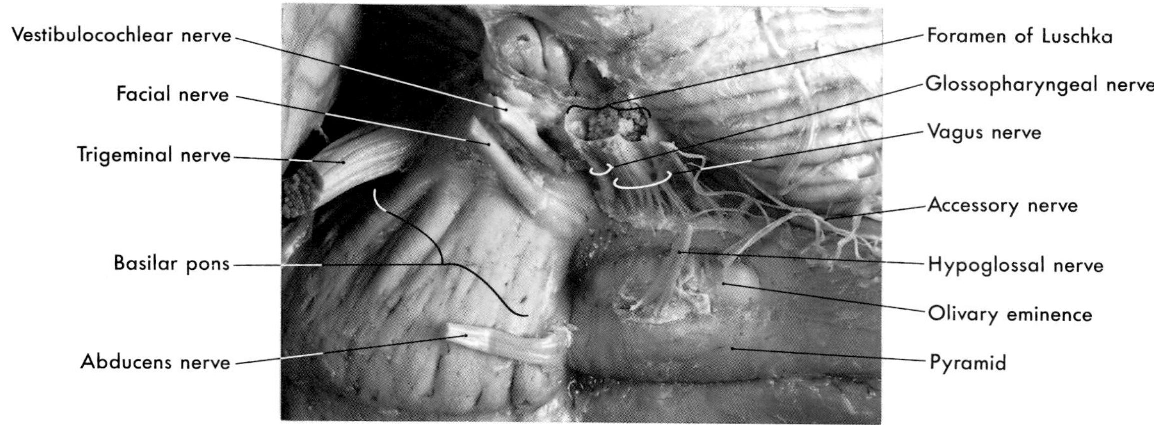

FIGURE 6-8 *Ventrolateral view of the brainstem at the pons-medulla junction showing the foramen of Luschka and the principal structures located in this area. Note the tuft of choroid plexus in the foramen. This area of the subarachnoid space, into which the foramen of Luschka opens, is the lateral cerebellomedullary cistern.*

the optic chiasm (Figs. 6-4A, 6-7). The floor of the third ventricle is formed by the *optic chiasm* and *infundibulum* and their corresponding recesses, plus a line extending caudally along the rostral aspect of the midbrain to the cerebral aqueduct. The caudal wall is formed by the *posterior commissure* and the recesses related to the pineal, whereas the roof is the *tela choroidae* from which the choroid plexus is suspended (Figs. 6-4A, 6-7).

Cerebral Aqueduct The *cerebral aqueduct*, the extension of the ventricle through the mesencephalon, communicates rostrally with the third ventricle and caudally with the fourth ventricle (Figs. 6-4A,B, 6-7). This midline channel is about 1.5 mm in diameter in adults and contains no choroid plexus. Its narrow diameter makes it especially susceptible to occlusion. For example, cellular debris in the ventricular system (from infections or *hemorrhage*) may clog the aqueduct. Tumors in the area of the midbrain (e.g., *pinealoma*) may compress the midbrain and occlude the aqueduct. The result is a blockage of CSF flow and enlargement of the third and lateral ventricles at the expense of the surrounding brain tissue. The cerebral aqueduct is surrounded on all sides by a sleeve of gray matter that contains primarily small neurons; this is the *periaqueductal gray* or *central gray*.

Fourth Ventricle The *fourth ventricle* is a somewhat pyramid-shaped space that forms the cavity of the metencephalon and myelencephalon (Figs. 6-4A,B, 6-7). The apex of this ventricle extends dorsally into the base of the cerebellum, and caudally it tapers to a narrow channel that continues into the cervical spinal cord as the central canal. Laterally the fourth ventricle extends over the surface of the medulla as the *lateral recesses*, to eventually open into the area of the pons-medulla junction through the *foramina of Luschka* (Figs. 6-4, 6-8). The irregularly shaped *foramen of Magendie* is located in the caudal sloping roof of the ventricle (Figs. 6-4, 6-9). Although the roof of the caudal part of the fourth ventricle and the lateral recesses is composed of *tela choroidea*, the rostral

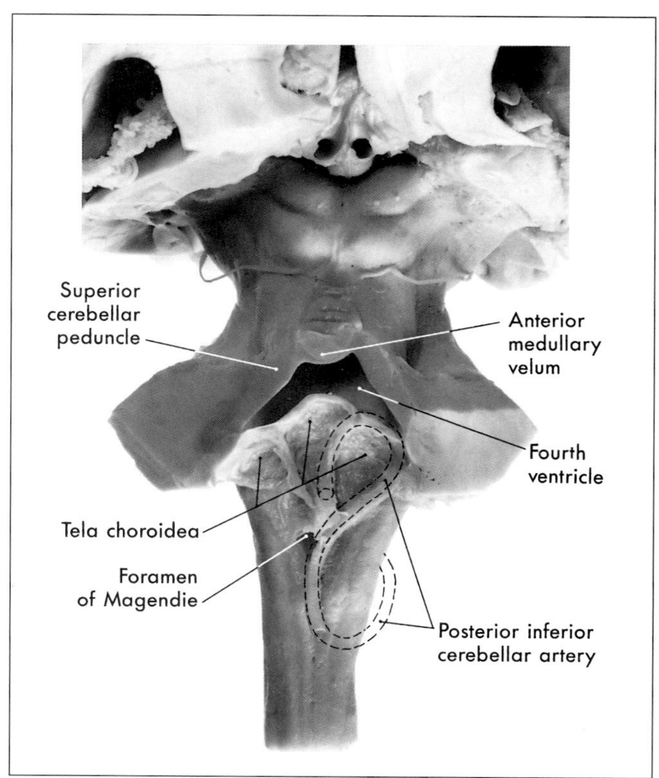

FIGURE 6-9 *Dorsal view of the brainstem with the cerebellum removed to expose the fourth ventricle, the tela choroidea of the caudal roof of the fourth ventricle, and the route of the posterior inferior cerebellar artery. The choroid plexus on the internal surface of the tela is served by this vessel.*

boundaries of this space are formed by brain structures. These include the cerebellum (covering about the middle third of the ventricle) and the superior cerebellar peduncles and anterior medullary velum (covering the rostral third of the ventricle). The floor is formed by the pons and medulla (Fig. 6-7). *The only openings between the ventricles of the brain and the subarachnoid space surrounding the brain are the foramina of Luschka and Magendie in the fourth ventricle.*

EPENDYMA, CHOROID PLEXUS, AND CSF

Ependyma The ventricles of the brain and the central canal of the spinal cord are lined by a simple cuboidal epithelium, the *ependyma*. Ependymal cells contain abundant mitochondria and are metabolically active. Their luminal surfaces are ciliated and have microvilli, and the bases contact the subependymal layer of astrocytic processes. There is no continuous basal lamina between ependymal cells and the subjacent glial cell processes (Fig. 6-10). Ependymal cells are attached to each other by *zonulae adherentes* (desmosomes).

In some regions, particularly the third ventricle, there are patches of specialized ependymal cells called *tanycytes* (Fig. 6-10). Tanycytes have basal processes that extend through the layer of astrocytic processes to form end feet on blood vessels and in the neuropil. They may function to transport substances between the ventricles and the blood. In contrast to ependymal cells, tanycytes are attached to each other and to immediately adjacent ependymal cells by tight junctions. Desmosomes are also present between tanycytes.

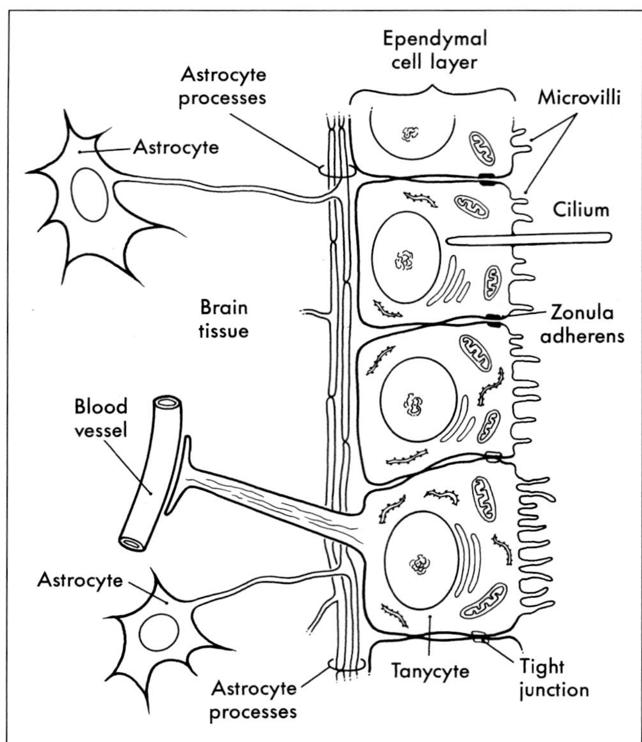

FIGURE 6-10 *The ependyma and its relationship to the layer of subependymal astrocytic processes, and a representation of a tanycyte.*

Choroid Plexus The choroid plexus in each ventricle is thrown into a series of folds called *villi* (sing. *villus*). These are covered on their ventricular (luminal) surfaces by a continuum of dome-shaped structures, each with numerous microvilli (Fig. 6-11A,B). Each dome represents the luminal surface of one *choroidal epithelial cell*, and the shallow grooves between domes are the points of contact between adjacent choroidal cells (Fig. 6-11B,C). Each villus consists of a core of highly vascularized connective tissue derived from the pia mater and a simple cuboidal covering (the choroidal epithelial cell layer), which is derived from ependymal cells (Figs. 6-3, 6-11B,C). The abundant capillaries in the connective tissue core of each villus are surrounded by a *basal lamina*. The endothelial cells of these capillaries have numerous *fenestrations*, which allow a free exchange of molecules between blood plasma and the extracellular fluid in the connective tissue core (Fig. 6-12). The connective tissue core itself consists of fibroblasts and collagen fibrils. Another basal lamina is formed at the interface between the connective tissue core and the choroidal epithelial cells that form the surface of each villus (Figs. 6-11C, 6-12). These choroidal cells have microvilli on their apical (ventricular) surface, interdigitating cell membranes on their sides, and irregular bases. Each is attached to its neighbor by continuous *tight junctions* (*zonulae occludentes*) that seal off the subjacent extracellular space from the ventricular space (Figs. 6-11C, 6-12). This represents the *blood-CSF barrier*. Choroidal epithelial cells contain a nucleus, numerous mitochondria, rough endoplasmic reticulum, and a small Golgi apparatus (Fig. 6-11C, 6-12). Thus they are specialized to control the flow of ions and metabolites into the CSF.

Although choroidal epithelial cells are joined by tight junctions, ependymal cells are not. Therefore fluid exchange occurs between CSF and the extracellular fluid of the brain parenchyma. The composition of CSF can thus sometimes reflect disease processes occurring in brain tissue. For example, catabolites of catecholamines are reduced in the CSF of patients with Parkinson's disease, a neurodegenerative disorder involving loss of dopaminergic neurons.

In humans the blood supply to the choroid plexuses is via the *choroidal arteries* and the *posterior cerebellar arteries*. Choroid plexus in the inferior horn, collateral trigone, and body of the lateral ventricle is served by the *anterior choroidal artery* (a branch of the internal carotid) and the *lateral posterior choroidal artery*. The *medial posterior choroidal artery* serves the choroid plexus of the third ventricle. The choroid plexus located inside the fourth ventricle is served by branches of the *posterior inferior cerebellar artery* (Figs. 6-7, 6-9), and the tuft that extends out of the foramen of Luschka into the subarachnoid space (Fig. 6-8) is served by the *anterior inferior cerebellar artery*.

CSF Choroidal epithelial cells secrete CSF by selective transport of materials from the connective tissue extracellular space (Fig. 6-12). Sodium chloride is actively transported into the ventricles, and water passively follows the concentration gradient thus established. Other materials, including large molecules, are transported in pinocytotic vesicles from the basal to the apical surface of the epithelium and exocytosed into the CSF. Compared to blood plasma, CSF has higher concentrations of chloride, magnesium, and sodium and lower concentrations of potassium, calcium, glucose, and proteins.

Normal CSF contains very little protein (15 to 45 mg/dl), little immunoglobulin, and only one to five cells (mononuclear) per milliliter. Changes from these normal values are useful in the diagnosis of a variety of disease processes. *Lumbar puncture* is used to collect a sample of CSF for analysis and to measure CSF pressure. A needle is inserted between the third and fourth (or fourth and fifth) lumbar vertebrae into the dural sac, and a few milliliters of fluid withdrawn. Because the average volume of CSF in the adult is about 120 ml, and the rate of production is about 450 to 500 ml/day, the sample removed is quickly replaced.

The numbers and types of cells found in CSF vary according to the type of disease. In bacterial meningitis or brain abscesses, neutrophils predominate and may reach concentrations of 1,000 to 10,000/ml. In meningeal syphilis, by contrast, 200 to 300 cells/ml would be typical, and most of these would be lymphocytes. Lymphocytes are also the predominant cell type found in active multiple sclerosis, even though there are usually fewer than 50 cells/ml of CSF. The diagnosis of multiple sclerosis also rests on specific changes in the immunoglobulin G content of CSF; immunoglobulin G is both derived from the blood and produced by lymphocytes in the CSF, where it is released during an autoimmune reaction attack.

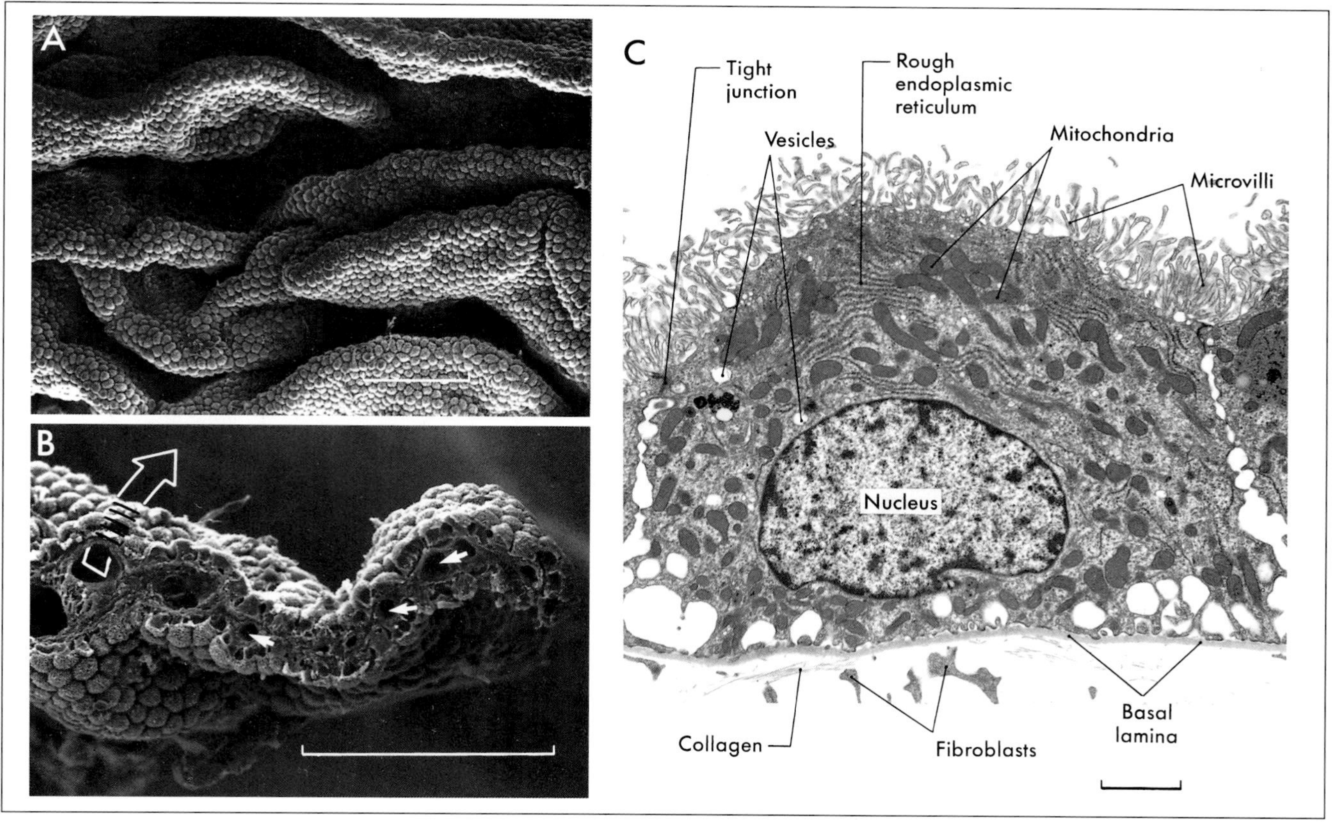

FIGURE 6-11 *Elements of the choroid plexus. Scanning electron micrographs of the surface (A) of several villi of the choroid plexus and the cut surface (B) of one villus. The characteristic appearance of the luminal surface of the choroidal cells is seen in both micrographs. The arrows in B mark vessels in the core of the villus (solid arrow) and the route of fluid movement from the vessels into the ventricular space (broken arrow). Transmission electron micrograph (C) showing the internal structure of one choroidal epithelial cell. Primate, scale = 100 μm for A, B; 2 μm for C.*

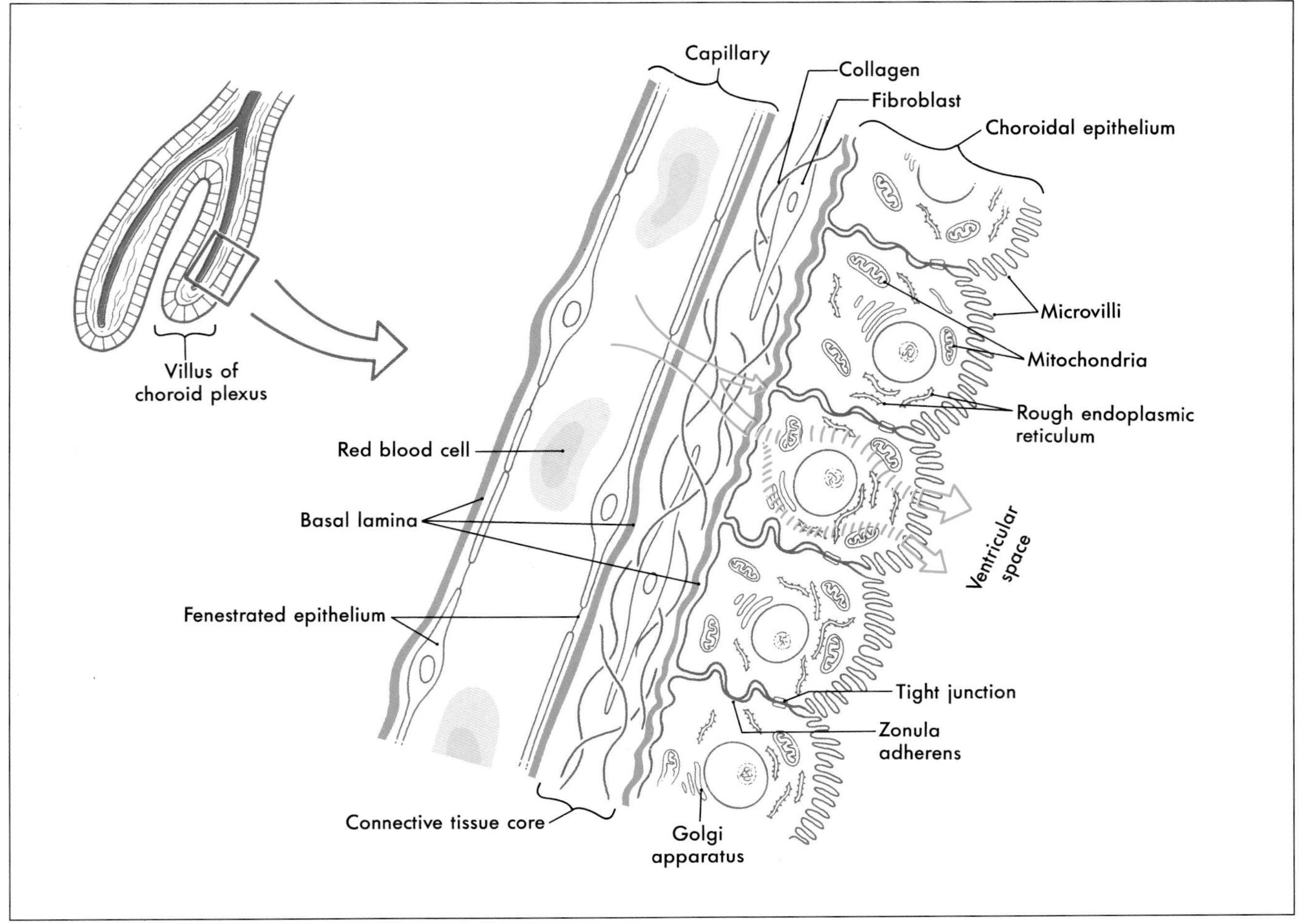

FIGURE 6-12 *The basic structure of the choroid plexus and the route of fluid transport (shown in green) through the choroidal epithelium to produce CSF.*

In marked contrast to the elevated numbers of white blood cells seen in central nervous system (CNS) infections, numerous red blood cells are present in the CSF of patients who have bleeding into the *subarachnoid space* (*subarachnoid hemorrhage*). For example, this condition may result from rupture of an intracranial aneurysm or arteriovenous malformation. Patients with subarachnoid hemorrhage or individuals with primary CNS tumors may have elevated protein levels in their CSF. Elevated CSF protein is also seen in patients with syphilis or meningitis, and in cancer patients in whom the disease has metastasized into the CNS. The CSF of cancer patients may also contain malignant cells characteristic of their primary lesions.

The CSF produced by the choroid plexuses passes through the ventricular system to exit the fourth ventricle through the foramina of Luschka and Magendie (Fig. 6-13). At this point, the CSF enters the subarachnoid space, which is continuous around the brain and spinal cord. The CSF in the subarachnoid space provides the buoyancy necessary to prevent the weight of the brain from crushing nerve roots and blood vessels against the internal surface of the skull. The weight of the brain, about 1400 g in air, is reduced to about 45 g when suspended in CSF. Consequently, the tethers formed by delicate connective tissue strands traversing the subarachnoid space, the *arachnoid trabeculae* (Fig. 6-13), are adequate to maintain the brain in a stable position within its CSF envelope.

The movement of CSF through the ventricular system and the subarachnoid space is influenced by two major factors. First, there is a subtle pressure gradient between the points of production of CSF (choroid plexuses in brain ventricles) and the points of transfer into the venous system (arachnoid villi). Because CSF is not compressible, it tends to move along this gradient. Second, CSF is also moved in the subarachnoid space by purely mechanical means. These include gentle movements of the brain on its arachnoid trabecular tethers during normal activities and the pulsations of the numerous arteries found in the subarachnoid space.

After passing through the subarachnoid space, the CSF reaches the arachnoid villi that extend into the superior sagittal sinus (Fig. 6-13). The subarachnoid space and the CSF it contains extend into the core of each villus. At this point CSF enters the venous circulation through two routes. A limited amount passes between the cells making up the arachnoid villus, whereas most is transported through these cells in membrane-bound vesicles (see also Chapter 7). About 330 to 380 ml of CSF enters the venous circulation per day.

HYDROCEPHALUS AND RELATED CONDITIONS

Blockage of CSF movement or a failure of the absorption mechanism will result in the accumulation of excessive fluid in or around the brain (Fig. 6-14). The results are commonly called *hydrocephalus*. Dilation of the cerebral ventricles may result from a blockage of CSF flow through the system, *obstructive hydrocephalus*, or from factors not related to impaired flow, *communicating hydrocephalus*.

AQUEDUCTAL STENOSIS—OBSTRUCTIVE HYDROCEPHALUS

Obstructive hydrocephalus, with enlargement of all ventricular spaces upstream to the blockage, may have a variety of causes. Although the obstruction can take place at any point, naturally narrow places such as the cerebral aqueduct or interventricular foramina are especially vulnerable.

Aqueductal stenosis may be caused by a tumor in the immediate vicinity of the midbrain (as in *pineoblastoma* or *meningioma*) that compresses the brain and occludes the aqueduct. This channel may also be occluded by the cel-

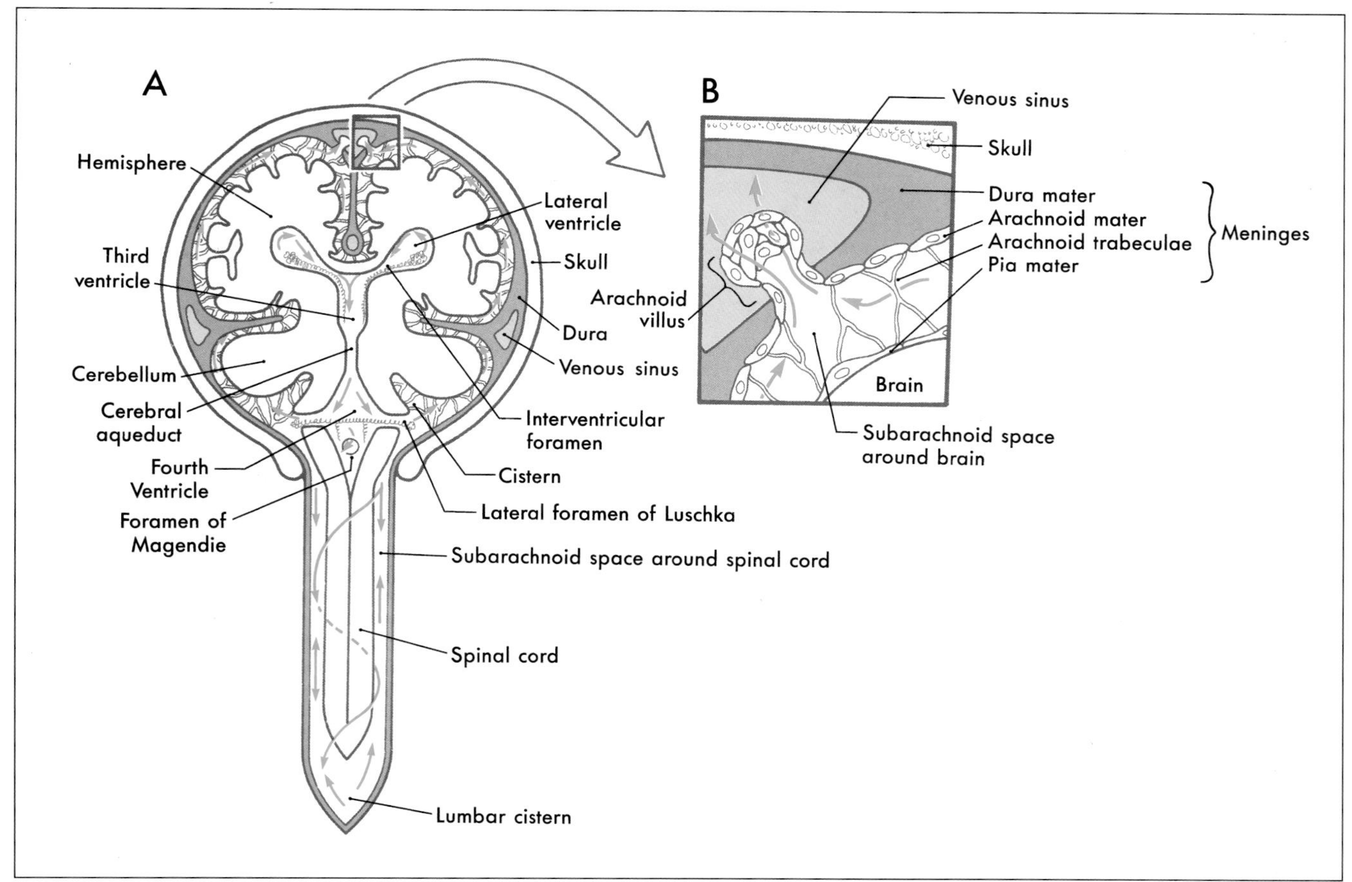

FIGURE 6-13 *Representation of the brain and spinal cord (A) showing the locations of choroid plexus (red) and the routes of flow taken by CSF (green) through the ventricles and the subarachnoid space around the CNS. The detail (B) shows the relationship of an arachnoid villus to the subarachnoid space and the venous sinus. Although not shown here, the venous sinus is lined by an endothelium. CSF enters the venous system primarily by transport through cells of the arachnoid villus (B, dashed green arrow), although some fluid moves between these cells (B, solid green arrow).*

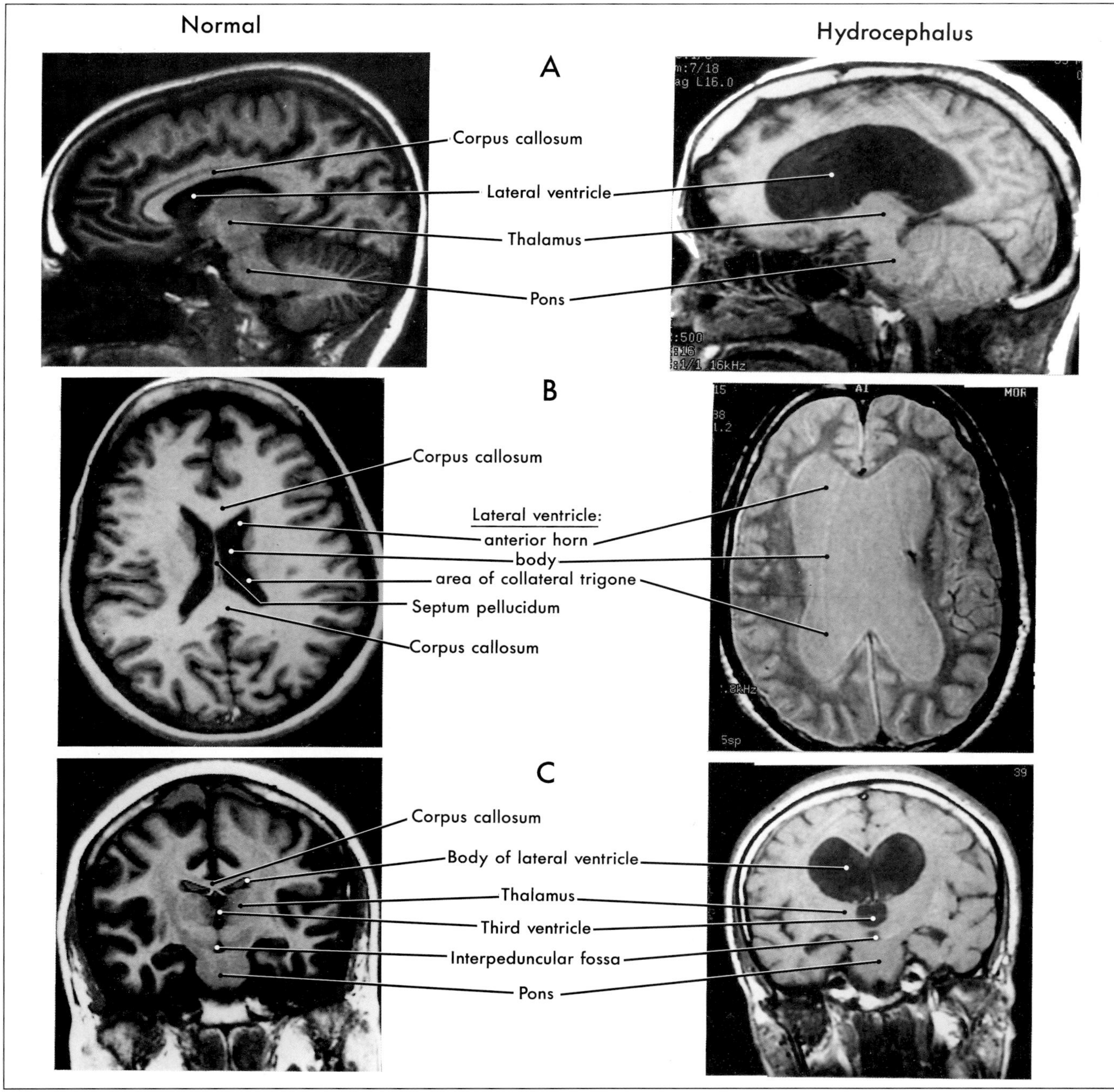

FIGURE 6-14 *Comparison of normal and hydrocephalic brains in sagittal (A), axial (B), and coronal (C) planes as seen in MRIs.*

lular debris seen following *intraventricular hemorrhage* or bacterial or fungal infections of the CNS. One major sequela of aqueductal blockade is enlargement of the third and both lateral ventricles (Fig. 6-14). Unilateral obstruction of one interventricular foramen, for example by a colloid cyst in one interventricular foramen, results in enlargement of the lateral ventricle on that side. Blockage of both interventricular foramina will produce enlargement of both lateral ventricles.

Communicating Hydrocephalus In *communicating hydrocephalus*, the flow of CSF through the ventricular system and into the subarachnoid space is not impaired. However, movement of CSF through the subarachnoid space and into the venous system is partially or totally blocked. This block may be caused by a congenital absence of arachnoid villi. Alternatively, these villi may be blocked by red blood cells subsequent to a subarachnoid hemorrhage. An exceedingly high level of protein in the CSF, as seen in patients with CNS tumors or infections, also contributes to communicating hydrocephalus. This high level is due partially to the sequestering of protein in the arachnoid villi and subsequent blockage of CSF transport into the venous system.

Additional causes of communicating hydrocephalus include the interruption of CSF movement through the subarachnoid space caused by either subarachnoid hemorrhage or a major CNS infection, and the subsequent inflammatory response. Overproduction of CSF in patients with *papilloma of the choroid plexus* may also be a factor. In all of these situations there is an enlargement of all parts of the ventricular system.

SOURCES AND ADDITIONAL READINGS

Davson H, Welch K, Segal MB: *Physiology and Pathophysiology of the Cerebrospinal Fluid.* Churchill Livingstone, Edinburgh, 1987.

Fishman RA: *Cerebrospinal Fluid in Diseases of the Nervous System.* WB Saunders, Philadelphia, 1992.

Kida S, Yamashima T, Kubota T, Ito H, Yamamoto S: A light and electron microscopic and immunohistochemical study of human arachnoid villi. J Neurosurg 69:429–435, 1988.

Peters A, Palay SL, Webster H deF: *The Fine Structure of the Nervous System, Neurons and Their Supporting Cells,* 3rd Ed. Oxford University Press, New York, 1991.

Segal MB (ed): *Barriers and Fluids of the Eye and Brain.* CRC Press, Boca Raton, 1992.

Upton ML, Weller RO: The morphology of cerebrospinal fluid drainage pathways in human arachnoid granulations. J Neurosurg 63:867–875, 1985.

Wood JH (ed): *Neurobiology of Cerebrospinal Fluid,* Vols. 1 and 2. Plenum Press, New York, 1980 and 1983.

Yamashima T: Functional ultrastructure of cerebrospinal fluid drainage channels in human arachnoid villi. Neurosurgery 22:633–641, 1988.

CHAPTER SEVEN

The Meninges

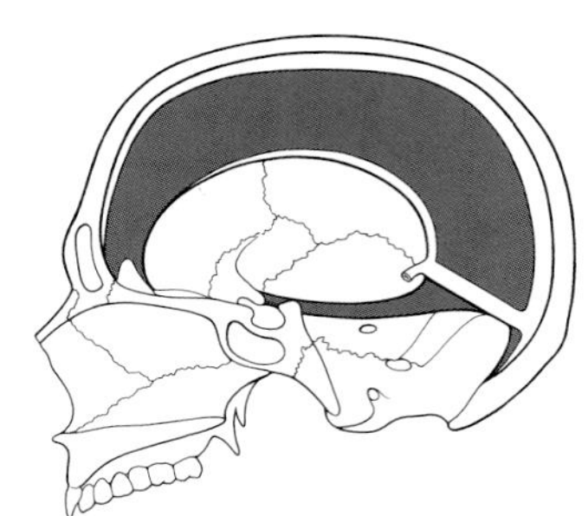

D. E. HAINES

The human nervous system is extremely delicate and lacks the internal connective tissue framework seen in most organs. For protection the brain and spinal cord are encased in bony shells, enveloped by fibrous coats, and delicately suspended within a fluid compartment. In the living state the nervous system has a gelatinous consistency but, when treated with fixatives, it becomes firm and easy to handle.

The brain and spinal cord are surrounded by the skull and vertebral column, respectively. With the exception of the intervertebral foramina, through which the spinal nerves and their associated vessels pass, and the foramina in the skull, which serve as conduits for arteries, veins, and cranial nerve roots, this bony encasement is complete. The membranous coverings of the central nervous system, the *meninges*, are located internal to the skull and vertebral column. The meninges (1) protect the underlying brain and spinal cord; (2) serve as a support framework for important arteries, veins, and sinuses; and (3) enclose a fluid-filled cavity, the *subarachnoid space*, which is vital to the survival and normal function of the brain and spinal cord.

This bony and meningeal encasement of the central nervous system is a double-edged sword. Although these structures offer maximum protection, in the case of trauma or in a disease process, they can be very unforgiving. For example, growth of a tumor creates a mass that will increase intracranial pressure and compress or displace various portions of the brain. Something has to give inside the skull when a space-occupying lesion develops, and it is the delicate tissue of the brain that gives. The neurologic deficits that result depend on the location of the mass, the rapidity with which it enlarges, and which parts of the brain are damaged.

DEVELOPMENT OF THE MENINGES

The meninges develop from cells of the *neural crest* and *mesenchyme* (mesoderm), which migrate to surround the developing central nervous system between 20 to 35 days of gestation (Fig. 7-1A–C). Collectively, these neural crest and mesodermal cells form the *primitive meninges* (*meninx primitiva*). At this stage no obvious spaces (venous sinuses, subarachnoid space) are present in the meninges. Between 34 and 48 days of gestation, the primitive meninges differentiate into an outer, more compact layer (*the ectomeninx*) and an inner, more reticulated layer (*the endomeninx*) (Fig. 7-1D). As development progresses (45 to 60 days' gestation), the ectomeninx becomes more compact, and spaces appear in this layer that correlate with the positions of the future venous sinuses. Concurrently, the endomeninx becomes more reticulated, and the spaces that appear in its inner part correspond to the subarachnoid spaces and cisterns of

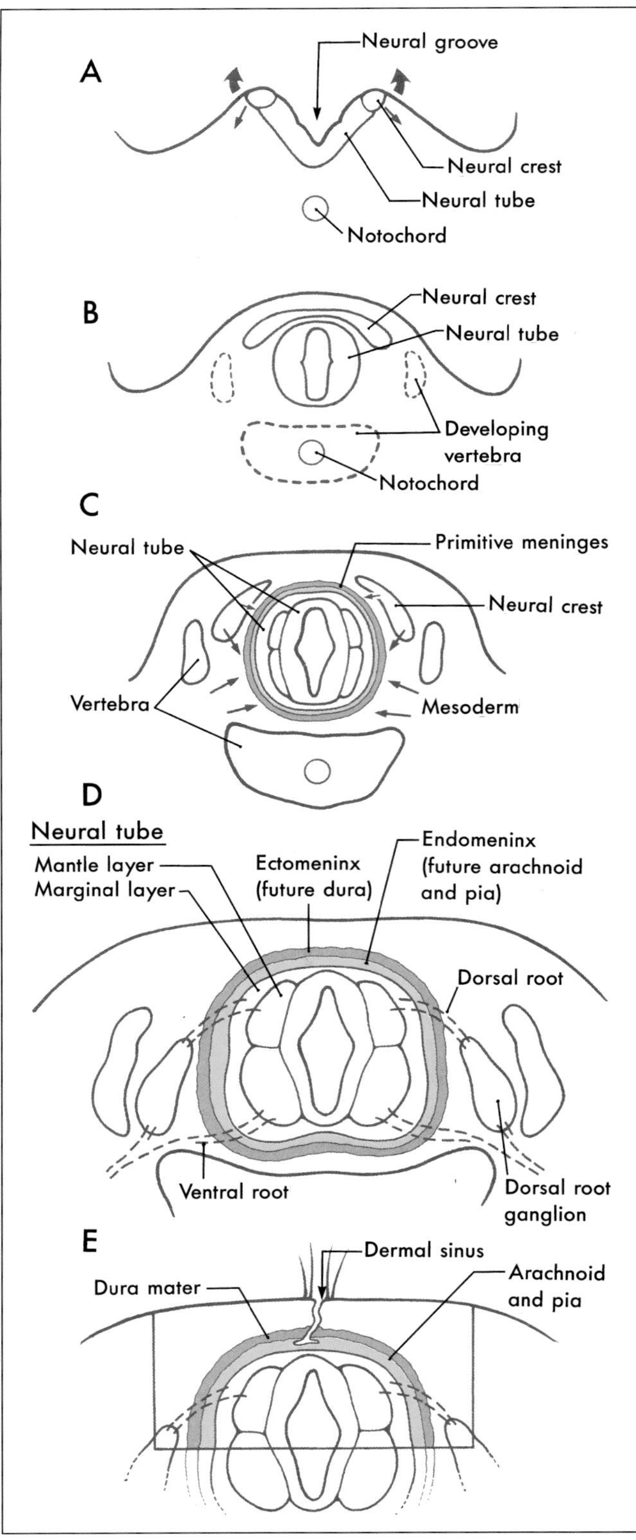

FIGURE 7-1 *Development of the meninges. After the neural tube closes (A,B), cells from the neural crest and mesoderm (C, arrows) migrate to surround the neural tube and form the primordia of the dura and of the arachnoid and pia (D). A dermal sinus (E) is a malformation in which there is a channel from the skin into the meninges.*

the adult. In general, the ectomeninx will become the *dura mater* of the adult, and the endomeninx will form the *arachnoid mater* and *pia mater* of the adult (Fig. 7-1D). By the end of the first trimester the meninges have reached the overall plan seen in the adult.

One developmental defect associated with closure of the neural tube and formation of the meninges in the lumbosacral area is the *congenital dermal sinus* (also called just dermal sinus) (Fig. 7-1E). This defect is caused by a failure of the ectoderm (future skin) to completely pinch off from the neuroectoderm and the primitive meninges that envelop it. As a result, the meninges are continuous with a narrow, epithelium-lined channel that extends to the skin surface (Fig. 7-1E). Dermal sinuses are sometimes discovered in young patients who have recurrent, unexplained bouts of meningitis. These lesions are surgically removed and recovery is usually complete.

The ectomeninx around the brain is continuous with the skeletogenous layer that forms the skull. This relationship is maintained in the adult, where the dura is intimately adherent to the inner surface of the skull. In the spinal column, the ectomeninx is also initially continuous with the developing vertebrae. However, as development proceeds the spinal ectomeninx dissociates from the vertebral bodies. A layer of cells remains on the vertebrae to form the periosteum, and the larger part of the ectomeninx condenses to form the spinal dura. The intervening space becomes the spinal *epidural space* (Fig. 7-2). This space is essential for the administration of *epidural anesthetics*.

OVERVIEW OF THE MENINGES

In general, the meninges consist of fibroblasts and varying amounts of extracellular connective tissue fibrils. The structural features of each meningeal layer reflect the fact that the fibroblasts of that particular layer are modified to serve a particular function.

The human meninges are composed of the *dura mater*, the *arachnoid mater*, and the *pia mater* (Figs. 7-2 and 7-3). The outermost portion, the *dura mater* (*pachymeninx*), is adherent to the inner surface of the skull, but is separated from the vertebrae by the epidural space (Fig. 7-2). Around the brain the inner portions of the dura give rise to infoldings or septa, such as the *falx cerebri* (Fig. 7-2), which separate brain regions from each other. Major venous sinuses are found at the points where these septa originate. Spinal and cranial nerves, as they enter or exit the central nervous system, must pass through a cuff of the dura that is continuous with the connective tissue of the peripheral nerve. Blood vessels traverse the dura in similar fashion.

Internal to the dura is the arachnoid mater (Figs. 7-2 and 7-3). The arachnoid is a thin cellular layer that is attached to the overlying dura, but, with the exception of

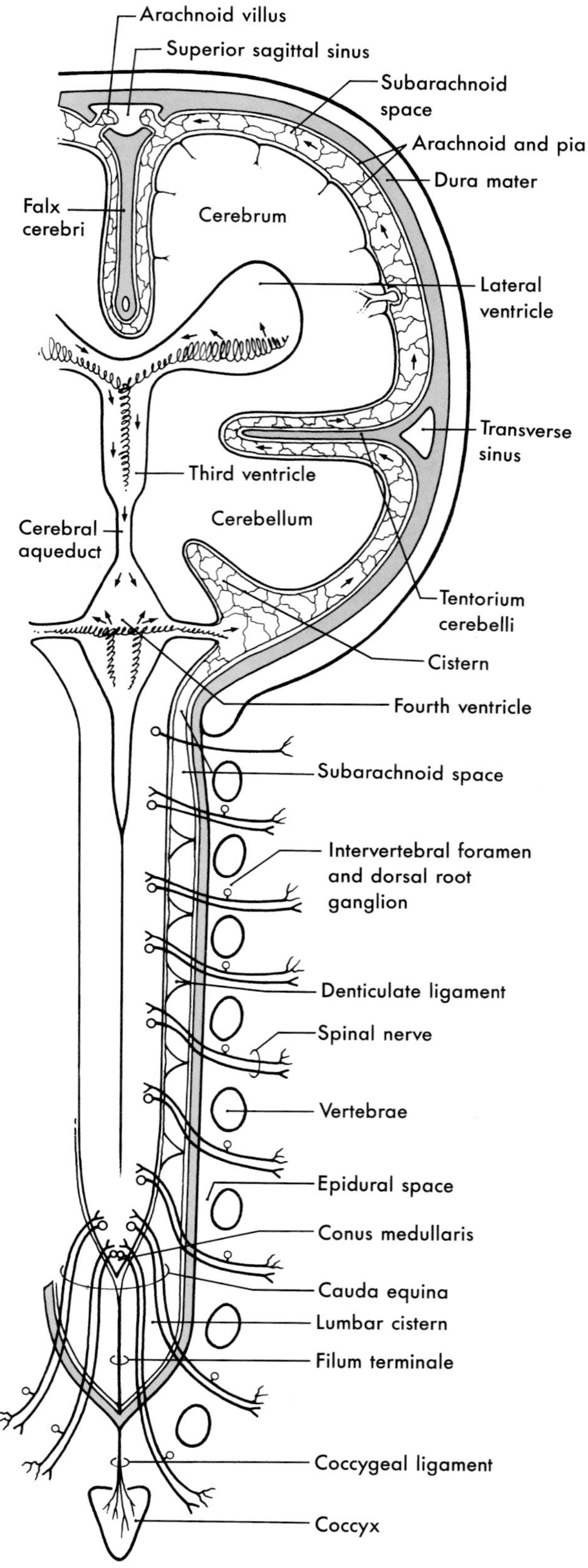

FIGURE 7-2 *The relation of the meninges to the brain and spinal cord and to their surrounding bony structures.*

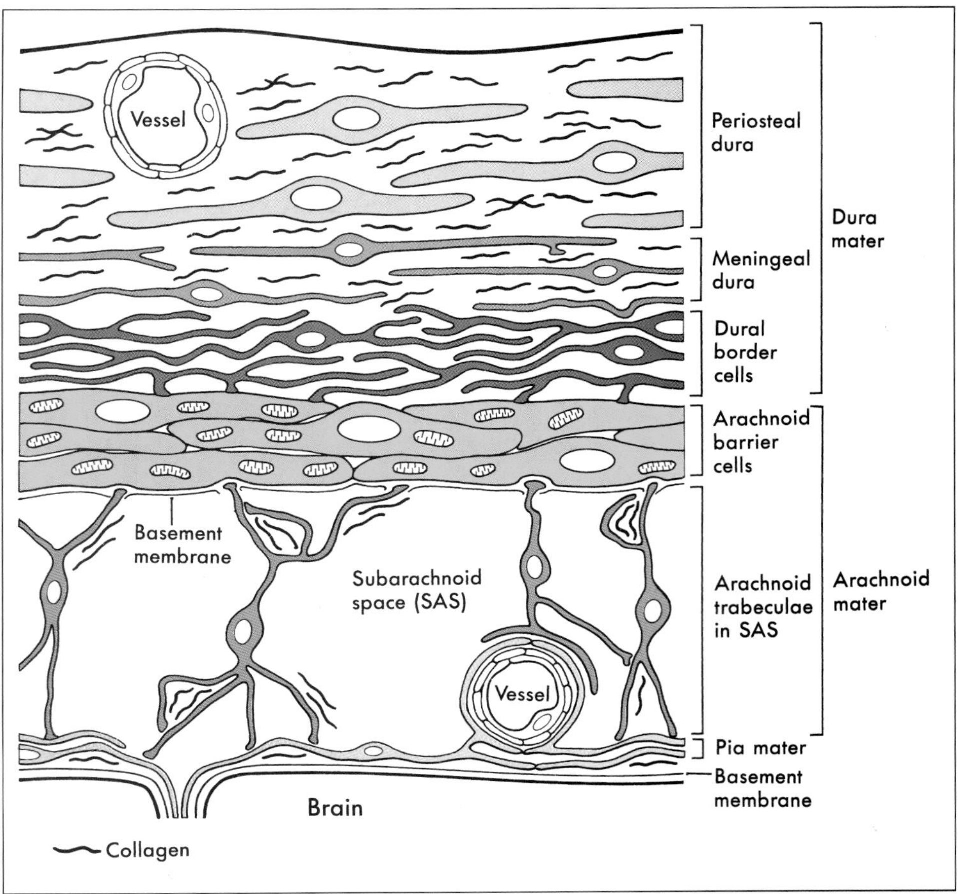

FIGURE 7-3 *The structure of the meninges. Layers of the dura are shown in shades of black, the arachnoid in shades of red, and the pia in green.*

the arachnoid trabeculae, is separated from the pia mater by the *subarachnoid space*. The arachnoid around the brain is directly continuous with the arachnoid lining the inner surface of the spinal dura (Fig. 7-2). Consequently, the spinal and cerebral subarachnoid spaces are also directly continuous with each other at the foramen magnum. The *subarachnoid space* contains cerebrospinal fluid and vessels and is bridged by fibroblasts of varying sizes and shapes that collectively form the *arachnoid trabeculae*. The arachnoid is avascular and does not contain nerve fibers.

The *pia mater* is located on the surface of the brain and spinal cord and closely follows all their various grooves and elevations (Figs. 7-2 and 7-3). Around the spinal cord the pia mater contributes to the formation of the *denticulate ligaments* and the *filum terminale* (*filum terminale internum*). It is common to refer to the arachnoid and pia as the *leptomeninges*.

THE DURA MATER

The *dura mater* (*pachymeninx*) is composed of elongated fibroblasts and copious amounts of collagen fibrils (Fig. 7-3). This membrane contains blood vessels and nerves and is generally divided into outer (*periosteal*), inner (*meningeal*), and *border cell* portions. There is no distinct border between periosteal and *meningeal* portions of the dura (Fig. 7-3). Fibroblasts of the *periosteal dura* are larger and slightly less elongated than other dural cells. This portion of the dura is adherent to the inner surface of the skull, and its attachment is particularly tenacious along suture lines and in the cranial base. In contrast, the fibroblasts of the *meningeal dura* are more flattened and elongate, their nuclei are smaller, and their cytoplasm may be darker than in periosteal cells. Although cell junctions are rarely seen between dural fibroblasts, the large amounts of interlacing collagen in periosteal and meningeal portions of the dura give these layers of the meninges great strength.

Dural Border Cell Layer The innermost part of the dura is composed of flattened fibroblasts that have sinuous processes. Collectively, these cells form the *dural border cell layer* (Fig. 7-3). The extracellular spaces between the flattened cell processes of dural border cells contain an amorphous substance but no collagen or elastic fibers. Cell junctions (desmosomes, gap junctions) are occasionally seen between dural border cells and cells of the underlying arachnoid.

Because of its loose arrangement, enlarged extracellular spaces, and lack of extracellular connective tissue fibrils, *the dural border cell layer constitutes a plane of structural weakness at the dura-arachnoid junction*. This layer is externally continuous with the meningeal dura and internally continuous with the arachnoid. Consequently, *bleeding into this area of the meninges will disrupt the dural border cell layer* rather than invade the overlying dura or the underlying arachnoid. We will consider meningeal hemorrhages after discussing the arachnoid.

Blood Supply The arterial supply to the dura of the anterior cranial fossa originates from the *cavernous portion of the internal carotid*, the *ethmoidal arteries* (via the ethmoidal foramina), and branches of the ascending pharyngeal artery (via the foramen lacerum). The *middle meningeal artery* serves the dura of the middle cranial fossa and may be compromised when there is trauma to the skull. It is a branch of the maxillary artery and enters the skull through the foramen spinosum. The accessory meningeal artery (via the foramen ovale) and small branches from the lacrimal artery (via the superior orbital fissure) also serve the dura of the middle fossa. The dura of the posterior fossa is served by small meningeal branches of ascending pharyngeal and occipital arteries and by minute branches of the vertebral arteries.

The spinal dura is served by branches of major arteries (e.g., vertebral, intercostal, and lumbosacral) that are located close to the vertebral column. These small meningeal arteries enter the vertebral canal via the intervertebral foramina to serve the dura and adjacent structures.

Nerve Supply The nerve supply to the dura of the anterior and middle fossae is from branches of the *trigeminal nerve*. *Ethmoidal nerves* and branches of the *maxillary* and *mandibular nerves* innervate the dura of the anterior fossa, whereas that of the middle fossa is served mainly by branches from the *maxillary* and *mandibular nerves*. The dura of the posterior fossa receives sensory branches from dorsal roots C1 to C3 and may have some innervation from the vagus nerve. The *tentorial nerve*, a branch of the ophthalmic nerve, courses caudally to serve the tentorium cerebelli. Autonomic fibers to the vessels of the dura originate from the superior cervical ganglia and gain access to the cranial cavity by simply following the progressive branching patterns of the vessels on which they lie.

Nerves to the spinal dura originate as recurrent branches of the spinal nerve located at that level. These delicate strands pass through the intervertebral foramina and distribute to the spinal dura and some adjacent structures.

Dural Infoldings and Sinuses As noted above, the dura has *periosteal* and *meningeal* parts. The *periosteal dura* lines the inner surface of the skull and functions as its periosteum. The meningeal dura is continuous with the periosteal dura, but draws away from it at specific locations to form the *dural infoldings* (or *reflections*). The largest of these is the *falx cerebri* (Fig. 7-4). It is attached to the crista galli rostrally, to the midline of the inner surface of the skull dorsally, and to the surface of the tentorium cerebelli caudally. The falx cerebri separates the right hemisphere from the left. The *superior sagittal sinus* is found where the falx cerebri attaches to the skull, the *straight sinus* where it fuses with the *tentorium cerebelli*, and the *inferior sagittal sinus* in its free edge (Fig. 7-4). Many large superficial veins located on the surface of the cerebral hemispheres empty into the superior sagittal sinus.

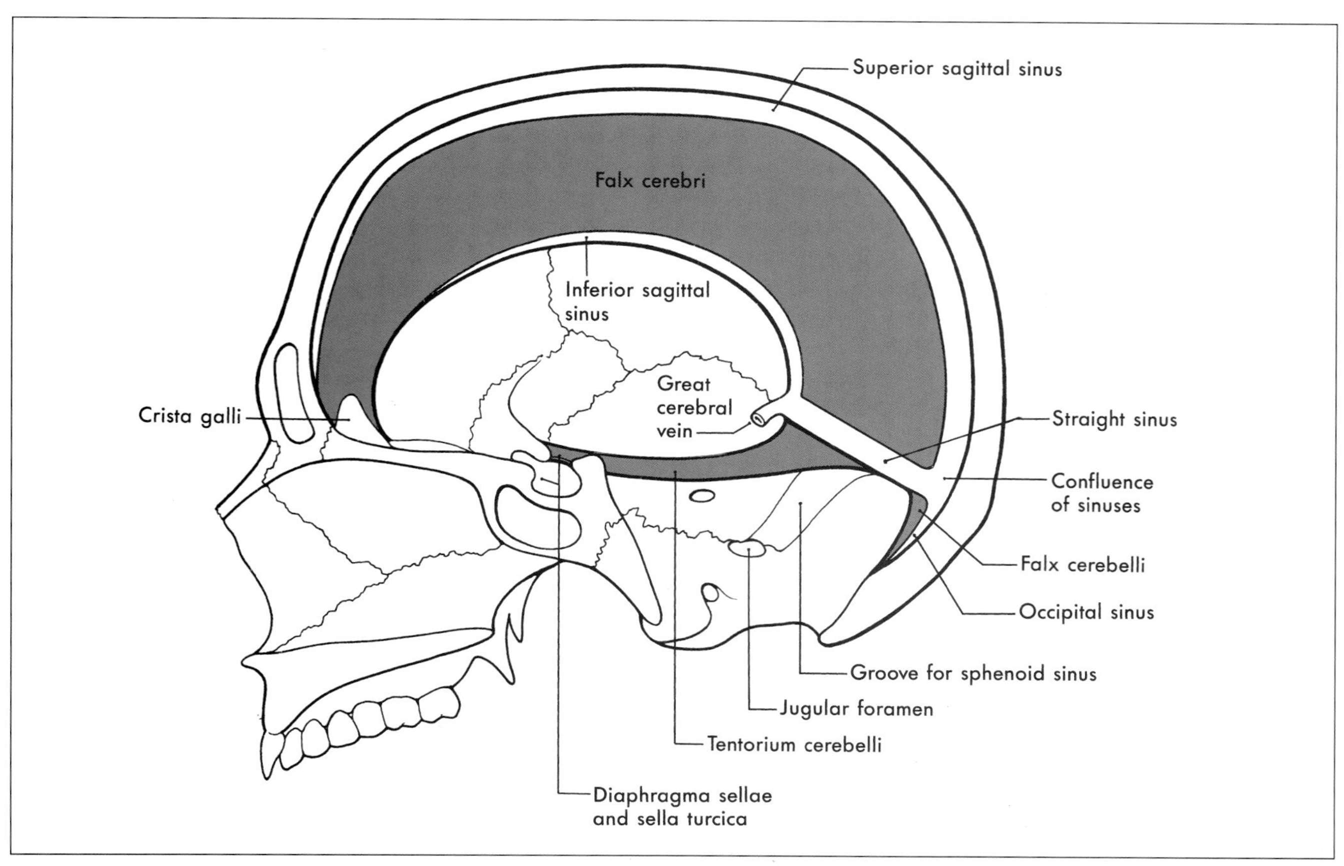

FIGURE 7-4 *Mid-sagittal view of the skull showing the dural infoldings (reflections) and venous sinuses associated with each.*

The *tentorium cerebelli* is the second largest of the dural infoldings (Figs. 7-4, 7-5). Rostrally, it attaches to the clinoid processes, rostrolaterally to the petrous portion of the temporal bone (location of the *superior petrosal sinus*), and caudolaterally to the inner surface of the occipital bone and a small part of the parietal bone (location of the *transverse sinus*) (Fig. 7-5). The tent shape of the tentorium divides the cranial cavity into *infratentorial* (below) and *supratentorial* (above) compartments. The supratentorial compartment is divided into right and left halves by the falx cerebri. The sweeping edges of the right and left tentoria, as they arch from the clinoid processes to join at the straight sinus, form the *tentorial notch* (Fig. 7-5). The occipital lobe is above the tentorium, the cerebellum below it, and the midbrain passes through the tentorial notch.

The restrictive nature of the tentorial notch and the tenacious nature of the falx cerebri and tentorium cerebelli become important factors in situations of increased intracranial pressure or space-occupying lesions. For example, lesions in supratentorial locations may force brain structures across the midline below the falx cerebri or down through the tentorial notch. Such lesions in infratentorial locations may result in the extrusion of brain structures either up through the tentorial notch or down through the foramen magnum.

Located below the tentorium cerebelli on the midline of the occipital bone is the *falx cerebelli* (Fig. 7-4). This small dural infolding extends into the space found between the cerebellar hemispheres and usually contains a small *occipital sinus* that communicates with the *confluence of sinuses*.

The smallest of the dural infoldings, the *diaphragma sella* (Figs. 7-4, 7-5), forms the roof of the hypophyseal fossa and encircles the stalk of the pituitary. The *cavernous sinuses* are found on either side of the sella turcica, and the *anterior* and *posterior intercavernous sinuses* are found in their respective edges of the diaphragma sella.

The relationships of venous sinuses are discussed in Chapter 8. It should be emphasized, however, that *venous sinuses are endothelium-lined spaces* that communicate with each other. In addition, large veins from the surface of the brain empty into the venous sinuses. As they enter the sinus, these veins are attached to a cuff of dura. Consequently, a blow to the head (or a minor bump to the head in an aged individual) may cause the brain to shift just enough in the subarachnoid space to tear a vein at the point where it enters the sinus. This tear may result in venous blood in the subarachnoid space or may create a hematoma at the dura-arachnoid interface (see Fig. 7-8).

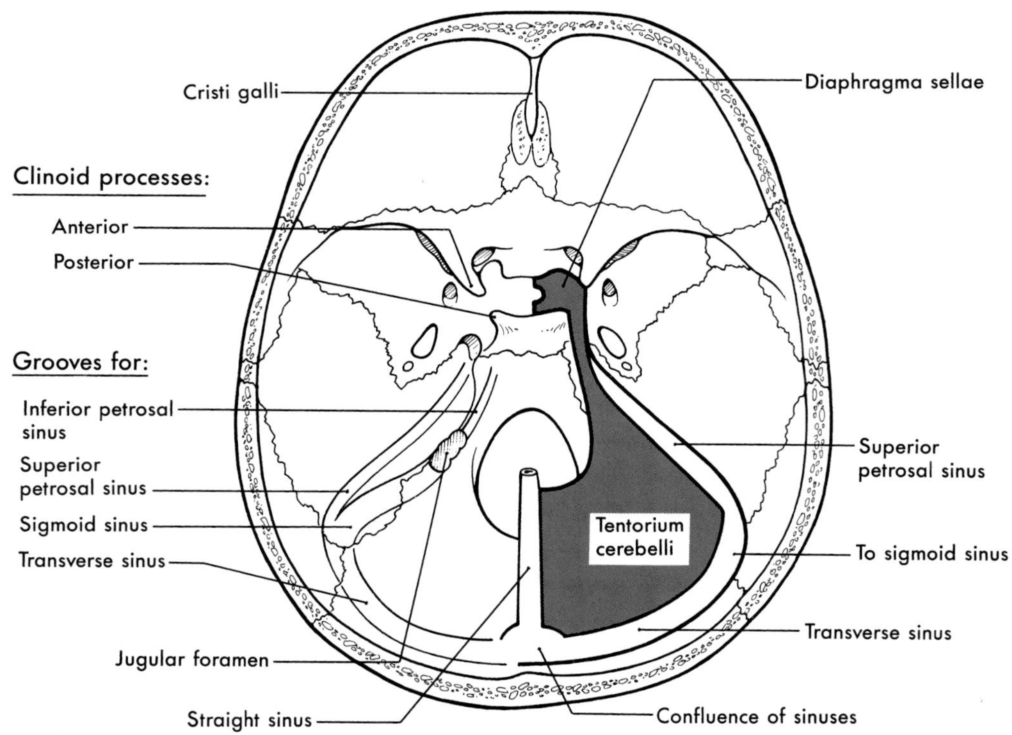

FIGURE 7-5 *View of the cranial base from the dorsal aspect showing the tentorium cerebelli (and its associated sinuses) and the diaphragma sellae. Also indicated are the positions of grooves formed by some of the major sinuses.*

Cranial Versus Spinal Dura At the margin of the foramen magnum, the *periosteal dura* essentially stops, but the *meningeal dura* continues caudally in the vertebral canal to eventually attach to the inner aspect of the sacrum as the *coccygeal ligament* (*filum terminale externum*) (Fig. 7-2). The *spinal dural sac* is anchored rostrally and caudally and is separated from the adjacent vertebrae by an *epidural space* that contains venous channels, some lymphatics, and fat deposits. There are no dural infoldings around the cord; consequently, there are no venous sinuses in the spinal dura.

ARACHNOID MATER

The *arachnoid mater* is located internal to the dural border cell layer and is regarded as having two parts (Fig. 7-3). The portion of the arachnoid directly apposed to the dural border cells is the *arachnoid barrier cell layer*, and the spindly cells that traverse the subarachnoid space constitute the *arachnoid trabeculae*.

The *subarachnoid space* is located between the arachnoid barrier cell layer and the pial cells located on the surface of the brain or spinal cord. This space contains cerebrospinal fluid and many superficial vessels. Enlarged regions of the subarachnoid space are called *subarachnoid cisterns*; these will be discussed later.

Arachnoid Barrier Cell Layer Fibroblasts of this layer are more plump than the flattened cells of the dura (Fig. 7-3). The arachnoid barrier cell layer is tenuously attached to the dural border cell layer by occasional cell junctions. In contrast, arachnoid barrier cells have closely apposed cell membranes and are joined to each other by *numerous tight* (occluding) *junctions*; hence the "barrier" characteristic of this layer. This close apposition of cell membranes excludes any significant extracellular space; consequently, no collagen is found in this layer of the meninges. The tight junctions between these arachnoid cells not only serve as a barrier against the movement of fluids or other substances, but also impart strength to the membrane. In the human, a basement membrane (basal lamina) is found on that surface of the barrier cell layer that faces the subarachnoid space.

Arachnoid Trabeculae and the Subarachnoid Space The *arachnoid trabeculae* are composed of flattened, irregularly shaped fibroblasts that bridge the subarachnoid space in a random fashion (Fig. 7-3). Trabecular cells attach to the barrier layer and may attach to each other, to pial cells, or to blood vessels in the subarachnoid space. Although much of the extracellular collagen associated with trabecular cells is confined in the folded processes of these cells, some may be found free in the subarachnoid space. The attachments of the trabecular cells and their framework of collagen fibrils give added strength to the arachnoid mater.

The *subarachnoid space* is located internal to the barrier cell layer and external to the pia mater (Figs. 7-2, 7-3). It contains cerebrospinal fluid, trabecular cells and collagen fibrils, arteries, and veins. Although some vessels may lie free in the subarachnoid space, most are covered by a thin layer of the leptomeninges (Fig. 7-3). These large vessels in the subarachnoid space may rupture, resulting in the spread of blood around the brain; this event is a *subarachnoid hemorrhage*. *Cerebrospinal fluid* is produced by the *choroid plexi* of the lateral, third, and fourth ventricles. It exits the ventricular system via the foramina of Magendie and Luschka to enter the subarachnoid space (see arrows on Fig. 7-2). After circulating around the brain and spinal cord, it reenters the vascular system primarily through the *arachnoid villi*. The subarachnoid space around the spinal cord is the route used to administer *spinal anesthesia*.

Although it is common to refer to the brain as "floating" in the cerebrospinal fluid of the subarachnoid space, it is actually *suspended within this space*. The structural basis for this fact is as follows. The dura is adherent to the skull, the arachnoid to the dura, the arachnoid trabeculae to the pia, and the pia to the surface of the brain. Consequently, the brain is suspended, through this chain, within the fluid milieu of the subarachnoid space by the numerous delicate strands of the arachnoid trabeculae. This structuring is possible because the brain looses about 97% of its weight when it is suspended in cerebrospinal fluid. For example, a brain that weighs about 1,400 g in air will weigh only about 45 to 50 g in fluid.

Because the arachnoid trabeculae are not rigid, the brain may move within the fluid-filled subarachnoid space. In a closed head injury, the brain may move on its trabecular tethers in response to a sudden blow and be subjected to minor damage (*concussion/contusion*). This injury may result in no, or only momentary, loss of consciousness. Such a minor injury may be found at the point of the blow or at a site opposite the contact (*contrecoup* injury).

Arachnoid Villi The small specialized portions of the arachnoid that protrude into the superior sagittal sinus through openings in the dura form the *arachnoid villi* or *granulations* (Fig. 7-6; also see Fig. 7-2). If they are especially large or calcified (as in older individuals), they may be called *pacchionian bodies*.

Arachnoid villi extend into the sinus through tight cuffs in the meningeal dura and are found just off the midline or in cul-de-sacs (the *lateral* or *venous lacunae*) of the sinus. The space in the center of each villus is continuous with the subarachnoid space around the brain. This space is enclosed in a layer of cells that are markedly similar to arachnoid barrier cells, and these arachnoid cells, in turn, are surrounded by a capsule of

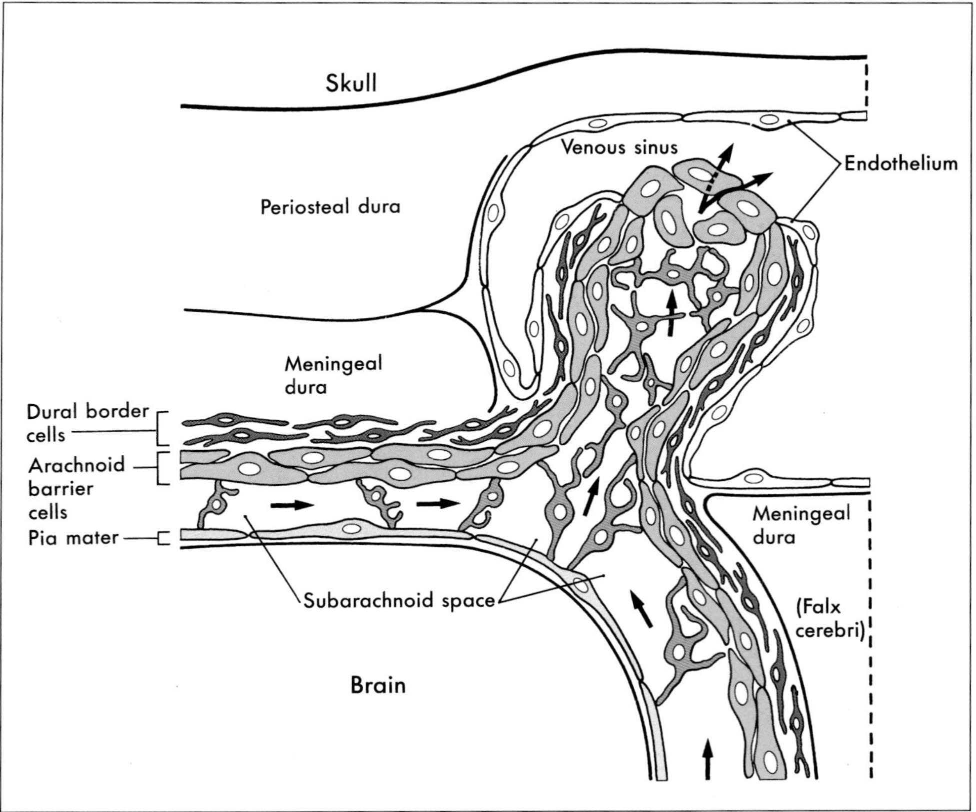

FIGURE 7-6 *Structure of the arachnoid villi. Note the continuity of the cell layers of the villus with those of the meninges. Cerebrospinal fluid (arrows) passes from the subarachnoid space into the villus and then into the venous sinus.*

cells that are essentially the same as dural border cells. These two layers are continuous with their respective meningeal layers through the stalk of the villus (Fig. 7-6). The *endothelial lining of the sinus* is reflected onto the villus and may cover this structure entirely or may leave a few arachnoid cells exposed; the exposed cells are called *arachnoid cap cells.* The endothelium covering the villus sits on a basement membrane beneath which some extracellular collagen may be found.

Arachnoid villi are structurally adapted for the transport of cerebrospinal fluid from the subarachnoid space into the venous circulation (Fig. 7-6). Cerebrospinal fluid moves only from the villus into the sinus. The two routes of fluid movement are through small intercellular channels located between cells and by way of a vacuole-mediated transport of fluid and other elements (bacteria, blood cells) through villus cells. As cerebrospinal fluid traverses the villus, it moves down a pressure gradient from a point of higher pressure (the subarachnoid space) to a point of lower pressure (the venous sinus). If the pressure on the venous side exceeds that on the subarachnoid space side, the flow of cerebrospinal will slow or stop. Venous blood, however, never flows from the sinus into the subarachnoid space.

MENINGIOMAS AND MENINGEAL HEMORRHAGES

At this point, it is appropriate to briefly consider meningiomas and meningeal hemorrhages because they are *specifically related to the dura and arachnoid portions of the meninges.* Tumors of the meninges, collectively called *meningiomas*, originate primarily from clusters of arachnoid cells found in the villi and at points where cranial nerves and blood vessels traverse the dura. These tumors are most often seen in patients between 40 and 50 years of age. Meningiomas are located outside the parenchyma of the brain and may invade the surrounding skull but almost never penetrate the brain or spinal cord (Fig. 7-7). Consequently, dysfunction is caused by compression of underlying regions of the brain or spinal cord. The treatment of choice for meningiomas is surgical removal.

If we exclude, for the moment, subarachnoid hemorrhages, *meningeal hemorrhages* can be generally described as *extravasated blood that strips the dura from the skull or dissects open the dural border cell layer* (Fig. 7-8). The most common cause in both situations is an injury to the head, with or without skull fracture. In a head injury, the periosteal dura may be loosened from the skull with consequent damage to a major artery; the middle and accessory meningeal arteries are common victims. Extravascular blood dissects the periosteal dura from the skull and collects to form an *extradural (epidural) hematoma* (Fig. 7-8). The neurologic deficits seen in patients with epidural hemorrhage are usually those characteristic of increased intracranial pressure. These deficits, in order of occurrence, are headache, confusion and disorientation, lethargy, and finally a state of unresponsiveness.

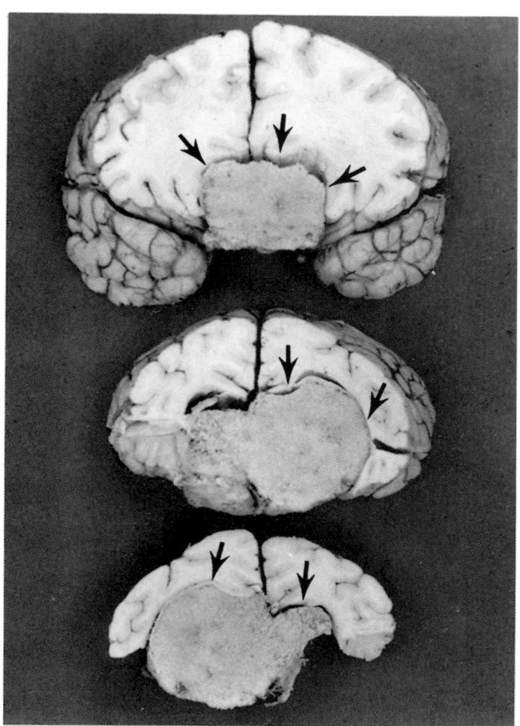

FIGURE 7-7 *A large olfactory groove meningioma. Note that this tumor has significantly compressed (arrows) but not invaded the brain in these sequential slices through the frontal lobe. (Courtesy of Dr. Jonathan Fratkin, University of Mississippi Medical Center.)*

In contrast to extradural hemorrhages, bleeding into the meninges at the approximate junction of the arachnoid with the dura originates mainly from venous structures. A common cause is the tearing of "bridging veins" as they pass through the subarachnoid space and enter a dural venous sinus (Fig. 7-8). Although these lesions are commonly called "subdural," as noted above, there is *no naturally occurring space at the arachnoid-dura junction*. Hematomas at this junction are usually caused by extravasated blood that splits open the dural border cell layer (Fig. 7-8). This extravascular blood does not collect within preexisting space, but rather creates a space at the dura-arachnoid junction. Because these so-called *subdural hematomas* are usually found *within a specific layer of cells*, they actually constitute *"dural border" hematomas*. These lesions generally contain blood in their central area and myofibroblasts, fibroblasts, mast cells, proliferating blood vessels, and dural border cells in the surrounding capsule.

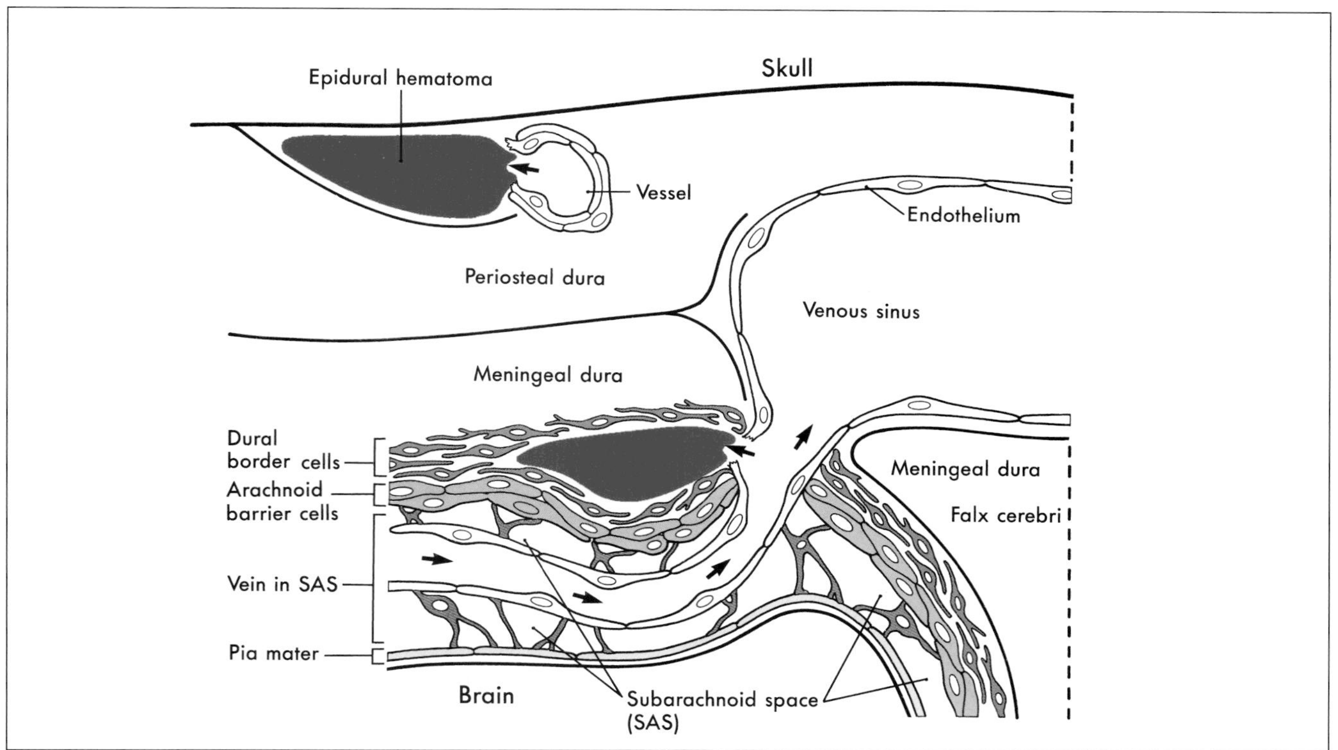

FIGURE 7-8 *The relationship of extravasated blood to the meninges. The epidural hematoma is located between the dura and the skull. Bleeding into the dura-arachnoid interface, classically called a "subdural" hematoma, is actually into a structurally weak cell layer at this juncture.*

Trauma to the skull may also result in tearing of the arachnoid membrane. In such instances, cerebrospinal fluid, which is under pressure (about 100–150 mmH_2O in a recumbent position), also may dissect open and collect with the dural border cell layer. These lesions are called *hygromas*.

PIA MATER

The *pia mater* consists of flattened cells with long, equally flattened processes that closely follow all the surface features of the brain and spinal cord (Fig. 7-3). The pia and arachnoid together constitute the *leptomeninges*. Vessels in the subarachnoid space (Fig. 7-3) may be covered by a single layer of pial cells, may be enveloped by several layers of leptomeningeal cells, or may lie free in this space. The pia is separated from the brain surface by a *glial basement membrane* and by occasional places where pial cells pull away from the brain to form a small *subpial space*. Pial cells at the brain surface may be arranged as a single layer or as several layers. Single pial cell processes and their subjacent collagen correspond to the *pia intima*; these closely follow surface features of the brain and spinal cord. When there are several tiers of pial cell processes, the outer layers correspond to the *epipial layer*. In general, the pia is thicker on the spinal cord than on the brain.

Where small vessels penetrate the surface of the brain and spinal cord, they pull along a small envelope of pial cell processes and extracellular space. These *perivascular spaces* (*Virchow-Robin spaces*) extend for varying distances into the parenchyma of the nervous system and may serve as a conduit for the movement of extracellular fluid between the minute spaces around neurons and glial cells and the subarachnoid space.

The spinal cord is anchored in the subarachnoid space by three structures: two pial modifications plus a reticulated septum of arachnoid cell processes that attaches to the dorsal midline of the cord. The first of the pial structures, the *denticulate ligaments*, run longitudinally along each side of the spinal cord about midway between the dorsal and ventral roots and attach to the inner surface of the arachnoid-lined dural sac (see Fig. 7-2). From each ligament a series of about 20 to 22 structures, shaped much like a shark's tooth, extend laterally to attach to the inner surface of the arachnoid-lined dural sac. Second, extending caudally from the conus medullaris is a tough strand composed primarily of pia; this is the *filum terminale* (*filum terminale internum*). The filum terminale attaches to the caudal end of the dural sac which, in turn, attaches to the coccyx as the *coccygeal ligament* (*filum terminale externum*) (see Fig. 7-2). Together, these anchoring structures serve a function analogous to that of the arachnoid trabeculae around the brain.

The large space caudal to the conus medullaris, which contains cerebrospinal fluid, dorsal and ventral roots (constituting the *cauda equina*), and the filum terminale, is the *lumbar cistern* (see Fig. 7-2). The retrieval of cerebrospinal fluid is an important diagnostic tool for evaluating a variety of central nervous system disorders. A needle introduced into the lumbar cistern (*spinal tap* or *lumbar puncture*) through the 3rd-4th or 4th-5th lumbar interspaces is the primary method used to collect a sample of cerebrospinal fluid from this cistern.

CISTERNS AND SUBARACHNOID HEMORRHAGES

The subarachnoid space around the brain has a number of naturally enlarged regions called *subarachnoid cisterns*, which contain cerebrospinal fluid, arteries, veins, and, in some cases, cranial nerve roots (Fig. 7-9). Cisterns occur where the brain draws away from the skull as part of its natural variation in shape, thus enlarging the subarachnoid space.

Cisterns are usually named according to the structures on which they border. For example, the *interpeduncular cistern* is found in the interpeduncular fossa, the *dorsal cerebellomedullary cistern* (*cisterna magna*) is found between the cerebellum and the medulla, and so on (Fig. 7-9). Typically, the shapes of cisterns, as seen in magnetic resonance imaging (MRI) and computed tomography (CT) scans, are determined by the corresponding shapes of surrounding brain structures (Fig. 7-10); this characteristic relationship is useful in diagnosis. The *cisterna magna* is a potential source of cerebrospinal fluid if the lumbar cistern is not accessible. In a *cisternal puncture*, a needle is carefully introduced into the cisterna magna through the atlantooccipital membrane and a sample of fluid withdrawn.

Cisterns are bordered by particular brain structures, contain segments of major vessels, and may also contain cranial nerve roots or other structures (Table 7-1). Consequently, a progressively enlarging aneurysm or a slow bleed into a particular cistern may result in signs or symptoms related to the structures found in or next to the cistern. For example, an aneurysm protruding into the interpeduncular cistern may affect the oculomotor nerve (Table 7-1) and, consequently, eye movements or pupil size.

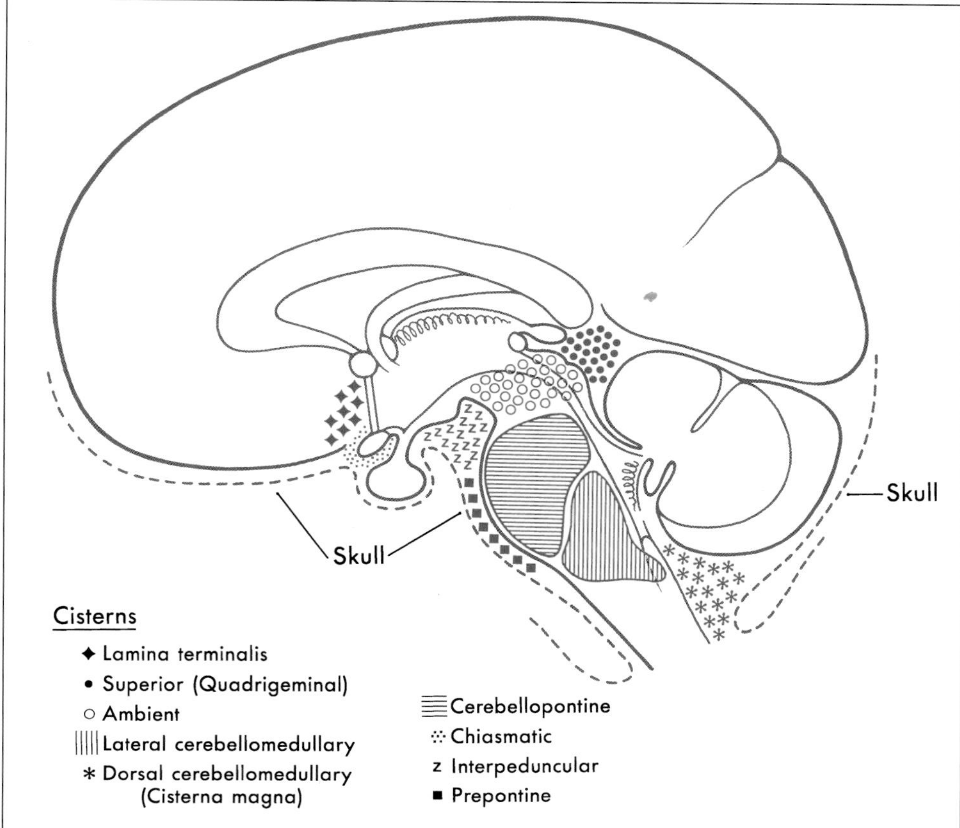

FIGURE 7-9 *The locations of the major subarachnoid cisterns in relation to brain structures. Although the cerebellopontine, lateral cerebellomedullary, and ambient cisterns are located on the lateral aspect of the brainstem, their approximate positions are indicated on this midsagittal view. Compare with Table 7-1.*

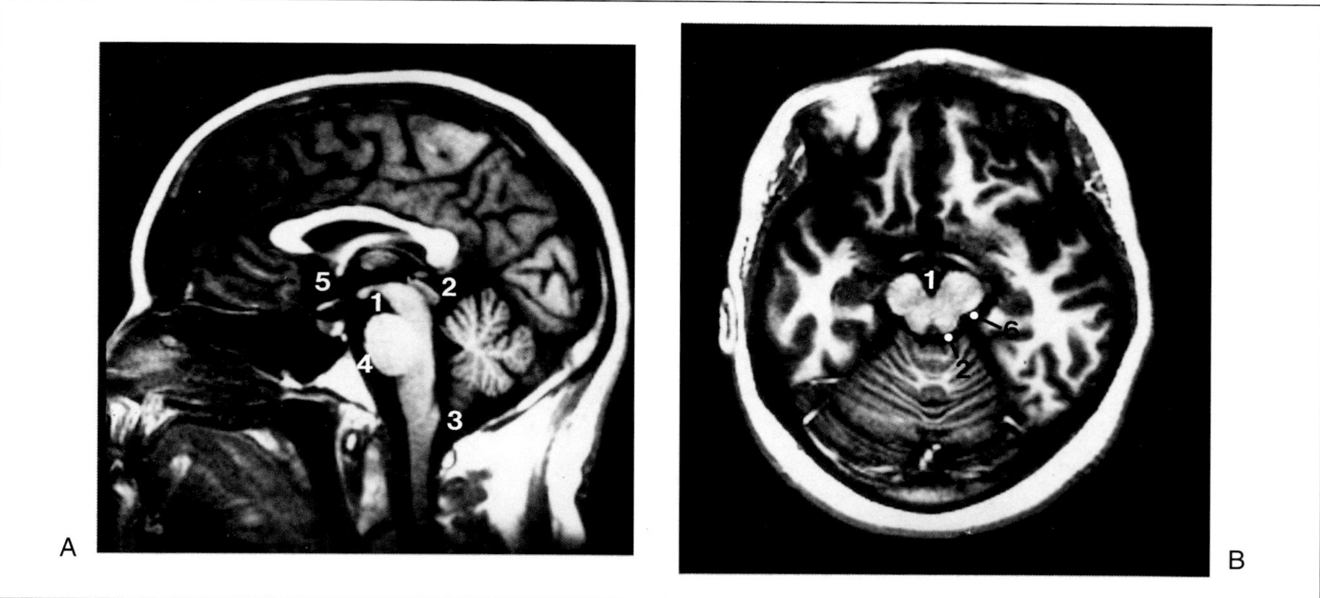

FIGURE 7-10 *Magnetic resonance images (MRI) in sagittal (A) and axial (horizontal, B) planes with some of the major cisterns indicated. 1, interpeduncular; 2, superior (quadrigeminal); 3, cisterna magna (dorsal cerebellomedullary); 4, prepontine; 5, of the lamina terminalis; 6, ambient.*

TABLE 7-1

SOME PRINCIPAL CISTERNS AND THE MAIN ARTERIES, VEINS, CRANIAL NERVES, AND OTHER STRUCTURES ASSOCIATED WITH THEM

CISTERN	ARTERY	VEIN	CRANIAL NERVE(S)	STRUCTURE
Ambient	Portions of posterior cerebral, quadrigeminal, and superior cerebellar aa.	Basal v. (of Rosenthal)	Trochlear	Lateral aspect of crus cerebri
Cerebellopontine (inferior—also called lateral cerebellomedullary)	Vertebral a. and proximal branches of PICA	Retroolivary and lateral medullary vv.	Glossopharyngeal, vagus, spinal accessory, and hypoglossal	Pyramid, inferior olivary eminence, and choroid plexus
Cerebellopontine (superior)	Distal branches of anterior inferior cerebellar, labyrinthine, and basilar aa.	Pontomesencephalic and petrosal vv.	Trigeminal, facial, and vestibulocochlear	—
Chiasmatic	Ophthalmic a. and small branches to chiasm and hypophysis	—	Optic nerve and optic chiasm	—
Cisterna magna (also called dorsal cerebellomedullary)	Distal branches of PICA, posterior spinal a., and branches to choroid plexus of fourth ventricle	Tonsillar and dorsal medullary vv.	—	Roots of C1, C2
Interpeduncular	Rostral end of basilar a. and portions of posterior cerebral, choroidal, and thalamogeniculate aa.	Portions of basal v. (of Rosenthal)	Oculomotor root	Mammillary body, medial edge of crus cerebri
Prepontine	Basilar a. and its branches	Pontine veins	Abducens	—
Quadrigeminal	Portions of posterior cerebral, quadrigeminal, and choroidal aa.	Great cerebral v. (of Galen)	Trochlear root	Pineal, superior and inferior colliculi

PICA = posterior inferior cerebellar artery.
(Data from Yasargil, 1984.)

A *subarachnoid hemorrhage* is an extravasation of blood (usually arterial) into the subarachnoid space. Some subarachnoid hemorrhages are caused by trauma, but most (about 70%) are caused by rupture of an aneurysm. *Aneurysms* are clearly defined dilations in the walls of arteries. Although many aneurysms are thought to be congenital in nature, they may also be caused by an ongoing pathologic process or by trauma, or may be secondary to a general systemic problem such as hypertension. Subarachnoid hemorrhage from a ruptured aneurysm is most common in individuals between 40 and 65 years of age.

The occurrence of a subarachnoid hemorrhage may be signaled by a sudden severe headache, vomiting and/or nausea, and a depression or loss of consciousness. Bloody cerebrospinal fluid obtained by lumbar or cisternal puncture is diagnostic for subarachnoid hemorrhage. This type of vascular accident is an especially catastrophic event with a poor prognosis; about 40% of spontaneous hemorrhages result in death. In individuals with neurologic signs that can be traced to an aneurysm, the treatment of choice is to clip the aneurysm or its stalk, thereby separating it from the cerebral circulation.

SOURCES AND ADDITIONAL READING

Alcolado R, Weller RO, Parrish EP, Garrod, D: The cranial arachnoid and pia mater in man: Anatomical and ultrastructural observations. Neuropathol Appl Neurobiol 14:1–17, 1988.

Al-Mefty O: *Meningiomas.* Raven Press, New York, 1991.

Frederickson RG: The subdural space interpreted as a cellular layer of meninges. Anat Rec 230:38–51, 1991.

Haines DE: On the question of a subdural space. Anat Rec 230:3–21, 1991.

Nabeshima S, Reese TS, Landis DMD, Brightman MW: Junctions in the meninges and marginal glia. J Comp Neurol 164:127–170, 1975.

Nicholas DS, Weller RO: The fine anatomy of the human spinal meninges. J Neurosurg 69:276–282, 1988.

Orlin JR, Osen K, Hovig T: Subdural compartment in pig: A morphologic study with blood and horseradish peroxidase infused subdurally. Anat Rec 230:22–37, 1991.

Peters A, Palay SL, Webster HD: *The Fine Structure of the Nervous System: The Neurons and Supporting Cells,* 3rd Ed. WB Saunders, Philadelphia, 1991.

Schachenmayr W, Friede RL: The origin of subdural neomembranes. I. Fine structure of the dura-arachnoid interface in man. Am J Pathol 92:53–68, 1978.

Williams PL, (ed): *Gray's Anatomy*, 38th Ed. Churchill Livingstone, New York, 1995.

Yasargil MG: *Microneurosurgery, I. Microsurgical Anatomy of the Basal Cisterns and Vessels of the Brain, Diagnostic Studies, General Operative Techniques and Pathological Considerations of the Intracranial Aneurysms.* Georg Theime Verlag, Stuttgart, 1984.

CHAPTER EIGHT

A Survey of the Cerebrovascular System

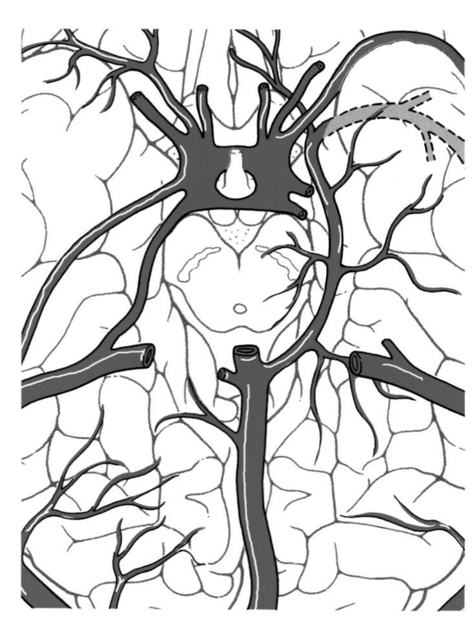

D. E. HAINES

About 50% of the problems that occur inside the cranial cavity and result in neurologic deficits are vascular in origin. Consequently, a good understanding of cerebral vascular patterns is essential to establish an accurate diagnosis. The brain is a voracious consumer of oxygen and, therefore, requires a great deal of oxygenated blood. Although it makes up only about 2% of total body weight in adults, the brain receives about 15–17% of the total cardiac output and consumes about 20% of the oxygen used by the entire body!

An ongoing flow of oxygenated blood is essential for continued brain function. The average individual will lose consciousness if the brain is deprived of blood for 10 to 12 seconds; after 3 to 5 minutes, irreparable brain damage or death may result. There are exceptions, however. Individuals who become hypothermic with a subsequent decrease in arterial blood flow to the brain, as in a winter drowning, may be revived after 10, 15, or even 20 minutes with little or no permanent damage. In these cases, the reduction in body temperature protects the brain against the consequences of reduced blood flow.

OVERVIEW

Blood is supplied to the brain by the *internal carotid* and *vertebral arteries*. The internal carotid arteries enter the skull and then divide into the *anterior* and *middle cerebral arteries*. The vertebral arteries pass through the foramen magnum and join to form the *basilar artery*; hence, the term *vertebrobasilar* is frequently applied to this part of the cerebral circulation. The basilar artery branches into right and left *posterior cerebral arteries*.

Venous outflow from the brain travels through superficial and deep veins, which drain into the dural venous sinuses. Blood in the sinuses, in turn, enters the internal jugular vein. Superficial veins in the scalp and veins in the orbit also may communicate with the dural sinuses, but these are not major conduits for venous drainage from the brain.

This chapter is an *overview of the cerebrovascular system, with emphasis on the distribution pattern of vessels on the surface of the brain*. Details of the distribution of blood vessels to internal structures are covered when we consider nuclei and tracts within the central nervous system.

CAUSES OF VASCULAR COMPROMISE

Intracranial hemorrhage originates from arteries or veins, and may result from diseases, trauma, developmental defects, or infections. Such bleeding is classified according to its location. *Meningeal hemorrhages* are found in relation to the coverings of the brain (see also Chapter 7), whereas bleeding into the subarachnoid space is called *subarachnoid hemorrhage*. Hemorrhage may occur into the ventricular spaces (*intraventricular*) or into the substance of the brain (*parenchymatous*). Although many events can lead to cerebral vascular problems with resultant dysfunction, only three examples—aneurysms, cerebral embolism, and *arteriovenous malformations* (Fig. 8-1)—are considered here.

An *aneurysm* is a dilation of a vessel wall, usually an artery. The cavity of the aneurysm is continuous with the lumen of the vessel from which it originates (Fig. 8-1A). Cerebral aneurysms may range from small (berry) to very large (giant aneurysms > 2 cm in diameter); the latter may cause damage by compression. Most intracranial aneurysms (about 85%) are found on branches of the internal carotid artery. Regardless of where they occur, *intracranial aneurysms are frequently located at the branch-points of vessels*. The treatment of choice is to clip the stalk of the aneurysm so as to separate its friable sac from the cerebral circulation.

A *cerebral embolism* is the occlusion of a cerebral vessel by some extraneous material (e.g., clot, tumor cells, plaque fragments). This occlusion leads to *ischemia* (a localized anemia) and, if prolonged, ultimately to *infarction* (a localized vascular insufficiency resulting in necrosis) of the area served by the vessel (Fig. 8-1B). An embolus made up exclusively of blood products is called

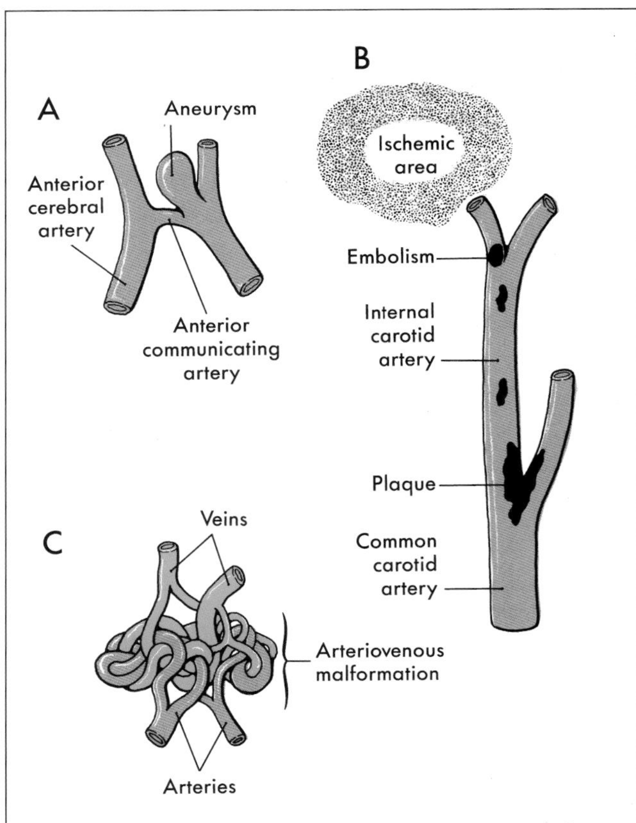

FIGURE 8-1 *Representation of an aneurysm (A), an embolism (B), and an arteriovenous malformation (C).*

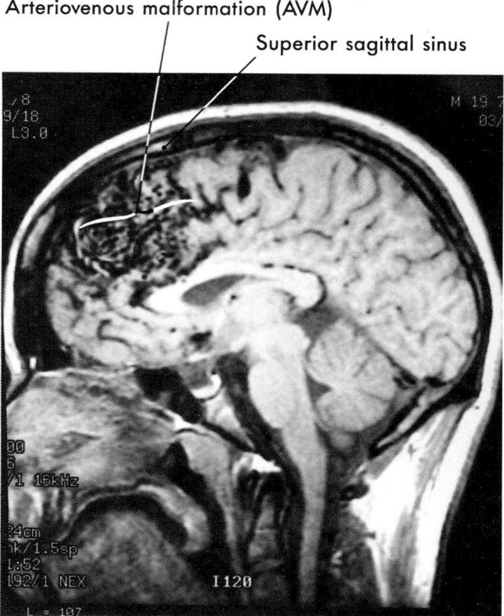

FIGURE 8-2 *Sagittal magnetic resonance imaging near the midline showing an AVM in the frontal lobe. This lesion includes branches of the anterior cerebral artery and drains into the superior sagittal sinus.*

a *thrombus*. The size of the embolus determines where it lodges. Very small emboli may temporarily occlude small cerebral vessels and give rise to a *transient ischemic attack*, a sudden loss of neurologic function that usually resolves within a day or so. On the other hand, large emboli that suddenly occlude major vessels result in catastrophic neurologic problems and even death.

An *arteriovenous malformation* (AVM) results when the communications between major arteries and veins do not develop normally (Fig. 8-1C). These lesions consist of masses of tortuous, interconnecting channels composed of large arteries and veins that communicate directly with each other. An AVM may be located on the surface of the brain or within the substance of the hemisphere or brainstem. Subarachnoid hemorrhage is most commonly the result of bleeding from an aneurysm or AVM. Superficially located AVMs can be surgically removed.

AVMs range from small and difficult to detect to quite large; large AVMs may cause damage to adjacent structures (Fig. 8-2). These lesions are commonly identified in the second or third decades of life, although symptoms (focal seizures, hemiparesis) may be seen earlier. Bleeding from AVMs is common and may be "silent" or may result in obvious neurologic deficits.

INTERNAL CAROTID SYSTEM

Internal Carotid Artery Distal to its point of entry into the skull bone, the *internal carotid artery* consists of three segments, *petrous*, *cavernous*, and *cerebral* parts. The *petrous part* is located in the carotid canal and has no branches of consequence. The *cavernous part* passes through the cavernous sinus and gives rise to hypophysial and meningeal branches.

The *cerebral part* of the internal carotid begins where this vessel penetrates the dura just ventral to the optic nerve. Its branches are the *ophthalmic*, *posterior communicating*, and *anterior choroidal arteries* (Fig. 8-3). As the ophthalmic artery enters the orbit, it gives rise to the *central artery of the retina*, a prime source of arterial blood to the retina. Occlusion of the ophthalmic artery results in significant visual loss in the ipsilateral eye. Also, aneurysms at the ophthalmic-carotid intersection may cause visual loss because of direct pressure on the optic nerve. The posterior communicating artery joins the *posterior cerebral artery*, and the anterior choroidal artery follows caudolaterally along the optic tract (Fig. 8-3). The internal carotid artery ends by dividing into the anterior and middle cerebral arteries.

Anterior Cerebral Artery The *anterior cerebral artery* passes superiorly over the optic chiasm and is joined to its counterpart by the *anterior communicating artery* (Fig. 8-3). That part of the anterior cerebral artery between its origin from the internal carotid and the anterior communicating artery is the A_1 segment. The anterior communicating artery and the distal parts of the A_1 segments are located in the *cistern of the lamina terminalis* and give rise to small branches that serve structures in the immediate area.

About 20–25% of all intracranial aneurysms are found either on the anterior communicating artery or where this vessel joins the anterior cerebral artery (Fig. 8-1A). Patients with aneurysms of the anterior communicating artery may have visual deficits because of the close proximity of this vessel to the optic chiasm.

Distal to the anterior communicator, the anterior cerebral artery branches over the medial surface of the hemisphere to about the level of the parieto-occipital sulcus; collectively these branches form the A_2 segment (Fig. 8-4. The main branches of A_2 lie within the *callosal cistern*. Figure 8-4 shows the distribution pattern of the major named branches of the anterior cerebral artery on the medial aspect of the hemisphere. Aneurysms of A_2, are usually located at the branch points of the frontopolar and callosomarginal arteries.

Middle Cerebral Artery The *middle cerebral artery* is usually (70% of the time) the larger of the terminal branches of the internal carotid artery (Fig. 8-3). The part of this vessel located between its origin from the internal carotid and the point where it branches in the Sylvian fissure is the M_1 segment. Branches from M_1 serve adjacent medial and rostral aspects of the temporal

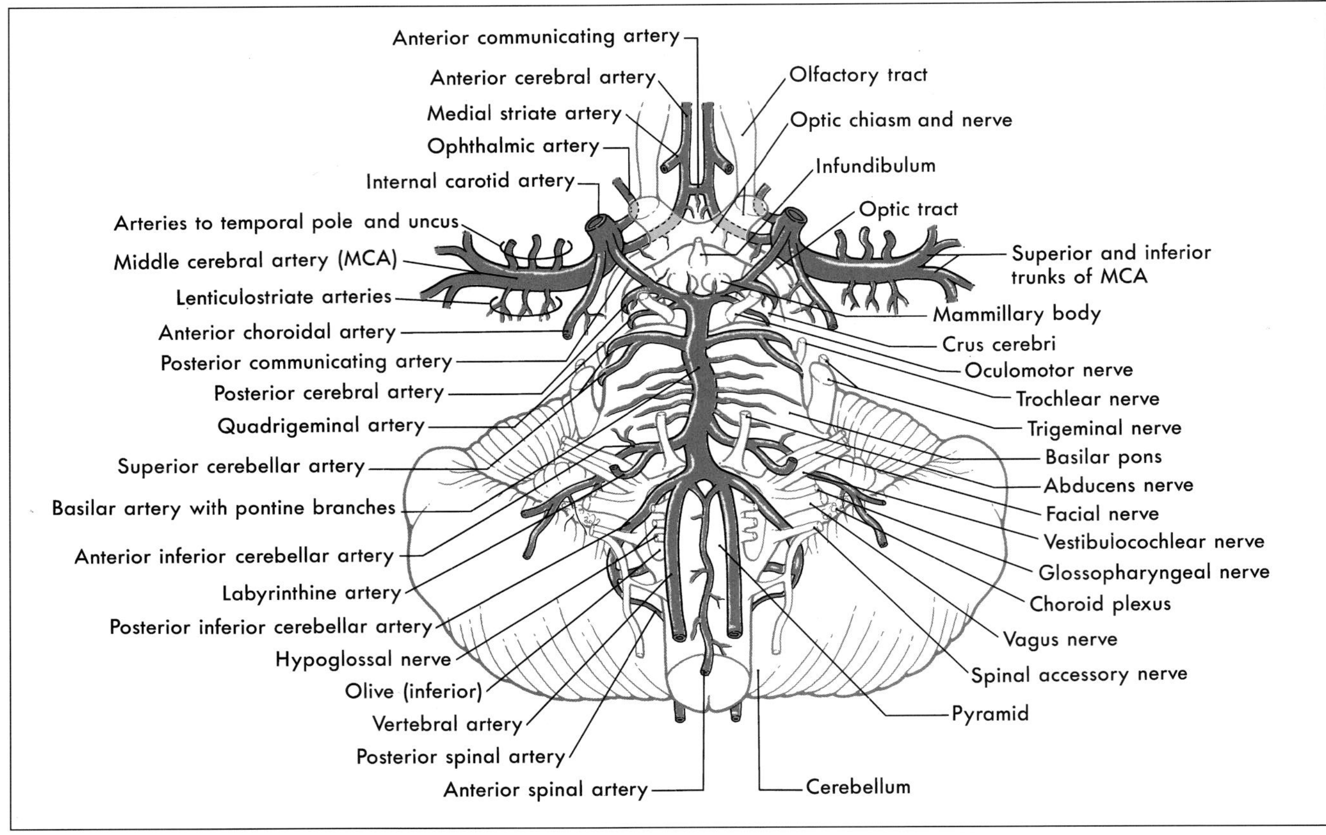

FIGURE 8-3 *Arteries on the base of the brain showing the relationship of vessels to structures and the arrangement of the circle of Willis (see also Fig. 8-9).*

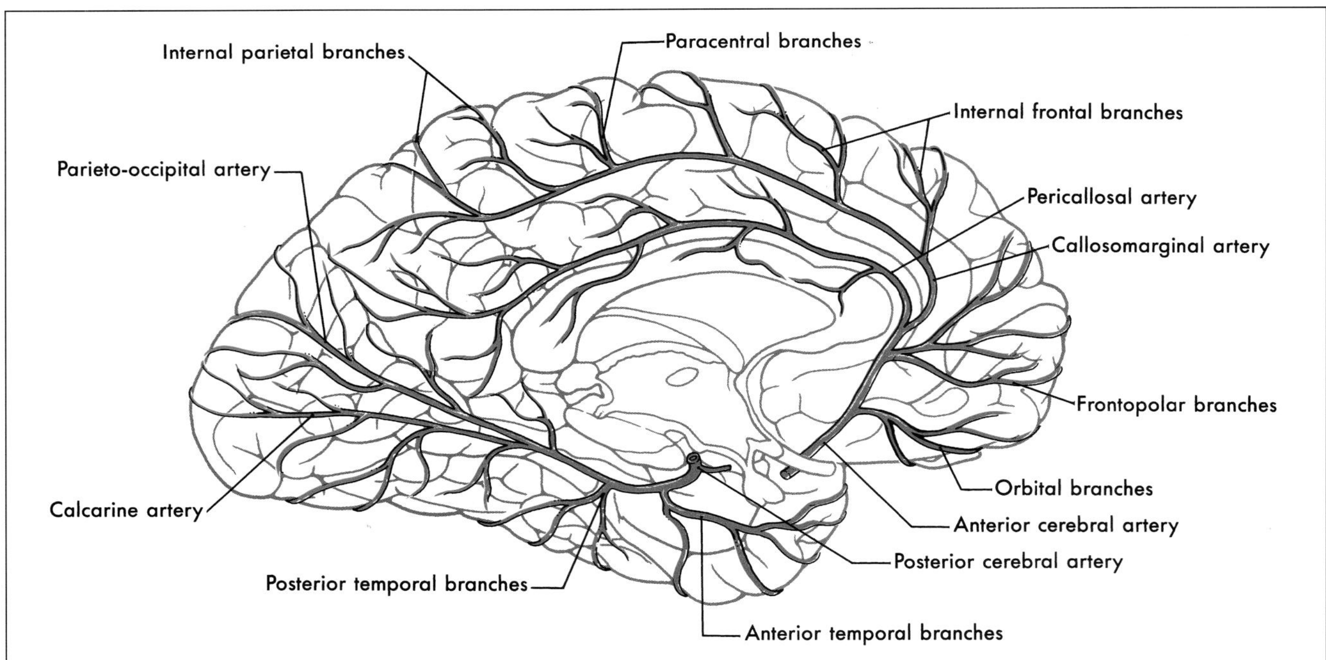

FIGURE 8-4 *Arteries on the medial surface of the cerebral hemisphere and on the inferior surface of the temporal lobe.*

lobe and via *lenticulostriate arteries*, structures located inside the hemisphere. The M_1 segment is located in medial portions of the *Sylvian cistern*.

On the ventromedial aspect of the insular cortex, the M_1 segment usually bifurcates into *superior* and *inferior trunks* (Fig. 8-3). These trunks and their distal branches collectively form the M_2 segment. As these vessels pass through the *Sylvian cistern*, they give rise to branches that serve the insular cortex. Distal branches of the superior and inferior trunks exit the lateral fissure and serve, respectively, *cortical areas located above and below this fissure*. Most of these distal branches are named according to the general area, or structure, they serve; their distribution patterns are shown in Figure 8-5. Aneurysms of the middle cerebral artery occur most frequently at the bifurcation of the M_1 segment into superior and inferior trunks.

VERTEBROBASILAR SYSTEM

Vertebral Artery The *vertebral artery* leaves the transverse foramen of C1, loops caudally and medially around the lateral mass of the atlas, and then pierces the atlanto-occipital membrane, to which it is anchored. This circuitous portion of the vertebral artery is vulnerable to injury. For example, hyperextension of the head may compress the vertebral artery between the occipital bone and the posterior arch of the atlas, and extreme rotation of the head may put torsion on this artery and restrict blood flow. The deficits seen in these patients fall under a general classification referred to as *vertebrobasilar insufficiency*. Once inside the subarachnoid space, the vertebral artery is located in the *lateral cerebellomedullary cistern*.

Branches of the vertebral artery supply the medulla, parts of the cerebellum, and the dura of the posterior fossa (Figs. 8-3, 8-6). Its first major branch, the *posterior inferior cerebellar artery* (PICA), arches around the dorsolateral medulla and sends branches to this part of the brainstem. Dorsally, the PICA is located in the *cisterna magna*. It serves the choroid plexus of the fourth ventricle and then branches over medial parts of the inferior cerebellar surface (Fig. 8-6). In about 75% of brains, the *posterior spinal artery* is a branch of the PICA; in the other 25%, it arises from the vertebral artery. The posterior spinal artery serves dorsolateral regions of the medulla *caudal to the area served by the PICA* (Fig. 8-6). The *vertebral artery* supplies the ventrolateral medulla and, just before joining its counterpart on the opposite side, gives rise to the *anterior spinal artery*. This vessel usually (85% of cases) originates as two small trunks, which join to form a single artery that courses caudally in the ventral median fissure of the medulla and continues into the spinal cord (Fig. 8-3; see also Fig. 8-18). The anterior spinal artery is found in the *premedullary cistern*.

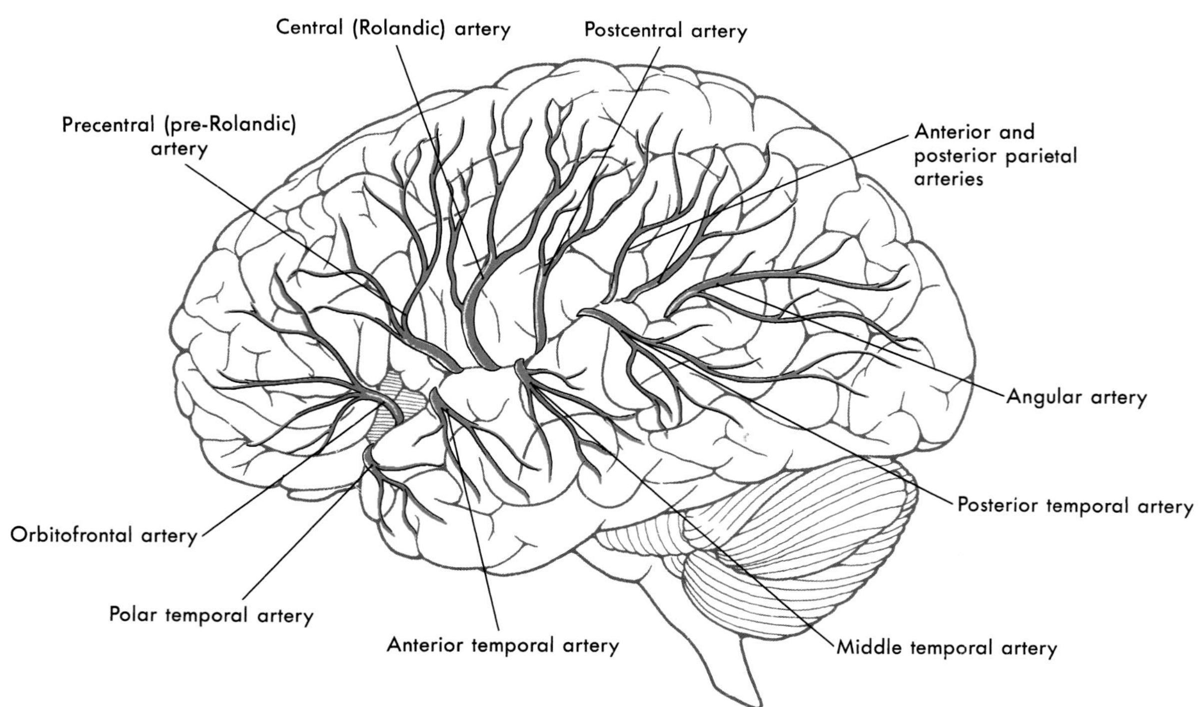

FIGURE 8-5 *Branches of the middle cerebral artery on the lateral surface of the hemisphere. Polar temporal and anterior temporal arteries are branches of M_1; the remaining arteries arise from M_2.*

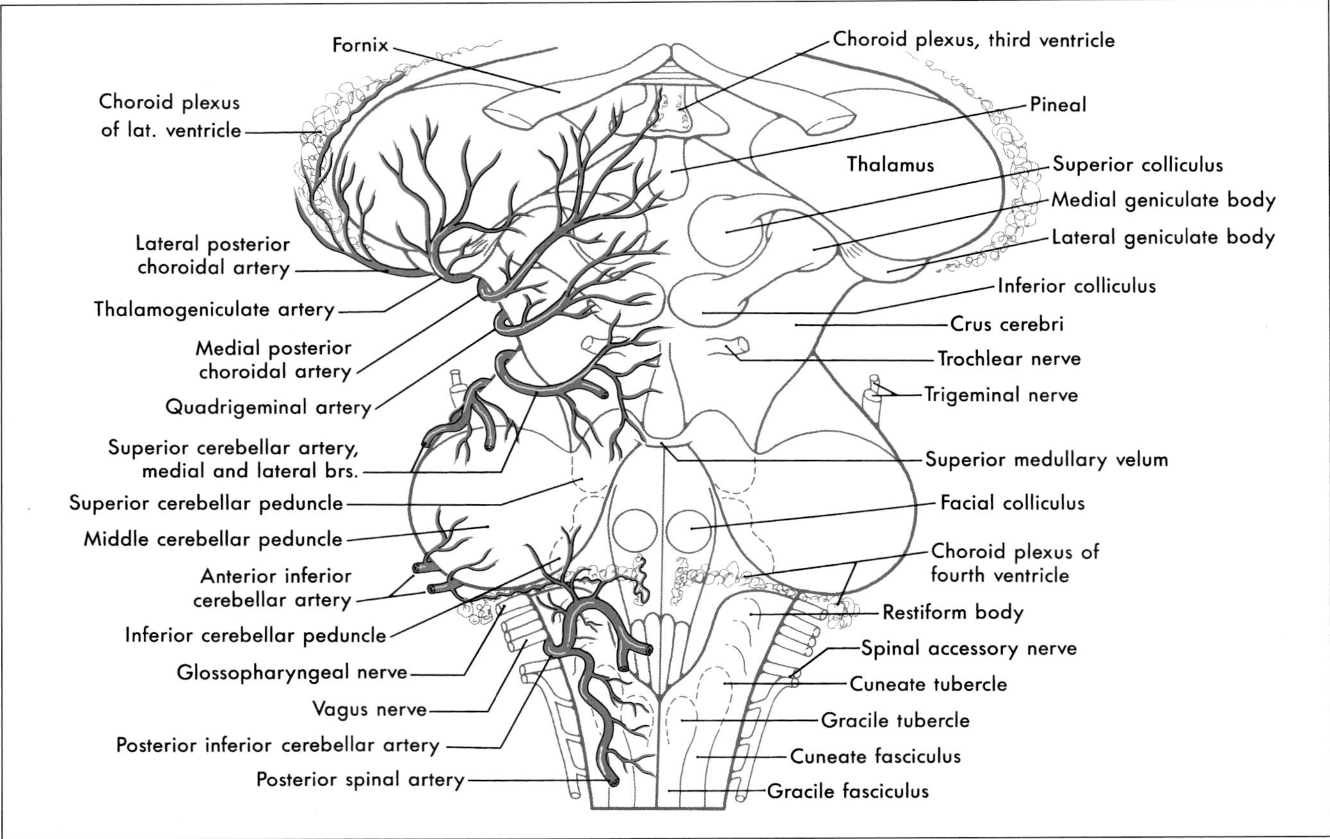

FIGURE 8-6 *Major arteries on the dorsal aspect of the brainstem, arteries to the caudal thalamus, and arteries to the choroid plexuses of the lateral, third, and fourth ventricles.*

Aneurysms of the vertebral artery or its major branches are not common. When present, they are usually found where the PICA branches from the vertebral artery.

Basilar Artery The *basilar artery* lies in a shallow groove on the ventral surface of the pons in the *prepontine cistern* (Fig. 8-3). Its first large branch, the *anterior inferior cerebellar artery* (*AICA*), arises from the lower one-third of the basilar artery and passes through the *cerebellopontine cistern* as it wraps around the caudal aspect of the middle cerebellar peduncle. The AICA serves ventral and lateral surfaces of the cerebellum, parts of the pons, and the portion of choroid plexus that extends out of the foramen of Luschka into the cerebellopontine angle (Fig. 8-6). The *labyrinthine artery* is usually a branch of the AICA. It arises close to the origins of the facial and vestibulocochlear nerves (Fig. 8-3) and enters the internal acoustic meatus along with these nerves.

The basilar artery gives rise to numerous *pontine arteries* (Fig. 8-3). These arteries may penetrate the pons immediately as *paramedian branches*, travel for a short distance around the pons as *short circumferential branches*, or pass for longer distances as *long circumferential branches*.

The last major branches of the basilar artery are the *superior cerebellar arteries*. These vessels pass laterally *just caudal to the root of the oculomotor nerve* and wrap around the brainstem in the *ambient cistern* to ultimately serve caudal parts of the midbrain and the entire superior surface of the cerebellum (Figs. 8-3, 8-6).

About 15% of all intracranial aneurysms occur in the vertebrobasilar system. Most are found in relation to the basilar bifurcation and, therefore, may involve the oculomotor nerve. In similar fashion, an aneurysm on the AICA may produce symptoms of facial or vestibulocochlear nerve involvement, as this vessel travels adjacent to these nerves.

Posterior Cerebral Artery At the pons-midbrain junction the *basilar artery* bifurcates in the *interpeduncular cistern* and gives rise to the *posterior cerebral arteries*. Each posterior cerebral artery passes laterally *just rostral to the root of the oculomotor nerve* (Fig. 8-3), wraps around the midbrain in the *ambient cistern*, and then joins the ventral and medial surface of the temporal lobes (Figs. 8-4, 8-7). The *posterior cerebral artery* sends branches to the midbrain and thalamus and to the ventral and medial surfaces of the temporal and occipital lobes as far as the level of the parieto-occipital sulcus (Figs. 8-4, 8-7).

The posterior cerebral artery is divided into *segments* called P_1 to P_4. The P_1 segment is located between the basilar bifurcation and the posterior communicating artery. It gives rise to small perforating vessels and to *quadrigeminal* and *thalamoperforating arteries* (Figs. 8-3, 8-6). The portion of the posterior cerebral artery between the posterior communicator and the inferior temporal branches is the P_2 segment. *Medial* and *lateral posterior choroidal* and *thalamogeniculate arteries*, as well as small perforating branches to the midbrain, originate from P_2. The *P_3 segment* is the portion of the artery that gives rise to its *temporal branches*, and the *parieto-occipital* and *calcarine arteries* form the *P_4 segment* (Figs. 8-4, 8-7).

CEREBRAL ARTERIES AND WATERSHED INFARCTS

The territories supplied by branches of the *anterior, middle*, and *posterior cerebral* arteries are summarized in Figure 8-8. Each of the functional regions of the cerebral cortex that we will discuss later lies within the distribution of a particular cerebral artery.

On the lateral surface of the hemisphere, the terminal branches of anterior and middle and middle and posterior cerebral arteries overlap, forming *border zones* between the areas served by these arteries (Fig. 8-8). Smaller border zones are located between the territories of the anterior and posterior cerebral arteries at the parieto-occipital sulcus and between the territories of the cerebellar arteries (Fig. 8-8). The brain tissue located in these *border zones* is particularly susceptible to damage under conditions of sudden *systemic hypotension* or when there is *hypoperfusion* of the distal vascular bed of a major cerebral artery.

In the case of the cerebral arteries, inadequate perfusion of the *border zones* may result in *watershed infarcts* (Fig. 8-8). Such lesions represent about 10% of all brain infarcts and may be caused by, for example, hypotension or embolic showers. Damage to the anterior cerebral-middle cerebral border zone in one hemisphere (*anterior watershed infarct*, Fig. 8-8) causes a contralateral hemiparesis of the leg and expressive language or behavioral changes. A *posterior watershed infarct* (damage at the middle cerebral-posterior cerebral border zone) commonly produces a partial visual loss accompanied by a variety of language problems.

CIRCLE OF WILLIS

The *circle of Willis* is actually a roughly shaped heptagon of arteries located on the ventral surface of the brain (Figs. 8-3, 8-9, 8-10). This loop of vessels passes around the optic chiasm and the optic tract, crosses the

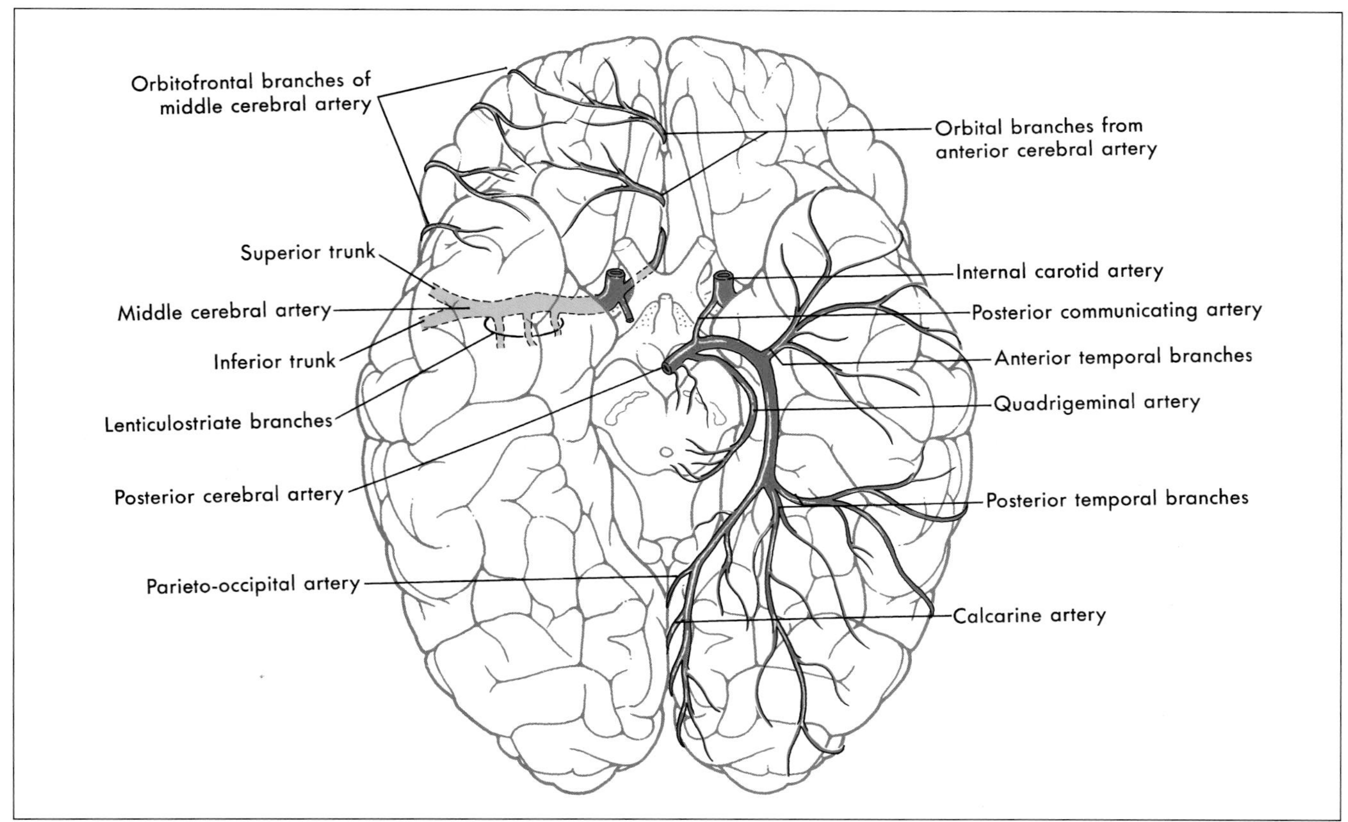

FIGURE 8-7 *Distribution pattern of the major branches of the posterior cerebral artery.*

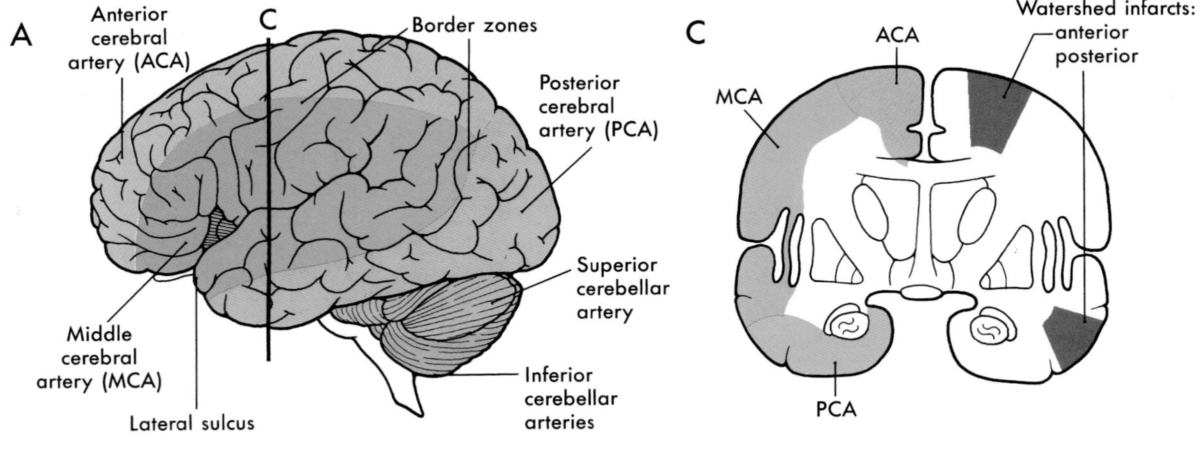

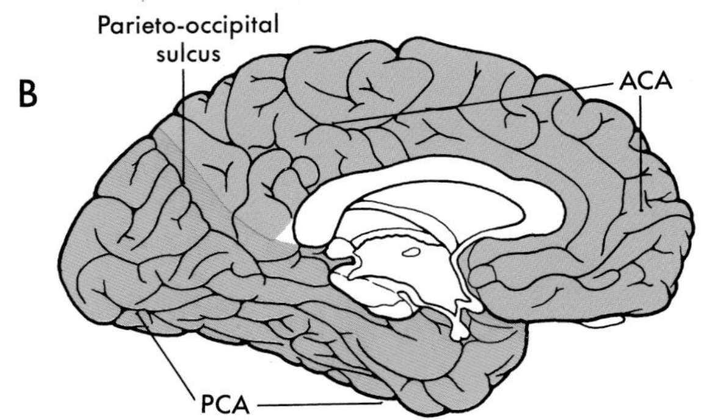

FIGURE 8-8 *Lateral (A), medial (B), and cross-sectional (C) views of the hemisphere showing the regions served by the anterior cerebral (green), middle cerebral (blue), and posterior cerebral (red) arteries. The distal territories of these vessels overlap at their peripheries and create border zones. These zones are susceptible to infarcts (C) in cases of hypoperfusion of the vascular bed. Small border zones also exist (A) between superior (green) and inferior (blue) cerebellar arteries.*

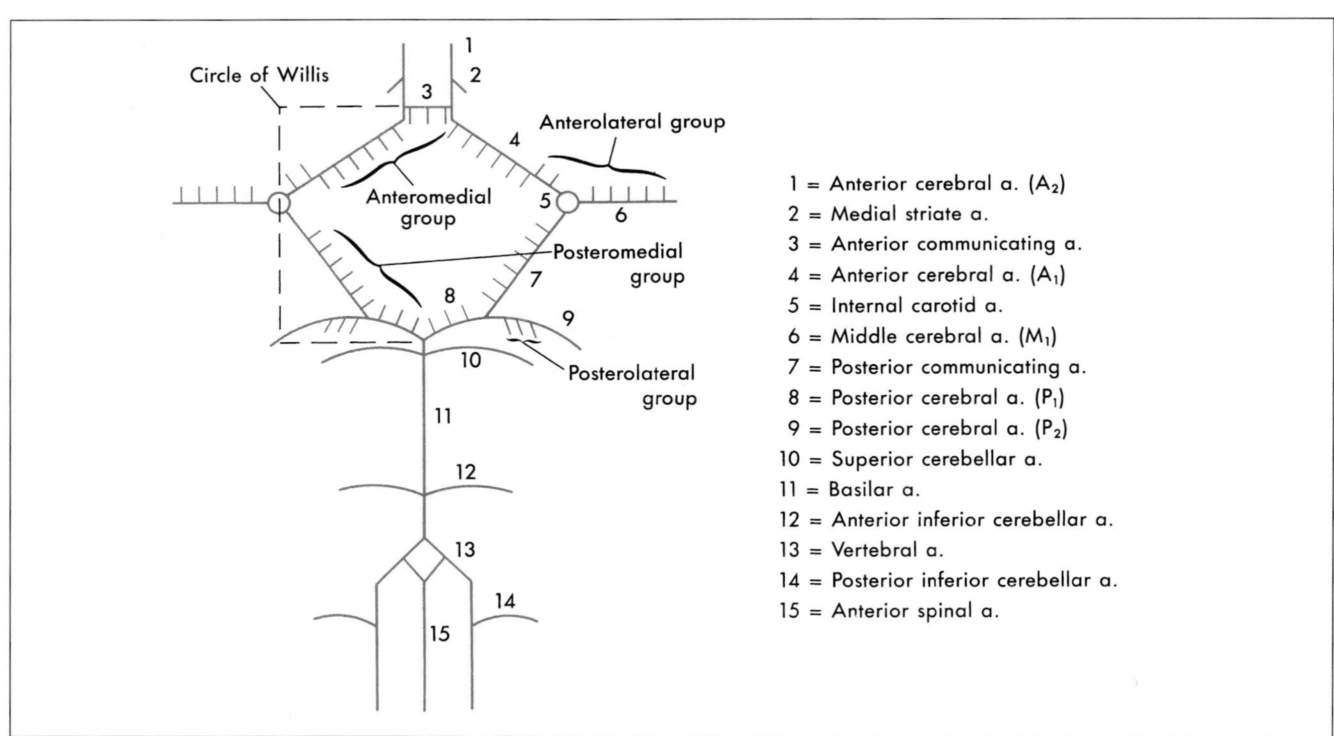

FIGURE 8-9 *Stick drawing of the vertebrobasilar and internal carotid systems showing the configuration of the circle of Willis. The four groups of ganglionic (or central) arteries are shown.*

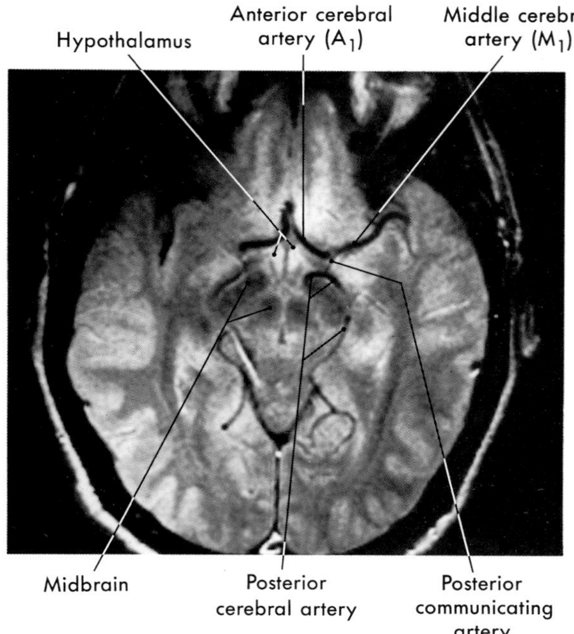

FIGURE 8-10 *An axial (horizontal) magnetic resonance image through the ventral aspect of the hemisphere showing most of the vessels forming the circle of Willis (compare with Figs. 8-3 and 8-9).*

crus cerebri of the midbrain, and joins at the pons-midbrain junction. Important structures located inside this circle include the optic chiasm and tracts, infundibulum and tuber cinereum, the mammillary bodies, the hypothalamus, and structures of the interpeduncular fossa (Figs. 8-3, 8-10). Arteries forming the circle of Willis give rise to numerous *perforating* (*central or ganglionic*) *branches*, which serve structures located deep to their origin (Figs. 8-3, 8-9) and to the large cortical branches (*anterior*, *middle*, and *posterior cerebral arteries*) discussed above (Fig. 8-10).

Central Branches The perforating branches of the circle of Willis are divided into four groups (Fig. 8-9). The *anteromedial group* originates from A_1 and from the anterior communicating artery. These vessels serve structures in the area of the optic chiasm and anterior parts of the hypothalamus. The *anterolateral group* arises from M_1, with A_1 also sending some branches into this area. Included in this group are the *lenticulostriate arteries*, which serve the interior of the hemisphere. Vessels of the anterolateral group enter the hemisphere via the *anterior perforated substance*. The *posteromedial group* originates from P_1 and from the posterior communicating artery. These vessels supply the crus cerebri, middle and caudal portions of the hypothalamus, and as they enter the interpeduncular fossa, they form the *posterior perforated substance*. The *thalamoperforating arteries* are part of the posteromedial group and, as their name implies, they serve the thalamus. *The posterolateral group* arises from P_2 and is composed of the *thalamogeniculate* and *posterior choroidal arteries* and some small penetrating branches that enter the midbrain. These vessels serve parts of the thalamus and the choroid plexus.

VEINS AND VENOUS SINUSES OF THE BRAIN

In contrast to the arterial supply of the brain, which comes from *two major sources*, the venous drainage of the brain exits the skull through *one major vessel*. Venous blood from superficial and deep veins enters the dural sinuses, which, in turn, drain into the *internal jugular vein*. The major venous sinuses are endothelium-lined channels closely associated with the meningeal reflections (see Chapter 7). The *superior* and *inferior sagittal sinuses* are located in the attached and free edges of the falx cerebri, respectively. The *straight sinus* is found where the falx cerebri attaches to the tentorium cerebelli. The other venous sinuses are located adjacent to the inner surface of the skull at specific locations.

Cerebral Hemispheres The *cerebral veins* on the lateral surface of the hemisphere drain into the superior sagittal and transverse sinuses and into the *superior anastomotic vein* (*of Trolard*) and the *inferior anastomotic vein* (*of Labbé*) (Fig. 8-11). These large *anastomotic veins* form channels between the superior sagittal and transverse sinuses with the *superficial middle cerebral vein*. The latter vessel courses medially around the temporal pole to end in the cavernous sinus (see Fig. 8-12).

Small vessels on the mid-sagittal surface of the hemisphere drain into the sagittal sinus and, from the medial region of the temporal lobe, into the *basal vein* (*of Rosenthal*) (Fig. 8-13). The venous blood in these channels, and from the corpus callosum and the interior of the hemisphere (*internal cerebral veins*), drains into the straight sinus. The *confluence of sinuses* (*confluens sinuum*) is formed by the junction of the straight sinus, the superior sagittal sinus, and both transverse sinuses (Figs. 8-12, 8-13). Rather than a true confluence, the superior sagittal sinus usually drains into the right transverse sinus and the straight sinus into the left.

Basal Aspect of the Brain Figure 8-12 shows the venous structures on the ventral surface of the hemisphere. The *basal vein* (*of Rosenthal*) begins on the orbital cortex as the *anterior cerebral vein* and in the Sylvian fissure as the *deep middle cerebral vein* and proceeds around the medial edge of the temporal lobe to join the straight sinus. It receives venous blood from the midbrain and medial areas of the temporal lobe (Figs. 8-12, 8-13). The *transverse* and *sigmoid sinuses* form a shallow

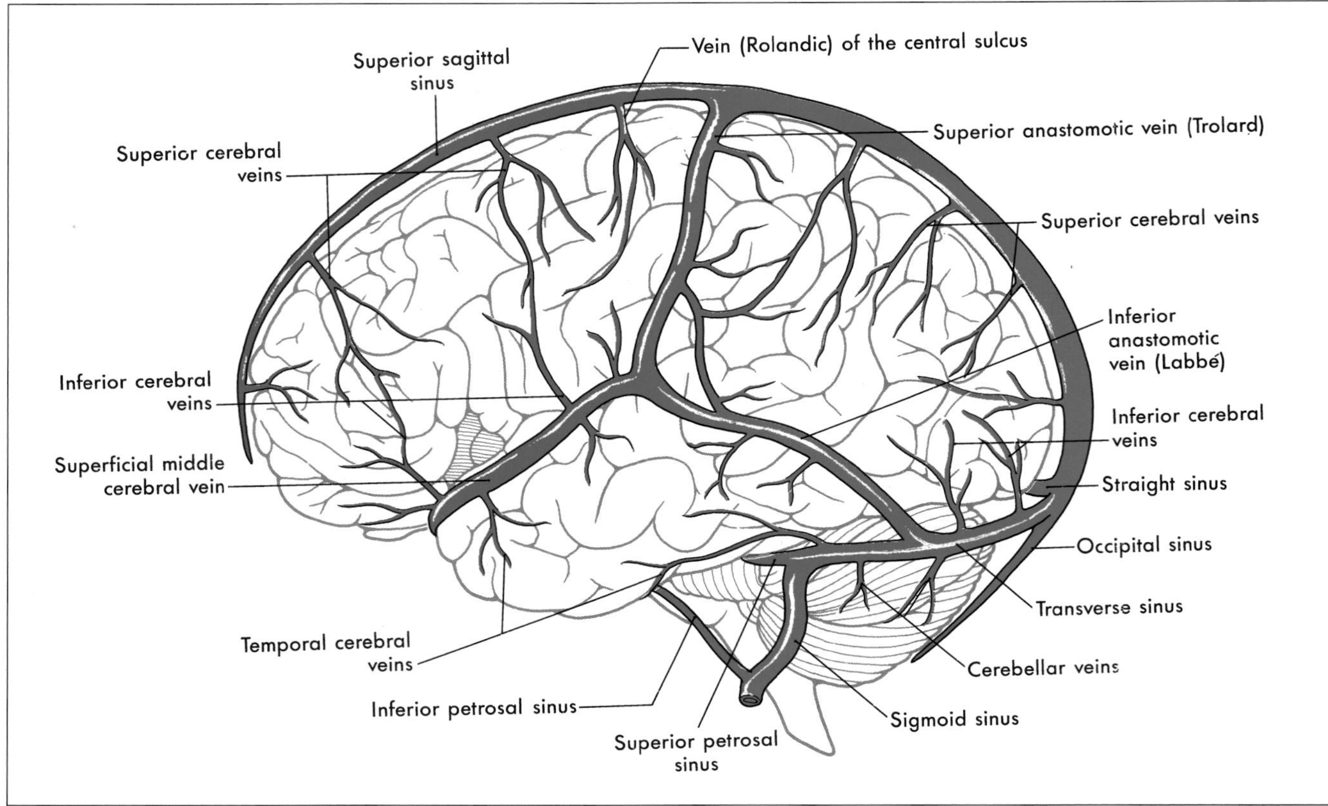

FIGURE 8-11 *Veins and sinuses of the brain from the lateral aspect.*

groove on the internal surface of the occipital and temporal bones, respectively, and receive several tributaries. Venous blood travels from the transverse and petrosal sinuses into the *sigmoid sinus* and then into the *internal jugular vein* at the jugular foramen (Figs. 8-11, 8-12).

The *cavernous sinus* is located on either side of the body of the sphenoid bone. It contains the *cavernous part of the internal carotid artery*; the *abducens*, *oculomotor*, and *trochlear nerves*; and the *ophthalmic* and *maxillary branches* of the trigeminal nerve (Fig. 8-14). These nerves are found internal to the dura surrounding the sinus but are external to its endothelial lining.

The main tributaries of the cavernous sinus are illustrated in Figure 8-12. Each cavernous sinus communicates with its counterpart through the *intercavernous sinuses*. Caudally, the cavernous sinus drains into the *superior* and *inferior petrosal sinuses* and the *basilar plexus* on the ventral aspect of the brainstem.

The neurologic deficits seen in patients with an aneurysm of the cavernous part of the carotid artery are related to the compact nature of the sinus and the close apposition of cranial nerves III, IV, VI, and V_1 and V_2 to the carotid artery (Fig. 8-14A,B). An expanding aneurysm will affect the adjacent nerves and result in a partial or complete paralysis of eye movement, loss of the corneal reflex, and paresthesias or pain within the distribution of the ophthalmic and maxillary nerves. The direct shunting of blood from the internal carotid artery into the cavernous sinus, a *carotid-cavernous fistula*, is rarely the result of a ruptured aneurysm but may occur secondary to trauma (Fig. 8-15).

Internal Veins of the Hemisphere The main venous channels draining internal structures of the hemisphere are the *internal cerebral veins* (Figs. 8-12, 8-16). They course along the dorsomedial edge of the thalamus and are located in the tela choroidea of the third ventricle. The principal tributaries of the internal cerebral veins and their relationships to adjacent structures are shown in Figure 8-16. Of these, the *thalamostriate vein* (also called the *terminal vein*) merits comment. It is found in association with the stria terminalis and drains the caudate nucleus (*transverse caudate veins*) and internal regions of the hemisphere dorsal and lateral to the caudate nucleus.

The two internal cerebral veins join to form the *great cerebral vein* (*of Galen*). This large venous channel has several notable tributaries (Figs. 8-12, 8-16) and is caudally continuous with the *straight sinus*.

Cerebral and spinal veins and dural sinuses lack valves. Consequently, pathologic processes may alter normal venous flow patterns and result in the transport of material *into* the brain. For example, a tumor or infec-

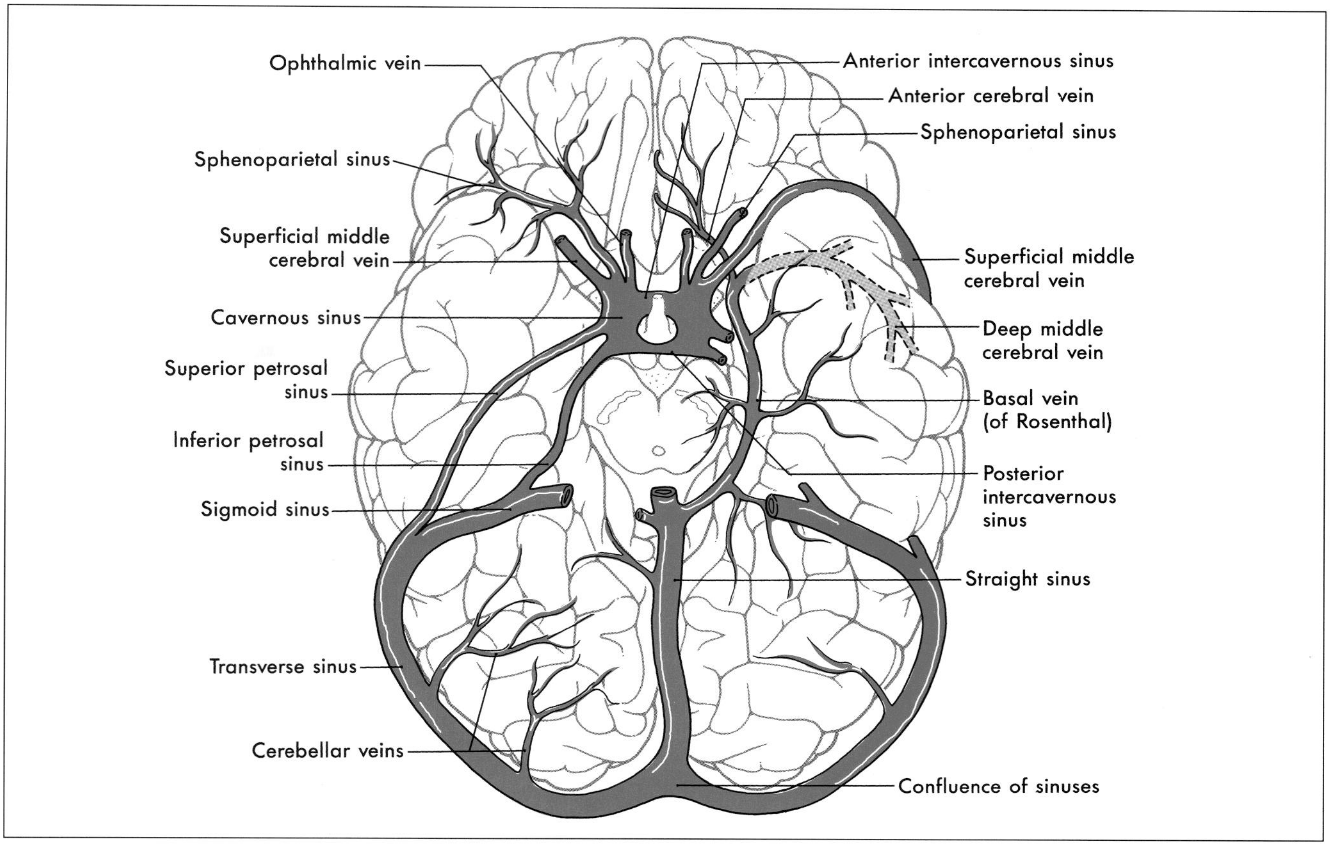

FIGURE 8-12 *Veins and sinuses on the ventral surface of the hemisphere. The cerebellum, pons, and caudal midbrain are removed. For clarity, the petrosal sinuses are shown only on the left and the basal vein (of Rosenthal) only on the right.*

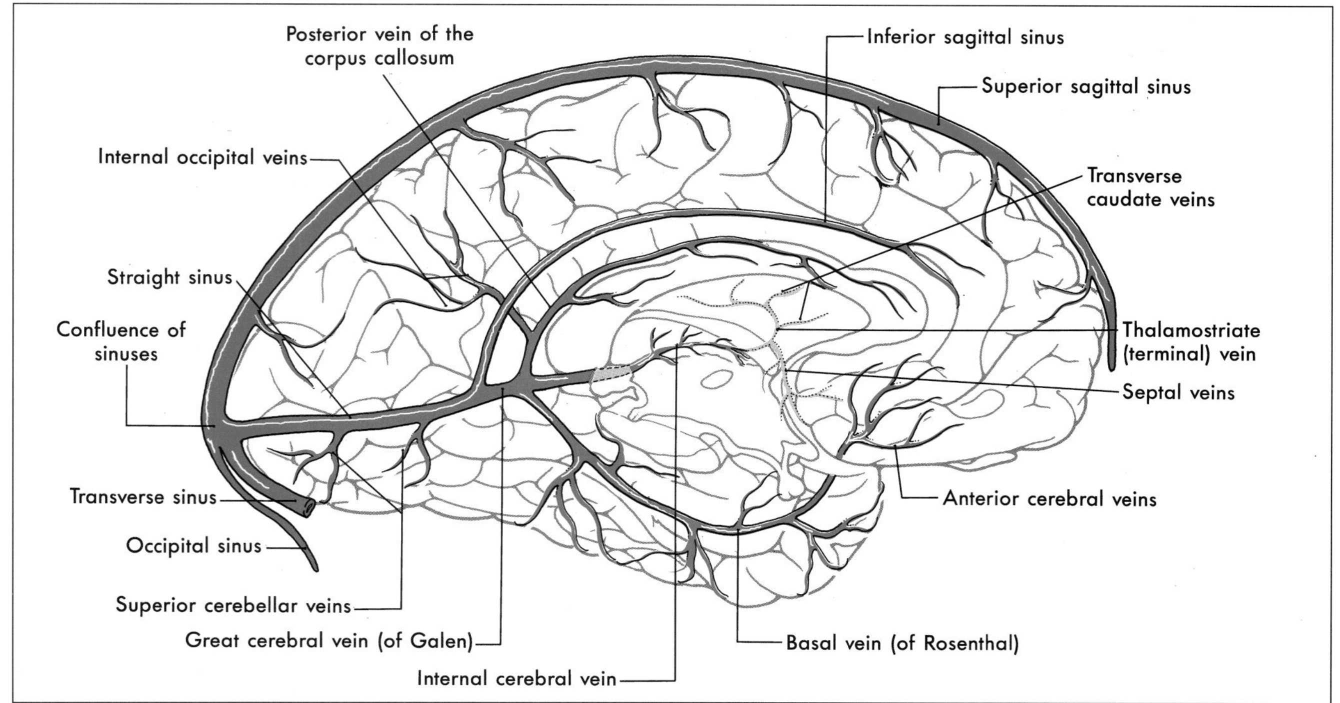

FIGURE 8-13 *Veins and sinuses on the medial surface of the hemisphere and ventral part of the temporal lobe.*

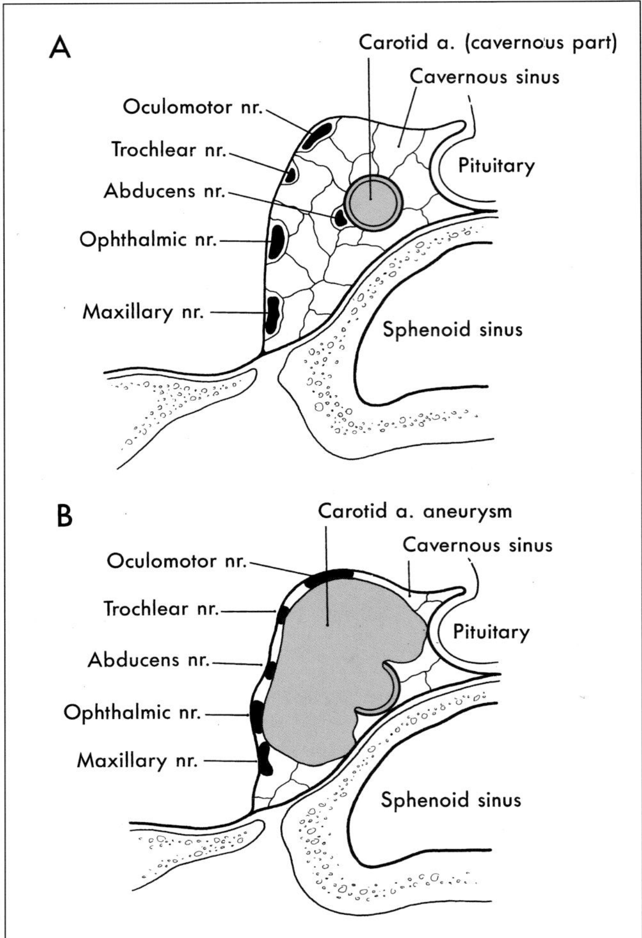

FIGURE 8-14 *The carotid artery and cranial nerves of the cavernous sinus (A) in cross section. An aneurysm of the cavernous part of the carotid artery (B) will damage some, or all, of the cranial nerves passing through the sinus.*

tion in the orbit may cause venous blood to flow toward the cavernous sinus rather than away from it. In this way, infectious material or tumor cells would pass from the orbit into the cavernous sinus and, through its connecting channels, to other parts of the brain.

Malformations of the great cerebral vein of Galen are sometimes described as a special type of AVM, or as an aneurysm. This lesion is usually seen in newborns or infants (Fig. 8-17). In these cases, the great cerebral vein is grossly enlarged and fed by large and abnormal branches of the cerebral and cerebellar arteries. Bulging fontanelles, progressive hydrocephalus (resulting from occlusion of the cerebral aqueduct), and dilated veins in the face and scalp are characteristic findings.

Brainstem and Cerebellum The brainstem is drained by a loosely organized network of venous channels located on its surface. In general these enter larger veins or venous sinuses located in the immediate vicinity. For example, veins of the midbrain enter the great cerebral and basal veins, whereas those of the pons and medulla enter the petrosal sinuses, the cerebellar veins, and (from the medulla) the venous channels on the surface of the spinal cord.

Venous drainage from the cerebellum is quite straightforward. The *superior cerebellar veins* enter the straight, transverse, or superior petrosal sinuses. The inferior cerebellar surface is drained by *inferior cerebellar veins*, which enter the inferior petrosal, transverse, or straight sinuses.

ARTERIES OF THE SPINAL CORD

The blood supply to the spinal cord comes from the *anterior* and *posterior spinal arteries* and from *spinal branches* of segmental arteries (Fig. 8-18). The anterior spinal artery gives off *central* (or *sulcal*) *branches*, which pass alternately to the right and left to serve central regions of the spinal cord. The posterior spinal arteries course on the surface

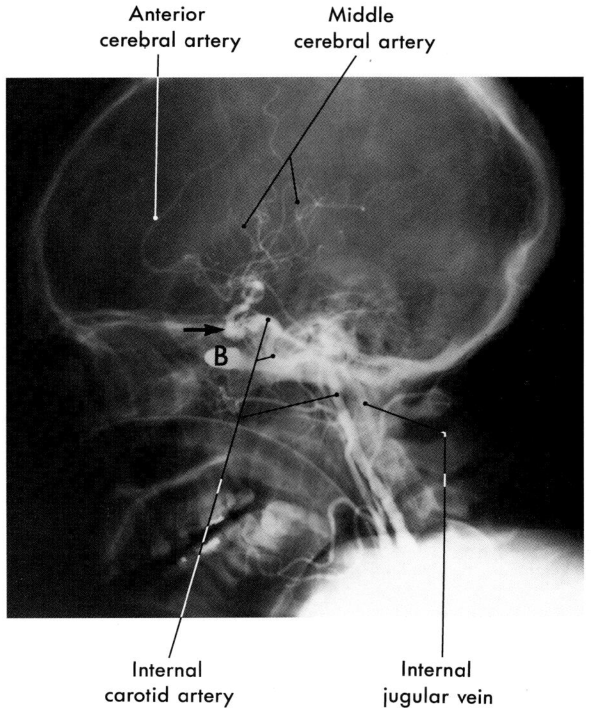

FIGURE 8-15 *Traumatic carotid-cavernous fistula (arrow). This patient was shot in the face, the bullet (B) entering the orbit and damaging the internal carotid artery in the cavernous sinus. Note that the radiopaque substance injected into the common carotid artery appears in the anterior and middle cerebral arteries* and *internal jugular vein* before *appearing in the veins and sinuses of the head. This means that some blood is passing from the internal carotid into the cavernous sinus and then directly into the internal jugular vein.*

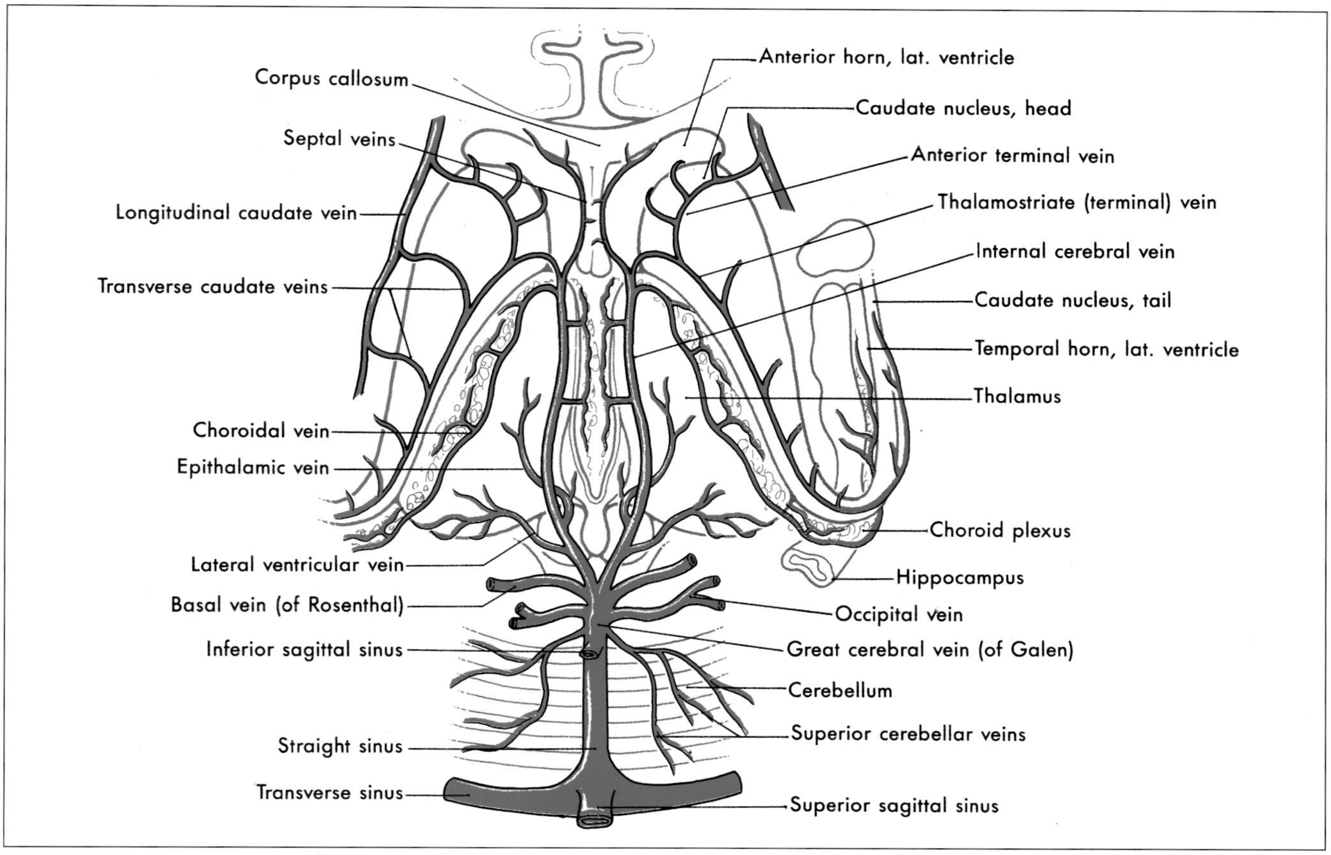

FIGURE 8-16 *Veins draining internal areas of the hemisphere and the tributaries of the great cerebral vein and straight sinus.*

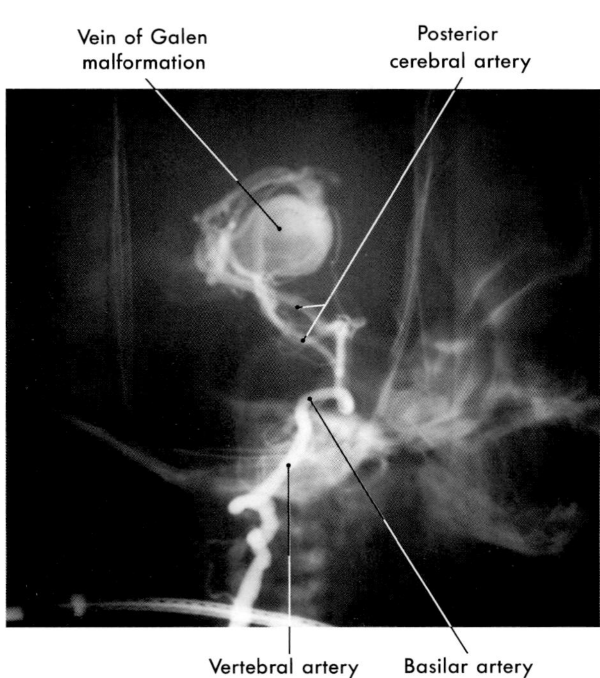

FIGURE 8-17 *Vein of Galen malformation in an infant. This lesion is served by the posterior cerebral arteries.*

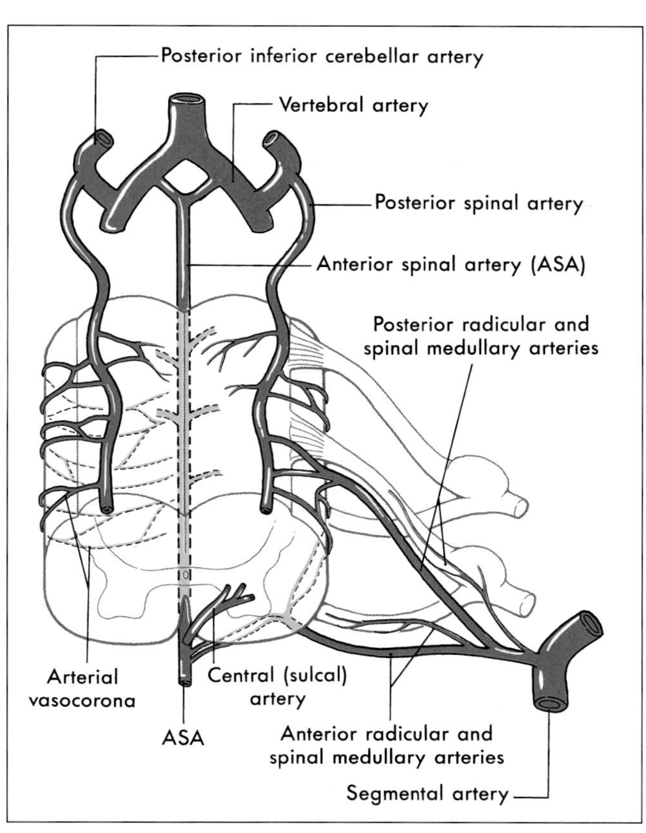

FIGURE 8-18 *Arteries serving the spinal cord.*

of the spinal cord medial to the dorsal root entry zone. These arteries serve dorsal parts of the spinal cord and contribute to the *arterial vasocorona* located on the surface of the spinal cord (Fig. 8-18).

The blood supply to the spinal cord is supplemented at most spinal levels by branches of segmental arteries. These *spinal branches* enter the intervertebral foramina and divide into *dorsal* (*posterior*) and *ventral* (*anterior*) *radicular* and *medullary arteries* (Fig. 8-18). The terminal branches of the medullary arteries contribute to the formation of the *arterial vasocorona*. At level T12, L1, or L2, one anterior radicular (or medullary) artery, usually on the left, is especially large. This is the *artery of Adamkiewicz*. Surgery in this area of the lower back must avoid compromise of this vessel, as it is a major source of blood to lower thoracic and upper lumbar cord levels.

VEINS OF THE SPINAL CORD

In general the venous drainage of the spinal cord mirrors its arterial supply. The location of *anterior* and *posterior spinal veins* and their relationship to other spinal venous structures is shown on Figure 8-19. It is important to note that there is extensive communication between spinal veins and the internal and external venous plexuses found adjacent to the dural sac and vertebral bodies. The veins forming these plexuses apparently lack valves, and the flow in these channels is easily reversed. This represents an important conduit through which metastases from the pelvis, kidney, or lung may spread to the vertebral bodies or into the central nervous system.

Spinal AVMs, although comparable to those of the cranial cavity, do have some unique features. In adults these lesions are usually served by branches of one segmental artery, whereas in children spinal AVMs are usually much larger and have several feeding arteries. Spinal AVMs bleed less frequently than their cranial counterparts and may give rise to more localizing symptoms. For example, in addition to low back pain and occasional sensory or motor problems, patients frequently experience impaired micturition.

BLOOD-BRAIN BARRIER

Although this chapter is primarily concerned with the distribution of vessels on the surface of the central nervous system, it is appropriate to mention briefly the *blood-brain barrier* (Fig. 8-20; see Chapter 2 for details). Although this is a physiologic barrier to the movement of many substances into or out of the brain, the blood-brain barrier has anatomic features that correlate with its function.

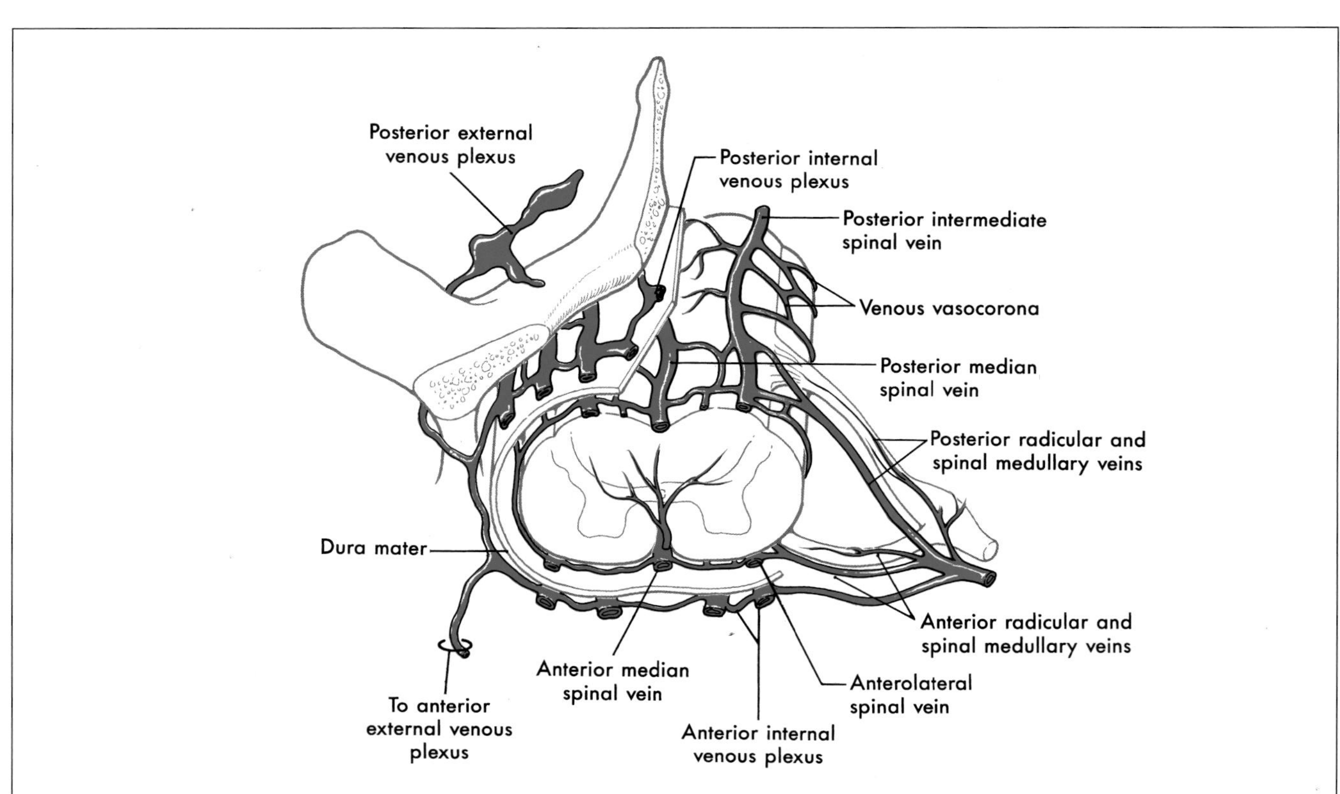

FIGURE 8-19 *Veins draining the spinal cord and the general relationships of internal and external venous plexuses.*

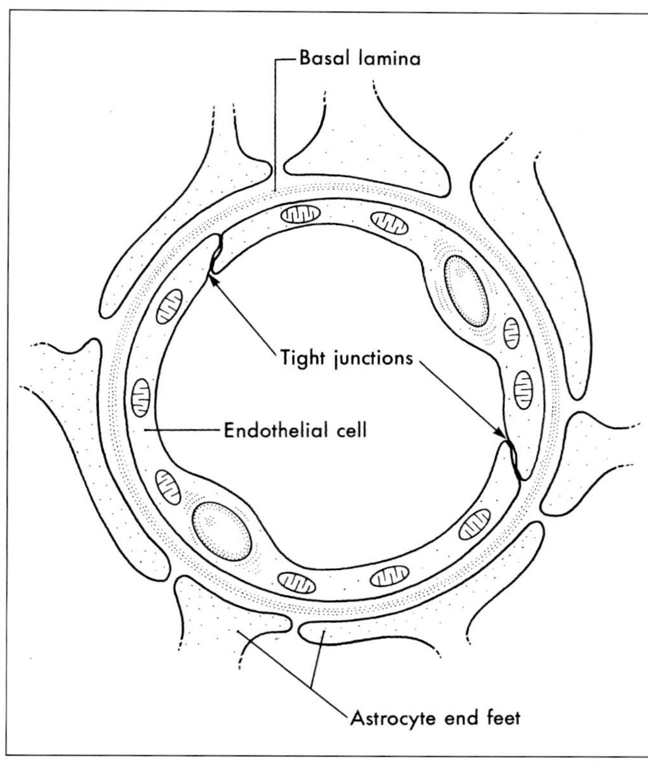

FIGURE 8-20 *Basic structure of the blood-brain barrier.*

The endothelial cells of brain capillaries form a continuous lining membrane; they are joined by numerous tight (occluding) junctions and have no intercellular pores or fenestrations (Fig. 8-20). In contrast, capillaries of the general circulation have fenestrations and intercellular pores. Endothelial cells of brain capillaries rest on a continuous basement membrane (basal lamina), which, in turn, is surrounded by the end feet of astrocytes (Fig. 8-18).

Under normal (healthy) conditions, the blood-brain barrier prohibits the movement of high molecular weight substances (such as proteins) and vital dyes into the brain. In some disease states, however, the barrier breaks down. In the case of a brain tumor, the new capillaries that proliferate into the lesion do not have a close apposition to astrocytes. As a result, the endothelium of these tumor capillaries develops fenestrations and intercellular pores. These can provide a mechanism for diagnosis. For example, intravascular injection of a radioactive amino acid in such a patient will exit the capillaries in the tumor and will localize therein but will not be found in other (normal) parts of the brain.

SOURCES AND ADDITIONAL READINGS

Crosby EC, Humphrey T, Lauer EW: *Correlative Anatomy of the Nervous System.* Macmillan Publishing, New York, 1962.

Duvernoy HM: *Human Brainstem Vessels.* Springer-Verlag, Berlin, 1978.

Gillilan L: The correlation of the blood supply to the human brain stem with clinical brain stem lesions. J Neuropath Exp Neurol 23:78–108, 1964.

Gillilan L: The arterial and venous blood supplies to the forebrain (including the internal capsule) of primates. Neurology 18:653–670, 1968.

Gillilan LA: Blood supply of vertebrate brains. In Crosby EC, Schnitzlein HN (eds): *Comparative Correlative Neuroanatomy of the Vertebrate Telencephalon.* Macmillan Publishing, New York, pp 266–314, 1982.

Hassler O: Deep cerebral venous system in man: A microangiographic study on its areas of drainage and its anastomoses with superficial cerebral veins. Neurology 16:505–511, 1966.

Hassler O: Blood supply to the human spinal cord. Arch Neurol 15:302–307, 1966.

Hassler O: Venous anatomy of human hindbrain. Arch Neurol 16:404–409, 1967.

Hassler O: Arterial pattern of human brain stem: Normal appearance and deformation in expanding supratentorial conditions. Neurology 17:368–375, 1967.

Nieuwenhuys R, Voogd J, van Huijzen CHR: *The Human Central Nervous System, A Synopsis and Atlas.* Springer-Verlag, Berlin, 1988.

Platzer W: *Pernkopf Anatomy, Atlas of Topographic and Applied Human Anatomy,* Vol. I, Head and Neck. Urban & Schwarzenberg, Baltimore, 1989.

Swash M, Oxbury J: Section 19 Cerebral Vascular Disease. In: *Clinical Neurology,* Churchill Livingstone, Edinburgh, pp 924–1020, 1991.

Yasargil MG: *Microneurosurgery, I. Microsurgical Anatomy of the Basal Cisterns and Vessels of the Brain, Diagnostic Studies, General Operative Techniques and Pathological Considerations of the Intracranial Aneurysms.* Georg Theime Verlag, Stuttgart, 1984.

CHAPTER NINE

The Spinal Cord

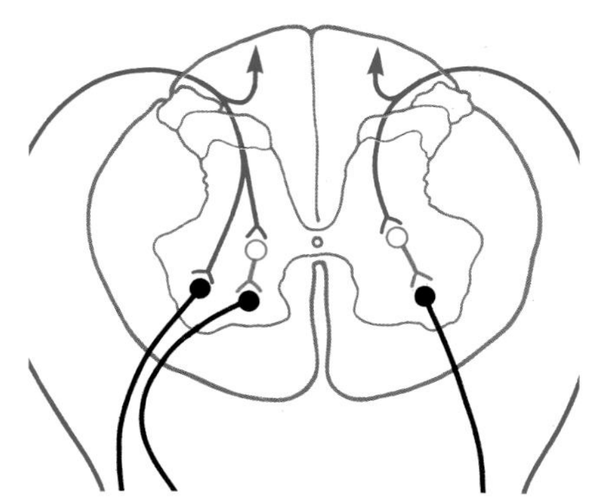

D. E. HAINES • G. A. MIHAILOFF • R. P. YEZIERSKI

Although small in diameter, the spinal cord is the most important conduit between the body and the brain. It conveys sensory input from the arms, trunk, legs, and most of the viscera and contains fibers and cells that control the motor elements found in these structures. Consequently, injury to the spinal cord, especially at cervical levels, may cause permanent and catastrophic deficits, or even death.

OVERVIEW

The spinal cord participates in four essential functions. First, it receives primary sensory input from receptors in skin, skeletal muscles and tendons (*somatosensory fibers*), and from receptors in thoracic, abdominal, and pelvic viscera (*viscerosensory fibers*). Through multi-synaptic relays in the spinal cord, much of this sensory input is conveyed to higher levels of the neuraxis.

Second, the spinal cord contains *somatic motor neurons* that innervate striated muscles and *visceral motor neurons* that, after synapsing in peripheral ganglia, influence smooth and/or cardiac muscle and glandular epithelium. Any disease process that damages the somatic motor neuron (as in *poliomyelitis*) or compromises its ability to elicit a response in the skeletal muscle (as in *myasthenia gravis*) will result in weakness or paralysis.

Third, somatosensory fibers enter the spinal cord and influence ventral horn motor neurons either directly, or indirectly through interneurons. These activated motor neurons, in turn, produce rapid involuntary contractions of skeletal muscles. The sensory fiber, the associated motor neuron, and the resultant involuntary muscle contraction constitute the circuit of the *spinal reflex*. Reflexes are essential to normal function and can be used as diagnostic tools to assess the functional integrity of the spinal cord.

Fourth, the spinal cord contains descending fibers that influence the activity of spinal neurons. These fibers originate in the cerebral cortex and brainstem, and damage to them adversely influences the activity of spinal motor and sensory neurons.

DEVELOPMENT

Neural Plate As explained in more detail in Chapter 5, the spinal cord arises from the caudal portion of the embryonic *neural plate* and from the *caudal eminence*. The neural plate gives rise to the cervical, thoracic, and lumbar levels, whereas the caudal eminence gives rise to the sacral and coccygeal levels. The neural plate appears as a specialized area of ectoderm (*neuroectoderm*, *neuroepithelial cells*) dorsal to the notochord at about 18 days (Fig. 9-1A,B). By 20 days of gestation, the neural plate is an oblong structure that is larger at its rostral area (future brain) and tapered caudally (future spinal cord).

Beginning on day 21, the edges of the neural plate (*neural folds*) rotate dorsomedially to meet on the midline (Fig. 9-1B,C). The initial apposition of the neural folds to form the *neural tube* takes place at what will become, in the adult, cervical levels of the spinal cord. This closure simultaneously proceeds in rostral and caudal directions, ultimately creating small openings at either end, between the *cavity of the neural tube* and the surrounding amniotic cavity. These openings are the *anterior* and *posterior neuropores*. The anterior and posterior neuropores close at 25 and 27 days of gestation, respectively.

A variety of defects result from a failure of the neural tube to close (Fig. 9-1D–F). *Rachischisis* occurs when the neural folds do not join and the undifferentiated neuroectoderm remains exposed. In its most extreme form, this deficit results in *anencephaly*. In this situation, the cephalic part of the neural tube (brain) does not form, and there is no skull. In other cases the neural tube may form normally, but the surrounding vertebrae may not, resulting in *spina bifida occulta*, *meningocele*, or *meningomyelocele*.

Neural Tube The neural tube consists of precursor cells forming the *ventricular layer* (or zone). Proliferating cells of this layer give rise to *ependymal cells* lining the neural tube and to *neuroblasts* and *glioblasts* that form the *mantle layer* (the intermediate layer). The *marginal layer* contains glioblasts and the out-growing processes of mantle layer neuroblasts. The *neural crests* detach from the lateral edge of the neural plate and migrate to various locations lateral and ventral to the neural tube.

Structures of the developing neural tube can be correlated with their adult counterparts (Fig. 9-1C). *Neural crest cells* differentiate into cells of the *dorsal* (*posterior*) *root ganglia*, among other structures. The mantle layer consists of four rostrocaudally oriented columns of neuroblasts, forming the paired *alar plates dorsally* and the paired *basal plates ventrally*. The alar plate and basal plate are separated from each other by the *sulcus limitans*. Neuroblasts of the alar plate differentiate into the tract neurons and interneurons of the *dorsal horn* of the adult, and those of the basal plate become the motor neurons and interneurons of the *ventral horn*. The axons of basal plate neuroblasts, which become somatic motor neurons, extend distally as parts of peripheral nerves. The intermediate zone (also called intermediate gray), an important region of the spinal cord insinuated between the dorsal and ventral horns of the adult, originates from portions of both alar and basal plates.

The marginal layer is invaded by processes of neuroblasts located in the mantle layer (Fig. 9-1C) and by the descending axons of neuroblasts found in the devel-

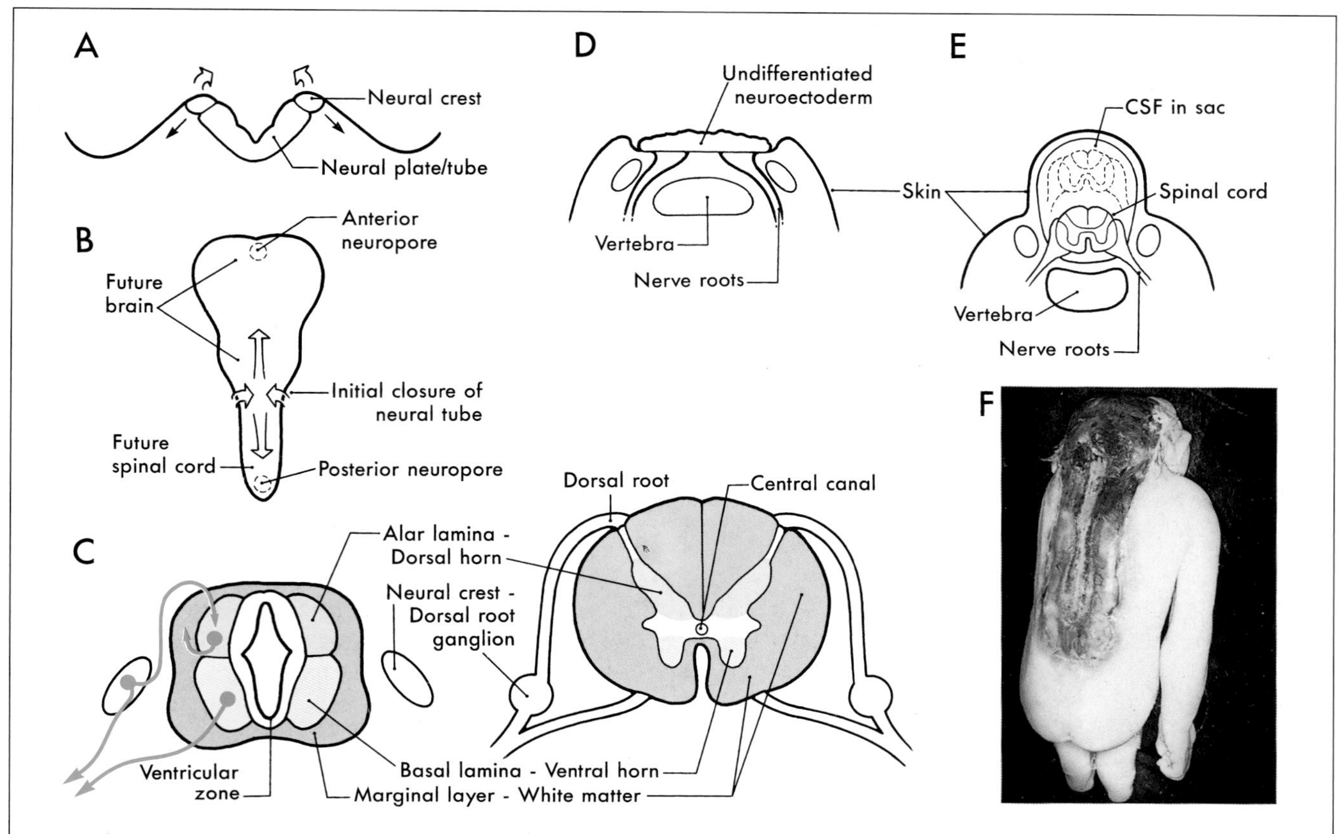

FIGURE 9-1 *Development of the spinal cord. Cross-sectional (A) and dorsal (B) views of the neural plate, and the correlation of neural tube structures with the adult cord (C). Malformations involving defects of the nerve tissue and/or surrounding bone include rachischis (D), meningocele (E, solid cord) or meningomyelocele (E, dashed cord), and anencephaly with rachischisis (F). (Photograph courtesy of Dr. Jonathan Fratkin.)*

oping brainstem or cerebral cortex. These axons, many of which become myelinated, form the various tracts of the white matter of the adult spinal cord.

SPINAL CORD STRUCTURE

The adult spinal cord is composed of a butterfly-shaped central area of neuron cell bodies, the *gray matter*, and a surround of myelinated fibers, the *white matter*. Although the cavity of the neural tube was prominent during development, this space is reduced to a small ependymal lined *central canal* in the adult (Fig. 9-1C).

Surface Features The human spinal cord extends from the *foramen magnum* to the level of the 1st or 2nd lumbar vertebra. It consists of 8 cervical, 12 thoracic, 5 lumbar, 5 sacral, and 1 coccygeal levels. *Each level (or segment) of the spinal cord is specified by the intervertebral foramina through which the dorsal and ventral roots attached to that segment exit the vertebral canal* (Fig. 9-2). Although generally cylindrical, the cord has *cervical* (C4 to T1) and *lumbosacral* (L1 to S2) *enlargements*, which serve, respectively, the upper and lower extremities.

There are few superficial markings on the spinal cord (Fig. 9-3). The *posterior median sulcus* separates the dorsal portion of the cord into two halves and contains a delicate layer of pia, the *posterior median septum*. The *posterolateral sulcus*, which runs the full length of the cord, represents the entry point of dorsal root (sensory) fibers. This area is frequently called the *dorsal root entry zone*. In cervical and upper thoracic regions, a *posterior intermediate sulcus* and *septum* are found between the posterolateral and posterior median sulci.

On the ventrolateral surface of the spinal cord, the *anterolateral sulcus* is the exit point for ventral root (motor) fibers (Fig. 9-3). Because the ventral roots exit in a somewhat irregular pattern, however, this sulcus is not distinct.

The *anterior median fissure* is a prominent space dividing the ventral part of the cord into halves (Fig. 9-3). This fissure contains delicate strands of pia and, more important, the *sulcal branches* of the *anterior spinal artery*.

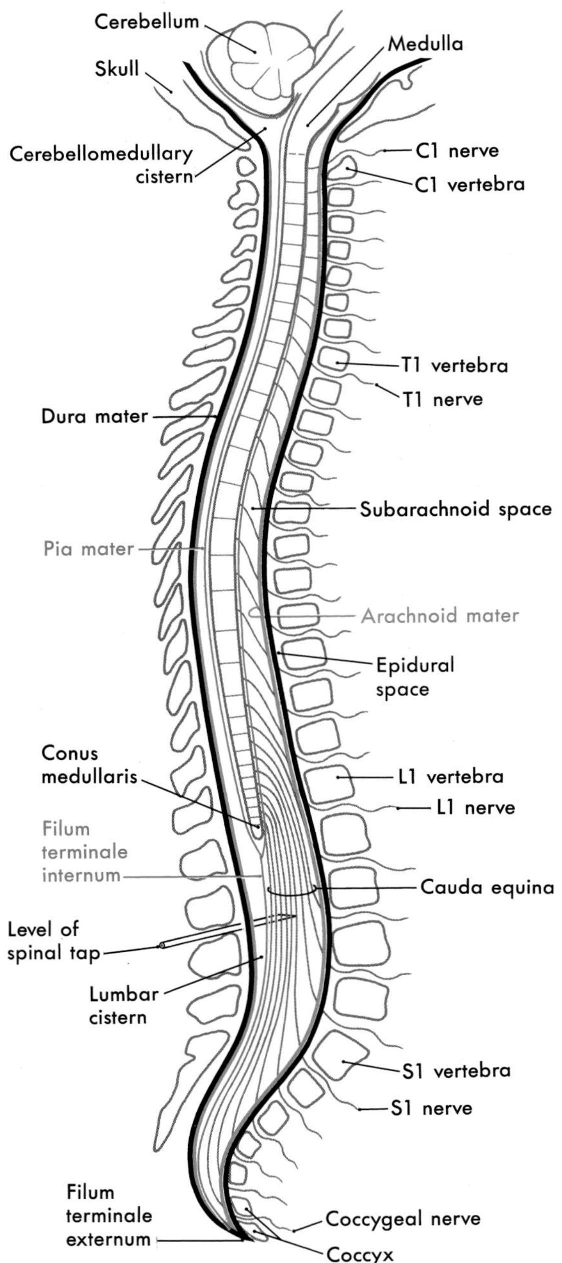

FIGURE 9-2 *Diagrammatic representation of the spinal cord, the meninges, and other adjacent structures. Note the location for a lumbar puncture (spinal tap).*

Spinal Meninges The tubular dural sac that encloses the spinal cord is attached cranially to the rim of the foramen magnum, and its closed caudal end is anchored to the coccyx by the *filum terminale externum* (Figs. 9-2, 9-3). This dural sac is separated from the vertebrae by the epidural space. The spinal cord, in turn, is attached to the dural sac by the laterally placed *denticulate ligaments* and the *filum terminale internum*. This latter structure extends caudally from the end of the spinal cord, the *conus medullaris*, and terminates in the dura. The *arachnoid mater* adheres to the inner surface of the *dura mater*, and the *pia mater* is intimately attached to the surface of the cord. The *subarachnoid space between these layers is continuous with the subarachnoid space around the brain and is likewise filled with cerebrospinal fluid.* In adults the conus medullaris is located at the level of the L1 or L2 vertebral body. Extending caudally from this point to the end of the dural sac is an enlarged part of the spinal subarachnoid space, the *lumbar cistern* (Fig. 9-2). This cistern contains the dorsal and ventral roots from spinal segments L2 to Coc1 as they sweep caudally. Collectively, these roots form the *cauda equina*. The method of choice for obtaining a sample of cerebrospinal fluid for diagnostic purposes is the *lumbar puncture* (*spinal tap*), in which a large-bore needle is introduced between the L3 and L4 vertebral arches into the lumbar cistern.

White Matter The white matter of the spinal cord is divided into three large regions; each is composed of individual *tracts* or *fasciculi*. The *posterior* (*dorsal*) *funiculus* is located between the posterior median septum and the medial edge of the dorsal horn (Fig. 9-3). At cervical levels this area consists of the *gracile* and *cuneate fasciculi*; collectively, these are commonly referred to as the *dorsal columns*.

The *lateral funiculus* is the area of white matter located between the posterolateral and anterolateral sulci (Fig. 9-3). This region of the cord contains clinically important ascending and descending tracts, the locations of which are shown in Figure 9-11. Those most important in diagnosing the neurologically impaired patient are the *lateral corticospinal tract* and the *anterolateral system* (ALS).

Located between the anterolateral sulcus and the ventral median fissure is a comparatively small region, the *anterior* (*ventral*) *funiculus* (Fig. 9-3). This area contains *reticulospinal* and *vestibulospinal fibers*, portions of the ALS, the *anterior* (*ventral*) *corticospinal tract*, and a composite bundle called the *medial longitudinal fasciculus* (MLF).

Two small but important components of the white matter are the *anterior* (*ventral*) *white commissure* and the *dorsolateral fasciculus* (Fig. 9-3). The former is located on the ventral midline and is separated from the central canal by a narrow band of small cells. The dorsolateral fasciculus is frequently called the *tract of Lissauer*. It is a small bundle of lightly myelinated and unmyelinated fibers capping the dorsal horn.

Gray Matter The gray matter of the spinal cord is composed of neuron cell bodies, their dendrites and the initial part of the axon, the axon terminals of fibers synapsing in this area, and glial cells. Because this area has few myelinated fibers, it appears distinctly light and has a characteristic shape in myelin-stained sections (Fig. 9-3; see also Fig. 9-5).

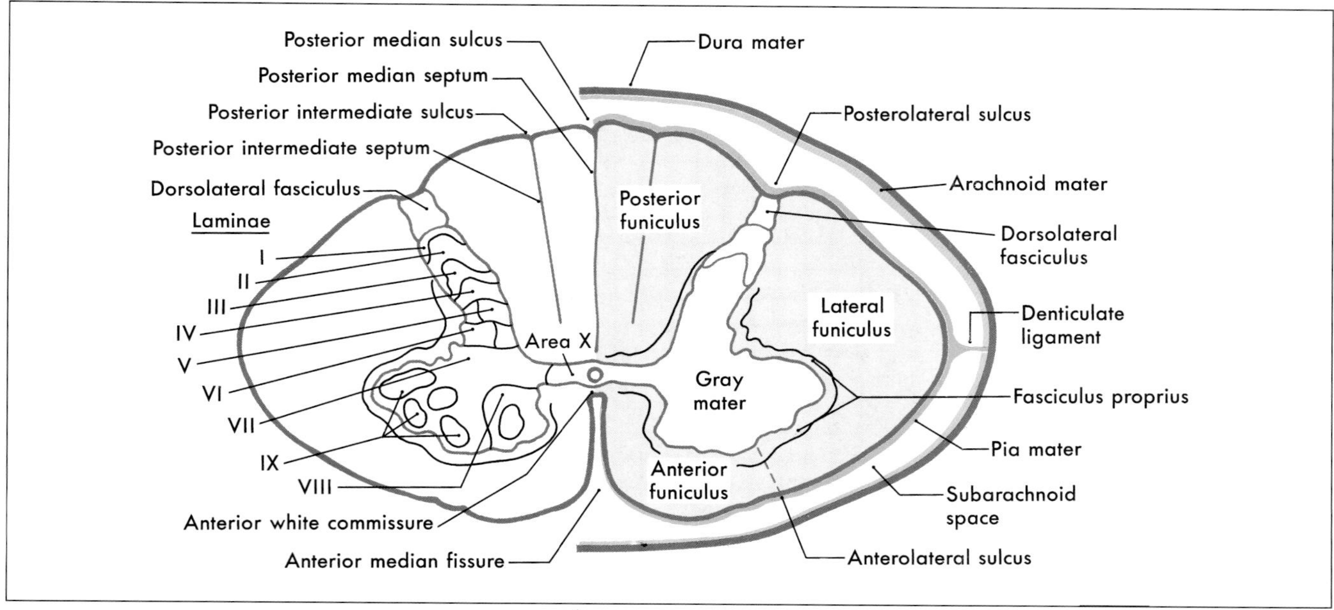

FIGURE 9-3 *The spinal cord at C7 showing the basic organization of the gray and white matter and the meninges (shown only on the right). The lamination pattern of the gray matter is shown only on the left.*

The spinal gray is divisible into a *dorsal* (*posterior*) *horn*, a *ventral* (*anterior*) *horn*, and the region where these meet, commonly called the *intermediate zone* (or *intermediate gray*). Based on the shape, size, and distribution of neurons located in these areas, the gray matter is divided into *laminae* (*Rexed's laminae*) *I to IX* and an *area X* around the central canal (Fig. 9-3). These laminae are also characterized by the input they receive and the trajectory of axons arising therein.

The dorsal horn is composed of laminae I to VI (Fig. 9-3). The most distinct structure in the dorsal horn, the *substantia gelatinosa* (lamina II), is capped by cells of the *posteromarginal nucleus* (lamina I). Laminae III to VI are arranged in a series ventral to the substantia gelatinosa. Laminae III and IV may also be called the *dorsal proper sensory nucleus*; their cells have elaborate dendrites that extend into lamina II. Laminae V and VI, which form the base of the dorsal horn, are usually divided into medial and lateral portions.

The intermediate zone, lamina VII, extends from the area of the central canal to the lateral edge of the spinal gray and varies in shape at different levels. Particularly characteristic of lamina VII in thoracic levels are the *dorsal nucleus of Clarke* (*Clarke's nucleus*) and the *intermediolateral nucleus*; the latter is frequently called the *intermediolateral cell column* (see Fig. 9-5).

The ventral horn is made up of laminae VIII and IX (Fig. 9-3). The former contains a population of smaller cells that are interneurons and tract cells. The latter consist of several distinct clusters of large motor neurons whose axons directly innervate skeletal muscle.

Blood Supply The blood supply to the spinal cord is derived from the *anterior* and *posterior spinal arteries* and from branches of segmental arteries (Fig. 9-4). The segmental branches that serve the dorsal and ventral roots and the dorsal root ganglia are called *radicular arteries*,

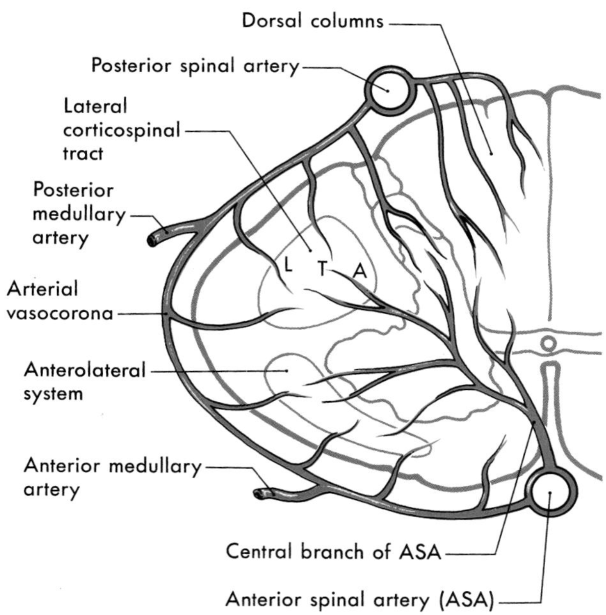

FIGURE 9-4 *Blood supply to the spinal cord. Note that the lateral corticospinal tract and the anterolateral system receive a dual blood supply; also note the topographical arrangement of corticospinal fibers (L, leg; T, trunk; A, arm).*

and the branches that largely bypass the roots to supplement the blood supply to the cord are called *medullary arteries*. One especially large medullary artery, the *artery of Adamkiewicz*, is most often seen at L2 on the left. This vessel is an important source of blood supply to the cord and must be preserved during surgery in this area; damage to this artery may result in an infarct of lower thoracic and upper lumbar levels of the cord. At each level, terminal branches of the spinal and medullary arteries join together to form an arterial network, the *arterial vasocorona*, on the surface of the spinal cord (Fig. 9-4).

The dorsal columns and peripheral parts of the lateral and anterior funiculi are served by the posterior spinal arteries and arterial vasocorona. Most of the gray matter and the adjacent parts of the white matter are served by the *central branches* of the *anterior spinal artery*. These central branches tend to alternate: one serves the left side of the cord; the next serves the right side. Trauma, as in hyperextension of the cervical spine, may cause occlusion or spasm of the anterior spinal artery or mechanical injury to the cord. The result is bilateral damage to the cervical cord (*central cord syndrome*).

REGIONAL CHARACTERISTICS

Although all spinal levels have dorsal, lateral, and ventral funiculi and dorsal and ventral horns, their shapes and proportion vary between major spinal regions (Fig. 9-5). For example, cervical (C4 to T1) and lumbosacral (L1 to S2) cord levels have prominent dorsal and ventral horns because of the extensive sensory input from, and motor outflow to, the arms and legs. In contrast, the dorsal and ventral horns in thoracic levels are small; sensory input is less dense, and there is no appendicular musculature at these levels.

Cervical Levels The cervical cord is oval in shape (Fig. 9-5). There is a large amount of white matter because a full complement of ascending and descending fiber tracts are present. The dorsal and ventral horns at levels C1 to C3 are comparatively small, whereas those at C4 to C8 are large.

Thoracic Levels In general the thoracic cord is round, and the dorsal and ventral horns are small (Fig. 9-5). From upper to lower thoracic levels there is a progressive decrease in the amount of white matter. However, the small size of the dorsal and ventral horns makes the white matter in thoracic levels appear proportionately large. Two structures especially obvious in the gray matter at thoracic levels are the *dorsal nucleus of Clarke* and the *lateral horn*. The former, a prominent cell group in medial parts of lamina VII, contains neurons whose axons project to the cerebellum. The latter, also part of

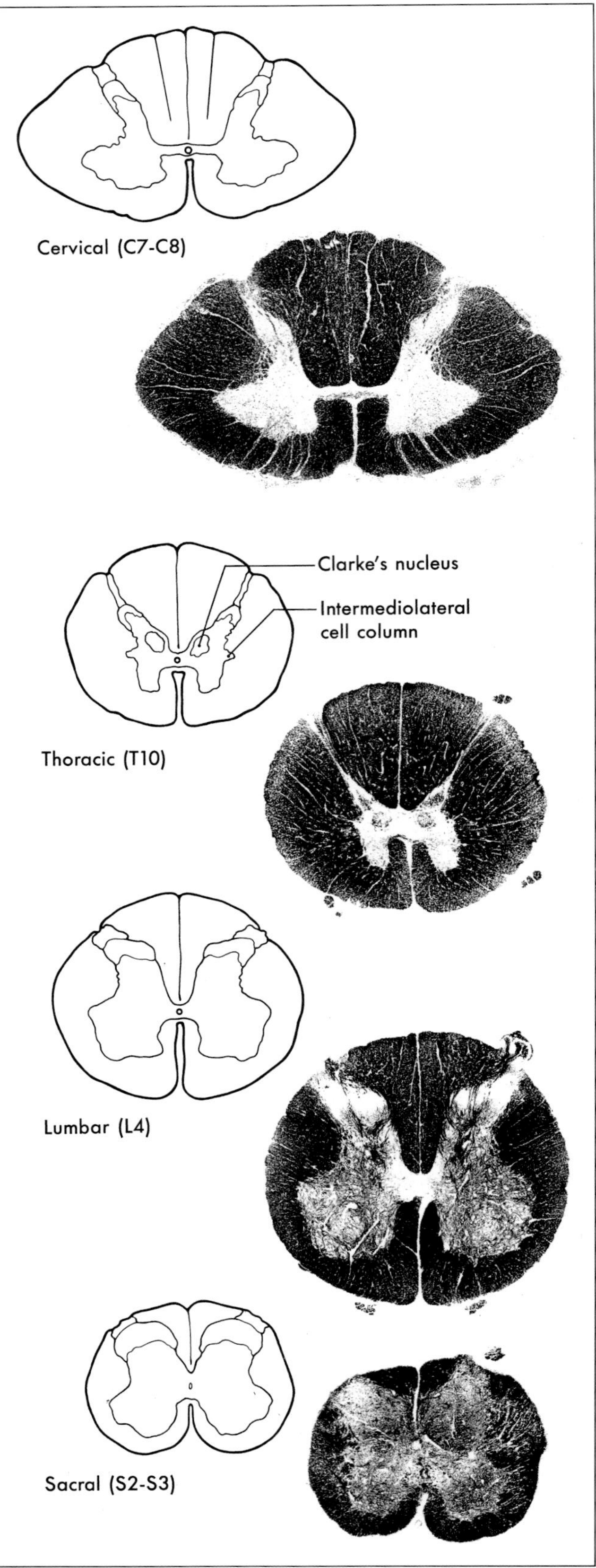

FIGURE 9-5 *Representative levels of the spinal cord.*

lamina VII, is a protrusion into the lateral funiculus formed by the *intermediolateral cell column*. These cells are preganglionic sympathetic neurons whose axons will terminate in peripheral ganglia.

Lumbar Levels At lumbar levels the cord is also round (Fig. 9-5). The dorsal and ventral horns are quite large, and there is less white matter than at higher levels. Therefore, the dorsal and ventral horns appear proportionately large, the reverse of the situation at thoracic levels.

Sacral Levels At sacral levels, the spinal cord is round and is smaller than at lumbar levels (Fig. 9-6). It consists mainly of gray matter, with the white matter forming a relatively thin shell. The intermediate gray matter of levels S2, S3, and S4 contains preganglionic parasympathetic cell bodies (the *sacral visceromotor nucleus*).

SPINAL NERVES

The spinal nerves are formed by the junction of the dorsal and ventral roots of the spinal cord (Fig. 9-6). As there are 31 spinal cord levels (8 cervical, 12 thoracic, 5 lumbar, 5 sacral, 1 coccygeal), so there are 31 corresponding pairs of spinal nerves. Each spinal nerve contains afferent fibers that convey sensory input from the periphery and efferent fibers arising from spinal motor neurons. These fibers, plus circuits in the spinal gray, are the structural basis for the *spinal reflexes* routinely tested in the neurologic examination.

The spinal nerve may contain up to four types of fibers. Two of these are sensory and have their cell bodies in the dorsal root ganglion, and two are motor and have their cell bodies in the spinal cord gray matter (Fig. 9-6).

Sensory Components of the Spinal Nerve Sensory information is brought to the spinal cord by neuronal processes whose cell bodies reside in the dorsal root ganglia. The central processes of these neurons penetrate the spinal cord, and the peripheral processes pass outward in the spinal nerves to innervate body structures. Sensory input originates from (1) the body surface; (2) deep structures such as muscles, tendons, and joints; and (3) internal organs. Fibers conveying input from the first two areas are classified as *general somatic afferent* (GSA), whereas sensory fibers from the gut and other visceral structures are classified as *general visceral afferent* (GVA). The GSA fibers are further classified as either *exteroceptive* or *proprioceptive*.

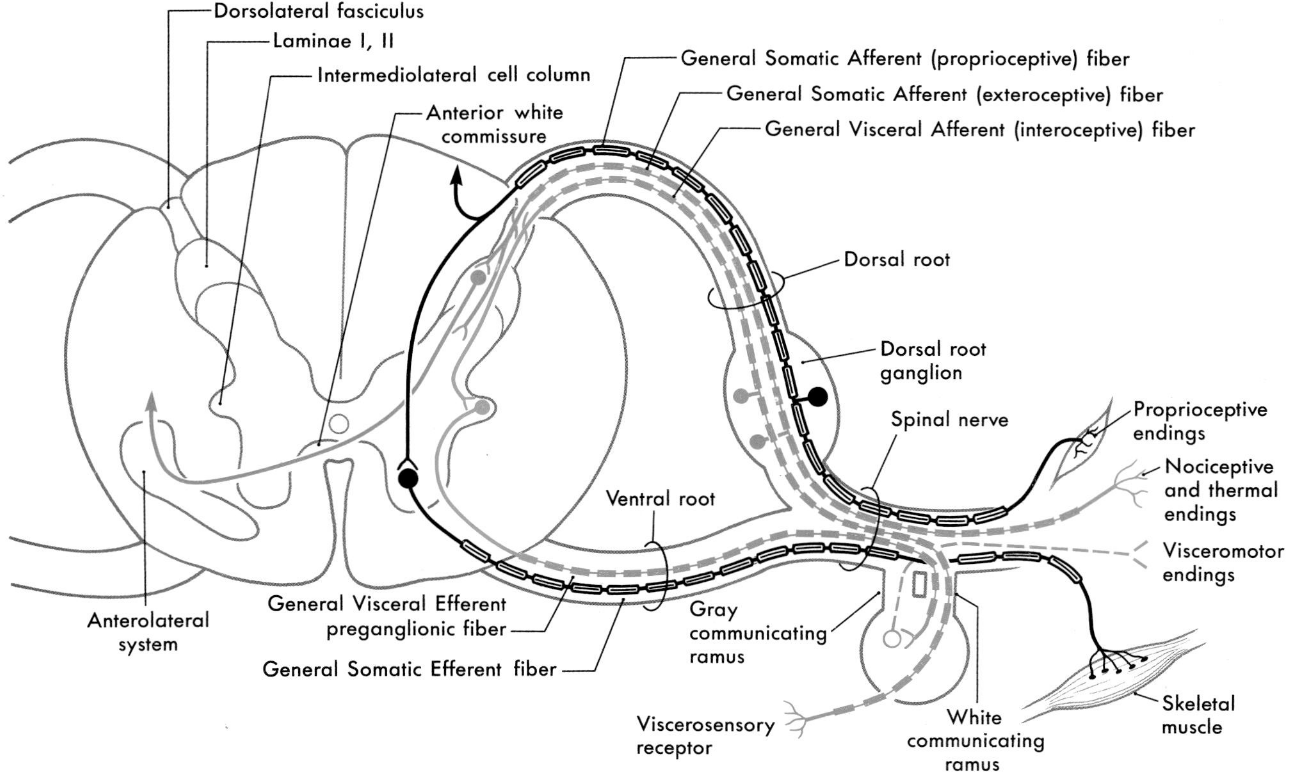

FIGURE 9-6 *The spinal nerve shown on a representative cord level. The relative thickness of the various types of fibers is indicated. The narrow-diameter fiber passing through the gray communicating ramus and terminating in visceromotor endings represents a GVE postganglionic fiber.*

Exteroceptive (GSA) fibers arise from receptors that are sensitive to (1) mechanical, thermal or chemical stimuli that may cause tissue damage, or (2) from receptors sensitive to discriminative touch or vibratory stimuli. The former fibers (A-δ and C) are slowly conducting (0.5–30 meters/sec), unmyelinated or lightly myelinated, and they enter the cord via the *lateral division of the dorsal root*. These fibers may ascend and/or descend in the *dorsolateral fasciculus* before entering the dorsal horn to terminate primarily in laminae I to V. The latter fibers (A-β) are rapidly conducting (30–70 meters/sec), heavily myelinated, and they enter the cord through the *medial division of the dorsal root*. After entering the dorsal funiculus, these fibers may give rise to ascending and/or descending collaterals.

Proprioceptive (GSA) fibers originate from receptors located in muscles, tendons, or joints that are sensitive to stretch or pressure; some vibratory sense is also conveyed by these fibers. These are rapidly conducting (70–120 meters/sec; Ia, Ib, A-β), heavily myelinated fibers that also enter the medial division of the dorsal root. The central processes of these proprioceptive fibers (and of the heavily myelinated exteroceptive fibers) may directly enter, and ascend in, the dorsal columns, or they may branch into the spinal gray to synapse in relay nuclei (such as the dorsal nucleus of Clarke) or on cells in the ventral horn that participate in spinal reflexes.

The spinal nerve also conveys sensory information from thoracic, abdominal, and pelvic viscera. This *interoceptive* input originates primarily from receptors that are sensitive to nociceptive stimuli and is conveyed via GVA fibers. These fibers travel through (for example) the *splanchnic nerves* and traverse the sympathetic chain and *white communicating ramus* to enter the spinal nerve (Fig. 9-6). Their central processes enter the lateral division of the dorsal root and terminate in laminae I and V to VII. These GVA fibers are also lightly myelinated and slowly conducting (1–20 meters/sec).

Neurotransmitters of Primary Sensory Neurons Although several neuroactive substances have been implicated as transmitters in primary afferent fibers, those having an important role are *substance P* (SP), *calcitonin gene-related peptide* (CGRP), and *glutamate*. Small-diameter (A-δ and C) fibers arising from visceral and somatic structures—that is, GVA and small-diameter GSA fibers—use one or more of these three neurotransmitters, and it is probable that some large-diameter, heavily myelinated GSA fibers use glutamate. Specifically, SP and/or CGRP can be found in small-diameter GVA and GSA fibers, and these peptides plus glutamate are also found in the smaller cell bodies of the dorsal root ganglion, from which these fibers arise. Centrally, fibers and terminals containing these three neurotransmitters can be found in laminae I, II, and V, where the small-diameter axons synapse with cells that relay the information to higher levels of the neuraxis. Conversely, some of the large bodies in dorsal root ganglia, which give rise to large-diameter GSA fibers, contain glutamate. This transmitter is also found in the dorsal columns in large-diameter, heavily myelinated fibers, indicating that it may function in the relay of proprioceptive information.

Motor Components of the Spinal Nerve The spinal cord gives rise to two types of motor fibers: (1) those that directly innervate skeletal (striated) muscle, and (2) autonomic fibers that synapse on a second neuron, usually located in an autonomic ganglion. The latter (or postganglionic) neurons innervate smooth muscle, cardiac muscle, or glandular epithelium (Fig. 9-6).

The motor cells that innervate skeletal muscle are located in the ventral horn; these cells and their peripheral processes are classified as *general somatic efferent* (GSE). GSE cells from the ventral horn also supply motor innervation to the specialized *intrafusal* muscle fibers of the *muscle spindles* (neuromuscular spindles), sensory structures in muscles that detect muscle length and various aspects of contraction dynamics. Large motor neurons in the ventral horn are organized in two general, but overlapping, patterns (Fig. 9-7). First, cells innervating proximal muscles are located medially, and cells innervating more distal muscles are located progressively more laterally. This explains why the ventral horn is smaller and narrower at thoracic than at cervical and lumbar levels. At thoracic levels the ventral horn contains motor neurons for only the axial muscles of the trunk, whereas at cervical and lumbar levels it also contains the more lateral groups of motor neurons that innervate the limbs. Second, within the ventral horn at C4 to T1 and L1 to S2,

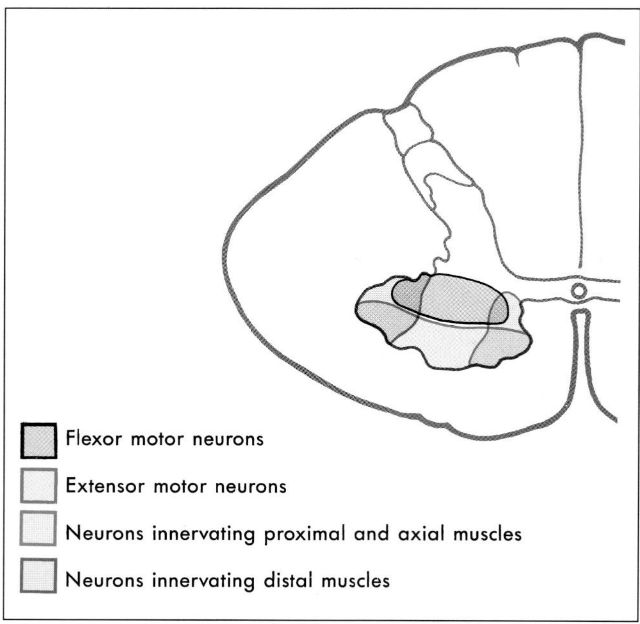

FIGURE 9-7 *Representation of the general organization of motor neurons in the ventral horn.*

motor neurons innervating extensors tend to be more ventrally located, whereas those innervating flexors tend to be found more dorsally.

The autonomic motor neurons of the spinal cord are classified as *general visceral efferent* (GVE) and have their cell bodies in lamina VII. At cord levels T1 to L2, these cells belong to the *intermediolateral cell column* (sympathetic cells), whereas at sacral levels S2 to S4, they belong to the parasympathetic system and form the *sacral visceromotor nucleus* located in the lateral part of lamina VII. Unlike the single-neuron GSE projection, autonomic pathways consist of two neurons in series (Fig. 9-6). The spinal cord neuron projects to an autonomic ganglion and is therefore classified as *GVE-preganglionic*. In the ganglion, it synapses with a *GVE-postganglionic neuron, which innervates the target structure*.

Motor fibers (GSE and GVE) exit in the ventral root and pass into the spinal nerve. GSE fibers continue through the spinal nerve and are conveyed by the progressive branching of peripheral nerves to the skeletal muscles of the body. In contrast, GVE-preganglionic fibers leave the spinal nerve to join the sympathetic trunk via the *white communicating ramus* (Fig. 9-6). Once they have entered the sympathetic trunk, these preganglionic fibers follow any of several routes, which are considered in detail in Chapter 28. Suffice it to say, GVE-postganglionic fibers from cells of the sympathetic chain ganglia rejoin the spinal nerves via the *gray communicating ramus*, whereas those of the prevertebral ganglia distribute only to the gut.

Neurotransmitters of Spinal Motor Neurons The three populations of spinal motor neurons are (1) large ventral horn cells (α-motor neurons) that innervate extrafusal skeletal muscle cells, (2) smaller cells (*γ-motor neurons*) that innervate only the intrafusal fibers of the muscle spindles, and (3) cells that give rise to preganglionic sympathetic (T1 to L1) or parasympathetic (S2 to S4) fibers, which terminate in peripheral motor (autonomic) ganglia. All three of these cell populations use *acetylcholine* as their neurotransmitter. Consequently, acetylcholine is abundant in axon terminals at the *neuromuscular junction*, and numerous nicotinic acetylcholine receptors are present on the postsynaptic junctional folds of the muscle membrane. *Myasthenia gravis*, a neurologic disease characterized by moderate to profound muscle weakness, is closely correlated with the presence of circulating antibodies directed against nicotinic receptor sites on the postsynaptic membrane. The result is a blockage of transmission at the neuromuscular junction.

SPINAL REFLEXES

Afferent fibers in spinal nerves may synapse on tract cells that relay information to higher levels of the neuraxis, or they may terminate on motor neurons or interneurons, both of which may participate in reflex circuits. Reflexes require an afferent fiber, interneurons and/or motor neurons, and a target tissue, usually skeletal muscle. Reflexes may be relatively simple and confined to a single cord level (*intrasegmental*), or complex, involving multiple cord segments (*intersegmental*). Certain diseases or central nervous system lesions can affect spinal reflexes, resulting in reflexes that are greatly exaggerated (*hyperreflexia*), diminished (*hyporeflexia*), or absent (*areflexia*). Numerous reflexes are part of the standard neurologic examination; only a few examples are given here.

Tendon Reflex Although a *tendon reflex* may be elicited by tapping any large tendon (e.g., triceps or Achilles), a common example is the *knee-jerk* or *quadriceps stretch reflex* (Fig. 9-8). A brisk tap on the patellar tendon stretches the primary sensory endings in muscle spindles located in the quadriceps muscle, sending an impulse toward the dorsal root ganglion via heavily myelinated, rapidly conducting group Ia fibers. The central processes of these afferent axons synapse on and excite motor neurons in the ventral horn that innervate the quadriceps muscle. The result is a

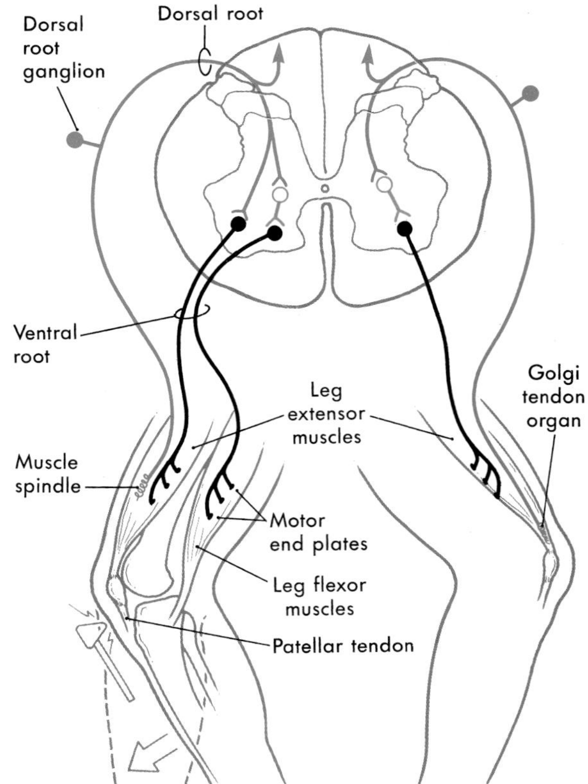

FIGURE 9-8 *Pathway for the patellar/tendon reflex and reciprocal inhibition (left) and autogenic inhibition (right). The inhibitory glycinergic interneurons are represented by the red open cell bodies.*

sudden contraction of these muscles and a twitch of the leg at the knee. Because this reflex requires only one synapse and is a response to muscle stretch, it also may be called a *monosynaptic stretch reflex* or a *myotatic reflex*.

An extension of the simple stretch reflex is seen in *reciprocal inhibition* and *autogenic inhibition* (also called the *inverse myotatic reflex*). In reciprocal inhibition, one group of muscles is excited and the antagonistic group is inhibited (Fig. 9-8). In this situation, the muscle spindle is stretched by a tap on the patellar tendon, and the impulse enters the spinal cord via a group Ia primary sensory fiber. This fiber branches and has excitatory terminations on quadriceps motor neurons and on group Ia inhibitory (glycinergic) interneurons. As a result, the quadriceps (extensor) contracts, whereas the interneurons inhibit spinal motor neurons innervating the hamstring (flexor) muscles, which remain passive. This action enhances the effectiveness of the reflex.

The muscle receptor involved in autogenic inhibition is the *Golgi tendon organ* (Fig. 9-8). This receptor responds to relatively high tension (higher than that needed to activate the muscle spindles). Activation causes an increase in the rate of firing of the group Ib sensory fibers that arise from this receptor. In the spinal cord, these fibers terminate on group Ib inhibitory (glycinergic) interneurons, which inhibit motor neurons that innervate the muscle attached to the tendon from which the afferent volley originated.

Flexor Reflex A further level of complexity in spinal reflexes is seen in the flexor reflex (withdrawal reflex or nociceptive reflex) (Fig. 9-9). This type of reflex is initiated by cutaneous input, is frequently a response to nociceptive stimuli, and represents an attempt to protect a body part by extricating it from the source of injury. Lightly myelinated or unmyelinated primary sensory fibers (A-δ or C fibers) conveying nociceptive input enter the dorsolateral fasciculus and branch. Many of these fibers enter the spinal gray where they form excitatory synaptic contacts with ascending tract cells and with both excitatory and inhibitory interneurons (Fig. 9-9). While tract neurons relay this nociceptive information to higher levels of the neuraxis, the excitatory glutaminergic interneurons synapse on flexor motor neurons, resulting in activation of the ipsilateral flexor muscles of the thigh, leg, and foot and withdrawal of the extremity. This action is enhanced by the synapse of inhibitory interneurons on extensor (antagonistic) motor neurons and the resultant decreased activity (inhibition) of extensor muscles. The flexor reflex, considering its afferent and efferent limbs, involves several spinal segments.

Crossed Extension Reflex The ***crossed extension reflex*** builds on the basic circuits of the flexor reflex, but also involves musculature of the contralateral side of the body (Fig. 9-10). By way of interneurons, nociceptive input on A-δ or C fibers excites ipsilateral leg flexor motor neurons and inhibits ipsilateral leg extensor motor neurons. Consequently, the flexors contract, the extensors relax, and the extremity is withdrawn from the painful stimulus. If the patient is standing or walking, however, the opposite leg must participate in the response to keep the individual from falling. The same nociceptive input that resulted in withdrawal on the ipsilateral side is conveyed to interneurons that project to the contralateral ventral horn (Fig. 9-10). These fibers excite motor neurons polysynaptically, innervating contralateral extensor muscles and inhibiting motor neurons that innervate contralateral flexor muscles. Thus, there is an ipsilateral flexion and withdrawal from the stimuli accompanied by an extension of the contralateral leg to support the body.

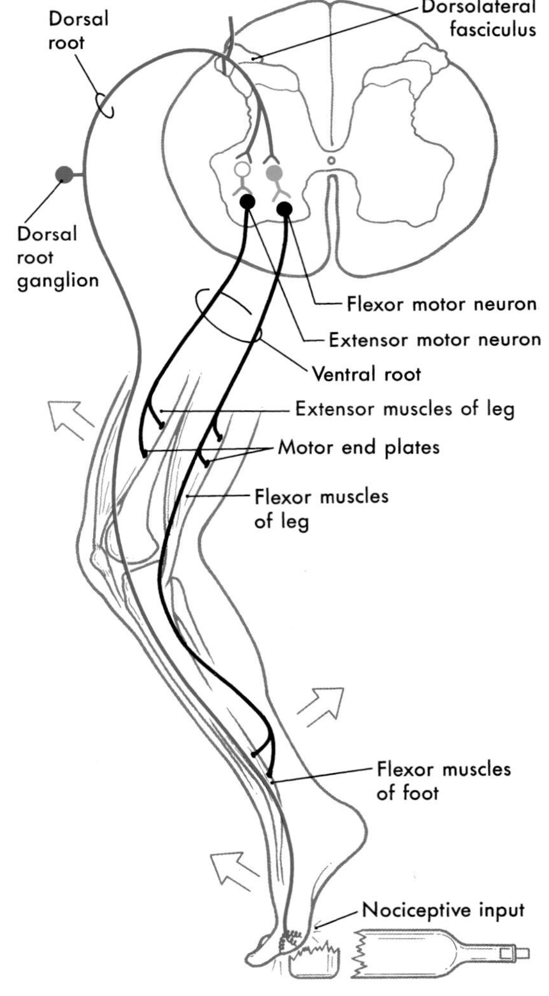

FIGURE 9-9 *Pathway for the flexor reflex. The inhibitory glycinergic interneuron is represented by the red open cell and the excitatory glutaminergic interneuron by the green closed cell.*

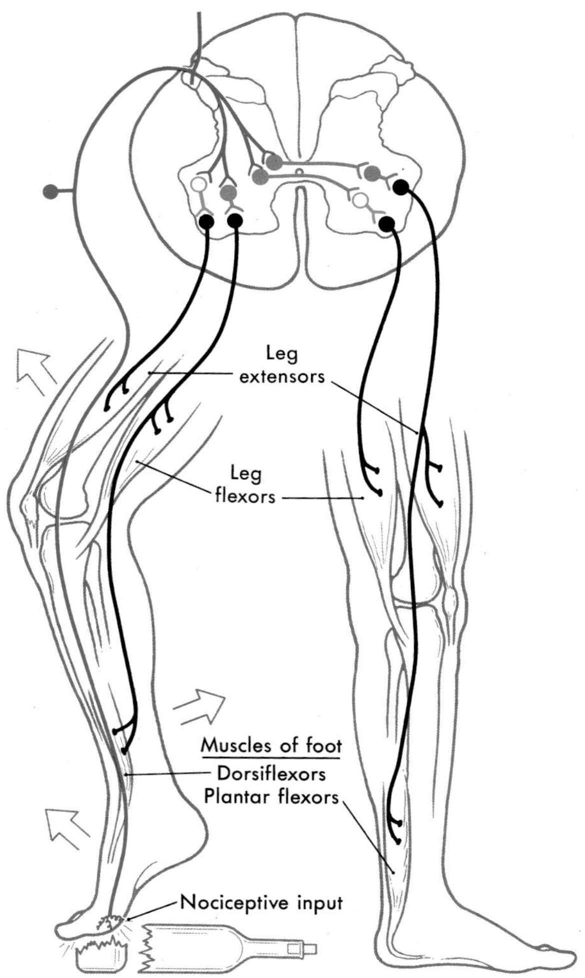

FIGURE 9-10 *Pathway for the crossed extension reflex. Glycinergic interneurons (inhibitory) are represented by the open red cell bodies and glutaminergic interneurons (excitatory) by the closed green cells.*

PATHWAYS AND TRACTS OF THE SPINAL CORD

The spinal cord white matter consists of (1) long *ascending* and *descending fibers/tracts*, which link the spinal cord to higher levels of the neuraxis, and (2) *propriospinal fibers* that project from one spinal level to another (Fig. 9-11). Ascending fibers convey information to higher levels of the neuraxis. Some descending fibers modulate the transmission of nociceptive information in the dorsal horn, whereas others influence the activity of motor neurons. Propriospinal fibers form the basis for a wide variety of intraspinal reflexes.

Many tracts/fibers in the nervous system are *named according to their origin and termination*. For example, the *corticospinal* fibers originate in the cerebral cortex (*cortico-*) and end in the spinal cord (*-spinal*). These are descending fibers because the *cortex* is a more rostral part of the neuraxis than is the *spinal* cord. Likewise, the term *spinothalamic fiber* indicates that the cell of origin is in the spinal cord, the termination in the thalamus, and these are ascending fibers. Keeping this basic principal in mind will expedite learning many tracts and pathways.

Ascending Tracts The *gracile* and *cuneate fasciculi*, collectively called the *dorsal columns*, are composed of the central processes of heavily myelinated primary sensory fibers that convey proprioceptive, tactile, and vibratory information from the ipsilateral side of the body (Fig. 9-11). Fibers in the gracile fasciculus originate from sacral, lumbar, and lower thoracic (below T6) levels; those in the cuneate fasciculus originate from upper thoracic (above T6) and cervical levels. Injury to the dorsal columns on one side results in an *ipsilateral loss of proprioception, discriminative touch, and vibratory sense below the level of the lesion.*

The posterior (*dorsal*), *spinocerebellar*, and *anterior* (*ventral*) *spinocerebellar tracts* are located on the lateral surface of the cord, meeting at approximately the level of the denticulate ligament (Figs. 9-3, 9-11). The fibers of the former tract arise from *Clarke's nucleus* (in lamina VII at T1 to L2), and the fibers of the latter tract arise primarily from cells of laminae V to VIII and from large ventral horn neurons called *spinal border cells*, both at lumbosacral levels.

In the ventrolateral area of the spinal cord is a large composite bundle called the *anterolateral system* (ALS) (Fig. 9-11). This system encompasses those regions of the white matter that were classically divided into anterior and lateral spinothalamic tracts. The ALS contains *spinothalamic*, *spinomesencephalic* (spinotectal, spinoperiaqueductal), *spinohypothalamic*, and *spinoreticular fibers*. The fibers of the ALS originate primarily from dorsal horn cells, but it is now known that some arise from neurons in the ventral horn. Those fibers that coalesce to form the ALS cross in the ventral white commissure, ascending one to two levels as they do so.

In general, fibers of the ALS convey nociceptive, thermal, and touch information to higher levels of the neuraxis. Consequently, *injury to the spinal cord that involves ALS fibers will result in a loss of pain, temperature, and crude touch sensations on the contralateral side of the body beginning about one to two segments below the lesion.* The ALS is somatotopically organized; this means that in the spinal cord, lower portions of the body are represented dorsolaterally and upper levels are represented ventromedially.

Nociceptive input and some discriminative touch are also carried by *postsynaptic dorsal column fibers* and by the *spinocervicothalamic tract*. The former originate from laminae III to VIII (mainly IV) and ascend ipsilaterally in the dorsal columns. The latter fibers arise from the same laminae but ascend as a diffuse population in the dorsal part of the lateral funiculus to end in the lateral cervical

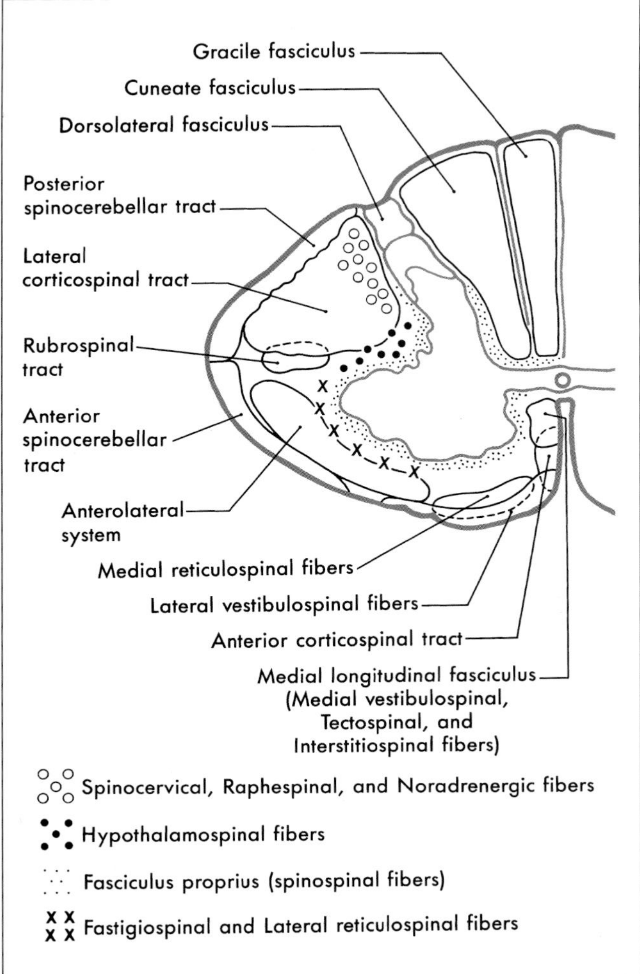

FIGURE 9-11 *Ascending and descending pathways of the spinal cord shown in cross section as they are organized at cervical levels.*

nucleus at levels C1 to C3. The existence of these minor fiber populations in humans may explain the recurrence of pain perception in some patients who have had an *anterolateral cordotomy* for intractable pain.

Other, more diffusely arranged, ascending fibers include *spino-olivary*, *spinovestibular*, and *spinoreticular fibers*. These are discussed in later chapters in relation to the functional systems they serve.

Descending Tracts The lateral funiculus (Fig. 9-11) contains the *lateral corticospinal* and *rubrospinal* tracts, as well as other fiber populations that are more diffuse in their distribution (*reticulospinal*, *fastigiospinal*, *raphespinal*, *hypothalamospinal*). Corticospinal fibers arise from the cerebral cortex and descend through the brainstem. At the medulla-spinal cord junction, most cross to form the *lateral corticospinal tract*, but some remain uncrossed as the *anterior corticospinal tract*. Lateral corticospinal fibers are somatotopically arranged; fibers that originate from leg areas of the cerebral cortex and project to lumbosacral levels are lateral, whereas those traveling to cervical levels from arm areas of the cortex are medial (Fig. 9-11). One important function of this tract is to influence spinal motor neurons, *especially those controlling fine movements of the distal musculature*. Consequently, lesions of lateral corticospinal fibers on one side of the cervical cord result in paralysis of the arm and leg on that side (*hemiplegia*).

Rubrospinal fibers arise from the red nucleus of the midbrain, cross at that level, and descend in the spinal cord with lateral corticospinal fibers (Fig. 9-11). In general, rubrospinal fibers, as well as lateral corticospinal fibers, excite flexor motor neurons and inhibit extensor motor neurons.

Although diffusely arranged, other descending fibers in the lateral funiculus serve important functions (Fig. 9-11). *Reticulospinal fibers* in this area originate from the medullary reticular formation, and *fastigiospinal fibers* from the fastigial nucleus of the cerebellum. At spinal levels, the former are uncrossed and the latter are crossed. Because their function is to help maintain posture, these fibers tend to excite extensor motor neurons and inhibit flexor motor neurons. *Raphespinal fibers* originate from the raphe nuclei of the brainstem, descend bilaterally in dorsal areas of the lateral funiculus, and function to modulate the transmission of nociceptive information at spinal levels. The activity of GVE motor neurons of the intermediolateral cell column is influenced by *hypothalamospinal fibers*, which descend through lateral areas of the brainstem and spinal cord. Lesions in the brainstem or cervical spinal cord that interrupt these fibers result in ipsilateral *ptosis*, *miosis*, *anhidrosis*, and *enophthalmos* (*Horner's syndrome*).

The *anterior funiculus* (Fig. 9-11) contains *reticulospinal* and *vestibulospinal* fibers, the *anterior corticospinal tract*, and the *medial longitudinal fasciculus* (MLF). Reticulospinal fibers in this area arise in the pontine reticular formation of the brainstem, whereas vestibulospinal fibers originate from the vestibular nuclei. *Lateral vestibulospinal fibers* arise from the lateral vestibular nucleus, and *medial vestibulospinal fibers* originate primarily from the medial vestibular nucleus. Reticulospinal and vestibulospinal fibers of the anterior funiculus function in postural mechanisms through their general excitation of extensor motor neurons and inhibition of flexor motor neurons. Fibers of the *anterior corticospinal tract* are uncrossed, but most of these fibers cross in the ventral white commissure before terminating on medial motor neurons that innervate axial muscles.

The MLF, although quite small, is generally regarded as a composite bundle containing *medial vestibulospinal fibers*, *tectospinal fibers*, *interstitiospinal fibers* from the interstitial nucleus of the rostral midbrain, and some *reticulospinal fibers*. Tectospinal and vestibulospinal fibers are found only in cervical levels; the other fibers extend to lower cord levels. These fibers terminate primarily in laminae VII and VIII, but ultimately influence motor neurons innervating primarily axial and neck musculature.

The comparatively simple structure of the spinal cord belies its functional importance. Although the cord is no larger than the little finger, descending motor control of the body below the neck and all sensory input from the same areas must traverse it. Consequently, lesions in the spinal cord that would be considered of little consequence in larger parts of the brain may cause global deficits or death. As the cord merges into the brainstem, the organization and function of the central nervous system become progressively more complex.

SOURCES AND ADDITIONAL READINGS

Brown AG: *Organization in the Spinal Cord: The Anatomy and Physiology of Identified Neurons*. Springer-Verlag, Berlin, pp 1-238, 1981.

Dado RJ, Katter JT, Giesler GJ: Spinothalamic and spinohypothalamic tract neurons in the cervical enlargement of rats. I. Locations of antidromically identified axons in the thalamus and hypothalamus. J Neurophysiol 71:959–980, 1994.

Quencer RM, Bunge RP, Egnor M, Green BA, Puckett W, Naidich TP, Post MJD, Norenberg M: Acute traumatic central cord syndrome: MRI-pathological correlations. Neuroradiology 34:85–94, 1992.

Rexed B: The cytoarchitectonic organization of the spinal cord in the cat. J Comp Neurol 96:415–495, 1952.

Rexed B: A cytoarchitectonic atlas of the spinal cord in the cat. J Comp Neurol 100:297–379, 1954.

Schoenen J, Faull RLM: Spinal cord: Cytoarchitectural, dendroarchitectural, and myeloarchitectural organization. In Paxinos G (ed): *The Human Nervous System*. Academic Press, San Diego, pp 19–53, 1990.

Willis WD: *The Pain System, The Neural Basis of Nociceptive Transmission in the Mammalian Nervous System*, Vol. 8. In Gildenberg PL (ed): Pain and Headache. S Karger, Basel, pp 1–346, 1985.

Willis W, Coggeshall RE: *Sensory Mechanisms of the Spinal Cord*, 2nd Ed. Plenum Press, New York, pp 1–575, 1991.

Yezierski RP: Spinomesencephalic tract: Projections from the lumbosacral spinal cord of the rat, cat, and monkey. J Comp Neurol 267:131–146, 1988.

CHAPTER TEN

An Overview of the Brainstem

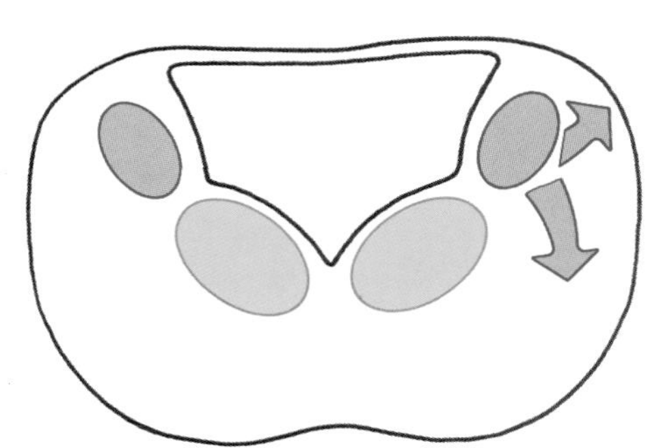

D. E. HAINES • G. A. MIHAILOFF

The term *brainstem* (sometimes written *brain stem*) is used in two ways: It can mean either the portion of the brain that consists of the medulla, pons, and midbrain, or the portion that consists of these structures plus the diencephalon. This book follows the former convention. For our purposes, therefore, *the brainstem consists of the rhombencephalon (excluding the cerebellum) and the mesencephalon*. These regions of the brainstem all share a basic organization, which is the topic of this chapter. The medulla, pons, and midbrain are discussed in detail in Chapters 11 to 13).

BASIC DIVISIONS OF THE BRAINSTEM

Medulla Oblongata At about the level of the foramen magnum, the spinal cord merges into the caudal-most portion of the brain, the *medulla oblongata* or *myelencephalon*, commonly called the medulla. The foramen magnum marks the approximate location of the *pyramidal decussation of the medulla* (Fig. 10-1A). The medulla is slightly cone-shaped and enlarges in diameter as it extends rostrally toward the pons. Dorsally, the pons-medulla junction is represented by the caudal edge of the middle and inferior cerebellar peduncles, whereas ventrally this border is formed by the caudal edge of the basilar pons (Fig. 10-1).

The cranial nerves associated with the medulla include the *hypoglossal* (XII, motor), parts of the *accessory* (XI, motor), *vagus* (X, mixed), and *glossopharyngeal* (IX, mixed) (Fig. 10-1A). The *abducens* (VI, motor), *facial* (VII, mixed), and *vestibulocochlear* (VIII, sensory) *nerves* are frequently called the *cranial nerves of the pons-medulla junction* because they exit the brainstem at this particular location (Fig. 10-1A).

Pons The pons (the ventral part of the *metencephalon*) extends from the pons-medulla junction to a line drawn from the exit of the trochlear nerve dorsally to the rostral edge of the basilar pons ventrally (Fig. 10-1A–C). This latter structure is bulbous and characterizes the ventral aspect of the pons. The cerebellum, although part of the metencephalon, is *not* part of the brainstem. It is joined to the brainstem by three large, paired bundles of fibers called the *cerebellar peduncles*. The *trigeminal nerve* (V, mixed) emerges from the lateral aspect of the pons (Fig. 10-1A).

Midbrain The *midbrain* (*mesencephalon*) extends rostrally from the pons-midbrain junction to join the diencephalon (thalamus). This latter interface is usually described as a line drawn from the posterior commissure dorsally to the caudal edge of the mammillary bodies ventrally (Fig. 10-1B). The *oculomotor nerve* (III, motor) exits the ventral aspect of the midbrain, whereas the *trochlear nerve* (IV, motor) exits its dorsal aspect (Fig. 10-1A,C).

Tegmental and Basilar Areas The central core of the midbrain and the pons is called the *tegmentum* and their ventral parts are the *basilar* areas. These regions are continuous with each other and with comparable areas of the medulla (Figs. 10-1B, 10-2). Although usually not considered part of the tegmentum *per se*, the central portion of the medulla shares structural and functional similarities with the former region. The *tegmentum of the pons and midbrain* and the *contiguous central portion of the medulla* contain ascending and descending tracts, many relay nuclei, and the nuclei of cranial nerves III to XII.

The basilar part of each brainstem division is ventral to the tegmentum (of the midbrain and pons) and to the central portion of the medulla (Fig. 10-2). Consequently, these basilar structures also form a rostrocaudal continuum. Basilar structures of the brainstem include the descending fibers of the *crus cerebri* (midbrain), *basilar pons*, and *pyramid* (medulla), and specific populations of neurons in the midbrain and pons that originate from the alar plate of the embryonic brain.

VENTRICULAR SPACES OF THE BRAINSTEM

The ventricular spaces of the brainstem are the cerebral aqueduct in the mesencephalon and the fourth ventricle in the rhombencephalon (Fig. 10-2). The *cerebral aqueduct* is a narrow channel, 1 to 3 mm in diameter, which connects the third ventricle (the cavity of the diencephalon) with the fourth ventricle. The cerebral aqueduct contains no choroid plexus; its walls are formed by a continuous mantle of cells collectively called the *periaqueductal gray*. The roof of the midbrain is the *tectum*.

The *fourth ventricle* is the cavity of the *rhombencephalon*. Its rostral portion lies between the pons and cerebellum, and its caudal part is located in the medulla (Fig. 10-2). The fourth ventricle is continuous rostrally with the cerebral aqueduct and caudally with the central canal of the cervical spinal cord. It also communicates with the subarachnoid space via three openings: the midline *foramen of Magendie* and the two lateral *foramina of Luschka*. The foramen of Magendie is located in the caudal roof of the ventricle and opens into the *dorsal cerebromedullary cistern* (*cisterna magna*) (Fig. 10-2). The foramina of Luschka are located at the ends of the lateral recesses of the fourth ventricle and open into the subarachnoid space at the cerebellopontine angles (see Fig. 6-8). The *lateral recesses* are horn-shaped widenings of the fourth ventricle that extend around the brainstem at the pons-medulla junction.

The roof of the fourth ventricle is mainly formed by the *anterior* (or *superior*) *medullary velum* rostrally, by the thin membranous *tela choroidea* caudally, and by a small part of the cerebellum in the middle (Figs. 10-1B, 10-2). From rostral to caudal, the walls of the fourth ventricle

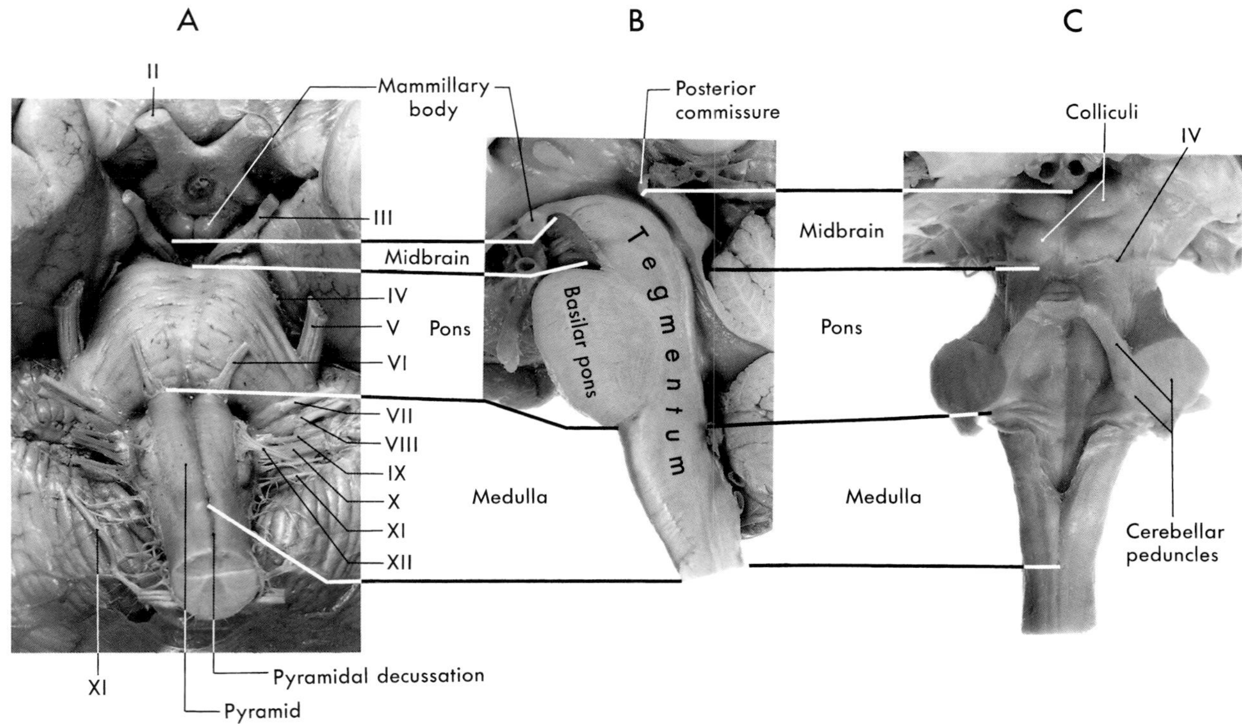

FIGURE 10-1 *Ventral (A), midsagittal (B), and dorsal (C) views of the brainstem. Cranial nerves are labeled by Roman numerals. In C, the cerebellum is removed to expose the dorsal surface of the brainstem and the fourth ventricle.*

are formed by the superior cerebellar peduncles, the middle and inferior cerebellar peduncles, and the attachment of the tela choroidea to the medulla (Fig. 10-3). The tela arises from the inferior surface of the cerebellum and sweeps caudally to attach to the V-shaped edges of the medullary portion of the ventricular space. The choroid plexus of the fourth ventricle is suspended from the inner surface of the tela, and parts of it protrude outward through the foramina of Luschka.

The floor of the fourth ventricle is called the *rhomboid fossa*. It is divided into two halves by a deep *median sulcus*, and each half is traversed rostrocaudally by a groove called the *sulcus limitans* (Fig. 10-3; see also Fig. 10-5). Several elevations can be distinguished in the rhomboid fossa that are related to the presence of underlying cranial nerve nuclei and associated fiber bundles (Fig. 10-3). As noted below, the cranial nerve nuclei that are located between the medial sulcus and the sulcus limitans are motor in function, whereas those located lateral to the sulcus are sensory in function. Medial to the sulcus limitans, the *hypoglossal* and *vagal trigones* represent

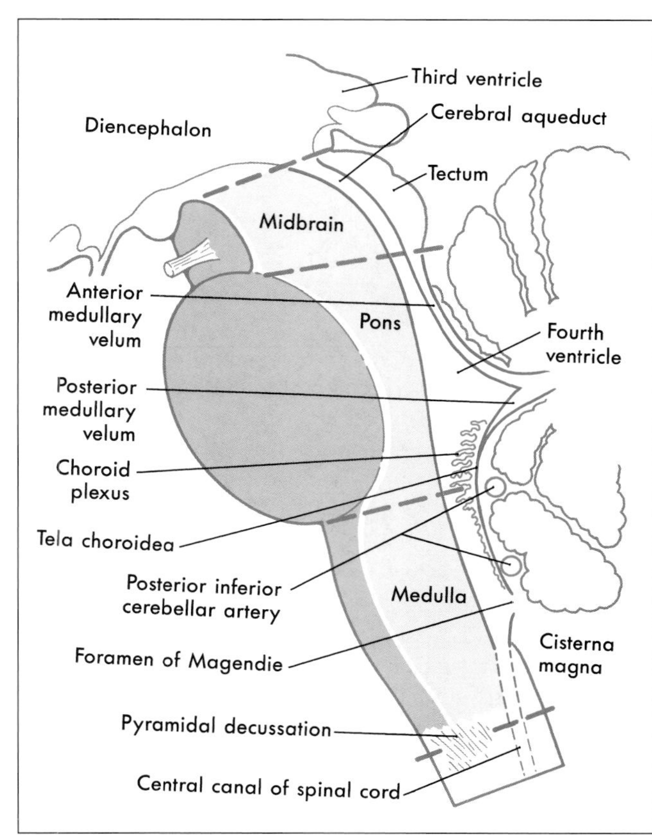

FIGURE 10-2 *Midsagittal drawing of the brainstem. Ventricular spaces of the brainstem are outlined in green. The* tegmental *and* basilar areas *and* contiguous areas of the medulla *are shown in light and dark gray, respectively. Compare with Fig. 10-1B.*

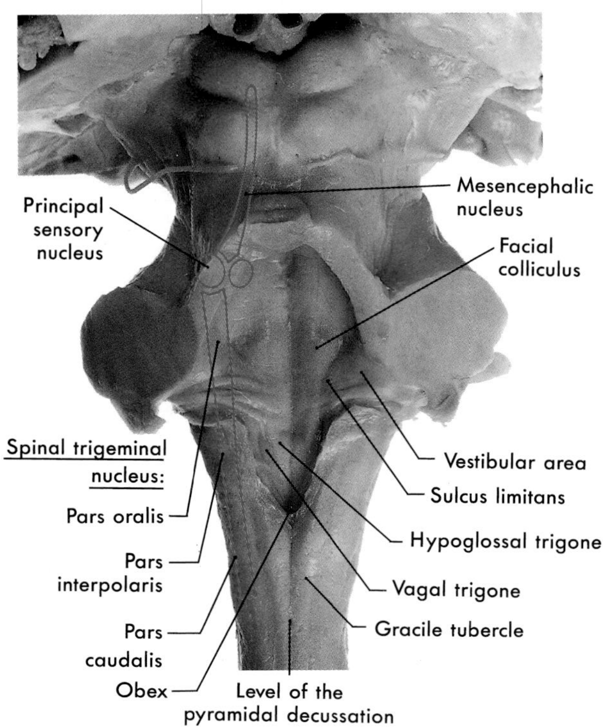

FIGURE 10-3 *Dorsal view of the brainstem. The approximate locations of the trigeminal nuclei are shown. The cerebellum is removed to expose the dorsal aspect of the medulla and midbrain and the rhomboid fossa. The trigeminal motor nucleus, located medial to the principal sensory nucleus, is not labeled.*

the underlying *hypoglossal* and *dorsal motor vagal nuclei.* In the caudal pontine region, the *facial colliculus* located medial to the sulcus limitans marks the location of the underlying *abducens motor nucleus* and the *internal genu* of the facial nerve (see Chapter 12). Lateral to the sulcus limitans in the medulla and caudal pons is a flattened region called the *vestibular area*, which marks the location of the *vestibular nuclei.*

CRANIAL NERVE NUCLEI AND THEIR FUNCTIONAL COMPONENTS

Cranial nerves, as is the case for spinal nerves, contain sensory or motor fibers or a combination of these fiber types. These various fibers are classified on the basis of their embryologic origin and/or common structural and functional characteristics. Primary sensory fibers, somatic motor neurons, and preganglionic and postganglionic visceromotor neurons that exhibit "... like anatomical and physiological characters so that they ... act in a common mode ..." (Herrick) are classified as having a specific *functional component.* For example, fibers conveying sharp pain, a specific type of input, from two widely separated body parts (the hand and the leg) have the same functional component. This principle, already introduced in relation to spinal nerves (see Chapter 9), is also directly applicable to cranial nerves.

Early in development, the rostrocaudally oriented cell columns forming the alar and basal plates essentially extend throughout the brainstem. As development progresses, neuroblasts in alar and basal plates begin to migrate to form their adult structures, and the caudocephalad continuity of the cell columns may be disrupted. In this respect, *the primitive cell column retains its relative position as it differentiates, but it may become discontinuous as the individual nuclei derived from the same column are formed.* Motor nuclei of cranial nerves arise from basal plate neuroblasts, whereas the nuclei that receive primary sensory input via cranial nerves originate from the alar plate.

In the caudal medulla, the rostral continuation of the central canal is small; therefore, basal and alar plates are located ventral and dorsal, respectively, to this space (Fig. 10-4). As the fourth ventricle flares open *at the level of the obex*, the alar plate shifts laterally and the basal plate retains a ventral (and now medial) position (Figs. 10-4, 10-5). Rostrally, as the fourth ventricle funnels into the cerebral aqueduct of the midbrain, the alar plate rotates back to a dorsal position, and the basal plate again assumes a ventral position (Fig. 10-4). The *sulcus limitans*, a persistent embryologic landmark in the medulla and pons, separates structures derived from the basal plate from those derived from the alar plate.

These points are clearly illustrated by first considering the basal plate. Some of these neuroblasts retain their position adjacent to the midline but become segmented into the *hypoglossal*, *abducens*, *trochlear*, or *oculomotor nuclei* (Figs. 10-5, 10-6). The skeletal muscles innervated by these motor neurons originate from occipital myotomes (tongue musculature) and from mesenchyme in the orbit (extraocular muscles); therefore their functional component is *general somatic efferent* (GSE). Lateral to the GSE cell groups, a second population of neuroblasts forms the *dorsal motor vagal nucleus*, *inferior salivatory nucleus*, *superior salivatory nucleus*, or the *Edinger-Westphal nucleus* (Figs. 10-5, 10-6). Because the axons of these cells synapse in peripheral ganglia which, in turn, innervate smooth muscle, cardiac muscle, or glandular epithelium, the functional component of these motor neurons is *general visceral efferent* (GVE). The third set of cranial nerve motor nuclei to originate from the basal plate is represented by neuroblasts that migrate ventrolaterally in the brainstem to form the *nucleus ambiguus*, the *facial nucleus*, or the *trigeminal motor nucleus* (Figs. 10-5, 10-6). These motor neurons innervate skeletal muscles that originate from the pharyngeal arches rather than from occipital myotomes. Consequently, their functional component is *special visceral efferent* (SVE).

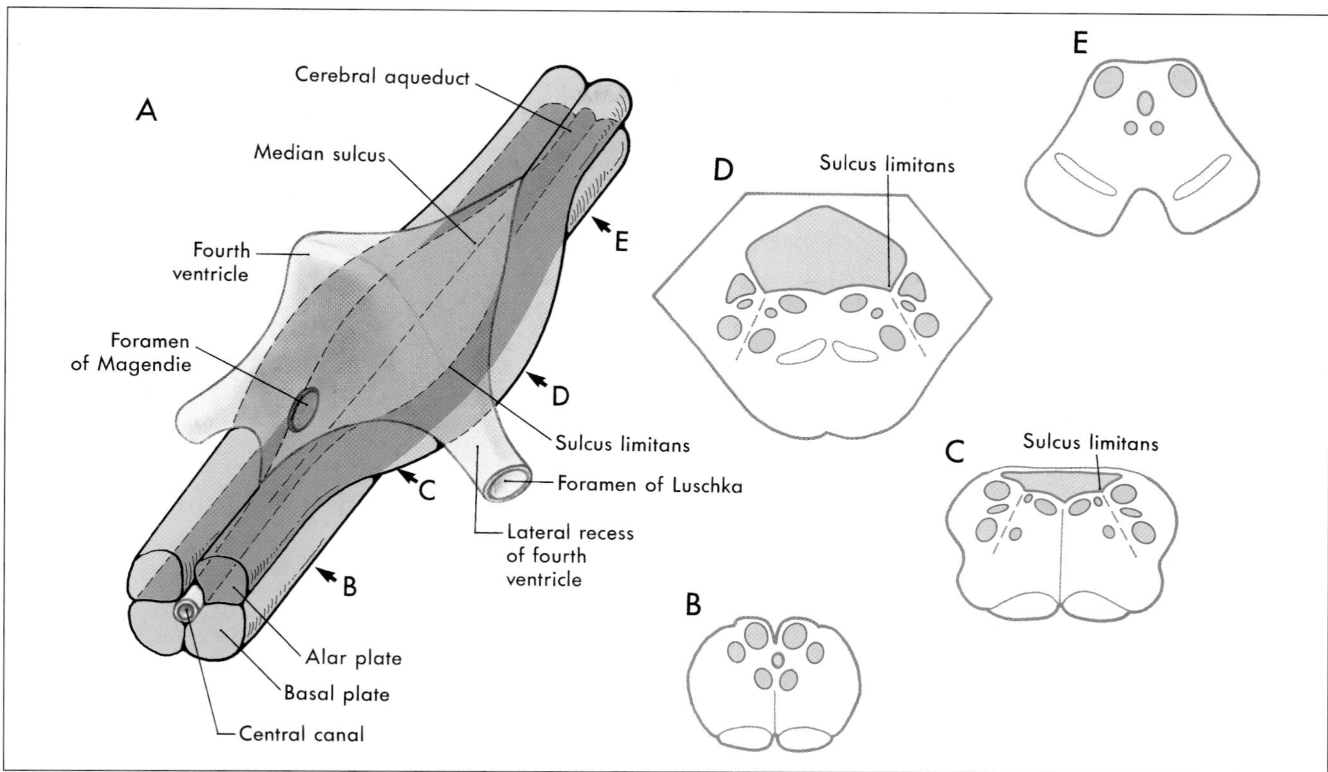

FIGURE 10-4 *Diagram showing the alar and basal plates in relation to the ventricular spaces. The alar plates shift laterally (A), where the fourth ventricle flares open at the obex, and then shift back to a dorsal position, where the ventricle funnels into the cerebral aqueduct. The position of structures derived from the alar and basal plates in relation to the sulcus limitans and the ventricular space is shown for the medulla (B, C), the pons (D), and the midbrain (E).*

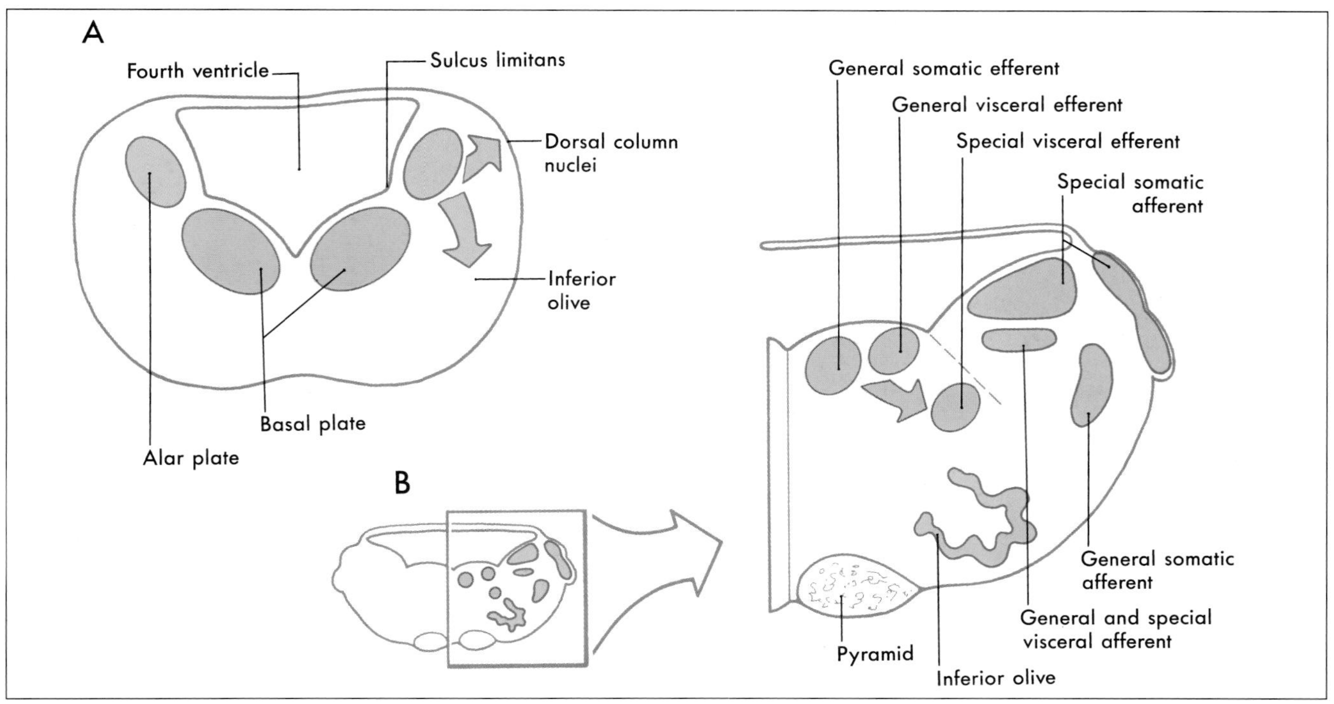

FIGURE 10-5 *Development of the alar and basal plates at early (A) and later (B) stages, showing their relation to functional components of the cranial nerve nuclei in the brainstem.*

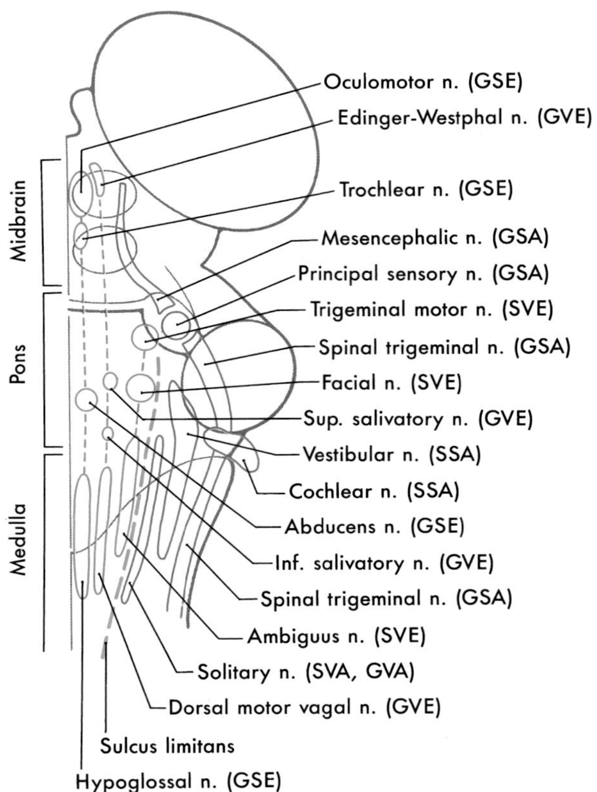

FIGURE 10-6 *Diagram showing the cranial nerve nuclei of the brainstem and their functional components in the brainstem. The various nuclei are unrolled onto a single plane (see also Fig. 10-5B) with basal plate derivatives (red) and alar plate derivatives (blue) located medial and lateral, respectively, to the sulcus limitans. (GSE, general somatic efferent; GVE, general visceral efferent; SVE, special visceral efferent; GSA, general somatic afferent; SSA, special somatic afferent; GVA, general visceral afferent; SVA, special visceral afferent.)*

Simultaneous with these developments in the basal plate, the alar plate gives rise to cell groups that will receive sensory input via cranial nerves. Taste (*special visceral afferent*, SVA) and *general visceral afferent* (GVA) sensations, such as pain from the gut, enter the brainstem with cranial nerves VII, IX, and X. The central processes of these sensory fibers form the *solitary tract* and end in the surrounding *solitary nucleus* (Fig. 10-6). Consequently, the functional components SVA and GVA are associated with these sensory fibers, which enter the solitary nucleus and tract. The eighth cranial nerve, the vestibulocochlear, transmits signals concerned with balance, equilibrium, and hearing. These sensory fibers end in the *vestibular* and *cochlear nuclei* of the medulla and pons, respectively (Figs. 10-5, 10-6). Because of the unique embryologic origin of the peripheral receptors of these nerves, the functional component associated with these fibers and nuclei is *special somatic afferent* (SSA) (Fig. 10-6). Sensory input from the face, oral cavity, and scalp to the apex of the head enters the brainstem via the trigeminal nerve. Centrally some of these fibers form the *spinal trigeminal tract* and synapse in the adjacent *spinal trigeminal nucleus*. Others end in the *principal* sensory nucleus or form the *mesencephalic tract*. In the latter case the cell bodies form the adjacent *mesencephalic nucleus*. Because these cell groups receive general sensory input, the functional component associated with these fibers and nuclei is *general somatic afferent* (GSA) (Figs. 10-5, 10-6). In addition, cranial nerves VII, IX, and X contribute GSA fibers to the spinal trigeminal tract and nucleus.

The spinal trigeminal nucleus extends caudally from about mid-pontine levels to the spinal cord-medulla junction. Based on its cytoarchitecture and on connections, the *spinal trigeminal nucleus* is divided into a *pars caudalis* (between the level of the cervical spinal cord and obex), a *pars interpolaris* (between the level of the obex and the rostral end of the hypoglossal nucleus), and a *pars oralis* (rostral to the level of the hypoglossal nucleus) (Fig. 10-3).

The blood supply to the brainstem is via branches of the *vertebral* and *basilar arteries*. As we will see in the next three chapters, branches of the vertebrobasilar system serve not only the medulla, pons, and most of the midbrain but also the entire cerebellum.

SOURCES AND ADDITIONAL READINGS

Readings for the brainstem chapters are listed at the end of Chapter 13.

CHAPTER ELEVEN

The Medulla Oblongata

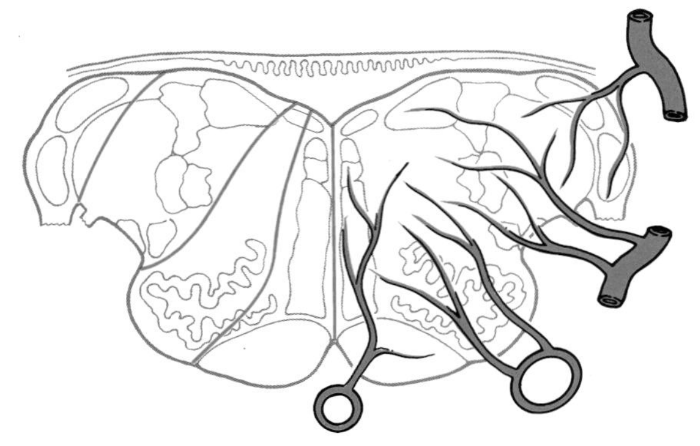

D. E. HAINES • G. A. MIHAILOFF

The medulla oblongata, or *myelencephalon*, is the most caudal segment of the brainstem. It extends rostrally from the level of the foramen magnum to the pons. The cavity of the medulla consists of a narrow, caudal part, which is the continuation of the central canal of the cervical spinal cord, and a flared, rostral portion, which is the medullary part of the *fourth ventricle*. The modest size of the medulla (0.5% of total brain weight) belies its importance. All the tracts passing to or from the spinal cord traverse the medulla and seven of the 12 cranial nerves (VI to XII) are associated with the medulla or the pons-medullary junction. Also, the medullary reticular formation contains cell groups that influence heart rate and respiration. The blood supply to the medulla is via branches of the *vertebral arteries*.

DEVELOPMENT

The basic structural plan of the medulla is an elaboration of that seen in the spinal cord (Figs. 11-1, 11-2). The basal and alar plates give rise to specific nuclei, and the surrounding mantle layer is invaded by axons originating from other levels. Beginning in the medulla, however, the basic derivatives of the primitive neural tube are augmented by the appearance of other structures that characterize each brainstem level.

Basal and Alar Plates ***Basal plate*** neuroblasts of the medulla give rise to the *hypoglossal nucleus* (general somatic efferent, GSE, cells), the *dorsal motor vagal nucleus* and the *inferior salivatory nucleus* (both contain general visceral efferent, GVE, cells), and the nucleus ambiguus (special visceral efferent, SVE, cells) (Fig. 11-2A,B). Caudal to the obex, the hypoglossal and dorsal motor vagal nuclei are quite small and are found in the central gray surrounding the central canal. Rostral to the obex, all of these nuclei are located medial to the sulcus limitans (Fig. 11-2B).

The cranial nerve nuclei derived from the *alar plate* in the medulla, and their corresponding functional components, include the *vestibular* and *cochlear nuclei* (special somatic afferent, SSA), the *solitary nucleus* (general visceral afferent, GVA, and special visceral afferent, SVA), and the *spinal trigeminal nucleus* (general somatic afferent, GSA) (Fig. 11-2). Alar plate neuroblasts caudal to the obex give rise to the *gracile* and *cuneate nuclei*. Rostral to the obex, some alar plate cells migrate ventromedially to form the *inferior olivary complex*.

Concurrent with these events, developing fibers traverse the medulla. An especially prominent bundle of axons collects on the ventral surface of the medulla to form the *pyramids* (Fig. 11-2B).

EXTERNAL FEATURES

Ventral Medulla The ventral aspect of the medulla is characterized by an *anterior (ventral) median fissure*, two laterally adjacent longitudinal ridges, the *pyramids*, and the *olive* (*inferior olivary eminence*) (Fig. 11-3). The pyramids issue from the basilar pons and extend caudally to the *pyramidal decussation*, where 90% of their fibers cross. Because many pyramidal fibers arise in the motor cortex, their crossing is frequently called the *motor decussation*. Rootlets of the *hypoglossal nerve* (XII) exit the medulla via the *preolivary sulcus*, a shallow groove located between the pyramid and the olive. The *abducens nerve* (VI) emerges at the pons-medullary junction, generally in line with the rootlets of XII.

Lateral Medulla On the lateral aspect of the medulla, a shallow trough, the *postolivary sulcus*, is located between the *restiform body* and the large eminence formed by the underlying *inferior olivary nucleus* (Fig. 11-4A,B). Cranial nerves IX (*glossopharyngeal*), X (*vagus*), and the medullary part of XI (*accessory*) emerge from the postolivary sulcus. The *facial nerve* (VII), along with the *intermediate root* of the facial nerve (VIIi; see Chapter 12), and the *vestibulocochlear nerve* (VIII) emerge from the dorsolateral medulla at the pons-medulla interface. On the lateral medullary surface caudal to the level of the obex, fibers of the spinal trigeminal nucleus and tract assume a superficial location and form the *trigeminal tubercle* (*tuberculum cinereum*) (Fig. 11-4B,C). Rostral to the obex, these trigeminal fibers are located internal to a progressively enlarging restiform body.

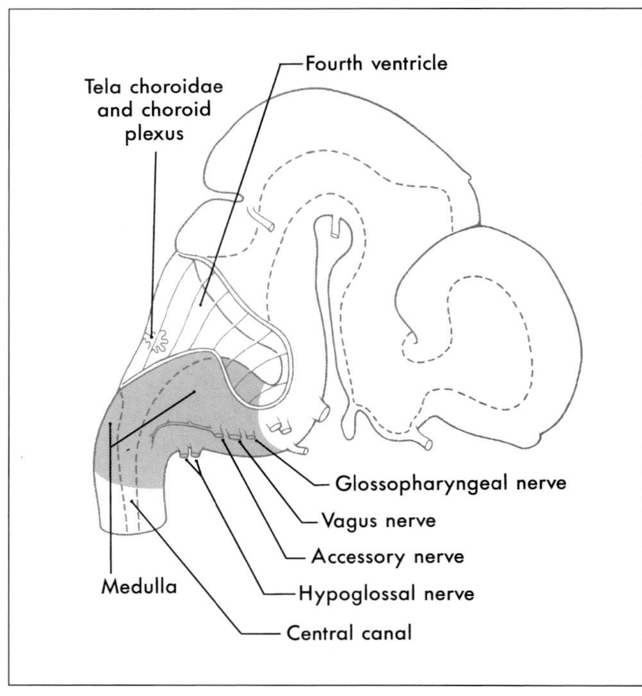

FIGURE 11-1 *Lateral view of the brain at about 7 weeks gestational age. The medulla is highlighted.*

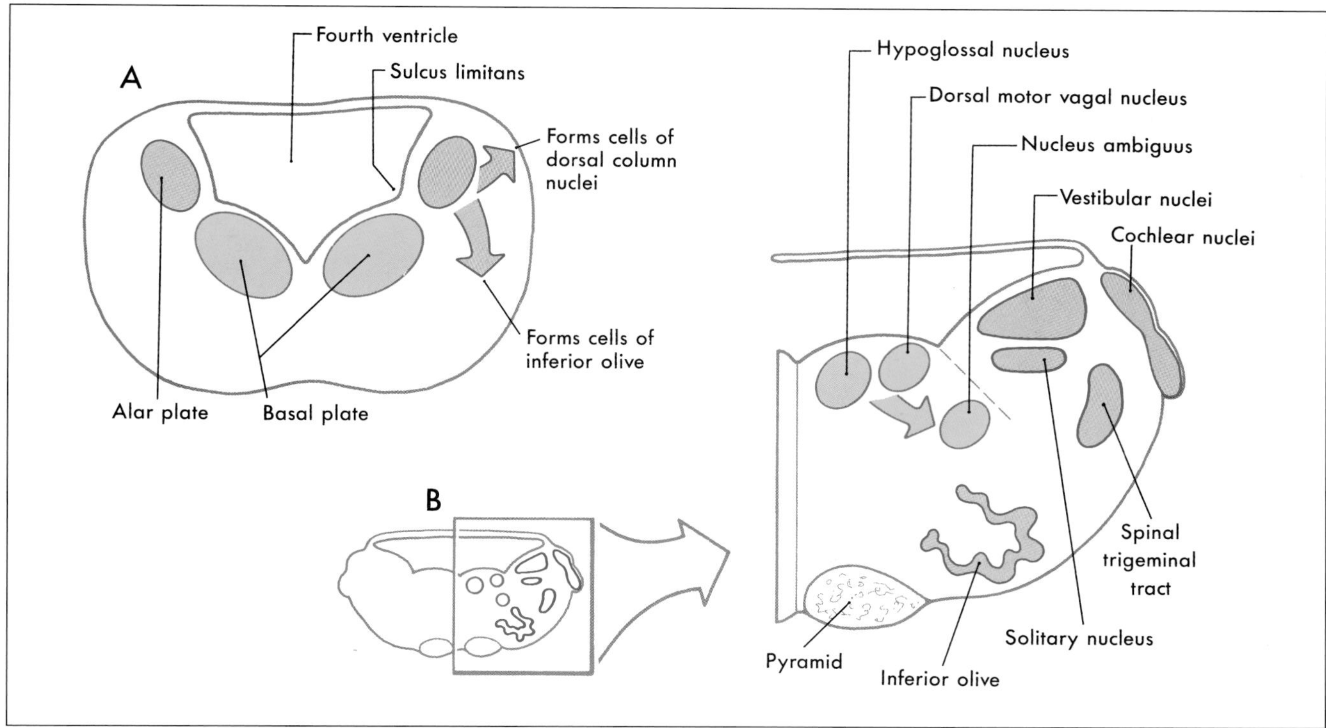

FIGURE 11-2 *Development of the medulla at early (A) and later (B) stages showing the relationships of alar and basal plates and their adult derivatives in the medulla.*

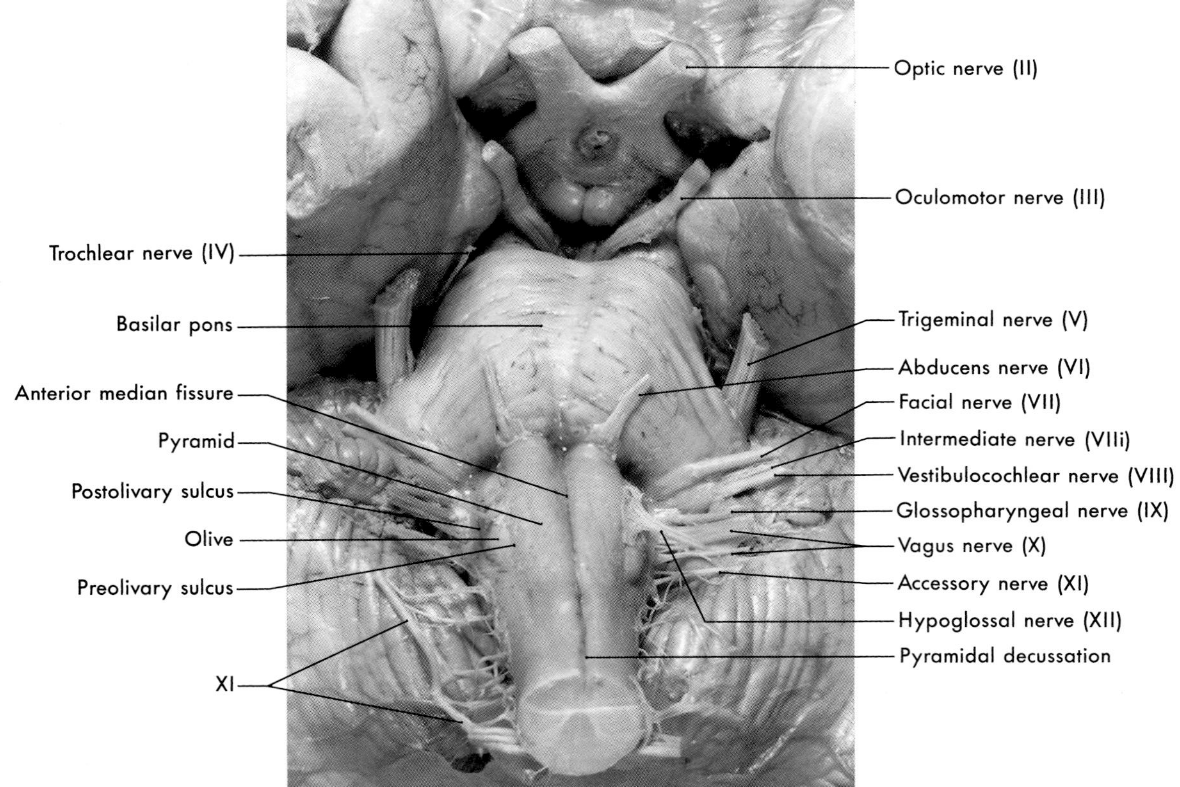

FIGURE 11-3 *Ventral view of the brainstem with emphasis on structures of the medulla.*

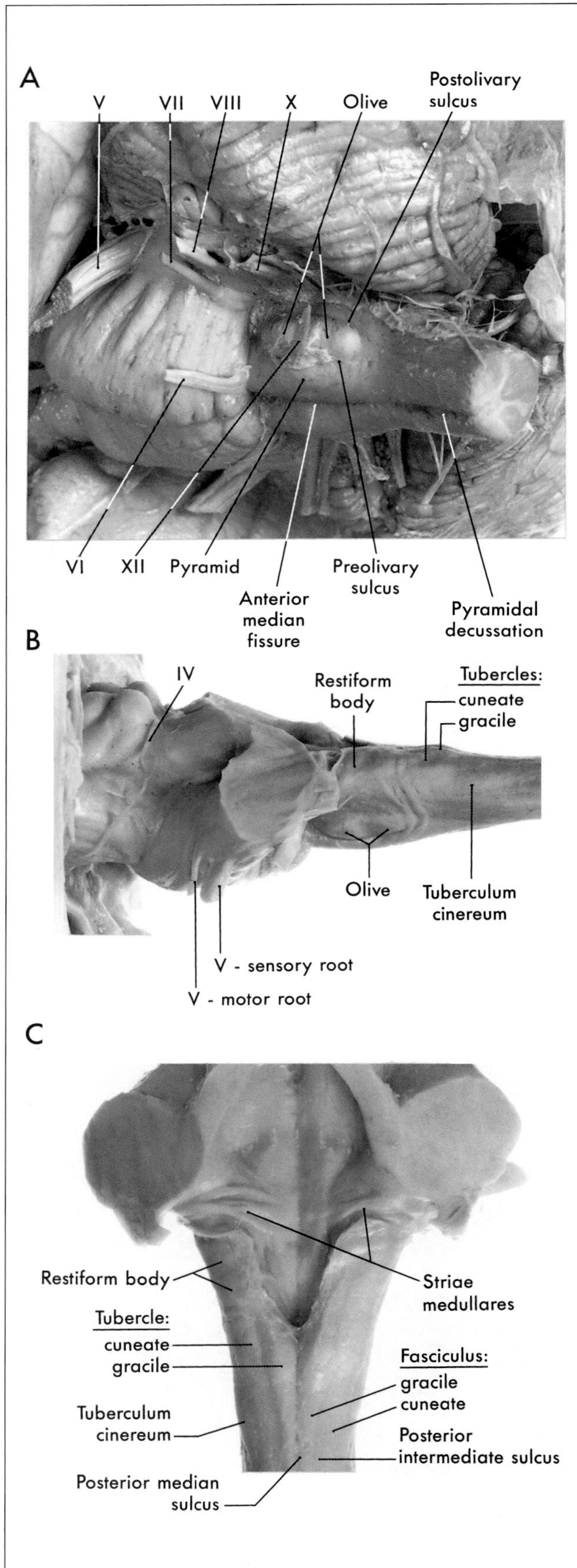

FIGURE 11-4 *Ventrolateral (A), lateral (B), and dorsal (C) views of the medulla. Cranial nerves are indicated by Roman numerals, and the cerebellum has been removed from B and C. The same specimen is used in B and C.*

Dorsal Medulla At and caudal to the level of the obex, the dorsal medulla is characterized by the *gracile* and *cuneate fasciculi* and their respective *tubercles* (Fig. 11-4C). These tubercles are formed by the underlying gracile and cuneate *nuclei*. Rostrolateral to the *gracile* and *cuneate* tubercles and forming a prominent elevation on the dorsolateral aspect of the medulla is the *restiform body*. This structure contains a variety of afferent cerebellar fibers and becomes progressively larger as it extends toward the pons-medulla junction. In the caudal pons, fibers of the restiform body join with a much smaller bundle, the *juxtarestiform body*, to form the *inferior cerebellar peduncle*.

Vasculature The blood supply to the entire medulla and to the choroid plexus of the fourth ventricle arises from branches of the *vertebral arteries* (see Fig. 11-16). In general, the medial medulla is served by the *anterior spinal artery*, the ventrolateral medulla by small branches from the *vertebral*, and the dorsolateral medulla rostral to the obex by the *posterior inferior cerebellar artery*. Caudal to the obex, the dorsal medulla is served by the *posterior spinal artery*. The internal distribution of these vessels is discussed below.

INTERNAL ANATOMY OF THE MEDULLA

Summary of Ascending Pathways The ascending tracts that originate from the spinal cord gray matter (*anterolateral system*, ALS; *dorsal* and *ventral spinocerebellar tracts*; etc.) and from dorsal root ganglion cells (*gracile* and *cuneate fasciculi*) continue into the medulla (Fig. 11-5). Some ALS fibers terminate in the medulla, as *spinoreticular fibers*, and others convey pain and temperature input to more rostral levels, including the thalamus. Dorsal column fibers synapse in the medulla, but the tactile and vibratory information carried by these fibers continues rostrally via the *medial lemniscus* (Fig. 11-5). Spinocerebellar axons enter the cerebellum through the restiform body (dorsal tract) or the superior cerebellar peduncle (ventral tract). Other ascending bundles, such as *spino-olivary* and *spinovestibular* fibers, terminate in the medulla.

Summary of Descending Pathways The descending tracts that originate from the cerebral cortex (*corticospinal*, Fig. 11-5) and from the midbrain (*rubrospinal*, *tectobulbospinal*), and pons (*reticulospinal*, *vestibulospinal*) traverse the medulla en route to the spinal cord. The medulla contributes additional fibers to the latter two fiber systems. At this level, the *medial longitudinal fasciculus* contains only

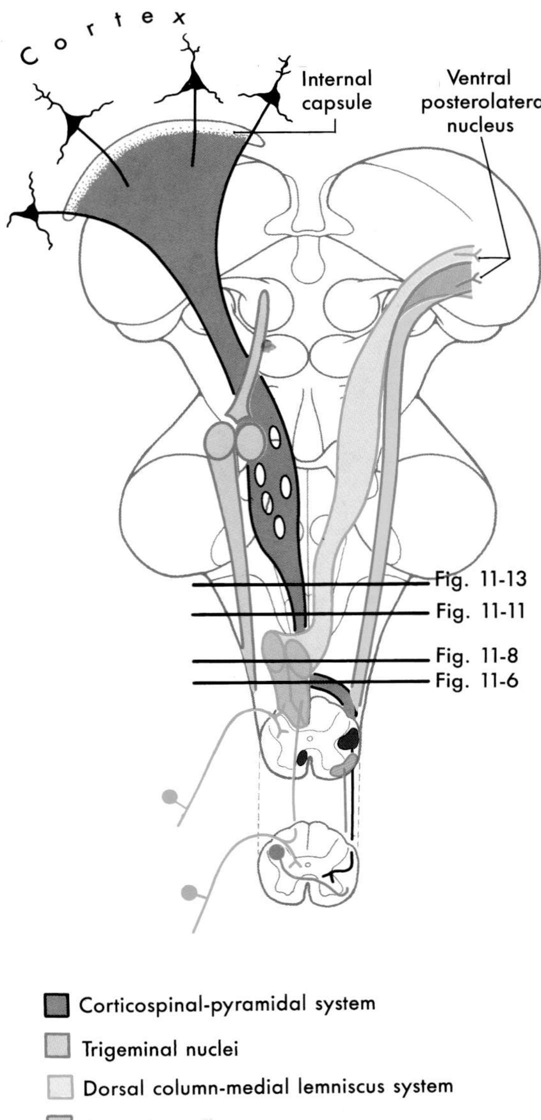

FIGURE 11-5 *Diagram of the brain showing the location and trajectory of three important pathways and the trigeminal nuclei. The color coding for each is continued in Figures 11-6, 11-8, 11-11, and 11-13.*

descending fibers. The majority of these descending axons influence, either directly or indirectly, the discharge patterns of motor neurons in the spinal gray matter.

Spinal Cord–Medulla Transition The spinal cord–medulla transition is characterized by changes that begin at the caudal level of the *pyramidal decussation* (Figs. 11-6, 11-7). The spinal cord gray matter is replaced by the pyramidal decussation; the central gray matter enlarges; the dorsolateral fasciculus and substantia gelatinosa of the spinal cord merge, respectively, into the spinal trigeminal tract and nucleus; and nuclei characteristic of the medulla appear. The caudal medulla is described in the following sections at the levels of the motor and sensory decussations.

Caudal Medulla—Level of the Motor Decussation At the level of the *motor decussation* (*pyramidal decussation*), about 90% of corticospinal fibers cross the ventral midline to form the contralateral lateral corticospinal tract of the cord (Figs. 11-5 to 11-7). Dorsally, at this level the *gracile* and *cuneate nuclei* first appear in their respective fasciculi (Figs. 11-6, 11-7). Because the *gracile* and *cuneate fasciculi* are collectively called the *dorsal columns*, their respective nuclei are frequently referred to as the *dorsal column nuclei*. Laterally, the *spinal trigeminal tract* (visible on the surface of the medulla as the *trigeminal tubercle* or *tuberculum cinereum*) is located on the medullary surface. Internal to the spinal tract is the *spinal trigeminal nucleus, pars caudalis* (Fig. 11-6).

The spinal trigeminal tract is composed of central processes of primary sensory fibers that enter the brain mainly in the trigeminal nerve. These fibers terminate on cells of the spinal trigeminal nucleus, which, in turn, projects to the contralateral thalamus as the *ventral trigeminothalamic tract*.

In the lateral medulla, the ALS and *rubrospinal tract* are found medial to the superficially located *dorsal* and *ventral spinocerebellar tracts* (Figs. 11-6, 11-7). It is appropriate to emphasize that ALS fibers (conveying pain and temperature input from the contralateral side of the body) and spinal trigeminal tract fibers (conveying pain and temperature from the ipsilateral face) are located adjacent to each other throughout the ventrolateral medulla.

The ventral medulla contains the most rostral part of the *accessory nucleus* (cranial nerve XI), remnants of the medial motor cell column of C1, and the *medial longitudinal fasciculus* (MLF) and *tectobulbospinal system* (TBS) (Figs. 11-6, 11-7). At this level, the tectospinal fibers in the TBS are incorporated into the MLF. These small bundles are displaced laterally by the pyramidal decussation compared to their position at more rostral levels.

The *central gray* surrounds the central canal of the medulla and contains the caudal extremes of the hypoglossal (XII) and dorsal motor vagal nuclei (X) (Fig. 11-6). When the ventricle flares open at the level of the obex, these nuclei occupy the medial floor of the ventricular space.

Caudal Medulla—Level of the Sensory Decussation Immediately rostral to the motor decussation are axons that originate from cells of the dorsal column (gracile and cuneate) nuclei and swing ventromedially to cross the

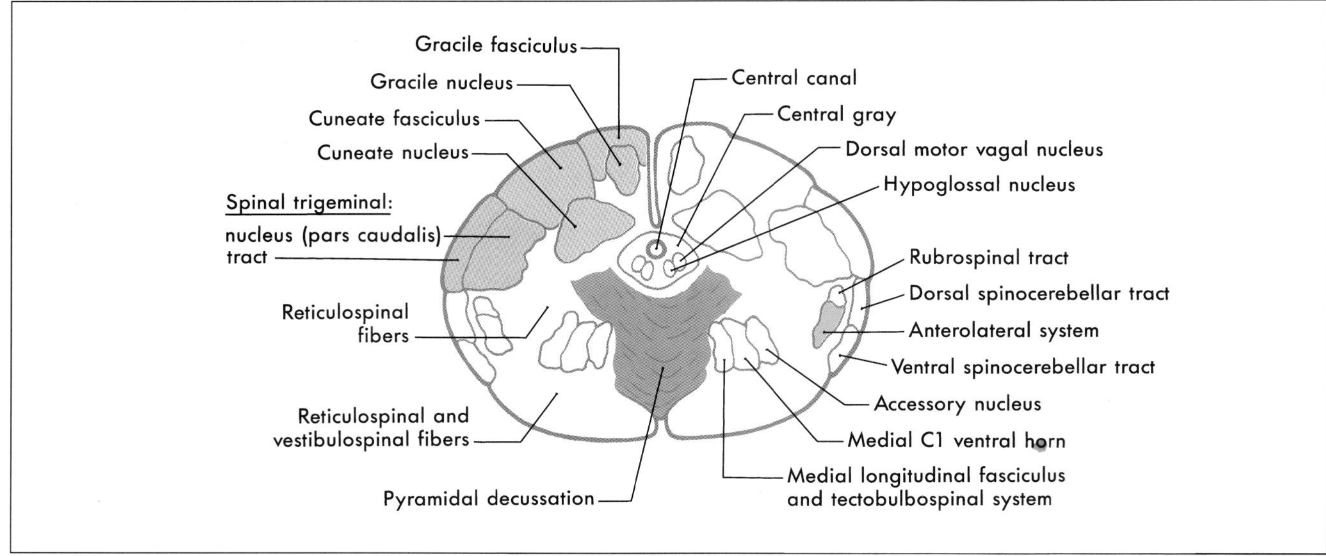

FIGURE 11-6 *Cross section of the medulla at the level of the pyramidal decussation. Correlate with Figure 11-5.*

midline (Fig. 11-5). This constitutes the *sensory decussation*, so named because it is the point at which a major ascending sensory pathway (dorsal column-medial lemniscus) crosses the midline.

At this level the dorsal columns (gracile and cuneate fasciculi) are largely replaced by the *gracile* and *cuneate nuclei* (Figs. 11-8, 11-9). Fibers conveying tactile and vibratory sensations from lower and upper levels of the body terminate, respectively, in the gracile and cuneate nuclei. The axons of these cells, in turn, form the *internal arcuate fibers*, which cross the midline as the sensory decussation and collect as the *medial lemniscus* on the contralateral side (Figs. 11-5, 11-8, 11-9). Information from lower extremities (gracile cell axons) is conveyed in the ventral part of the medial lemniscus, and information from the upper extremities (cuneate cell axons) is conveyed in the dorsal part of the medial lemniscus. The *accessory cuneate nucleus* is located lateral to the cuneate nucleus (Fig. 11-8). Its cells receive primary sensory input via cervical spinal nerves and project to the cerebellum as *cuneocerebellar fibers*. In doing so, they represent the upper extremity equivalent of the dorsal spinocerebellar tract.

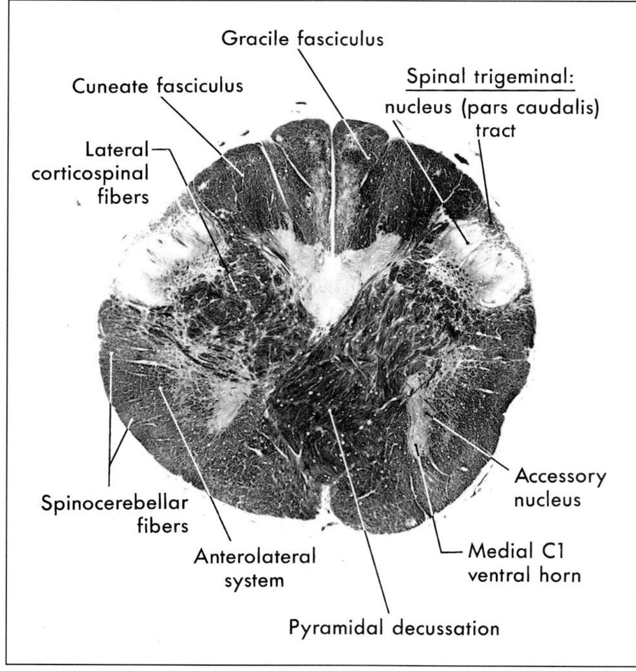

FIGURE 11-7 *A fiber (myelin) stained cross section of the medulla at the level of the pyramidal decussation. Compare with Figure 11-6.*

The *spinal trigeminal tract* and *nucleus* (*pars caudalis*) maintain their position in the lateral medulla. At this level, however, dorsal spinocerebellar fibers have migrated dorsally to cover the spinal tract, heralding the beginnings of the *restiform body* (Figs. 11-8, 11-9). Just medial to the spinal trigeminal nucleus, a small column of motor neurons, the *nucleus ambiguus*, appears (Fig. 11-8). The axons of these special visceral efferent (SVE) cells travel in the glossopharyngeal (IX) and vagus (X) nerves. Fibers of the *ALS* and *rubrospinal tract* are located in the ventrolateral medulla (Fig. 11-8). The *lateral reticular nucleus*, a distinct cell group adjacent to the ALS, receives spinal input and projects to the cerebellum.

Structures characteristic of the ventral medulla at this level include the *pyramid*, fibers of the *hypoglossal nerve*, and the caudal end of the inferior olivary complex (Figs. 11-8, 11-9). The inferior olive, which becomes larger at more rostral levels, receives input from a variety of areas and projects primarily to the cerebellum. Internal to the pyramid, and along the midline from ventral to dorsal, are the *medial lemniscus*, *tectobulbospinal systems*, and

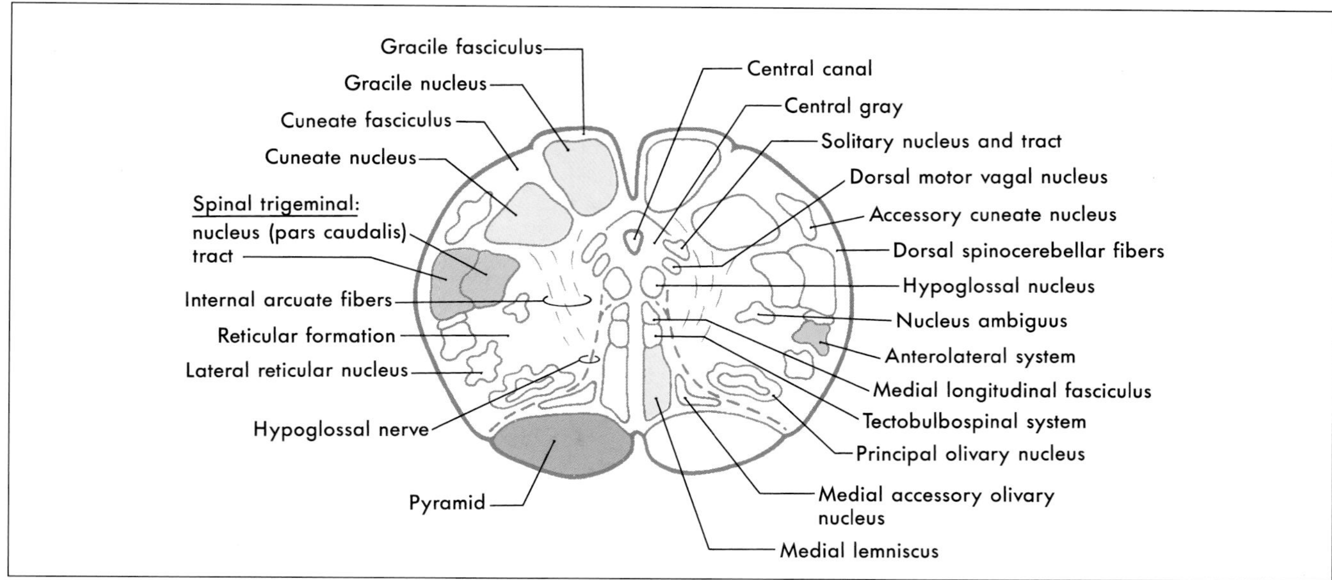

FIGURE 11-8 *Cross section of the medulla at the level of the sensory decussation. Correlate with Figure 11-5.*

medial longitudinal fasciculus (MLF) (Fig. 11-8). At this level, MLF fibers are characteristically found adjacent to the midline and ventral to structures of the central gray.

The *central gray* is larger than at the level of the motor decussation, and caudal parts of the *hypoglossal* and *dorsal motor vagal nuclei* and the *solitary nucleus* and *tract* can be clearly identified along its perimeter (Figs. 11-8, 11-9). Hypoglossal (GSE) motor neurons innervate the ipsilateral half of the tongue. These fibers course ventrally along the lateral edge of the medial lemniscus and pyramid and share a common blood supply with these structures. The GVE cells of the dorsal motor vagal nucleus send their axons to autonomic ganglia associated with viscera in the thorax and abdomen. The *solitary tract* and *nucleus* receive GVA and SVA (taste) input from cranial nerves VII, IX, and X. At this level of the medulla, GVA input to the solitary nucleus comes mainly from thoracic and abdominal viscera (via nerve X) and the carotid sinus (via nerve IX).

The fourth ventricle flares open at the level of the *obex* (Fig. 11-10). The *area postrema* is an emetic (vomiting) center located in the wall of the ventricle at this

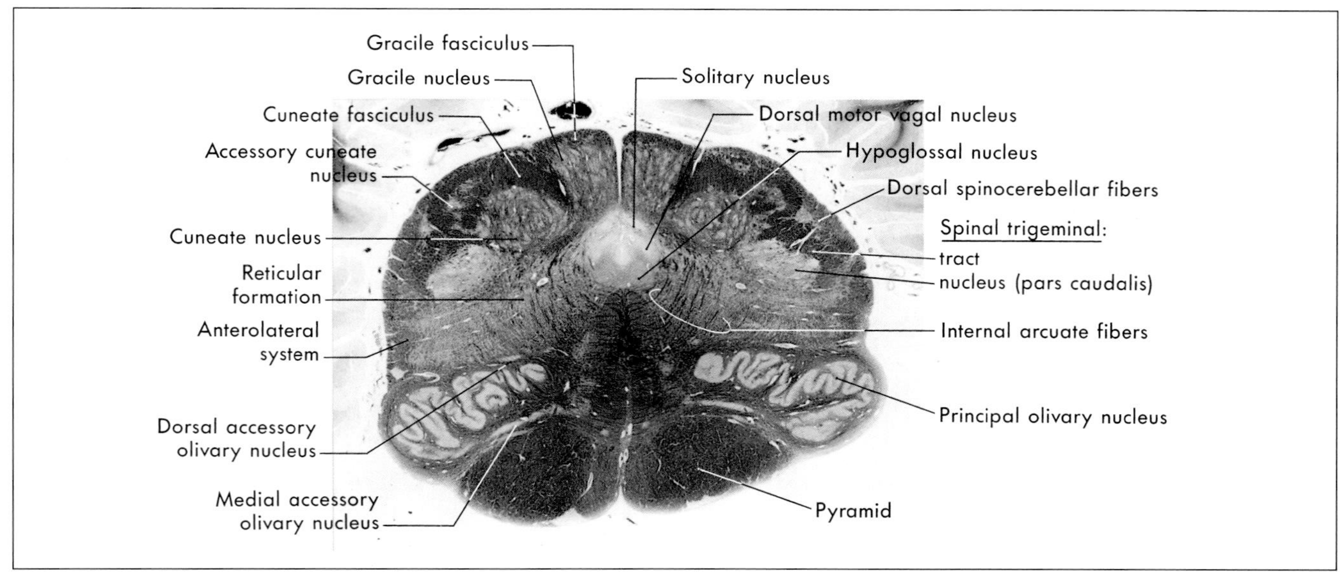

FIGURE 11-9 *A fiber (myelin) stained cross section of the medulla at the level of the sensory decussation. Compare with Figure 11-8.*

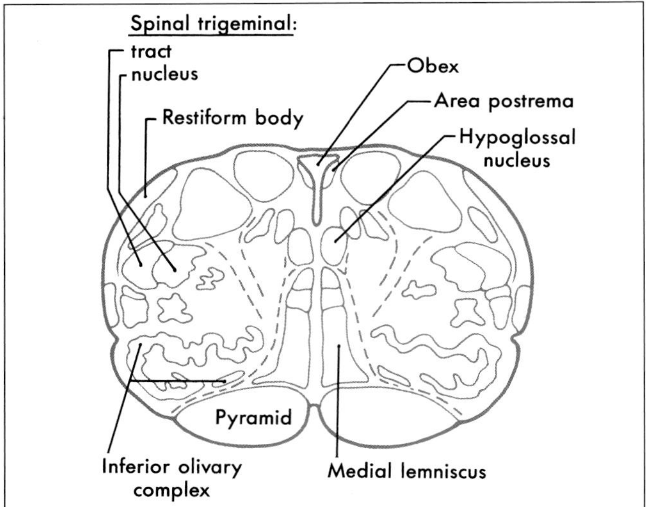

FIGURE 11-10 *Cross section of the medulla at the level of the obex.*

level. Especially noticeable changes at this point compared to more caudal levels include enlargement of the inferior olivary complex and restiform body.

Mid-Medullary Level Rostral to the obex, the structures in the medial floor of the fourth ventricle are the *hypoglossal* and *dorsal motor vagal nuclei* and, lateral to the *sulcus limitans*, the vestibular nuclei (Figs. 11-11, 11-12). The latter cell groups consist at this level of *medial* and *inferior* (or spinal) *vestibular nuclei*. They receive input from nerve VIII and interconnect with areas of the brain concerned with balance and eye movement. The *solitary tract* and *nucleus* occupy their characteristic position ventral to the vestibular nuclei.

Laterally, the *restiform body* forms a prominent elevation on the dorsolateral aspect of the medulla (Figs. 11-11, 11-12). This structure contains *dorsal spinocerebellar*, *cuneocerebellar*, *olivocerebellar*, *reticulocerebellar*, and other cerebellar afferents. In the base of the cerebellum, these fibers join with the *juxtarestiform body* to form the *inferior cerebellar peduncle*.

The *spinal trigeminal tract* and *nucleus* (*pars interpolaris*) are internal to the restiform body (Figs. 11-11, 11-12). Other structures in the lateral medulla are comparable to those seen more caudally. These include the *nucleus ambiguus* and the *lateral reticular nucleus* as well as the *anterolateral system*, *ventral spinocerebellar tract*, and *rubrospinal tract* (Fig. 11-11). At all medullary levels, neurons of the nucleus ambiguus contribute axons to cranial nerves IX and X, which innervate pharyngeal and laryngeal muscles, including those of the vocal folds.

Ventrally, the inferior olivary complex is prominent at mid-medullary levels and is composed of a large, saccular *principal olivary nucleus* and diminutive *medial* and *dorsal accessory olivary nuclei* (Figs. 11-11, 11-12). These cell groups receive input from a variety of central nervous system nuclei and project primarily to the contralateral cerebellum (as *olivocerebellar fibers*) through the restiform body. Ventrally and medially, the orientation of the *pyramid*, *medial lemniscus*, *medial longitudinal fasciculus*, and *tectobulbospinal system* remains essentially the same as at more caudal levels (Figs. 11-11, 11-12).

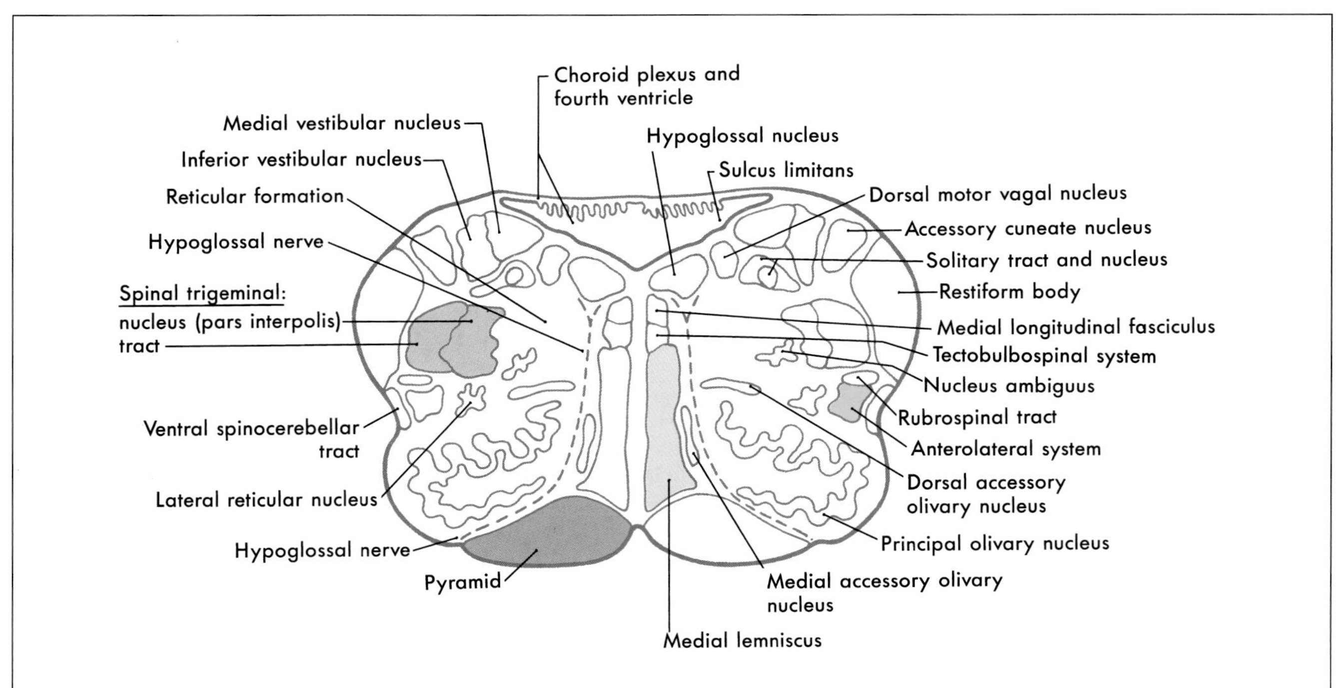

FIGURE 11-11 *Cross section of the medulla at mid-olivary levels. Correlate with Figure 11-5.*

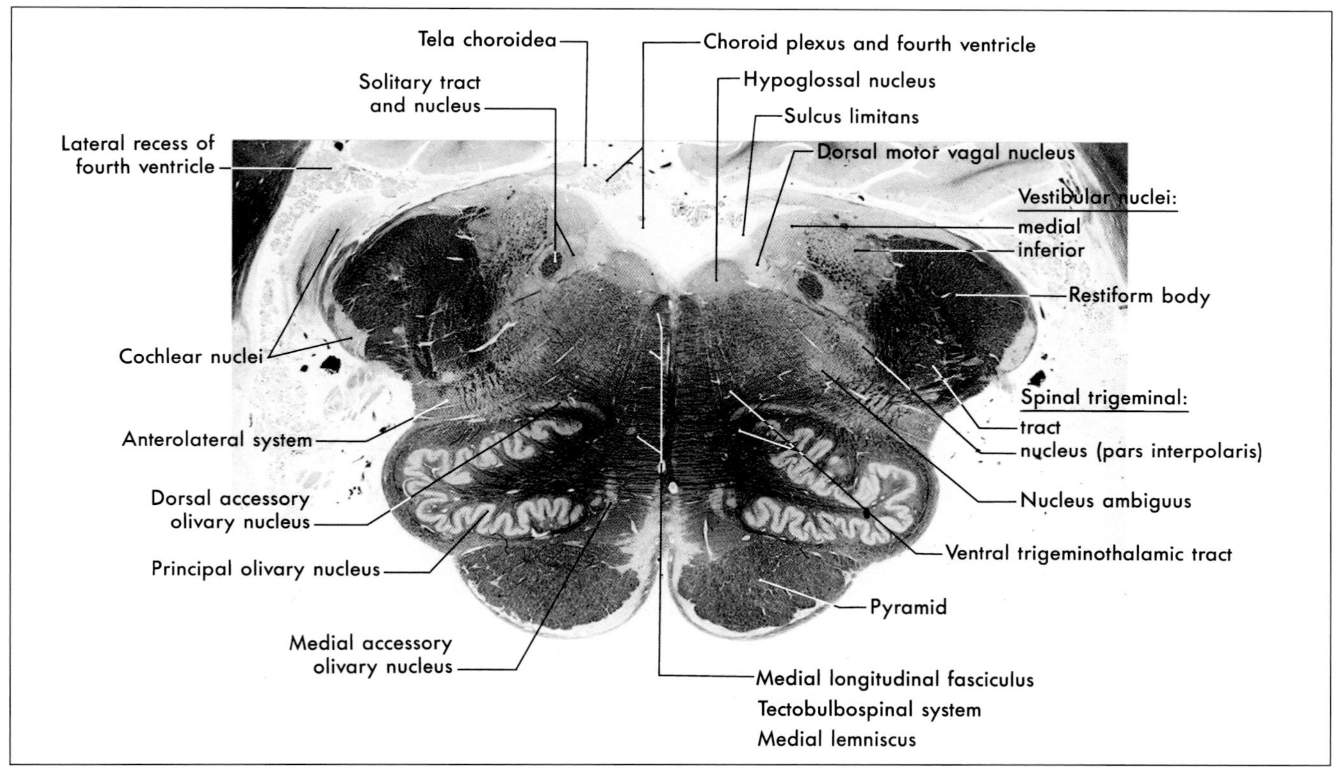

FIGURE 11-12 *A fiber (myelin) stained cross section of the medulla at mid-olivary levels. This section is between the levels represented in Figures 11-11 and 11-13.*

Rostral Medulla and Pons-Medulla Junction Comparison of Figures 11-11 and 11-13 shows that many of the structures seen in the mid-medulla are present in essentially the same locations in the rostral medulla. Therefore, we will emphasize the features that are different in the rostral medulla.

In the floor of the fourth ventricle, the positions occupied by the hypoglossal and dorsal motor vagal nuclei at more caudal levels are taken by the *prepositus hypoglossal nucleus* and the *inferior salivatory nucleus* (Fig. 11-13). The prepositus nucleus is a small, somewhat flattened cell group that is easily distinguished from the hypoglossal nucleus. The GVE cells of the inferior salivatory nucleus are located ventral to the medial vestibular nucleus and medial to the solitary tract and nucleus. Axons of these cells distribute to the otic ganglion via peripheral branches of the glossopharyngeal nerve.

The *medial* and *inferior* (or spinal) *vestibular nuclei* are prominent at this level and are joined, in this plane of section, by the *dorsal* and *ventral cochlear nuclei* (Fig. 11-13; see also Fig. 11-12). The latter nuclei are located on the dorsal and lateral aspects of the restiform body at the pons-medullary junction. The medial vestibular nucleus appears homogeneous in fiber-stained sections, and the inferior vestibular nucleus has a salt-and-pepper appearance (Fig. 11-12). This appearance results because small descending bundles of myelinated fibers (pepper) are intermingled with cells (salt) in the inferior nucleus. Medial to the restiform body is the spinal trigeminal tract and the *pars oralis* of the *spinal trigeminal nucleus* (Fig. 11-13; see also Fig. 10-3).

Although structures in ventral and medial areas of the medulla are unchanged from mid-medullary levels, some changes take place at the pons-medullary junction that merit comment (Fig. 11-14). Fibers of the *restiform body* arch dorsally to enter the cerebellum. The *facial motor nucleus* (SVE cells) appears ventrolaterally, and the *trapezoid body* and *superior olivary nucleus* (both conveying auditory information) appear adjacent to the facial nucleus and the spinal trigeminal tract and nucleus (Fig. 11-14). The inferior olivary complex disappears, and the *central tegmental tract*, one source of input to the inferior olive, appears about where the latter cell group was located. Finally, the *medial lemniscus* begins to shift ventrolaterally and to rotate from the dorsoventral orientation, which it exhibits in the medulla, toward a horizontal orientation more characteristic of the pons (Fig. 11-14). At the pons-medulla junction, the cross section of the medial lemniscus is oriented obliquely (dorsomedial to ventrolateral); by the level of the mid-pons, it is horizontal.

Reticular and Raphe Nuclei The word *reticulum* comes from the Latin for "little net" (diminutive of *rete*, net) and denotes meshlike structures. The *reticular nuclei* of the brainstem are diffuse and ill-defined and have little apparent internal organization. Collectively, they make up the *reticular formation*, which may be thought of simply as including all the cells that are interspersed among the more compact and named structures of the brainstem.

Raphe is a Greek word for "suture" or "seam." Thus, the *raphe nuclei* are bilaterally symmetric cell groups in the brainstem that are located directly adjacent to the midline.

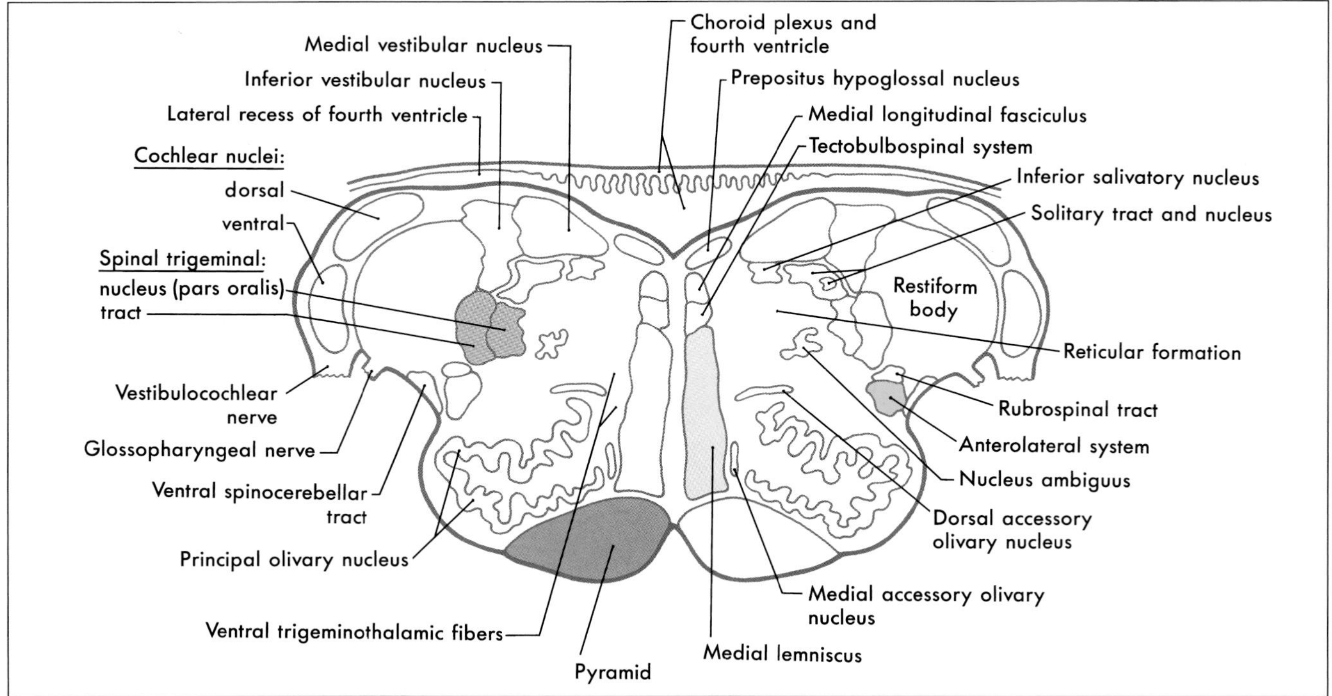

FIGURE 11-13 *Cross section of the medulla at rostral olivary (and medullary) levels. Correlate with Figure 11-5.*

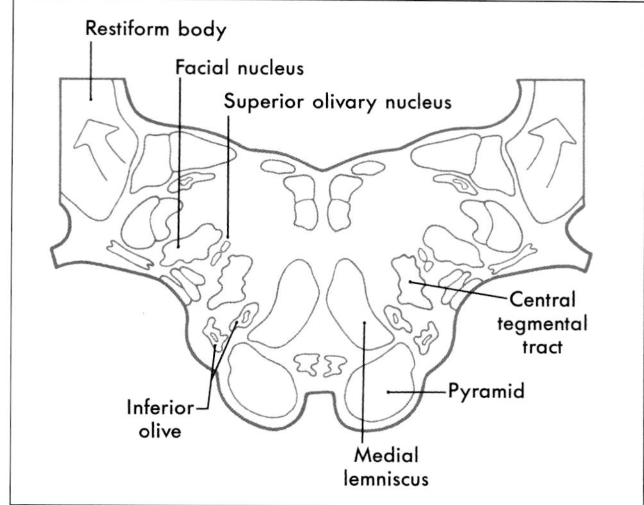

FIGURE 11-14 *Cross section of the medulla at the pontomedullary junction. Fibers of the restiform body sweep up (arrows) into the cerebellum at this level.*

The *medial medullary reticular area* consists of the *central nucleus of the medulla* at caudal medullary levels and the *gigantocellular reticular nucleus* rostrally; the latter cell group extends into the pons (Fig. 11-15). The *lateral medullary reticular area* contains a compact column of cells, the *lateral reticular nucleus*, and a diffuse population of cells that forms the *parvocellular nucleus* and the *ventrolateral reticular area* (area reticularis superficialis ventrolateralis) (Fig. 11-15). The latter cells function in the control of heart rate and respiration. Consequently, a sudden onset of *central apnea*, indicating damage to these respiratory areas, is often a prime early sign of medullary compression. For example, apnea may occur when an increase in intracranial pressure forces the tonsils of the cerebellum downward into the foramen magnum (*tonsillar herniation*) so that they press against the medulla.

The raphe nuclei of the medulla are the *nuclei raphe pallidus* and *obscurus* and, at rostral levels, the *nucleus raphe magnus* (Fig. 11-15). The nuclei pallidus and obscurus are located at mid to rostral medullary levels along the dorsal and ventral midline, respectively. The nucleus raphe magnus begins in the rostral medulla and extends into the caudal pons (Fig. 11-15). Cells of these raphe nuclei receive input from several areas, including the central gray of the mesencephalon, and project to the spinal cord. *Raphespinal* fibers from raphe magnus are especially important for the inhibition of pain transmission in the dorsal horn of the spinal cord.

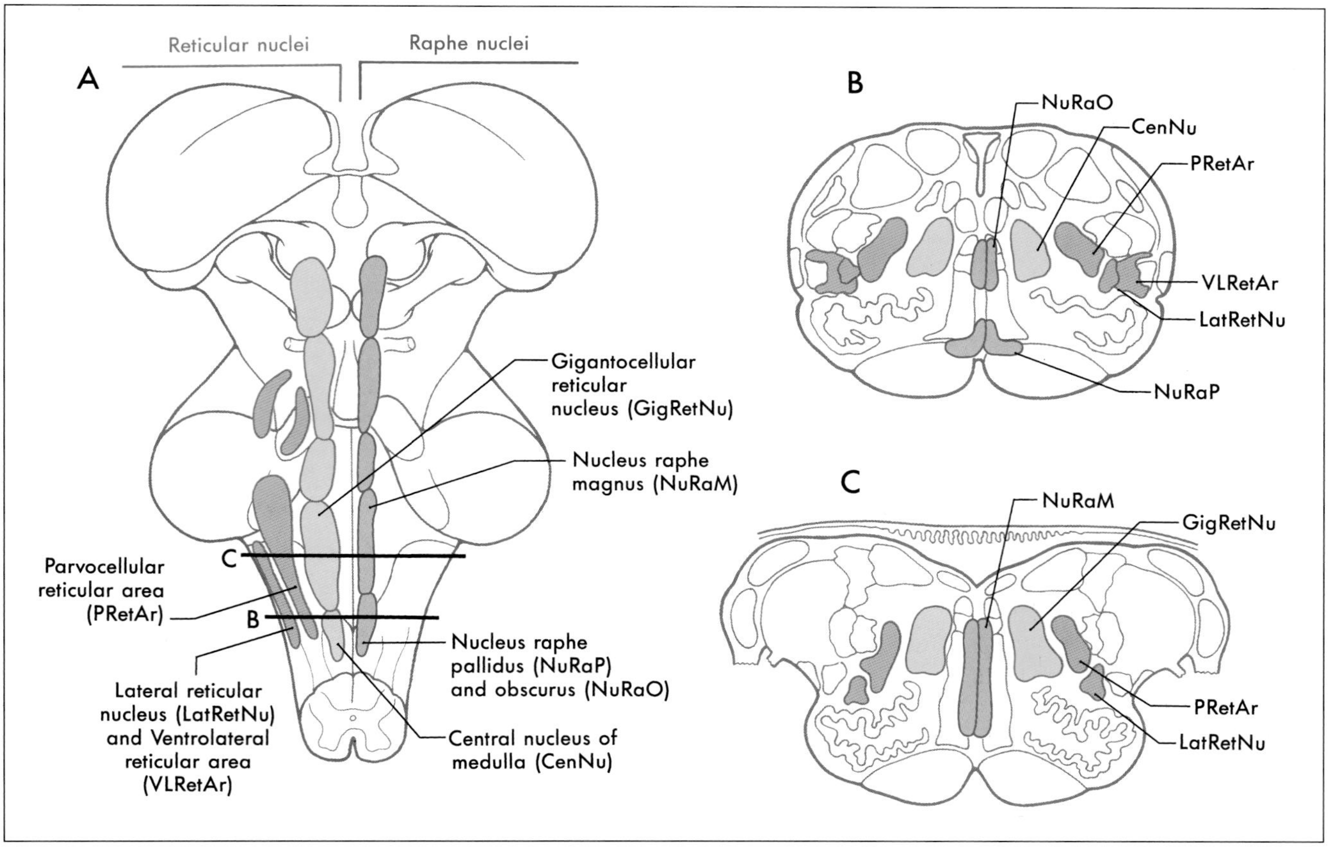

FIGURE 11-15 *Dorsal (A) view of the brainstem and caudal (B) and rostral (C) cross sections showing the raphe and reticular nuclei of the medulla.*

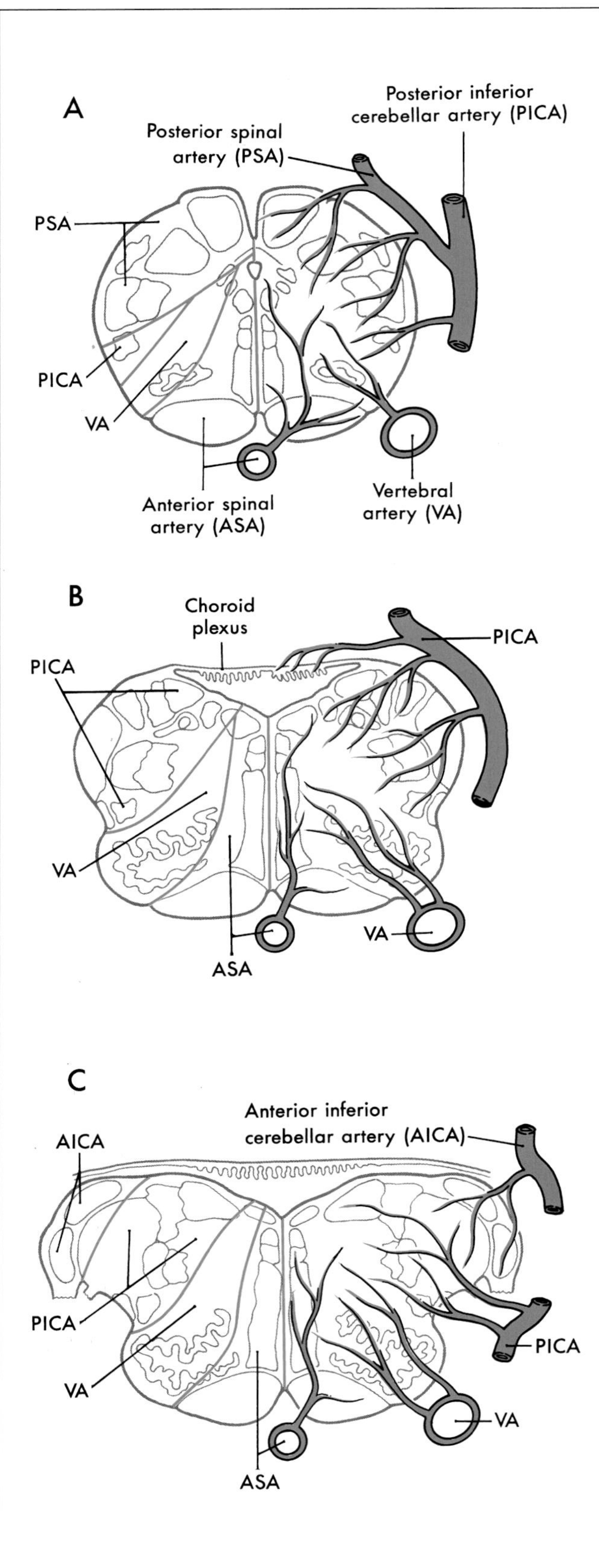

Vasculature The blood supply to the medulla is via branches of the *vertebral arteries* (Fig. 11-16). These are the *anterior spinal artery* and the *posterior inferior cerebellar artery* (PICA). The *posterior spinal artery* is usually a branch of the PICA.

Medial structures of the medulla at all levels, including the pyramid, medial lemniscus, and hypoglossal nucleus and roots, are served by penetrating branches of the *anterior spinal artery* (Fig. 11-16). Vascular insufficiency of these branches may result in a contralateral hemiparesis (pyramidal/corticospinal damage), contralateral loss of proprioception and vibratory sense (medial lemniscus), and a deviation of the tongue to the ipsilateral side when protruded (hypoglossal root or nucleus injury).

The dorsal medulla caudal to the obex is served by branches of the posterior spinal artery (Fig. 11-16A). Major structures in this area include the dorsal column (gracile and cuneate) nuclei and the spinal trigeminal tract and nucleus. Although vascular lesions of the posterior spinal artery are rare, they produce an ipsilateral loss of proprioception and vibratory sense on the body (damage to dorsal columns and nuclei) coupled with an ipsilateral loss of pain and temperature sensation from the face (spinal trigeminal tract).

Rostral to the obex, the entire dorsolateral medulla is served by branches of the *posterior inferior cerebellar artery* (PICA) (Figs. 11-16B,C; 11-17). Included in the territory served by this vessel are the anterolateral system, spinal trigeminal tract and nucleus, vestibular nuclei, solitary tract and nucleus, and the nucleus ambiguus. Vascular insufficiency of the PICA gives rise to a characteristic set of sensory and motor deficits commonly called the *lateral medullary syndrome* or *Wallenberg's syndrome* (Fig. 11-17). The deficits seen and the corresponding structures involved are (1) contralateral loss of pain and temperature sensation from the body (anterolateral system), (2) ipsilateral loss of pain and temperature sensation from the face (spinal trigeminal tract and nucleus), (3) some vertigo and nystagmus (vestibular nuclei), (4) loss of taste from the ipsilateral half of the tongue (solitary tract and nucleus), and (5) hoarseness and dysphagia (nucleus ambiguus and/or roots of IX and X) (Fig. 11-17C). In addition to this broad expanse of medulla, branches of the PICA also serve the choroid plexus of the fourth ventricle. At the pons-medulla junction, the cochlear nuclei and a small adjacent part of the restiform body are served by branches of the anterior inferior cerebellar artery (Fig. 11-16C).

FIGURE 11-16 *Blood supply of the medulla caudal to the obex (A), at mid-medullary (B), and at rostral medullary (C) levels. Arteries are shown on the right, and the territories served by each is indicated on the left.*

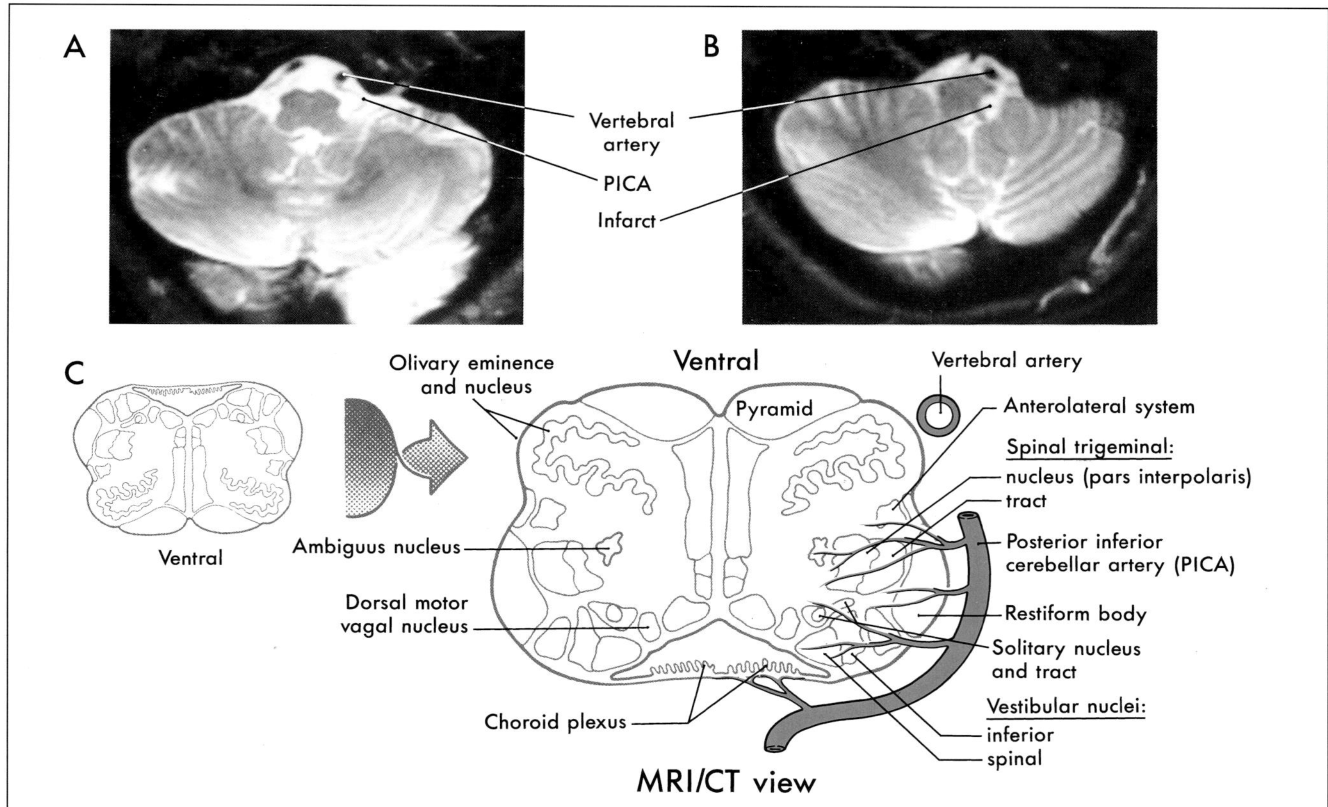

FIGURE 11-17 *Lateral medullary (Wallenberg's) syndrome. A normal magnetic resonance image (A) shows the vertebral and posterior inferior cerebellar (PICA) arteries in relation to the medulla. The patient shown in B had an occlusion of the PICA, which resulted in an infarct of the territory of the medulla served by this vessel. The structures damaged in this lesion are shown in C. Compare with Figure 11-16.*

SOURCES AND ADDITIONAL READINGS

Readings for the brainstem chapters are listed at the end of Chapter 13.

CHAPTER TWELVE

The Pons and Cerebellum

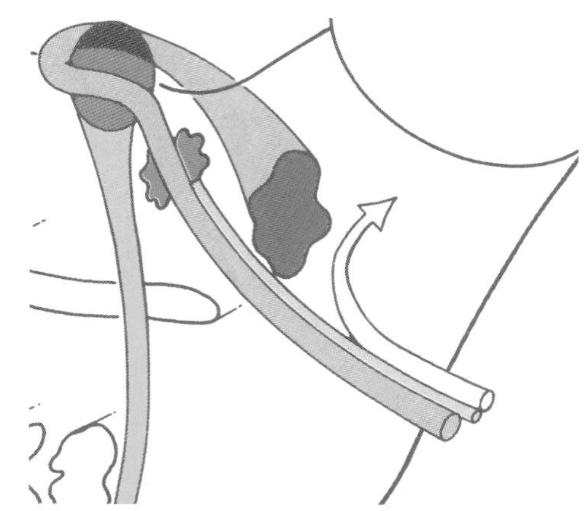

G. A. MIHAILOFF • D. E. HAINES

The *metencephalon* consists of the pons and cerebellum. The pons is the middle segment of the brainstem, the caudal part being the medulla and the rostral portion being the midbrain. The motor and sensory nuclei and the exit points of cranial nerves V to VIII are associated with the pons. The *cerebellum is not part of the brainstem* but rather is considered a suprasegmental structure because it is located dorsal to the brainstem. Functionally, the cerebellum is part of the motor system. The blood supply to the pons and cerebellum arises from branches of the basilar and cerebellar arteries.

DEVELOPMENT

The pons and cerebellum are considered together in this chapter because they arise from the same region of the developing neural tube. The metencephalon is rostral to the myelencephalon and extends from the pontine flexure to the mesencephalic isthmus (Fig. 12-1). At this level, the cavity of the neural tube is enlarged, forming the parts of the fourth ventricle associated with the pons and cerebellum.

Basal and Alar Plates The *basal* and *alar plates* of the brainstem extend from the medulla rostrally into the developing pons. The cranial nerve *motor* nuclei found in the pons (trigeminal, abducens, facial, and superior salivatory) originate from the basal plate and are located medial to the sulcus limitans (Fig. 12-2A,B). The functional components of these motor neurons include *special visceral efferent* (SVE, trigeminal and facial), *general*

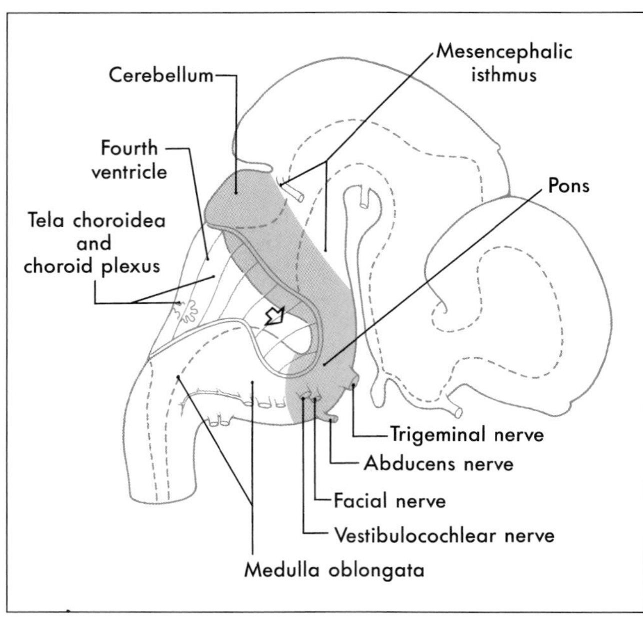

FIGURE 12-1 *Lateral view of the brain at about 7 weeks' gestational age. The pons and cerebellum are highlighted.*

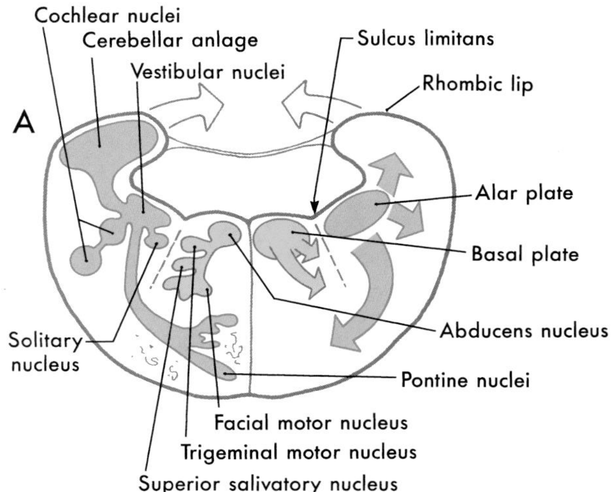

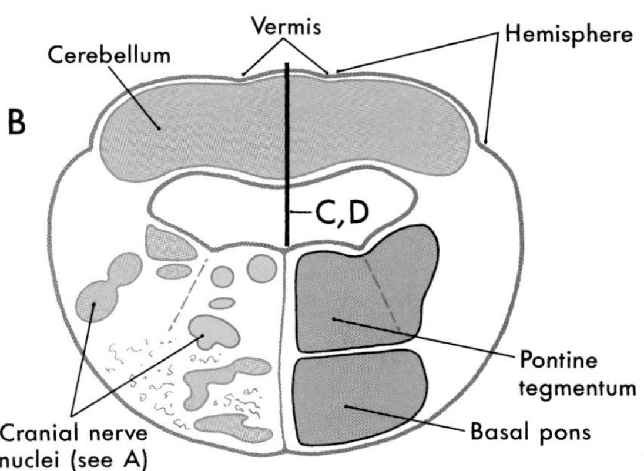

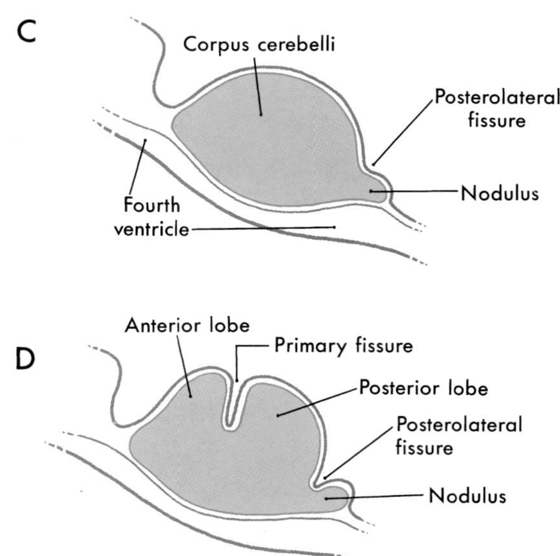

FIGURE 12-2 *Development of the pons and cerebellum. The pontine alar and basal plates give rise to cranial nerve nuclei and the pontine nuclei, and the cerebellum originates from the rhombic lips (A,B). Sagittal views of the cerebellum (C,D; plane of section from B) show the relationships of posterolateral and primary fissures.*

somatic efferent (GSE, abducens), and *general visceral efferent* (GVE, superior salivatory).

The cranial nerve *sensory* nuclei located in the pons include portions of the trigeminal and vestibulocochlear nuclei, and the rostral tip of the solitary nucleus. These nuclei originate from the alar plate and are found lateral to the sulcus limitans (Fig. 12-2A,B). Their functional components are *general somatic afferent* (GSA) for the trigeminal, *special somatic afferent* (SSA) for vestibular and cochlear nuclei, and *special* and *general visceral afferent* (SVA, GVA) for the solitary nucleus. The portion of the dorsal pons that contains these motor and sensory nuclei, as well as the reticular formation and several ascending and descending tracts, is the *pontine tegmentum* (Fig. 12-2B).

The ventral area of the developing pons is invaded by large numbers of fibers. Although some will terminate here, others pass through to more caudal targets. *Alar plate* neuroblasts also migrate into this ventral pontine region to form the *basilar pontine nuclei*. These nuclei, their axons, and the descending fibers passing to and through this area collectively form the *basilar pons* (Fig. 12-2B).

Cerebellum The cerebellum develops from the *rhombic lips* of the pontine alar plates. These lips expand dorsomedially toward each other until they meet at the midline and fuse to form the *cerebellar plate* (Fig. 12-2A,B), which is the rudiment of the cerebellum. As development progresses, the cerebellum is divided by transverse fissures into lobes and lobules. The first of these fissures to appear is the *posterolateral fissure*, which separates the *flocculonodular lobe* caudally from the *corpus cerebelli* rostrally. The *primary fissure*, the second groove to appear, divides the corpus cerebelli into the *anterior* and *posterior lobes* (Fig. 12-2B–D). Internal changes, such as development of the cerebellar cortex and nuclei, take place concurrently with these external events.

EXTERNAL FEATURES

Basilar Pons The portion of the brainstem lying between the midbrain rostrally and the medulla caudally is the pons (*pons* is the Latin word for "bridge"). Ventrally (Fig. 12-3), the pons consists of a massive bundle of transversely oriented fibers that enter the cerebellum as the *middle cerebellar peduncle* (*brachium pontis*). The exit of the trigeminal nerve marks the transition from the basilar pons, which is ventral to the trigeminal root, to the middle cerebellar peduncle, which lies dorsal to the exit of the trigeminal nerve (Figs. 12-3, 12-4A). Rostrally, the large bundles forming the *crus cerebri* of the midbrain extend into the basilar pons. Caudally, some of these descending axons emerge to form the *pyramids* of the medulla (Fig. 12-3A).

The cranial nerves that emerge from the pons are the *trigeminal* (V), *abducens* (VI), *facial* (VII), and *vestibulocochlear* (VIII) nerves. The trigeminal nerve exits laterally and is composed of a large sensory root and a small motor root. The portion of the trigeminal nerve that traverses the subarachnoid space between the pons and the trigeminal ganglion forms a landmark that is visible on magnetic resonance imaging taken at this level (Fig. 12-4B). The

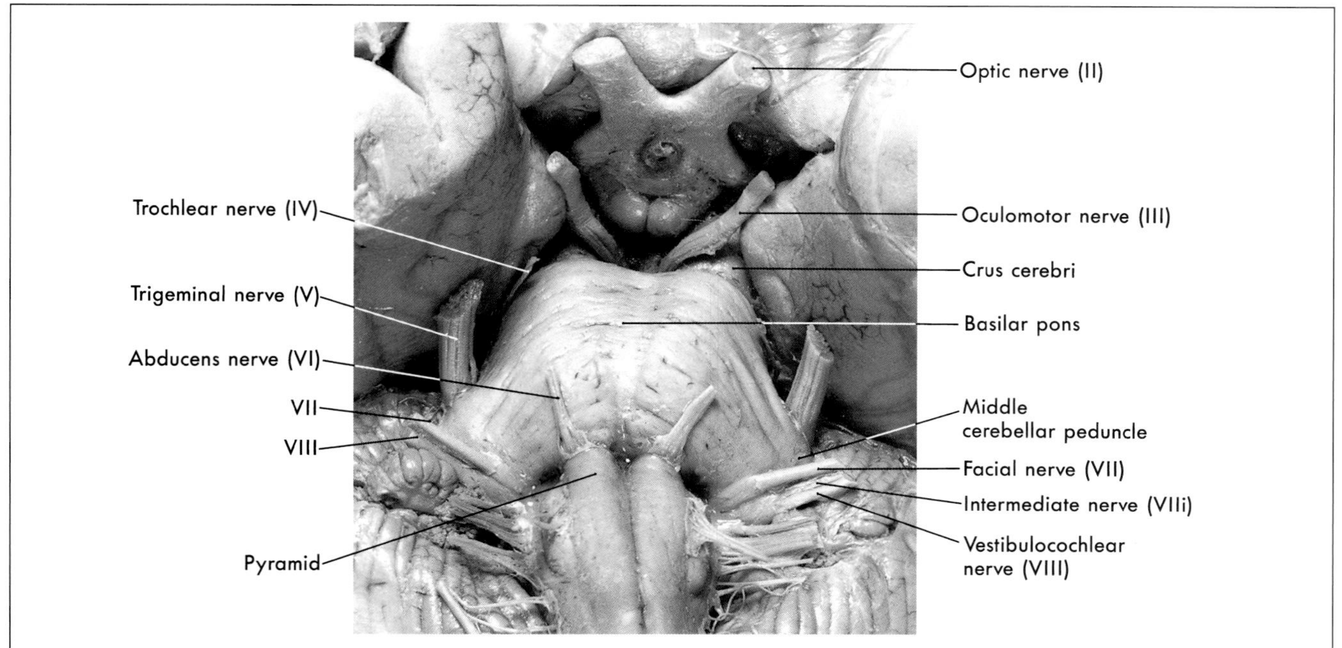

FIGURE 12-3 *Ventral view of the brainstem with emphasis on the pons.*

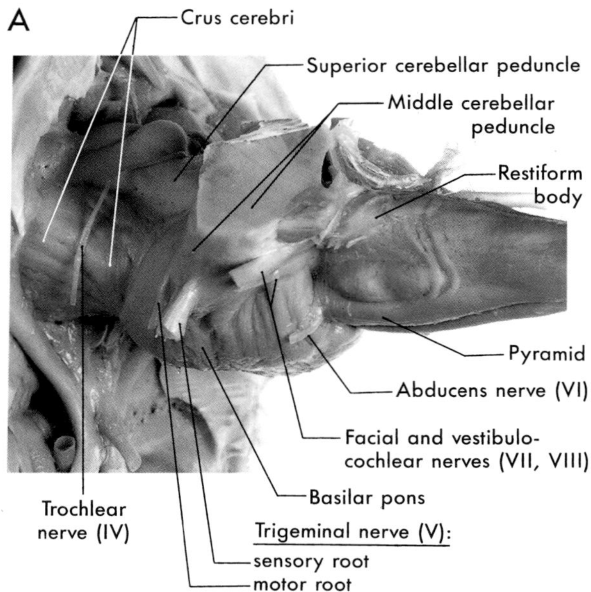

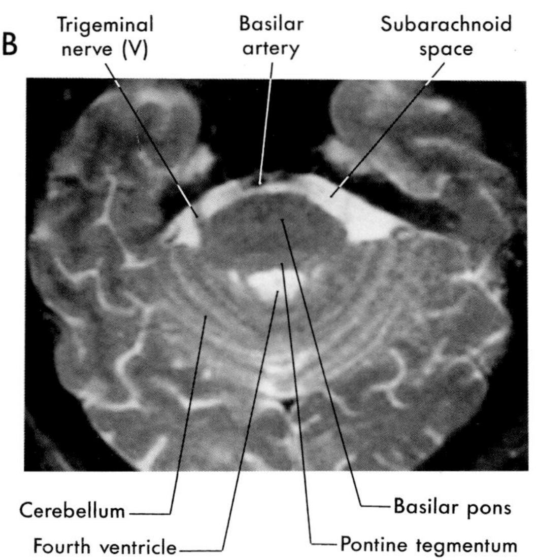

FIGURE 12-4 *A dissection with the cerebellum removed showing the cranial nerves and relationships of the pons (A) and a magnetic resonance image of the pons and root of the trigeminal nerve (B).*

abducens, facial, and vestibulocochlear nerves emerge in medial to lateral sequence along the pons-medulla junction (Fig. 12-3). Although cranial nerve VII is commonly called the facial nerve, it is actually composed of two roots, the *facial nerve* (special visceral efferent fibers, SVE) and the *intermediate nerve* (special visceral afferent, SVA; general visceral efferent, GVE; and general somatic afferent, GSA). The *vestibulocochlear nerve* (special somatic afferent, SSA fibers) emerges dorsolaterally and, with the facial and intermediate nerves and *labyrinthine artery*, occupies the internal acoustic meatus.

Rhomboid Fossa of the Pons The *rhomboid fossa* forms the floor of the fourth ventricle. Its caudal portion is located in the medulla, and its larger, more rostral, area is in the pons. The dorsal surface of the *pontine tegmentum*, which forms the floor of the fourth ventricle, is visible only when the cerebellum is detached from the brainstem (Fig. 12-5). This part of the ventricular floor is characterized by an elevation called the *facial colliculus* located between the median fissure and the *sulcus limitans* and by an area called the *vestibular area* located lateral to the sulcus limitans. The facial colliculus is formed by the underlying abducens nucleus and internal genu of the facial nerve (see below), and the vestibular area marks the location of the vestibular nuclei. The *brachium pontis* and the *brachium conjunctivum* form the lateral walls of the fourth ventricle in the pons; the roof is formed by the anterior medullary velum, by a small part of the cerebellum, and by a portion of the tela choroidea (Fig. 12-6).

Cerebellum The cerebellum is located dorsal to the brainstem and fills much of the posterior fossa. It is attached to the brainstem by three pairs of cerebellar peduncles (*superior*, *middle*, and *inferior*). In sagittal section, the human cerebellum appears wedge-shaped (Fig. 12-6), with its superior surface apposed to the tentorium cerebelli and its inferior surface curving toward the foramen magnum.

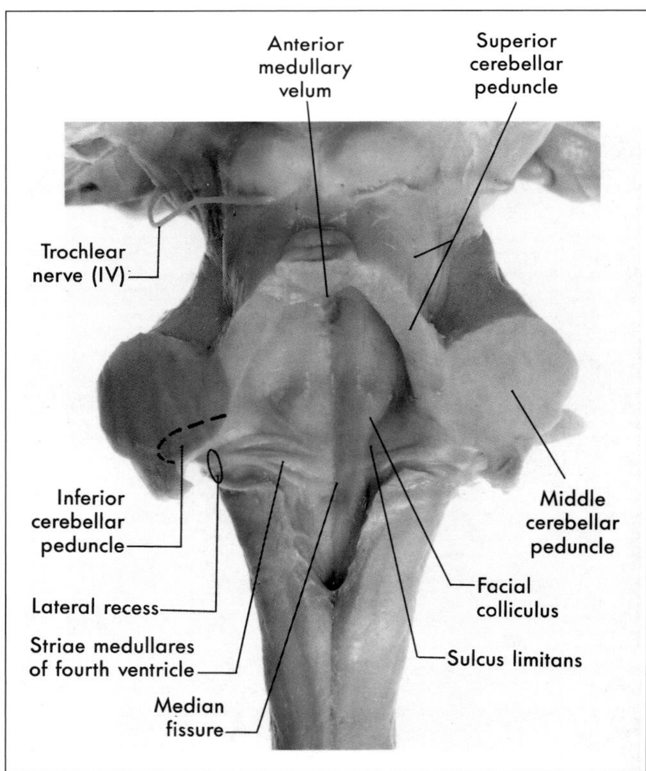

FIGURE 12-5 *Pontine part of the fourth ventricle and rhomboid fossa.*

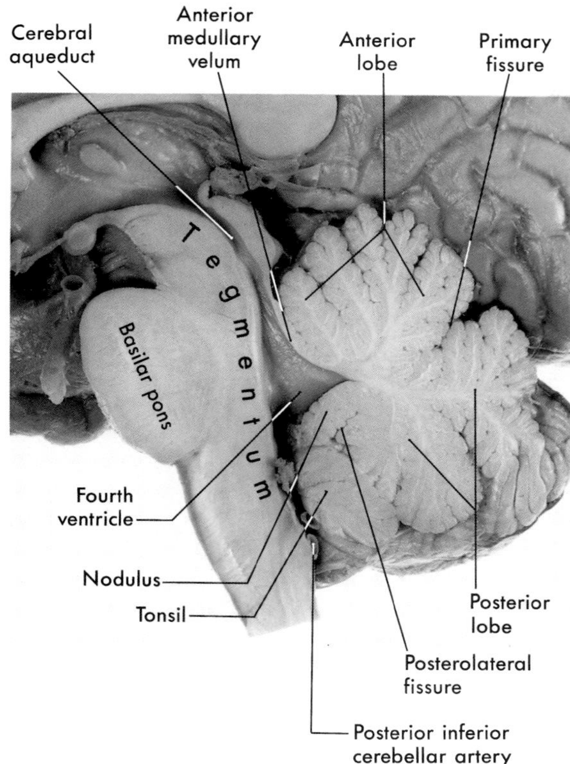

FIGURE 12-6 *Sagittal view of the brainstem (emphasis on pons) and cerebellum.*

The cerebellum consists of *anterior*, *posterior*, and *flocculonodular lobes*; each lobe, in turn, is composed of lobules (Figs. 12-6, 12-7). Lobes and lobules are separated from each other by *fissures*. Lobules are made up of yet smaller folds of cerebellar cortex called *folia* (singular, *folium*). Cerebellar folia, lobules, and lobes can often be followed across the midline from one side of the cerebellum to the other.

Each cerebellar lobe (and lobule) is also divided into rostrocaudally oriented *vermis* (medial), *intermediate* (paravermis), and *hemisphere* (lateral) areas of cortex (Fig. 12-7). The vermal cortex is approximately 1.0 cm at its widest point. The hemisphere is expansive in the human cerebellum and is separated from the vermis by a somewhat ill-defined intermediate cortex.

Four *cerebellar nuclei* are located in the white matter core of each hemisphere. From medial to lateral, they are the *fastigial*, *globose*, *emboliform*, and *dentate* nuclei (Fig. 12-7). These cells receive input from branches of cerebellar afferent fibers and from Purkinje cells located in the cerebellar cortex. In turn, axons of cerebellar nuclear cells provide the main output signals of the cerebellum. The structure, function, and connec-

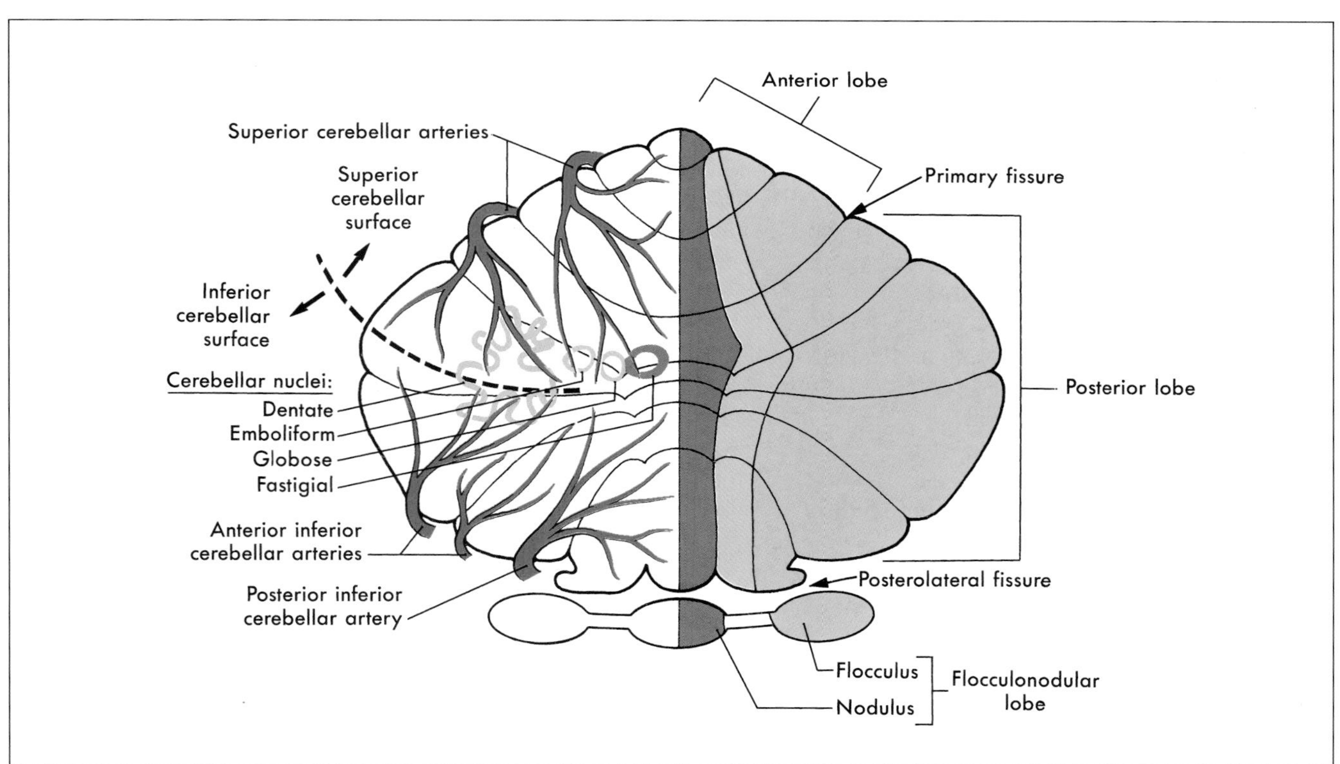

FIGURE 12-7 *Unfolded dorsal view of the cerebellum with lobes and fissures labeled on the right side and the blood supply to the various lobes and their underlying nuclei indicated on the left. The colors of the rostrocaudal regions (zones) of the lobules correlate with the cerebellar nuclei to which they are related; gray, medial (vermis) area; green, intermediate (paravermis) area; blue, lateral (hemisphere) area.*

tions of the cerebellar cortex and nuclei are considered in greater detail in Chapter 26.

Vasculature The *basilar artery* and its branches serve basilar and tegmental areas of the pons. The internal distribution of the basilar artery and its branches is discussed later in this chapter. The *superior cerebellar artery* distributes to the superior surface of the cerebellum and most of the cerebellar nuclei; the inferior surface is served by *anterior* and *posterior inferior cerebellar arteries* (Fig. 12-7).

INTERNAL ANATOMY OF THE PONS

Summary of Ascending Pathways The major ascending pathways seen in the medulla continue into the pons (Fig. 12-8). These include the *medial lemniscus*, *anterolateral system* (ALS), *ventral trigeminothalamic fibers*, and the *ventral spinocerebellar tract*. Although most of these fibers continue through the pons, some ALS fibers terminate in the pontine reticular formation (as *spinoreticular* fibers), and ventral spinocerebellar axons enter the cerebellum. The *restiform body*, a prominent structure in the rostral medulla, sweeps dorsally into the cerebellum in the caudal pons.

Summary of Descending Pathways The most prominent groups of descending fibers arise from cells located in the midbrain or forebrain and therefore traverse the pons (Fig. 12-8). These include the *corticospinal fibers*, the *central tegmental* and *rubrospinal tracts*, and the *tectobulbospinal system*. The *medial longitudinal fasciculus* occupies a characteristic position near the midline in the floor of the fourth ventricle. At the pons-medullary junction, this bundle contains mainly descending fibers; in the rostral pons it is made up of ascending fibers.

Caudal Pontine Level As noted earlier, the pons is divided into a dorsal part, the *tegmentum*, and a ventral region, the *basilar pons*. The following sections describe the anatomy of the pons at three levels: caudal pontine, mid-pontine, and rostral pontine. Each level is described dorsoventrally, beginning with the tegmentum and proceeding to the basilar pons.

At caudal pontine levels, the *facial colliculus* is formed by the underlying *abducens nucleus* and fibers comprising the *internal genu* of the facial nerve (see Figs. 12-10 to 12-12). Axons from the *general somatic efferent* (GSE) cells of the abducens nucleus course ventrally through the tegmentum, pass adjacent to the corticospinal fibers in the basilar pons, and exit the brainstem at the pons-medullary junction as the abducens nerve. The internal genu of VII is composed of the axons of *special visceral efferent* (SVE) cells from the facial nucleus. These axons loop around the abducens nucleus from caudal to rostral, as the internal genu, then course ventrolaterally to exit the brainstem (see Fig. 12-12). Ventrolateral to the abducens nucleus, these SVE fibers are surrounded by cells of the *superior salivatory nucleus*, the axons of which exit the brainstem as the GVE component of the intermediate nerve (see Fig. 12-12).

Medial to the abducens nucleus is the *medial longitudinal fasciculus* and the *tectobulbospinal system* (see Fig. 12-11). As in the medulla, these bundles are internal to the ventricular space and adjacent to the midline.

The dorsolateral tegmentum contains the vestibular nuclei and the solitary tract and nucleus (Fig. 12-9; see Fig. 12-11). The *lateral*, *medial*, and *inferior vestibular*

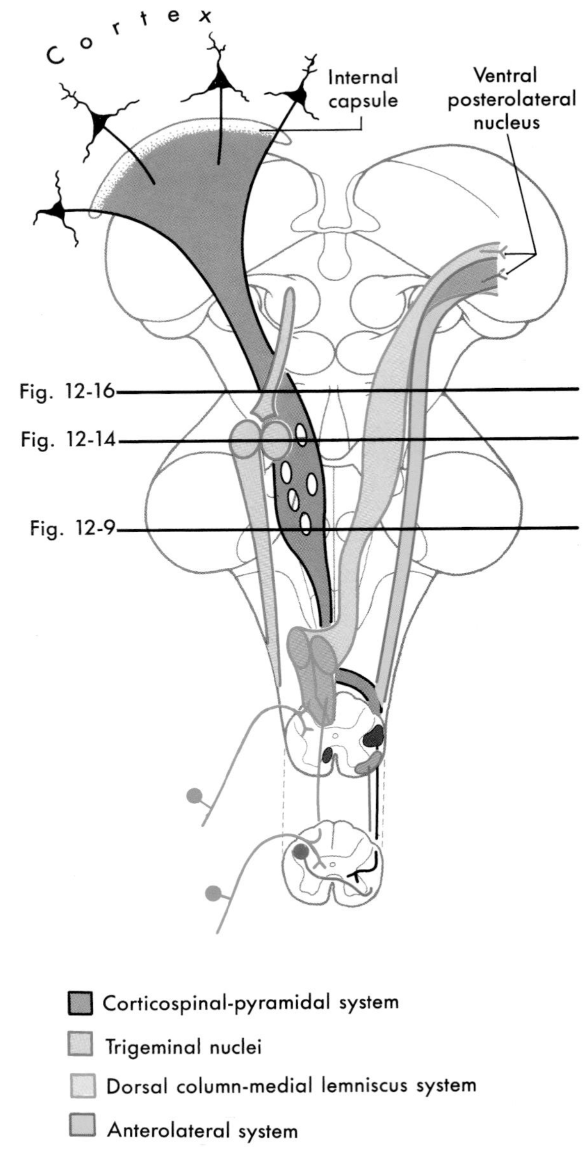

FIGURE 12-8 *Diagrammatic representation of the brain showing the location and trajectory of three important pathways and the trigeminal nuclei. The color coding for each is continued in Figures 12-11, 12-14, and 12-16.*

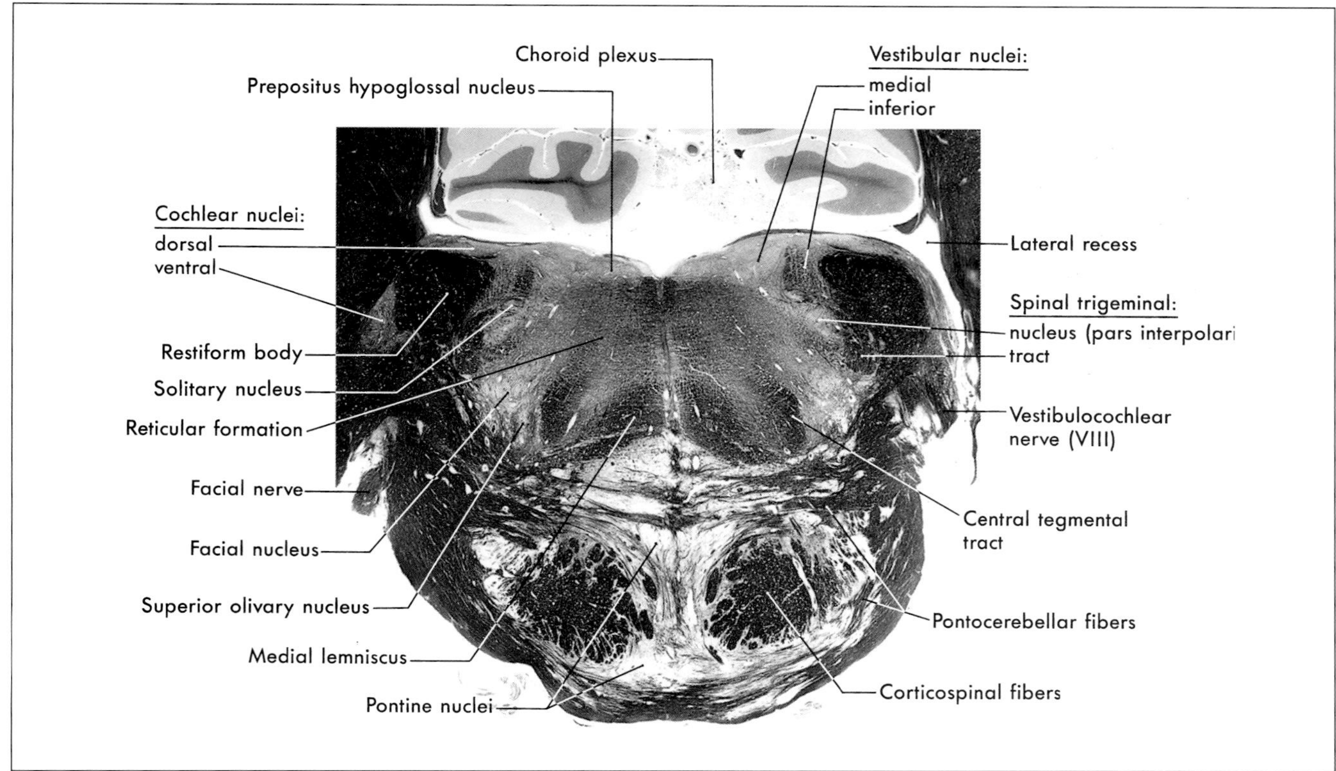

FIGURE 12-9 *A fiber (myelin) stained cross section at the level of the facial motor nucleus (caudal pons).*

nuclei are present at this level, whereas the *superior vestibular nucleus* becomes prominent more rostrally. The small bundles of fibers coursing between the vestibular nuclei and the cerebellum in the wall of the fourth ventricle form the *juxtarestiform body* (Figs. 12-10, 12-11). This structure is composed of vestibulocerebellar and cerebellovestibular fibers and, along with *the laterally adjacent restiform body*, comprises the *inferior cerebellar peduncle*. The rostral portions of the *solitary tract* and *nucleus* are located ventral to the vestibular nuclei and consist of a core of primary sensory fibers (tract) surrounded by cell bodies (nucleus). This part of the solitary complex receives mainly taste input (*special visceral afferent*, SVA fibers) and is sometimes called the *gustatory nucleus*.

The central portion of the pontine tegmentum at caudal levels contains, from medial to lateral, the *central tegmental tract*, the *superior olivary nucleus*, the *facial motor nucleus*, and the *spinal trigeminal tract* and *nucleus* (Figs. 12-9 to 12-12). A major part of the central tegmental tract includes fibers from the red nucleus of the midbrain to the inferior olive of the medulla (rubro-olivary fibers). Cells of the superior olive receive input from the ventral cochlear nucleus and send their axons into the lateral lemniscus on both sides. The route followed by motor facial fibers is shown in Figure 12-12; these axons innervate the ipsilateral muscles of facial expression. The spinal trigeminal tract is composed of general sensory (*general somatic afferent*, GSA) fibers from

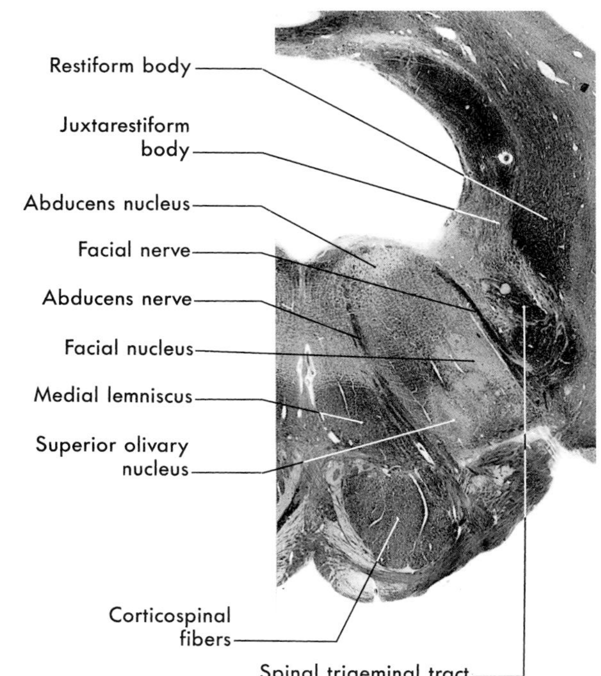

FIGURE 12-10 *A fiber (myelin) stained cross section of the pons at the level of the facial colliculus. Compare with Figure 2-11.*

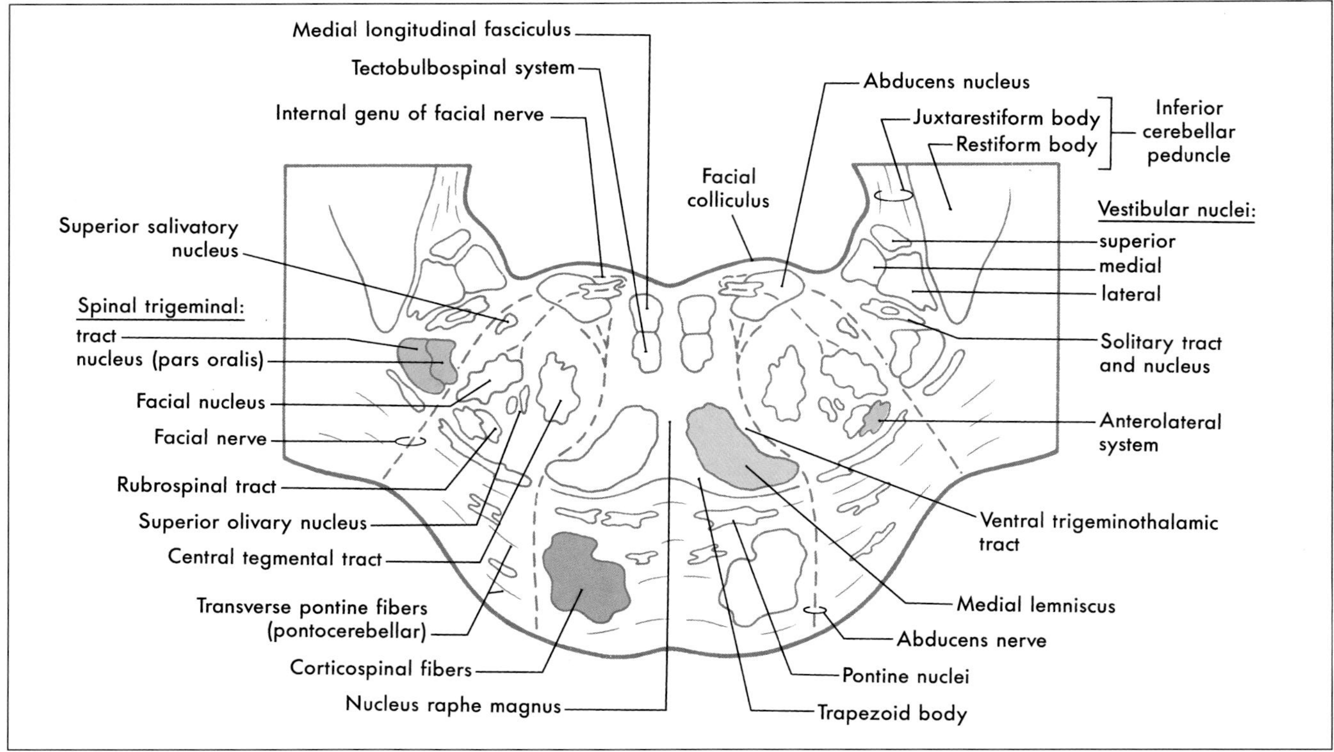

FIGURE 12-11 *Cross section of the caudal pons at the level of the facial colliculus. Correlate with Figure 12-8.*

the ipsilateral half of the face, oral cavity, and much of the scalp. Although most of this input is via the trigeminal nerve (hence the name of the tract), cranial nerves VII, IX, and X also make modest contributions to the spinal trigeminal tract and nucleus. Axons of the spinal trigeminal tract synapse in the spinal trigeminal nucleus, the cells of which project to the contralateral thalamus as *ventral trigeminothalamic fibers*.

The *anterolateral system*, *rubrospinal tract*, and *trapezoid body* are located in the ventrolateral tegmentum (Fig. 12-11). Although pain and temperature signals from the contralateral side of the body are conveyed by ALS fibers, some of these axons end in the pontine reticular formation as *spinoreticular fibers*. The trapezoid body is composed of decussating axons from the cochlear nuclei. After crossing, these fibers form the *lateral lemniscus* and convey auditory signals to the midbrain.

The *medial lemniscus* was oriented vertically in the medulla, but in the caudal pons it begins to shift to a horizontal position (Figs. 12-9, 12-13). At this level, the ventral part of the medial lemniscus (lumbosacral representation) shifts somewhat laterally, and its dorsal portion (brachial representation) assumes a more medial location. The ventral surface of the medial lemniscus lies along the border between the tegmental and basilar pons.

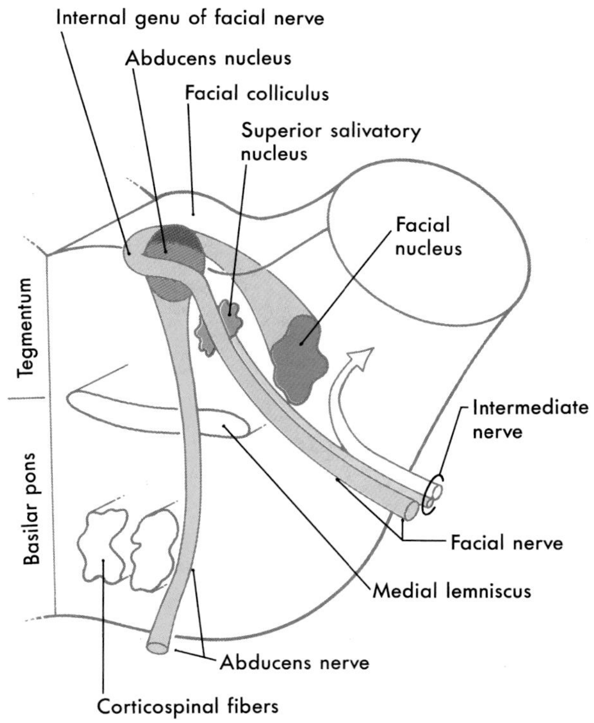

FIGURE 12-12 *Diagrammatic representation of the left side of the pons, as viewed from rostral to caudal, showing the relationships of the abducens and facial nerves. The open arrow indicates the GSA and SVA parts of the intermediate nerve; these fibers course caudally to enter the spinal trigeminal and solitary tract and nuclei, respectively.*

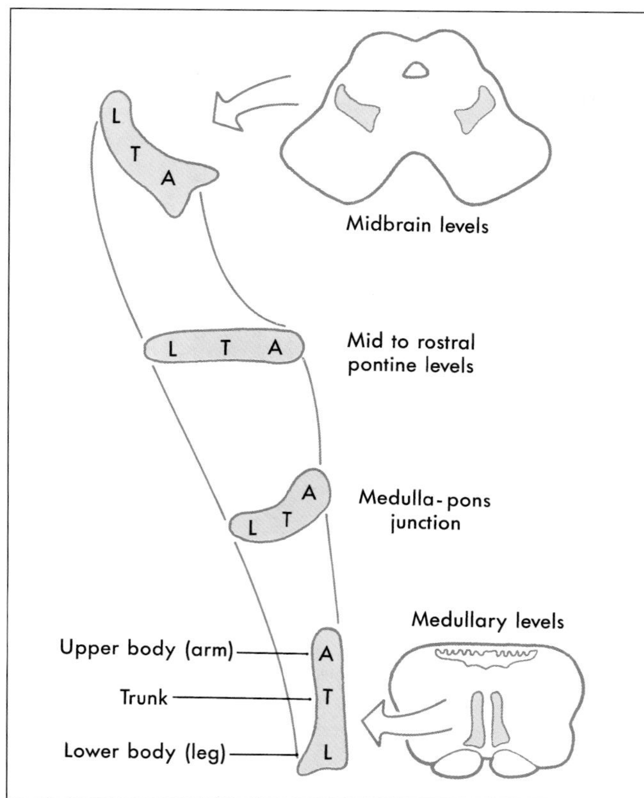

FIGURE 12-13 *The orientation of the medial lemniscus at all brainstem levels.*

The ventral pons contains the *basilar pontine nuclei*, longitudinally running *corticospinal* and *corticopontine fibers*, and transversely oriented *pontocerebellar fibers* (Figs. 12-9, 12-11). On each side of the midline, the corticospinal fibers located in the pyramid of the medulla are now completely surrounded by the *basilar pontine nuclei*. These pontine cells receive input from diverse regions of the neuraxis. In turn, most of their axons cross the midline and enter the cerebellum via the *middle cerebellar peduncle* (brachium pontis) as pontocerebellar fibers.

Mid-Pontine Level Prominent features of the tegmentum at this level are the *principal* (or *chief*) *sensory trigeminal nucleus*, the *trigeminal motor nucleus*, and the *mesencephalic tract* and *nucleus* (Figs. 12-14, 12-15). The principal sensory and motor trigeminal nuclei are located in the lateral tegmentum, and the mesencephalic tract and nucleus extend rostrally in the lateral wall of the central gray. Cells of the principal sensory nucleus receive *general somatic afferent* (GSA) input from the ipsilateral trigeminal nerve and project to the thalamus via *dorsal trigeminothalamic* (uncrossed) and *ventral trigeminothalamic* (crossed) *fibers*. The *special visceral efferent* (SVE) cells of the trigeminal motor nucleus innervate the masticatory muscles on the ipsilateral side. Lastly, the unipolar cell bodies of the mesencephalic nucleus and their processes (the mesencephalic tract) convey proprioceptive input to a variety of nuclei, including the trigeminal motor nucleus.

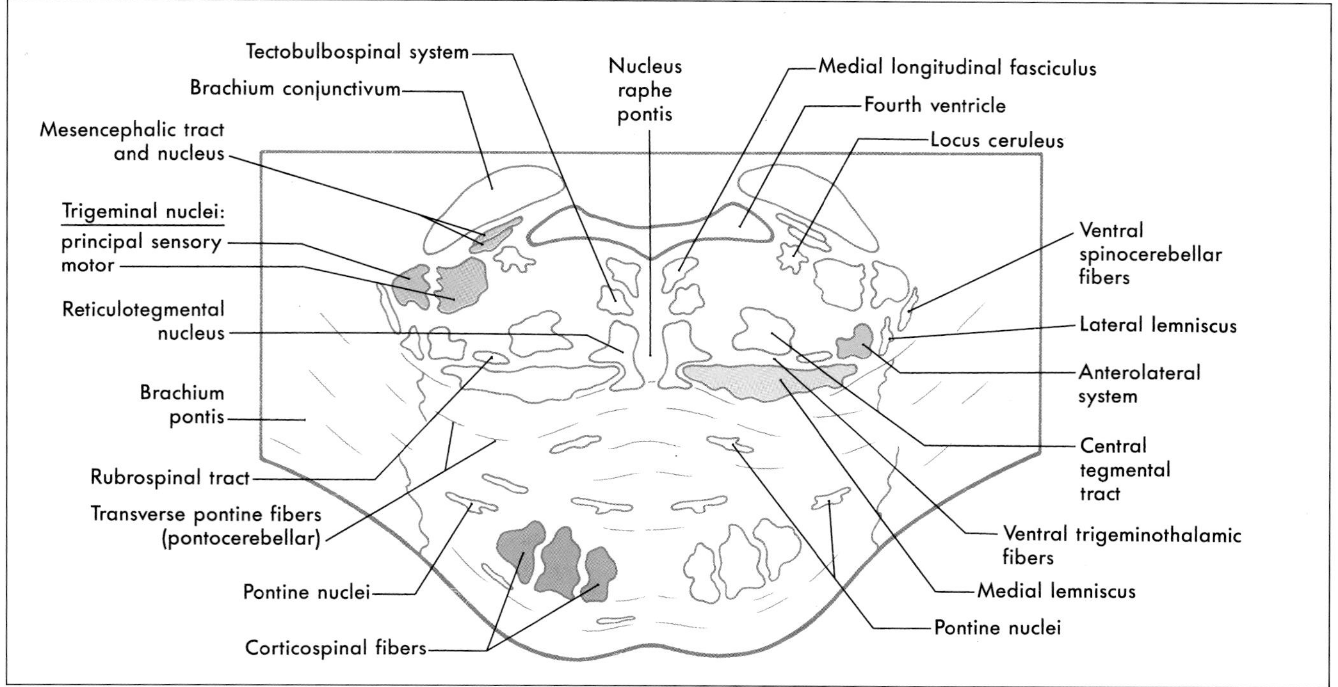

FIGURE 12-14 *Cross section of the mid-pons at the level of the principal sensory and motor trigeminal nuclei. Correlate with Figure 12-8.*

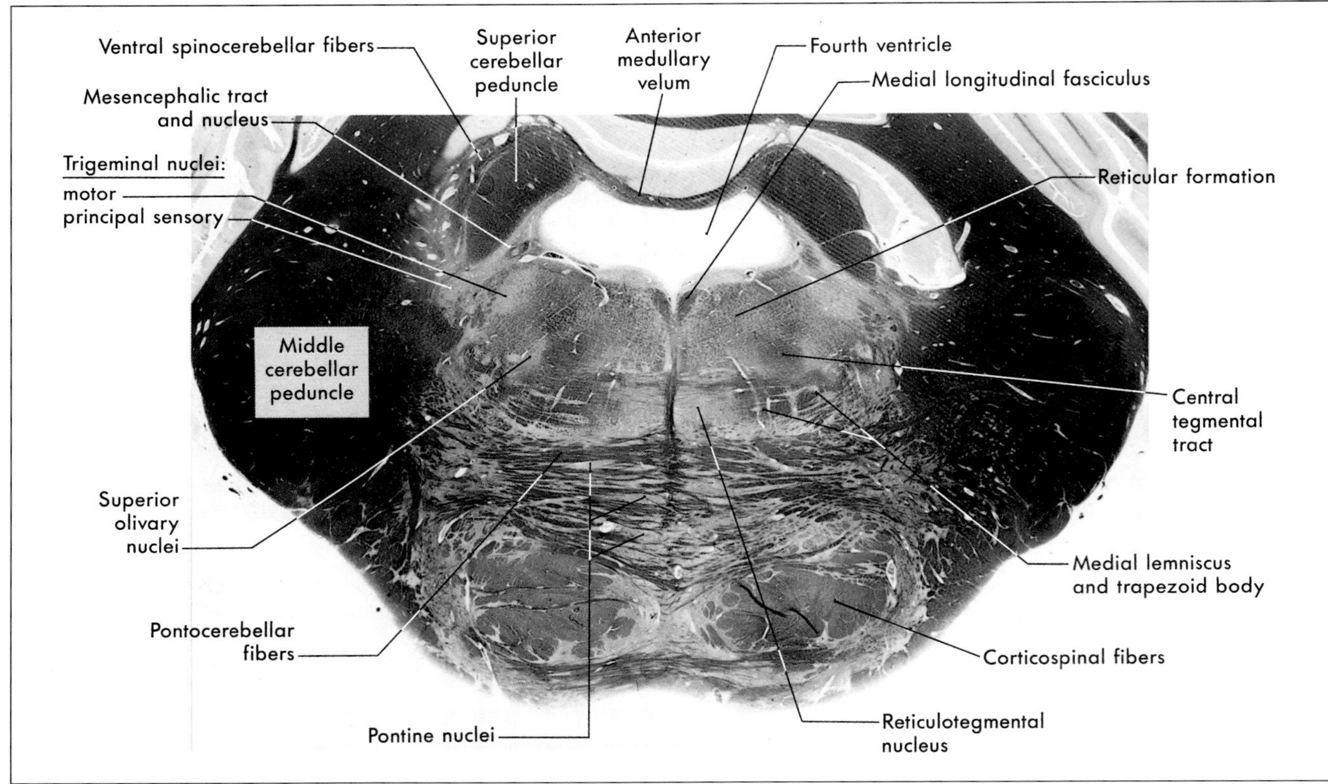

FIGURE 12-15 *A fiber (myelin) stained cross section at the level of the principal sensory and motor trigeminal nuclei. Compare with Figure 12-14.*

The *locus (nucleus) ceruleus* is located in the lateral floor of the fourth ventricle at this level (Fig. 12-14). Ceruleus neurons contain pigment (hence the name *nucleus pigmentosus pontis*) and represent the largest single source of noradrenergic axons in the central nervous system.

A comparison of Figures 12-11 and 12-14 reveals that most major tracts in the pontine tegmentum (*medial longitudinal fasciculus, tectobulbospinal system, medial lemniscus, anterolateral system*, and *ventral trigeminothalamic fibers*) occupy positions comparable to those seen at more caudal levels. Consequently, this section emphasizes only the features that are new at this level. The medial lemniscus is oriented horizontally at this level (Figs. 12-13, 12-15), and the *rubrospinal fibers* have shifted medial to the anterolateral system. Most auditory fibers are now concentrated in the *lateral lemniscus*, and rostral parts of the *superior olivary nucleus* appear just lateral to the central tegmental tract. *Ventral spinocerebellar fibers* migrate dorsally and enter the cerebellum by coursing over the surface of the superior cerebellar peduncle. The *brachium conjunctivum (superior cerebellar peduncle)* arises from the cerebellar nuclei, sweeps rostrally, forming the lateral wall of the fourth ventricle (Figs. 12-14, 12-15), and enters the caudal midbrain tegmentum, where it decussates.

Neurons in the ventral tegmentum close to the midline and extending into dorsal portions of the basilar pons comprise the *reticulotegmental nucleus* (Figs. 12-14, 12-15). This cell group is continuous with the basilar pontine nuclei, and its axons enter the cerebellum through the contralateral *brachium pontis*. These cells also share similar cytologic features and afferent projections with neurons of the basilar pons. Although the *basilar pons* at mid-pontine levels is actually quite similar to that seen at caudal levels, it appears larger in comparison to the area of the tegmentum (Figs. 12-11, 12-14).

Rostral Pontine Level The only cranial nerve structures present in the pontine tegmentum at rostral levels are the *mesencephalic nucleus* and *tract*. These structures are located in the lateral aspect of the periaqueductal gray and remain in this position into the midbrain (Fig. 12-16). Ventral to the mesencephalic tract and nucleus is the *locus (nucleus) ceruleus*; this noradrenergic cell group also extends into the caudal midbrain.

The *brachium conjunctivum (superior cerebellar peduncle)* converges toward its decussation in the caudal midbrain, and most other major tracts in the tegmentum occupy positions comparable to those seen at mid-pontine levels (compare Fig. 12-14 with Fig. 12-16). The rubrospinal tract is shifted even more medially and the

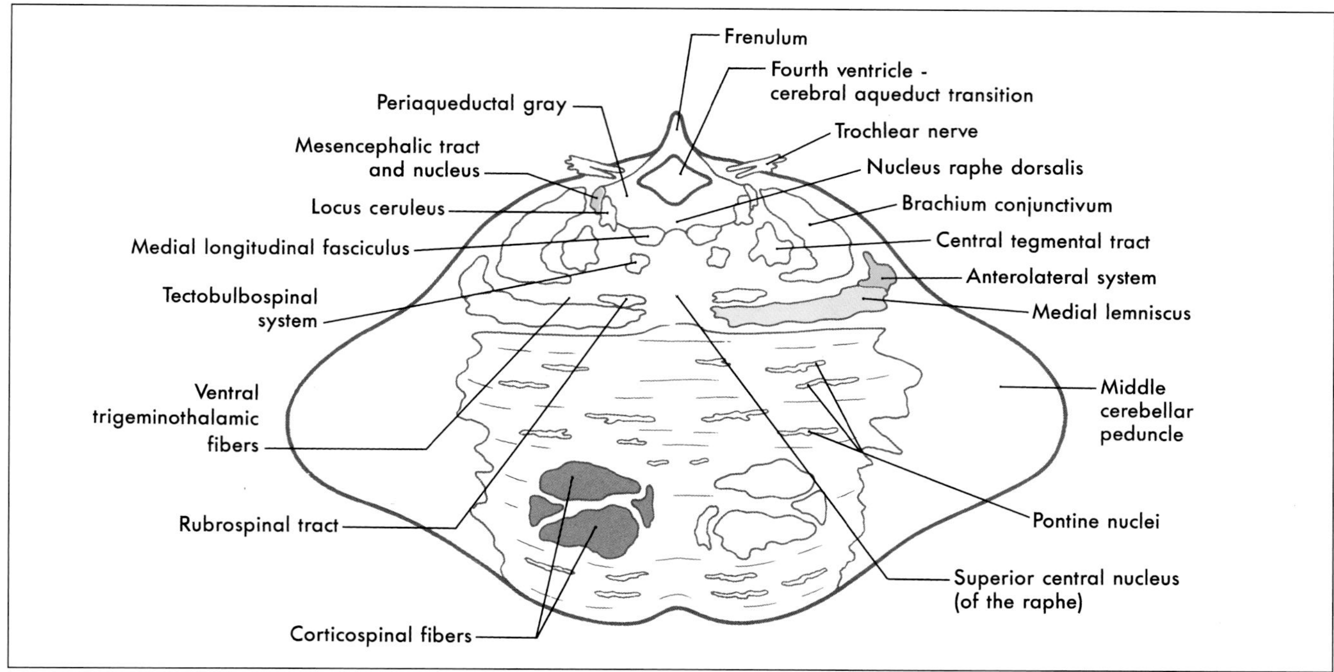

FIGURE 12-16 *Cross section of the rostral pons.*

lateral lemniscus is close to the dorsolateral surface of the brainstem at this point. Also, the composition of the *basilar pons* is essentially the same as that seen at mid-pontine levels (Figs. 12-14, 12-16).

Reticular and Raphe Nuclei Much of the pontine tegmentum is occupied by the reticular formation. This central core is generally divided into a medial area of primarily large neurons (magnocellular region) and a lateral area of mainly small neurons (parvicellular region) (Fig. 12-17). The magnocellular reticular nuclei of the pons are, from caudal to rostral, the *gigantocellular reticular nucleus* and the *caudal* and *oral pontine reticular nuclei*. The parvicellular area nuclei of the pons contain a diffuse *lateral reticular formation* at caudal and mid-pontine levels and include the *medial* and *lateral parabrachial nuclei* at rostral levels. The latter cell groups are located adjacent to the brachium conjunctivum.

The *raphe nuclei* are symmetrically distributed on either side of the midline (Fig. 12-17). Next to the medial lemniscus at caudal pontine levels is the *nucleus raphe magnus*. This cell cluster extends caudally into the rostral medulla and is an important synaptic station for signals involved in the inhibition of pain at medullary and spinal levels. In the caudal one-third of the tegmental pons, this cell group is replaced by the *nucleus raphe pontis*, which extends a little beyond mid-pontine levels. The *central superior nucleus* and *nucleus raphe dorsalis* are found in the rostral pons; the latter extends into the caudal midbrain (Fig. 13-16).

Vasculature Internal areas of the tegmental and basilar pons are served by branches of the *basilar artery* (Fig. 12-18). *Paramedian branches* distribute to medial areas of the basilar pons, including corticospinal fibers and the exiting fibers of the abducens nerve. The lateral part of the basilar pons is served by *short circumferential branches*, and the entire tegmental area plus a wedge of the middle cerebellar peduncle receives blood via the *long circumferential branches*. At caudal levels (levels of the facial colliculus), the long circumferential supply is supplemented by branches of the *anterior inferior cerebellar artery*. Rostrally, beginning at about the level of the principal sensory and motor trigeminal nuclei, the blood supply to the pontine tegmentum is supplemented by branches of the *superior cerebellar artery*.

INTERNAL ANATOMY OF THE CEREBELLUM

Cerebellar Cortex Each folium of the cerebellum is organized into three layers oriented parallel to the cortical surface. From external to internal, these are the *molecular*, *Purkinje cell*, and *granular layers* (Fig. 12-19). Internal to the granular layer, and forming the core of each folium, is a layer of subcortical white matter composed of all fibers arriving (afferents to the cortex) or leaving (efferents of the cortex) the cerebellar cortex.

The *molecular layer* contains *stellate cells*, *basket cells*, and the expansive dendrites of *Purkinje cells* (Fig. 12-19). In addition, the molecular layer contains two other major fiber populations: *climbing fibers* and *parallel fibers*.

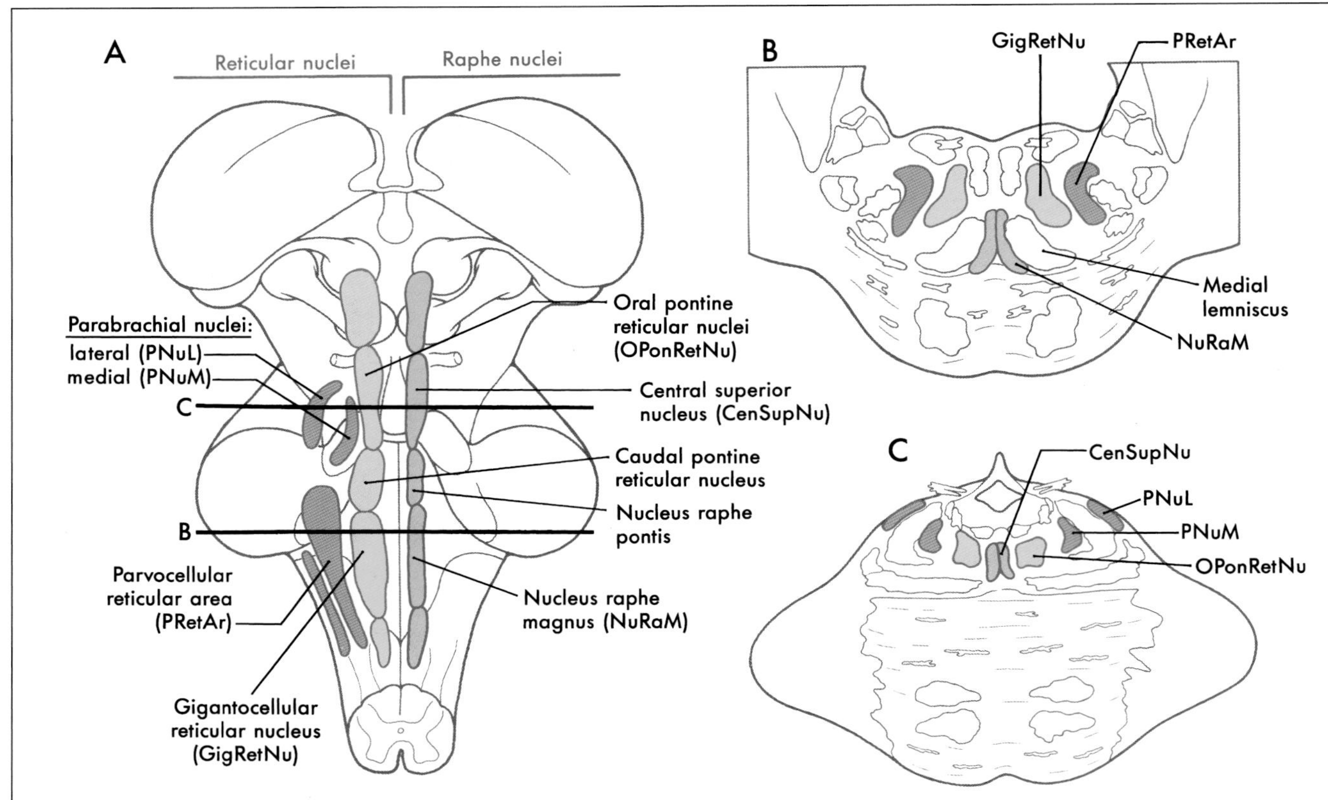

FIGURE 12-17 *Dorsal view of the brainstem (A) and caudal (B) and rostral (C) cross sections showing the raphe and reticular nuclei of the pons.*

The former originate from the contralateral inferior olive and the latter from the granule cells (Fig. 12-19).

Purkinje cells are the efferent neurons of the cerebellar cortex. Their large somata form a single layer at the molecular-granular interface (Fig. 12-19). Each Purkinje cell has a complex dendritic tree that extends into the molecular layer and exhibits a fanlike orientation that is perpendicular to the long axis of the folium. Axons of Purkinje cells pass through the subcortical white matter to end in the cerebellar and vestibular nuclei.

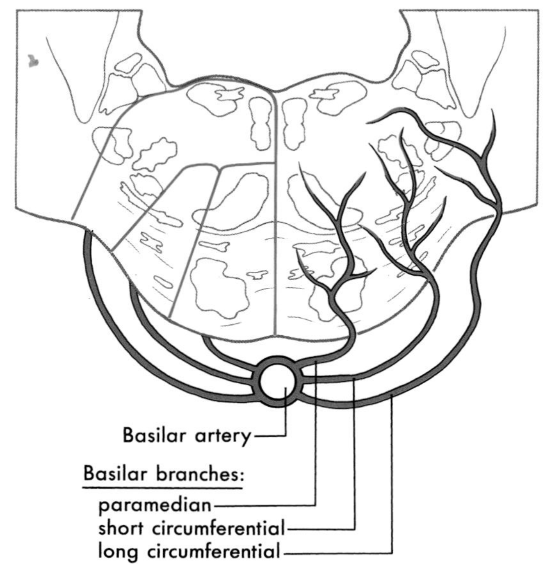

FIGURE 12-18 *Blood supply of the pons. Arteries are shown on the right and the territories served by each on the left.*

The granular layer contains *granule cells*, *Golgi cells*, and *mossy fibers* (Fig. 12-19). The latter originate from cells located in many nuclei throughout the brainstem and spinal cord. Traversing the granular layer are climbing fibers, en route to the molecular layer, and Purkinje cell axons leaving the cortex.

Cerebellar Nuclei The cerebellar nuclei are, from medial to lateral, the *fastigial*, *globose*, *emboliform*, and *dentate* (Fig. 12-7). They receive input from Purkinje cell axons and from collaterals of cerebellar afferent fibers. In general, Purkinje cells of the vermis relate to the fastigial nucleus, those of the intermediate cortex to the globose and emboliform nuclei, and those of the lateral cortex to the dentate nucleus (Fig. 12-7). Some Purkinje cells of the vermis and the flocculonodular lobe send axons directly to the vestibular nuclei. Cerebellar nuclear cells project to a variety of cell groups throughout the neuraxis, primarily via the brachium conjunctivum.

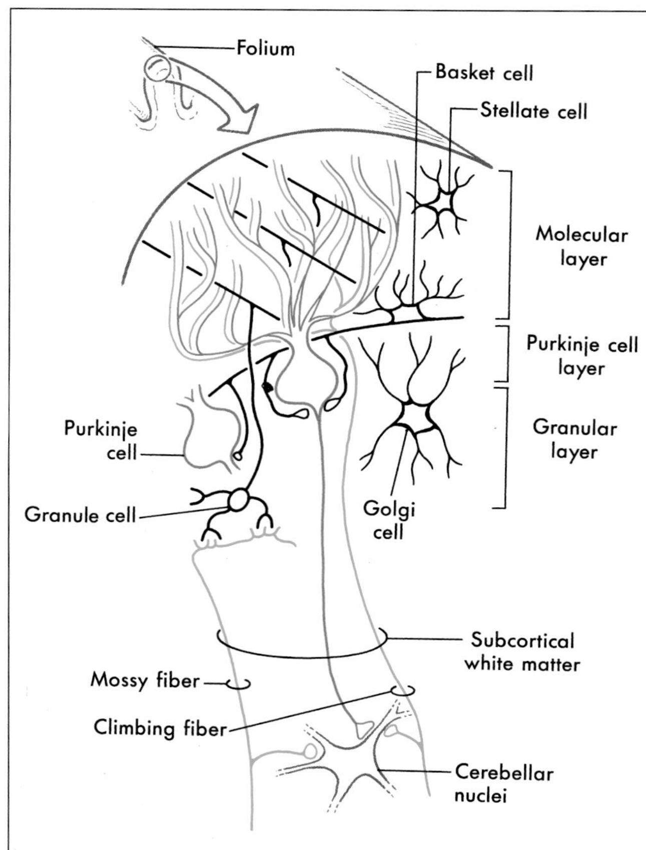

FIGURE 12-19 *Diagrammatic representation of the cells and fibers of the cerebellar cortex.*

Vasculature The blood supply to the cerebellar cortex and nuclei arises via branches of the *superior cerebellar artery* and the *anterior* and *posterior inferior cerebellar arteries* (Fig. 12-7). The superior cerebellar artery serves the superior surface of the cerebellum and, by way of penetrating branches, the fastigial, globose, and emboliform nuclei and rostrodorsal areas of the dentate nucleus. This vessel also distributes to the superior and middle cerebellar peduncles.

Branches of the anterior inferior cerebellar artery distribute to the more lateral areas of the inferior cerebellar surface, including the flocculus (Fig. 12-7). They also serve parts of the middle cerebellar peduncle, caudoventral regions of the dentate nucleus, and the choroid plexus in the subarachnoid space at the cerebellopontine angle.

The posterior inferior cerebellar artery (PICA) branches to more medial regions of the inferior cerebellar surface and to the nodulus (Fig. 12-7). The choroid plexus of the fourth ventricle receives branches from the PICA. This vessel is also an important source of blood to dorsolateral regions of the medulla.

SOURCES AND ADDITIONAL READINGS

Readings for the brainstem chapters are listed at the end of Chapter 13.

CHAPTER THIRTEEN

The Midbrain

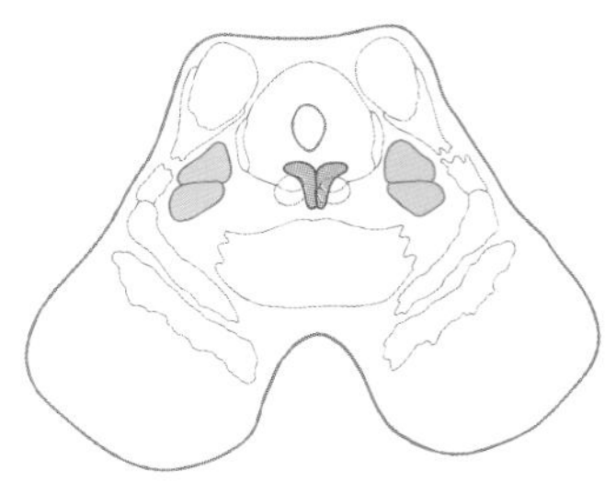

G. A. MIHAILOFF • D. E. HAINES • P. J. MAY

The *mesencephalon* or midbrain is the most rostral portion of the brainstem. It gives rise to cranial nerves III and IV, conducts ascending and descending tracts, and contains nuclei that are essential to motor function. Caudally, the midbrain is continuous with the pons, and rostrally it joins the diencephalon. The *cerebral aqueduct*, the cavity of the midbrain, is continuous rostrally with the third ventricle and caudally with the fourth ventricle. The blood supply to the mesencephalon is primarily from proximal branches of the posterior cerebral arteries (P_1 or P_2) and from small penetrating branches of the posterior communicating artery.

DEVELOPMENT

The mesencephalon (Fig. 13-1) arises early in development as one of the three primary brain vesicles. Neuroblasts of the *mantle layer* give rise to *alar* and *basal plates*, which are continuous with the alar and basal plates of the rhombencephalon (Fig. 13-2). These cell groups are the rostral continuations of the same primitive cell columns described for the metencephalon. The surrounding *marginal layer* contains the developing axons of cells located in other levels of the neuraxis. The *cerebral aqueduct* is narrow relative to the fourth ventricle; therefore, the basal and alar plates lie ventral and dorsal to it as in the spinal cord and caudal medulla.

Basal and Alar Plates The primitive neurons of the alar plate give rise to the quadrigeminal plate, from which the *superior* and *inferior colliculi* arise (Fig. 13-2). In humans, the superior colliculus consists of alternating layers of cells and fibers, whereas the inferior colliculus appears homogenous even though it consists of several subnuclei. Alar plate neuroblasts also migrate into ventral areas of the developing midbrain to form the red nucleus and the *substantia nigra* (Fig. 13-2B).

Basal plate neuroblasts give rise to the *general somatic efferent* (GSE) neurons of the *oculomotor* and *trochlear nuclei*. In addition, the *general visceral efferent* (GVE) cells of the *Edinger-Westphal nucleus*, a visceral motor cell group associated with the oculomotor complex, also arise from the basal plate (Fig. 13-2).

As the basal and alar plates differentiate, the marginal layer is invaded by axons originating from cells located outside the midbrain. These fibers collect in the ventrolateral area of the developing mesencephalon to form an especially prominent bundle, the *crus cerebri* (plural, *crura cerebri*; Fig. 13-2B).

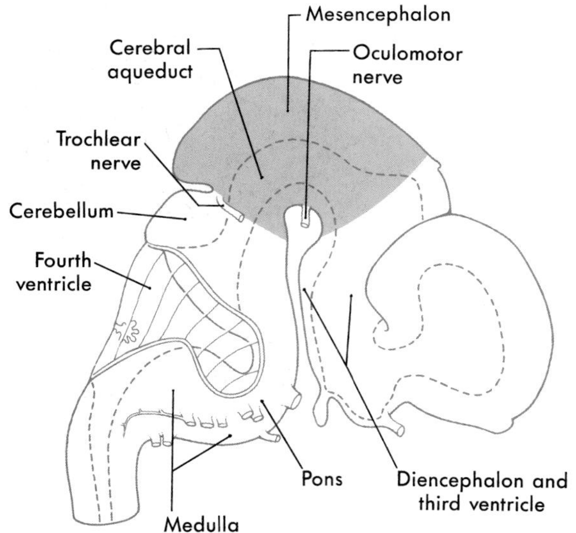

FIGURE 13-1 *Lateral view of the human brain at about 7 weeks' gestational age. The midbrain is highlighted.*

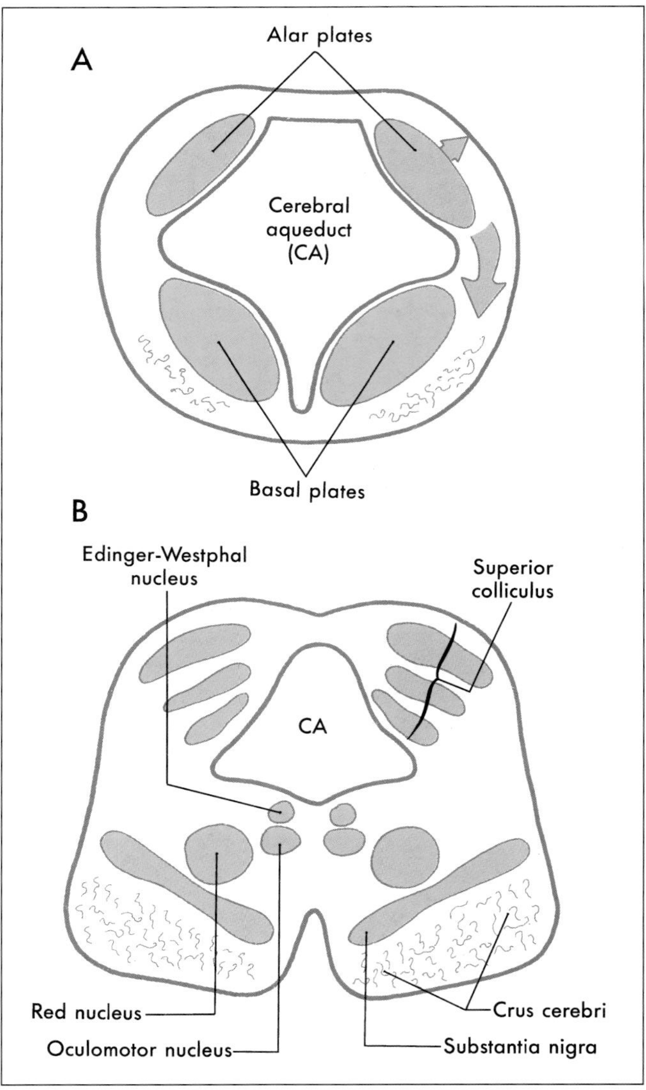

FIGURE 13-2 *Development of the midbrain at early (A) and later (B) stages showing the alar and basal plates and the structures derived from each. The level shown at B is diagrammatic of the rostral (superior colliculus) midbrain.*

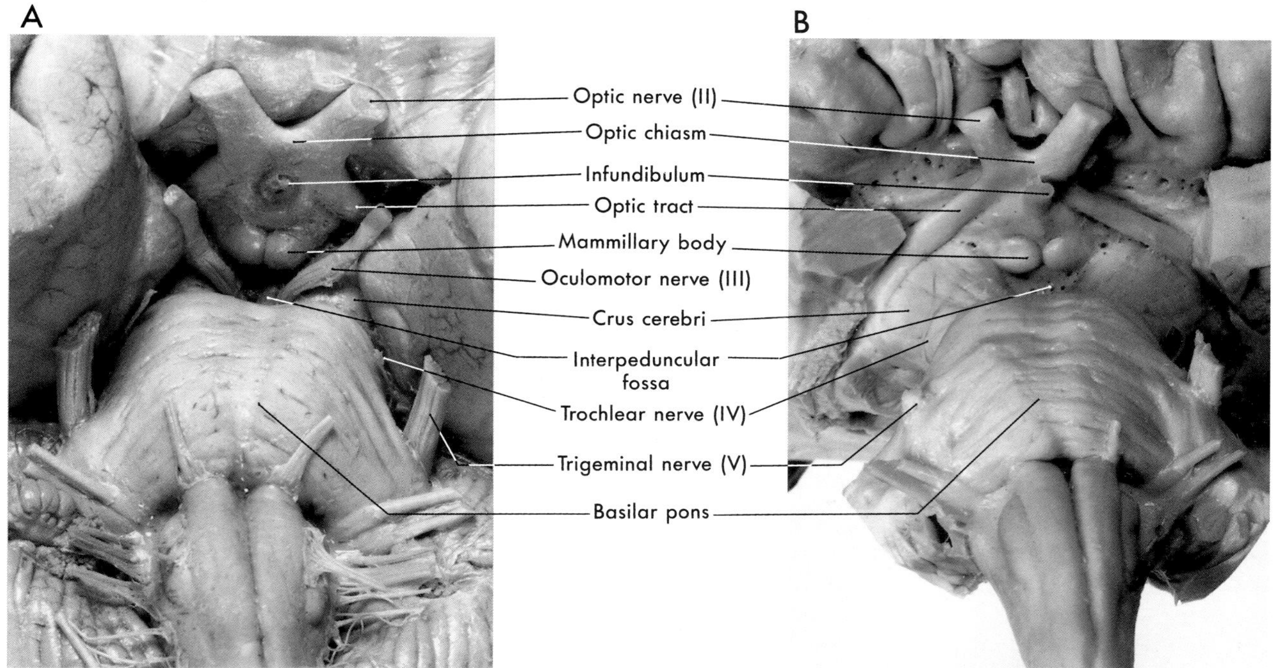

FIGURE 13-3 *Ventral views (A, undissected; B, dissected) of the brainstem with emphasis on the midbrain.*

EXTERNAL FEATURES

Ventral Midbrain The presence of a pair of large axon bundles, the crura cerebri, is a characteristic feature of the ventral midbrain. These bundles emerge from the cerebral hemispheres caudal to the optic tracts, converge slightly toward the midline as they course through the midbrain, and disappear into the basilar pons (Fig. 13-3). The *oculomotor nerves* exit the medial edge of each crus and pass through the space between the crura, which is called the *interpeduncular fossa* (Fig. 13-3). Ventrally, the rostral limit of the midbrain is marked by the exit of the crura cerebri from the cerebral hemispheres and by the caudal edge of the mammillary bodies. The caudal border of the midbrain is formed where each crus enters the basilar pons.

The subarachnoid space of the interpeduncular fossa is called the *interpeduncular cistern*. This cistern contains the oculomotor nerves and the upper part of the basilar artery, including its bifurcation and proximal branches. Numerous vessels penetrate the floor of this fossa and create many small perforations (Fig. 13-3B). Consequently, this area is frequently called the *posterior perforated substance*.

Dorsal Midbrain The dorsal surface of the adult midbrain is characterized by four elevations collectively called the *corpora quadrigemina* (Fig. 13-4). The rostral two elevations are the *superior colliculi*, and the caudal two are the *inferior colliculi*. Just caudal to the inferior colliculus, the exit of the trochlear nerve marks the pons-midbrain junction on the dorsal surface of the brainstem, whereas the midbrain-diencephalic boundary is formed by the posterior commissure (Fig. 13-5).

Rostrolaterally, the inferior colliculus is joined to the *medial geniculate body* of the diencephalon by a fiber bundle called the *brachium of the inferior colliculus* (Fig. 13-4). The inferior colliculus and the medial geniculate body

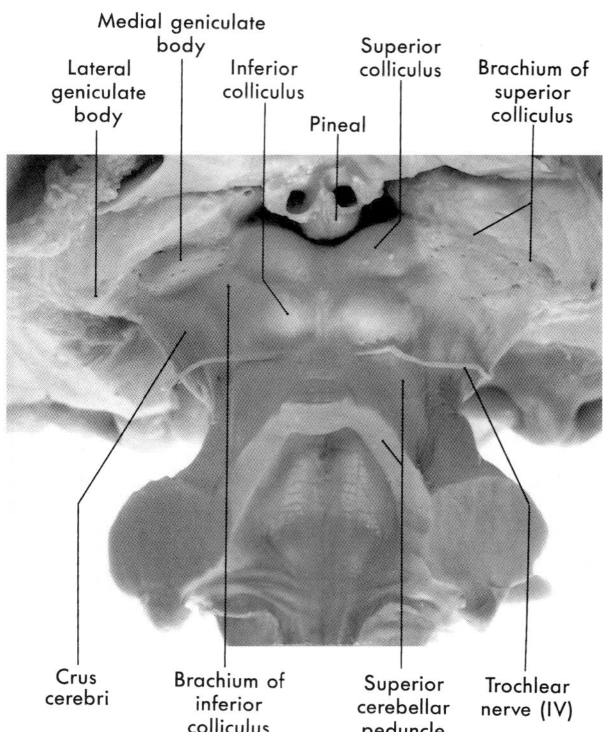

FIGURE 13-4 *Dorsal view (dissected) of the brainstem with emphasis on the midbrain and its junction with the diencephalon.*

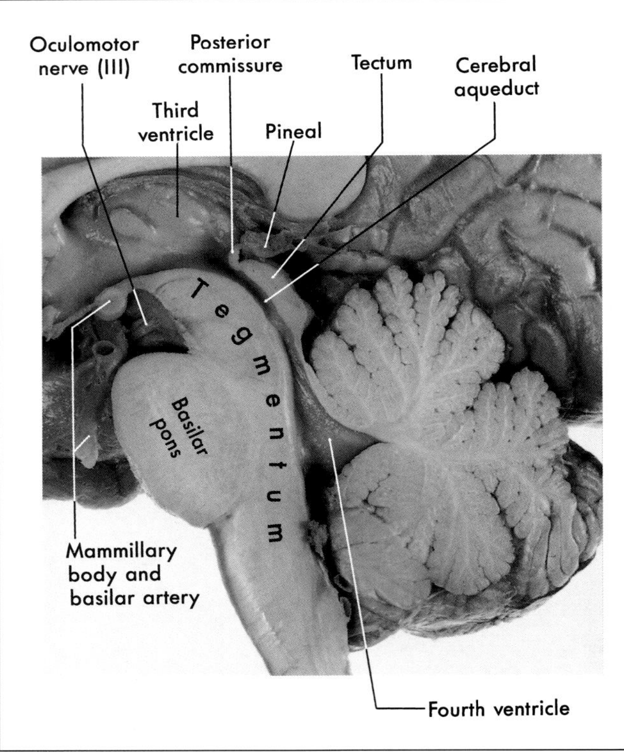

FIGURE 13-5 *Sagittal view of the brainstem with emphasis on midbrain structures.*

are part of the auditory system. The *brachium of the superior colliculus* extends from the optic tract to the superior colliculus in a groove located between the *medial geniculate body* and *pulvinar* of the diencephalon (Fig. 13-4; see also Fig. 13-14). The superior colliculus, pulvinar, and lateral geniculate body are parts of the visual and visual-motor systems.

On the midline, the *pineal gland*, a diencephalic structure, extends posteriorly above and between the superior colliculi (Fig. 13-5). Tumors of the pineal produce noncommunicating (obstructive) hydrocephalus because of compression of the dorsal midbrain and resulting occlusion of the cerebral aqueduct.

The subarachnoid space immediately dorsal to the colliculi is the *quadrigeminal cistern*. This cistern contains the exiting trochlear nerves, the great vein of Galen, and distal branches of the posterior cerebral arteries.

Vasculature The primary blood supply to the mesencephalon arises via branches of the *basilar artery*, with smaller branches from the *superior cerebellar*, *anterior choroidal*, and *posterior communicating arteries*. An important source of blood to the dorsal portion of the midbrain is the *quadrigeminal artery*, a branch of the posterior cerebral artery (P_1 segment). Also, the *superior cerebellar artery* gives rise to branches that serve caudal parts of the dorsal midbrain and adjacent regions of the pons.

INTERNAL ANATOMY OF THE MIDBRAIN

General Regions (Tectum, Tegmentum, and Basis Pedunculi) The midbrain is divisible into three regions, which can be appreciated best in cross section. Dorsal to the cerebral aqueduct is the *tectum* (roof) of the midbrain (Figs. 13-5, 13-6). The characteristic structures of this area are the superior and inferior colliculi. The *tegmentum of the midbrain* extends from the base of the tectum to, but does not include, the substantia nigra. The ventrolateral portion of the midbrain on either side is formed by the *basis pedunculi*, which consists of the *substantia nigra* and the *crus cerebri* (Fig. 13-6). In turn, the *crus cerebri* is composed primarily of corticopontine and corticospinal fibers. The term *cerebral peduncle* is sometimes used for the crus cerebri but actually represents the entire midbrain below the tectum (tegmentum + basis pedunculi).

Summary of Ascending Pathways The long ascending pathways that traverse the pons and medulla continue through the midbrain (Fig. 13-7). These include the *medial lemniscus*, *anterolateral system*, and the *dorsal* and *ventral trigeminothalamic tracts*. Other, smaller bundles, such as the *medial longitudinal fasciculus*, are also present and occupy positions comparable to those seen in lower levels of brainstem.

Summary of Descending Pathways Descending fibers from the cerebral cortex pass through the midbrain and, as we have seen, are also prominent features of the pons, medulla, and spinal cord (Fig. 13-7). At midbrain levels these *corticospinal*, *corticobulbar*, and *corticopontine* fibers are parts of the crus cerebri. Some of the descending

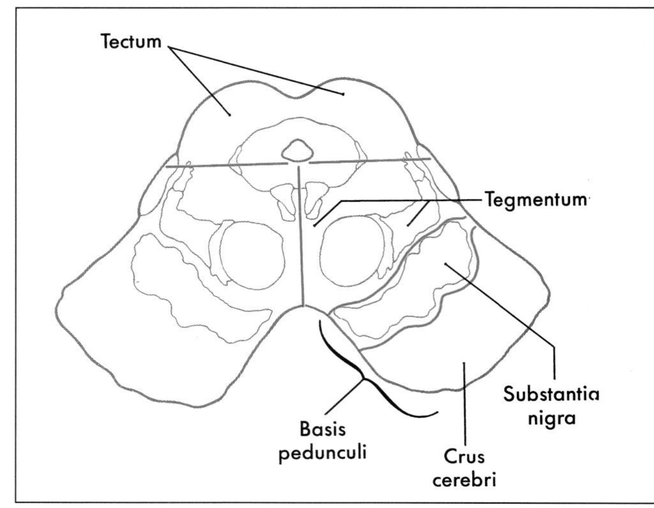

FIGURE 13-6 *Major subdivisions of the midbrain.*

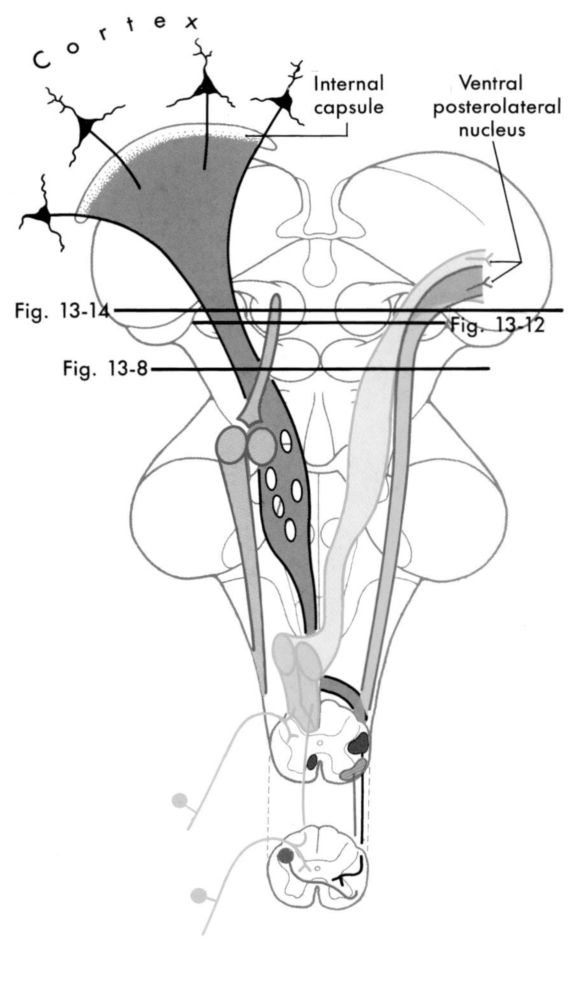

FIGURE 13-7 *Diagrammatic representation of the brain showing the location and trajectory of three important pathways and the trigeminal nuclei. The color coding for each is continued in Figures 13-8, 13-12, and 13-14.*

fiber bundles that were discussed in the pons and medulla, such as the *tectobulbospinal system* and the *rubrospinal* and *central tegmental tracts*, originate from nuclei of the midbrain.

The following sections describe the anatomy of the midbrain at caudal and rostral brain levels and at the level of the midbrain-diencephalon junction. The anatomy is presented dorsoventrally at each level, starting with the tectum and proceeding through the tegmentum to the basis pedunculi.

Caudal Midbrain Levels Transverse sections through the caudal midbrain are characterized by the presence of the *inferior colliculus*, *trochlear nucleus*, and *decussation of the superior cerebellar peduncle* (Figs. 13-8 to 13-10). The inferior colliculus is composed of a large *central nucleus* bordered dorsally by a smaller *pericentral* (*dorsal*) *nucleus* and laterally by an *external* (*lateral*) *nucleus*. Lateral lemniscus fibers enter the inferior colliculus from a ventral direction and give this area a gobletlike appearance (Figs. 13-8, 13-9). The lateral lemniscus provides auditory input to the nuclei of the inferior colliculus, which then transmit this information to the medial geniculate body via fibers of the *brachium of the inferior colliculus*. A complete tonotopic representation (frequency map) of the cochlea is found in several of the central auditory nuclei (see Chapter 20).

The *periaqueductal gray*, or *central gray*, is a prominent collection of small neurons surrounding the cerebral aqueduct at all midbrain levels (Figs. 13-8 to 13-10; see also Figs. 13-12, 13-13). It receives somatosensory input, is interconnected with the hypothalamus and thalamus, and projects caudally to brainstem nuclei. Its connections and the presence of a high level of opiate receptor binding activity indicate that the periaqueductal gray plays an important role in the brain mechanisms responsible for the suppression and modulation of pain (analgesia). The *mesencephalic tract* and *nucleus* are located in the lateral edge of the periaqueductal gray, and the *nucleus raphe dorsalis* is located ventrally on the midline, adjacent to the trochlear nucleus and medial longitudinal fasciculus (Fig. 13-8).

The central area of the tegmentum at the level of the inferior colliculus is occupied by the decussating fibers of the *superior cerebellar peduncle* (Figs. 13-8, 13-9). From this point, the majority of these cerebellar efferent axons pass rostrally to targets in the midbrain and thalamus, although some turn caudally and enter the pons and medulla. Just dorsal to this decussation, the *general somatic efferent* (GSE) cell bodies of the *trochlear nucleus* form an oval-shaped cell group nestled in the fibers of the *medial longitudinal fasciculus* (Figs. 13-8, 13-10). The axons of trochlear motor neurons pass laterally and dorsally around the periaqueductal gray to cross the midline before exiting the brainstem just caudal to the inferior colliculus. They innervate the superior oblique muscle. *Tectobulbospinal fibers* are ventral to the *medial longitudinal fasciculus*, and, as their name states, the fibers of the *central tegmental tract* occupy the center of the tegmentum (Figs. 13-8–13-10).

Immediately ventral to the decussation of the superior cerebellar peduncle are the *interpeduncular nucleus* and *rubrospinal fibers* (Fig. 13-8). The interpeduncular nucleus is related to the limbic system, a part of the brain that functions in the control of emotional behavior.

Located in ventrolateral and lateral portions of the midbrain tegmentum are fibers of the *anterolateral system* and *medial lemniscus* (Fig. 13-8). The anterolateral system

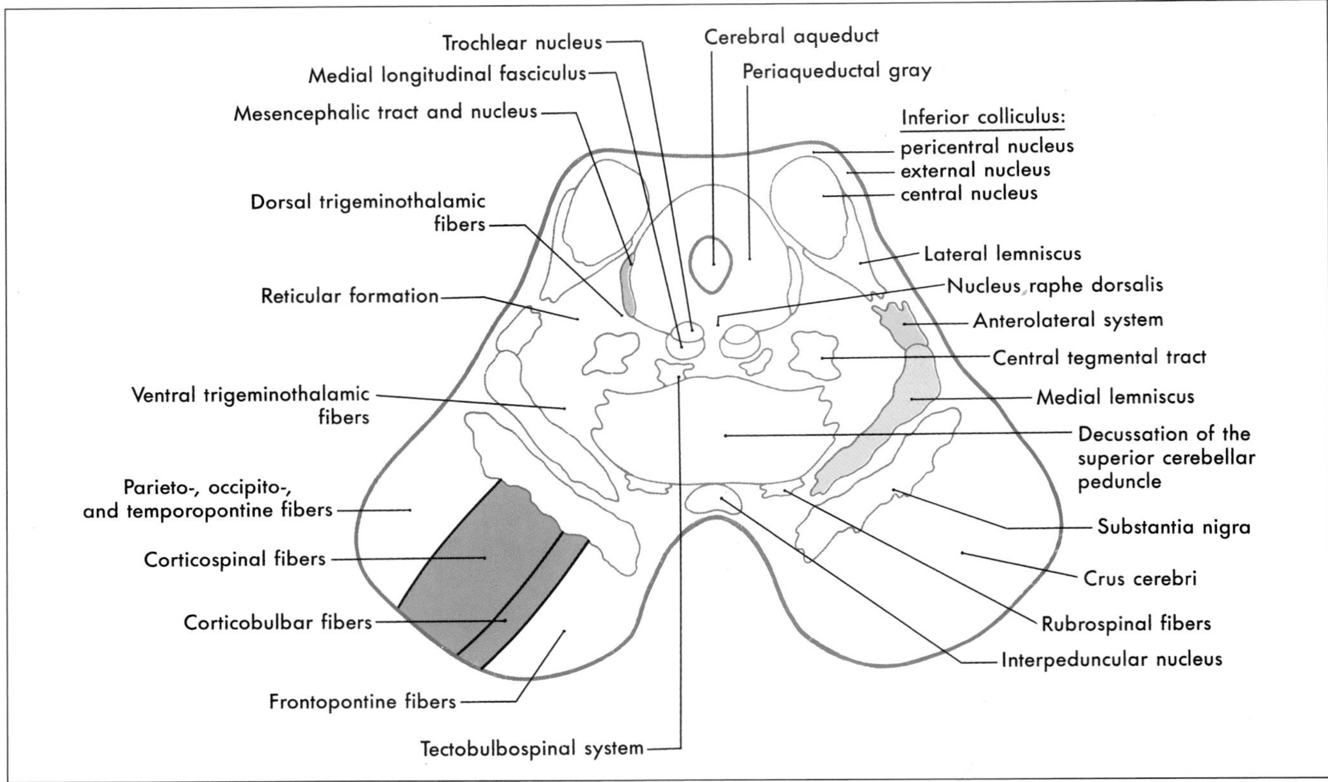

FIGURE 13-8 *Cross section of the midbrain at the level of the inferior colliculus. Correlate with Figure 13-7.*

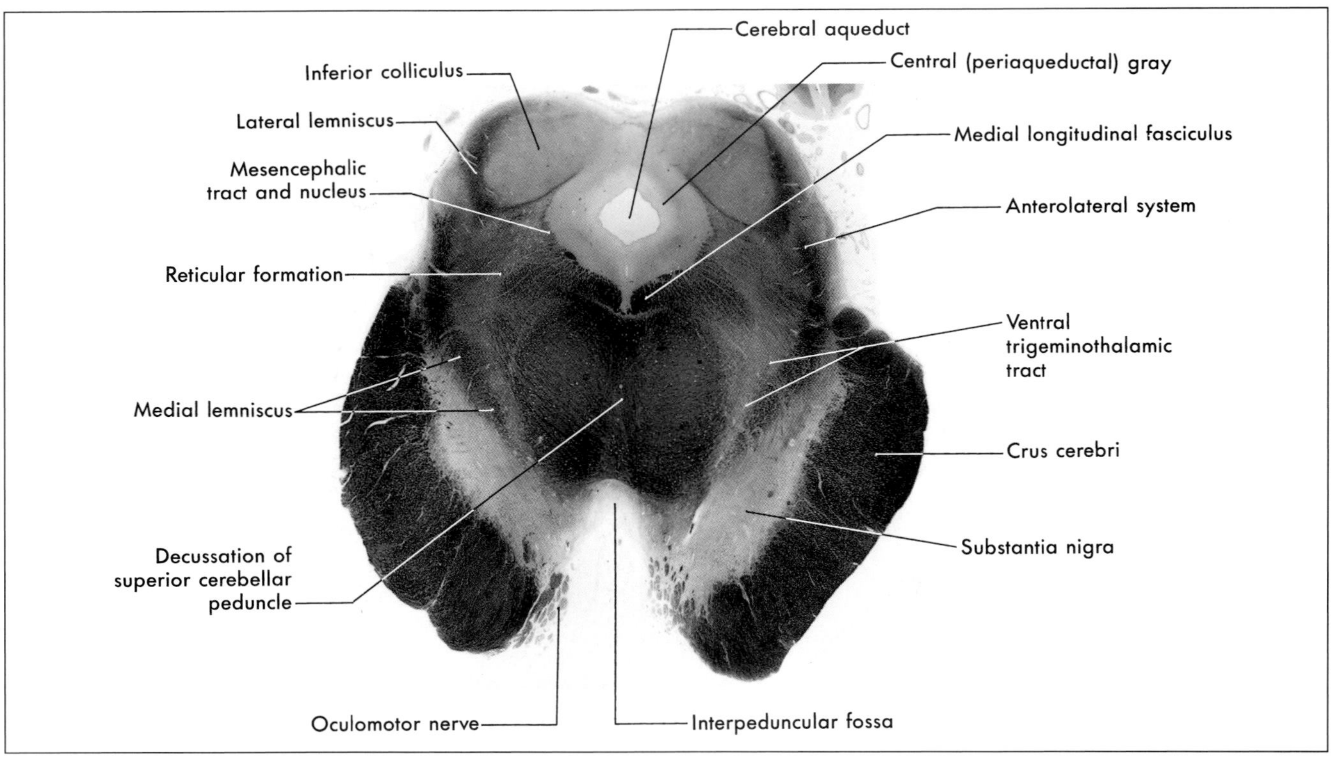

FIGURE 13-9 *A fiber (myelin) stained cross section of the midbrain at the level of the inferior colliculus. Compare with Figure 13-8.*

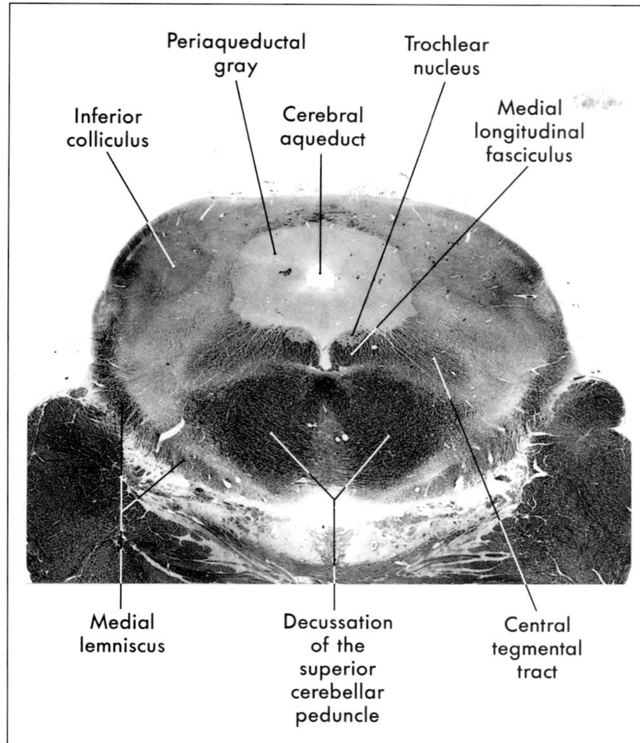

FIGURE 13-10 *A fiber (myelin) stained cross section of the midbrain at an intercollicular level showing the trochlear nucleus. Compare with Figure 13-8.*

conveys pain and temperature signals from the contralateral side of the body. Some of its fibers terminate in the dorsal thalamus and therefore pass through the midbrain, but it also contains *spinomesencephalic fibers* that end in the midbrain. Some of these fibers synapse in the reticular nuclei of the midbrain and pons (*spinoreticular fibers*), some synapse in the tectum (*spinotectal fibers*), and some synapse in the periaqueductal gray. The latter fibers represent an important link in the pathways involved in the inhibition of pain at medullary and spinal levels. Medial lemniscal fibers have shifted from a horizontal position, characteristic of the rostral pons, to the ventromedial to dorsolateral orientation of the midbrain. The somatotopy of fibers in the medial lemniscus follows accordingly (Fig. 13-11).

Dorsal and *ventral trigeminothalamic tracts* are located adjacent to the central tegmental tract and the medial lemniscus, respectively (Figs. 13-8, 13-9). These bundles are somewhat diffusely arranged and convey crossed (ventral) and uncrossed (dorsal) somatosensory information from the face.

The base of the midbrain on either side is formed by the *basis pedunculi*, which consists of the *substantia nigra* and the *crus cerebri* (Fig. 13-8). The substantia nigra is related to motor function and is considered in the next section of this chapter. The crus cerebri contains *corticospinal*, *corticobulbar*, and *corticopontine* fibers. The former two fiber populations are found in the middle third of the crus cerebri. Corticospinal fibers projecting to lower cord levels (leg representation) are found laterally and corticobulbar fibers (projections to cranial nerve nuclei; face representation) are located more medially. Corticopontine fibers in the medial third of the crus arise from the frontal lobe (*frontopontine*), whereas those in the lateral third originate from parietal, occipital, and temporal lobes (*parietopontine*, *occipitopontine*, and *temporopontine*).

Rostral Midbrain Levels Transverse sections through the rostral midbrain are characterized by the presence of the *superior colliculus*, *red nucleus*, and *oculomotor nuclei* (Figs. 13-12, 13-13). The paired *superior colliculi* are composed of alternating layers of gray matter (cells) and white matter (fibers). The "superficial" three layers (I to III) receive input from the retina and visual cortices and project to the thalamus. The "deep" four layers (IV to VII) subserve gaze changes including eye movements and project to the thalamus, brainstem, and spinal cord. Descending crossed projections from the tectum pass to a variety of brainstem areas (for example, as *tectoreticular*, *tecto-olivary*, and *tectofacial fibers*) and—although this element is minor in humans—to the cervical spinal cord as *tectospinal fibers*. Together, these fibers constitute the *tectobulbospinal system* of the brainstem; the name "tectobulbospinal" reflects the diversity of the fibers in this bun-

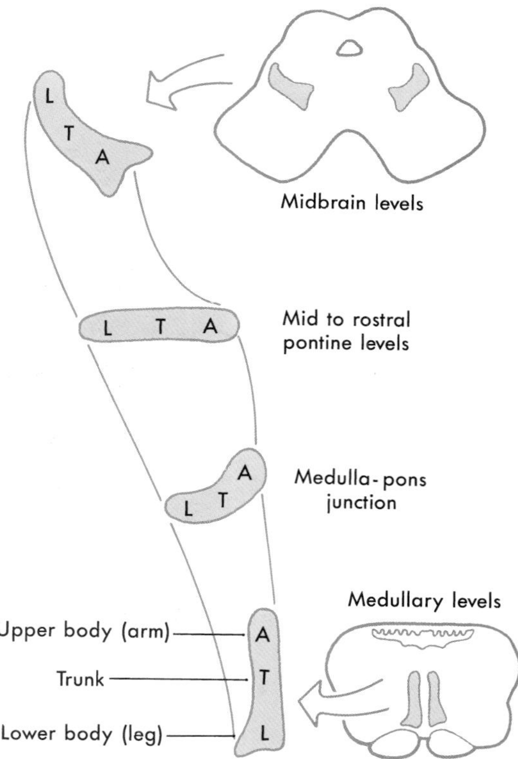

FIGURE 13-11 *The orientation of the medial lemniscus in the midbrain as compared to that in the pons and medulla.*

dle. *Rubrospinal fibers* originate from the contralateral red nucleus; these fibers descend to medullary and spinal levels, where they influence the activity of motor neurons innervating skeletal muscles.

Ventral to the periaqueductal gray, the oculomotor complex forms a V-shaped region between the medial longitudinal fasciculi (Figs. 13-12 to 13-14). This complex consists of the *general somatic efferent* (GSE) cells of the oculomotor nucleus and the *general visceral efferent* (GVE) cells of the Edinger-Westphal nucleus. Oculomotor fibers arch through and medial to the red nucleus, emerge from the brainstem at the medial edge of the basis pedunculi, and innervate four of the six extraocular muscles. The *Edinger-Westphal nucleus* (Figs. 13-13, 13-14) lies dorsal to the oculomotor nucleus and provides the preganglionic parasympathetic fibers that travel to the ciliary ganglion via the oculomotor nerve. These parasympathetic fibers are located on the perimeter of the oculomotor nerve and, consequently, are the first to be affected in a compression injury to this nerve. The postganglionic fibers from the ciliary ganglion innervate the sphincter pupillae and ciliary muscles.

Several other cell groups are located close to the main oculomotor nucleus (Figs. 13-13, 13-15). These include (1) the *interstitial nucleus of Cajal* located adjacent to the fibers of the medial longitudinal fasciculus (MLF), (2) the *nucleus of Darkschewitsch* situated within the ventrolateral border of the periaqueductal gray, (3) the *nuclei of the posterior commissure*, and (4) the *rostral interstitial nucleus of the MLF*. Most of these nuclei are involved in the control of eye movements.

The *red nucleus*, a prominent structure in the midbrain tegmentum at this level (Figs. 13-12 to 13-15), is so named because in the unfixed brain the dense vascularity of the region gives it a pink color. It is composed of a caudal magnocellular and a rostral parvocellular region, but these subdivisions are less distinct in primates and humans than in most other animals. The red nucleus is involved in motor function and has extensive connections throughout the neuraxis. Its efferents include the *rubrospinal tract*, which travels to the contralateral spinal cord, and *rubro-olivary fibers*, which descend in the *central tegmental tract* to the ipsilateral inferior olivary complex. Afferents to the red nucleus arise from the contralateral cerebellar nuclei and the ipsilateral cerebral cortex.

The major tracts of the caudal midbrain tegmentum are present in similar locations at rostral midbrain levels (Fig. 13-12). *Dorsal* and *ventral tegmental decussations* cross the midline at the levels of the dorsal and ventral limits of the red nuclei, respectively. The dorsal decussation carries tectobulbospinal fibers, and the ventral decussation carries rubrospinal fibers. Immediately lateral to the red nucleus are *cerebellorubral* and *cerebel-*

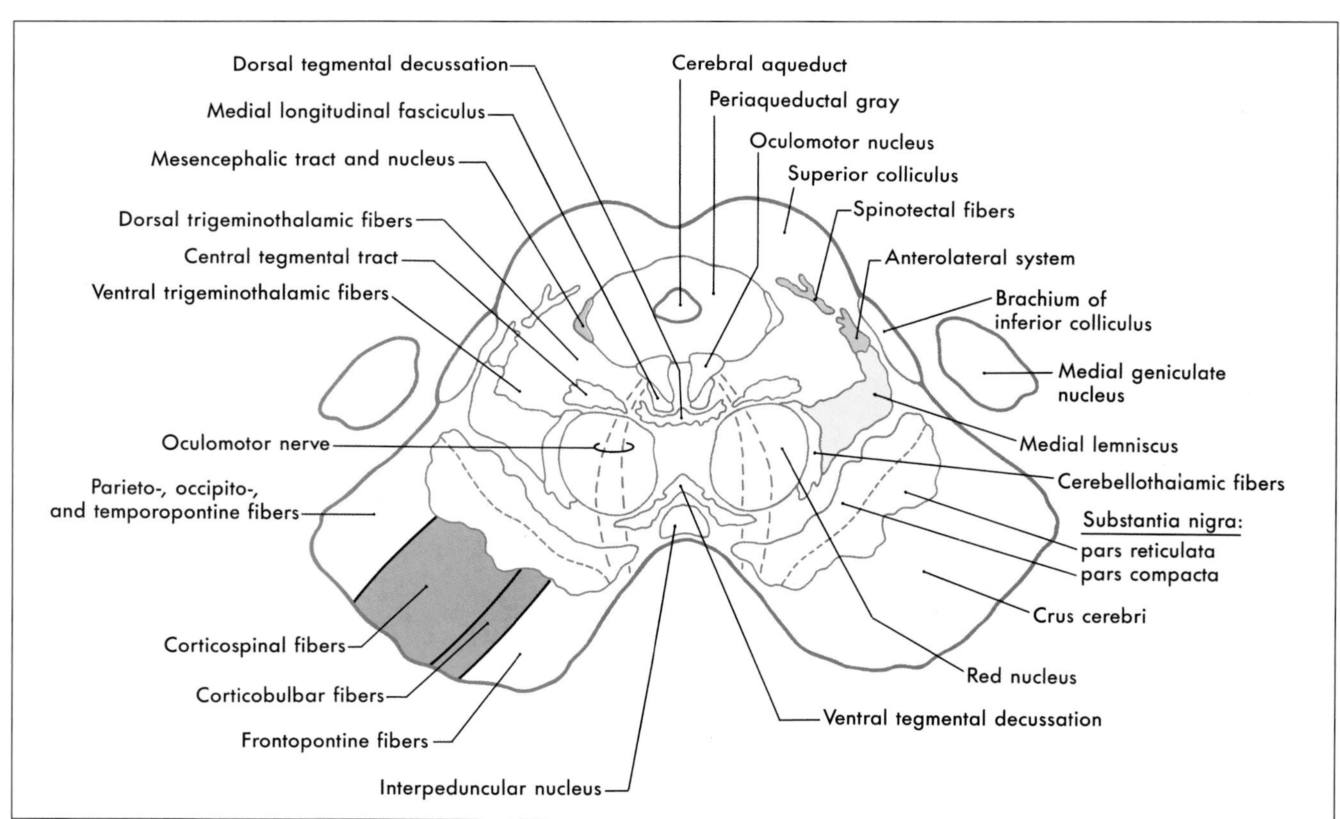

FIGURE 13-12 *Cross section of the midbrain at the level of the superior colliculus. Correlate with Figure 13-7.*

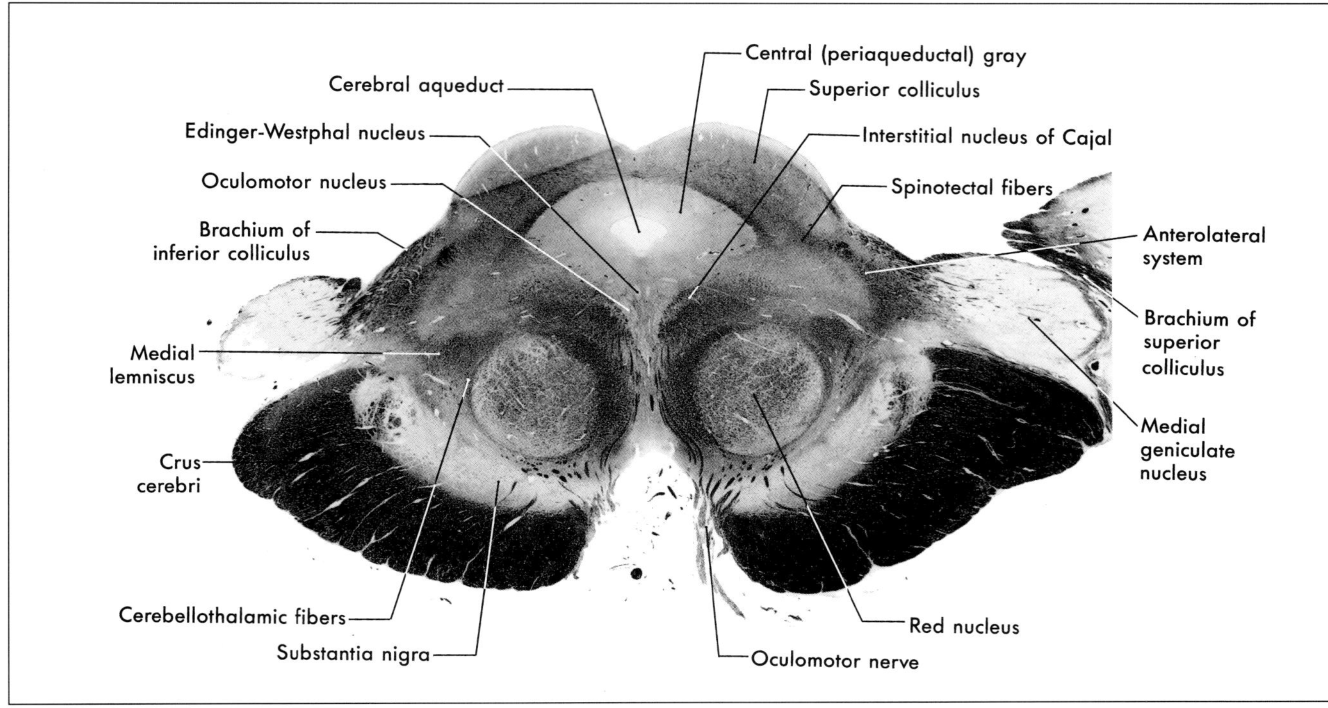

FIGURE 13-13 *A fiber (myelin) stained cross section of the midbrain at the level of the superior colliculus. This plane of section is between those illustrated in Figures 13-12 and 13-14.*

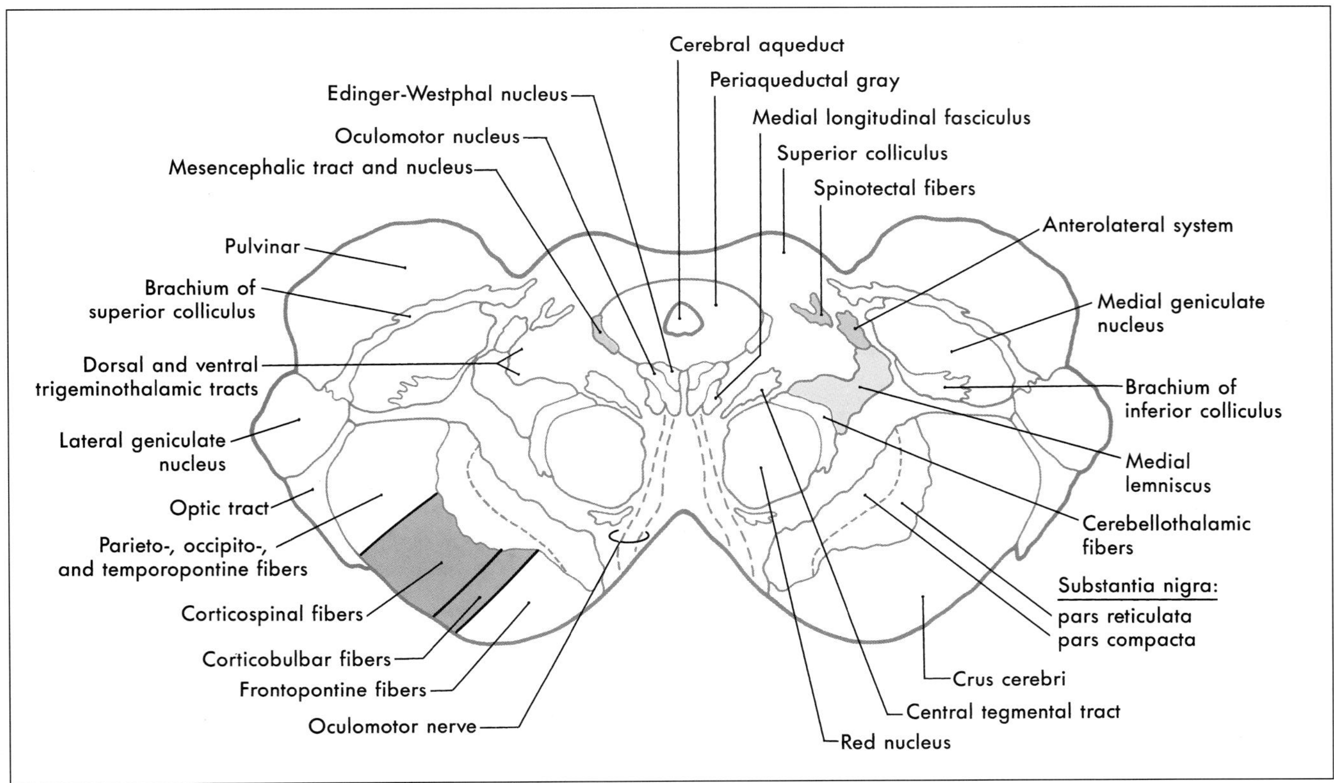

FIGURE 13-14 *Cross section of the midbrain at the mesencephalon-diencephalon junction. Correlate with Figure 13-7.*

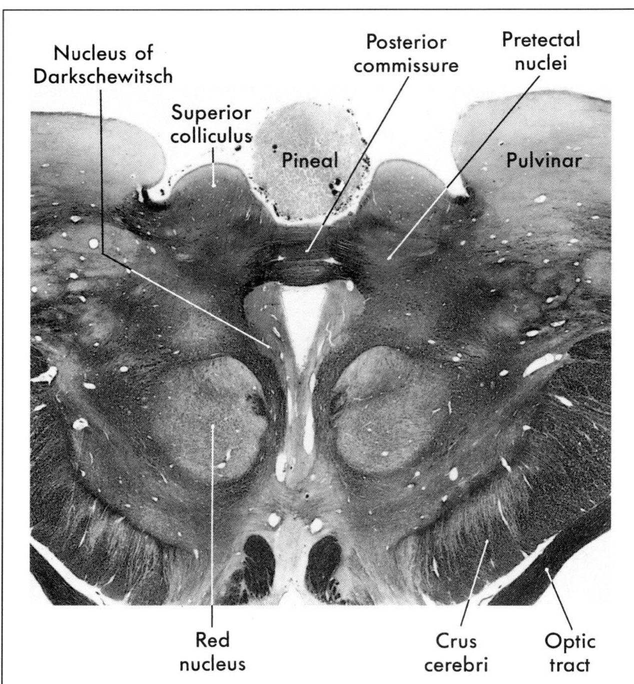

FIGURE 13-15 *A fiber (myelin) stained cross section of the midbrain at the mesencephalon-diencephalon junction, showing the pretectal area.*

lothalamic fibers. These are crossed ascending fibers from the decussation of the superior cerebellar peduncle. *Dorsal* and *ventral trigeminothalamic fibers* and the *medial lemniscus* occupy central and more lateral areas of the tegmentum (Fig. 13-12). Fibers of the *anterolateral system* are located at the lateral extreme of the medial lemniscus and, at this level, *spinotectal fibers* enter the deep layers of the superior colliculus (Fig. 13-13).

The structure of the *basis pedunculi* and the organization of fibers in the *crus cerebri* are the same in the rostral midbrain as more caudally (Fig. 13-12). The *substantia nigra* is functionally associated with the basal ganglia and is commonly divided into a *compact part* (pars compacta) and a *reticular part* (pars reticulata) (Figs. 13-12, 13-14). Cells of the reticular part project to the superior colliculus, thalamus, and pontine reticular formation, whereas those of the compact part project diffusely to the *caudate nucleus* and *putamen* where their synaptic terminals release dopamine. Parkinson's disease, a deficit characterized by tremor and difficulty in initiating or terminating movement, is associated with the loss of dopamine-containing cells in the compact part.

Midbrain-Diencephalon Junction The major tracts and nuclei seen at the level of the superior colliculus are still present in their same locations, at the mesencephalic-diencephalic interface. A comparison of Figures 13-12 and 13-14 reveals these similarities.

Groups of cells related primarily to the visual system and referred to as the *pretectal nuclei* are located at the rostral extent of the superior colliculus (Fig. 13-15). The pretectal nuclei include a controlling center for the *pupillary light reflex*. Neurons involved in this reflex receive bilateral retinal inputs and project bilaterally to visceral motor neurons in the Edinger-Westphal nucleus.

Several structures appear at this junction that signal the transition from midbrain to thalamus (Fig. 13-14). Laterally, a small part of the *pulvinar* is present, and portions of the *medial* and *lateral geniculate nuclei* also appear in the same plane. Fibers of the *brachium of the inferior colliculus* convey auditory signals to the *medial geniculate nucleus*. The optic tract contains visual fibers, some of which terminate in the lateral geniculate nucleus while others continue into the superior colliculus and pretectal area via the *brachium of the superior colliculus* (Figs. 13-14, 13-15; see also Fig. 13-4). The *ventral tegmental area* (*of Tsai*) is a diffuse cell group located ventromedial to the red nucleus and rostrally continuous with the lateral hypothalamic area. These neurons receive input from and project to hypothalamic and limbic structures that function in emotional behavior.

Reticular and Raphe Nuclei The reticular formation of the midbrain tegmentum is composed of the *cuneiform* and *subcuneiform nuclei* (Fig. 13-16). The midbrain reticular formation participates in the ascending systems that regulate states of consciousness. Many of these neurons project to the thalamus, especially the thalamic reticular nucleus, and the hypothalamus. This ascending fiber system is largely responsible for maintaining an alert, wakeful state and thus forms part of the *ascending reticular activating system*. Lesions involving the midbrain reticular formation can result in *hypersomnia*, which is characterized by slow respiration and an EEG pattern (large amplitude slow waves) indicative of a sleep state.

Located in ventral parts of the periaqueductal gray, the *nucleus raphe dorsalis* extends from the rostral pons into the caudal midbrain (Fig. 13-16). The serotonergic cells of this nucleus project to wide areas of the cerebral cortex, where they modulate neuronal activity involved in sleep/dream cycles.

Vasculature The blood supply to the midbrain originates from the *basilar artery* and its major branches (the *quadrigeminal* and *superior cerebellar arteries*) and, to a limited degree, from the *anterior choroidal artery*, a branch of the internal carotid (Fig. 13-17). Medial regions of the midbrain receive numerous small branches from the P_1 segment of the posterior cerebral artery and from the *posterior communicating artery*. These paramedian branches constitute the *posteromedial group* of branches from the circle of Willis (see Fig. 8-9). Included in their territory are the oculomotor, trochlear, and Edinger-Westphal nuclei; the exiting

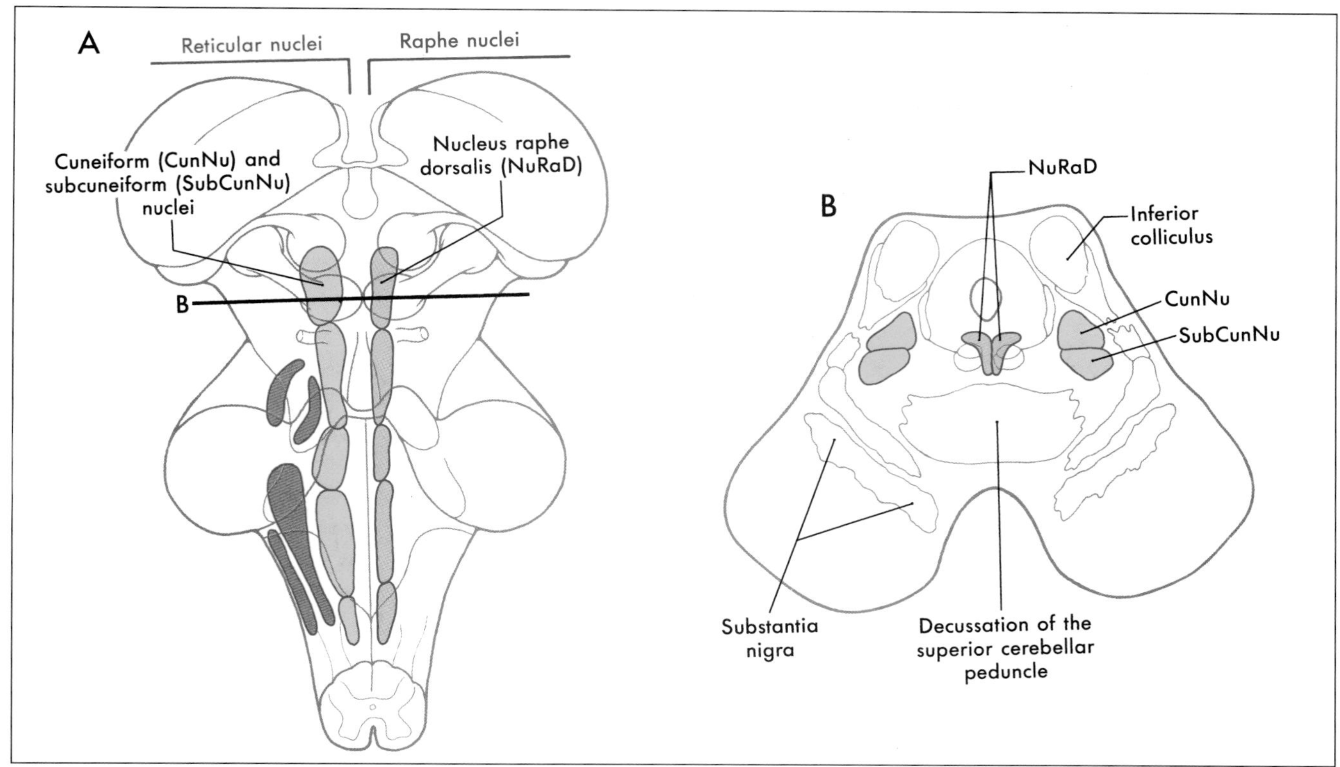

FIGURE 13-16 *Dorsal (A) view of the brainstem and a cross section (B) at the level of the inferior colliculus showing the raphe and reticular nuclei of the brainstem.*

oculomotor fibers; the red nucleus; and medial aspects of the substantia nigra and crus cerebri (Fig. 13-17). Occlusion of the paramedian branches that serve only the crus cerebri and oculomotor roots results in a contralateral hemiplegia (corticospinal involvement) and an ipsilateral oculomotor palsy with pupillary dilation (oculomotor involvement). This combination of deficits is *Weber's syndrome*.

Ventrolateral regions of the midbrain are served by penetrating branches of the *quadrigeminal artery* and by branches of the *anterior choroidal artery* (Fig. 13-17). The region served by these branches includes the lateral parts of the crus and substantia nigra and the medial lemniscus.

The dorsal midbrain is served primarily by the *quadrigeminal artery* (*collicular artery*), which typically arises from P_1 (Fig. 13-17). Much of the periaqueductal gray, the nuclei of the superior and inferior colliculi, the anterolateral system, and the brachium of the inferior colliculus are served by quadrigeminal branches. Additional blood supply to the area surrounding the exit of the trochlear nerve and the inferior colliculus arises from medial branches of the *superior cerebellar artery*.

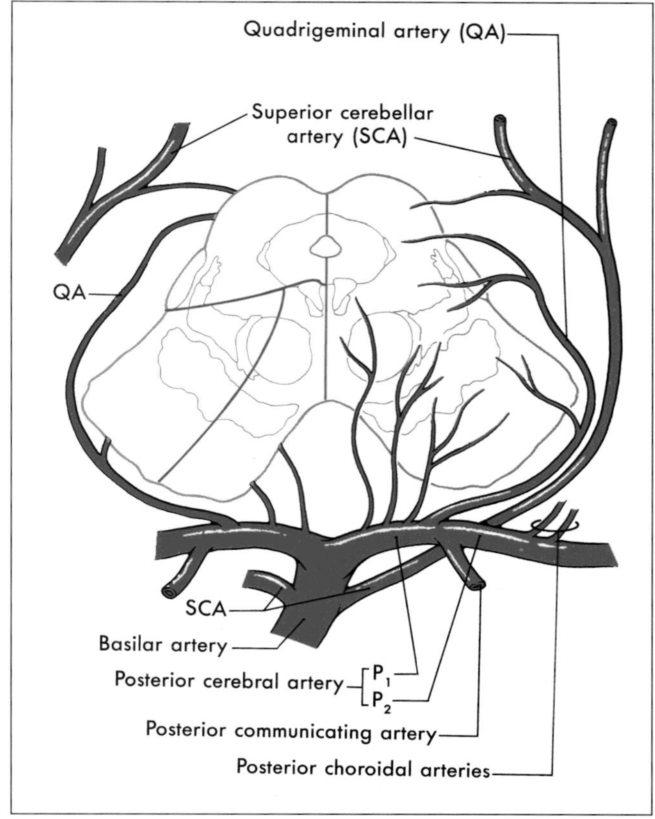

FIGURE 13-17 *Blood supply of the midbrain. Arteries are shown mainly on the right and the territories served by each on the left.*

SOURCES AND ADDITIONAL READINGS

Bobillier P, Seguin S, Petitjean F, Salvert D, Tovret M, Jouvet M: The raphe nuclei of the cat brain stem: a topographical atlas of their efferent projections as revealed by autoradiography. Brain Res 113:449–486, 1976.

Bogerts B: A brainstem atlas of catecholaminergic neurons in man, using melanin as a natural marker. J Comp Neurol 197:63–80, 1981.

Brodal A: *The Reticular Formation of the Brainstem, Anatomical Aspects and Functional Correlations*. Charles C Thomas, Springfield, IL, 1958.

Brodal A: *Neurological Anatomy*, 3rd Ed. Oxford University Press, New York, 1981.

Carpenter MB, Sutin J: *Human Neuroanatomy*, 8th Ed. Williams & Wilkins, Baltimore, 1983.

Crosby EC, Humphrey T, Lauer EW: *Correlative Anatomy of the Nervous System*. Macmillan Publishing, New York, 1962.

Duvernoy HM: *Human Brainstem Vessels*. Springer-Verlag, Berlin, 1978.

Haines DE: *Neuroanatomy, An Atlas of Structures, Sections, and Systems*, 4th Ed. Williams & Wilkins, Baltimore, 1995.

Hobson JA, Brazier MAB (eds): *The Reticular Formation Revisited: Specifying Function for a Nonspecific System*. Int. Brain Res. Organization Monogram Series Vol 6. Raven Press, New York, 1980.

Hubbard JE, DiCarlo V: Fluorescence histochemistry of monoamine-containing cell bodies in the brain stem of the squirrel monkey (*Saimiri sciureus*) III. Serotonin-containing groups. J Comp Neurol 153:385–398, 1974.

Larsell O, Jansen J: *The Comparative Anatomy and Histology of the Cerebellum, The Human Cerebellum, Cerebellar Connections, and Cerebellar Cortex*. The University of Minnesota Press, Minneapolis, 1972.

Nieuwenhuys R: *Chemoarchitecture of the Brain*. Springer-Verlag, Berlin, 1985.

Nieuwenhuys R, Voogd J, van Huijzen CHR: *The Human Central Nervous System, A Synopsis and Atlas*, 3rd Ed. Springer-Verlag, Berlin, 1988.

Olszewski J, Baxter D: *Cytoarchitecture of the Human Brain Stem*, 2nd Ed. S. Karger, Basel, 1982.

Palay SL, Chan-Palay V: *Cerebellar Cortex, Cytology and Organization*, Springer-Verlag, New York, 1974.

Paxinos G (ed): *The Human Nervous System*, Academic Press, San Diego, 1990. Chapters 7-14.

Taber E, Brodal A, Walberg F: The raphe nuclei of the brain stem in the cat. I. Normal topography and cytoarchitecture and general discussion. J Comp Neurol 114:161–187, 1960.

Weber JT, Martin GF, Behan M, Huerta MF, Harting JK: The precise origin of the tectospinal pathway in three common laboratory animals: a study using the horseradish peroxidase method. Neurosci Lett 11:121–127, 1979.

CHAPTER FOURTEEN

The Diencephalon

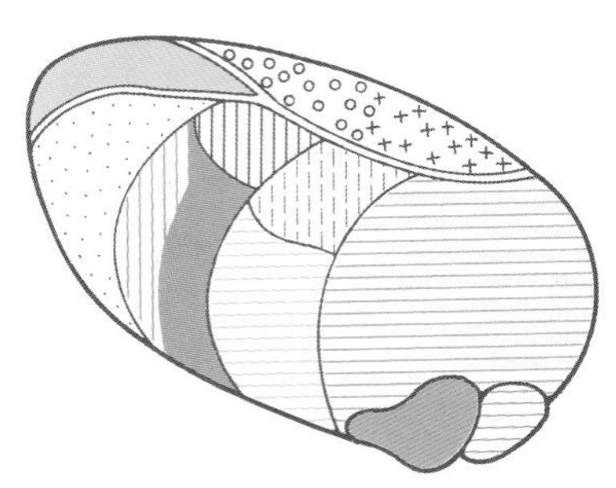

G. A. MIHAILOFF • D. E. HAINES

Although considered by some to be part of the brainstem, the diencephalon is treated here as a portion of the forebrain. The diencephalon includes the *dorsal thalamus*, *hypothalamus*, *ventral thalamus*, and the *epithalamus*, and it is situated between the telencephalon and the brainstem. In general, the diencephalon is the main processing center for information destined to reach the cerebral cortex from all ascending sensory pathways (except those related to olfaction) and numerous other subcortical cell groups. The right and left halves of the diencephalon, for the most part, contain symmetrically distributed cell groups separated by the space of the third ventricle.

OVERVIEW

The *dorsal thalamus*, or *thalamus* as it is commonly called, is the largest of the four principal subdivisions of the diencephalon and consists of pools of neurons that collectively project to all areas of the cerebral cortex. Some of the thalamic nuclei receive somatosensory, visual, or auditory input and transmit this information to the appropriate area of the cerebral cortex. Other thalamic nuclei receive input from subcortical motor areas and project to those parts of the overlying cortex that influence the successful execution of a motor act. A few thalamic nuclei receive a more diffuse input and, in turn, relate in a more diffuse way to widespread areas of cortex.

The *hypothalamus* is also composed of multiple nuclear subdivisions and is connected primarily to the forebrain, brainstem, and spinal cord. This part of the diencephalon is involved in the control of visceromotor (autonomic) functions. In this respect, the hypothalamus regulates functions that are "automatically" adjusted (such as blood pressure, temperature, etc.) without our being aware of the change. In contrast, conscious sensation and voluntary motor control are mediated by the dorsal thalamus.

The *ventral thalamus* and *epithalamus* are the smallest subdivisions of the diencephalon. The former includes the subthalamic nucleus, which is linked to the basal ganglia of the forebrain and functions in the motor sphere; lesions in the subthalamus give rise to very characteristic involuntary movement disorders. The epithalamus is functionally related to the limbic system.

DEVELOPMENT OF THE DIENCEPHALON

The cell groups that give rise to the diencephalon form in the caudomedial portion of the prosencephalon, bordering on the space that will become the third ventricle. The developing brain at this level consists initially of a roof plate and the two alar plates; it lacks a well-defined floor plate and basal plates.

A shallow groove appears in the wall of the third ventricle and extends rostrally from the developing cerebral aqueduct to the ventral edge of the interventricular foramen (Fig. 14-1A,B). This groove, the *hypothalamic sulcus*, divides the alar plate into a dorsal area, the future *dorsal thalamus*, and a ventral portion, the future *hypothalamus*. The third ventricle, which is the vertically oriented, midline space of the diencephalon, is continuous with the paired lateral ventricles via the *interventricular foramina* (Fig. 14-1B). The dorsal thalamus on each side of the third ventricle increases rapidly in size and, in many brains, will partially fuse across the space of the third ventricle to form the *massa intermedia* or *interthalamic adhesion*. This structure is present in about 80% of the general population.

The epithalamus develops from the caudal portion of the roof plate (Fig. 14-1B). By the seventh week, a small thickening of the roof plate forms. It gradually increases in size, and evaginates to form the *epiphysis*, which develops into the *pineal gland* of the adult. The portion of the roof plate immediately rostral to the epiphysis gives rise to the *habenula*, a small thickening in which the habenular nuclei will develop.

Just anterior to the habenular region, the roof plate epithelium and adjacent pia mater give rise to the choroid plexus of the third ventricle which, in the adult, remains suspended from the roof of this space (Fig. 14-1B–D). This choroid plexus is continuous through the interventricular foramina with that of the lateral ventricles. Elsewhere, in locations around the perimeter of the third ventricle, specialized patches of ependyma lie on the midline and form unpaired structures called the *circumventricular organs*. These structures include the subfornical organ, the organum vasculosum of the lamina terminalis, the subcommissural organ, and the pineal gland (Fig. 14-2). These cellular regions are characterized by the presence of fenestrated capillaries, which implies an absence of the blood-brain barrier. These structures are thought to release metabolites and neuropeptides into the cerebrospinal fluid or into the cerebrovascular system.

The development of the pituitary gland during the third week is linked to that of the diencephalon (Fig. 14-1B,C). A downward extension of the floor of the third ventricle, the *infundibulum*, meets *Rathke's pouch*, an upward outpocketing of the stomodeum, the primitive oral cavity. By the end of the second month, Rathke's pouch loses its connection with the developing oral cavity but maintains its attachment to the infundibulum. As development continues, Rathke's pouch gives rise to the *anterior lobe* (*adenohypophysis*) and *pars intermedia* of the pituitary gland, whereas the infundibulum differentiates

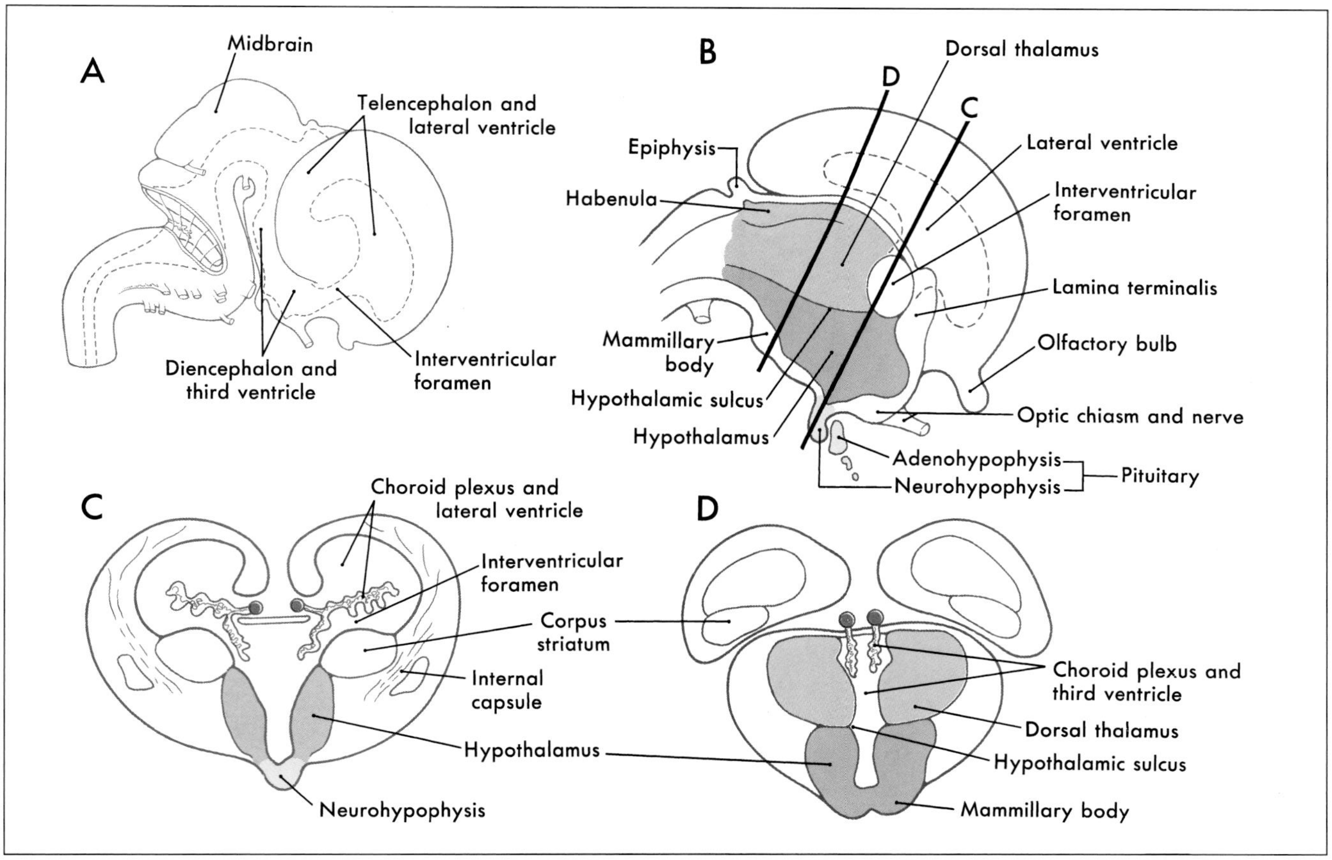

FIGURE 14-1 *Development of the diencephalon. Lateral (A) and mid-sagittal (B) views of the forebrain at about 8 to 9 weeks of gestational age. The cross-sectional views (C,D) are taken from the planes shown in B and emphasize diencephalic structures.*

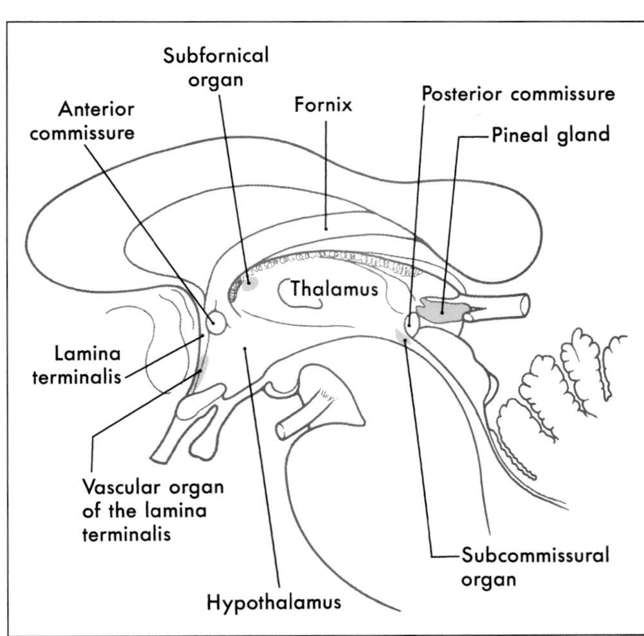

FIGURE 14-2 *Mid-sagittal view showing the locations of circumventricular organs.*

into the *posterior lobe* of the pituitary gland, or *neurohypophysis* (Fig. 14-1B). A *craniopharyngioma* (Rathke's pouch tumor) can arise from a portion of Rathke's pouch that fails to undergo proper migration and apposition to the infundibulum. These tumors mimic lesions of the pituitary and may cause visual problems, diabetes insipidus, and increased intracranial pressure.

BASIC ORGANIZATION

The junction between the diencephalon and midbrain lies along a line extending from the posterior commissure to the caudal edge of the mammillary body on the medial aspect of the hemisphere (Fig. 14-3). On the ventral surface of the hemisphere, this interface is represented by a line following the caudal edge of the optic tract (Fig. 14-4). The boundary between the diencephalon and surrounding telencephalon is less distinct and is represented laterally by the internal capsule (discussed later) and rostrally by the interventricular foramen, lamina terminalis, and optic chiasm (Fig. 14-3).

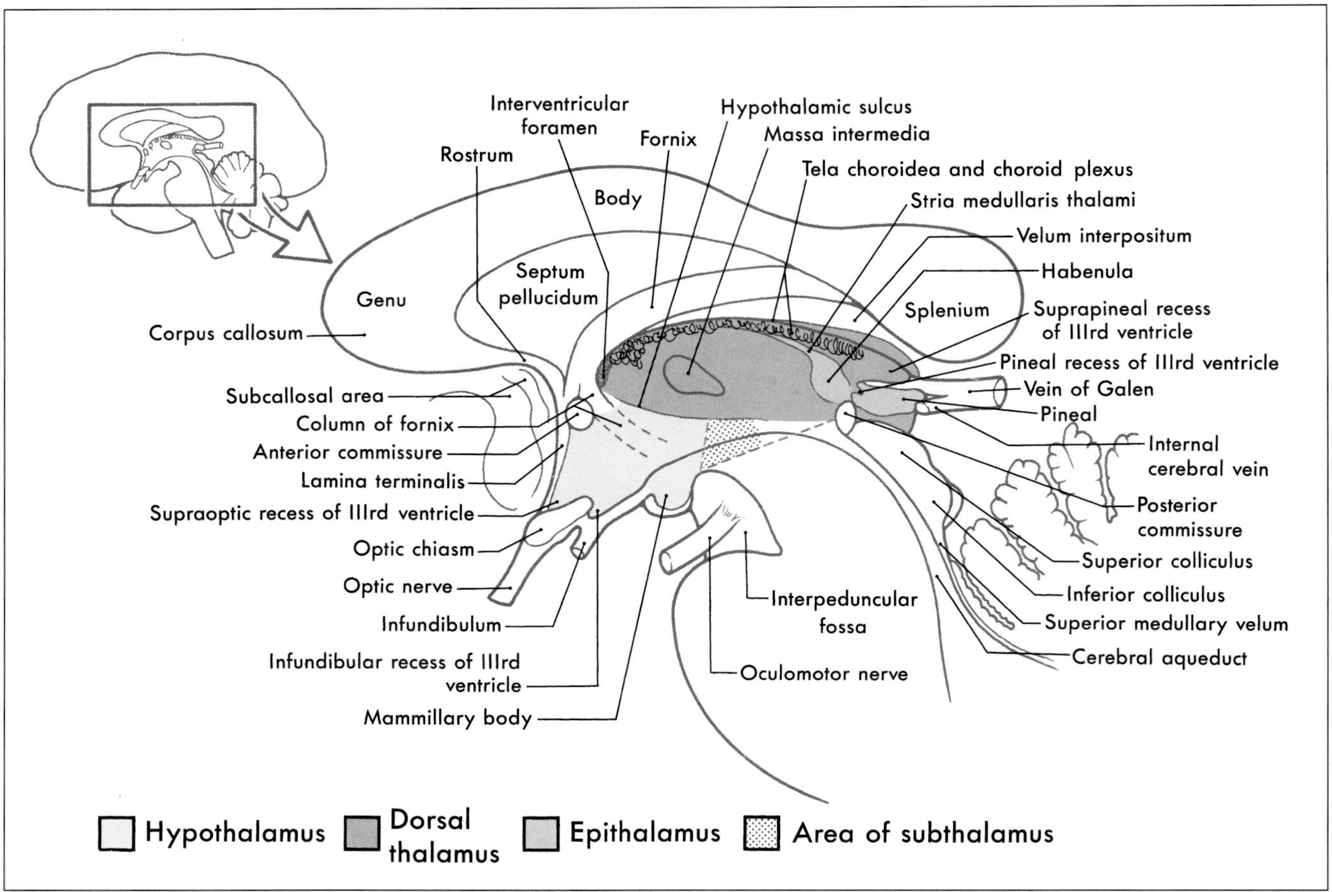

Interventricular foramen
Hypothalamic sulcus
Massa intermedia
Fornix
Rostrum
Tela choroidea and choroid plexus
Stria medullaris thalami
Body
Velum interpositum
Habenula
Septum pellucidum
Genu
Splenium
Suprapineal recess of IIIrd ventricle
Corpus callosum
Pineal recess of IIIrd ventricle
Vein of Galen
Pineal
Subcallosal area
Column of fornix
Anterior commissure
Lamina terminalis
Internal cerebral vein
Posterior commissure
Supraoptic recess of IIIrd ventricle
Optic chiasm
Superior colliculus
Optic nerve
Inferior colliculus
Interpeduncular fossa
Infundibulum
Superior medullary velum
Cerebral aqueduct
Infundibular recess of IIIrd ventricle
Oculomotor nerve
Mammillary body
Hypothalamus
Dorsal thalamus
Epithalamus
Area of subthalamus

FIGURE 14-3 *Mid-sagittal view of the diencephalon and closely related structures. This is a drawing of the specimen shown in Figure 14-5.*

Optic chiasm
Anterior perforated substance
Optic tract
Crus cerebri
Posterior perforated substance
Infundibulum
Lateral eminence
Median eminence
Mammillary body

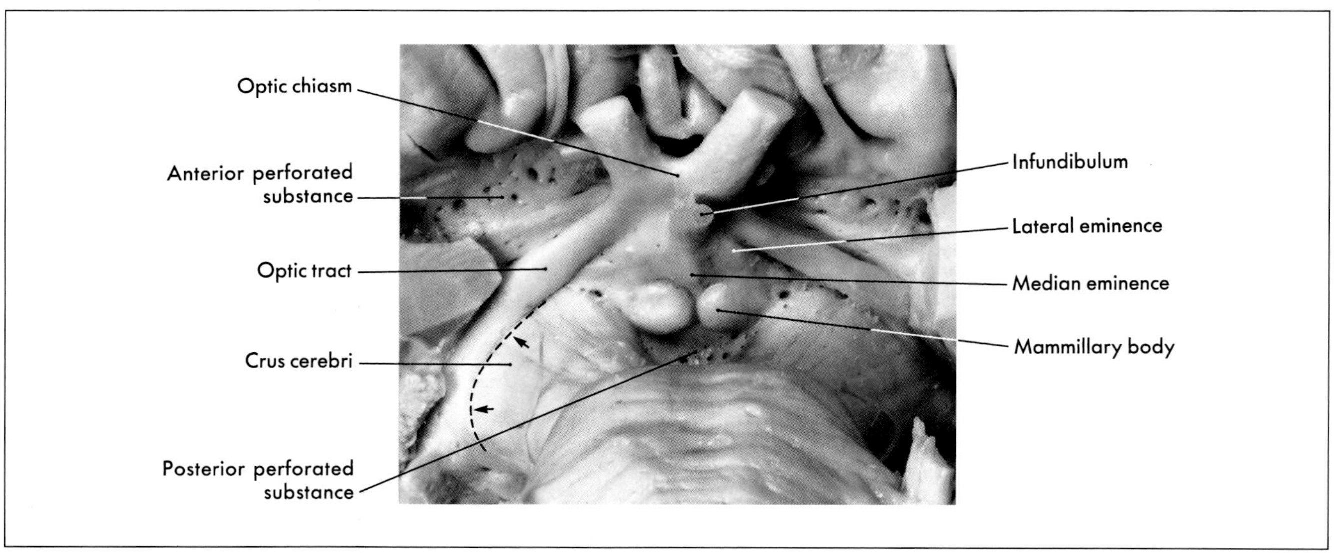

FIGURE 14-4 *Ventral view of the hemisphere emphasizing diencephalic structures visible on the surface and showing the diencephalic-mesencephalic interface as represented by the caudal edge of the optic tract (arrows).*

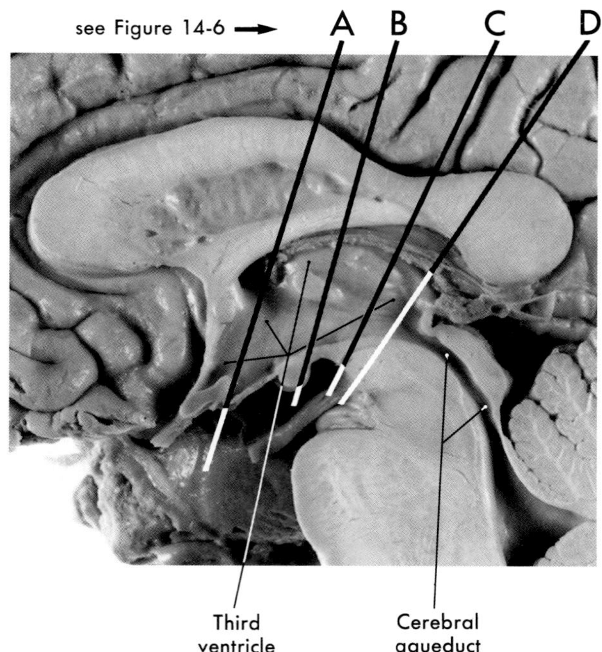

FIGURE 14-5 *Mid-sagittal view of the diencephalon. This view correlates with the drawing in Figure 14-3. The four lines indicate the planes of the stained sections in Figure 14-6.*

The cavity of the diencephalon, the *third ventricle*, is a narrow, vertically oriented midline space located between the paired dorsal thalami and hypothalami of the two sides (Figs. 14-5, 14-6). In addition to its connections with the lateral ventricles and the cerebral aqueduct, the third ventricle has small evaginations or *recesses* associated with the optic chiasm, the infundibulum, and the pineal gland (Figs. 14-1, 14-3).

All four diencephalic subdivisions can be approximated in a mid-sagittal section of the forebrain (Figs. 14-3, 14-5). The *dorsal thalamus* is located dorsal to the hypothalamic sulcus and extends from the interventricular foramen caudally to the level of the splenium of the corpus callosum. The *hypothalamus* lies ventral to the hypothalamic sulcus and is bordered anteriorly by the lamina terminalis and caudally by a line that extends from the posterior aspect of the mammillary body dorsally to intersect with the hypothalamic sulcus. The only diencephalic structures visible on the ventral surface of the hemisphere are those related to the hypothalamus, including the optic chiasm, infundibulum, medial and lateral eminences, and mammillary bodies (Fig. 14-4). The *ventral thalamus* (subthalamus) does not border on the ventricle; rather it occupies a position caudal to the hypothalamus, rostral to the diencephalon-midbrain junction, and *lateral to the midline* (Figs. 14-3, 14-6B). Epithalamic structures are located posteriorly and dorsally, in close apposition to the posterior commissure, and include the pineal gland, the habenular nuclei, and the main afferent bundle of these nuclei, the stria medullaris thalami.

DORSAL THALAMUS (THALAMUS)

The *dorsal thalamus* (or *thalamus*) is a massive collection of neuronal cell groups that participate in a widely diverse array of functions involving motor, sensory, and limbic systems. Typically, thalamic output neurons project to the cerebral cortex; in fact, very little information reaches the cerebral cortex without first being processed by thalamic neurons. As a result, the thalamus is often regarded as the functional "gateway" to the cerebral cortex. In turn, nearly all regions of the cerebral cortex give rise to reciprocal projections that return to the thalamic region from which they originally received input.

The thalamus is covered on its lateral aspect by a layer of myelinated axons, the *external medullary lamina*, which includes fibers that enter or leave the subcortical white matter (Fig. 14-6B,C). Within the external medullary lamina are clusters of neurons that form the *thalamic reticular nucleus*. The medial surface of the thalamus borders the third ventricle, and the external medullary lamina and thalamic reticular nucleus blend with the thalamic fasciculus and zona incerta, respectively, to form an interface between dorsal and ventral thalami (Fig. 14-6B).

An *internal medullary lamina*, also consisting of myelinated fibers, extends into the substance of the thalamus, where it forms partitions or boundaries that divide the thalamus into its principal cell groups (Fig. 14-7): the *anterior*, *medial*, *lateral*, and *intralaminar nuclear groups*. The latter are located in the portion of the internal medullary lamina that separates the lateral and medial nuclear groups. In addition, there are *midline thalamic nuclei*, and a diffuse population of neurons located in the external medullary lamina, the *thalamic reticular nucleus*.

Finally, attached to the caudolateral portion of the thalamus are the medial and lateral geniculate bodies (Figs. 14-6D, 14-7). Although considered here as components of the lateral nuclear group, the geniculate nuclei are sometimes considered as a separate part of the thalamus, the *metathalamus*.

Anterior Thalamic Nuclei This group of cells consists of a large principal nucleus and two smaller nuclei; collectively, these cell groups form the *anterior nucleus of the thalamus* (Figs. 14-6A, 14-7). The anterior nucleus forms a prominent wedge on the rostral aspect of the dorsal thalamus just caudolateral to the interventricular foramen; this wedge is the *anterior thalamic tubercle*. Rostrally, the inter-

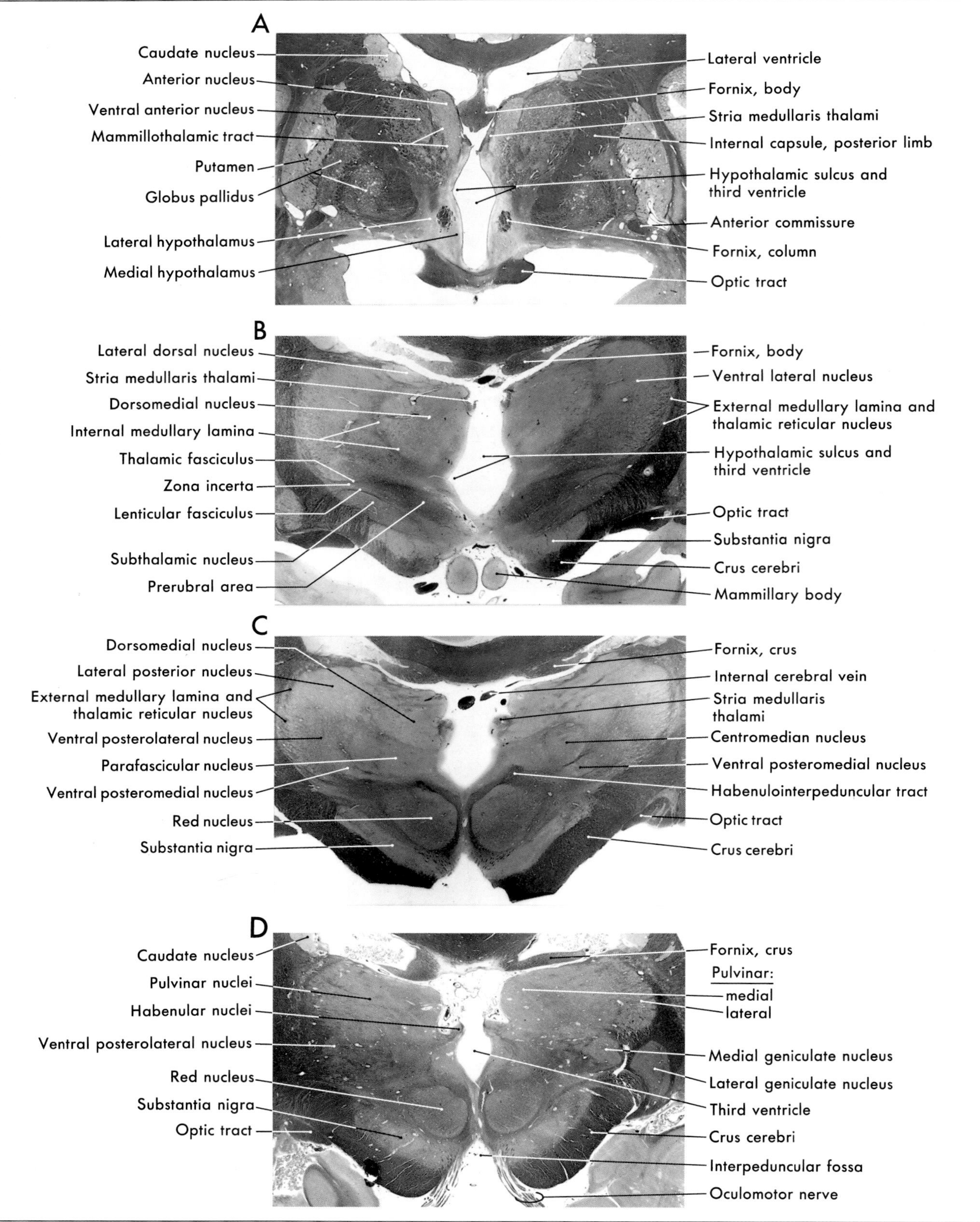

FIGURE 14-6 *Four levels of the forebrain from rostral (A) to caudal (D) showing the internal structure of the hemisphere with emphasis on the diencephalon. These levels correlate with those shown in Figure 14-5, and with the planes represented in the exploded view in Figure 14-7. Weil stain.*

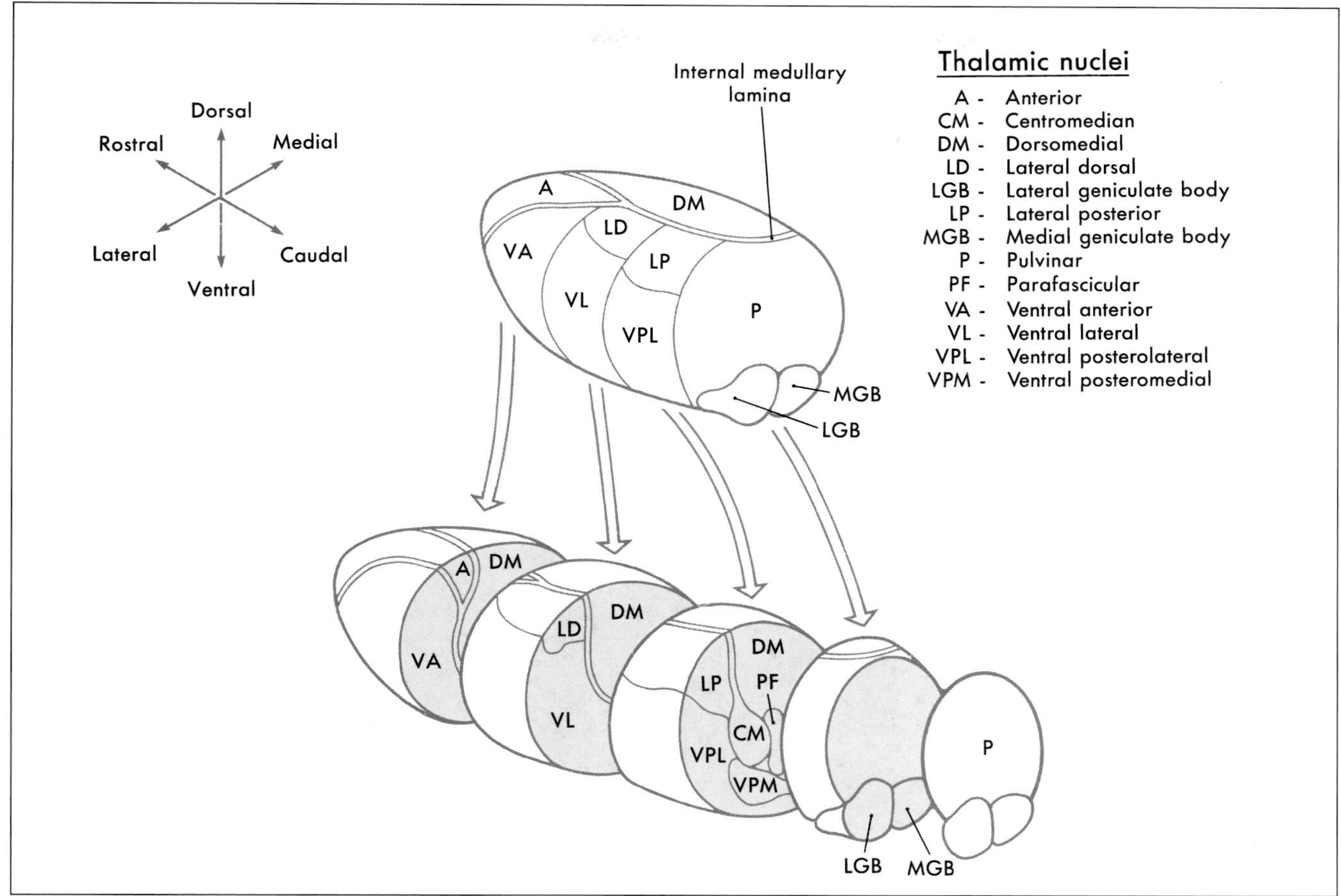

FIGURE 14-7 *Exploded view of the dorsal thalamus illustrating the organization of thalamic nuclei. Compare with Figure 14-6A-D.*

nal medullary lamina divides to partially encapsulate the anterior nucleus. The cells of this nucleus receive dense limbic-related projections from (1) the mammillary nuclei via the mammillothalamic tract and (2) the medial temporal lobe via the fornix. The output of this nucleus is primarily directed to the cingulate gyrus through the anterior limb of the internal capsule (Fig. 14-8).

Medial Thalamic Nuclei This region of the dorsal thalamus comprises the *dorsomedial nucleus*. This expansive group of neuronal cell bodies is composed of large parvicellular (located caudally) and magnocellular (located rostrally) parts, and of a small paralaminar part adjacent to the internal medullary lamina (Figs. 14-6B–D, 14-7). The two larger portions are linked to parts of the frontal and temporal lobes and to the amygdaloid complex (Fig. 14-8). Cells of the paralaminar subdivision receive input from the frontal lobe and substantia nigra and may play a role in the control of eye movement.

Lateral Thalamic Nuclei This large collection of thalamic neurons is grouped into *dorsal* and *ventral tiers*. The relatively small group of *dorsal tier* nuclei includes the *lateral dorsal* and *lateral posterior nuclei* along with the much larger *pulvinar nucleus* (pulvinar) (Figs. 14-6B–D, 14-7). The connections of the lateral dorsal and lateral posterior nuclei are formed with the cingulate gyrus and parietal lobe respectively (Fig. 14-8). The large *pulvinar nucleus* consists of anterior, medial, lateral, and inferior subdivisions. The inferior division receives input from the superior colliculus and projects to visual association cortex. Other portions of the pulvinar project to areas of the temporal, parietal, and frontal lobes that are especially concerned with visual function and eye movements (Fig. 14-8).

The large *ventral tier* of the lateral group consists of three separate nuclei (Figs. 14-6A–D, 14-7). The *ventral anterior nucleus* (VA) and the slightly more caudal *ventral lateral nucleus* (VL) are important motor-related nuclei, whereas the *ventral posterior nucleus* (*consisting of ventral posterolateral and ventral posteromedial nuclei*) conveys somatosensory information to the cerebral cortex. The *VA* is composed of a large parvicellular portion and a small magnocellular part. The former receives input from the medial segment of the globus pallidus and the latter receive afferents from the reticular portion of sub-

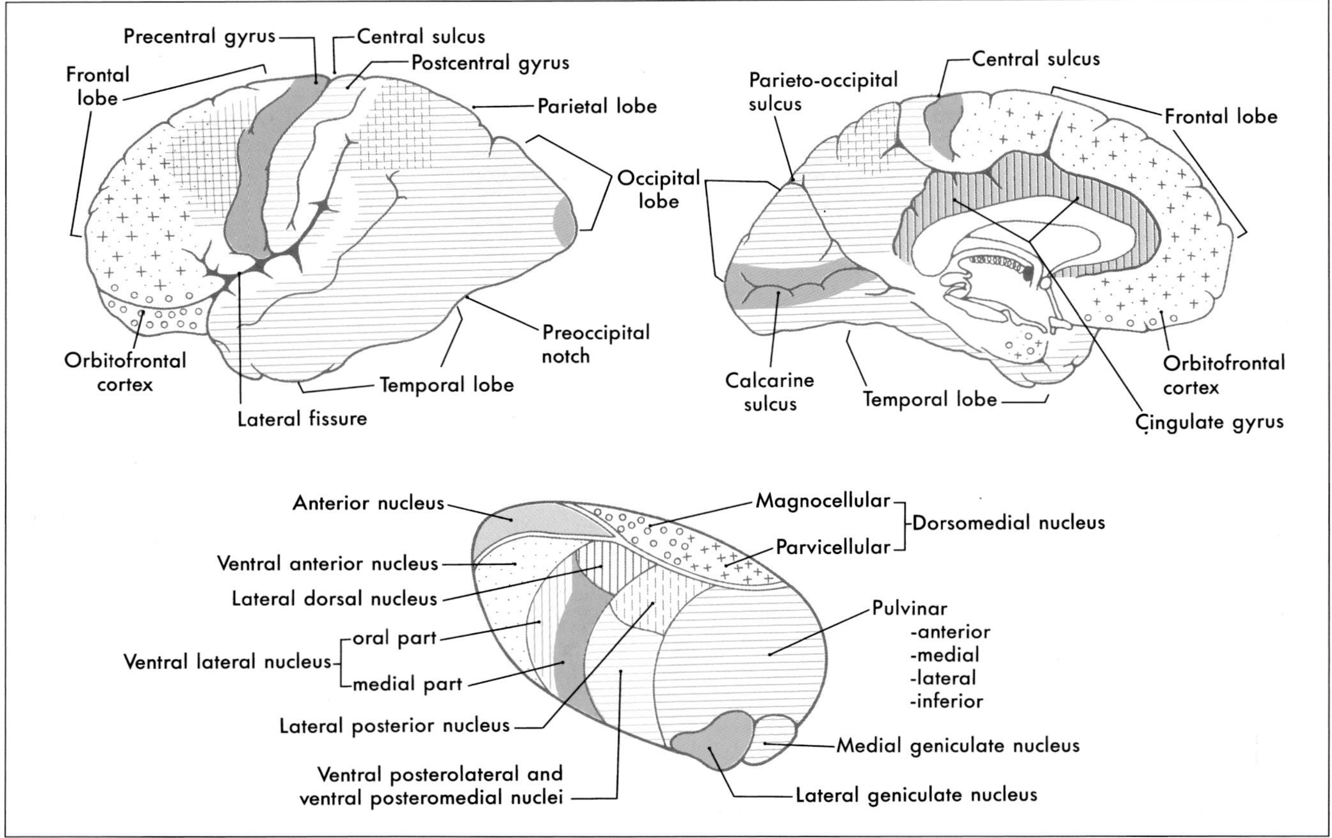

FIGURE 14-8 *Relationship of the thalamic nuclei with the cerebral cortex as seen by the patterns of thalamocortical connections. Each thalamic nucleus is pattern-coded or color-coded to match its target area in the cerebral cortex.*

stantia nigra. The efferent projections from *VA* are diffuse and appear to include selected parts of the frontal lobe (Fig. 14-8).

The *VL* (Figs. 14-6B, 14-7) is also composed of three subdivisions, a pars oralis, a pars medialis, and a pars caudalis. The largest of these, the pars oralis, receives a dense projection from the internal segment of the globus pallidus; some of these afferents enter the caudal subdivision. In contrast, the medial subdivision receives its main input from the cerebellar nuclei. Consequently, pallidal and cerebellar projections are largely *segregated* within this nucleus. The output of *VL* reflects its segregated input in that the oral and medial parts project to separate areas of the frontal lobe (Fig. 14-8).

The *ventral posterior nucleus* is divided into a larger *ventral posterolateral nucleus* (*VPL*) and a smaller *ventral posteromedial nucleus* (*VPM*), both of which receive somatosensory input (Figs. 14-6C, 14-7). The medial lemniscus and spinothalamic fibers terminate in a somatotopic manner (cervical fibers medial, sacral fibers lateral) within *VPL*, whereas trigeminothalamic fibers from the spinal trigeminal nucleus and the principal trigeminal sensory nucleus terminate in *VPM*. Both *VPL* and *VPM* project to the postcentral gyrus of the frontal lobe (Fig. 14-8).

A small group of cells called the *ventral posterior inferior nucleus* is situated ventrally between VPL and VPM. These cells process vestibular input and project to lateral areas of the postcentral gyrus that are located in the depths of the central sulcus. Similarly, a small group of cells forming the rostral (oral) portion of VPL receives cerebellar input and projects to the precentral gyrus of the frontal lobe; this nucleus probably represents a few cells that have been displaced from the slightly more rostrally located VL. This cell group is also called the *ventral intermediate nucleus* because of its location between VL and VPL.

The *lateral* and *medial geniculate nuclei* are considered parts of the lateral nuclear group (Figs. 14-6D, 14-7). The medial geniculate nucleus receives ascending *auditory* input via the brachium of the inferior colliculus and projects to the primary auditory cortex in the temporal lobe. The lateral geniculate nucleus receives *visual* input from the retina via the optic tract and in turn projects to the primary visual cortex on the medial surface of the occipital lobe (Fig. 14-8).

Located in the posterior thalamus at about the level of the pulvinar and geniculate nuclei is a cluster of cell groups collectively called the *posterior nuclear complex*. This complex consists of the suprageniculate nucleus, nucleus limitans, and the posterior nucleus. These nuclei are positioned dorsal to the medial geniculate and medial to the rostral pulvinar. The posterior nuclear complex receives and sends to the cortex nociceptive cutaneous input transmitted over somatosensory pathways.

Intralaminar Nuclei Embedded within the internal medullary lamina are the discontinuous groups of neurons that form the *intralaminar nuclei*. These cells are characterized by their projections to the striatum and to other thalamic nuclei, along with diffuse projections to the cerebral cortex. Two of the most prominent are the *centromedian* and *parafascicular nuclei* (Fig. 14-7). The centromedian nucleus projects to the striatum and to motor areas of the cerebral cortex, whereas the parafascicular nucleus projects to rostral and lateral areas of the frontal lobe. Other intralaminar nuclei receive input from ascending pain pathways and project to somatosensory and parietal cortex.

Midline Nuclei The midline nuclei are the least understood components of the thalamus. The largest is the *paratenial nucleus*, which is located just ventral to the rostral portion of the stria medullaris thalami; other cells are associated with the interthalamic adhesion (massa intermedia). Although inputs are poorly defined, efferent fibers reach the amygdaloid complex and the anterior cingulate cortex, thus suggesting a role in the regulation of visceral function.

Thalamic Reticular Nucleus The cells of this nucleus are situated within the external medullary lamina and between this lamina and the internal capsule (Fig. 14-6B,C). Axons of these cells project medially into the nuclei of the dorsal thalamus or to other parts of the reticular nucleus, but not into the cerebral cortex. Afferents are received from the cortex and from nuclei of the dorsal thalamus via collaterals of thalamocortical and corticothalamic axons. It appears that thalamic reticular neurons modulate, or gate, the responses of thalamic neurons to incoming cerebral cortical input.

Summary of Thalamic Organization Each thalamic nucleus (with a few exceptions) gives rise to efferent projections (thalamocortical axons) that target some portion of the cerebral cortex. That region of cortex then provides a reciprocal projection (corticothalamic axons) that returns to the original thalamic nucleus. Figure 14-8 reviews the major thalamocortical relationships.

The nuclei of the thalamus can also be divided into *specific* and *nonspecific* categories. The medial and lateral geniculate nuclei and VPL and VPM are classically described as *specific* because their input is derived from a singular, restricted sensory system. Accordingly, all other thalamic nuclei are considered to be *nonspecific*. Another and perhaps more useful scheme divides the thalamic nuclei into *relay nuclei* or *association nuclei*. The *relay nuclei* are those that project to a single, functionally uniform region of cortex; they include the VL, VPL, VPM, and geniculate nuclei. The *association nuclei*, by contrast, project to association cortex (i.e., nonmotor, nonsensory cortex); they include the pulvinar complex and the VA nucleus. Unfortunately, neither of these two schemes deals unequivocally with all thalamic nuclei. For example, VL is described by some authors as a specific nucleus, whereas others regard it as nonspecific. Similarly, VA is regarded by some as a relay nucleus, but others consider it to be an association nucleus. Chapter 31 contains additional information on these topics.

THE INTERNAL CAPSULE

Axons pass between the diencephalon, particularly the dorsal thalamus, and the cerebral cortex in a fan-shaped mass of fibers, the *internal capsule*, that passes from the central core of the hemisphere into the brainstem (Figs. 14-6A–D, 14-9). Even though this structure consists mostly of axons that reciprocally link the thalamus and cerebral cortex, it also contains cortical efferent fibers that project to the brainstem (corticorubral, corticoreticular, corticobulbar) or spinal cord (corticospinal).

Although the internal capsule is described in detail in Chapter 15, it is summarized here because of its important relationship to the thalamus. As seen in axial section (Fig. 14-9) the internal capsule consists of an *anterior limb*, *genu*, and *posterior limb*. The genu is located immediately lateral to the anterior thalamic nucleus, at about the same level as the interventricular foramen. The anterior limb extends rostrolateral from the genu and is insinuated between the caudate and lenticular nuclei. The posterior limb extends caudolateral from the genu and consists of a large part separating the thalamus from the globus pallidus and smaller portions that link temporal and occipital lobes with thalamic nuclei.

HYPOTHALAMUS

Unlike the thalamus, which is primarily related to somatic functions, the hypothalamus is mainly involved in *visceromotor*, *viscerosensory*, and *endocrine* activities. The hypothalamus and related limbic structures receive sensory input regarding the internal environment and, in turn, regulate through four mechanisms the motor systems that modify the internal environment. *First*, the hypothalamus is a principal modulator of autonomic

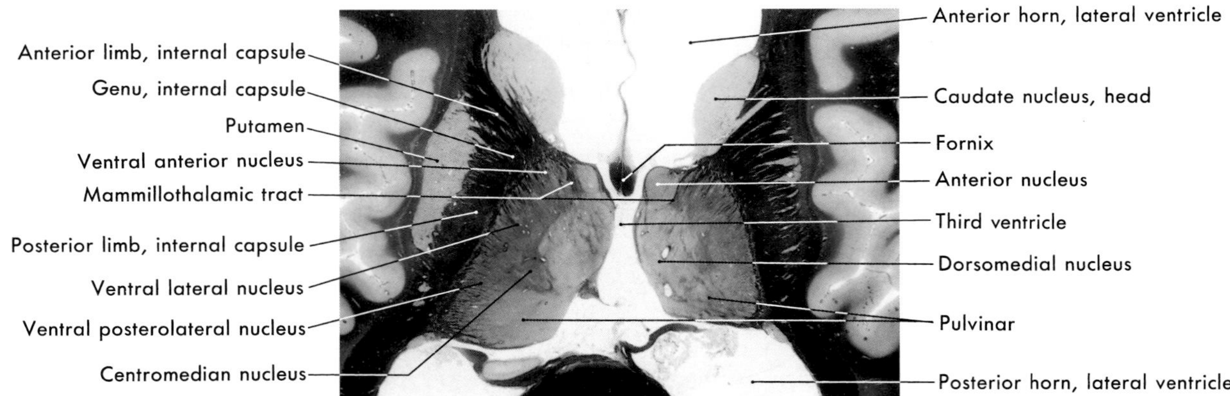

FIGURE 14-9 *Axial view of the forebrain showing the relationship of the thalamus to the limbs of the internal capsule. Weil stain.*

nervous system function. *Second*, it is a viscerosensory transducer, containing neurons with specialized receptors capable of responding to changes in the temperature or osmolality of blood, as well as specific hormonal levels in the general circulation. *Third*, the hypothalamus regulates the activity of the anterior pituitary through the production of releasing factors (hormone-releasing hormones), and, *fourth*, it performs an endocrine function by producing and releasing oxytocin and vasopressin into the general circulation within the posterior pituitary.

The hypothalamus can be divided into lateral, medial, and periventricular zones (Figs. 14-6A, 14-10). The *lateral zone*, often called the *lateral hypothalamic area*, extends the full rostrocaudal length of the hypothalamus and is separated from the medial zone by a line drawn through the fornix in the sagittal plane (Fig. 14-11C). The *medial zone* is divided from rostral to caudal into three regions: the chiasmatic, the tuberal, and the mammillary regions (Figs. 14-10, 14-11A–D). The *periventricular zone* includes the neurons that border the ependymal surfaces of the third ventricle.

Lateral Hypothalamic Zone The *lateral hypothalamic area* (Fig. 14-11A–D) is composed of diffuse clusters of neurons intermingled with longitudinally oriented axon bundles. The latter, which form the *medial forebrain bundle*, are diffusely organized in the human brain. No discrete named nuclei are present in this lateral area, although the supraoptic nucleus (see below) is partially located in it. Cells of the lateral hypothalamic area are involved in cardiovascular function and in the regulation of food and water intake.

Medial Hypothalamic Zone In contrast to the lateral zone, the medial hypothalamus contains well-defined groups of neurons whose function and connections are established. Within the *chiasmatic* (anterior) *region* are five nuclei: the *preoptic*, *supraoptic*, *paraventricular*, *anterior*, and *suprachiasmatic nuclei* (Fig. 14-11A,B). Nuclei in the chiasmatic region are generally involved in regulating hormone release (preoptic, supraoptic, periventricular), cardiovascular function (anterior), circadian rhythms (suprachiasmatic), and body temperature and heat loss mechanisms (preoptic). In the *tuberal region* are the *dorsomedial*, *ventromedial*, and *arcuate nuclei* (Fig. 14-11A,C). The *ventromedial nucleus* is regarded as the food

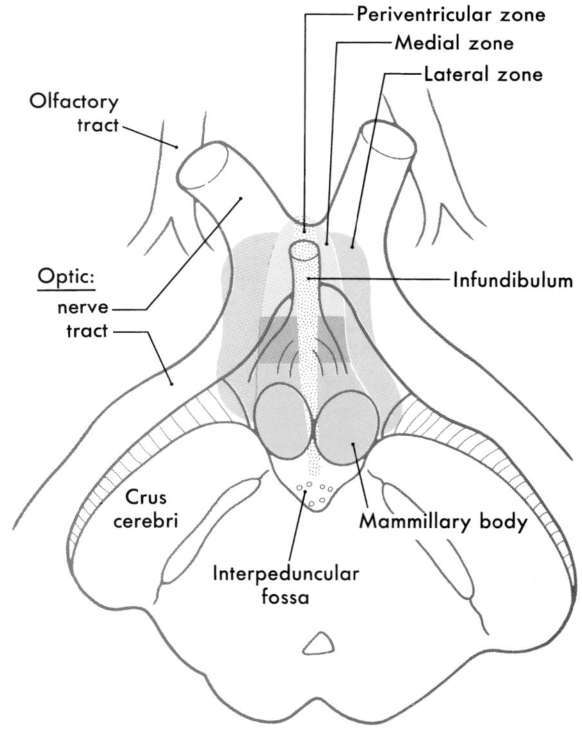

FIGURE 14-10 *Ventral view of the diencephalon illustrating the three zones of the hypothalamus as superimposed on external structures. The colors used for medial and lateral zones correlate with those in Figure 14-11.*

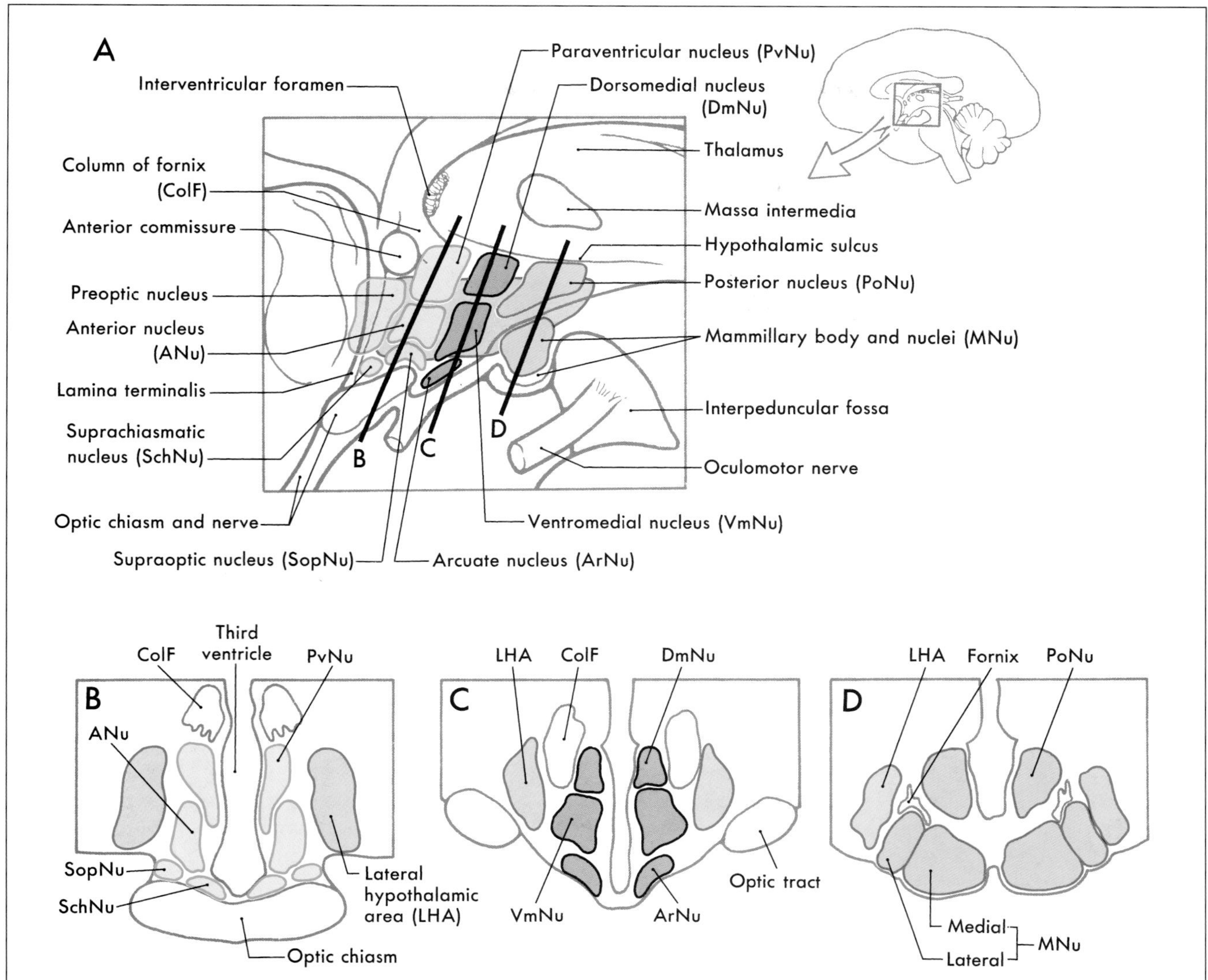

FIGURE 14-11 *Mid-sagittal (A) and cross-sectional (B-D) views illustrating the nuclei of medial and lateral hypothalamic zones and the nuclei associated with chiasmatic (B), tuberal (C), and mammillary (D) regions. The colors used here correlate with those in Figure 14-10. (A is adapted from Haymaker W, Anderson E, Nauta WJH:* The Hypothalamus. *Charles C Thomas, Springfield, IL, 1969, with permission.)*

intake (satiety) center. Bilateral lesions of this hypothalamic region produce *hyperphagia*, a greatly increased food intake with resultant obesity. Cells of the *arcuate nucleus* deliver peptides to the portal vessels and, through these channels, to the anterior pituitary. Some of these peptides are *releasing factors*, which cause an increase in the secretion of specific hormones by the anterior pituitary, and some are *inhibiting factors*, which inhibit the secretion of specific hormones by the anterior pituitary.

At caudal levels, the *mammillary region* is composed of the *posterior nucleus* and the *mammillary nuclei* (Fig. 14-11A,D). In humans, the mammillary nuclei consist of a large medial and a small lateral nucleus. Although both of these nuclei receive input via the fornix, only the medial nucleus projects to the anterior nucleus through the mammillothalamic tract. This latter bundle traverses the internal medullary lamina as it enters the anterior nucleus (Figs. 14-6A, 14-9). The neurons of the posterior nucleus are involved in activities that include elevation of blood pressure, pupillary dilation, and shivering or body heat conservation. The mammillary nuclei are involved in the control of various reflexes associated with feeding, as well as mechanisms relating to memory formation.

Afferent Fiber Systems Although many axonal systems extend into the hypothalamus, only four inputs are mentioned here; the entire group is discussed in Chapter 29. The fornix and the stria terminalis are two major afferent fiber bundles that reach the hypothalamus (Fig. 14-6A–D). The *fornix* consists of axons that largely originate in the hippocampus, and the *stria terminalis* arises from neurons in the amygdaloid complex (see Fig. 15-15).

Fibers comprising the *ventral amygdalofugal bundle* exit the amygdala and course through the substantia innominata to enter the hypothalamus and thalamus (see Fig. 15-15). As mentioned earlier, the *medial forebrain bundle* passes bidirectionally through the lateral hypothalamic region. This composite fiber bundle consists of ascending axons that originate in areas throughout the neuraxis and terminate in the hypothalamus and other axons that exit the hypothalamus to reach forebrain and brainstem targets.

Efferent Fibers The hypothalamus is the source of a diverse array of efferent fibers (see Chapter 29). Several nuclei give rise to descending fibers that contribute to the dorsal longitudinal fasciculus, the medial forebrain bundle, and to diffuse projections that pass into the tegmentum. These fiber systems project directly to numerous brainstem nuclei, as well as to preganglionic sympathetic and parasympathetic neurons in the spinal cord. Other projections reach the thalamus and frontal cortex, and still others extend to the posterior pituitary or to the tuberohypophysial portal system for delivery of substances to the anterior pituitary.

VENTRAL THALAMUS (SUBTHALAMUS)

The *ventral thalamus* (also called *subthalamus*) includes the large *subthalamic nucleus*, the medially adjacent *prerubral area* (*field H of Forel*), and dorsally the *zona incerta* (Figs. 14-3, 14-6B). As the term *ventral thalamus* (or subthalamus) implies, these cell groups are located ventral to the large expanse of the dorsal thalamus. The *subthalamic nucleus* is a lens-shaped cell group situated rostral and dorsal to the substantia nigra and immediately ventral to a distinct myelinated fiber bundle, the *lenticular fasciculus* (Fig. 14-6B). The cells of the subthalamic nucleus receive input from motor areas of the cerebral cortex, project to the substantia nigra, and are reciprocally connected with the globus pallidus. The subthalamic nucleus can be affected by vascular lesions involving posteromedial branches of the posterior cerebral or posterior communicating arteries, which results in a characteristic clinical condition known as *hemiballismus*. Patients with this involuntary movement disorder exhibit rapid and forceful flailing movements, which usually involve the contralateral upper extremity. These movements can be very debilitating because the patient has no control over their initiation or duration.

The *zona incerta* is located dorsal to the subthalamic nucleus and is separated from it by the *lenticular fasciculus* (Fig. 14-6B). Dorsal to the zona incerta are the myelinated axons of the thalamic fasciculus. The zona incerta contains output neurons that project to a variety of locations, including the cerebral cortex, superior colliculus, pretectal region, and the basilar pons. Afferent projections arise from the motor cortex and as collaterals from the medial lemniscus.

The *prerubral area* (*Forel's field H*) is located just rostral to the red nucleus and medial to the subthalamic nucleus (Fig. 14-6B). There are scattered neurons in this region, and traversing the prerubral area are fibers from the *lenticular fasciculus* (Forel's field H_2) that enter the *thalamic fasciculus* (Forel's field H_1).

EPITHALAMUS

The *pineal gland*, *habenular nuclei*, and *stria medullaris thalami* are the principal components of the epithalamus (Figs. 14-3, 14-12). The pineal gland consists of richly vascularized connective tissue containing glial cells and pinealocytes but no true neurons. Mammalian pinealocytes are related to the photoreceptor elements found in this gland in lower forms, such as amphibians. In humans, however, they remain only indirectly light sensitive and receive information concerning photic stimuli through a neural circuit.

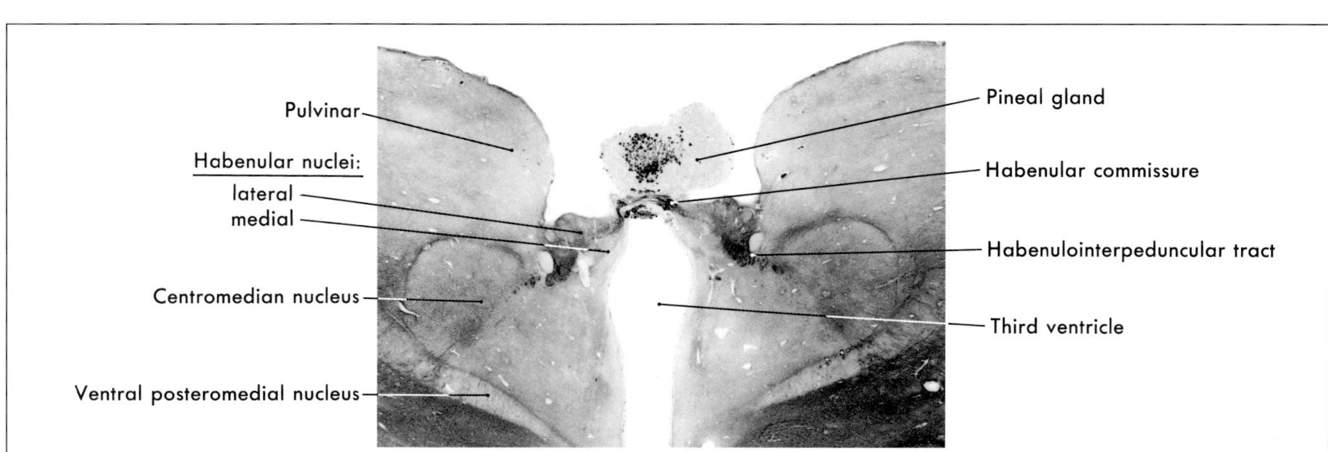

FIGURE 14-12 *Caudal level of the diencephalon showing the habenular nuclei, pineal gland, and related structures. The stria medullaris thalami has disappeared at this level because its fibers have dispersed to end in the habenular nuclei.*

Pinealocytes have clublike processes that are apposed to blood vessels but do not have direct synaptic contacts with central nervous system neurons. These cells synthesize melatonin from serotonin via enzymes that are sensitive to diurnal fluctuations in light. Levels of serotonin-*N*-acetyltransferase increase during the night (in the absence of photic stimulation), and the synthesis of melatonin is enhanced. Exposure to light turns off the enzymatic activity and melatonin production is diminished. Thus the production of melatonin by pinealocytes is rhythmic and calibrated to the 24-hour cycle of photic input to the retina. This is called a *circadian rhythm*.

Photic stimulation of pinealocytes occurs through an indirect route. Retinal ganglion cells project to the suprachiasmatic nucleus of the hypothalamus, which, in turn, influences neurons of the intermediolateral cell column in the spinal cord through descending connections. These preganglionic sympathetic neurons project to the superior cervical ganglion, which, in turn innervates, the pineal gland via postganglionic fibers that travel on branches of the internal carotid artery.

Pinealocytes also produce serotonin, norepinephrine, and neuroactive peptides, such as thyrotropin-releasing hormone (TRH), which are normally associated with the hypothalamus. These secretory products are released into the general circulation or the cerebrospinal fluid.

Pinealomas (tumors with large numbers of pinealocytes) are accompanied by depression of gonadal function and delayed puberty, whereas lesions that lead to the *loss* of pineal cells are associated with precocious puberty. This indicates that pineal secretory products exert an inhibitory influence on gonadal formation.

The habenular nuclei are located just anterior to the pineal gland and consist of a large lateral nucleus and a small medial nucleus (Fig. 14-12). Both nuclei contribute axons to the *habenulo-interpeduncular tract* (fasciculus retroflexus), which terminates in the midbrain interpeduncular nucleus. The *stria medullaris thalami*, which arches over the dorsal surface of the thalamus near the midline, conveys input to both habenular nuclei. The *habenular commissure*, a small bundle of fibers riding on the dorsal edge of the posterior commissure, connects the habenular regions of the two sides.

VASCULATURE OF THE DIENCEPHALON

The diencephalon is supplied by smaller vessels that branch from the various arteries making up the circle of Willis and by larger arteries that originate from the proximal parts of the posterior cerebral artery (Figs. 14-13, 14-14A,B). The hypothalamus and subthalamus are supplied by *central* (*perforating* or *ganglionic*) *branches* of the circle of Willis. Anterior parts of the hypothalamus are served by central branches (*anteromedial group*) arising from the anterior communicating artery and the A_1 segment of the anterior cerebral artery and from branches of the proximal part of the posterior communicating artery. Caudal hypothalamic regions and the ventral thalamus are supplied by branches of the *posteromedial group*; these branches arise from the posterior communicating artery and the P_1 segment of the posterior cerebral artery.

Arising from the P_1 segment near the basilar bifurcation are the *thalamoperforating arteries*. These large vessels (of which there may be more than one on each side) penetrate deeply to supply anterior areas of the thalamus (Figs. 14-13, 14-14A,B). If these vessels are occluded during surgery in this region, as can occur, for example, when an aneurysm of the basilar bifurcation is clipped, the patient can be rendered permanently comatose. Slightly more distal branches, which usually arise from the P_2 segment, are the *posterior choroidal* and *thalamogeniculate* arteries. These arteries also supply portions of the diencephalon (Figs. 14-13, 14-14A,B). A narrow portion of the caudal

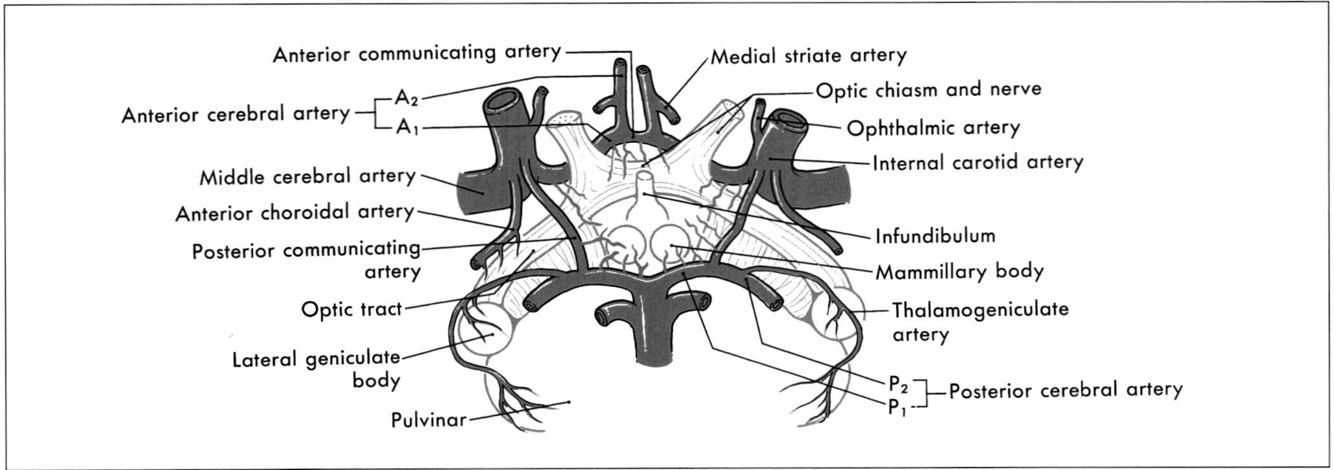

FIGURE 14-13 *Ventral aspect of the diencephalic region showing the arterial circle of Willis and the distribution of central branches to hypothalamic structures.*

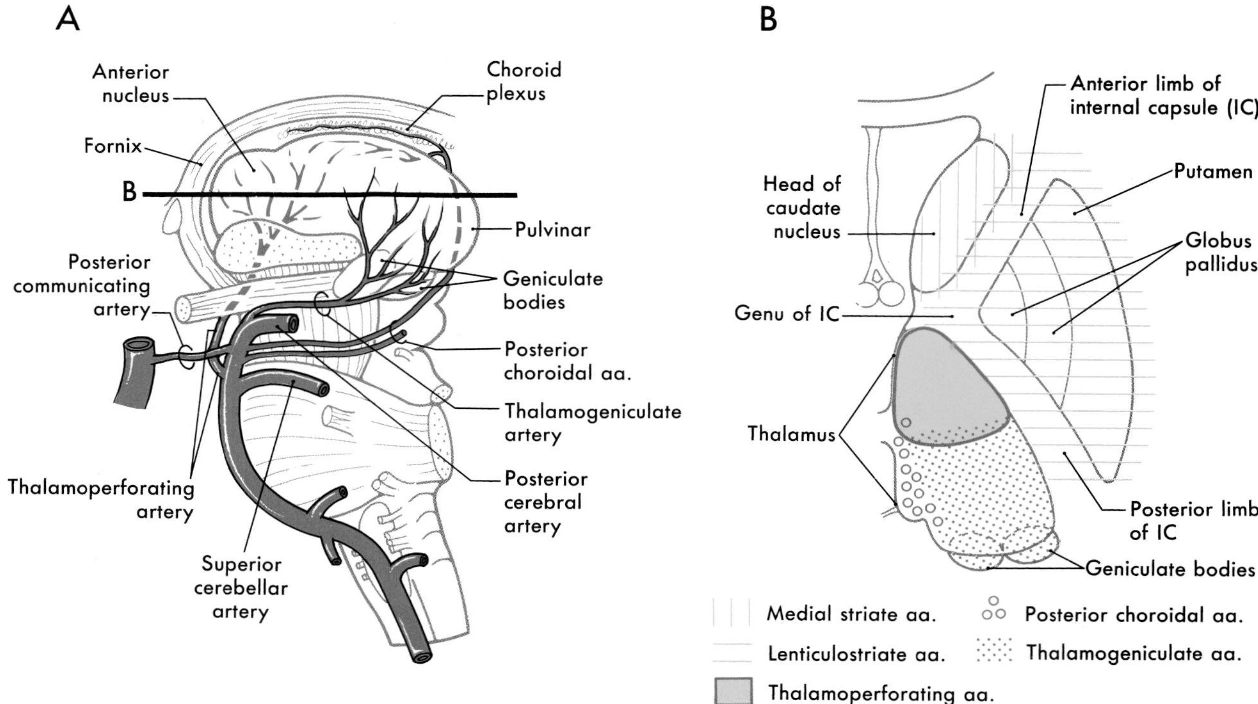

FIGURE 14-14 *Blood supply to the dorsal thalamus. The lateral aspect (A) shows the general distribution of the main arteries. An axial plane (B) through the hemisphere shows the internal territories served. The distribution of the main striate arteries, especially to the internal capsule, is also shown.*

and medial thalamus bordering on the third ventricle is supplied by the *medial posterior choroidal artery*, whereas the thalamogeniculate branches irrigate the posterior thalamus, including the pulvinar and the geniculate nuclei (Figs. 14-13, 14-14B). In addition, branches of the medial posterior choroidal artery also serve the choroid plexus of the third ventricle.

Although the thalamus receives a blood supply largely separate from that of the internal capsule (Fig. 14-14B), vascular lesions in the thalamus may extend into the internal capsule or vice versa. Ischemic or hemorrhagic strokes in the hemisphere may result in *contralateral hemiparesis* in combination with *hemianesthesia*. These losses correlate with damage to corticospinal and thalamocortical fibers in the internal capsule. On the other hand, strokes involving the larger thalamic arteries may result in total or dissociated sensory losses. These patients may subsequently experience persistent, intense pain (*thalamic pain, Dejerine-Roussy syndrome*) as the lesion in the thalamus resolves with time.

SOURCES AND ADDITIONAL READINGS

Alexander GE, Crutcher MD, DeLong MR: Basal ganglia-thalamocortical circuits: parallel substrates for motor, oculomotor, "prefrontal" and "limbic" functions. In Uylings HBM, Van Eden CG, DeBruin JPC, Corner MA, Feenstra MGP (eds): *Progress in Brain Research*, Vol. 85. *The Prefrontal Cortex: Its Structure, Function, and Pathology*. Elsevier Biomedical Press, Amsterdam, 1990.

Burt AM: *Textbook of Neuroanatomy*. W.B. Saunders, Philadelphia, 1993.

Carpenter MB, Sutin J: *Human Neuroanatomy*, 8th Ed. Williams & Wilkins, Baltimore, 1983.

Hirai T, Jones EG: A new parcellation of the human thalamus on the basis of histochemical staining. Brain Res Rev 14:1–34, 1989.

Jones EG: *The Thalamus*. Plenum Press, New York, 1986.

Nieuwenhuys R, Voogd J, van Huijzen Chr: *The Central Nervous System, A Synopsis and Atlas*, 3rd Ed. Springer-Verlag, Berlin, 1988.

Schell GR, Strick PL: The origin of thalamic inputs to the arcuate premotor and supplementary motor areas. J Neurosci 4:539–560, 1984.

Ungerleider LG, Galkin TW, Mishkin M: Visuotopic organization of projections from striate cortex to inferior and lateral pulvinar in rhesus monkey. J Comp Neurol 217:137–157, 1983.

Walker AE: *The Primate Thalamus*. University of Chicago Press, Chicago, 1938.

CHAPTER FIFTEEN

The Telencephalon

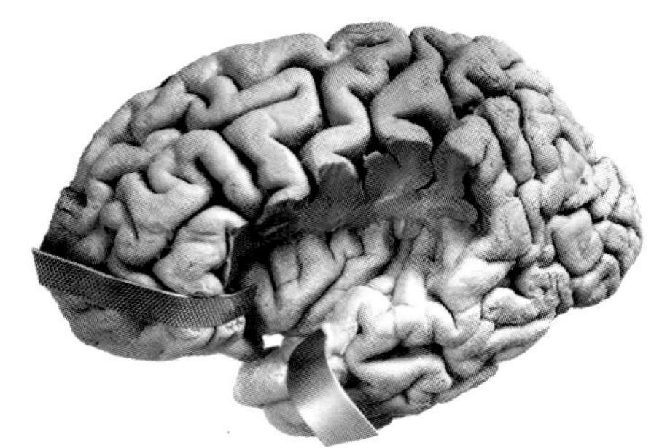

D. E. HAINES • G. A. MIHAILOFF

The telencephalon is the largest part of the human brain and is that portion in which all modalities are represented. Various sensory inputs (e.g., vision and hearing) are localized in some areas, whereas motor functions are represented in other regions and are modulated by subcortical nuclei. The telencephalon contains circuits that interrelate regions that have specific functions, such as motor or visual, with other regions called *association areas*. Seeing a familiar image may precipitate a cascade of neural events having olfactory, emotional, sensory, and/or motor components. Damage to association areas results in complex neurologic deficits. The patient is not blind or paralyzed but may be unable to recognize sensory input (*agnosia*), express ideas or thoughts (*aphasia*), or perform complex goal-directed movements (*apraxia*).

OVERVIEW

The telencephalon consists of two large *hemispheres* separated from each other by a deep *longitudinal cerebral fissure*. Each hemisphere has an outer surface, the *cerebral cortex*, which is composed of layers of cells. The cortex is thrown into elevations called *gyri* (singular, *gyrus*) separated by grooves called *sulci* (*sulcus*). Internal to the cortex are large amounts of *subcortical white matter*, along with aggregates of gray matter that form the *basal ganglia* and the *amygdala*. Although not parts of either the telencephalon or the basal ganglia, the *subthalamic nucleus* (of the diencephalon) and the *substantia nigra* (of the mesencephalon) have important connections that functionally link them with the basal ganglia.

Information passing into or out of the cerebral cortex must traverse the subcortical white matter. The myelinated fibers forming the white matter are organized into (1) association bundles that connect adjacent or distant gyri in one hemisphere; (2) commissural bundles that connect the two hemispheres, the largest of these being the *corpus callosum*; and (3) the *internal capsule*. The latter contains axons projecting to numerous downstream nuclei (*corticofugal fibers*) and axons conveying information to the cerebral cortex (*corticopetal fibers*).

The *hippocampal complex* and the *amygdala* are located in the walls of the temporal horn of the lateral ventricle. The axons of cells in these structures coalesce to form the *fornix* and *stria terminalis*, respectively.

DEVELOPMENT

Enlargements of the prosencephalon, the *telencephalic* (*cerebral*) *vesicles*, appear at about 5 weeks of gestational age. As the cerebral vesicles enlarge in all directions, they pull along portions of the neural canal that will form the cavities of the telencephalon, the *lateral ventricles* (Fig. 15-1A,B). The primitive lateral ventricles extend into frontal, parietal, temporal, and occipital areas as they develop and form that portion of the ventricle found in each of these lobes in the adult. The *interventricular foramina*, which connect each lateral ventricle to the midline *third ventricle* (cavity of the diencephalon), are initially large but become smaller as development progresses.

Cells forming the *corpus striatum* appear in the floor of the developing lateral ventricle at the time when primordial cell groups in the wall of the third ventricle are giving rise to diencephalic structures (Fig. 15-1C,D). As development progresses, the corpus striatum is bisected by axons growing to and from the cerebral cortex. These axons form the *internal capsule* of the adult and divide the corpus striatum into a medially located *caudate nucleus* and a laterally located *putamen*. As the diencephalon enlarges, it gives rise to the thalamus and hypothalamus and to cells that migrate across the developing internal capsule to assume a position medial to the putamen (Fig. 15-1D). These cells become the *globus pallidus* of the adult and, in combination with the putamen, form the *lenticular nucleus*.

The initial development of the major commissural bundles and of the hippocampus takes place along the medial aspect of the hemisphere (Fig. 15-1C–E). In the adult, there are three major interhemispheric commissures: the *anterior commissure*, the *hippocampal commissure*, and the *corpus callosum*. The first of these to appear, the *anterior commissure*, arises within the lamina terminalis. The latter is a membrane-like structure that extends from the anterior commissure ventrally to the rostral edge of the optic chiasm (see Fig. 14-3). The second to form, the *hippocampal commissure*, develops along with the hippocampal primordium. As growth occurs, the hippocampus, which originates in the dorsomedial part of the hemisphere, is displaced into the temporal lobe where it assumes a position characteristic of the adult (Fig. 15-1E). In the process, fibers from one side cross to the other side, as the *hippocampal commissure*, just ventral to the area that will be occupied by the corpus callosum. The third commissure to develop, the *corpus callosum*, originates from the area of the lamina terminalis as a structure initially composed of astrocytic processes. Axons from developing neurons in each hemisphere traverse this glial structure to access the contralateral side. As this takes place, the corpus callosum enlarges in a caudal direction to form the prominent structure found in the adult (Figs. 15-1E, 15-2A).

Failure of the corpus callosum to develop (*agenesis of the corpus callosum*) may be accompanied by an absence of the anterior and hippocampal commissures (Fig. 15-2A,B). Although some individuals with this condition experience focal seizures and have mental retardation, others live for many years with few or no obvious deficits. These individuals frequently have developmental abnormalities in other parts of the nervous system.

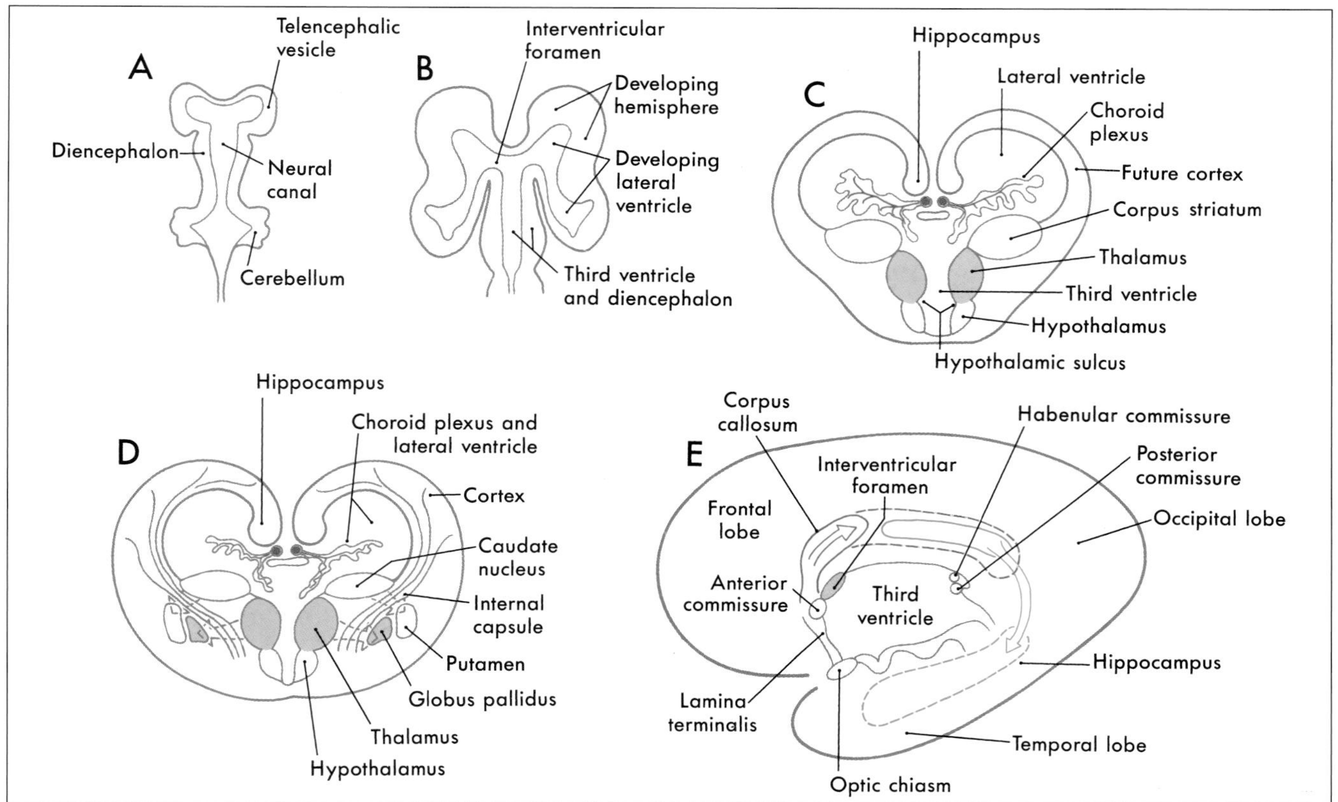

FIGURE 15-1 *Development of the telencephalon. Enlargement of the ventricles is shown in longitudinal view (A, B), differentiation of the basal ganglia and internal capsule in cross section (C, D), and growth of the corpus callosum (in red) and hippocampus (in green) in the sagittal plane (E).*

LOBES OF THE CEREBRAL CORTEX

Based on the arrangement of major sulci, the cerebral cortex is divided into five lobes plus an area called the *insula*. Four of these lobes are named according to the overlying bones of the skull.

On the lateral surface of the hemisphere the major sulci are the *central sulcus*, the *lateral* (*Sylvian*) *sulcus*, and a small lateral end of the *parieto-occipital sulcus* (Fig. 15-3). The *preoccipital notch* is the high point along the ventrolateral margin of the hemisphere. An imaginary line connecting the dorsal terminus of the parieto-occipital sulcus with the preoccipital notch intersects with another line drawn caudally from the lateral sulcus. These lines, along with the central and lateral sulci, divide the lateral surface into *frontal*, *parietal*, *temporal*, and *occipital lobes*.

On the medial surface of the hemisphere, the major sulci separating lobes are the *cingulate*, *parieto-occipital*, and *collateral* (Fig. 15-4). Two imaginary lines also sepa-

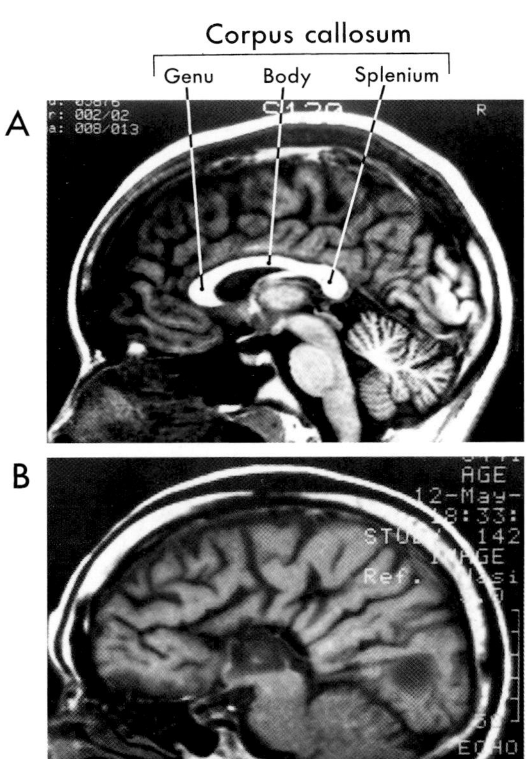

FIGURE 15-2 *Magnetic resonance image in the sagittal plane of a normal adult (A) and of an individual, also adult, with agenesis of the corpus callosum (B). A comparison of B with A reveals an absence of the corpus callosum, aberrant gyri on the medial surface of the hemisphere, and other structural defects.*

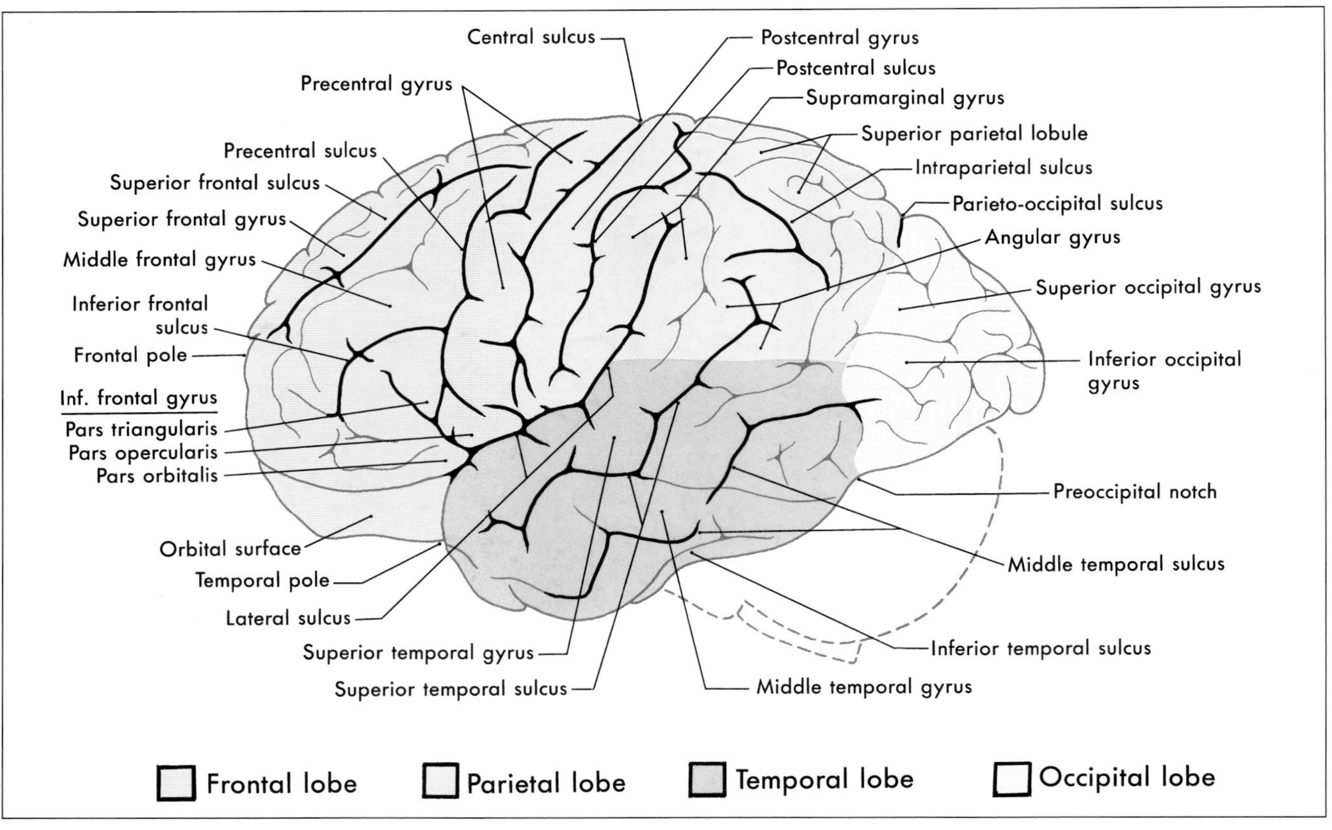

FIGURE 15-3 *Lateral aspect of the left cerebral hemisphere showing lobes and their associated gyri and sulci.*

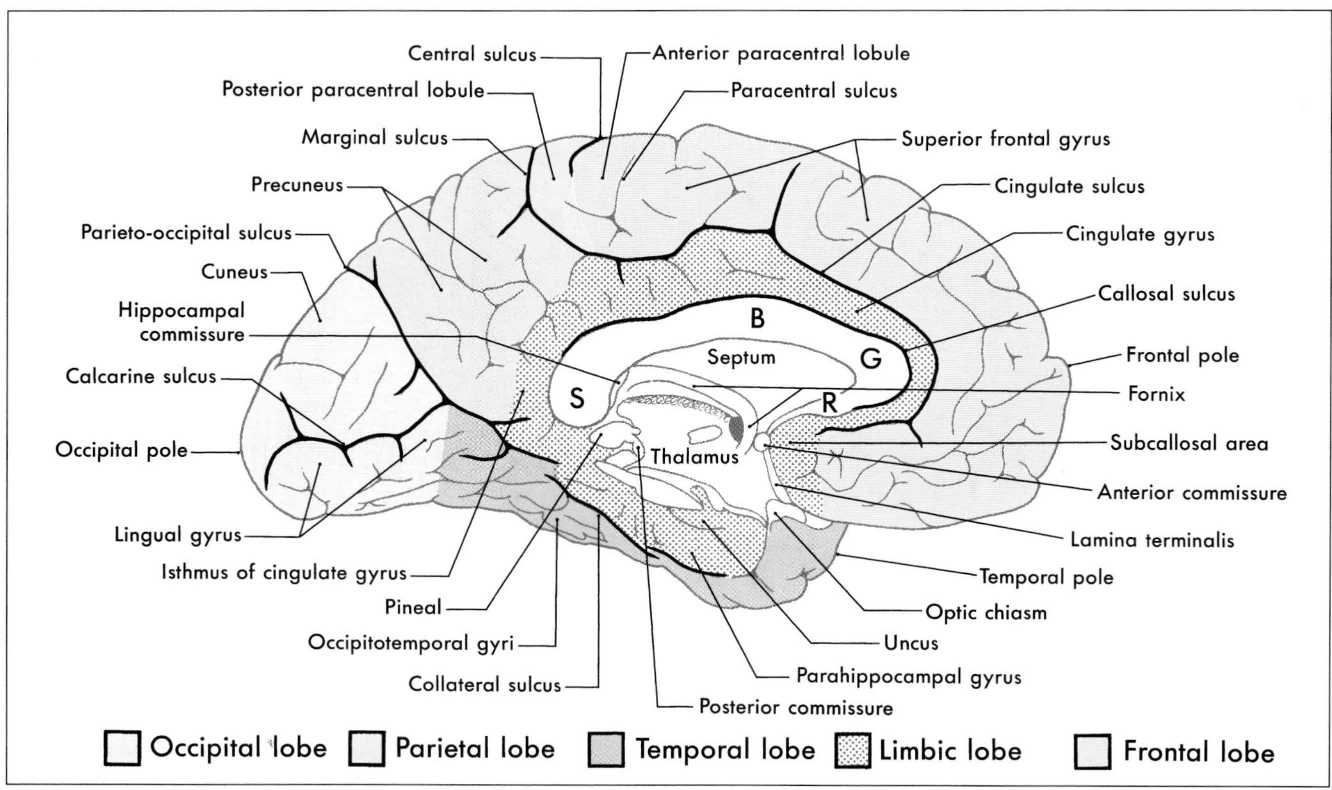

FIGURE 15-4 *Medial aspect of the left cerebral hemisphere showing lobes and their associated gyri and sulci. The parts of the corpus callosum are R = rostrum, G = genu, B = body (or trunk), and S = splenium.*

rate lobes on the medial surface. One connects the medial end of the central sulcus with the cingulate sulcus; the other joins the parieto-occipital sulcus with the preoccipital notch. This combination of sulci and lines separate the four lobes noted previously, plus the limbic lobe, on the medial surface of the hemisphere.

Frontal Lobe The lateral surface of the frontal lobe is divided by *inferior* and *superior frontal sulci* into *inferior*, *middle*, and *superior frontal gyri* (Fig. 15-3), the latter folding onto the medial aspect of the hemisphere. The inferior frontal gyrus is divided into a *pars opercularis*, *pars triangularis*, and *pars orbitalis*. The ventral surface of the frontal lobe is composed of the *gyrus rectus*, the *olfactory sulcus*, and a series of *orbital gyri* (Fig. 15-5). The rostral-most point of this lobe is the *frontal pole* of the brain.

The *olfactory bulb* and *tract*, which relay sensory information, lie on the ventral surface of the frontal lobe in the olfactory sulcus (Fig. 15-5). At the point where the olfactory tract attaches to the hemisphere, it bifurcates into medial and lateral striae (Fig. 15-6). The triangle formed by this bifurcation is called the *olfactory trigone*.

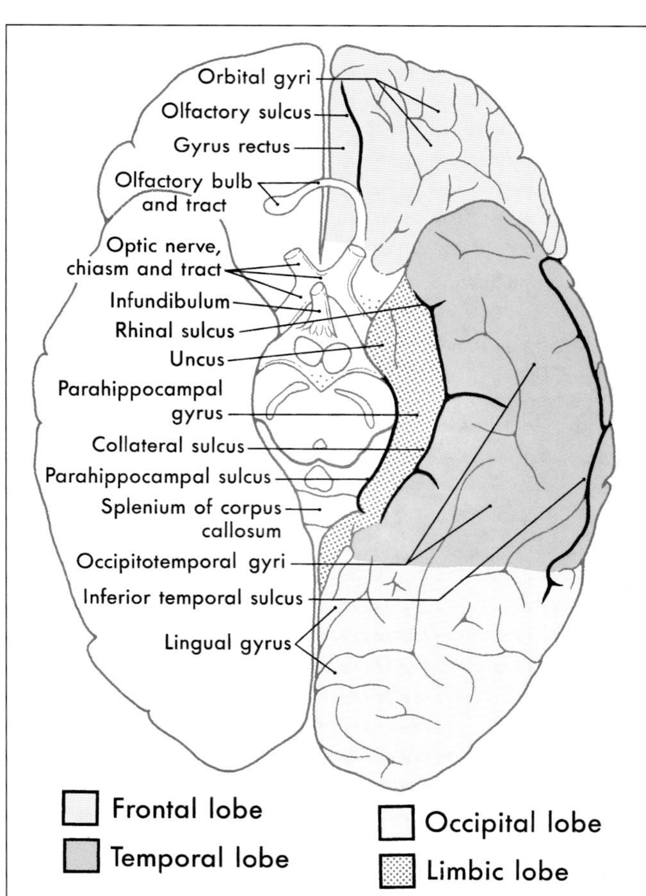

FIGURE 15-5 *Ventral aspect of the left cerebral hemisphere showing lobes and their associated gyri and sulci.*

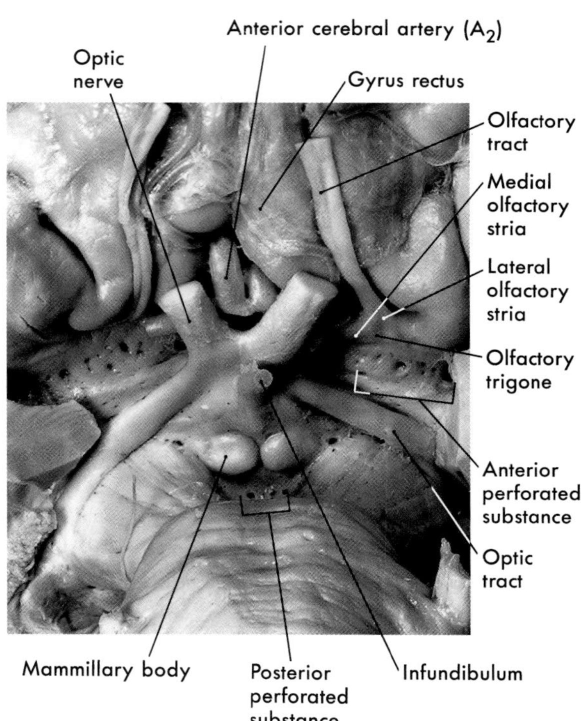

FIGURE 15-6 *Ventral aspect of the hemisphere in the area of the anterior perforated substance.*

Immediately caudal to this trigone, the surface of the hemisphere is characterized by numerous small holes formed by vessels (lenticulostriate arteries) as they enter the brain; this is the *anterior perforated substance* (Fig. 15-6). The olfactory tract and striae and the cell groups associated with the anterior perforated substance are functionally related to the limbic system.

The *precentral gyrus* is continuous on the medial surface of the hemisphere with the *anterior paracentral lobule*; the latter is separated from the superior frontal gyrus by the *paracentral sulcus* (Fig. 15-4). These two gyri collectively form the *primary motor cortex*.

Parietal Lobe The parietal lobe consists of the *postcentral gyrus*, located between the *central* and *postcentral sulci*, and the *superior* and *inferior parietal lobules*, which are separated by the *intraparietal sulcus* (Fig. 15-3). Gyri forming the superior parietal lobule extend onto the medial surface of the hemisphere as the *precuneus*, whereas the inferior lobule is made up of the *angular* and *supramarginal gyri*. The latter is a crescent-shaped ridge of cortex around the caudal terminus of the lateral sulcus.

As the *postcentral gyrus* extends onto the medial surface of the hemisphere, it is continuous with the *posterior paracentral lobule* (Fig. 15-5). This cortical area is bordered rostrally by an imaginary line that connects the central sulcus to the cingulate sulcus and caudally by the *marginal ramus of the cingulate sulcus*. The latter is frequently called the *marginal sulcus*. Taken together, the postcentral gyrus and posterior paracentral lobule constitute the *primary somatosensory cortex*.

Temporal Lobe The gyri that form the temporal lobe are found on lateral and ventral aspects of the hemisphere between the *lateral sulcus* and the *collateral sulcus* (Figs. 15-3 to 15-5). These gyri are, beginning at the lateral sulcus, the *superior*, *middle*, and *inferior temporal gyri* and a broad area of cortex, the *occipitotemporal gyri*, extending from the *temporal pole* to the occipital lobe. The *superior temporal sulcus* ends in the loop of cortex forming the angular gyrus of the inferior parietal lobule. An *inferior temporal sulcus* may be found between the inferior temporal and occipitotemporal gyri, or it may be absent, in which case these gyri blend around the inferior margin of the hemisphere.

On the upper edge of the temporal lobe, and extending into the depths of the lateral fissure, are the *transverse temporal gyri* (of Heschl). These gyri (Fig. 15-7) form the *primary auditory cortex*.

Insula The oval region of cortex located deep inside the lateral fissure is the *insula* (Fig. 15-7). This area is characterized by a set of long gyri in its caudal part (the *gyri longi*) and a set of short gyri in its rostral part (the *gyri breves*). The insular cortex is continuous with that of the adjacent frontal, parietal, and temporal lobes. This continuity forms lips on each lobe that overlie the insula to form the *frontal*, *parietal*, and *temporal opercula* (Fig. 15-7; see also Fig. 15-9). The *limen insulae* (threshold to the insula) is the area where the ventral surface of the hemisphere is continuous with the insular cortex (Fig. 15-7). Although the function of the insula is still somewhat unclear, it is known that the insular cortex receives *nociceptive* and *viscerosensory input*.

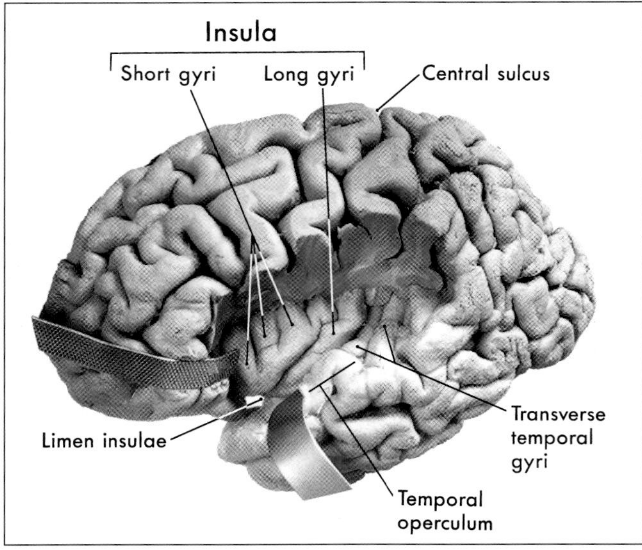

FIGURE 15-7 *View of the left hemisphere from anterior and lateral. The frontal and parietal opercula are removed and the temporal operculum retracted to expose the insula and transverse temporal gyri.*

Occipital Lobe The occipital lobe forms the caudal end of the hemisphere; its caudal extreme is the *occipital pole* of the brain (Figs. 15-3, 15-4). The irregular collection of gyri on the lateral surface of the occipital lobe form the *occipital gyri*. An important landmark on the medial aspect of the occipital lobe is the *calcarine sulcus*. This sulcus separates the cuneus, which is dorsal to it, from the *lingual gyrus*, which is ventrally located. The *primary visual cortex* is located in the portions of these gyri that border directly on the calcarine sulcus.

Limbic Lobe The *limbic lobe* is part of a more complex entity commonly called the *limbic system*. As discussed in Chapter 30, the limbic system includes this lobe plus its afferent and efferent connections with other telencephalic, diencephalic, and brainstem nuclei.

The limbic lobe is a ring of cortex that makes up the medial-most rim of the hemisphere. Although not specified as a separate lobe in some texts, it is designated here as such because of its unique functional characteristics. Beginning anteriorly (and just ventral to the rostrum of the corpus callosum) this ring of cortex consists of the *subcallosal area*, *cingulate gyrus*, *isthmus of the cingulate gyrus*, *parahippocampal gyrus*, and the *uncus* (Figs. 15-4, 15-5). The limbic cortex is separated from adjacent cortical areas by the *cingulate sulcus* and the *collateral sulcus* and from the *corpus callosum* by the *callosal sulcus*. On the ventral surface of the temporal lobe, the *rhinal sulcus* is located between the rostral extreme of the parahippocampal gyrus and the laterally adjacent occipitotemporal gyri (Fig. 15-5).

The function of the limbic lobe is complex and cannot be designated as, for example, primarily sensory or motor. Rather, it is linked to circuits that influence complex functions such as memory, learning, and behavior.

Vasculature of the Cerebral Cortex The lobes comprising the cerebral cortex receive their blood supply via terminal branches of the *anterior*, *middle*, and *posterior cerebral arteries*. Details of individual branches are given in Chapter 8; only general points are reviewed here.

The *anterior cerebral artery* branches from the internal carotid and is joined to its counterpart by the *anterior communicating artery*. The anterior cerebral artery arches dorsally and caudally to serve the medial surface of the hemisphere to about the level of the parieto-occipital sulcus (Fig. 15-8). The part of the anterior cerebral artery between the internal carotid and the anterior communicator is called the A_1 segment. Branches of A_1 serve structures in the immediate vicinity, including the optic chiasm and anterior parts of the hypothalamus. The branches of the anterior cerebral distal to the anterior communicator collectively form the A_2 segment. These serve the medial surface of the frontal and parietal lobes, including the *anterior* (lower extremity part of motor cortex) and *posterior* (lower extremity part of somatosensory cortex) *paracentral lobules*.

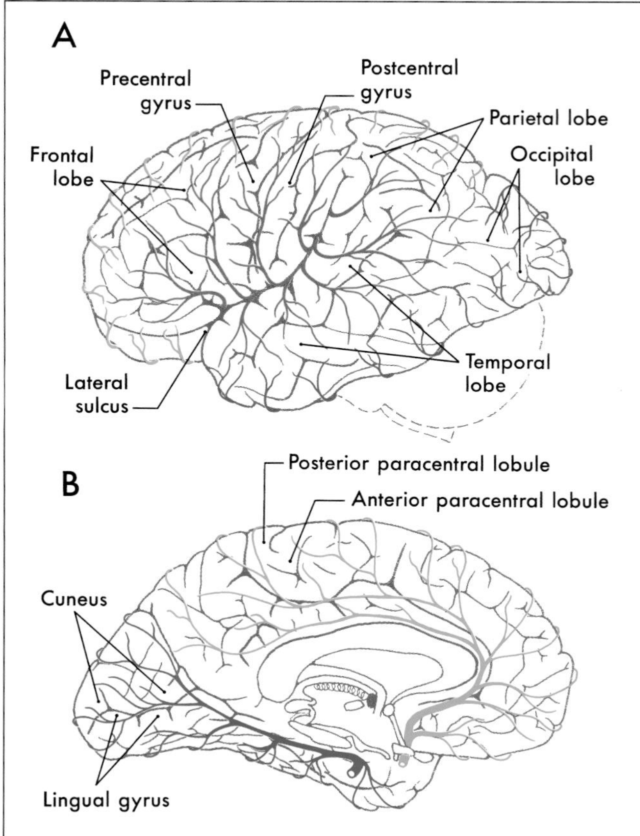

FIGURE 15-8 *Lateral (A) and medial (B) views of the left cerebral hemisphere showing the distribution of the anterior (in green), middle (in red), and posterior (in blue) cerebral arteries.*

The *middle cerebral artery* originates from the internal carotid and passes laterally to the *limen insula*, where it generally branches into *superior* and *inferior trunks* (Figs. 15-8, 15-9). This initial part of the middle cerebral, the M_1 segment, gives rise to the *lenticulostriate* (*lateral striate*) *arteries*. The superior and inferior trunks and their distal branches collectively form the M_2 segment. Whereas the insular cortex itself is served by both trunks, the branches of the superior and inferior trunks exit the lateral sulcus and distribute, respectively, to cortex above and below this sulcus (Fig. 15-9). Important cortical regions served by M_2 include the trunk, upper extremity, and head areas of the motor cortex (*precentral gyrus*), somatosensory cortex (*postcentral gyrus*), auditory cortex (*transverse temporal gyri*), parietal association cortex (*parietal lobules*), and large parts of the dorsolateral surface of the frontal lobe.

The *posterior cerebral artery* begins at the basilar bifurcation (Fig. 15-8). The first and second parts of this vessel (P_1 and P_2 segments) are located, respectively, between the basilar bifurcation and the *posterior communicating artery* and just distal to the latter vessel. Midbrain and diencephalic structures are the main targets of branches of P_1 and P_2 segments. The P_3 segment of the posterior cerebral artery is the part from which the *temporal branches* originate, and P_4 is the segment that gives rise to *parieto-occipital* and *calcarine arteries*. The latter vessel is the source of blood to the *primary visual cortex*, which borders on the calcarine sulcus. Other important telencephalic structures in the domains of P_3 and P_4 include the *parahippocampal gyrus* and the *precuneus*.

WHITE MATTER OF THE CEREBRAL HEMISPHERE

All information entering or leaving the cerebral cortex or connecting one part of the cortex with another must pass through the subcortical white matter. In general, the white matter core of the hemisphere contains *association fibers*, *commissural fibers*, and *projection fibers*.

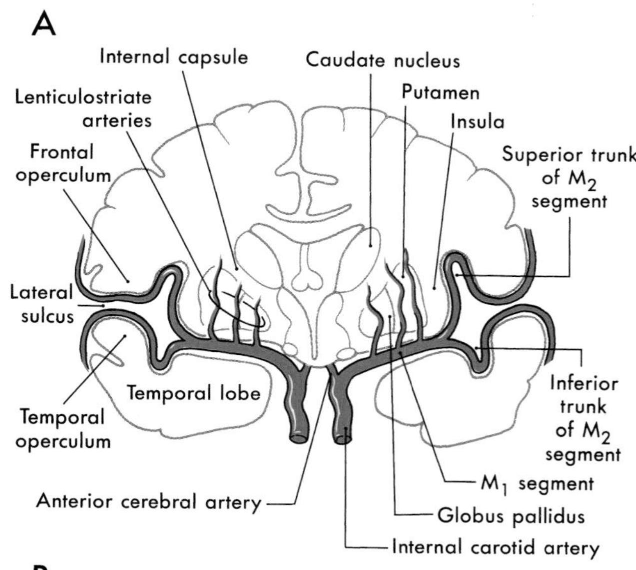

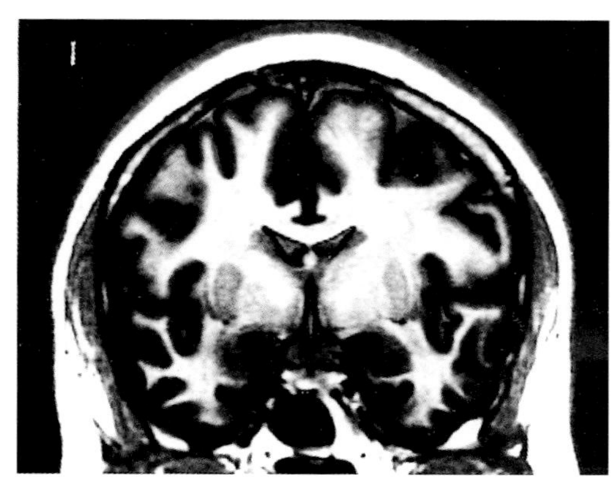

FIGURE 15-9 *Cross section of the cerebral hemisphere showing the main branches of the middle cerebral artery (A) and a magnetic resonance image (B) in the coronal plane at a comparable level.*

Association Fibers The *association fibers* interconnect different areas of cortex within the same hemisphere. These may be *short association fibers* that connect the cortices of adjacent gyri or *long association fibers* that interconnect more distant areas of cortex (Fig. 15-10). Important examples of the latter are the *cingulum* located internal to the cingulate gyrus and continuing into the parahippocampal gyrus, the *inferior longitudinal fasciculus* (temporal-occipital interconnections), and the *uncinate fasciculus* (frontal-temporal interconnections). The *superior longitudinal fasciculus*, located in the core of the hemisphere, interconnects frontal, parietal, and occipital cortices, whereas the *arcuate fasciculus* interconnects frontal and temporal lobes (Fig. 15-10). In the white matter of the temporal lobe, fibers passing between the frontal and occipital areas make up the *inferior fronto-occipital fasciculus*.

The *claustrum*, a thin layer of neuron cell bodies located internal to the insular cortex, is sandwiched between two small association bundles (Fig. 15-11). The *external capsule* is insinuated between the claustrum and putamen, and the *extreme capsule* is located between the claustrum and the insular cortex.

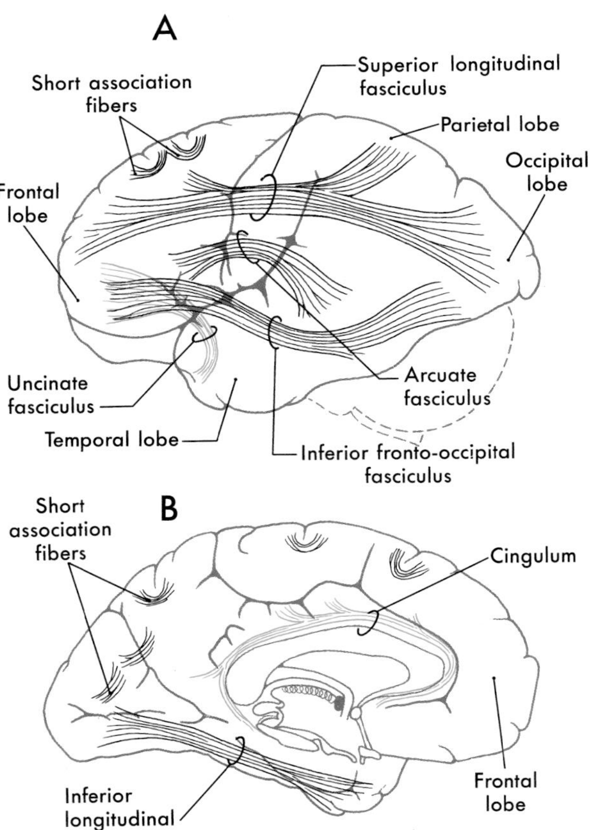

FIGURE 15-10 *Main association bundles as visualized from lateral (A) and medial (B) aspects of the left hemisphere.*

Commissural Fibers—The Corpus Callosum In general commissural fibers interconnect corresponding structures on either side of the neuraxis. The largest bundle of commissural fibers is the *corpus callosum* (Figs. 15-2, 15-4). This huge bundle is located dorsal to the diencephalon and forms the roof of much of the lateral ventricles. The corpus callosum consists of, from rostral to caudal, a *rostrum*, *genu*, *body* (also called *trunk*), and *splenium* (Fig. 15-4). Many of the fibers passing through the genu arch rostrally to interconnect the frontal lobes; these form the *frontal* (or *minor*) *forceps*. The fibers interconnecting the occipital lobes loop through the splenium of the corpus callosum, forming the *occipital* (or *major*) *forceps*. The *tapetum*, which is located in the lateral wall of the collateral trigone and posterior horn of the lateral ventricle is also composed of fiber bundles that cross in the splenium.

Smaller commissural bundles are the *anterior commissure* and the *hippocampal commissure* (Fig. 15-4). In sagittal view, the anterior commissure is located caudal to the rostrum of the corpus callosum but rostral to the main part of the fornix. This bundle interconnects various parts of the frontal and temporal lobes. The *hippocampal commissure* is formed by fibers that originate in the hippocampal formations and cross the midline as a thin layer ventral to the splenium of the corpus callosum.

The *posterior commissure* and the *habenular commissure* are caudal parts of the diencephalon (Figs. 15-1, 15-4). The former crosses the midline at the base of the pineal gland and just dorsal to the cerebral aqueduct. The latter is a small fascicle running along the dorsal aspect of the posterior commissure and interconnecting the habenular nuclei.

Projection Fibers—The Internal Capsule The projection fibers of the hemispheres include both the axons that originate outside the telencephalon and project to the cerebral cortex (*corticopetal*) and the axons that arise from cerebral cortical cells and project to downstream targets (*corticofugal*). A prime example of the former are projections from the thalamus to the cerebral cortex (*thalamocortical fibers*); examples of the latter are *corticospinal*, *corticopontine*, and *corticothalamic fibers*. Projection fibers are organized into a large, compact bundle called the *internal capsule* (Fig. 15-11), which has intimate structural associations with the diencephalon and basal ganglia. Consequently, to divide the internal capsule into its constituent parts, reference must be made to these adjacent cell groups.

In an axial plane through the hemisphere, the internal capsule appears as a prominent V-shaped structure (Figs. 15-11, 15-12). It is divided into three parts: (1) an *anterior limb* insinuated between the head of the

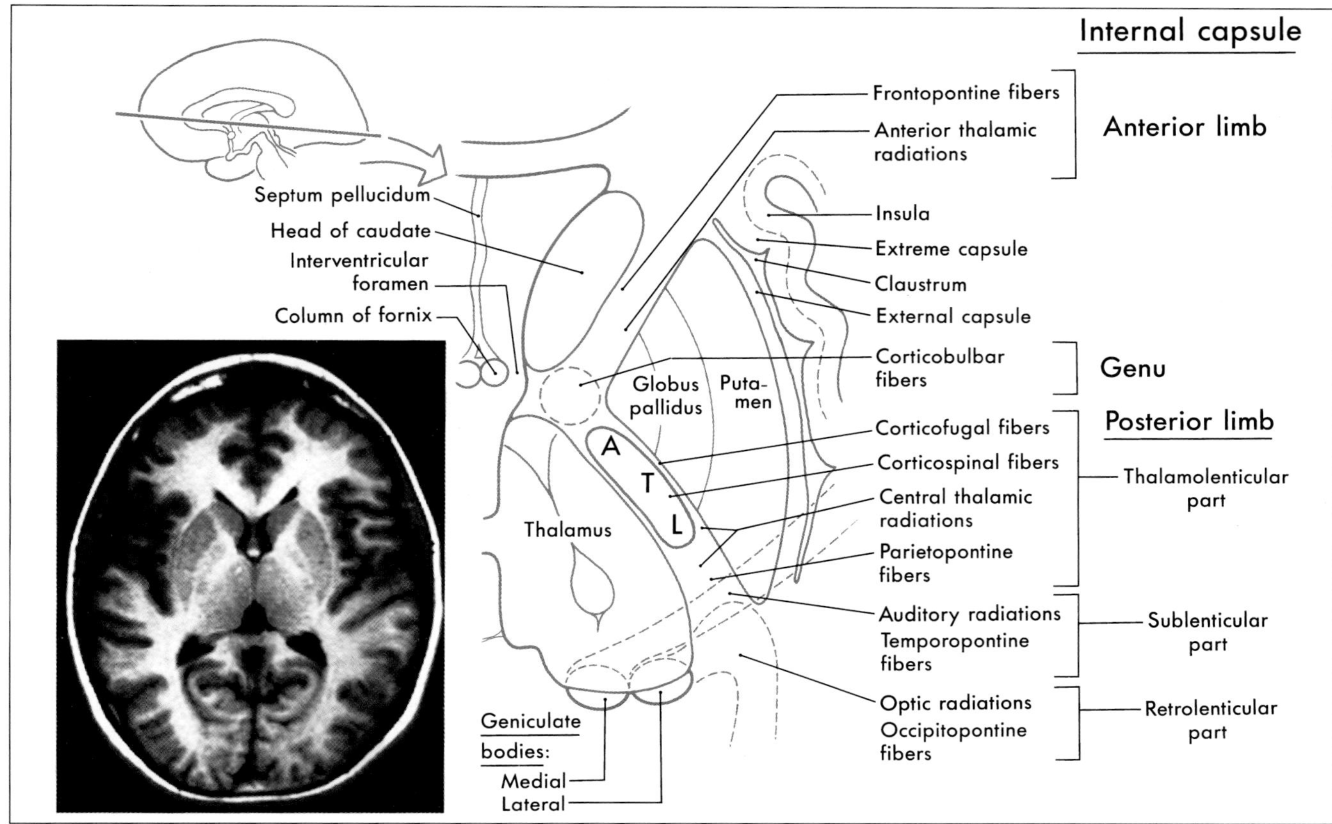

FIGURE 15-11 *The internal capsule in the axial plane with a magnetic resonance image at a comparable level. The various parts of the internal capsule labeled in the drawing can be identified.* Corticofugal *is an umbrella term under which specific populations of descending fibers, such as* corticoreticular, corticorubral, *and* corticotectal *are included. Although not labeled here, specific types of corticofugal fibers are also present in the sublenticular and retrolenticular parts of the posterior limb.*

caudate nucleus and the lenticular nucleus, (2) a *posterior limb* located between the dorsal thalamus and the lenticular nucleus, and (3) a *genu* located at the intersection of the anterior and posterior limbs, which is approximately at the level of the interventricular foramen (Fig. 15-11). The *anterior limb of the internal capsule* contains *thalamocortical/corticothalamic fibers* (collectively called the *anterior thalamic radiations*), which interconnect the dorsomedial and anterior thalamic nuclei with areas of the frontal lobe and the cingulate gyrus. *Frontopontine fibers*, especially those from the prefrontal areas, also pass through this structure. The *genu of the internal capsule* contains *corticobulbar fibers* that arise in the frontal cortex just anterior to the precentral sulcus and from the precentral gyrus (primary motor cortex).

The *posterior limb of the internal capsule* is larger and more complex (Fig. 15-11). It is frequently divided into a *thalamolenticular part* (located between the thalamus and

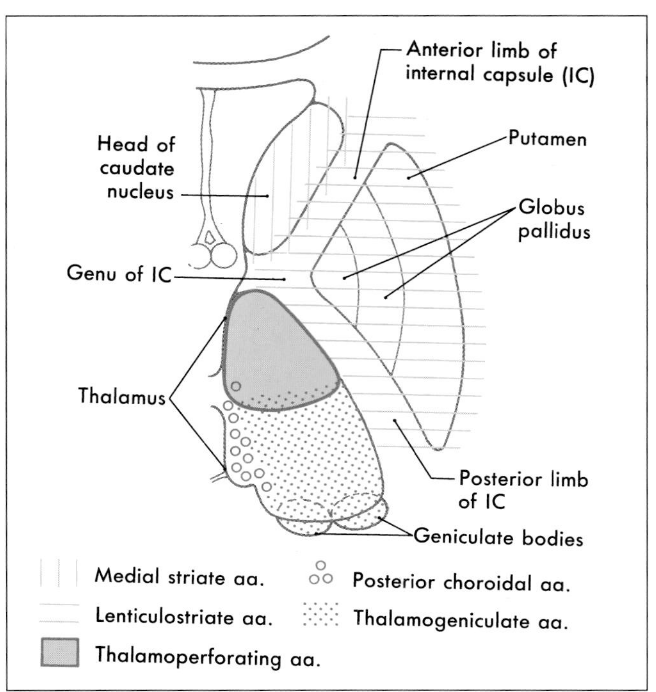

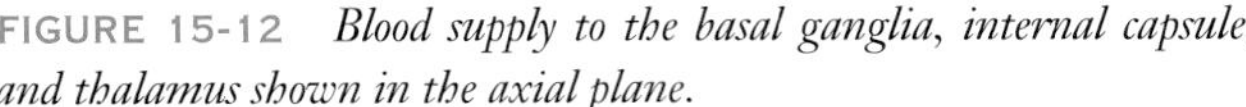
FIGURE 15-12 *Blood supply to the basal ganglia, internal capsule, and thalamus shown in the axial plane.*

the lenticular nucleus), a *sublenticular part* (fibers passing ventral to the lenticular nucleus), and a *retrolenticular part* (fibers located caudal to the lenticular nucleus). However, it is also common to refer to the "thalamolenticular part" as the *posterior limb*, the "sublenticular part" as the *sublenticular limb*, and the "retrolenticular part" as the *retrolenticular limb*. This latter terminology is considerably less cumbersome and much easier to remember.

The major fiber populations passing through the posterior limb of the internal capsule are summarized in Figure 15-11. Included are *corticospinal fibers* arising from the motor cortex and projecting to the contralateral spinal cord and *thalamocortical/corticothalamic fibers* (*central thalamic radiations*) that interconnect nuclei of the dorsal thalamus with the overlying cortex. *Geniculotemporal radiations* (*auditory radiations*) convey auditory information from the medial geniculate nucleus to the transverse temporal gyri through the sublenticular part (or limb). Visual input from the lateral geniculate body to the occipital cortex is conveyed via *geniculocalcarine radiations* through the retrolenticular part (or limb) (Fig. 15-11). The geniculocalcarine radiations, *optic radiations*, form a distinct lamina of fibers immediately lateral to the tapetum as they course caudally into the occipital lobe.

Fibers of the internal capsule flare out into the hemisphere as they pass distal to the caudate and putamen. This abrupt divergence of internal capsule fibers forms the *corona radiata* ("radiating crown"), which contains *converging corticofugal fibers*, as well as diverging corticopetal fibers.

Vasculature of the Internal Capsule The blood supply to the genu and most of the posterior limb of the internal capsule is via the *lenticulostriate arteries*; these are branches of the M_1 segment of the middle cerebral artery (Figs. 15-9, 15-12). Branches of the anterior choroidal artery supply the ventral region of the posterior limb and the immediately adjacent retrolenticular limb. The anterior limb receives somewhat of a dual blood supply, in that lenticulostriate arteries and branches of the *medial striate artery* (usually a branch of A_2) serve this area. Occlusion (ischemic stroke) or rupture (hemorrhagic stroke) of the vessels serving the posterior limb may result in motor and sensory deficits. Most commonly seen is a contralateral hemiparesis, indicating damage to corticospinal fibers. Hemiparesis may be accompanied by contralateral somatosensory losses, which result from the interruption of thalamocortical fibers relaying sensory information from the thalamus to the cortex.

BASAL GANGLIA (NUCLEI)

The term "basal ganglia" is somewhat of a misnomer. The cells forming these structures are not "ganglia," a term usually reserved to describe an aggregation of neuronal somata in the peripheral nervous system, but are "nuclei" in the central nervous system. In addition, the definition of what cell groups make up the basal ganglia has been revised over the years, with contemporary views focusing on the functional characteristics of these nuclei. In this respect, *the basal ganglia consist of (1) the caudate and lenticular nuclei (together forming the dorsal basal ganglia), (2) the nucleus accumbens plus parts of the adjacent olfactory tubercle (the ventral striatum), and (3) the substantia innominata (ventral pallidum)* (Fig. 15-13). As described above, the *subthalamic nucleus* and *substantia nigra* are not components of the basal ganglia, but they are included in this discussion because of their important structural/functional relationship with the caudate and lenticular nuclei.

The basal ganglia function primarily in the motor sphere. Damage to these nuclei, as in vascular lesions, degenerative genetic disorders, or problems of unknown etiology, results in a variety of motor deficits, some of which are recognized as characteristic involuntary movements.

Caudate and Lenticular Nuclei Collectively, the caudate and lenticular nuclei form the *corpus striatum*. The corpus striatum, in turn, is divided into the *neostriatum*, consisting of the *caudate nucleus* and the *putamen*, and the *paleostriatum* or *globus pallidus* (Fig. 15-13). Collectively, the globus pallidus and putamen form the *lenticular nucleus*.

The caudate nucleus is characteristically located in the lateral wall of the lateral ventricle and consists of three parts, a *head*, *body*, and *tail* (Figs. 15-14, 15-15). The *head of the caudate* nucleus forms a prominent bulge in the anterior horn of the lateral ventricle. At about the level of the interventricular foramen, the caudate diminishes in size, but continues caudally as the *body of the caudate nucleus* in the lateral wall of the body of the lateral ventricle. In the lateral wall of the collateral trigone, the body of the caudate nucleus turns ventrally then anteriorly to continue rostrally as the *tail of*

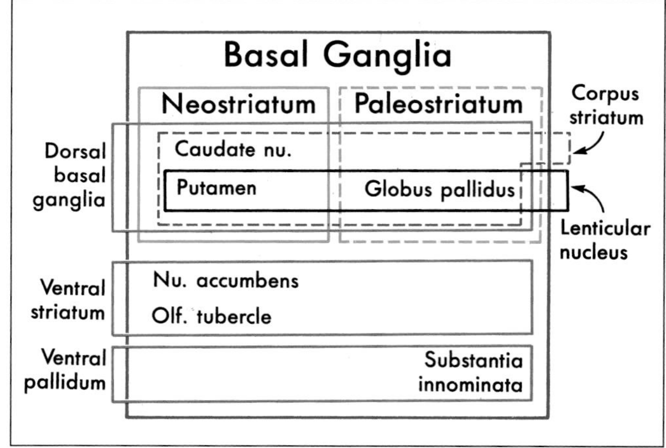

FIGURE 15-13 *A series of stacked boxes showing which nuclei form the various parts of the basal ganglia and the terms used to describe them.*

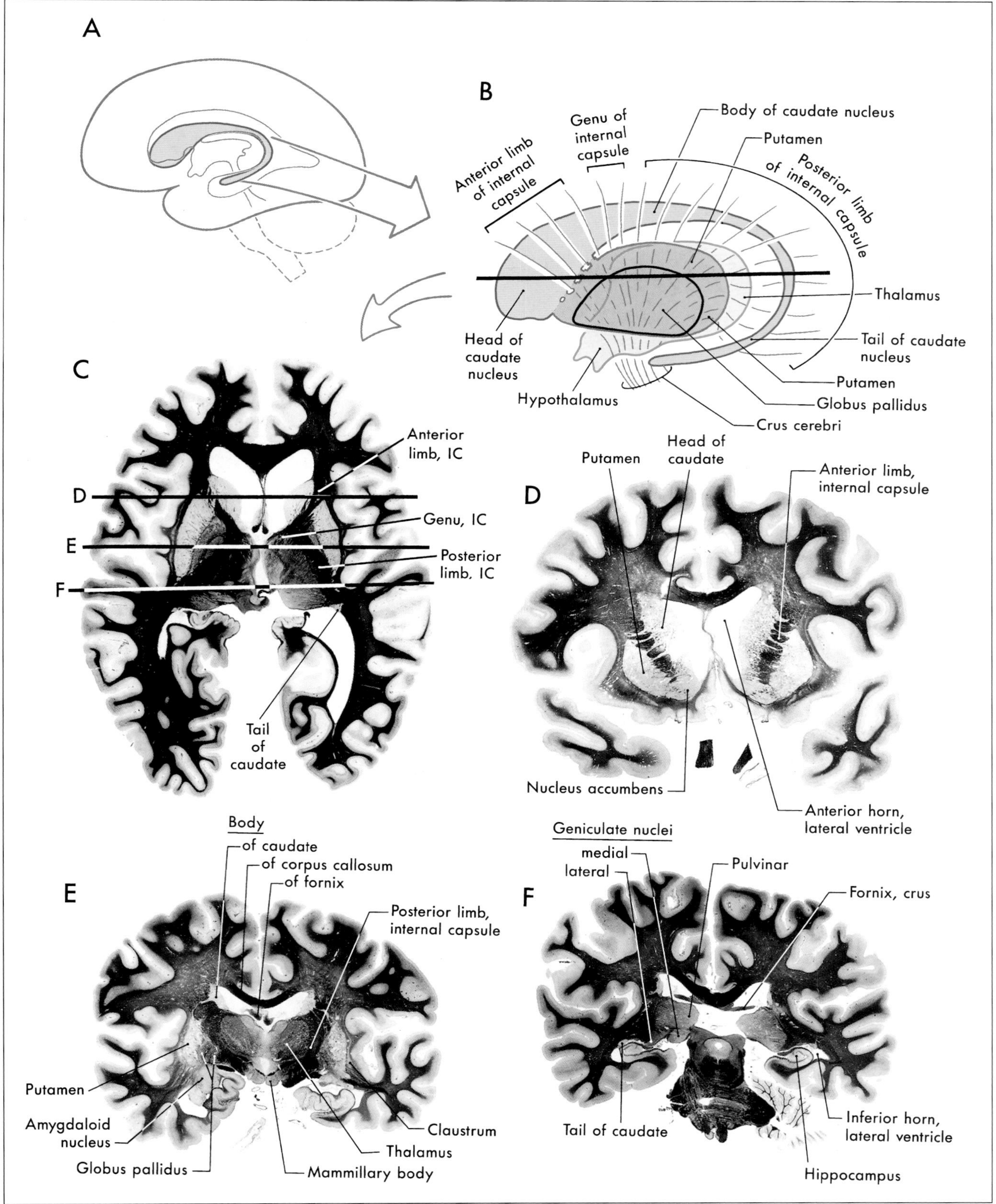

FIGURE 15-14 *Three-dimensional drawing (A,B) of the relationships of the internal capsule, basal ganglia, and thalamus. The axial section (C) represents the approximate plane shown in (B); the coronal sections (D, E, F) are taken from the three levels indicated in C. C–F, Weil stains.*

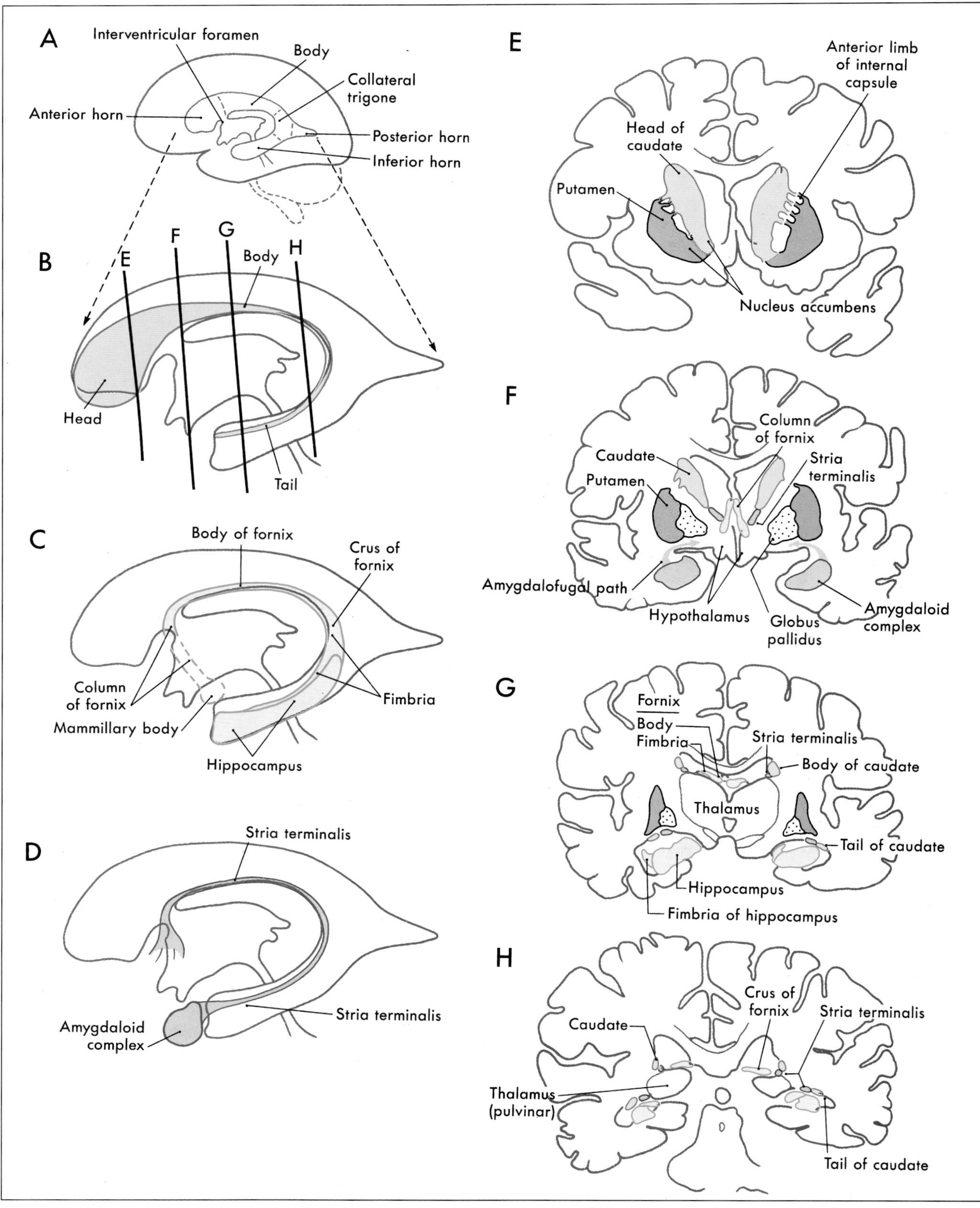

FIGURE 15-15 *The lateral ventricle (A) and its relationship to the caudate nucleus (B), the hippocampus and fornix (C), and the amygdala and stria terminalis (D). The coronal levels in E through H correlate with the planes indicated in B and are color-coded to match the corresponding structure in B, C, or D.*

the caudate nucleus in the lateral wall of the temporal horn. This part of the caudate is located in the dorsolateral wall of the temporal horn. Thus, the C-shape of the caudate nucleus faithfully follows the C-shape of the lateral ventricle (excluding the posterior horn) (Figs. 15-14A,B, 15-15A,B,E–H).

The *lenticular nucleus* is located within the base of the hemisphere and is surrounded by white matter (Fig. 15-14, 15-15). The internal capsule borders the lenticular nucleus medially, and the external capsule separates it from the claustrum laterally. The lenticular nucleus consists of a larger lateral part, the *putamen*, and a smaller medial portion, the *globus pallidus* (or *pallidum*). The putamen extends more dorsal, more anterior, and more caudal than does the globus pallidus and is clearly the larger part of the lenticular nucleus when viewed in axial or coronal planes (Figs. 15-14B–E). The *globus pallidus* is located internal to the putamen and is smaller in all dimensions (Figs. 15-14, 15-15). It is divided into *medial* (*internal*) and *lateral* (*external*) *parts* by thin sheets of dorsoventrally oriented white matter. The globus pallidus is also separated from the putamen by a thin lamina of white matter.

Nucleus Accumbens and Substantia Innominata The *nucleus accumbens* is located rostrally and ventrally in the hemisphere where the putamen is continuous with the head of the caudate nucleus (Fig. 15-16). This cell group is also closely apposed to the septal nuclei and the nucleus of the diagonal band, both of which extend into the base of the septum pellucidum. At a slightly more caudal level, the *substantia innominata* (basal nucleus of Meynert) is located internal to the anterior perforated substance in the area ventral to the anterior commissure (Fig. 15-16). This area contains clusters of small and large cells and diffuse bundles of fibers. Although many areas of the nervous system are affected in Alzheimer disease, there is an especially noticeable loss of larger neurons in the substantia innominata.

Subthalamic Nucleus and Substantia Nigra Although not a part of the telencephalon either developmentally or geographically, the subthalamic nucleus and the substantia nigra are intimately allied with the basal ganglia based on their connections (Fig. 15-17). The *subthalamic nucleus*, a component of the diencephalon, is a flattened, lens-shaped cell group located rostral to the substantia nigra. It is medial to the internal capsule and is capped by a thin sheet of fibers called the *lenticular fasciculus*. The *substantia nigra*, a part of the midbrain, is found dorsomedial to the crus cerebri and immediately caudal to the subthalamic nucleus. It is divided into a reticulated part (*pars reticulata*) and a compact part (*pars compacta*); the latter is characterized by numerous melanin-containing neuron cell bodies.

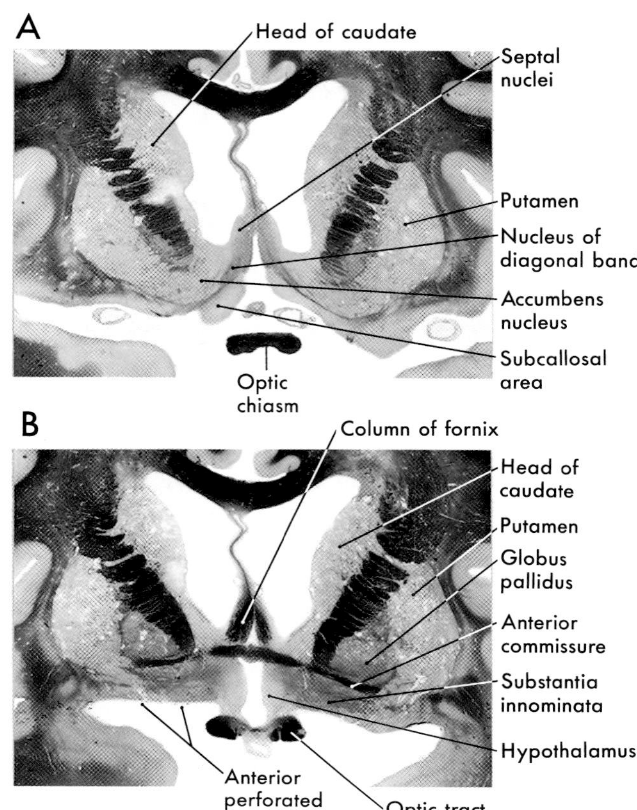

FIGURE 15-16 *Cross sections of the telencephalon at rostral levels showing the nucleus accumbens (A) and the substantia innominata (B) and adjacent structures. Weil stain.*

Major Connections of the Basal Ganglia The connections of the basal ganglia are discussed in their entirety in Chapter 25. It is appropriate, however, to briefly review these relationships here, with particular emphasis on the major fiber bundles of the basal ganglia (Fig. 15-17).

The two largest bundles of efferent fibers exiting the basal ganglia are the *lenticular fasciculus* and the *ansa lenticularis*. The former leaves the globus pallidus, traverses the posterior limb of the internal capsule, and forms a thin sheet of fibers insinuated between the subthalamic nucleus and the zona incerta. These fibers loop around the medial aspect of the zona incerta and pass laterally (and dorsally) as the *thalamic fasciculus* (Fig. 15-17). The fibers of the *ansa lenticularis* originate at a slightly more rostral level, arch around the anteroventral aspect of the internal capsule, and pass caudally to join with the fibers of the lenticular fasciculus as they enter the thalamic fasciculus (Fig. 15-17).

Two smaller but equally important bundles are the *subthalamic fasciculus* and connections between the substantia nigra and neostriatum (Fig. 15-17). The subthalamic fasciculus is composed of bidirectional connections between the globus pallidus and the subthalamic

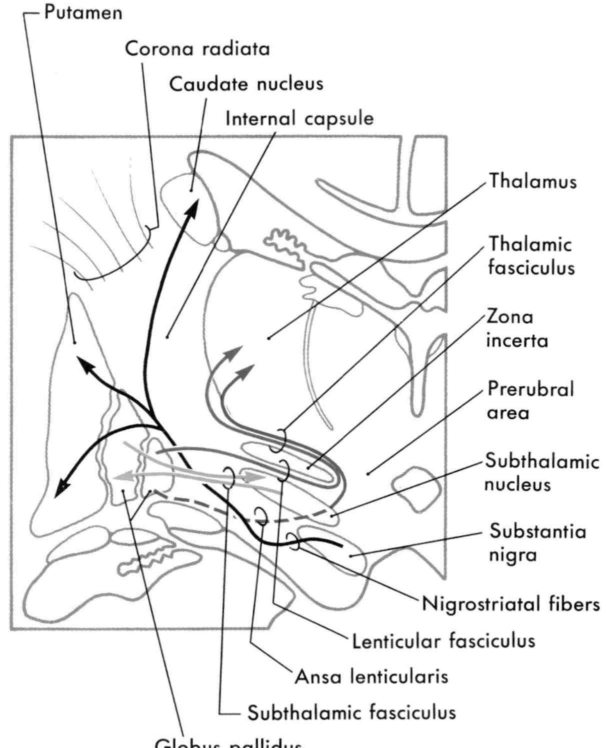

FIGURE 15-17 *Major fiber bundles associated with the basal ganglia and with the subthalamic nucleus and substantia nigra.*

nucleus; these fibers also traverse the internal capsule. The connections between the substantia nigra and the neostriatum are named according to the origin and termination of the fiber. Axons that project from nigral cells to the neostriatum are *nigrostriatal* projections, and fibers that project from striatal cells to the substantia nigra are *striatonigral* fibers.

Vasculature of the Basal Ganglia and Related Structures
The blood supply to the caudate and putamen is provided by branches of the *medial striate artery*, *lenticulostriate branches* of M_1, and the *anterior choroidal artery* (Figs. 15-9, 15-12). The medial striate artery, usually a branch of A_2, serves much of the head of the caudate nucleus. Most of the lenticular nucleus and the surrounding internal and external capsules are supplied by the lenticulostriate branches of M_1. Ventral and medial portions of the head of the caudate, as well as the body of the caudate, are also served by these arteries. The tail of the caudate, ventrocaudal portions of the lenticular nucleus, and adjacent temporal lobe structures (hippocampus, choroid plexus) receive their blood supply via the anterior choroidal artery, a branch of the internal carotid artery. The blood supply to the subthalamic nucleus and the substantia nigra arises from the *posteromedial branches* of the P_1 segment and branches of the posterior communicating artery. These vessels penetrate the brain at the midbrain-diencephalon junction.

HIPPOCAMPUS AND AMYGDALA

The *hippocampal formation* and the *amygdaloid complex* are located in the temporal lobe. The former lies in the ventromedial floor of the temporal horn of the lateral ventricle and the latter in the rostral end of this space. Through a variety of pathways (see Chapter 30), these structures interconnect with numerous telencephalic and diencephalic centers.

Developmentally, the hippocampus is formed by a rolling-in of primitive cortex to form the curved, multilayered structure characteristic of the adult (Fig. 15-14F). The hippocampal formation is found internal to the parahippocampal gyrus and is composed of the *subiculum*, the *hippocampus* (also called *Ammon's horn*), and the *dentate gyrus*. The cortex of the parahippocampal gyrus is continuous with the subiculum, which, in turn, is continuous with the hippocampus. The dentate gyrus forms a reverse loop around the hippocampus and, in doing so, presents a serrated surface that is medially exposed to the subarachnoid space. Details of the internal structure of the hippocampal formation are discussed in Chapter 30 and summarized in Figure 30-4.

Axons of hippocampal neurons converge to form a prominent bundle that arches around caudal, dorsal, and anterior aspects of the thalamus. This bundle, the *fornix*, is a major efferent path of the hippocampal formation (Figs. 15-15A–C, F–H). It is composed of a flattened caudal part, the *crus*, a compact dorsal portion, the *body*, and a part that arches around the anterior part of the thalamus and passes through the hypothalamus to terminate in the mammillary body; this is the *column of the fornix*. Located along the edge of the dentate gyrus and continuing on the lateral edge of the crus and body of the fornix is a thin fringe of fibers called the *fimbria*.

The *amygdaloid complex* (commonly called the *amygdala*) is located internal to the cortex of the uncus (Figs. 15-4, 15-15D,F). Two major efferent bundles are related to the amygdala. First, the *stria terminalis* follows a looping trajectory that shadows, in a reverse direction, the orientation of the caudate nucleus (Fig. 15-15D). In the temporal horn, the stria terminalis is located just medial to the tail of the caudate nucleus. As the stria terminalis arches rostrally, it assumes a position in the shallow groove between the caudate nucleus and the dorsal thalamus (Fig. 15-15G,H) where it is accompanied by the terminal vein. At about the level of the interventricular foramen, the fibers of the stria terminalis fan out to enter and terminate in the hypothalamus, septal area, and the neostriatum.

The second major efferent bundle of the amygdala is the diffusely arranged *ventral amygdalofugal pathway*. These fibers leave the amygdaloid complex, pass medially through the *substantia innominata*, and continue medially to enter hypothalamic and septal nuclei, or turn caudally and distribute to the brainstem (Fig. 15-15F).

Cell groups located internal to the subcallosal area collectively form the *septal nuclei* (Fig. 15-16). Consequently, the subcallosal area and a small strip of cortex located adjacent to the lamina terminalis, the *paraterminal gyrus*, are commonly called the *septal area*. These septal nuclei are medially adjacent to the nucleus accumbens and continuous with sheets of neuronal cell bodies that extend into the *septum pellucidum*. The latter structure extends, in general, from the fornix to the inner surface of the corpus callosum. It forms the medial wall of the anterior horns, and a small part of the bodies of the lateral ventricles (Figs. 15-11, 15-14). In general, the septal nuclei have complex interconnections with hippocampal, amygdaloid, and limbic structures.

Injuries to the temporal lobe, especially bilateral damage, almost always involves the hippocampus and amygdala. Deficits most directly linked to trauma to these structures include profound changes in eating and sexual behavior, in aggression levels, and in memory function. Concerning the latter, the patient demonstrates a loss of recent memory or shows an inability to acquire new memory (learn new tasks), but memory of events that took place in the distant past remains intact.

Vasculature of the Hippocampus and Amygdala The blood supply to the hippocampal formation and amygdaloid complex is primarily via the *anterior choroidal artery*. This vessel arises from the internal carotid, passes along the medial edge of the temporal horn, and sends branches into the hippocampus and amygdala. It also serves the tail of the caudate, the choroid plexus of the temporal horn, and ventral regions of the lenticular nucleus. The cortex of the uncus and of the parahippocampal gyrus are served by superficial branches of the *middle cerebral* and *posterior cerebral arteries*, respectively.

SOURCES AND ADDITIONAL READINGS

Bailey P, von Bonin G: *The Isocortex of Man*. University of Illinois Press, Urbana, 1951.

Carpenter MB, Sutin J: *Human Neuroanatomy*, 8th Ed. Williams & Wilkins, Baltimore, 1983.

Crosby EC, Humphrey T, Lauer EW: *Correlative Anatomy of the Nervous System*. Macmillan Publishing, New York, 1962.

Kuhlenbeck H: *The Central Nervous System of Vertebrates: A General Survey of Its Comparative Anatomy With an Introduction to the Pertinent Fundamental Biologic and Logical Concepts*, Vol 5, Part I: *Derivatives of the Prosencephalon: Diencephalon and Telencephalon*. S. Karger, Basel, 1977.

Kuhlenbeck H: *The Central Nervous System of Vertebrates: A General Survey of Its Comparative Anatomy With an Introduction to the Pertinent Fundamental Biologic and Logical Concepts*, Vol 5, Part II: *Mammalian Telencephalon: Surface Morphology and Cerebral Cortex*. S. Karger, Basel, 1978.

Nieuwenhuys R, Voogd J, van Huijzen Chr: *The Human Central Nervous System, A Synopsis and Atlas*, 3rd Ed., Springer-Verlag, Berlin, 1988.

Paxinos G (ed): *The Human Nervous System*. Academic Press, San Diego, pp 439–755, 1990.

Zola-Morgan S, Squire LR: Neuroanatomy of memory. Annu Rev Neurosci 16:547–563, 1993.

CHAPTER SIXTEEN

The Somatosensory System I: Discriminative Touch and Position Sense

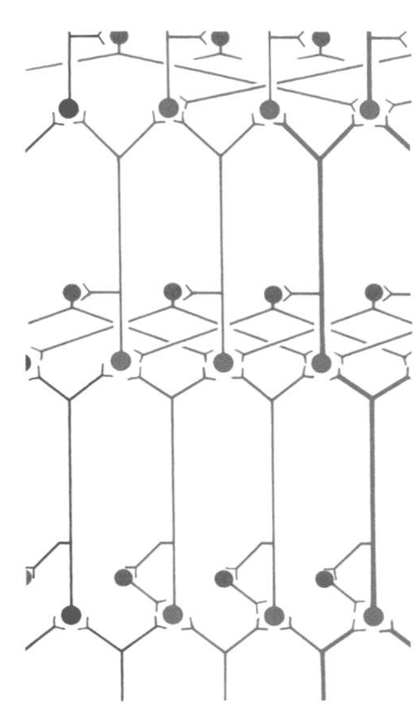

S. WARREN • R. P. YEZIERSKI • N. F. CAPRA

If you reach into your pocket to determine the types of coins present, you are gathering information through the activation of specialized receptors of the *somatosensory system*. Specifically, the size of a coin is determined by noting the joint angles when the coin is held between the forefinger and thumb. "Heads and tails" may be identified using *slowly adapting* receptors sensitive to stimuli that indent the skin. Dimes can be distinguished from pennies by stroking their edges with the fingertips and activating *rapidly adapting* receptors. This information is transmitted to the cerebral cortex by a multi-synaptic pathway called the *dorsal column-medial lemniscal system*. At the same time, much of this information, along with information concerning muscle tension and length, is also transmitted to the cerebellar cortex, where it is used to regulate muscle activity that allows manipulation of the coins. The *spinocerebellar pathways* are among those that subserve these nonconscious somatosensory functions.

OVERVIEW

In general, the somatosensory system transmits and analyzes *touch/tactile* information from external and internal locations on the body and head. The result of these processes leads to the appreciation of somatic sensations, which can be subdivided into the submodalities *discriminative touch*, *flutter-vibration*, *proprioception* (*position sense*), *crude* (*nondiscriminative*) *touch*, *thermal* (*hot and cold*) *sensation*, and *nociception* (*pain*). The following anatomically and functionally discrete pathways transmit these signals: (1) the *dorsal column-medial lemniscal* pathway, (2) the *trigeminothalamic* pathways, (3) the *spinocerebellar* pathways, and (4) the *anterolateral system*.

This chapter describes pathways that transmit discriminative touch, flutter-vibration, and proprioceptive information. These pathways are the *dorsal column-medial lemniscal pathway*, portions of the trigeminothalamic pathways originating in the *principal trigeminal sensory nucleus*, and the *spinocerebellar* pathways. The pathways subserving the submodalities of pain, temperature and crude touch, itch, and tickle, comprise the anterolateral system. These, and portions of the trigeminothalamic pathways are described in Chapter 17.

DORSAL COLUMN-MEDIAL LEMNISCAL SYSTEM (DCMLS)

The dorsal column-medial lemniscal pathway (see Figs. 16-6 and 16-8) is involved with the perception and appreciation of mechanical stimuli. It underlies the capacity for fine form and texture discrimination, form recognition of three-dimensional shape (*stereognosis*), and motion detection. This pathway is also involved in transmitting information related to conscious awareness of body position (*proprioception*) in space.

Characteristic features of the DCMLS include transmission in general somatic afferent (GSA) fibers that have fast conduction velocities, a limited number of synaptic relays where processing of the signal occurs, and a precise somatotopic organization. These features provide the basis for the accurate localization of the body region touched. There is only limited convergence along the pathway; consequently, the signal is transmitted with high fidelity and a high degree of spatial and temporal resolution. This pathway signals somatic sensations using *frequency* and *population codes*. In frequency coding, a cell's firing rate signals stimulus intensity or temporal aspects of the tactile stimulus. In population coding, the distribution in time and space of activated cells in the central nervous system signals location of the stimulus, as well as its motion or direction if any.

The high degree of resolution in the DCMLS is the result of inhibitory mechanisms such as *lateral* (*surround*) *inhibition*. This mechanism, which sharpens and enhances the discrimination between separate points on the skin, is critical for *two point discrimination*. The ability to discriminate between two points simultaneously applied varies widely over different parts of the body.

Peripheral Mechanoreceptors The first step in evoking somatic sensations is the activation of peripheral mechanoreceptors. Mechanical pressure, such as skin deformation, is *transduced* into an electrical signal in the peripheral process of a primary afferent neuron. This leads to a depolarizing *graded membrane potential* across the membrane of the neuron. If this potential depolarizes the *trigger zone*, located at the first myelin segment of the axon, to *threshold*, an *action potential* is produced (see Chapter 3). In most receptors, transduction occurs between the mechanoreceptor and subjacent primary afferent membrane. In contrast, Merkel cells undergo depolarization in response to a stimulus. They influence the associated primary afferent axon by vesicular release of a transmitter substance. Depolarization of the Merkel cell changes the rate at which the transmitter is released, leading to depolarization of the axon.

Each morphologic type of receptor responds to different tactile stimuli. *Cutaneous tactile receptors* (Table 16-1; Fig. 16-1) are located in the basal epidermis and dermis of *glabrous* (palms, sole, lips) and *hairy* skin. These low threshold mechanoreceptors may be encapsulated, such as *Meissner's* and *Pacinian corpuscles*, or unencapsulated, such as *Merkel cell neurite complexes* and *hair follicle receptors*. Meissner's corpuscles, hair follicle receptors, and Pacinian corpuscles respond to transient, phasic, or vibratory stimuli. These receptors respond to each initial application or removal of a stimulus but fail to respond during maintained stimulation. Consequently, they are

TABLE 16-1

CUTANEOUS MECHANORECEPTORS AND THEIR ASSOCIATED FIBER TYPES AND SENSATIONS

RECEPTOR TYPE (ADAPTATION RATE)	SENSATION	FUNCTION-SIGNAL	FIBER TYPE
Meissner's corpuscles (Rapidly Adapting RA)	tap and flutter	texture; velocity	A-beta (Aβ)
Hair follicle receptors (Rapidly Adapting) (Slowly Adapting SA)	motion, direction	velocity	A-beta (Aβ)
Pacinian corpuscle (Rapidly Adapting RA)	vibration, 100–300 Hz	vibration; transients	A-beta (Aβ)
Merkel's cell (Slowly Adapting SA)	light pressure	edge detection; displacement velocity	A-beta (Aβ)
Ruffini complex (Slowly Adapting SA)	unknown	skin stretch, joint movement; displacement velocity	A-beta (Aβ)

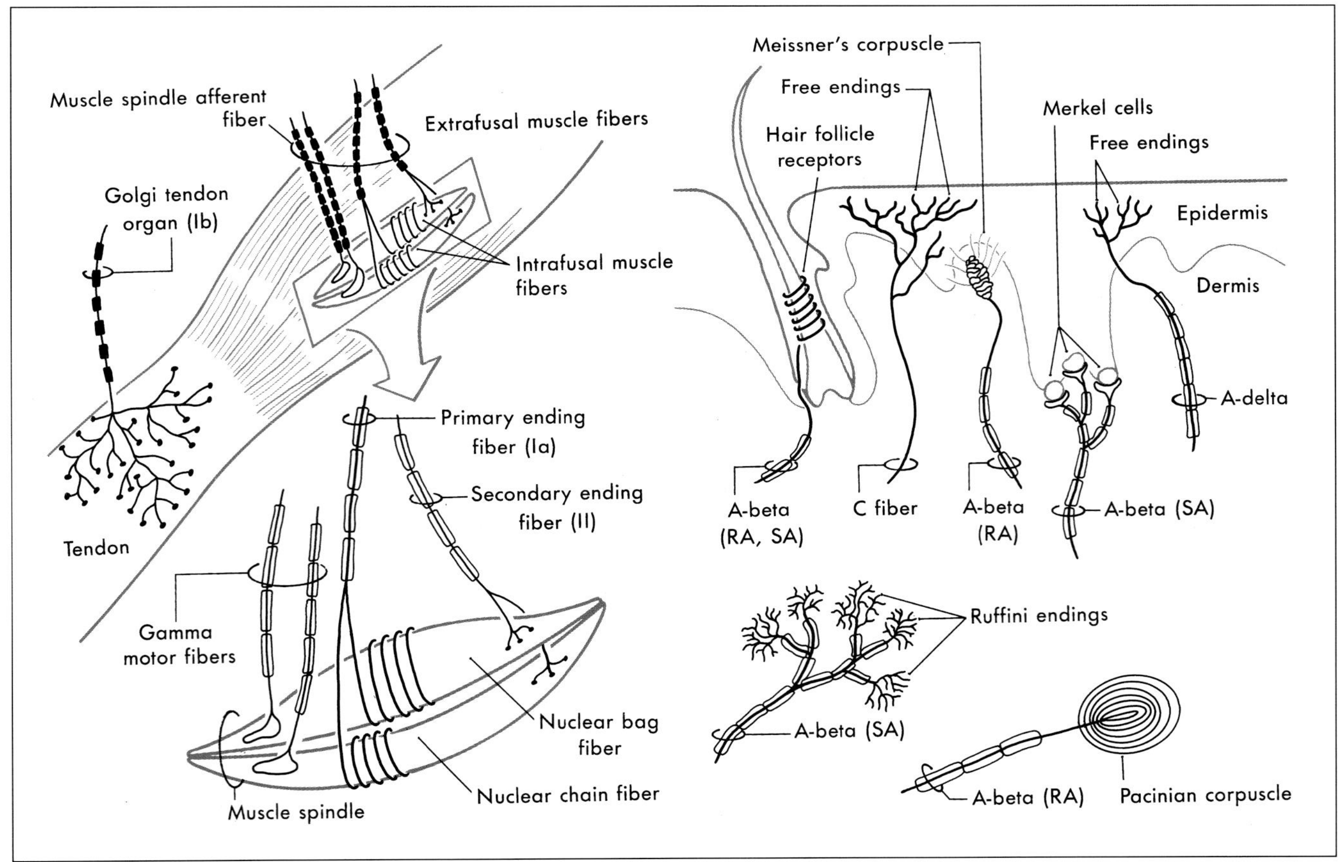

FIGUXRE 16-1 *Proprioceptive receptors and cutaneous mechanoreceptors and their afferent fibers. Cutaneous receptors are either rapidly adapting (RA) or slowly adapting (SA).*

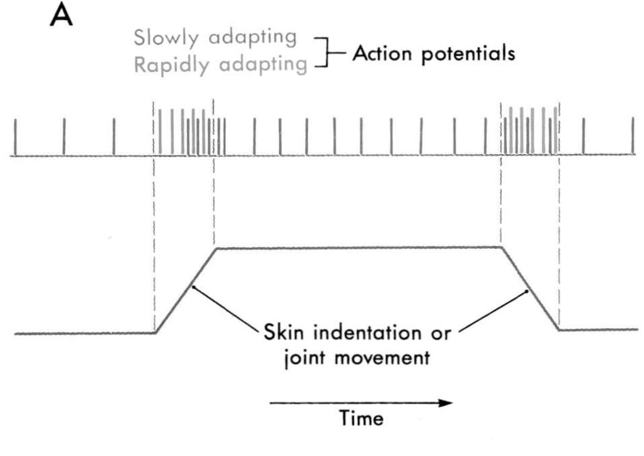

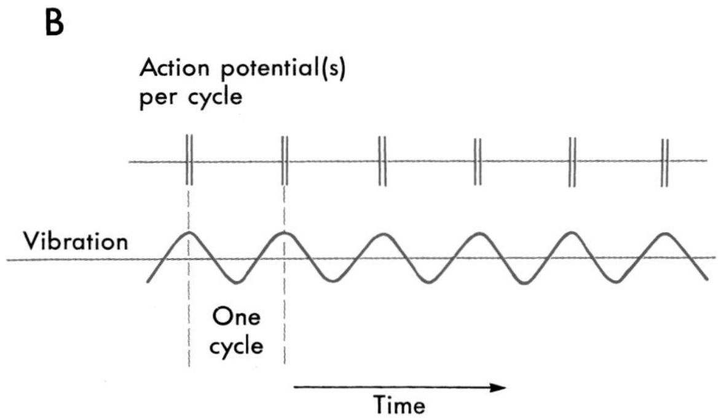

FIGURE 16-2 *Diagrammatic action potentials (top trace, A) evoked by skin indentation and removal of a cutaneous stimulus or of joint movement (bottom trace, A) in primary afferent fibers innervating slowly adapting (SA:Red) and rapidly adapting (RA:Green) cutaneous mechanoreceptors. Diagrammatic action potentials (blue B) evoked in a Pacini's corpuscle afferent fiber by sinusoidal stimulation of the skin surface (bottom trace).*

rapidly adapting (RA) *receptors* (Fig. 16-2A). Hair follicle receptors are also capable of signaling motion, its direction or orientation, and its velocity.

Merkel cell neurite complexes and some hair follicle receptors signal tonic events such as discrete small indentations in the skin. They provide input related to both the displacement and velocity of a stimulus. They are also capable of encoding stimulus intensity or duration because they are *slowly adapting* (SA) and are active as long as the stimulus is present (Fig. 16-2A). For example, Merkel cell complexes are crucial to reading Braille.

Deep tactile mechanoreceptors are found within the dermis of the skin and in the fascia surrounding muscles and bone. These receptors include *Pacinian corpuscles*, *Ruffini's endings*, and other encapsulated nerve endings located in the periosteum, the deep fascia, and the mesenteries. The receptors of this group respond to pressure, vibration (Fig. 16-2B; Table 16-1), or skin stretch and distention.

Proprioceptive receptors (Table 16-2; Fig. 16-1A) are located in muscles, tendons, and joint capsules. These receptors include the nuclear bag and nuclear chain intrafusal muscle fibers of *muscle spindles* and their associated nerve fibers. The *Golgi tendon organs* and their group Ib fibers and the encapsulated Ruffini-like joint receptors also function in this capacity. They respond to static limb and joint position or to the dynamic movement of the limb (*kinesthesia*) and are important sources of information for balance, posture, and limb movement.

The accuracy with which a tactile stimulus is detected depends on the density of receptors and the size of receptive fields (Fig. 16-3). The greatest density of cutaneous tactile receptors is found on the tips of the glabrous digits and in the perioral region. Other regions, like the back, have much lower density, thus creating a receptor density gradient between various body parts. The *receptive field* is the area of skin innervated by branches of a GSA fiber, the stimulation of which activates its receptors (Fig. 16-3). *Small receptive fields* are found in areas, such as the finger tips, where receptor density is high and each receptor serves an extremely small area of skin. In such regions the individual is able to discriminate small variations in a variety of sensory inputs. In other regions, receptor density is low and each receptor serves an expansive area of skin, creating *large receptive fields* with resultant reduction in discriminative ability.

At all levels of the tactile pathway, densely innervated body parts are represented by greater numbers of

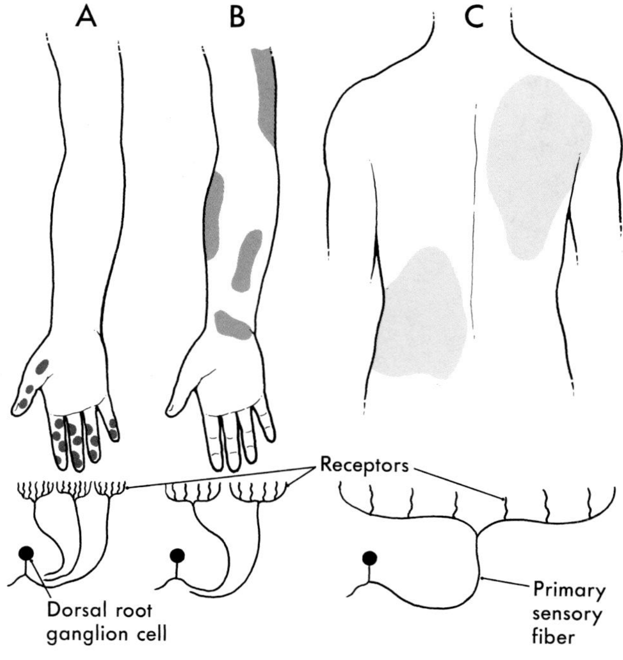

FIGURE 16-3 *Variation in the size of receptive fields as a function of peripheral innervation density. The greater the density of receptors, the smaller the receptive fields of individual afferent fibers.*

TABLE 16-2

CLASSIFICATION OF PROPRIOCEPTIVE RECEPTORS AND THEIR ASSOCIATED FIBER TYPES AND SENSATIONS

RECEPTOR TYPE (ADAPTATION RATE)	SENSATION	FUNCTION-SIGNAL	FIBER TYPE
Nuclear bag fiber (Slowly Adapting—primary annulospiral endings)	high dynamic sensitivity	length and rate of change; length and velocity	Ia
Nuclear chain fiber (Slowly Adapting—secondary flower spray ending)	low dynamic sensitivity	length; tension	II
Golgi tendon organ (Slowly Adapting)	tension	protect from overstretching; tension	Ib
Ruffini endings (Slowly Adapting)	limb position	joint movement and pressure	I

neurons and take up a disproportionately large part of the somatosensory system's body representation. As a result, the finger tips and lips provide the central nervous system with the most specific and detailed information about a tactile stimulus.

Primary Afferent Fibers As initially described in Chapter 9, primary afferent GSA fibers consist of (1) a *peripheral process* extending from the dorsal root ganglion to either contact peripheral mechanoreceptors or end as free nerve endings, (2) *a central process* extending from the dorsal root ganglion into the central nervous system, and (3) *a pseudounipolar cell body* in the dorsal root ganglion. The peripheral distribution of the afferent nerves associated with each spinal level delineates the segmental pattern of *dermatomes*. In clinical testing, these ribbon-like strips of skin are associated primarily with fibers and pathways that convey pain and thermal information; they are considered in Chapter 17.

Primary afferent peripheral processes are classified by two schemes. One is based on their contribution to a compound action potential (A, B, and C waves) recorded from an entire mixed peripheral nerve after electrical stimulation of that nerve. The other scheme is based on fiber diameter, myelin thickness, and conduction velocity (classes I, II, III, and IV) (Table 16-3; Fig. 16-4). The two schemes are related because conduction velocity determines a fiber's contribution to the compound action potential. Discriminative touch and position sense are transmitted by type Ia, Ib, and II fibers (Tables 16-1, 16-2).

Spinal Cord and Brainstem On the basis of cell and fiber diameter, primary sensory fibers are categorized as *large* and *small*. Large-diameter fibers subserve discriminative touch, flutter-vibration, and proprioception (groups Ia, Ib, II, and A-β; Tables 16-1, 16-2). They enter the spinal cord via the *medial division of the dorsal root* (see Chapter 9) and then branch (Fig. 16-5). One set of branches ter-

TABLE 16-3

FIBER GROUPS, DIAMETERS, AND CONDUCTION VELOCITIES FOUND IN PERIPHERAL MIXED AND CUTANEOUS NERVES

MIXED NERVE	CUTANEOUS NERVE	FIBER DIAMETER (μm)	CONDUCTION VELOCITY (m/sec)
I	N/A	13–20	80–120
II	A-beta (Aβ)	6–12	35–75
III	A-delta (Aδ)	1–5	5–30
IV	C	0.2–1.5	0.5–2

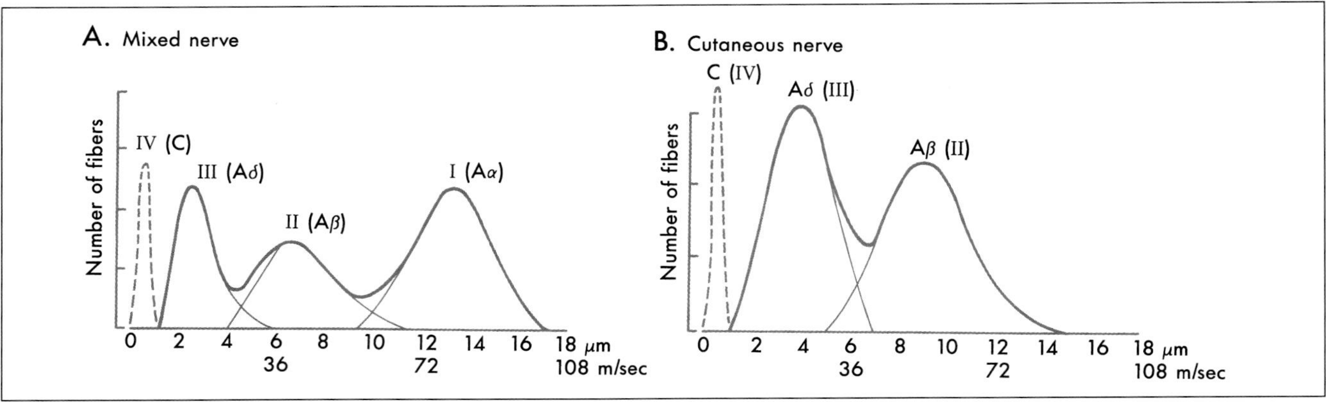

FIGURE 16-4 *Compound action potential evoked in a mixed nerve (A) and a cutaneous nerve (B) in response to electrical stimulation. Note the increase in the number of small diameter fibers and the absence of the A-α fibers in the cutaneous nerve (B).*

minates on second-order neurons in the spinal cord gray matter at, above, and below the level of entry. These branches contribute to a variety of spinal reflexes and to ascending projections such as *postsynaptic dorsal column fibers*. The largest set of branches ascends cranially and contributes to the formation of the *gracile* and *cuneate fasciculi*. These fiber bundles are collectively termed the *dorsal (posterior) columns* owing to their position in the spinal cord (Figs. 16-5, 16-6).

Within the dorsal columns, fibers from different dermatomes are organized topographically. Sacral level fibers assume a medial position, and fibers from progressively more rostral levels (up to thoracic level T6) are added laterally to form the *gracile fasciculus*. Thoracic fibers from above T_6 and cervical fibers form the laterally placed *cuneate fasciculus* in the same manner. Thus, the leg is represented medially and the arm laterally within the dorsal columns (Fig. 16-6). Compromise of blood flow in the posterior spinal artery, which supplies the dorsal funiculus, or mechanical injury to the dorsal columns (as in *Brown-Sequard syndrome*) results in an *ipsilateral reduction or loss of discriminative, positional, and vibratory tactile sensations at and below the segmental level of the injury*.

The *dorsal column nuclei*, the *gracile and cuneate nuclei*, are found in the dorsal medulla at the rostral end of their respective fasciculi. They are supplied by the posterior

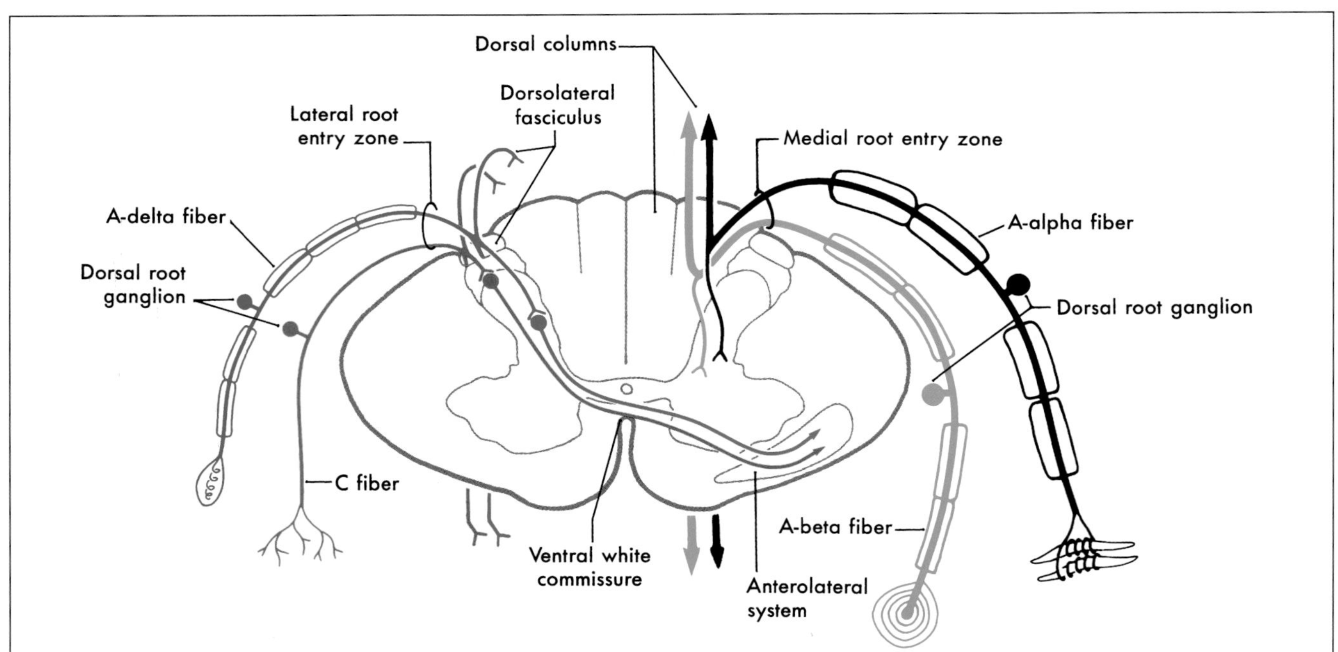

FIGURE 16-5 *A representative section of the cervical spinal cord showing large diameter A-α and A-β fibers on the right and small diameter A-δ and C fibers on the left.*

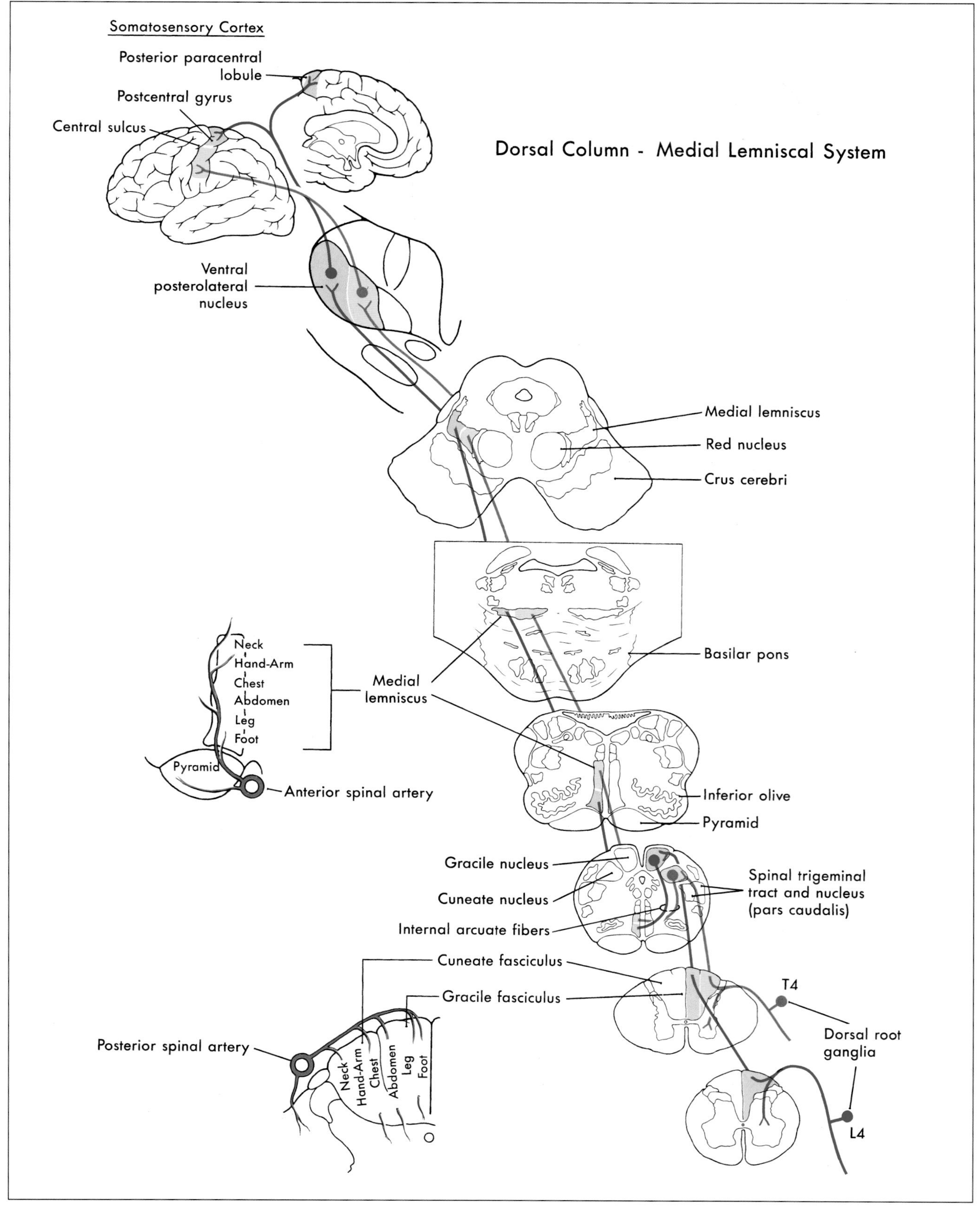

FIGURE 16-6 *The dorsal column medial lemniscal system. Note the somatotopic arrangement of body parts at each level of this pathway.*

spinal artery (Fig. 16-6). The cell bodies of the gracile and cuneate nuclei are the *second-order neurons* in the DCMLS. They receive input from *first-order neurons*, having cell bodies in the ipsilateral dorsal root ganglia. The gracile nucleus receives input from sacral, lumbar, and lower thoracic levels via the *gracile fasciculus*; the cuneate nucleus receives input from upper thoracic and cervical levels through the *cuneate fasciculus*.

In addition to the somatotopic organization of projections to the dorsal column nuclei, there is a submodality segregation of tactile inputs within these nuclei. The relay neurons that receive excitatory input from primary afferent fibers form submodality-specific rostrocaudal bands (Fig. 16-7). Rapidly adapting inputs terminate centrally and caudally within the nuclei. Slowly adapting cutaneous input and muscle spindle and joint inputs project to the rostral pole of the cuneate and gracile nuclei and to the rostrally adjacent *nucleus z*. The dorsal column nuclei also receive descending axons from the contralateral primary somatosensory cortex and from the medullary reticular formation (n. reticularis gigantocellularis) (Fig. 16-7).

The dorsal column nuclei have an "inner core" region of large projection neurons surrounded by a diffuse "shell" of small fusiform and radiating cells (Fig. 16-7). The latter are interneurons responsible for feedback inhibition in the dorsal column nuclei. This feedback alters activity of projection neurons of the inner core. In addition, the presence of non-dorsal column inputs to these projection cells suggests that information received by the dorsal column nuclei is not simply relayed but undergoes signal processing.

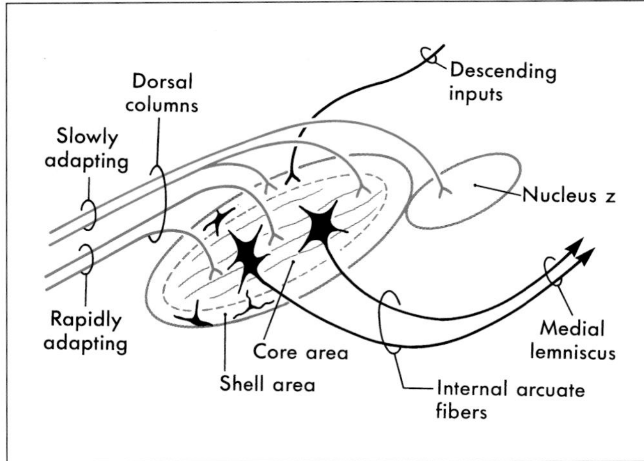

FIGURE 16-7 *Schematic representation of the dorsal column nuclei and nucleus z. Slowly adapting inputs, including those from joint and muscle receptors, terminate preferentially in the shell region. Rapidly adapting input, including Meissner's corpuscles and some hair follicle receptors, project to the core region. The output of the dorsal column nuclei is influenced by descending inputs arising in other brainstem and cortical areas.*

The *second-order cells* in the core regions of the dorsal column nuclei send their axons to the contralateral thalamus (Fig. 16-6). In the medulla, the *internal arcuate fibers*, axons of cells in the dorsal column nuclei, arc ventromedially toward the midline, decussate, and ascend as the *medial lemniscus* on the opposite side. Fibers in the medial lemniscus that arise in the cuneate nucleus are located dorsal to those that originate from the gracile nucleus (Fig. 16-6). The anterior spinal artery supplies the medial lemniscus in the medulla, and penetrating branches of the basilar artery supply it in the pons. Vascular damage at these brainstem levels leads to deficits in discriminative touch, vibratory, and positional sensibilities over the contralateral body. As the medial lemniscus moves rostrally through the brainstem, it rotates laterally so that the arm representation comes to lie medially and the leg laterally (Fig. 16-6). This somatotopic organization is maintained as the lemniscus ascends through the brainstem and terminates on cells in the *ventral posterolateral nucleus* (VPL) of the thalamus.

Two supplemental pathways, the *spinocervicothalamic pathway* and the *postsynaptic dorsal column pathway*, relay nociceptive and nondiscriminative tactile signals to supraspinal levels. Although these pathways are small in humans, they may provide the morphologic basis for the return of some tactile sensation following vascular accidents involving the DCMLS. The *spinocervicothalamic pathway* is considered in Chapter 17.

The *postsynaptic dorsal column pathway* consists of non-primary afferent axons carrying tactile signals in the dorsal columns (Fig. 16-8). The cells of origin of this pathway are located in laminae III and IV of the dorsal horn. Axons of these second-order *postsynaptic dorsal column fibers* travel in the dorsal columns and, together with other tactile primary afferent fibers, terminate in the dorsal column nuclei. Cells of these nuclei relay this postsynaptic dorsal column input to the contralateral thalamus via the medial lemniscus (Fig. 16-8).

Ventral Posterolateral Nucleus The *ventral posterior nucleus*, sometimes called the *ventrobasal complex*, is a wedge-shaped cell group located caudally in the thalamus. Its lateral border abuts the internal capsule, and ventrally it borders on the external medullary lamina. The ventral posterior nucleus is composed of the laterally located *ventral posterolateral nucleus* (VPL) and the medially located *ventral posteromedial nucleus* (VPM). In humans, these nuclei have also been termed the ventralis caudalis externus and ventralis caudalis internus, respectively. The VPL is separated from the VPM by fibers of the *arcuate lamina*. The ventral posterior nucleus is supplied by thalamogeniculate branches of the posterior cerebral artery, and compromise of these vessels can result in loss of all tactile sensation over the contralateral body and head (Fig. 16-9).

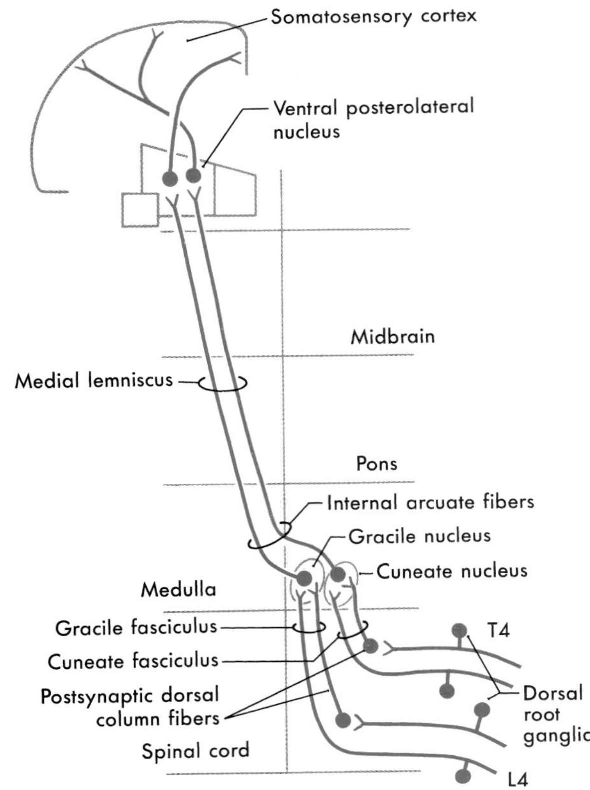

FIGURE 16-8 *Summary of the postsynaptic dorsal column pathway.*

The VPL receives ascending input from the medial lemniscus, and input to the VPM is from the trigeminothalamic tracts. Within VPL, medial lemniscal fibers from the contralateral cuneate nucleus terminate medial to those from the gracile nucleus. As a result, the representation of the lower limb and foot is lateral and the upper limb is medial in VPL (Fig 16-9). The representation of an individual body part is organized as a C-shaped lamina. Tactile signals are also represented in other thalamic nuclei receiving lemniscal input, including the ventral posterior inferior nucleus and the pulvinar and lateral posterior group.

In addition to their somatotopic organization, the medial lemniscal fibers that terminate in the ventral posterior nucleus are segregated on the basis of their functional properties. Rapidly and slowly adapting inputs terminate on different cell groups within the "core" region of VPL. Pacinian inputs and inputs arising from joints and muscles are confined to the more superficial dorsal, anterior, and ventral "shell" regions of the nucleus. Individual lemniscal axons arborize in the sagittal plane to terminate on longitudinal cell clusters, called *rods*, in the VPL. This arrangement of inputs and target cells creates isorepresentations consisting of neurons with similar receptive fields and submodalities arranged along a rostrocaudal axis.

The ventral posterior nucleus contains two populations of identified neurons. The first consists of large-diameter multipolar cells that give rise to axons that traverse the posterior limb of the internal capsule and terminate mainly in the primary (SI) and secondary (SII) somatosensory cortices. These *thalamocortical cells* and *fibers* are the *third-order neurons* in the DCMLS that provide excitatory input (glutaminergic) to the cortex. The second population consists of inhibitory (GABAergic) *local circuit interneurons*, which receive excitatory corticothalamic inputs and influence the firing rates of thalamocortical cells. In addition, these relay cells are also influenced by GABAergic input from the thalamic reticular nucleus and by excitatory (glutaminergic) corticothalamic fibers that arise in layer VI of the primary and secondary somatosensory cortices.

Primary Somatosensory (SI) Cortex Axons from third-order thalamic neurons terminate in the *primary somatosensory cortex* (SI) (Figs. 16-6, 16-9, 16-10). This cortical region is bordered anteriorly by the central sulcus and posteriorly by the postcentral sulcus, and com-

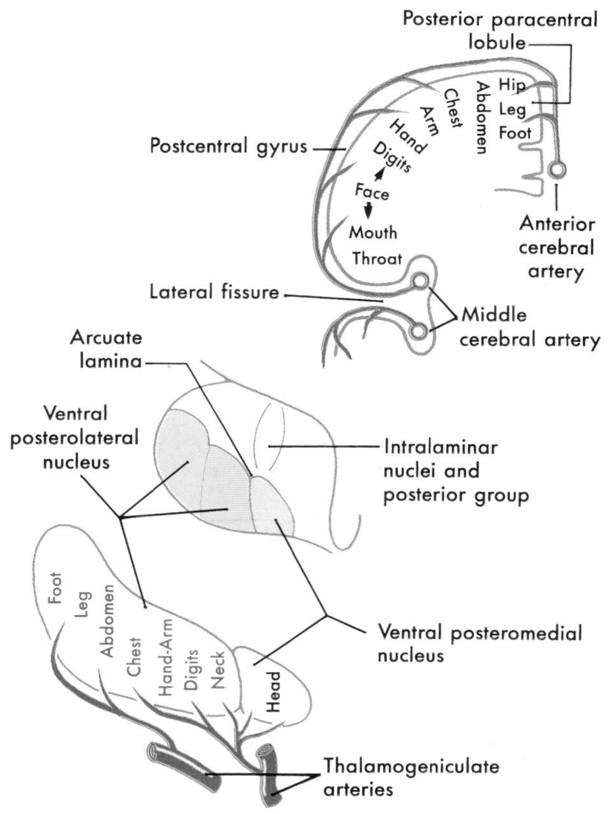

FIGURE 16-9 *Blood supply and somatotopic organization of the body in the ventral posterolateral and posteromedial nuclei and in the primary somatosensory (SI) cortex.*

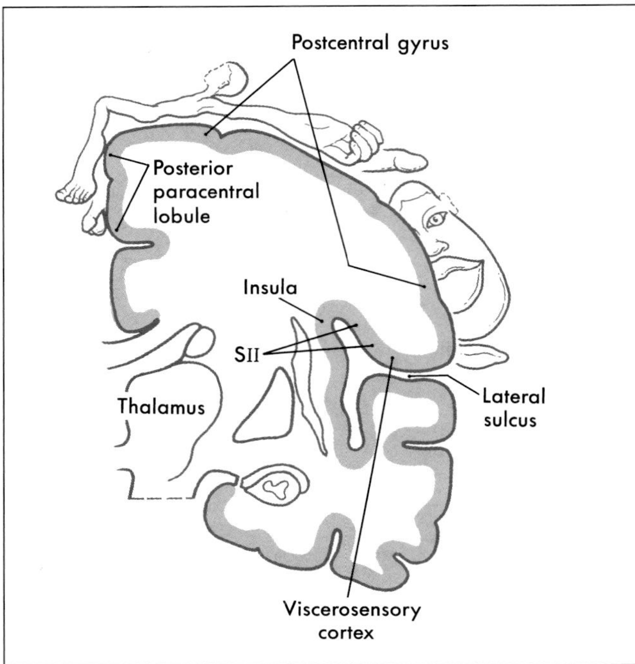

FIGURE 16-10 *The homunculus (body representation) of the primary somatosensory (SI) cortex. (Adapted from Penfield and Rasmussen, 1968, with permission.)*

prises the postcentral gyrus and the posterior paracentral lobule (Fig. 16-11). The cortex contains a somatotopic representation of the body surface (a *homunculus*, or "little man"), which is laid out in a "foot to tongue" pattern along the medial-to-lateral axis (Fig. 16-10). Body regions, such as the hand and the lips, with a high density of receptors have a disproportionately large amount of cortical tissue dedicated to their central representation. In contrast, regions, such as the back, with low receptor density have small cortical representations (Figs. 16-3, 16-10). Blood supply to the SI cortical areas is provided by the anterior and middle cerebral arteries. Vascular lesions involving the middle cerebral artery produce tactile loss over the contralateral upper body and face, and those involving the anterior cerebral artery affect the contralateral lower limb (Fig. 16-9).

Histologically, the primary somatosensory cortex is subdivided into four distinct areas; from anterior to posterior, these are *Brodmann's areas 3a, 3b, 1,* and *2* (Fig. 16-11). Area 3a is located in the depths of the central sulcus and abuts area 4 (primary motor cortex). Areas 3b and 1 extend up the side of the sulcus onto the shoulder of the postcentral gyrus, whereas area 2 lies on the gyral surface and abuts area 5 (somatosensory association cortex).

Each of these four cytoarchitectural areas of the SI cortex receives submodality-specific inputs. Areas 3a and 2 are primarily targeted by the shell region of VPL. They receive proprioceptive inputs arising from muscle spindle afferents (mainly area 3a), Golgi tendon organs, and joint afferents (mainly area 2). These two areas are capable of processing kinesthetic information related to muscle length and tension, as well as static and transient joint position. Areas 3b and 1 are mainly targeted by the core region of the VPL. They receive cutaneous afferents from receptors such as Meissner's corpuscles (RA) and Merkel cells (SA).

Damage to various parts of the somatosensory cortex results in certain types of sensory losses. Lesions involving area 1 produce a deficit in texture discrimination, whereas damage of area 2 results in loss of size/shape discrimination (*astereognosis*). Injury to area 3b has a more profound effect than damage to either area 1 or 2 alone, producing deficits in both texture and size/shape discrimination. This difference suggests that there is hierarchical processing of tactile information in SI cortex, with area 3b performing the initial processing and distributing the information to areas 1 and 2.

Additional Cortical Somatosensory Regions The second somatosensory cortex (SII) lies deep in the inner face of the upper bank of the lateral sulcus (Fig. 16-10). It, too, contains a somatotopically organized "representation" of the body surface. Inputs to SII cortex arise from the ventral posterior inferior nucleus of the thalamus and from the ipsilateral SI cortex. This cortical area is also supplied primarily by the middle cerebral artery, so it cannot substitute functionally for SI following vascular compromise of this artery (Fig. 16-9).

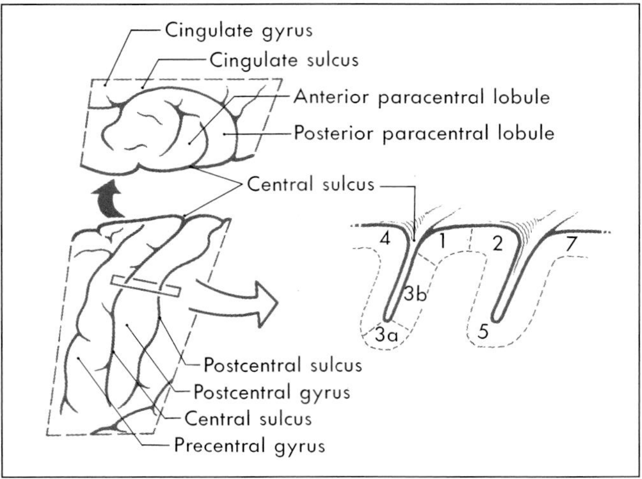

FIGURE 16-11 *The primary somatosensory cortex (SI) of the parietal lobe. The views on the left show the location of SI in the postcentral gyrus (lateral view, bottom) and as it extends medially into the posterior paracentral lobule (top view). See Figure 16-12 for an overview of the entire hemisphere. The cross section at right shows the subdivisions of the SI cortex into the four cytoarchitecturally distinct areas 3a, 3b, 1, and 2. Rostral to these is motor cortex (area 4), and caudal to them are association areas 5 and 7.*

Posterior to area 2, additional parietal cortical regions also receive tactile inputs. These regions include area 5 and lateral portions of area 7 (7b). The anterior pulvinar and lateral posterior group, which receive some medial lemniscal input, project to areas 5 and 7 (Fig. 16-11). In addition, they also receive input from primary somatosensory cortex. Lesions in the parietal association area can produce agnosia, in which contralateral body parts are lost from the personal body map. Sensation is not radically altered, but the limb is not dressed and is not recognized as part of the patient's own body.

TRIGEMINAL SYSTEM: TRIGEMINO-THALAMIC TRACTS

The trigeminal system is responsible for the relay and central processing of most somatosensory information from the head, especially the face, oral and nasal cavities, teeth, and cornea. Like spinal cord pathways, trigeminal pathways can be subdivided into those for the submodalities of pain and temperature (see Chapter 17) and those for tactile discrimination, flutter-vibration, and proprioception. The following discussion considers trigeminal structures involved with the latter submodalities.

Peripheral Receptors Tactile sensations originating in the head and neck derive from the same types of sensory receptors found in other parts of the body (Fig. 16-1). However, owing to their association with structures unique to this region, some of these receptors serve specialized functions. For example, receptors in the periodontal ligament (connective tissue surrounding each tooth) are exquisitely sensitive to tooth displacement and bite force. A large number of encapsulated receptors, particularly Meissner's corpuscles, are found beneath the surface of the lips and perioral skin. Sensory receptors of the face, the oral cavity as far as the pharyngeal arches (with the exception of the taste buds), the nasal cavity, the cerebral blood vessels, and the meninges, are also innervated by GSA fibers of the trigeminal nerve.

Trigeminal Nerve As its name implies, the trigeminal nerve has three peripheral divisions: *ophthalmic* (V_1), *maxillary* (V_2), and *mandibular* (V_3). A few nerve fibers in the *facial* (CN VII), the *glossopharyngeal* (CN IX), and the *vagus* (CN X) nerves transmit GSA innervation from a small cutaneous area around the ear. The peripheral distribution of these nerves delineates the facial "dermatomes" (see Chapter 17).

The pseudounipolar cell bodies of the trigeminal nerve are located in the *trigeminal* (*Gasserian* or *semilunar*) *ganglion* (Figs. 16-12, 16-13), the only exception being the cells of the mesencephalic trigeminal nucleus. The central processes of these trigeminal ganglion cells form the large *sensory root* of the trigeminal nerve (*portio major*) as they enter the lateral aspect of the pons. Within the brainstem, central processes of most trigeminal ganglion cells bifurcate into ascending and descending branches before terminating on *second-order neurons*. These neurons form a long column of cells that, based on functional and cytoarchitectural criteria, is divided into a *principal sensory nucleus*, located in the pons, and a *spinal nucleus of the trigeminal nerve*. This latter nucleus extends from the caudal pons into the upper cervical spinal cord and is subdivided into three regions oriented rostrocaudally (see Chapter 17). The central processes of the ganglion cells that terminate in the spinal nucleus form the *spinal tract of the trigeminal nerve* and convey information concerned mainly with pain, thermal sense, and nondiscriminative touch. The discussion in this chapter focuses on the *principal nucleus and its involvement in discriminative touch, proprioception, and kinesthesia* from the head.

Central Trigeminal Pathways The *principal* (*chief*) *sensory nucleus* is situated in the middle pons at the rostral pole of the spinal trigeminal nucleus (Figs. 16-12, 16-13). Most of the primary afferent fibers that terminate in the principal sensory nucleus are large diameter fibers (e.g., A-beta) concerned with discriminative sensation from the face and oral cavity. The principal sensory nucleus can be divided into dorsomedial and ventrolateral regions. The dorsomedial division receives most of its primary afferent input from the oral cavity, and the ventrolateral division receives input from all three components of the trigeminal nerve. Thus, the somatotopic pattern of the principal sensory nucleus is inverted, with V_1 being ventral, V_3 dorsal, and V_2 sandwiched in between (Fig. 16-12).

Second-order neurons in the principal sensory nucleus relay discriminative tactile information from the head to the *ventral posteromedial nucleus* (VPM). Neurons in the ventrolateral part of the principal sensory nucleus give rise to axons that project to the contralateral VPM along with fibers originating in the spinal nucleus. This combined ascending projection forms the *trigeminal lemniscus*, or *ventral trigeminothalamic tract* (Figs. 16-12, 16-13), which travels in close proximity to the medial lemniscus. Neurons in the dorsomedial division of the principal sensory nucleus project to the ipsilateral VPM by way of the *dorsal trigeminothalamic tract* (Fig. 16-12). This pathway ascends in the pontine tegmentum lateral to the periaqueductal gray in close association with the central tegmental tract. The afferent projections from the principal sensory nucleus terminate somatotopically within VPM so that the oral cavity is represented medially and the lateral facial structures more laterally (Figs. 16-9, 16-12). Third-order thalamocortical neurons in VPM project via the posterior limb of the internal capsule to the laterally placed face area of SI in the postcentral gyrus (Figs. 16-9, 16-10). Perioral regions have the highest peripheral innervation density, and consequently the largest representation along the postcentral gyrus (Fig. 16-10).

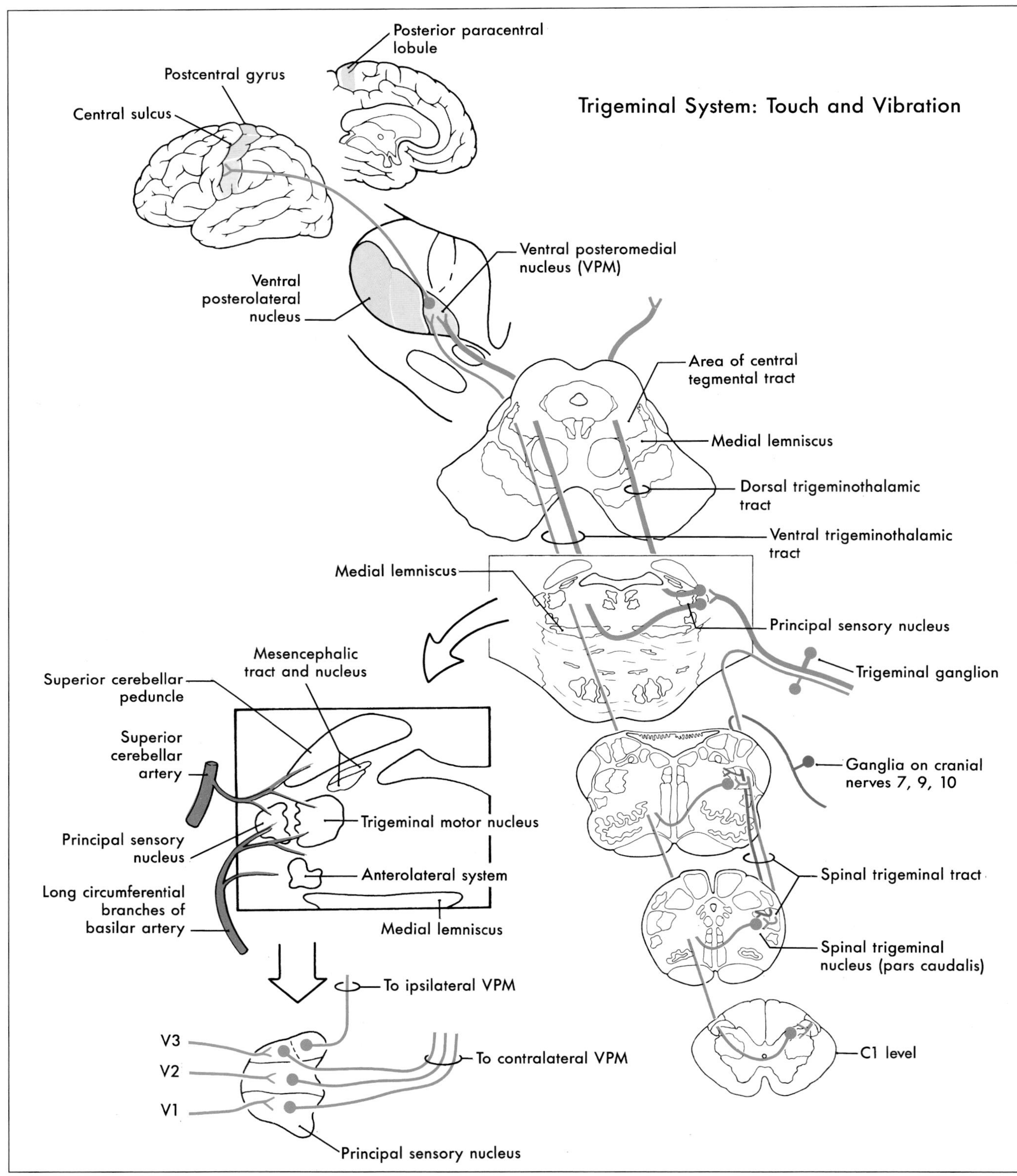

FIGURE 16-12 *The trigeminal pathways carrying discriminative touch, flutter-vibration, and proprioception. Thick green lines represent large-diameter primary afferent fibers and thin green lines represent smaller-diameter fibers conveying primarily thermal and nociceptive information. Note the inverted pattern of primary afferent fibers in the principal sensory nucleus.*

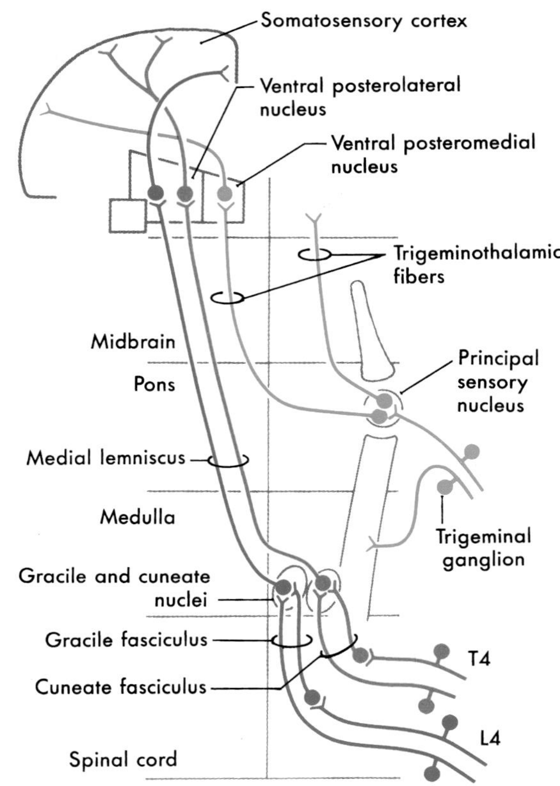

FIGURE 16-13 *Summary of dorsal column-medial lemniscal and trigeminothalamic pathways carrying discriminative touch, flutter-vibration, and proprioception to the contralateral primary somatosensory cortex.*

Proprioceptive endings in jaw muscles and many periodontal ligament receptors are innervated by primary afferent neurons located in the *trigeminal mesencephalic nucleus*. This brainstem nucleus, which consists of a slender column of pseudounipolar cells, represents a collection of neurons of neural crest origin that remain within the neural tube during development. Cells of the mesencephalic nucleus extend from the rostral pons to upper midbrain levels, where they form a thin band of neurons along the lateral edge of the periaqueductal gray. An important difference between the cell bodies of this nucleus and typical ganglion cells is that the former receive synaptic inputs from neurons in the brainstem. This synaptic influence on the neurons of the trigeminal mesencephalic nucleus provides a unique form of presynaptic modulation before central relay of the primary afferent information.

The peripheral processes of cells of the trigeminal *mesencephalic nucleus form the mesencephalic tract of the trigeminal nerve*, which is situated adjacent to the mesencephalic nucleus (Fig. 16-12; see also Fig. 16-16). These central processes send collaterals to brainstem nuclei, including the *trigeminal motor nucleus*, the reticular formation, the spinal trigeminal nucleus, and the cerebellum to form connections essential to normal function of the oral cavity. For example, the afferent limb of the *myotatic jaw-jerk reflex* consists of the processes of trigeminal mesencephalic nucleus neurons that originate as proprioceptive endings in jaw-closing muscles and terminate monosynaptically on trigeminal motor neurons. In turn, the axons of trigeminal motor neurons innervate muscles (i.e., temporalis) that elevate the jaw. A gentle tap on the jaw activates the afferent fibers of this reflex and initiates a contraction of the muscle from which the afferent volley originated. In a similar fashion, afferents from the periodontal ligament provide feedback to jaw muscle motor neurons during mastication. The central connections of the trigeminal mesencephalic nucleus are consistent with their broad participation in the modulation of oral motility patterns.

Proprioceptive endings are also innervated by trigeminal ganglion cells. Afferent fibers from the temporomandibular joint, the extraocular muscles, and some periodontal receptors originate from trigeminal ganglion cells. Since most trigeminal ganglion axons bifurcate when they enter the brainstem, both the principal sensory and the spinal trigeminal nucleus receive proprioceptive input. However, the principal sensory nucleus receives a disproportionate share of large-diameter, heavily myelinated fibers and may be considered the trigeminal homologue of the dorsal column nuclei.

RECEPTIVE FIELD PROPERTIES OF CORTICAL NEURONS

Neurons located in cortical areas representing the body and head are organized into functional units called *cortical columns* (see Fig. 31-9). These are distributed from the pial surface to the cortical white matter. Each column contains neurons responsive to one submodality, and the cells in a column all have similar peripheral receptive field loci. Thalamocortical inputs terminate as axodendritic or axosomatic synapses on stellate cells in layer IV and lower parts of layer III of the SI cortex. Axons of the stellate cells distribute information vertically to the pyramidal cells of layer V within individual columns.

The receptive field properties of cortical neurons are more complex than those at subcortical levels (Fig. 16-14). Cortical neurons respond to a specific stimulus orientation (edges) and to specific textures. They are also capable of coding the velocity, speed, and direction of moving stimuli. At least three distinct populations of neurons receive proprioceptive inputs. The first consists of simple neurons that receive input from a single joint or muscle group. These rapidly adapting cells signal movement. The second group consists of postural neurons that signal the final position of a joint once the movement is completed. The third is made of neurons that receive inputs from several joints and muscle groups (multi-joint) and signal complex joint-muscle interactions.

The functional properties of cortical neurons reflect the processing and integration of sensory information as it ascends from the dorsal column and ventral posterior nuclei to the final processing station in the cortical columns. Both processes can include (1) convergence of afferent input, which increases receptive field size while decreasing resolution; (2) divergence of output signal, which allows relay cells to amplify the sensory signal and supply it to multiple targets; (3) facilitation; and (4) inhibition. These processes act in concert to enhance the signal/noise ratio in terms of both space and time (Fig. 16-14).

In general, cortical neurons display larger receptive fields and more complex inhibitory surrounds than their subcortical inputs (Fig. 16-14). For example, a tactile stimulus in the center of a receptive field (Fig. 16-14, heavy green fibers) results in amplification of the sensory signal and increased activity in a restricted population of cortical cells. Conversely, stimulation at the edge of the receptive field (Fig. 16-14, blue fibers) suppresses the activity in these neurons. This mechanism provides the circuits active in two-point discrimination.

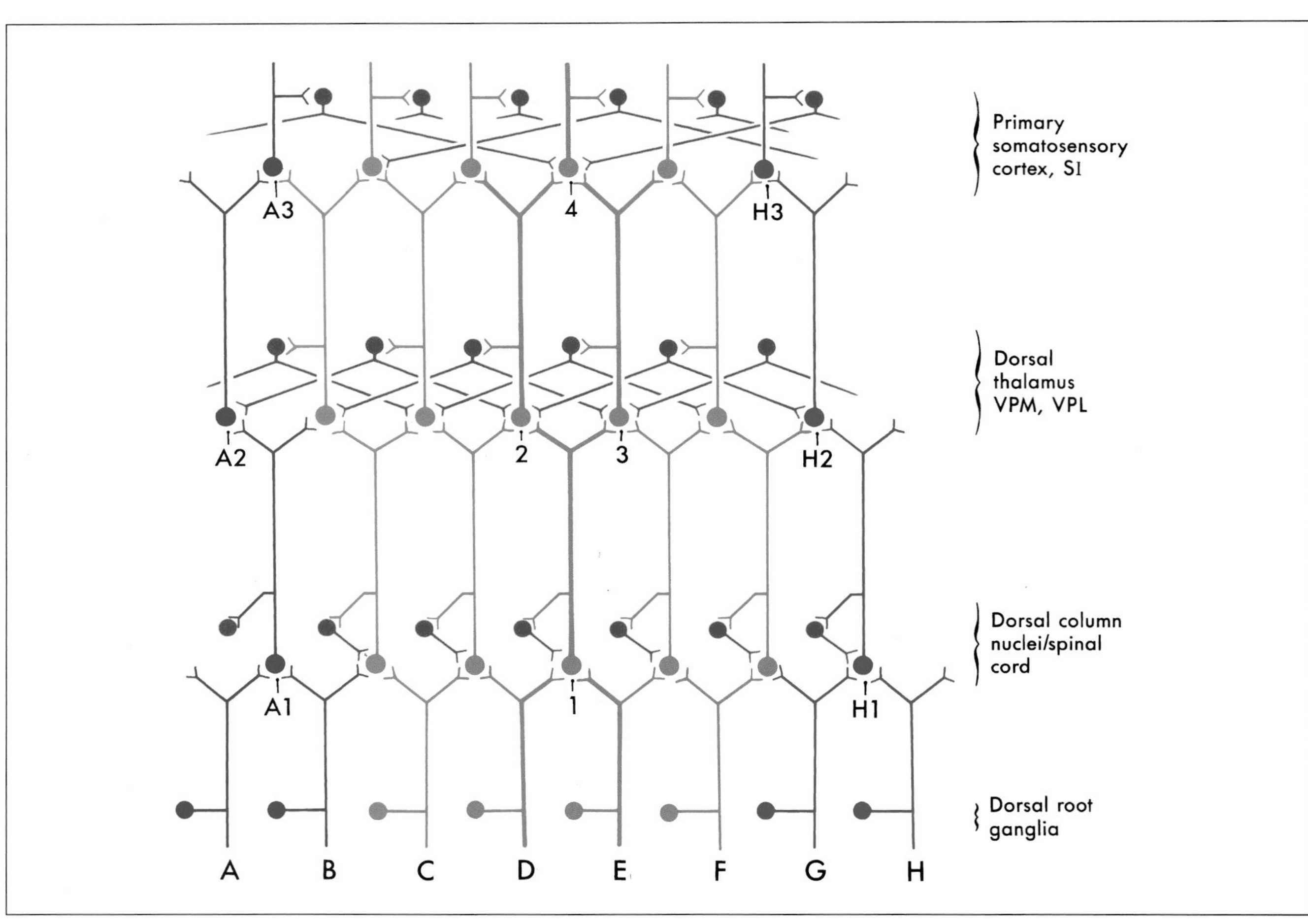

FIGURE 16-14 *Neuronal circuitry underlying excitation and inhibition in SI cortex. Indentation of receptors in the skin at point D or E activates primary afferent fibers that converge on cell 1 in the dorsal column nuclei and excite it. Cell 1, in turn, excites cells 2 and 3 in the thalamus via a divergent collateral branch. Collateral axons from thalamic cells 2 and 3 converge on cell 4 of the SI cortex and excite it. Divergent collateral axons of thalamic cells 2 and 3 excite SI cortical cells on either side of cell 4. Through the convergence and divergence of the peripheral excitatory signal, the receptive field of any SI cortical neuron is larger than the receptive field at lower levels of the pathway as can be seen following the thick green fibers. Red neurons are inhibitory. Cortical excitation is accompanied by inhibition (blue fibers). Stimulation of receptors in the skin at points A and H excites neurons A1 and H1 in the dorsal column nuclei via convergent collateral branches. Dorsal column neurons A1 and H1 excite thalamic neurons A2 and H2 via divergent collateral branches. Thalamic neurons A2 and H2 excite cortical neurons A3 and H3, respectively, and these neurons (A3, H3) excite interneurons that inhibit cortical neuron 4.*

NONCONSCIOUS PROPRIOCEPTION: SPINOCEREBELLAR PATHWAYS

Four spinocerebellar pathways transmit proprioceptive information and limited exteroceptive signals from cutaneous mechanoreceptors to the cerebellum (Fig. 16-15). These sensory signals include information about limb position, joint angles, and muscle tension and length. Input to the cerebellum plays an integral role in guiding cerebellar control of body muscle tone, movement, and posture.

Spinocerebellar tract axons terminate in the cerebellar nuclei and as mossy fibers in the vermis and paravermal region of the cerebellum. Sometimes these areas are

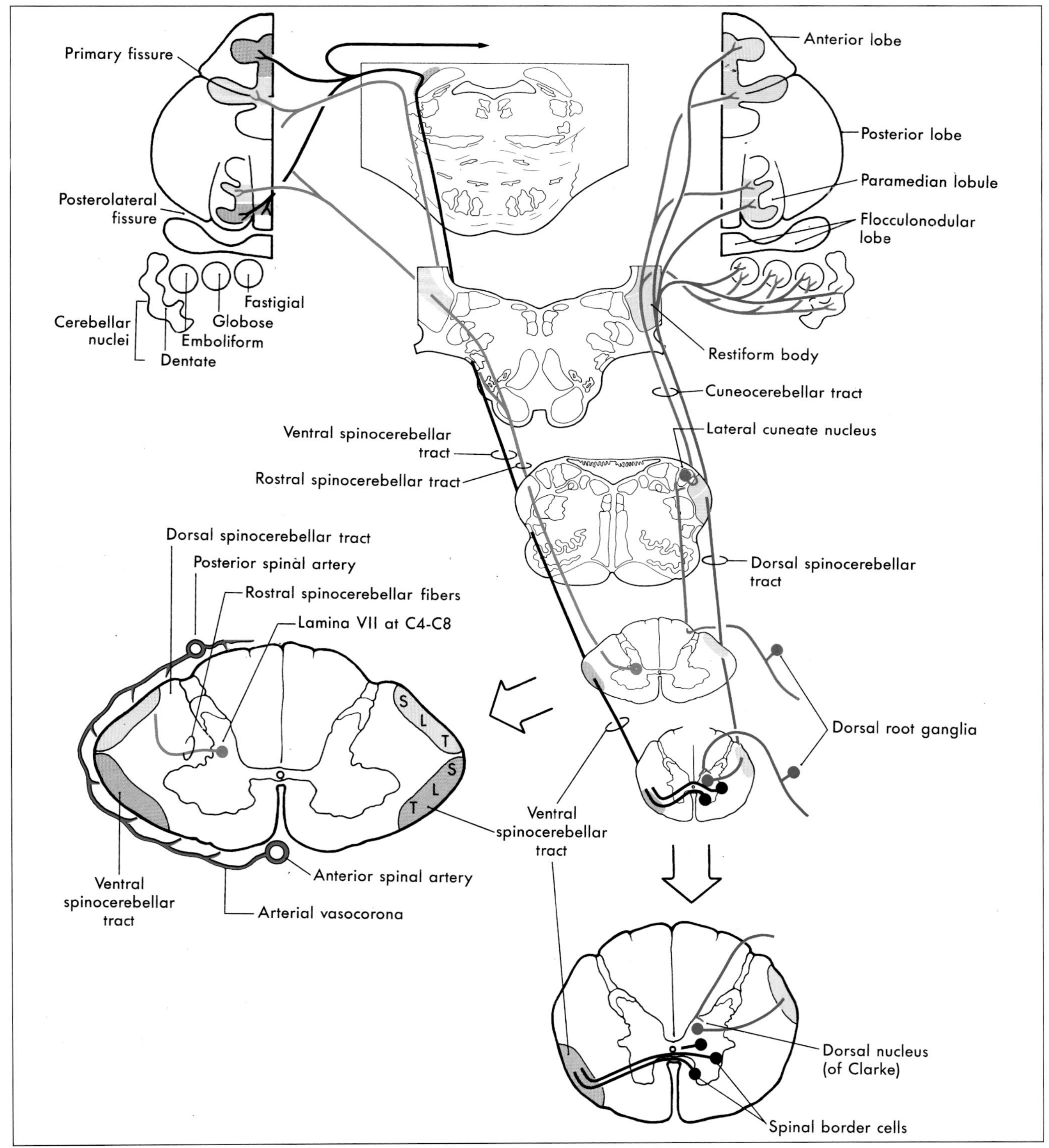

FIGURE 16-15 *Organization of dorsal, ventral, and rostral spinocerebellar tracts and of the cuneocerebellar tract.*

collectively called the *spinocerebellum*. This afferent input to the cerebellum forms a pair of somatotopic representations of the body surface in the anterior lobe and the paravermis of the posterior lobe (Fig. 16-15). Degeneration of the major spinocerebellar tracts occurs in diseases such as Friedreich's ataxia. The result is *cerebellar ataxia*—lack of coordination during walking and other movements that occurs because the cerebellum is not receiving the sensory feedback necessary to regulate movement.

Dorsal Spinocerebellar Tract Proprioceptive afferents and a limited number of exteroceptive (cutaneous) afferents from the lower limb and lower trunk travel the dorsal spinocerebellar tract to reach the ipsilateral cerebellar cortex (Fig. 16-15). Dorsal root fibers from the lower body and leg terminate on cells in the *dorsal nucleus of Clarke*, which is located in lamina VII of the intermediate zone in segments T1 to L2. Primary afferent fibers from the spinal cord levels caudal to L2 ascend in the dorsal funiculus to reach this nucleus. Axons from cells in the dorsal nucleus of Clarke traverse the ipsilateral lateral funiculus and collect on the surface of the spinal cord lateral to the corticospinal tract. These fibers ascend to reach the cerebellum via the restiform body.

Group I muscle spindle and Golgi tendon organ afferents monosynaptically activate tract cells in Clarke's nucleus. The discharge rate of these dorsal spinocerebellar tract cells shows a linear relationship to muscle length; therefore, their firing rate can encode muscle length as a frequency code. Group II and group III tactile fibers also terminate on cells in Clarke's nucleus. The cells in this nucleus that respond to proprioceptive input are different from the ones that respond to exteroceptive input.

Cuneocerebellar Tract The cuneocerebellar tract is the upper limb equivalent of the dorsal spinocerebellar tract (Fig. 16-15). Dorsal root fibers in spinal segments C2 to T4 carry muscle spindle and exteroceptive information in the ipsilateral cuneate fasciculus to the main cuneate nucleus. In the lower medulla, proprioceptive primary afferent fibers terminate somatotopically in the *lateral cuneate nucleus*. Cells of the lateral cuneate nucleus project as *cuneocerebellar fibers* to the cerebellum via the restiform body. Exteroceptive input arising from the rostral end of the main cuneate nucleus also ascends to the cerebellar cortex to terminate in the folia of the anterior lobe in lobule V.

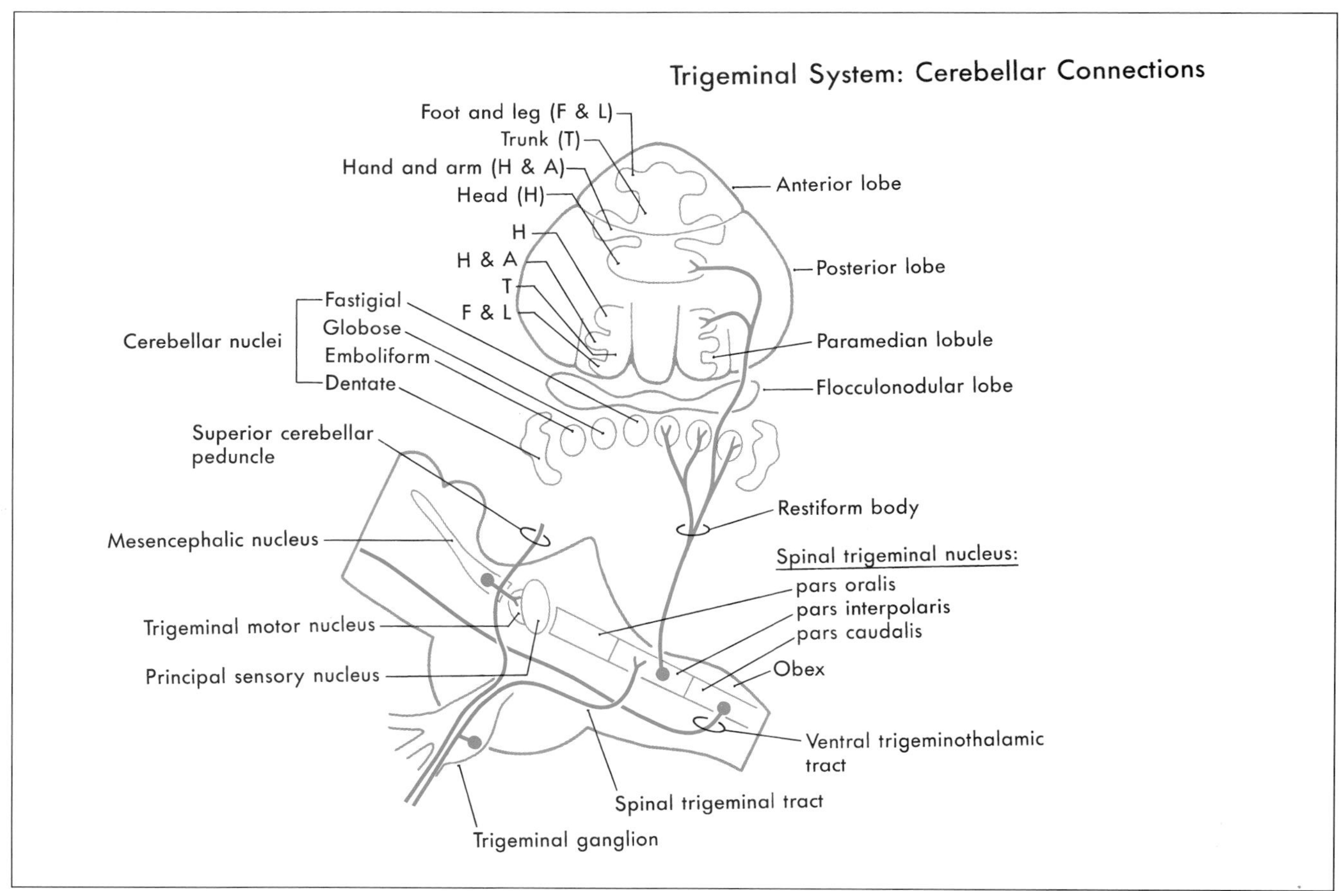

FIGURE 16-16 *Organization of trigeminocerebellar pathways.*

Ventral Spinocerebellar Tract This pathway relays information from group I and flexor reflex afferents arising in the lower limb. The cells of origin of this pathway are located in lumbar segments L3 to L6. They are located in the lateral part of Rexed's laminae V to VII, as well as along the ventrolateral border of the ventral horn, where they are called *spinal border cells* (Fig. 16-15). The axons of ventral spinocerebellar tract (VSCT) cells immediately *cross the midline* in the anterior white commissure and ascend in the lateral funiculus ventral to the dorsal spinocerebellar tract. In the pons, these fibers turn dorsolateral to enter the cerebellum via the *superior cerebellar peduncle*. Most fibers *recross* to terminate in the cerebellum ipsilateral to their side of origin. These VSCT fibers are distributed more laterally in the cerebellum than are those of the dorsal tract. Cells giving rise to VSCT fibers are strongly influenced by descending projections of the reticulospinal and corticospinal pathways. Reticulospinal input inhibits VSCT cells, and the corticospinal input facilitates VSCT cells. Vestibulospinal and rubrospinal projections also monosynaptically excite VSCT cells.

Rostral Spinocerebellar Tract This tract, the upper limb equivalent of the VSCT, arises from cell bodies located in lamina VII of the cervical enlargement (C4 to C8) (Fig. 16-15). The efferent projections from these neurons ascend uncrossed in the lateral funiculus of the spinal cord. Although most of these axons enter the cerebellum via the restiform body, some travel in the superior cerebellar peduncle. The rostral spinocerebellar tract from the upper limb and the ventral spinocerebellar tract from the lower limb relay cutaneous tactile information from Meissner's, Merkel's, and Pacinian mechanoreceptors (group II and group III afferents) to the cerebellum.

Trigeminocerebellar Connections The oral motor system requires continual feedback during mastication. As food is chewed its texture and consistency are altered, changing the demands on jaw muscles. In addition, adaptation is required for long-range functional changes. For example, there are modifications in jaw motility patterns during the transition from suckling to chewing in the newborn and from natural dentition to the use of dentures. It is probable that proprioceptive information reaching the cerebellum from jaw muscle spindles, periodontal afferents, and the temporomandibular joint are involved in these processes.

Branches of the central processes of the mesencephalic trigeminal neurons are distributed to the cerebellar hemispheres and nuclei via the superior cerebellar peduncle (Fig. 16-16). Additional proprioceptive signals from the spinal trigeminal nucleus *pars interpolaris* and *pars caudalis* enter the cerebellum by way of the restiform body.

SOURCES AND ADDITIONAL READINGS

Brodal A: *Neurological Anatomy in Relation to Clinical Medicine*, 3rd Ed. Oxford University Press, New York, 1981.

Burgess PR, Perl ER: Cutaneous mechanoreceptors and nociceptors. In A Iggo (ed): *Handbook of Sensory Physiology, Vol 2: Somatosensory System*. Springer-Verlag, New York, pp 30–78, 1973.

Carpenter MB: *Human Neuroanatomy*, 7th Ed. Williams & Wilkins Co, Baltimore, 1976.

Johansson RS, Vallbo ÅB: Tactile sensory coding in the glabrous skin of the human hand. Trends Neurosci 6:27–32, 1983.

Lenz FA, Dostrovsky JO, Tasker RR, Yamashiro K, Kwan HC, Murphy JT: Single-unit analysis of human ventral thalamic nuclear group: somatosensory responses. J Neurophysiol 59:299–316, 1988.

Mountcastle VB: Neural mechanisms in somesthesis. In Mountcastle VB (ed): *Medical Physiology*, 14th Ed. Vol I. CV Mosby, St Louis, pp 348–390, 1980.

Penfield W, Rasmussen T: *The Cerebral Cortex of Man: A Clinical Study of Localization of Function.* Hafner Publishing, New York, 1968 (Facsimile of 1950 ed).

Vallbo ÅB, Olsson KA, Westberg K-G, Clarke FJ: Microstimulation of single tactile afferents from the human hand: sensory attributes related to unit type and properties of receptive fields. Brain 107:727–749, 1984.

CHAPTER SEVENTEEN

The Somatosensory System II: Nondiscriminative Touch, Temperature, and Nociception

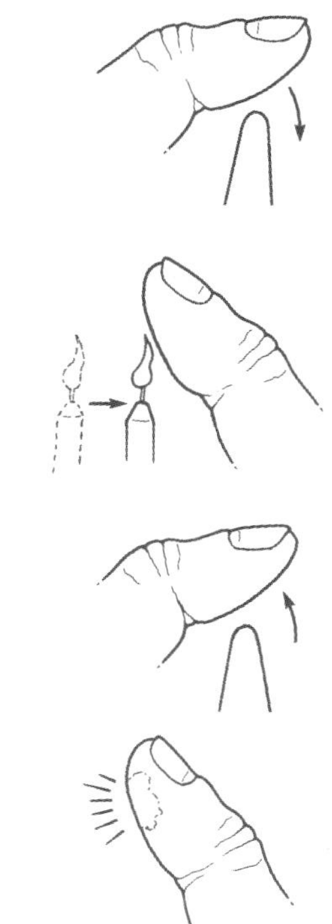

S. WARREN • N. F. CAPRA • R. P. YEZIERSKI

One crucial role of the somatosensory system is to supply the brain with information related to insults that could damage tissue. These signals ascend the neuraxis in a fiber bundle called the *anterolateral system* (ALS). Anyone who has used a hammer or hot skillet has had experience with this system. Hit your thumb with a hammer and, if you are lucky, only *high-threshold mechanoreceptors* that signal excess skin deformation will be activated. If you are unlucky, tissue is damaged and the result is pain (*nociception*). Specifically, *mechanonociceptors* have been stimulated. One common response is to gently rub the damaged area. This activates central nervous system (CNS) pathways that decrease the transmission of nociceptive signals and alter the perception of pain.

After the blow, damaged tissues release chemicals that activate another type of pain receptor, the *chemonociceptor*. These receptors are responsible for long-term pain and tenderness (*hyperalgesia*). Similarly, the temperature of a skillet is detected by *thermoreceptors* in the skin and transmitted through the ALS. If a burn is produced, the tissue damage is signalled by high frequency firing of *thermonociceptors*. ALS activation can lead to a variety of responses, including withdrawal reflexes, the conscious perception of pain, emotional effects such as suffering, and behavioral changes aimed at avoiding the cause of the pain.

OVERVIEW

Nondiscriminative (crude or light) touch, *innocuous thermal*, and *nociceptive (mechanical, chemical and thermal)* sensations are conveyed by pathways that collectively make up the *ALS*. This system transmits signals originating in peripheral receptors to spinal cord and brainstem neurons (Fig. 17-1). These signals are then forwarded to thalamic nuclei and from there to the trunk and extremity representations in the *somatosensory cortex*. The *ventral trigeminothalamic pathway* (Fig. 17-1; also see Fig. 17-11) carries similar signals that originate from receptors in the head. These are relayed through brainstem and thalamic nuclei to the face area of the somatosensory cortex. The touch fibers of the ALS differ from those described for the dorsal column-medial lemniscal system (DCML) (see Chapter 16) in several ways: (1) they yield a generalized feeling of being touched but do not give precise localization, (2) their receptive fields are larger, and (3) they are smaller in diameter and more slowly conducting. Disruption of the ALS can produce symptoms ranging from reduced sensibility (*hypesthesia*), to numbness, tingling, and prickling (*paresthesia*), to a complete loss of sensibility (*anesthesia*).

ANTEROLATERAL SYSTEM

Overview The ALS is a composite bundle that includes *spinothalamic*, *spinomesencephalic*, *spinoreticular*, *spinobulbar*, and *spinohypothalamic fibers*. Spinothalamic fibers project directly from the spinal cord to the ventral posterolateral (VPL) nucleus, the posterior nuclear group, and intralaminar nuclei of the thalamus. Collaterals to the reticular formation arise from some of these axons. Spinomesencephalic axons project to the periaqueductal gray and to the tectum; the latter are *spinotectal fibers*. Although spinoreticular fibers project to the reticular formation of the medulla, pons, and midbrain, collaterals may ascend to other targets such as the thalamus. Projections of less relevance to the somatosensory system, such as *spino-olivary fibers*, are grouped under the category of *spinobulbar fibers*. Spinohypothalamic fibers terminate in hypothalamic areas and nuclei, including some that give rise to hypothalamospinal axons.

Fibers classically described as comprising the lateral spinothalamic tract were considered to carry *only* pain and temperature information, whereas the ventral spinothalamic tract concerned *only* nondiscriminative touch. This older view of separate tracts conveying separate types of information is not used in this chapter. Current thinking holds that all parts of the ALS carry all modalities (pain, temperature, and touch), but that there are direct and indirect routes. The former is the

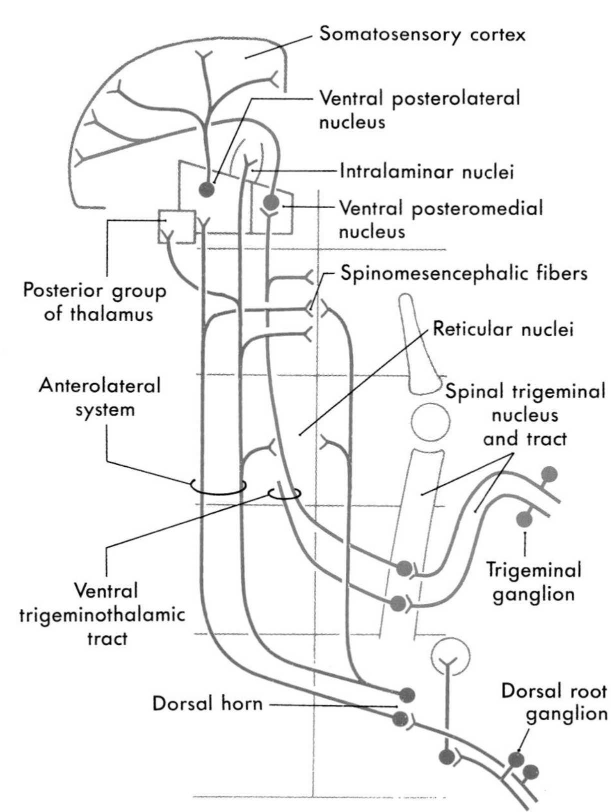

FIGURE 17-1 *Summary of anterolateral system and ventral trigeminothalamic tract fibers conveying nondiscriminative tactile, nonnociceptive thermal, and nociceptive inputs to the contralateral somatosensory cortex.*

TABLE 17-1

CLASSIFICATION OF THE CUTANEOUS MECHANICAL, INNOCUOUS THERMAL, AND NOCICEPTIVE RECEPTORS USING SMALL-DIAMETER FIBERS AND OF THEIR ADEQUATE STIMULI

RECEPTOR	ADEQUATE STIMULUS
Cutaneous mechanoreceptors	**Nondiscriminative tactile stimuli**
Group A-delta high-threshold mechanoreceptors	Rub, tap, stretch, squeeze, pinch
Group C high-threshold mechanoreceptors	Rub, tap, stretch, squeeze, pinch
Cutaneous thermoreceptors	**Innocuous warm and cold thermal stimuli**
Cold thermoreceptors	Transient reduction in temperature
Warm thermoreceptors	Transient increase in temperature
Cutaneous nociceptors	**Mechanical,thermal, and/or chemical stimuli**
Mechanonociceptors	
Group A-delta mechanonociceptors	Mechanical tissue damage
Group C mechanonociceptors	Mechanical tissue damage
Thermonociceptors	
A-delta heat thermonociceptors	Noxious heat and tissue damage
A-delta, C cold thermonociceptors	Noxious cold and tissue damage
Chemonociceptors (C)	Insect venom, bradykinin, histamine
Polymodal nociceptors (C)	Noxious heat/cold, tissue damage, algesic chemicals

neospinothalamic pathway (spinal cord → thalamus), whereas the latter is the polysynaptic *paleospinothalamic pathway* (spinal cord → reticular formation → thalamus). Both of these pathways, plus other fibers as defined above, collectively form the ALS.

Receptors and Primary Neurons The receptors for nondiscriminative touch, innocuous thermal stimuli, and nociceptive stimuli are distributed in glabrous and hairy skin, as well as in deep tissues including joints and muscles (Table 17-1). Morphologically, these receptors are all *free (naked) nerve endings* (see Fig. 16-1); that is, they lack specialized receptor cells or encapsulations. Because of this lack, the basis for their submodality specificity is unclear. These submodalities are transduced by activation of peripheral branches of either *thinly myelinated A-delta fibers* or *unmyelinated C fibers*. The density of free nerve endings and the corresponding size of receptive fields vary over the body surface in the same way as for other cutaneous receptors (see Fig. 16-3), being highest on the hands and in the perioral area. Regardless of size or location, however, each field is exquisitely sensitive to temperature, chemical, or mechanical stimuli.

Nondiscriminative touch results from the stimulation of free nerve endings that act as non-nociceptive (or non-noxious) *high-threshold mechanoreceptors* (Table 17-1). These receptors respond to any rough stimulus (tapping, squeezing, rubbing, stretching of the skin) that does not result in tissue damage (Fig. 17-2A). These fibers generally have no background activity when not stimulated, and they respond with a sustained discharge that signals stimulus duration (Fig. 17-2A). Mechanical nociception results from the activation of a separate type of receptor, called a *mechanonociceptor*, that responds to mechanical injury. These receptors are associated with fibers in the A-delta range. In addition, a class of C fibers, *C polymodal nociceptors*, respond to mechanical, thermal, and chemical stimuli (Table 17-1).

Non-nociceptive *thermoreceptors* fall into two classes, those activated by heat (35–45°C) and those activated by cold (17–35°C). They show a graded response to changes in ambient temperature (Fig. 17-2B). With repeated stimulation, these receptors become sensitized and show a decreased threshold and larger response to the application of a stimulus. Levels of heat (>45°C) or cold (<17°C) that burn the skin produce high-frequency firing in both A-delta and C *thermonociceptors* (Fig. 17-2B). C polymodal fibers also respond to painful thermal stimuli (Table 17-1).

Tissue damage causes the release of a number of chemical substances that activate a class of free nerve endings called *chemonociceptors* (Table 17-1). Specifically, chemonociceptors are activated by the release of endogenous substances such as bradykinin, H^+ ions, and

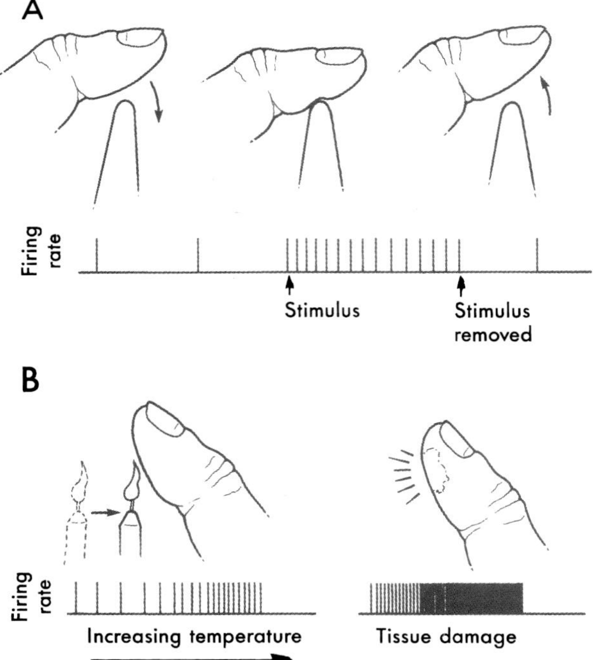

FIGURE 17-2 *Fibers conveying information from high-threshold mechanoreceptors (A) respond to the application of a punctate stimulus. Thermoreceptors show a graded response to increases in temperature (B, left), whereas burns produced by prolonged thermal stimulation evoke high frequency response in thermonociceptors (B, right).*

foreign irritants such as insect venoms. These free nerve endings are the peripheral processes of C fibers. As mentioned above, C polymodal nociceptors also respond to chemical stimuli.

Of the two primary afferent fiber types carrying nociceptive sensations, A-delta fibers have a slightly faster conduction velocity (5–30 m/sec) than that of C fibers. They carry well localized sensations, which do not evoke an affective component to the sensory experience. A pinprick, used clinically to test ALS function, is one stimulus that activates A-delta fibers. On the other hand, C fibers are smaller and conduct more slowly (0.5–2 m/sec). They transmit poorly localized sensations that produce a noticeable affective component. For example, the dull, persistent ache that follows a muscle pull results from activation of C fibers. Both A-delta and C fibers are considerably smaller and conduct more slowly than fibers of the DCML system. Nerve blocks or anoxia preferentially affect large-diameter, heavily myelinated fibers and, thus, usually result in loss of discriminative tactile, vibratory, and postural sensations to varying degrees. Local anesthetics, such as lidocaine, preferentially affect small-diameter A-delta and C fibers and thus result in loss of nociception (*analgesia*).

The cell bodies of C and A-delta fibers are generally small compared to other pseudounipolar neurons in the dorsal root ganglion. The central processes of these cells enter the spinal cord via the lateral division of the dorsal root. These smaller fibers may contain peptides such as substance P and calcitonin gene-related peptide (CGRP), which may serve as neurotransmitters. In addition to their normal trajectory into the dorsal horn, a small number of C fibers enter the spinal cord through the ventral root of a spinal nerve. It is possible that these fibers provide a basis for the return of pain after *dorsal rhizotomy*, a procedure in which dorsal roots are sectioned in an attempt to alleviate intractable pain.

The strip of skin that is innervated by the peripheral cutaneous branches of a given spinal nerve is called a *dermatome* (Figs. 17-3, 17-4). The central processes of these nerves that convey cutaneous input terminate in the dorsal horn. Clinically useful, dermatomes and their relation to landmarks on the body include, for example, C6 for the index finger, the T4-T5 border at the nipples, T10 at the navel, L1 along the pelvic rim, L5 for the big toe, and S4 and S5 for the genitalia and anus (Fig. 17-4). There is overlap between both the peripheral and central distribution of adjacent spinal nerves and consequently between their dermatomes (Fig. 17-3). This overlap reduces the effects of injury to a single spinal root.

Shingles (*herpes zoster*) is a disease of viral etiology that is noteworthy for its dermatomal distribution. Subsequent to a bout of chickenpox, viral DNA may infect and become latent in trigeminal and dorsal root ganglion cells. The virus may reactivate periodically, producing infectious virions that travel down the peripheral processes of the neurons to produce a painful skin irritation in the dermatomal distribution of the ganglion. When an injury or disease process affects a series of nerve roots, the result is diminished sensibility (*hypesthesia*) over the dermatomes served by those roots. The borders of the hypesthetic region correspond to dermatomal boundaries.

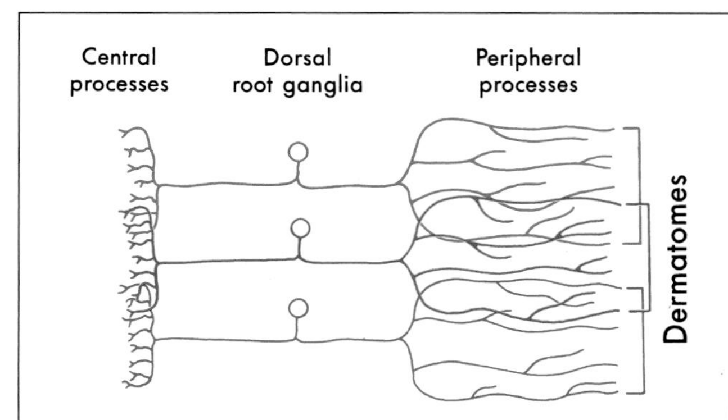

FIGURE 17-3 *The dermatomes formed by the peripheral processes of adjacent spinal nerves overlap on the body surface (see also Fig. 17-4). The central processes of these fibers also overlap in their spinal distribution.*

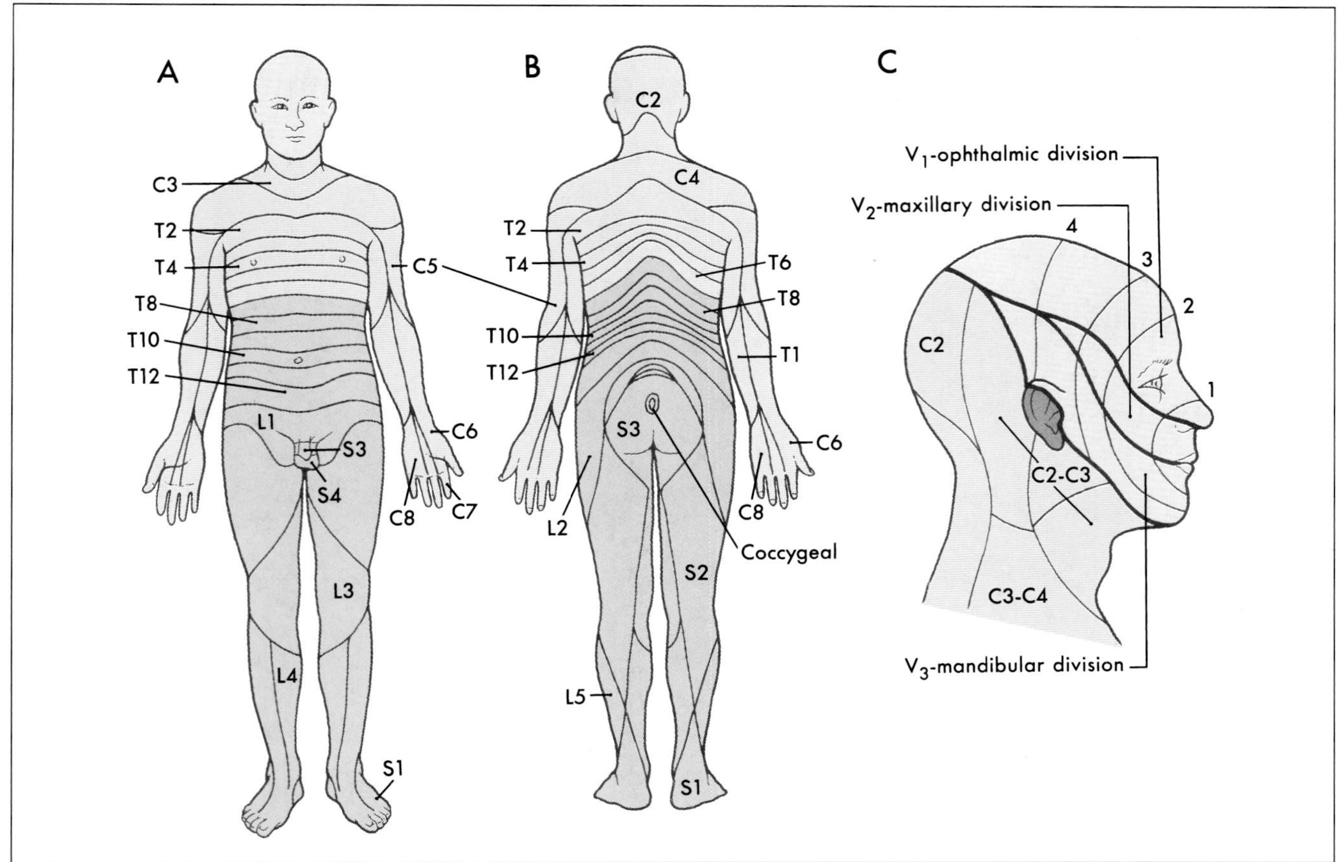

FIGURE 17-4 *Dermatomal maps of the peripheral distribution of spinal nerves (A and B) and trigeminal nerve (C).*

Clinically, it is important to test patients for intact ALS and DCML function. In testing the ALS, a single pin point applied to the skin should evoke a response (*perception of pain*) from the patient. Function of the intact DCML is tested by the simultaneous application of two points spaced at measured intervals. As the points are moved closer, the ability to identify them as separate stimuli (*two-point discrimination*) decreases, and eventually disappears.

Central Pathways As mentioned above, A-delta and C fibers enter the spinal cord via the *lateral division* of the *dorsal root entry zone*. The fibers enter the dorsolateral fasciculus (*Lissauer's tract*) and bifurcate into ascending and descending branches (see Fig. 16-5). Some collaterals terminate on interneurons in the spinal gray matter. These connections participate in the circuits that mediate spinal reflexes such as the *flexor withdrawal reflex* (see Chapter 9).

The functional properties of dorsal horn neurons reflect the type of primary afferent fiber input received. These neurons are classified as *low-threshold* (non-nociceptive), *nociceptive-specific* (noxious), *wide dynamic range* (non-nociceptive and noxious), or *deep* on the basis of their responses to different stimulus intensities.

The central target of nociceptive primary afferent fibers includes laminae I, II, and V of the dorsal horn (Fig. 17-5). Rexed's lamina I, the *posteromarginal nucleus*, receives mainly input from A-delta fibers (Fig. 17-6A). Neurons in this nucleus project to other spinal cord laminae, the brainstem reticular formation, and the thalamus (Fig. 17-6B). Lamina II, the *substantia gelatinosa*, is divided into outer (IIo) and inner (IIi) layers. Input to IIo and IIi is derived primarily from C fiber primary afferents, and IIi also receives input from collaterals of non-nociceptive afferent fibers. Lamina II contains excitatory and inhibitory interneurons that project to other laminae of the dorsal horn. In addition, some neurons in IIo are involved in relaying sensory information to supraspinal sites, including the thalamus (Fig. 17-6B).

Neurons in laminae III and IV, the *nucleus proprius* (also called the *dorsal proper sensory nucleus*), receive non-noxious inputs from the periphery. Cells in these laminae project to deeper laminae of the spinal cord, to the dorsal column nuclei, and to other supraspinal relay centers including the midbrain, thalamus, and hypothalamus (Fig. 17-6B).

Lamina V neurons receive both noxious and non-noxious (nociceptive and non-nociceptive) inputs and project to the medullary and mesencephalic reticular for-

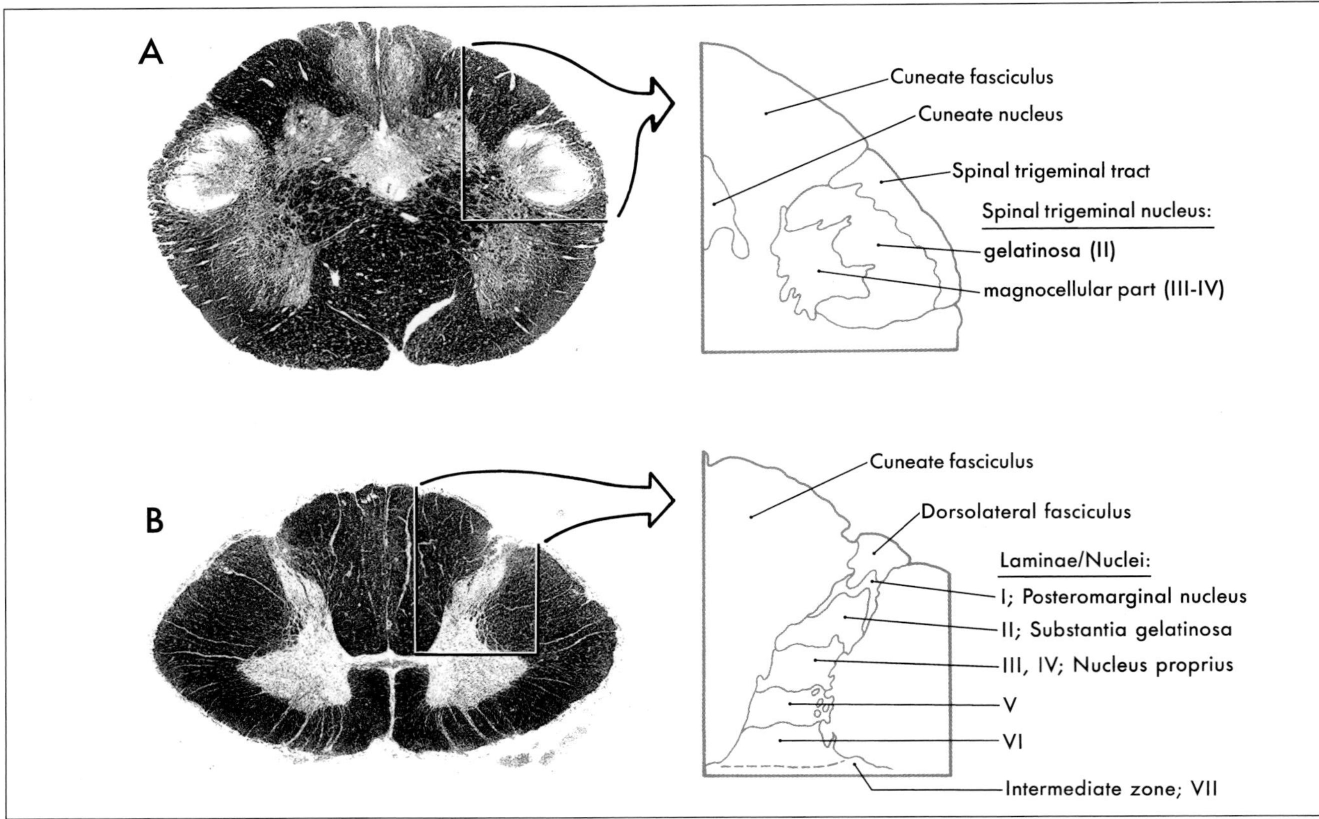

FIGURE 17-5 *Weil (myelin) stained section of the medullary dorsal horn (A) at the level of the pyramidal decussation and the spinal dorsal horn (B) at levels C7 to C8. The spinal laminae that correspond to medullary structures are labeled.*

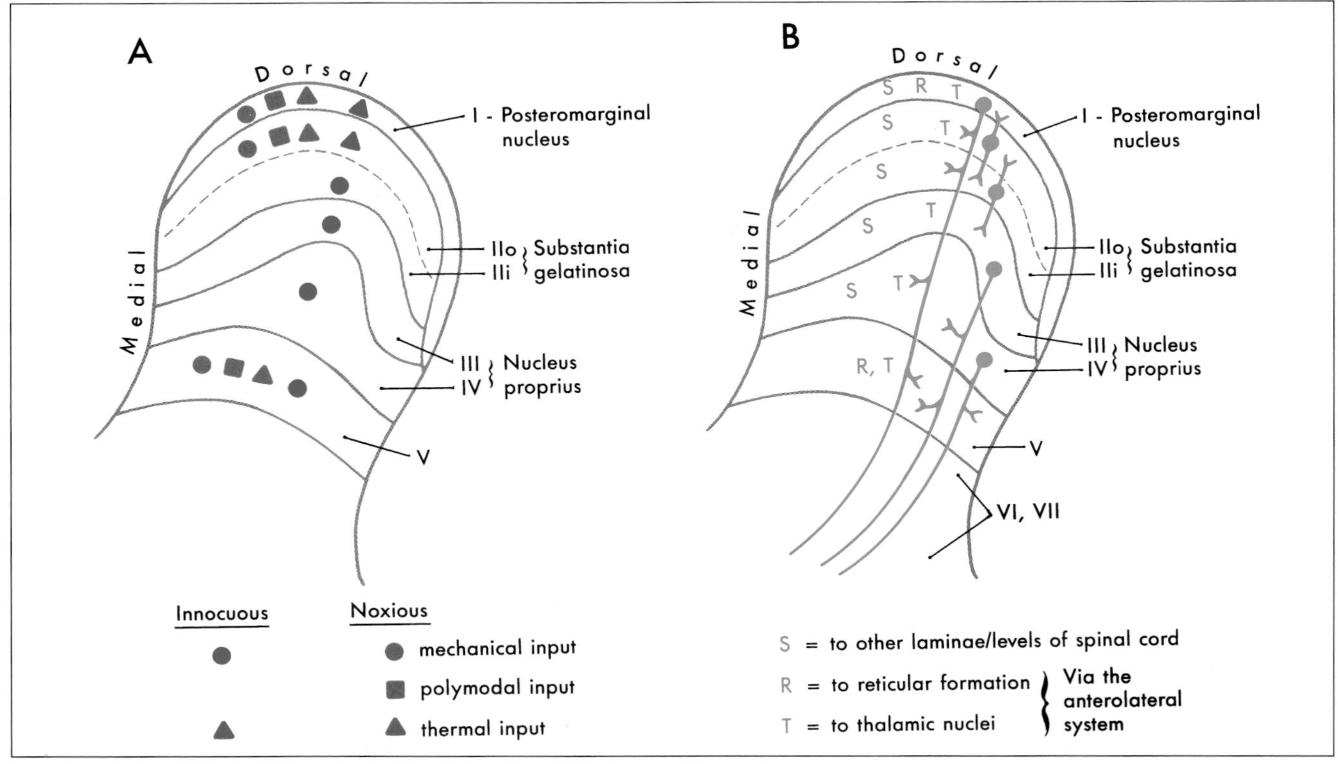

FIGURE 17-6 *Summary of dorsal horn laminae and their major sensory inputs (A) and major outputs (B).*

mation, thalamus, and hypothalamus (Fig. 17-6B). Neurons in deeper laminae of the spinal gray receive (directly and indirectly) noxious and non-noxious inputs and connect with neurons in other spinal cord levels (*propriospinal connections*).

As mentioned above, the fibers of the ALS participate in both direct and indirect spinothalamic pathways (Fig. 17-1). Most A-delta fibers participate in the *direct* (*neospinothalamic*) pathway, which carries nondiscriminative tactile, innocuous thermal, and nociceptive signals. When these A-delta fibers enter the dorsolateral fasciculus and bifurcate, the main branch ascends three to five spinal levels to terminate on a second-order neuron (tract cell) in lamina I of the dorsal horn (Fig. 17-6A). These tract cells, in turn, project to the thalamus. The great majority of their axons cross the midline of the spinal cord obliquely via the ventral (anterior) white commissure and ascend in the contralateral ALS. A few ascend in the *ipsilateral* ALS. The thalamic (third-order) neurons of these pathways are located mainly in the ventral posterolateral nucleus (VPL), the posterior nucleus, and the intralaminar nuclei.

The polysynaptic *indirect* (*paleospinothalamic*) component of the ALS relays noxious and innocuous mechanical and thermal information to the brainstem reticular formation. The input to this pathway originates chiefly from C fibers. Branches of these fibers ascend and descend by one or two levels in the dorsolateral fasciculus to synapse on interneurons in laminae II and III (Fig. 17-6B). These interneurons influence tract cells in laminae I to III and V to VIII, which send axons that cross obliquely through the ventral white commissure (over a distance of one to three segments) to join the contralateral ALS. These *spinoreticular fibers* terminate in the brainstem reticular formation, which in turn projects to the thalamus.

Fibers in the ALS are arranged *somatotopically* in the spinal cord. Axons from lower levels (coccygeal/sacral) of the body are found dorsolaterally, whereas those from more rostral levels of the cord are added in an orderly ventromedial sequence (Fig. 17-7). Because of its location, the ALS does not receive blood supply from a single vessel. Instead, its blood supply originates from the *arterial vasocorona*, and via *sulcal branches* of the *anterior spinal artery* (Fig. 17-7). Consequently, occlusion of either of these vessels results in a *patchy loss of nociceptive*, *thermal*, and *touch sensations over the contralateral side of the body* beginning about two segments below the lesion. In contrast, a *complete loss of these sensations* is seen in patients who have had an *anterolateral cordotomy* for relief of intractable pain.

The ALS may be involved in trauma or diseases of the spinal cord. For example, a hemisection of the spinal cord (as in the *Brown-Séquard syndrome*) results in a combination of sensory and motor losses. Sensory deficits include (1) *contralateral* loss of nociceptive and thermal sensations over the body below the level of the lesion (ALS damage), and (2) *ipsilateral* loss of discriminative tactile, vibratory, and position sense over the body below the level of the lesion (dorsal column damage) (Fig. 17-8A). The motor loss is manifest as an ipsilateral paralysis of the leg or leg and arm, depending on the level of the hemisection (see Chapter 23). *Syringomyelia*, a condition in which there is cystic cavitation of central regions of the spinal gray matter, may impinge on the ventral white commissure and decussating ALS fibers (Fig. 17-8B). When located at C3 to C4 levels of the spinal cord, this lesion produces *bilateral loss* of nondiscriminative tactile, nociceptive, and thermal sensations beginning several segments below the level where fibers are interrupted. The symptoms present as sensory losses in the configuration of a cape draped over the shoulders and extending down to the nipple.

In the medulla, ALS fibers retain their position near the ventrolateral surface. They are located ventral to the spinal trigeminal nucleus, dorsolateral to the inferior olive, and remain separated from the DCML as both course through the medulla and pons (Fig. 17-7). Therefore, vascular lesions or tumors in the lower brainstem can affect discriminative touch and nociception differentially. At the pontomesencephalic junction, the medial lemniscus rotates dorsolaterally so that ALS fibers are adjacent to its lateral margin. Thereafter, the DCML and ALS pathways course together to terminate in the thalamus (Fig. 17-7).

As the ALS ascends through the medulla, it decreases in size because of the departure of the *spinoreticular axons*, which originate in laminae V to VIII and terminate in the reticular formation. The reticular formation also receives *collaterals from lamina I spinothalamic axons* (Fig. 17-7). Several other pathways ascend in the ALS. For example, spinomesencephalic axons may terminate in the periaqueductal gray (PAG) or, as *spinotectal fibers*, in deep layers of the superior colliculus and anterior pretectum. Many tract cells with axons in the ALS project via collaterals to multiple targets as the primary axon ascends through the brainstem.

In addition to direct spinothalamic fibers, spinal cord neurons also project to brainstem targets that indirectly influence thalamic nuclei. Most notably, the reticular formation, which receives *spinoreticular fibers*, projects via *reticulothalamic fibers* to the intralaminar nuclei and posterior group (Fig. 17-7). The intralaminar nuclei project to the striatum and wide areas of cerebral cortex, and subserve the alerting response to painful stimuli. Nuclei of the posterior thalamic group project to secondary somatosensory (SII) cortex, and the retroinsular cortex. These polysynaptic pathways underlie the dull, poorly localized, but persistent, painful sensations that are perceived following localized thalamic lesions.

Somatosensory information, including nociceptive input from dorsal horn cells, also ascends directly to the hypothalamus via the *spinohypothalamic fibers* of the ALS. In addition, spinal input is indirectly conveyed to the

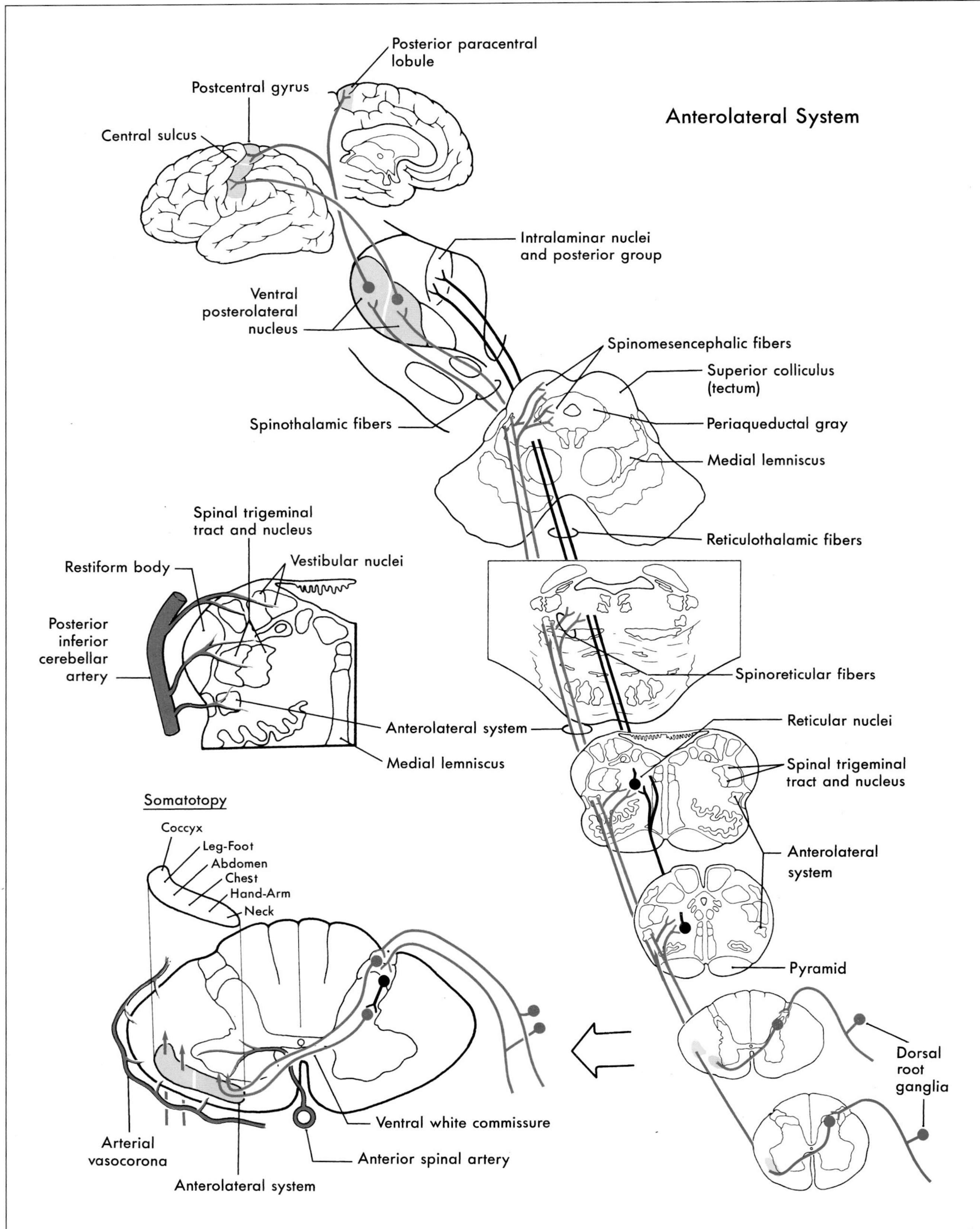

FIGURE 17-7 *The anterolateral system and blood supply to these fibers in the spinal cord and medulla.*

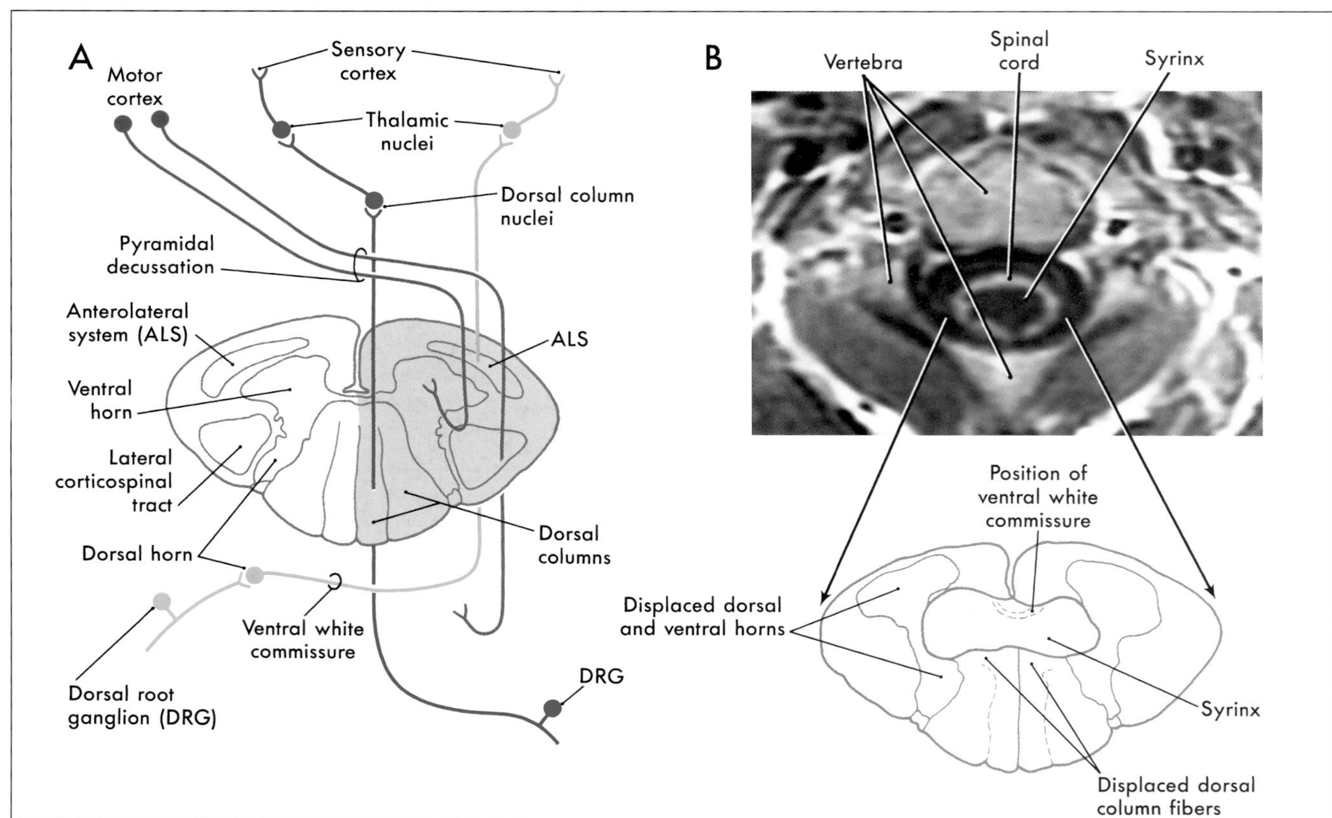

FIGURE 17-8 *Major tracts interrupted in a spinal cord hemisection (A, Brown-Séquard syndrome) that account for the characteristic sensory and motor losses. Magnetic resonance image of a cervical syringomyelia (B) with resultant expansion of the lesion into fibers of the ventral white commissure. In both A and B, the cross section of the spinal cord is shown in an orientation identical to that seen in the clinical setting.*

hypothalamus by way of synaptic relays in the reticular formation and in the periaqueductal gray (see Fig. 17-14). Through these ascending pathways, nociceptive information is transmitted to brain centers, such as the limbic system, that underlie emotional and autonomic responses to nociceptive stimuli.

Input to the VPL is somatotopically organized such that lower body areas are represented laterally, and upper body regions (exclusive of the head) are represented medially (Figs. 17-7, 17-9). Within the VPL, fibers of the ALS terminate on clusters of cells located in the periphery of the nucleus. Most of these cells are different from the ones targeted by DCML axons. However, some VPL neurons, called *multimodal cells*, receive input from both ALS and DCML pathways. The functional classes of cells found in VPL reflect the peripheral input received by tract cells of the spinal cord. These classes include, as in the spinal cord, *nociceptive-specific*, *wide dynamic range*, *low-threshold non-nociceptive*, and *deep neurons*.

Thalamocortical axons carrying nondiscriminative tactile, nociceptive, and thermal signals project via the posterior limb of the internal capsule to the somatosensory cortices (Fig. 17-7). Fibers originating in the VPL project mainly to SI cortex (areas 3, 1, 2), whereas those from the posterior nucleus terminate principally in the SII cortex. The somatotopy observed in the VPL is reflected in the cortex. Thalamocortical fibers from lateral areas of VPL project to the *posterior paracentral lobule* (leg/foot), whereas progressively more medial parts of the VPL project in an orderly manner to sequentially more lateral areas of the *postcentral gyrus* (Figs. 17-7, 17-9). These thalamocortical fibers terminate primarily at the 3b/1 border on specific physiologic classes of SI neurons: *low-threshold non-nociceptive*, *nociceptive-specific*, and *wide dynamic range cells*. Loss of nociceptive and thermal sensations over the contralateral body and face can result from vascular compromise of either the middle (for trunk, leg, and face) or the anterior (for leg and foot) cerebral artery (Fig. 17-9). Not only the sensation but also the ability to localize is lost.

Not all nociceptive information reaches the thalamus via the ALS. The *spinocervicothalamic pathway* is a multimodal tract that carries discriminative innocuous

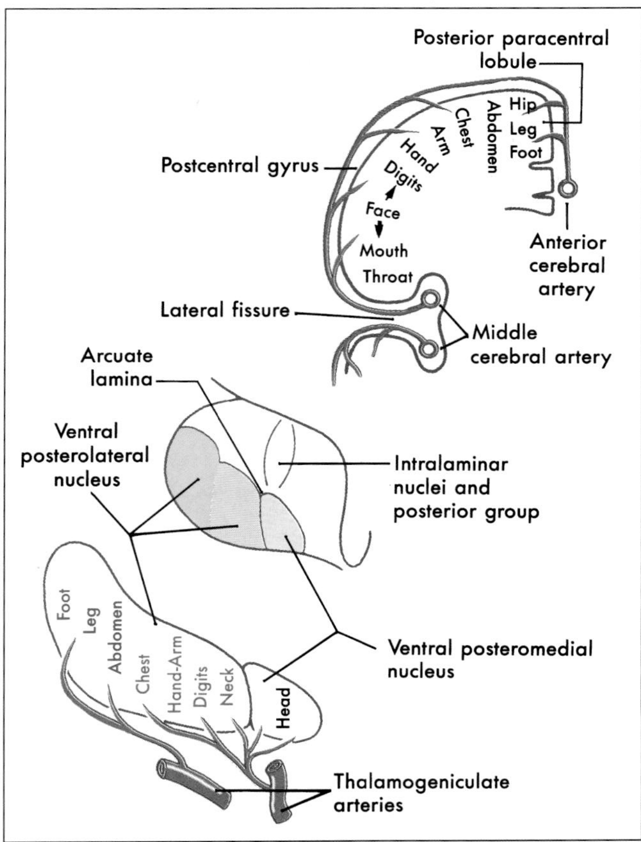

FIGURE 17-9 *Somatotopic organization of, and blood supply to, the ventral posteromedial and posterolateral nuclei and the SI somatosensory cortex.*

tactile information as well as nociceptive signals (Fig. 17-10). This pathway begins with afferent fibers that terminate on second-order cells in laminae III and IV of the dorsal horn. The axons of these second-order cells travel in the *ipsilateral lateral funiculus* to levels C1 and C2, where they terminate on third-order neurons in the *lateral cervical nucleus*. The axons of these cells decussate at the level of the spinomedullary junction and ascend in the medial lemniscus (Fig. 17-10). Like DCML axons, these *cervicothalamic axons* terminate in the VPL nucleus. This pathway is not essential for pain perception and is not especially prominent in humans. However, these fibers and the uncrossed axons in the ALS may be the basis for the retention of some nociceptive function after lesions involving the ALS, or for the return of pain perception after *anterolateral cordotomy*.

SPINAL TRIGEMINAL PATHWAY

Overview The transmission of nondiscriminative touch, thermal, and nociceptive signals from the head is provided by the *trigeminal nerve* (cranial nerve V), with smaller contributions from the area of the ear via cranial nerves VII, IX, and X. The primary sensory fibers of these nerves have their cell bodies in the *trigeminal ganglion*, the *geniculate ganglion* of cranial nerve VII, and the *superior ganglia* of cranial nerves IX and X. The central processes of all these cells coalesce to form the *spinal trigeminal tract*. Second-order neurons in the *spinal trigeminal nucleus* send axons to the contralateral side of the brainstem, where they collect to form the *ventral trigeminothalamic tract*. These fibers terminate on third-order neurons in the *ventral posteromedial* (*VPM*) nucleus, which, in turn, projects to face regions of the somatosensory cortex.

Primary Neurons Cranial nerves V, VII, IX, and X serve the cutaneous receptors of the face, the oral cavity, and the dorsum of the head except for the area served by the cervical nerves (Figs. 17-4, 17-11). A-delta and C fiber nociceptors are found throughout this region, and they are particularly prominent in the tooth pulp. Some of these fibers extend into the dentinal tubules, and carious lesions of the tooth expose these and other pulpal nerves to stimuli that result in dental pain. It is probable that the dull aching

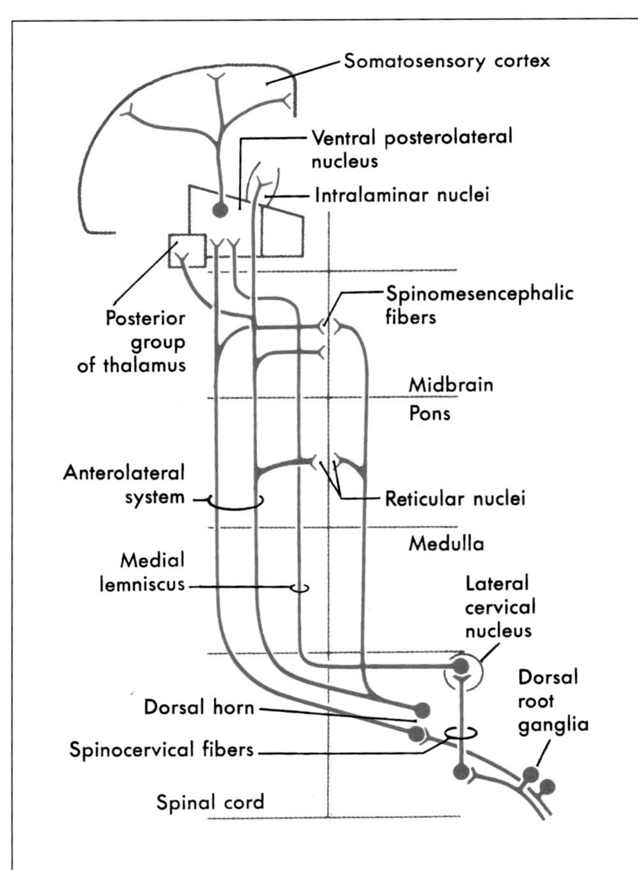

FIGURE 17-10 *Summary of the spinocervicothalamic tract that carries innocuous discriminative tactile, thermal, and nociceptive sensations.*

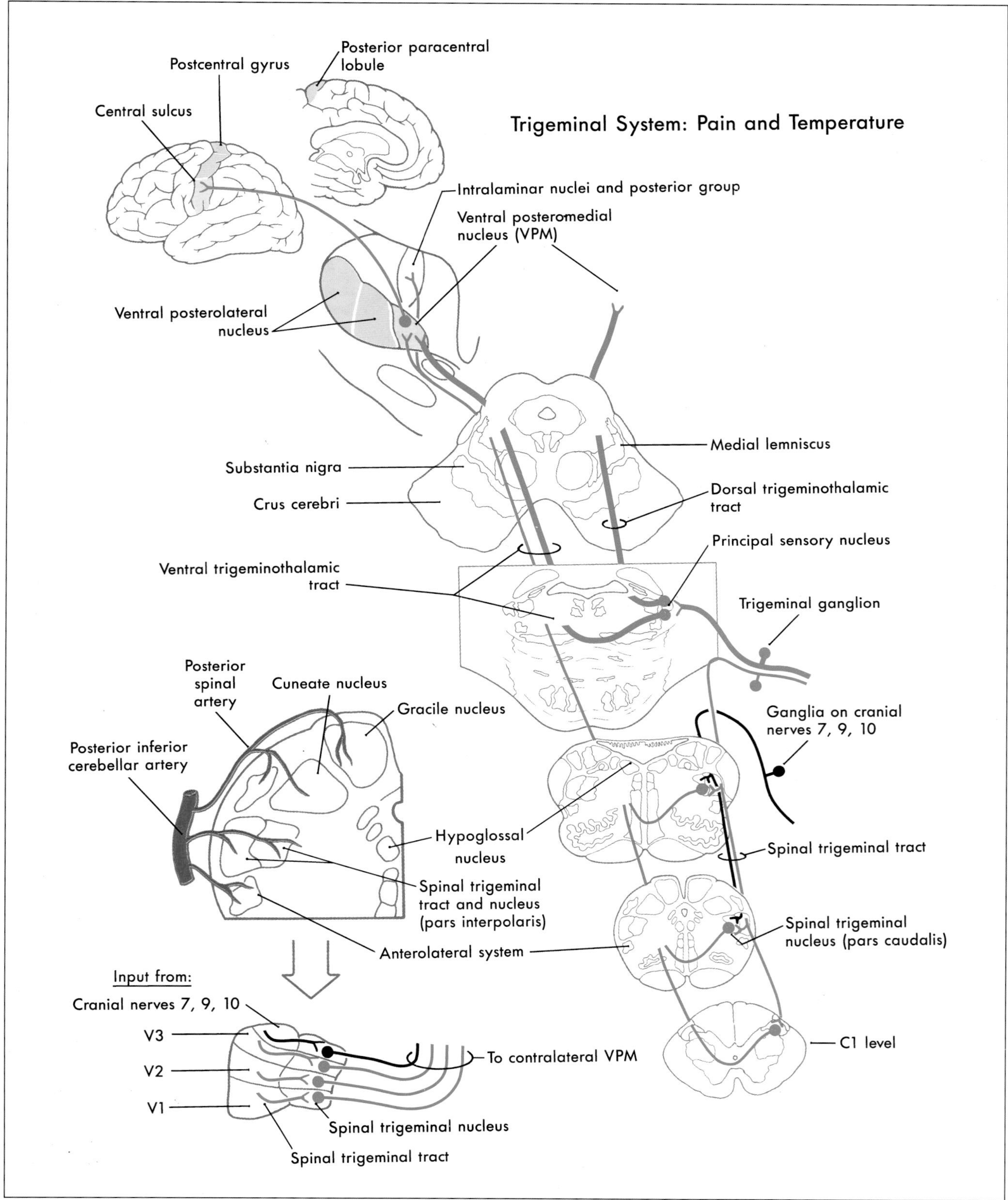

FIGURE 17-11 *The distribution of primary trigeminal fibers, of trigeminothalamic fibers to the ventral posteromedial nucleus, and the blood supply to trigeminal structures in the medulla. The large-diameter fibers convey discriminative touch, vibratory, and proprioceptive input, whereas the smaller-diameter fibers are the pathway for nondiscriminative tactile, innocuous thermal, and nociceptive signals.*

pain caused by pulp inflammation is the product of C fiber activity. Dental hypersensitivity, often characterized by sharp sensation, represents A-delta fiber activity. The cornea also receives a large number of nociceptive fibers, which form the afferent limb of the *corneal (blink) reflex*.

The central processes of small and large trigeminal ganglion cells form the *trigeminal sensory root* as they enter the pons (Fig. 17-11). As the small-diameter axons course dorsomedially into the pontine tegmentum, some bifurcate, sending an ascending branch to the *principal sensory nucleus*. The descending branch of these fibers and other unbranched small-diameter fibers form the *spinal trigeminal tract*. Through the caudal pons and the rostral medulla, this tract is internal to the restiform body. However, in the lower medulla caudal to the obex, it forms a superficial landmark lateral to the cuneate tubercle, known as the *tuberculum cinereum*.

The spinal trigeminal tract extends from the middle pons to the second or third cervical spinal cord segment, where its fibers interdigitate with those of the dorsolateral fasciculus (Lissauer's tract) (Figs. 17-5, 17-11; see also Fig. 16-5). Fibers of the mandibular division lie dorsal to those of the maxillary division, which, in turn, are dorsal to the ophthalmic division; this arrangement creates an inverted hemiface representation (Fig. 17-12). The primary afferent neurons associated with cranial nerves VII, IX, and X have cell bodies in their respective ganglia, enter the medulla, and take a position dorsal to those of the mandibular division.

The peripheral distribution of the branches of the trigeminal nerve (V_1, V_2, and V_3) delineates the facial dermatomes (Figs. 17-4, 17-12). Unlike the spinal segmental dermatomes, which partially overlap, the boundaries between adjacent facial dermatomes are sharply defined. This segregation of trigeminal branches is maintained by their central processes in the spinal trigeminal tract.

Injury to trigeminal nerve fibers produces a *paresthesia* restricted to specific regions of the face. The pain of *tic douloureux* (*trigeminal neuralgia*) produces episodic "paroxysmal" pain usually restricted to the peripheral distribution of the maxillary or mandibular divisions on one side. Trigeminal neuralgia is further characterized

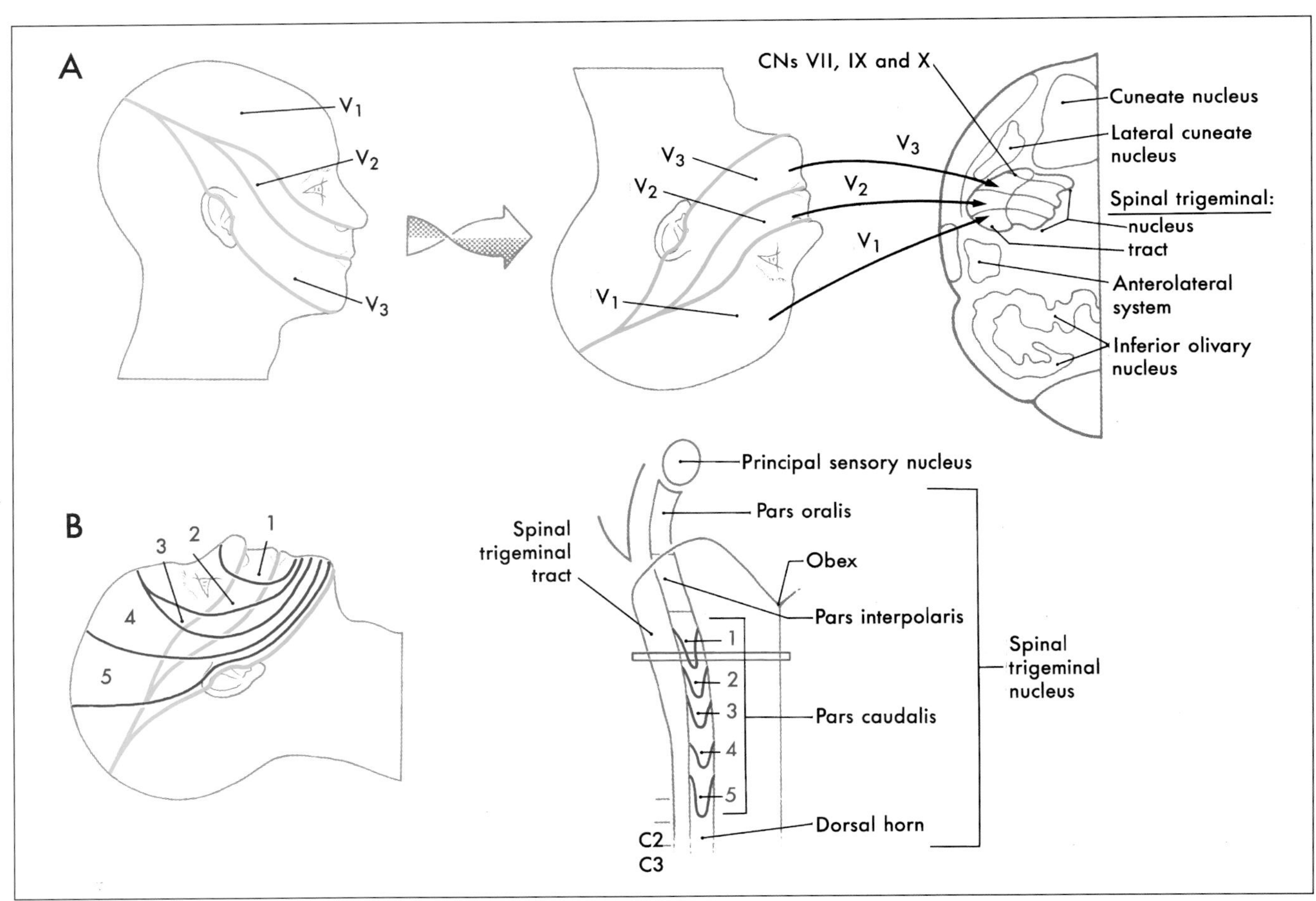

FIGURE 17-12 *Peripheral distribution of the trigeminal nerve and the inverted somatotopic arrangement of the hemiface within the spinal trigeminal nucleus pars caudalis (A). Functional onion-skin pattern of facial pain is superimposed along the caudal to rostral axis of pars caudalis (B). The red line through the long axis of pars caudalis (B) correlates with the approximate cross section represented in A. CN, cranial nerve.*

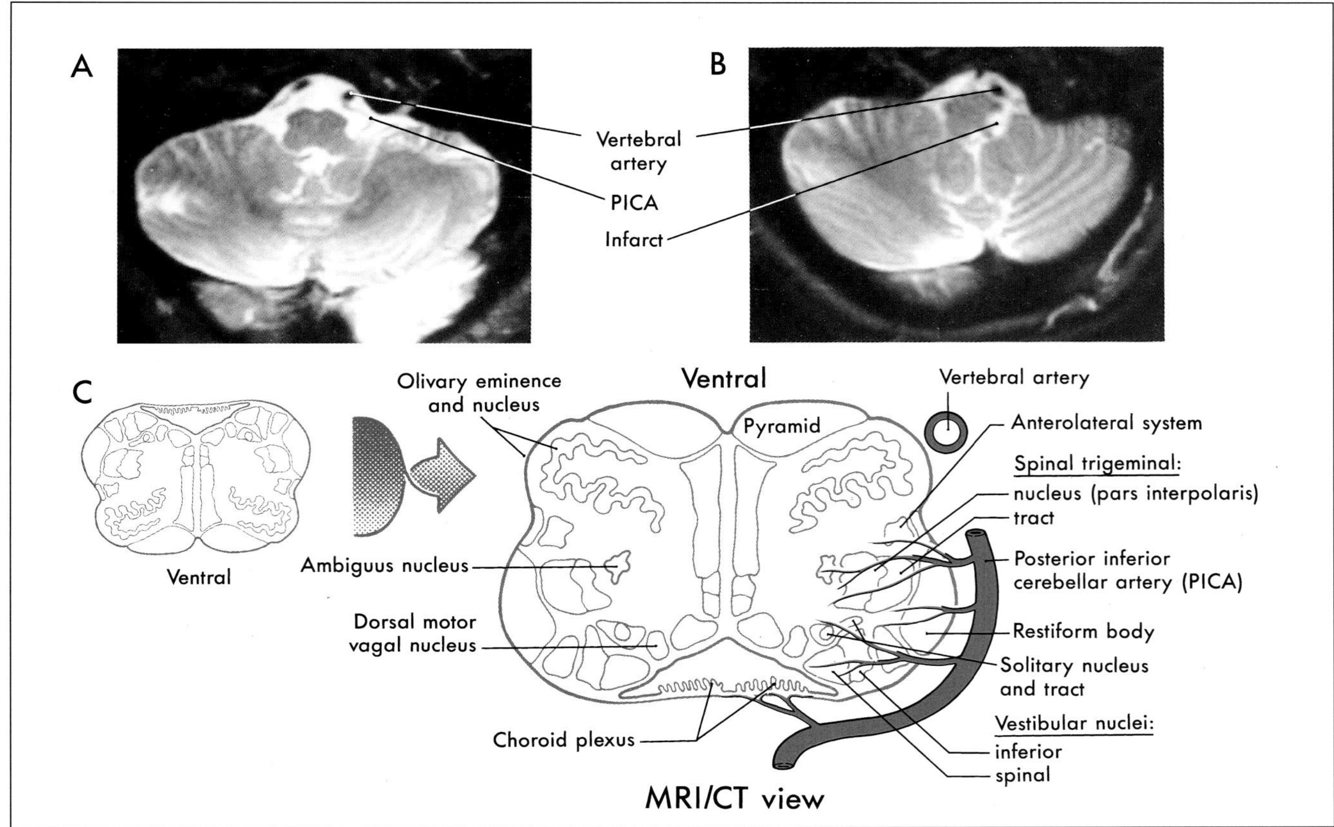

FIGURE 17-13 *Lateral medullary (Wallenberg's) syndrome. A normal magnetic resonance image (A) showing vertebral and posterior inferior cerebellar (PICA) arteries. A patient with an occlusion of PICA (B) resulting in an infarct of the dorsolateral medulla. The lesioned area contains trigeminal structures, the ALS, and other important nuclei and tracts (C).*

by the presence of "trigger zones," which, upon the most gentle stimulation (such as a light breeze or a brush with a wisp of cotton), produce stabbing pain on one side of the face. The precise etiology of this condition remains enigmatic, but vascular compression of the trigeminal nerve root or microneuromas are likely causes. The meninges are also supplied by fibers that terminate in the spinal trigeminal nucleus. These fibers are thought to be involved in the pain of migraine headaches.

Central Pathways The *spinal trigeminal nucleus*, located medial to the spinal tract, is the site of termination for fibers of the spinal trigeminal tract (Fig. 17-11). On the basis of cytoarchitecture, this nucleus is divided into a *pars caudalis*, a *pars interpolaris*, and a *pars oralis*. The caudal subnucleus (*pars caudalis*) (Figs. 17-5, 17-12) extends from C2 or C3 rostrally to the level of the obex. This part of the spinal nucleus shares many cytoarchitectural similarities with the dorsal horn. For this reason, it has been termed the *medullary dorsal horn* and has been divided into layers that correspond to Rexed's spinal cord laminae (Fig. 17-5). The *substantia gelatinosa* is largely continuous with lamina II of the spinal cord, and the *magnocellular* region is continuous with laminae III and IV. The pars caudalis and the dorsal horn also show homology in the distribution of neurotransmitters. For example, substance P and calcitonin gene-related peptide (CGRP) are localized in nociceptive C fibers that terminate in both of these areas.

At medullary levels, the posterior inferior cerebellar artery supplies the territory of the ALS fibers, as well as the spinal trigeminal nucleus and tract (Figs. 17-7, 17-11). Vascular accidents involving this vessel produce characteristic sensory disturbances, which include a contralateral loss of pain (*hemianalgesia*) and temperature (*hemithermoanesthesia*) sensibility over the body and ipsilateral loss of these modalities over the face. These symptoms are part of the *lateral medullary* (*Wallenberg's*) *syndrome* (Fig. 17-13).

The pars caudalis plays an important role in the transmission of nondiscriminative touch, nociceptive, and thermal sensations. This role is reflected by the fact that central processes of A-delta and C fibers terminate somatotopically in this subnucleus. In addition to the somatotopy within the pars caudalis, an *onion-skin pattern of facial pain* representation is oriented along the rostrocaudal axis of the subnucleus (Fig. 17-12). The nociceptive fibers that innervate circumoral and intraoral zones (teeth, gums, and lips) terminate rostrally, close to the obex in the caudal pars interpolaris. Fibers innervating progressively more caudal and lateral regions of the face terminate caudally within the spinal nucleus.

The interpolar subnucleus (*pars interpolaris*) is located between the level of the obex and the rostral pole of the hypoglossal (XII) nucleus. The most rostral subdivision is the oral subnucleus (*pars oralis*), which extends from the level of the rostral pole of the hypoglossal nucleus to the caudal end of the trigeminal motor nucleus (Figs. 17-11, 17-12). Some neurons in the pars interpolaris and pars oralis contribute to ascending somatosensory pathways, whereas others project to the cerebellum (see Fig. 16-16). In addition to projection neurons, the spinal trigeminal nucleus contains many local circuit neurons involved in brainstem reflexes and internuclear connections between the different parts of the trigeminal sensory nuclei.

The axons of *second-order trigeminothalamic neurons* in the spinal trigeminal nucleus decussate, coalesce to form the *ventral trigeminothalamic tract*, and ascend through the brainstem just dorsal to the medial lemniscus (Fig. 17-11). These fibers terminate in the ventral posteromedial, posterior, and intralaminar nuclei of the thalamus. As noted in Chapter 16, this pathway also carries crossed fibers from the principal trigeminal nucleus. The principal nucleus fibers terminate in the core of VPM, whereas spinal nucleus fibers terminate in its periphery. At the pontomesencephalic junction, ventral trigeminothalamic fibers intermingle with ALS fibers at the lateral margin of the medial lemniscus (Figs. 17-7, 17-11). Like ALS fibers, ascending ventral trigeminothalamic axons terminate in or give rise to collaterals that supply the reticular formation. In addition to regulating oral and facial reflexes, projections from the reticular formation terminate in the dorsal thalamus in the intralaminar nuclei and the medial region of the posterior nucleus. The intralaminar nuclei project widely to the striatum and cortex, especially frontal and somatosensory cortex. The medial region of the posterior nucleus projects to the head representation in secondary somatosensory cortex.

PAIN PERCEPTION AND CENTRAL PAIN SYNDROME

Vascular compromise of middle or anterior cerebral arteries (Fig. 17-9) produces a loss of sensibility (discriminative, nondiscriminative, innocuous thermal, and nociceptive) over contralateral regions of the body. Over time, however, appreciation of sensation returns. Pain sensations are first to return, followed by nondiscriminative tactile and thermal sensations. Discriminative tactile, vibratory, and proprioceptive sensations often fail to return to normal levels. If occlusion of the middle cerebral artery affects most of the postcentral gyrus, sensation begins returning first on the face and oral regions, then on the neck and trunk, and finally on the extremities and the distal parts of the limbs. This return of function indicates that other cortical areas may partially take over the appreciation of somatosensory stimuli, using the input they receive through non-lemniscal, non-ALS pathways.

At least some forms of somatosensory stimuli can be perceived at subcortical levels. In fact, electrical stimulation of the primary somatosensory cortex does not result in a complaint of pain, whereas thalamic stimulation may elicit paresthesia and sensations of dull pain and pressure. Furthermore, painful stimuli can be recognized and produce suffering without the presence of primary and secondary cortices, leading to the concept that *pain is perceived at subcortical levels*. However, damage to specific cortical regions eliminates the ability to precisely localize pain, suggesting that this is a function of somatosensory cortex and its lemniscal inputs (Fig. 17-14).

A second dissociation can occur in pain pathways. Pain perception and its affective component, suffering, are served by separate brain regions. The neospinothalamic pathway to primary somatosensory cortex is involved in the localization of painful stimuli (Fig. 17-14). Paleospinothalamic pathways that access the hypothalamus and limbic system via the reticular for-

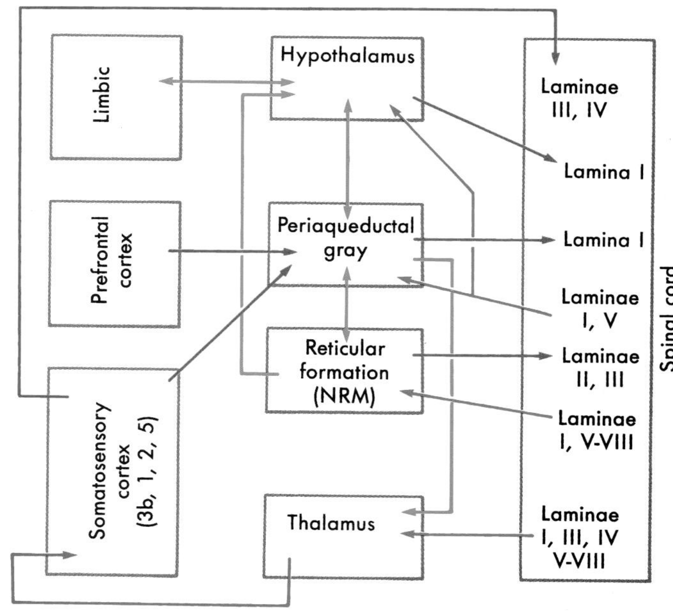

FIGURE 17-14 *Summary of pathways associated with pain perception. Ascending inputs are shown in blue, descending fibers in red, and interconnections in green. NRM, nucleus raphe magnus.*

mation and periaqueductal gray are involved in the suffering component of the pain experience (Fig. 17-14). This dissociation can be regulated pharmacologically, as some drugs eliminate suffering without affecting pain perception. Patients taking benzodiazepines report that the pain is still present, but its unpleasant nature is diminished.

Central or thalamic pain is a poorly understood sequela of natural or surgical lesions of structures involved in somatic sensibility. Central pain was originally observed following thalamic lesions, but it can occur following lesions of the ALS below the level of the thalamus. In the *central pain syndrome*, the analgesia that initially results from the lesion is replaced after a period of weeks, months, or years by spontaneous *paresthesia*, *dysesthesia*, or unusual painful responses. *Allodynia*, pain resulting from a stimulus that does not normally evoke pain, and *hyperalgesia*, an increased response to a stimulus that is normally painful, are common neurologic signs associated with the central pain syndrome. Patients often characterize central pain as burning, aching, pricking, or lacerating and as occurring in paroxysms that vary in intensity.

Central pain may last for years and is intractable to current analgesics. Pharmacologic agents, such as antidepressants and antiepileptic drugs, have been used with varying degrees of success to treat central pain. The time course and symptoms of central pain suggest that it could be due to the sprouting of inappropriate connections of non-nociceptive or nociceptive fibers, to increased excitability of central pain neurons, or to the removal of inhibitory influences on pain neurons.

Central pain syndrome can also result from vascular accidents. Patients surviving Wallenberg's syndrome (Fig. 17-13) may ultimately develop central pain, suggesting some sparing of alternate or parallel pain pathways. Although the etiology for this condition has not been proven, it is possible that this type of pain represents deafferentation phenomenon; that is, it results from removal of primary afferent influence on central neurons.

Patients experiencing central pain may obtain temporary relief from *transcutaneous electrical nerve stimulation* (*TENS*; electrical stimulation of nerves through the skin), from electrical stimulation of the dorsal columns, or from chronic stimulation of the periaqueductal or periventricular gray regions by stereotaxically positioned electrodes (*deep brain stimulation*). These stimulation techniques are discussed below. Neuroablative surgical procedures that have been used in the treatment of central pain include anterolateral cordotomy, trigeminal tractotomy, lesions of the dorsal root entry zone, thalamotomies, and cortical ablation. Unfortunately none of these procedures is successful in the long term.

PAIN TRANSMISSION AND CONTROL

The relaying of information from the spinal cord to supraspinal centers is an important event in the higher-order processing of nociceptive sensory signals. Based on the localization of putative neurotransmitters and secondary messengers in the dorsal horn, several candidates, such as *peptides* (calcitonin gene related peptide, substance P), *glutamate*, and *nitric oxide*, may be involved in this process. These and other chemical agents underlie the central pharmacology of nociceptive transmission and are responsible for the varied qualities associated with central pain pathways. Pain can be classified as acute or chronic, fast or slow, dull or sharp, or burning, throbbing, or aching. Because pain is such a complex sensory experience, reduction or elimination of nociceptive sensations is of obvious clinical importance. Effective clinical approaches that can be used for the purpose of controlling pain include *pharmacologic intervention* and *stimulation-produced analgesia*.

The CNS has neural circuits designed to modulate pain transmission (Figs. 17-14 to 17-16). These systems have components at all levels of the neuraxis, are capable of controlling nociceptive neurons, and are sensitive to opiates. Central structures implicated in the *descending control* of nociceptive transmission include (1) the somatosensory cortex, (2) the periventricular nucleus of the hypothalamus, (3) the pontine reticular formation, and (4) raphe nuclei and adjacent medullary reticular formation. Descending pathways originating in these structures are active during emergencies that could result in tissue damage.

Sites in the brainstem and hypothalamus modulate the processing of nociceptive information in the brainstem and spinal cord (Figs. 17-14, 17-15). The *periventricular gray* of the hypothalamus communicates with the *periaqueductal gray* (*PAG*) of the midbrain via an enkephalinergic pathway. Descending PAG fibers exert an excitatory influence on serotinergic neurons in the medullary *nucleus raphe magnus*, both directly and through interneurons in the lateral medullary reticular formation (Fig. 17-15). This projection uses *serotonin*, *neurotensin*, *somatostatin*, and *glutamate*. *Raphespinal neurons* project, in turn, to the dorsal horn and pars caudalis of the trigeminal nucleus (Figs. 17-15, 17-16). These serotonergic axons terminate on enkephalinergic interneurons in laminae II and III, which act pre- and post-synaptically to suppress the activity of pain transmission (Figs. 17-15, 17-16). In addition, both the hypothalamus (via *hypothalamospinal fibers*) and the PAG project directly to the medullary and spinal cord dorsal horns to act on incoming nociceptive signals (Fig. 17-14). *Cholecystokinin* and substance P are among the putative neurotransmitters used by PAG projection neurons.

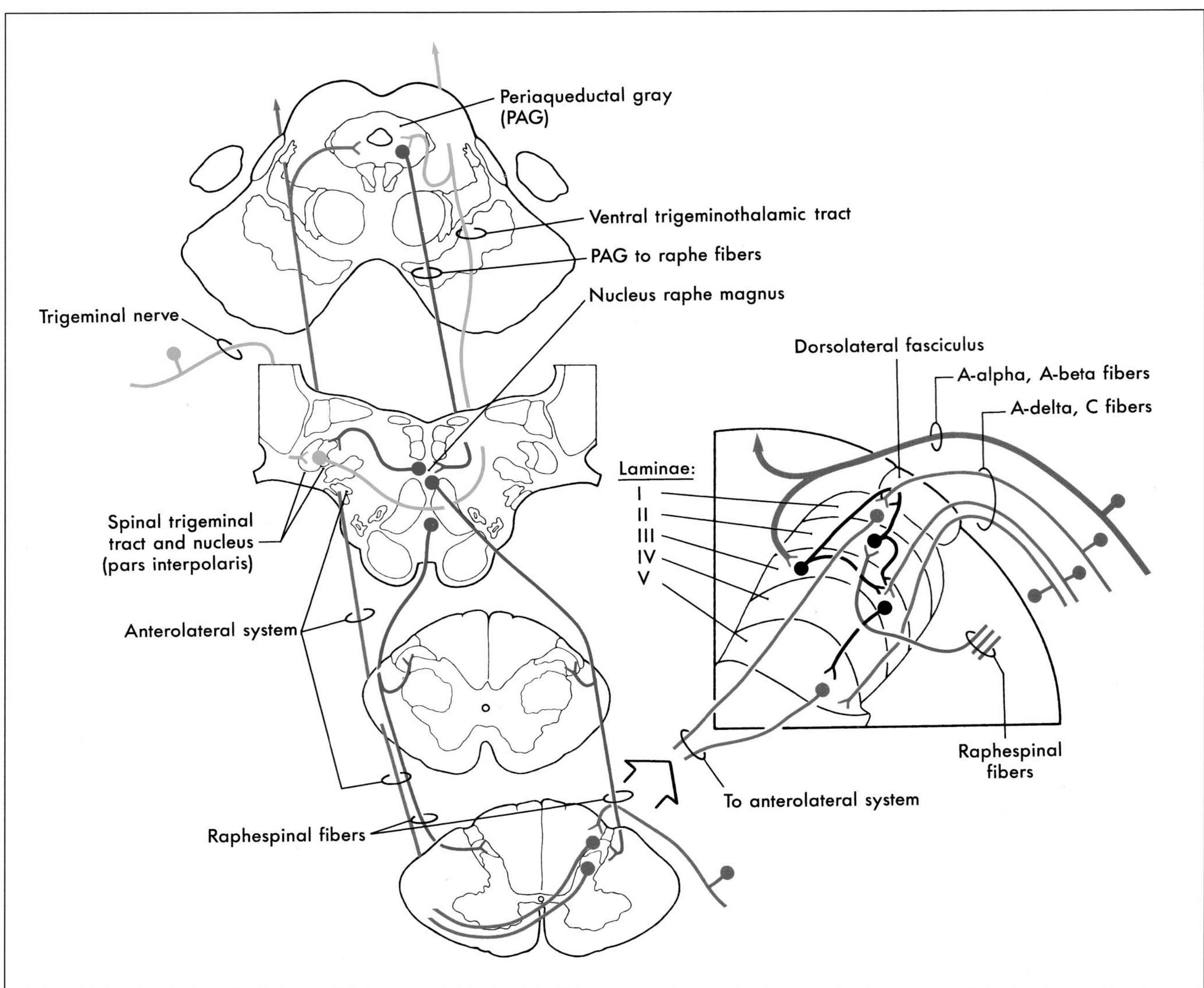

FIGURE 17-15 *Descending brainstem pathways that influence pain transmission and control within the brainstem (for trigeminal pathways) and in the spinal cord (for ALS projections).*

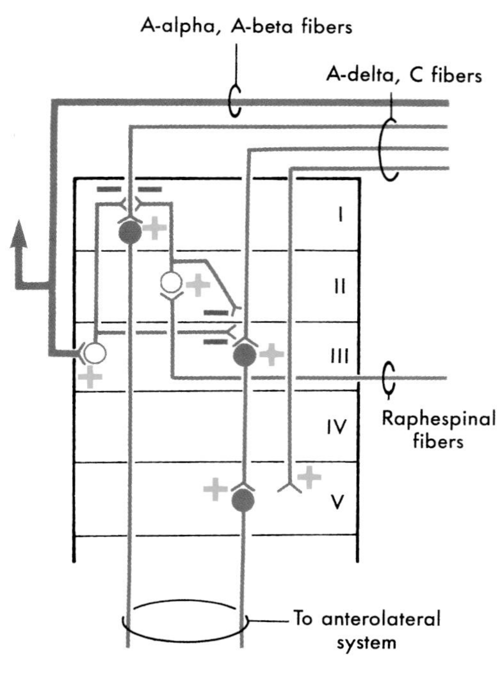

FIGURE 17-16 *Dorsal horn circuits that influence primary sensory fibers and ascending tract cells conveying nociceptive inputs. In addition to raphespinal fibers, some reticulospinal fibers also influence pain transmission in the dorsal horn.*

Stimulation-produced analgesia (*SPA*) relies on electrical stimulation of CNS structures to induce the release of endogenous chemicals, such as *enkephalin*, from cells in pain-control circuits. Endogenous opiates such as enkephalin inhibit pain transmission. Stimulation of periventricular gray, the PAG, or the nucleus raphe magnus results in the release of enkephalin or monoamines and in analgesia. Systemic administration of pharmacologic opiates, such as *morphine*, excites periventricular and periaqueductal neurons, supplementing their natural activity. This increase in activity suppresses neurons in the spinal and medullary dorsal horns that transmit painful information and produces analgesia. The delivery of opioids directly to the spinal cord (epidural) also is used to produce a powerful analgesia in a variety of postoperative and chronic pain states.

Current therapies for the control of pain transmission include transcutaneous electrical nerve stimulation and chronic stimulation of the dorsal columns by implanted electrodes. Dorsal column stimulation activates large-diameter myelinated fibers. Antidromic activation of these fibers discharges collaterals in the dorsal horn (Fig. 17-16). These collaterals stimulate the enkephalinergic interneurons in the dorsal horn that inhibit the transmission of pain signals. This stimulation also provides long-term diminution of pain for reasons that are poorly understood. *Acupuncture-like stimulation* also may produce local analgesia by stimulating these fibers. Future trends in pain management might include strategies such as receptor-specific pharmacologic agents, circumscribed surgical intervention, and behavioral modifications.

SOURCES AND ADDITIONAL READINGS

Brodal A: *Neurological Anatomy in Relation to Clinical Medicine*, 3rd Ed. Oxford University Press, New York, 1981.

Burgess PR, Perl ER: Cutaneous mechanoreceptors and nociceptors. In Iggo A (ed): *Handbook of Sensory Physiology*, Vol 2. *Somatosensory System*. Springer-Verlag, New York, 1973, pp 30–78.

Dubner R, Bennett GJ: Spinal and trigeminal mechanisms of nociception. Annu Rev Neurosci 6:381–418, 1983.

Dubner R, Sessle B, Storey A: *The Neural Basis of Oral and Facial Function*. Plenum Press, New York, 1978.

Light A: *The Initial Processing of Pain and Its Descending Control: Spinal and Trigeminal Systems*, Vol 12. *Pain and Headache*. Karger, New York, 1992.

Mayer DJ, Liebeskind JC: Pain reduction by focal electrical stimulation of the brain: An anatomical and behavioral analysis. Brain Res 68:73–93, 1974.

Poggio GF, Mountcastle VB: A study of the functional contributions of the lemniscal and spinothalamic systems to somatic sensibility. Central nervous mechanisms in pain. Johns Hopkins Hosp Bull 106:266–316, 1960.

Wall PD, Melzack R: *Textbook of Pain*, 3rd Ed. Churchill Livingstone, Edinburgh, 1994.

Willis WD: *The Pain System: The Neural Basis of Nociceptive Transmission in the Mammalian Nervous System*, Vol 8. *Pain and Headache*. Karger, New York, 1985.

Young RF: Effect of trigeminal tractotomy on dental sensation in humans. J Neurosurg 56:812–818, 1982.

CHAPTER EIGHTEEN

Viscerosensory Pathways

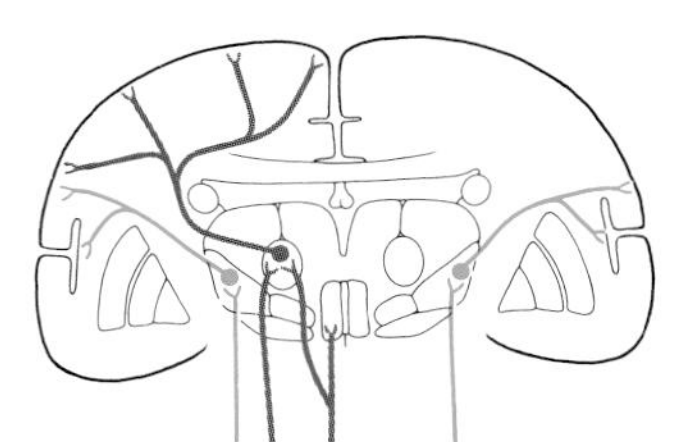

S. G. P. HARDY • J. P. NAFTEL

The somatosensory system conveys information from sensory receptors in the skin, joints, and skeletal muscles that allows one to perceive and respond to input from the external environment. Functioning in parallel with somatosensory pathways are fibers that convey information from visceral receptors. This input allows the body to make appropriate responses to changes in its internal environment.

VISCEROSENSORY RECEPTORS

Functionally, viscerosensory receptors may be categorized as nociceptors or physiologic receptors. *Nociceptors* are free nerve endings that respond to stimuli that have the potential to damage tissue or to stimuli resulting from the presence of damaged tissue. These receptors signal changes in visceral structures that result from pathologic processes (such as myocardial ischemia or appendicitis) or from benign conditions such as gastrointestinal cramping or bloating.

Physiologic receptors are responsive to innocuous stimuli, and they monitor the functions of visceral structures on a continuing basis. These receptors also mediate normal visceral reflexes such as the baroreceptor reflex. Examples of this type of receptor are (1) rapidly adapting mechanoreceptors, (2) slowly adapting mechanoreceptors, and (3) various types of specialized receptors.

Rapidly adapting mechanoreceptors signal the occurrence of dynamic events, such as movement or sudden changes in pressure. This type of receptor is present in the thoracic, abdominal, and pelvic cavities. In the thoracic cavity, it is represented by free nerve endings that exist in the epithelia of pulmonary airways. Because these nerve endings are sensitive to the presence of inhaled particles, they have been referred to as "cough receptors." Rapidly adapting mechanoreceptors in the abdominal and pelvic cavities vary greatly in size and location and may be either unencapsulated or encapsulated. The largest example of a rapidly adapting mechanoreceptor is the Pacinian corpuscle.

Slowly adapting mechanoreceptors signal the presence of stretch or tension within a visceral structure. These typically unencapsulated receptors are located in the smooth muscle layer of the pulmonary airways and in the smooth muscle layers of hollow abdominal and pelvic viscera. They are essential for the perception of a sense of fullness in certain viscera, such as the stomach or bladder.

Certain *specialized receptors* are unique to the viscerosensory system. These include *baroreceptors*, *chemoreceptors*, *osmoreceptors*, and *internal thermal receptors*. *Baroreceptors* (Fig. 18-1A) are found in the walls of the aortic arch and carotid sinus and respond to rapid increases or decreases in blood pressure. For baroreceptors to effectively perform this task, blood pressure must be in the range of about 30–150 mmHg. *Chemoreceptors* (Fig. 18-1B) are found in structures called *carotid bodies* (located at the bifurcation of the common carotid artery) and *aortic bodies* (located in the aortic arch) and are activated by changes in the composition of arterial blood. These include changes in oxygen and carbon dioxide tension, acidity, and the presence of certain drugs such as nicotine or cyanide.

In addition, specialized visceroreceptors also exist in the hypothalamus as *chemoreceptors*, *osmoreceptors*, and *internal thermal receptors*. These visceroreceptors are activated by changes in blood chemistry or osmolarity, or by changes in the temperature of blood circulating through the hypothalamus. Hypothalamic neurons that respond to these changes by altering their firing rates are considered to be the "receptor" cells.

VISCEROSENSORY FIBERS

Sympathetic and parasympathetic divisions of the autonomic nervous system (see Chapter 28) have been considered traditionally to consist only of visceromotor (GVE) fibers. These fibers travel through sympathetic

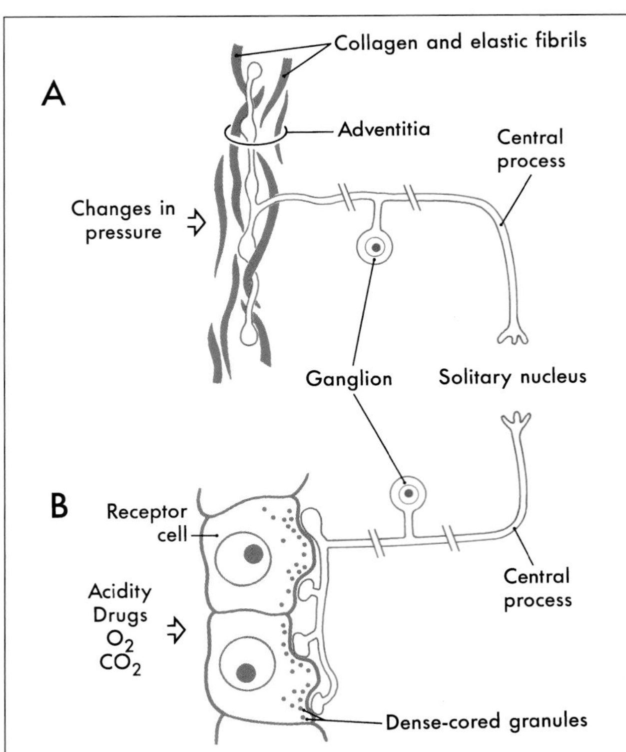

FIGURE 18-1 *Diagrammatic representation of a baroreceptor (A) and a chemoreceptor (B). The baroreceptor is located in the adventitia in apposition to collagen and elastic fibrils. This receptor alters its firing rate in response to changes in blood pressure. The chemoreceptor is composed of a receptor cell and the numerous afferent endings it contacts. Changes in blood chemistry trigger responses in the receptor cell that result in an action potential in the subjacent afferent ending.*

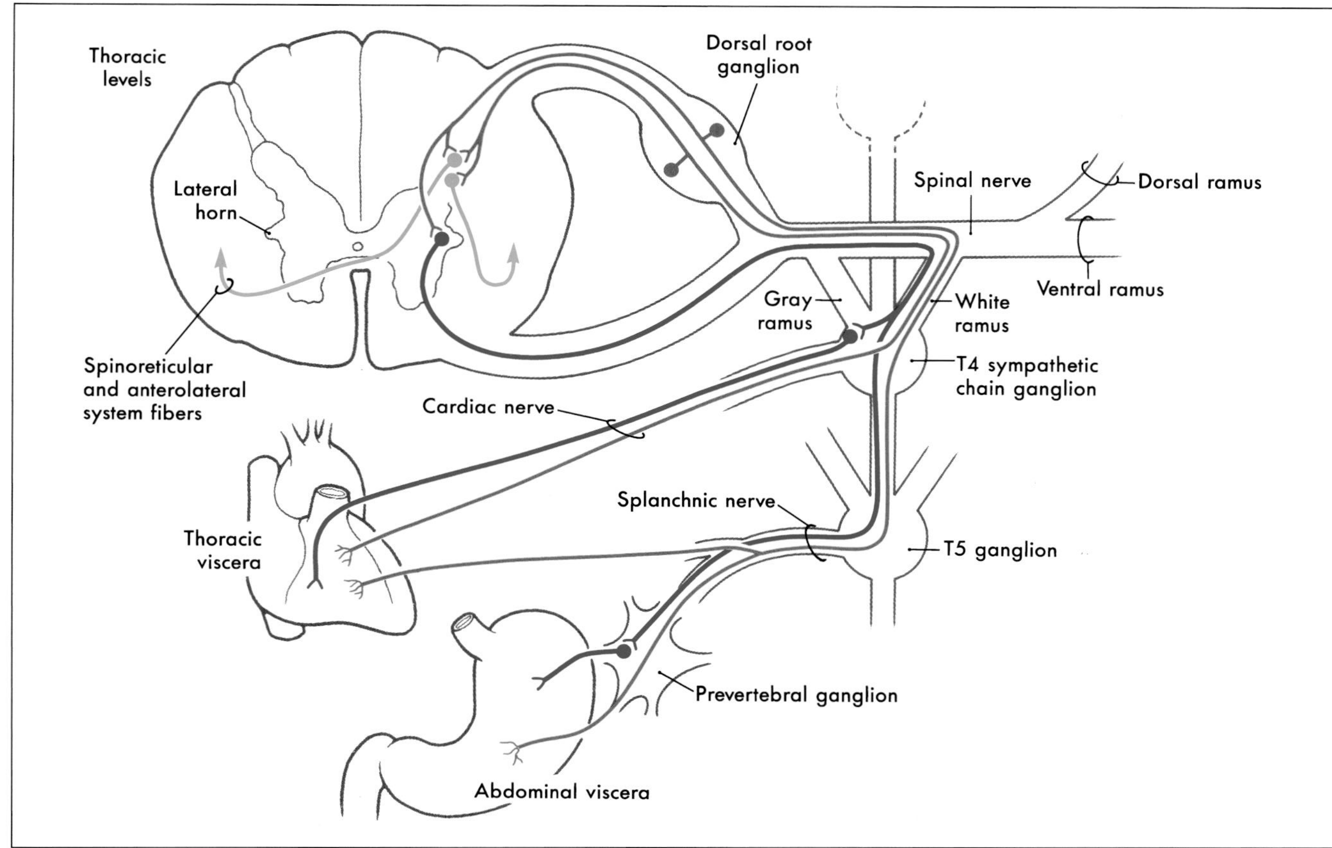

FIGURE 18-2 *Primary sensory sympathetic afferent fibers (red) shown in relation to dorsal horn tract cells (green) conveying visceral information to the thalamus and to GVE neurons (blue).*

nerves (such as splanchnic and cardiac nerves) or through parasympathetic nerves (such as vagus and pelvic nerves). However, these *sympathetic and parasympathetic nerves also contain viscerosensory (GVA) fibers* that serve many important functions. In this chapter, *the terms "sympathetic afferent" and "parasympathetic afferent" are used to describe viscerosensory fibers contained in sympathetic and parasympathetic nerves*, respectively. In addition to its conciseness, this usage complies with the terminology introduced by Langley, a pioneer in studies on the autonomic nervous system (see Cervero and Foreman).

Visceral afferents tend to predominate in parasympathetic nerves but are comparatively sparse in sympathetic nerves. For example, more than 80% of the fibers in the vagus nerve (a parasympathetic nerve) are viscerosensory, whereas less than 20% of the fibers in the greater splanchnic nerve (a sympathetic nerve) are visceral afferents. Most visceral afferents (90%; both sympathetic and parasympathetic) are either unmyelinated or thinly myelinated and, therefore, are slowly conducting fibers.

Information originating from *physiologic receptors* (innocuous input) travels primarily in parasympathetic nerves. Some input of this type may be conveyed by sympathetic nerves, but this is a very minor source of such information. In contrast, input originating from *nociceptors* is conducted almost exclusively by sympathetic nerves. Thus, there is a division of responsibility between parasympathetic and sympathetic nerves in terms of viscerosensory input.

ASCENDING PATHWAY FOR SYMPATHETIC AFFERENTS

Afferent fibers conveying nociceptive information from thoracic and abdominal viscera travel via the cardiac and splanchnic nerves (Fig. 18.2). For example, primary sensory fibers that originate from the stomach join the greater splanchnic nerve, enter the sympathetic trunk, and pass through a white ramus to join the spinal nerve. Nociceptive input from pelvic viscera such as the prostate and sigmoid colon is conveyed by viscerosensory fibers traveling through the hypogastric plexus and lumbar splanchnic nerves.

The cell bodies of origin of sympathetic afferent fibers are located in dorsal root ganglia at about levels T1 to L2 (Fig. 18-2). The central processes of these fibers enter the spinal cord via the lateral division of the dorsal root. They may ascend or descend one or two spinal levels in the dorsolateral fasciculus before terminating in laminae I and V and/or laminae VII and VIII. Cells in laminae I and V project mainly to the contralateral side as part of the *anterolateral system*, whereas the neurons in laminae VII and VIII project bilaterally as *spinoreticular fibers*. In addition, some primary viscerosensory fibers terminate on preganglionic sympathetic cell bodies located in the intermediolateral cell column at spinal levels T1 to L2 (Fig. 18-2). The axons of these cells, in turn, exit through the ventral root as GVE preganglionic sympathetic fibers.

In general, viscerosensory fibers that enter the spinal cord at a particular level originate from structures that receive GVE input from the same spinal level (Fig. 18-2). For example, visceral afferent fibers from the stomach enter the spinal cord over the dorsal roots of T5 to T9 and terminate in the same spinal segments that convey visceral efferent outflow to the stomach.

Projections to Thalamus Some neurons located in laminae I and V receive nociceptive input from sympathetic afferent fibers and send their axons rostrally via two routes in the anterolateral system (ALS) (Fig. 18-3). Some fibers cross in the ventral white commissure and ascend in the ALS, whereas others ascend in this bundle on the ipsilateral side. These ALS fibers terminate in the ventral posterolateral nucleus (VPL), which, in turn, projects to the inferolateral part of the postcentral gyrus (the parietal operculum) and to the insular cortex (Fig. 18-3). The location from which this visceral nociceptive information originated is encoded in these particular regions of the cerebral cortex. However, visceral pain is poorly localized (lacks detailed point-to-point representation) because receptor density is low and receptive fields correspondingly large and because this input converges in the pathway. Consequently, it is not possible to tell whether pain is coming from the stomach or the duodenum, *only* that it is coming from the general area of the upper abdomen.

Projections to Reticular Formation In addition to the direct path to thalamus and sensory cortex via the ALS and VPL, there are indirect routes via the reticular formation through which visceral nociceptive information can reach the cortex. The reticular formation receives spinoreticular inputs (mainly from laminae VII and VIII) and collaterals from the ALS (Fig. 18-3). In turn, cells of the reticular formation project to progressively higher levels of the neuraxis, thus relaying viscerosensory information in a multisynaptic fashion to progressively higher levels of the brain. Neurons located in the reticular formation and in the periaqueductal gray ultimately project to the hypothalamus and to the intralaminar nuclei of the thalamus (Fig. 18-3). These latter cell groups project to the cortex.

Reticulohypothalamic fibers travel via the *dorsal longitudinal fasciculus*, the *mammillary peduncle*, and the *medial forebrain bundle*. The first originates mainly from the periaqueductal gray, and the latter two originate mainly from the mesencephalic reticular formation. These midbrain

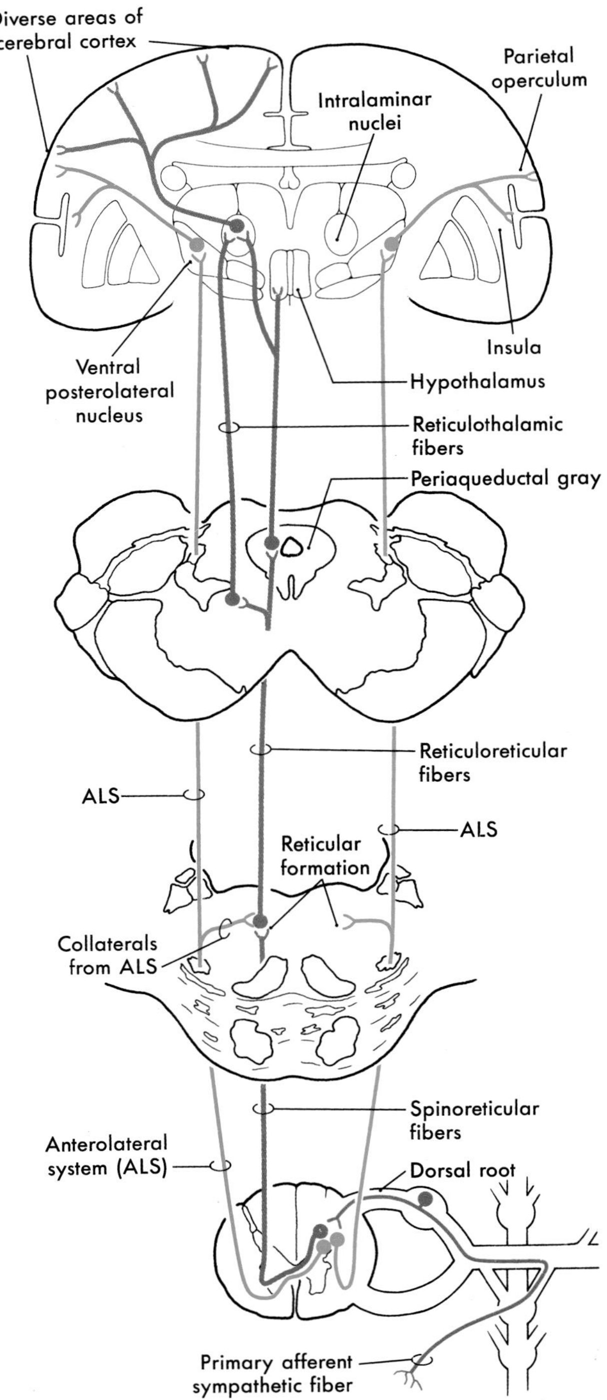

FIGURE 18-3 *Ascending visceral afferent input travels in the anterolateral system (green) and through multisynaptic circuits via the reticular formation of the brainstem (gray). These fibers influence specific and diverse areas of the cerebral cortex.*

centers receive both viscerosensory and somatosensory input and, through their projections, hypothalamic centers may be influenced by either system. For example, viscerosensory input resulting from distention of the bowel may result in increased heart rate or cutaneous flushing. On the other hand, somatosensory stimuli, such as that associated with coitus or suckling, may increase the release of the hypothalamic hormone oxytocin.

Referred Pain Referred pain is the phenomenon whereby noxious stimuli that originate in a visceral structure, such as the heart or the stomach, are perceived by the patient as pain arising from a superficial part of the body such as the skin, bones, or skeletal muscles (Fig. 18-4). Although referred pain may mask the true origin of the information, certain patterns of referred pain are clearly diagnostic of diseases in particular visceral locations. Visceral pain is transmitted by sympathetic sensory fibers and is typically referred to those somatic structures whose afferents enter the cord via the same dorsal roots.

The mechanism underlying referred pain is thought to involve a convergence of somatic and visceral afferent information onto pools of dorsal horn neurons, the axons of which ascend to higher levels of the neuraxis (Fig. 18-5). Normally, a visceral nociceptive fiber (e.g., from the heart) synapses on a spinothalamic tract cell whose axon will travel to the VPL, and from there the information is relayed to visceral parts of the sensory cortex. Consequently, this sensory input is perceived as arising from deep in the thorax (Fig. 18-5A). In some situations, however, collaterals of visceral afferent fibers may synapse on and excite dorsal horn tract cells that usually transmit only somatosensory information (Fig. 18-5B). In this case, the tract cell is activated by a visceral afferent collateral but sends information via the VPL to a part of the somatosensory cortex that represents the body wall (Fig. 18-5B). Consequently, the pain is "referred" to (interpreted as coming from) the surface of the body (Fig. 18-4) even though the stimulus actually originated from a visceral structure.

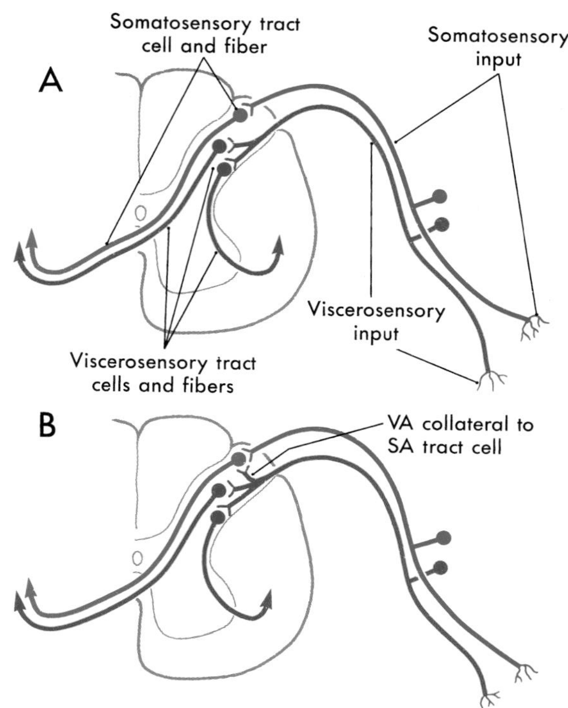

FIGURE 18-5 *The circuits involved in referred pain. Somatic afferent (blue) and visceral afferent (red) fibers terminate on tract cells that convey their respective types of information to the thalamus (A). Collaterals from visceral afferent fibers (VA) may activate tract cells that usually convey somatic (SA) data (B). The brain interprets this input as originating from the body wall.*

Angina Referred pain may occur in conjunction with any internal organ; however, it is frequently associated with diseases of the heart (Fig. 18-4). The pain resulting from heart disease is termed *angina*. In about 80% of patients, angina is initially perceived as an unpleasant, squeezing sensation originating from behind the sternum. This discomfort may also be perceived as pain radiating down the left arm or, more rarely, down both arms. The predilection of the pain for the left side of the chest, or extending down the left arm, reflects the predominance of myocardial disease in the left side of the heart. Consequently, nociception from the left side of the heart is referred to the left side of the body. Because angina is typically perceived as a pain of the chest,

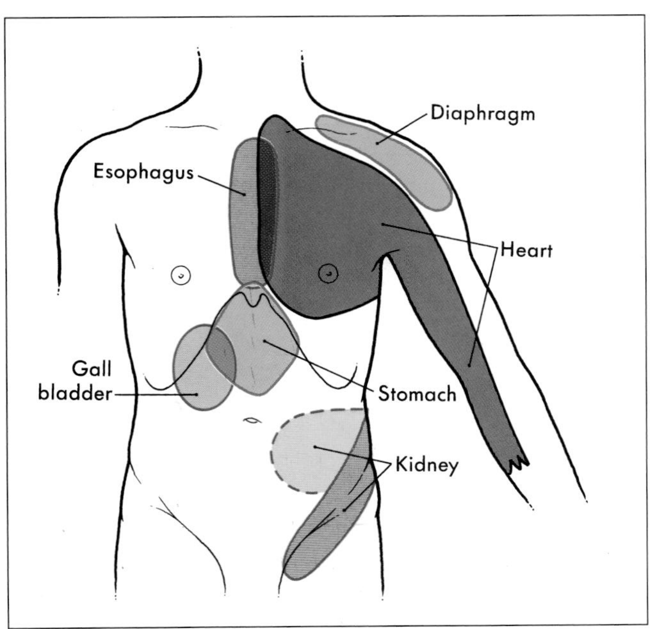

FIGURE 18-4 *Superficial areas to which pain is commonly referred from the corresponding deep structures.*

including the sternum and pectoral muscles, it is frequently called *angina pectoris*.

The pathways involved in angina are shown in Figure 18-6. Afferent fibers from the heart enter the sympathetic trunk through either the *cervical cardiac* or *thoracic cardiac nerves*. The former join the sympathetic chain at superior, middle, and inferior cervical ganglia, whereas the latter nerves join the sympathetic ganglia associated with spinal nerves T1 to T5 (Fig. 18-6). These primary viscerosensory fibers enter the spinal cord and terminate in laminae I and V of the dorsal horn. These same spinal segments also receive cutaneous somatosensory input from dermatomes of the chest wall and arm (Fig. 18-6). Tract cells in the dorsal horn that receive primarily somatosensory input may also be activated, as noted above, by collaterals of visceral afferent fibers from the heart. Consequently, the cerebral cortex interprets the pain as originating from the surface of the body (the upper chest and/or arm) when actually the stimulus that has produced the painful input is located in a visceral structure (the heart).

PATHWAYS FOR PARASYMPATHETIC AFFERENTS

Sacral Parasympathetic Afferents The *pelvic nerves* are parasympathetic and contain viscerosensory fibers passing to cord levels S2 to S4, and GVE preganglionic fibers originating from these levels. These primary sen-

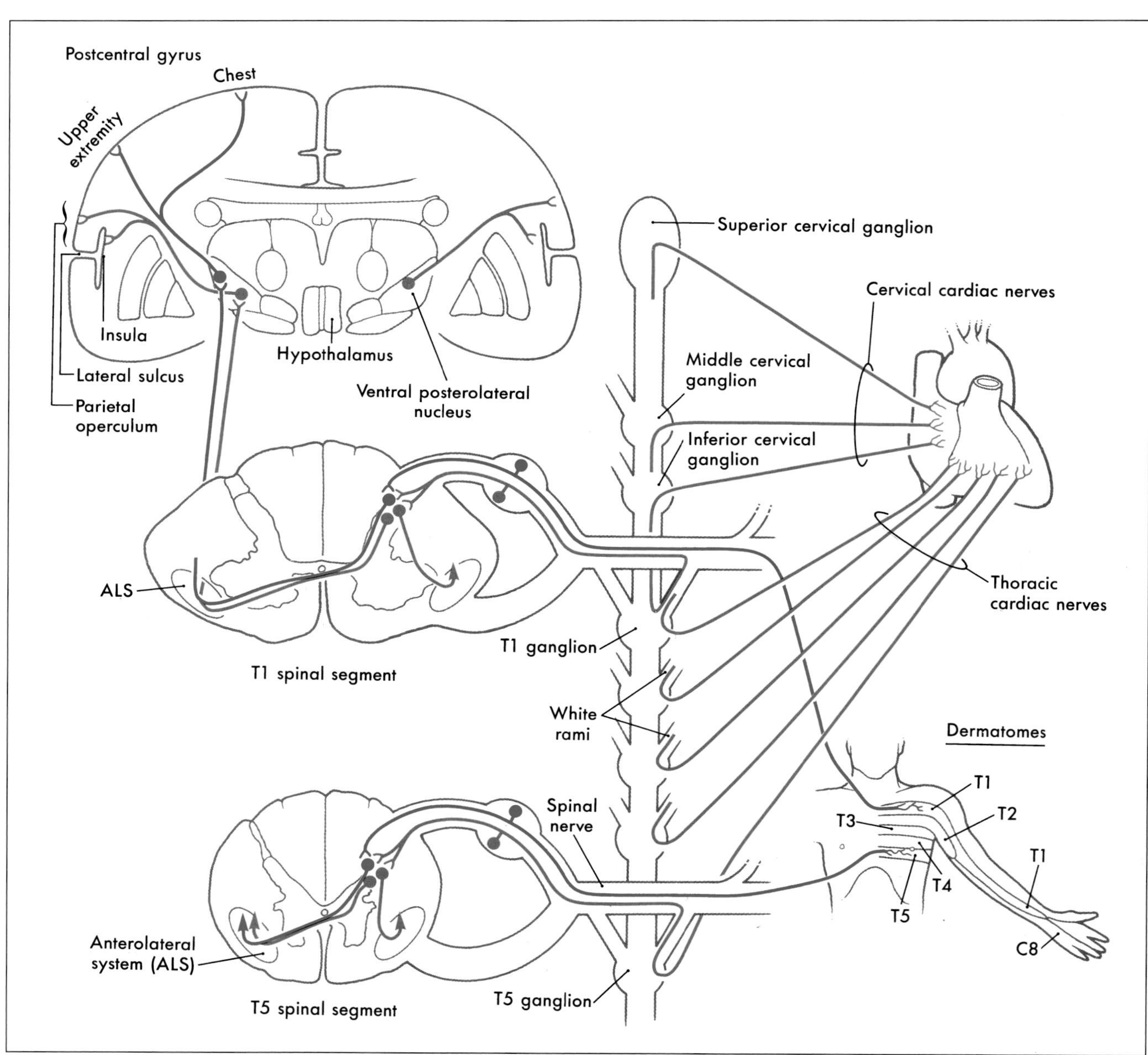

FIGURE 18-6 *The pathways mediating cardiac pain and the circuits through which cardiac pain may be "referred" to superficial parts of the body wall.*

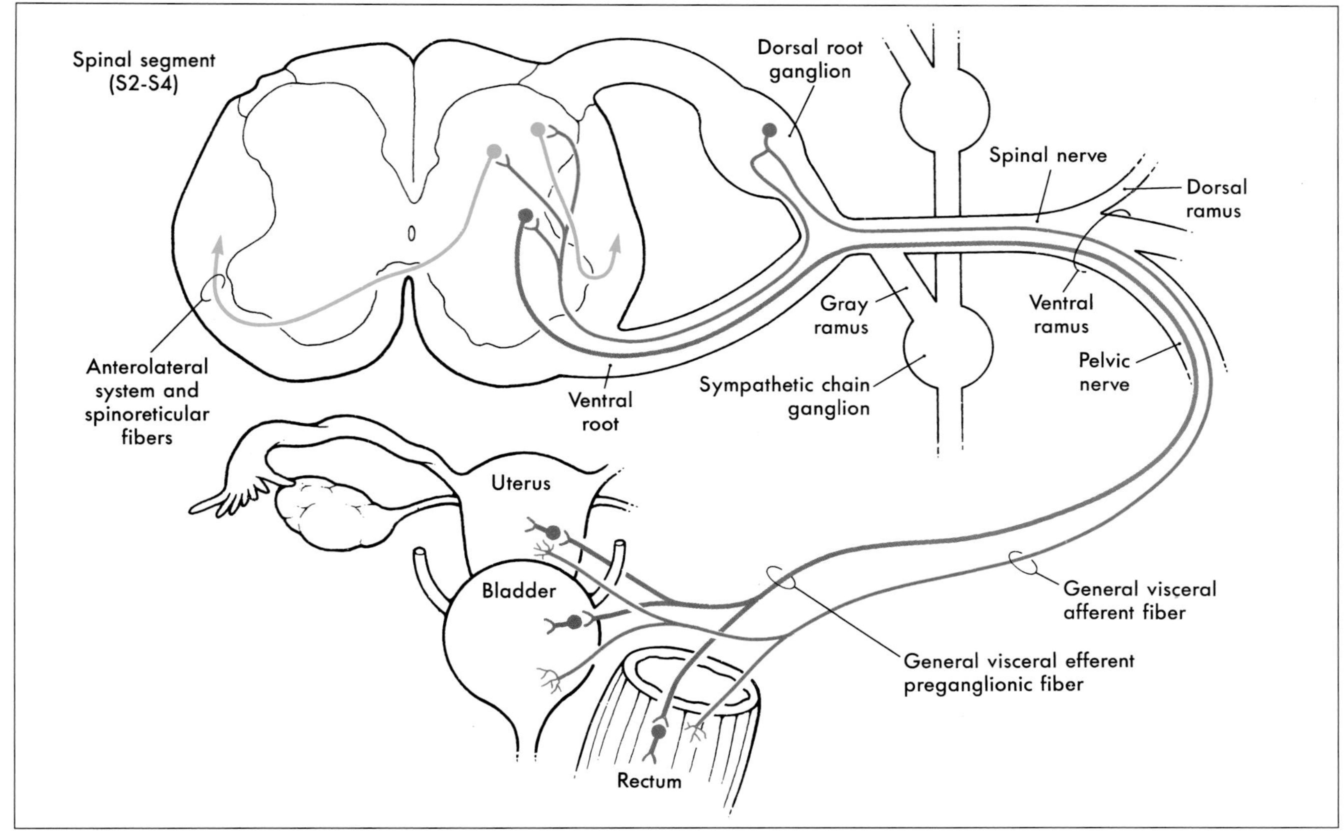

FIGURE 18-7 *Primary sensory parasympathetic fibers (red) shown throughout their trajectory in relation to dorsal horn tract cells (green), and GVE neurons (gray).*

sory parasympathetic fibers pass through the pelvic nerve, enter the spinal nerve, and have their cell bodies in dorsal root ganglia of S2 to S4 (Fig. 18-7). Rather than entering the dorsal root, however, their central processes course into the spinal cord via the ventral root.

Once in the cord, these viscerosensory fibers terminate in the dorsal horn and in the immediate vicinity of the visceral efferent preganglionic motor neurons. These dorsal horn cells relay information on bladder or bowel distention (a sense of "fullness") to the VPL via the ALS and spinoreticular pathways described above and then, through this thalamic nucleus, to the insular and parietal opercular cortices (Figs. 18-3, 18-6). In this way, a full bladder is perceived and interpreted as such. In addition, ascending input from the pelvic viscera is also shunted into the hypothalamus for the initiation of *supraspinal autonomic reflexes*. Those GVE cell groups in S2 to S4 that receive viscerosensory afferents give rise to parasympathetic preganglionic axons that synapse on postganglionic cells located in pelvic viscera (Fig. 18-7). This forms the basis for *spinal autonomic reflexes*.

Cranial Parasympathetic Afferents Cranial nerves III, VII, IX, and X contain fibers of GVE preganglionic parasympathetic motor neurons. However, only cranial nerves VII, IX, and X have sensory ganglia. Of these only cranial nerves IX (the *glossopharyngeal*) and X (the *vagus*) have significant numbers of parasympathetic afferent fibers.

Visceral afferent fibers traveling in the glossopharyngeal nerve originate primarily from *chemoreceptors* of the carotid body and *baroreceptors* of the carotid sinus wall (Fig. 18-8). In addition, nociceptive and tactile input from the oropharynx (the general area of the palatine tonsil) is also conveyed on the ninth cranial nerve. These sensory fibers form the afferent limb of the gag reflex.

The *carotid body* is composed of specialized neural elements, *chemoreceptors*, and is innervated by viscerosensory branches of the glossopharyngeal nerve. Within the carotid body, these chemoreceptors are located in close proximity to arterial sinusoids. As a result, they are responsive to changes in arterial oxygen and carbon dioxide tension, to the acidity of the blood, and to drugs. Carotid *baroreceptors* are located in the carotid sinus wall and respond to rapid changes in arterial blood pressure (Fig. 18-8). The *aortic arch* also contains chemoreceptors and baroreceptors that are similar in structure and function to those found in the carotid body and sinus. These specialized receptors, however, are innervated by aortic and/or cardiac branches of the vagus nerve.

Fibers of the vagus nerve transmit a wide variety of physiologic information from thoracic viscera and from

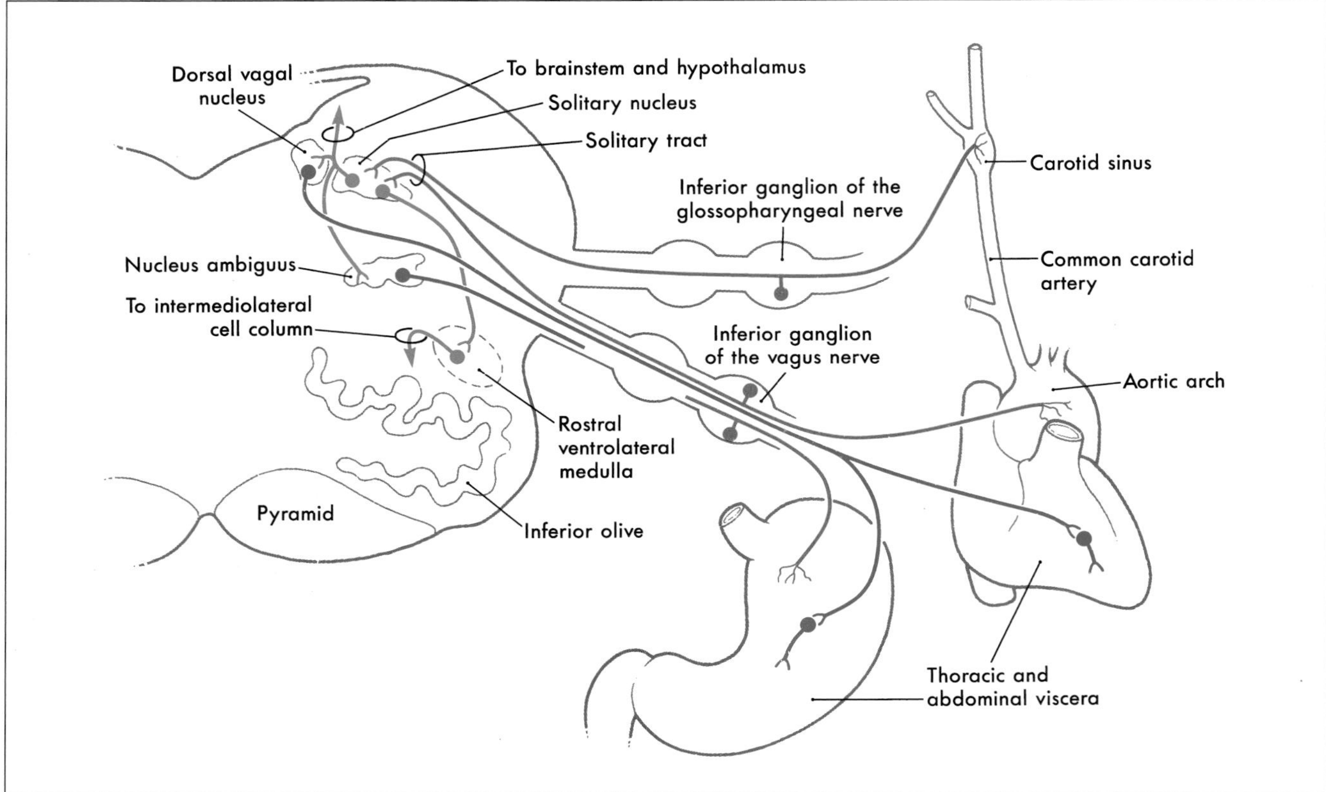

FIGURE 18-8 *The nucleus of the solitary tract receives viscerosensory fibers via the glossopharyngeal (IX) and vagus (X) nerves and projects to a variety of other nuclei.*

all viscera of the abdominal cavity above the level of the splenic flexure of the large colon. These vagal fibers convey information regarding the functional status of these structures but are not responsible for conveying information on pain.

The peripheral viscerosensory fibers, traveling in the glossopharyngeal and vagus nerves, enter the skull through the jugular foramen. Within this foramen there is a superior and inferior ganglion on each nerve (Fig. 18-8). Cell bodies of primary viscerosensory neurons (GVA) are found within the inferior ganglion, whereas the superior ganglion contains the cell bodies of primary somatosensory neurons (GSA).

The central processes of primary visceral fibers in cranial nerves IX and X enter the medulla, form the *solitary tract*, and synapse with neurons of the adjacent *solitary nucleus* (Fig. 18-8). Some of these fibers use substance P and cholecystokinin as their transmitters. Second-order neurons in the solitary nucleus project to and influence a variety of neurons in the brainstem and hypothalamus. These targets include the dorsal vagal nucleus, the nucleus ambiguus, rostral areas of the ventrolateral medulla, and the parabrachial nuclei. The *dorsal vagal nucleus* is the primary source of preganglionic parasympathetic neurons that project to thoracic and abdominal viscera. In addition, the *nucleus ambiguus* also contains some parasympathetic visceromotor cells (GVE, autonomic) that innervate cardiac ganglia. However, the targets of most nucleus ambiguus cells (these are somatic visceral efferent, SVE, motor neurons) are muscles of the larynx and pharynx.

Visceral efferent (GVE) motor neurons of the nucleus ambiguus and the dorsal vagal nucleus receive input from the solitary nucleus and project, via the vagus nerve, to parasympathetic ganglia of the heart. Activation of this pathway causes a decrease in heart rate and a corresponding decrease in blood pressure, a "vasodepressor" response. Conversely, neurons located in rostral parts of the ventrolateral medulla receive solitary input and project to the spinal cord, where they influence the activity of preganglionic sympathetic motor neurons in the intermediolateral cell column. In doing so, this medullary center causes an increase in blood pressure and thereby serves a "vasopressor" function.

Baroreceptor Reflex Projections from the solitary nucleus to the dorsal vagal nucleus, to cells associated with the nucleus ambiguus, and to rostral parts of the ventrolateral medulla are essential to the normal operation of the *baroreceptor reflex* (Fig. 18-8). In this reflex, afferent input from carotid and aortic baroreceptors enters the medulla on cranial nerves IX and X and terminates in the solitary nucleus. Increases in blood pressure cause the baroreceptors to increase their discharge frequency, whereas decreases in blood pressure result in a lower rate of baroreceptor discharge. In this manner, blood pressure is continuously monitored and the resulting information is forwarded to the solitary nucleus. Within this nucleus, neurons projecting to the dorsal vagal nucleus and the nucleus ambiguus respond in a manner opposite to those neurons projecting to the rostral parts of the ventrolateral medulla. For example, during a period of acute *hypertension*, solitary neurons excite "vasodepressor" cells of the dorsal vagal nucleus and nucleus ambiguus and inhibit "vasopressor" neurons of the rostral ventrolateral medulla. The inhibitory solitary neurons may be GABAergic, whereas the excitatory neurons presumably use one of the excitatory amino acids. As a result of this dual influence from solitary neurons, blood pressure is lowered and the hypertension is diminished. Conversely, during a period of acute *hypotension*, projections from the solitary nucleus inhibit "vasodepressor" cells and excite "vasopressor" neurons, leading to an elevation of blood pressure. Consequently, blood pressure is elevated and the hypotension relieved.

In addition to projections from the solitary nucleus to ambiguus, dorsal vagal, and medullary nuclei, solitary neurons also project into the reticular formation. Consequently, visceral afferent information entering the medulla on the vagal and glossopharyngeal nerves may also influence the hypothalamus and diverse areas of the cerebral cortex. The solitary nucleus does not relay any appreciable amount of general visceral sensation to the dorsal thalamus. Consequently, most of the general visceral afferent information conveyed by the glossopharyngeal and vagal nerves does not reach a level of consciousness. On the other hand, the solitary nucleus does relay taste information to the cerebral cortex via the ventral posteromedial nucleus.

VISCERAL INPUT TO THE RETICULAR ACTIVATING SYSTEM

The reticular formation of the brainstem receives a wide range of inputs and projects to, among other targets, the intralaminar nuclei of the thalamus. These cell groups, in turn, send their axons to broad areas of the cerebral cortex, with the largest number terminating in the frontal lobe. These reticulothalamic and thalamocortical pathways "alert" or "activate" the cerebral cortex as a whole and are one important part of the *ascending reticular activating* system (ARAS, see also Chapters 17 and 31).

As discussed above, the reticular formation receives viscerosensory input via spinoreticular fibers and collaterals from the ALS. Some of these ascending fibers to the reticular formation convey nociceptive visceral afferent information originating from the gut on sympathetic afferent fibers. Other ascending fibers convey a sense of bladder (or bowel) fullness originating from pelvic viscera on parasympathetic afferents. Both types of viscerosensory input feed into the reticular formation and participate in the "arousal" of the cerebral cortex through the ARAS. For example, either sudden pain from the stomach or small intestine or the stimulus of a full bladder will excite the reticulothalamocortical circuit and wake a person from a deep sleep. The initial sensation is not one of specific information (full bladder, stomach pain) but rather the sense of just being awakened. However, once the cortex has been "alerted," the conscious/perceptive part of the brain takes over (recognizes the source of the arousal) and addresses the problem.

SOURCES AND ADDITIONAL READINGS

Ammons WS: Cardiopulmonary sympathetic afferent excitation of lower thoracic spinoreticular and spinothalamic neurons. J Neurophysiol 64:1907–1916, 1990.

Bieger D, Hopkins DA: Viscerotopic representation of the upper alimentary tract in the medulla oblongata in the rat: the nucleus ambiguus. J Comp Neurol 262:546–562, 1987.

Brody MJ: Central nervous system mechanisms of arterial pressure regulation. Fed Proc 45:2700–2706, 1986.

Cechetto DF, Saper CB: Evidence for a viscerotopic sensory representation in the cortex and thalamus in the rat. J Comp Neurol 262:27–45, 1987.

Cervero F, Foreman RD: Sensory innervation of the viscera. In Loewy AD, Spyer KM (eds): *Central Regulation of Autonomic Functions*. Oxford University Press, New York, 1990, pp 104–125.

Coggeshall RE, Applebaum ML, Frazen M, Stubbs TB III, Sykes MT: Unmyelinated axons in human ventral roots, a possible explanation for the failure of dorsal rhizotomy to relieve pain. Brain 98:157–166, 1975.

Garrison DW, Chandler MJ, Foreman RD: Viscerosomatic convergence onto feline spinal neurons from esophagus, heart and somatic fields: effects of inflammation. Pain 49:373–382, 1992.

Procacci P, Zoppi M: Heart pain. In Wall PD, Melzack R (eds): *Textbook of Pain*. Churchill Livingstone, Edinburgh, 1989, pp 410–419.

Reis DJ, Granata AR, Joh TH, Ross CA, Ruggiero DA, Park DH: Brain stem catecholamine mechanisms in tonic and reflex control of blood pressure. Hypertension Suppl II 6:7–15, 1984.

Willis W: Visceral pain. In Brooks FP, Evers PW (eds): *Nerves and the Gut*. Charles B Slack, Thorofare, NJ, 1977, pp 350–364.

CHAPTER NINETEEN

The Visual System

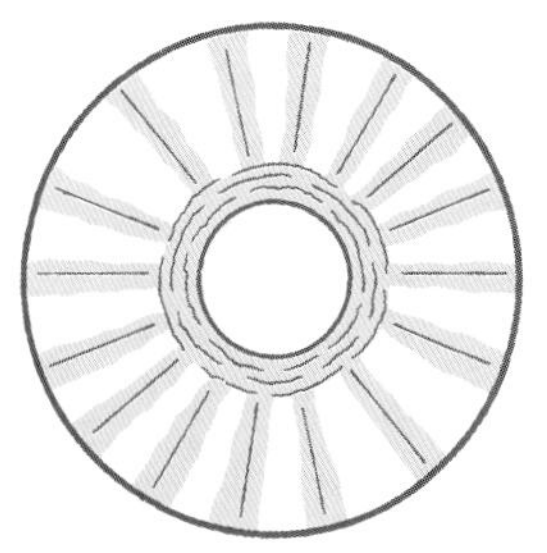

J. B. HUTCHINS • J. J. CORBETT

Of all the senses, humans rely most on vision. Consequently, more is known about vision than about any of the other senses, and a great deal of basic science and clinical research effort has been spent examining its function. The visual system is enormously important in clinical diagnosis. About 40% of the fibers in the brain carry information related to visual function; therefore, damage to the brain is often reflected by some type of visual dysfunction.

OVERVIEW

Like other sensory systems, the visual system creates a somatotopic "map" of its sensory field (the visual world) that is preserved at all levels. Light information is received by the retinal photoreceptors, and the initial processing of the visual signal occurs in the retina. Although the retina projects to several diencephalon and midbrain structures, most retinal axons terminate in a thalamic relay nucleus, which in turn innervates a large region of cortex in the occipital lobe. From there, visual input is sent to several cortical regions, which govern much of the temporal and parietal lobes.

ANATOMY OF THE EYE

The human eye is an optical instrument specialized for the collection of light and the initial processing of visual information. Its anatomy is considered only briefly here.

Cornea The *cornea* provides a transparent protective coating for the optical structures of the eye (Fig. 19-1). Its lateral margin is continuous with the *conjunctiva*, a specialized epithelium covering the "white" (sclera) of the eye. Although the conjunctiva and sclera have a blood supply (most evident when the eyes are irritated), the cornea is normally not vascularized. Several conditions, such as vascularization or opacities, may degrade the optical quality of this structure, requiring transplantation of a donor cornea.

Chambers of the Eye Just behind the cornea is the fluid-filled *anterior chamber*, which is bounded posteriorly by the *iris* and the opening of the *pupil* (Fig. 19-1). A second fluid-filled space, the *posterior chamber*, is bounded anteriorly by the iris and posteriorly by the lens and its encircling suspensory ligament (zonule fibers). Fluid is continuously produced by the epithelium over the *ciliary body* around the rim of the posterior chamber and flows through the pupillary opening into the anterior chamber. It then drains into a set of modified veins, the *canals of Schlemm*, that are located around the rim of the anterior chamber in the *angle* where the iris meets the cornea. Because the suspensory ligament encircling the lens consists of discrete strands, the fluid in the posterior chamber is in contact with the *vitreous body*, the gelatinous mass that fills the main space of the eyeball between the lens and the retina. Therefore, any condition that obstructs the drainage of fluid via the canals of Schlemm can lead to *glaucoma*, a buildup of pressure in the entire eyeball that may result in blindness.

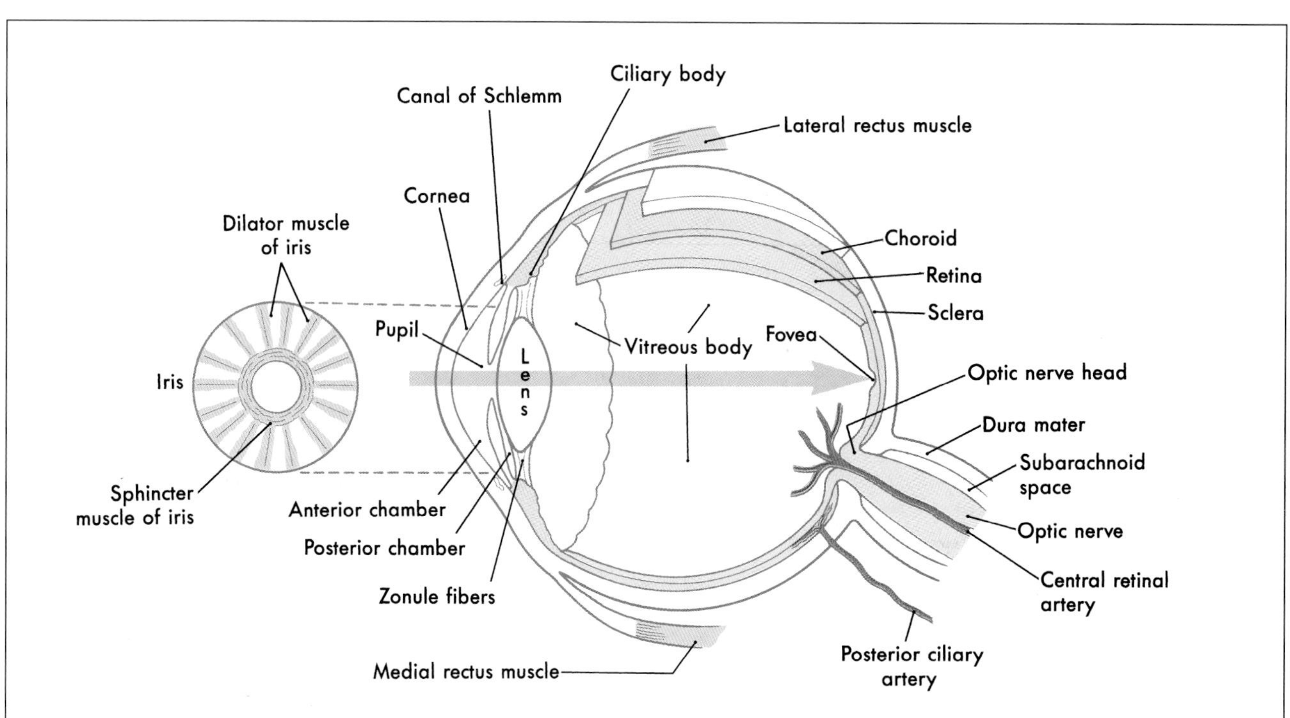

FIGURE 19-1 *Cross section of the human eye. The gray arrow shows the path of light through the optical apparatus.*

Iris The *iris* is a pigmented structure lying directly anterior to the lens (Fig. 19-1). The *stroma* of the iris contains melanocytes that reflect or absorb light to give the iris its characteristic color. Also embedded in the stroma are the circumferentially organized *sphincter muscle* of the iris and the radially arranged *dilator muscle* (Fig. 19-1).

The *iris sphincter* receives parasympathetic input from the *Edinger-Westphal nucleus* through a synaptic relay in the *ciliary ganglion*. The axons of these postganglionic parasympathetic neurons release acetylcholine onto the muscarinic receptors of the sphincter muscle cells. Subsequent contraction of these muscle cells results in a decrease in diameter of the pupil (*miosis*).

The action of the iris sphincter muscle is opposed by the dilator muscle. The preganglionic fibers in the pathway to the dilator arise in the *intermediolateral cell column* at upper thoracic levels. These preganglionic fibers terminate on postganglionic sympathetic fibers in the *superior cervical ganglion* that innervates the dilator. Within the iris, sympathetic nerve endings release norepinephrine onto the radially arranged muscles, and their contraction results in pupillary enlargement (*mydriasis*). This phenomenon is a measure of the general state of the sympathetic tone. Anger, pain, or fear may result in an enlargement of the pupil in the absence of a change in lighting conditions. The *pupillary light reflex*, a contraction of the pupil in response to light, is used to assess the function of the nervous system at midbrain levels (see Chapter 27).

The circumference of the pupillary margin changes by a factor of six, a change in muscle length greater than any other in the human body. To accomplish this change, acetylcholine is released onto *both* the sphincter and dilator muscles. The effect is to activate muscarinic receptors that *depolarize* sphincter muscle cells and cause contraction. Additionally, acetylcholine released by collaterals onto the dilator muscle mediates presynaptic inhibition of norepinephrine release and *blocks* dilator contraction. Thus, as the sphincter contracts, the dilator relaxes, strengthening the pupillary response to light.

Lens The *lens* is a clear structure that *focuses light on the retina* (Fig. 19-1). Mechanisms that change the curvature of the lens are discussed in detail in Chapter 27.

Beginning at about age 40, the lens begins to lose its elasticity, so that the shape it adopts when relaxed is more flattened than in a younger person. This change reduces the individual's ability to focus on near objects, a condition called *presbyopia*. Bifocal corrective lenses are prescribed to aid the patient in performing tasks requiring close, detailed vision.

Opacities in the lens, known as *cataracts*, are relatively common and can be seen as a cloudiness of the lens. Cataracts may be caused by *congenital defects* (e.g., secondary to maternal infection with *rubella*), persistent exposure to *ultraviolet light*, or poorly understood mechanisms that occur in aging. Current therapy consists of replacement of the lens with an inert plastic prosthesis, restoring sight but with a concomitant loss of accommodation.

Uvea The iris, ciliary body, and choroid make up the *vascular tunic* of the eye, also called the *uvea*. The *choroid* is a highly vascularized, pigmented tissue layer lying between the *retinal pigment epithelium* and the *sclera*, the tough outer coating of the eye. *Uveitis* is an inflammation of these structures, often secondary to eye injury.

THE NEURAL RETINA AND PIGMENT EPITHELIUM

The inner surface of the posterior aspect of the eye is covered by the *retina*, which is composed of the *neural retina* and the *retinal pigment epithelium* (Fig. 19-2). When describing the layers and cells of the retina, it is common to use the terms *inner* and *outer*. Inner refers to structures located toward the vitreous and the center of the eyeball, whereas outer is used in reference to structures located toward the pigment epithelium and choroid.

The *retinal pigment epithelium* is a continuous sheet of pigmented cuboidal cells bound together by tight junctions that block the flow of plasma or ions. It (1) supplies the neural retina with *nutrition* in the form of glucose and *essential ions*, (2) *protects* retinal photoreceptors from potentially damaging levels of light, and (3) plays a key role in the maintenance of photoreceptor anatomy.

The *neural retina* contains the photoreceptors and associated neurons of the eye and is specialized for *sensing* light and *processing* the resultant information. The *photoreceptors* absorb quanta of light (photons) and convert this input to an electrical signal. The signal is then processed by retinal neurons as discussed below. Finally, the retinal neurons called *ganglion cells* send the processed signal to the brain via axons that travel in the optic nerve.

The contact between the neural retina and the pigment epithelium is the adult remnant of the ventricular space of the developing eyecup. As such, it is mechanically unstable. This instability is clearly demonstrated in a *retinal detachment* when the neural retina tears away from the pigment epithelium. Because photoreceptors are metabolically dependent on their contact with pigment epithelial cells, a detached retina must be repaired to avoid further damage. The detached part of the neural retina is welded to the pigment epithelium using surgical procedures. Although this repair prevents an increase in the area of detachment, the detached portion of the retina does not regain function.

The neural retina has seven characteristic layers (Fig. 19-2). From outer to inner they are (1) a layer containing the *photoreceptor cell outer and inner segments*, (2) an *outer nuclear layer* consisting of the nuclei of photoreceptor cells, (3) the *outer plexiform layer* consisting of the synaptic connections of photoreceptors with second-order retinal cells, (4) the *inner nuclear layer* containing somata of sec-

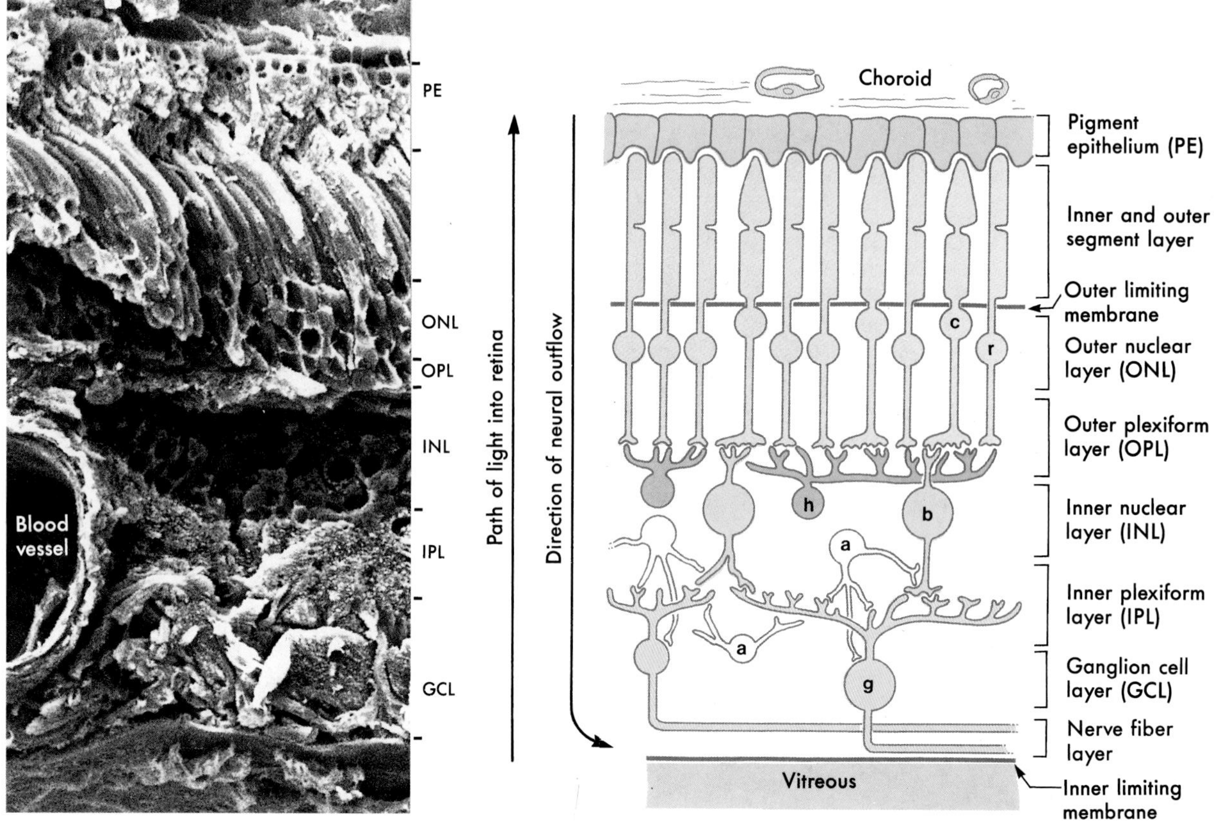

FIGURE 19-2 *Cells and layers of the retina. A scanning electron micrograph of the primate retina correlated with a schematic drawing showing the main retinal cell types. Photoreceptors (rods, r, and cones, c) are shown in green. Horizontal cells (h, gray) and bipolar cells (b, blue) receive input from photoreceptors; the bipolar cells in turn synapse onto amacrine cells (a, white) and ganglion cells (g, red).*

ond-order and some third-order retinal cells, (5) the *inner plexiform layer*, another area of synaptic contact, (6) the *ganglion cell layer* containing the cell bodies of the *ganglion cells*, and (7) the *nerve fiber layer* (or *optic fiber layer*) composed of the axons of the ganglion cells. These axons converge at the *optic disc* to form the *optic nerve*. Layers 2 through 7 are flanked by a pair of *limiting membranes*, which consist of glial cell processes joined by tight junctions. The *outer limiting membrane* is located between layers 1 and 2, and the *inner limiting membrane* is located between the nerve layer and the vitreous.

The *photoreceptor outer segments* interdigitate with the melanin-filled processes of pigment epithelial cells (Fig. 19-2). These processes are mobile, and they elongate into the pigmented layer when the light entering the eye is bright (*photopic* conditions) and retract when the light is dim (*scotopic* conditions). This mechanism combines with contractions of the iris to protect the retina from light conditions, which would otherwise damage the photoreceptors. The iris, pigment epithelium, and circuitry of the retina all contribute to the eye's ability to resolve the visual world over a wide range of light conditions.

The blood supply of the neural retina arises from branches of the *ophthalmic artery*: the *central artery of the retina* and the *ciliary arteries*. The former branches out from the optic nerve head to serve inner portions of the neural retina. The latter vessels penetrate the sclera around the exit of the optic nerve and feed the *choriocapillaris* (a portion of the choroid), which, in turn, provides nutrients to the outer portions of the neural retina.

PHOTORECEPTOR CELLS

The rods and cones of the retina are responsible for *photoreception*, the process by which photons are detected and the information is transduced into an electrochemical signal. There are two basic types of photoreceptors: *rods* and *cones* (Figs. 19-3 and 19-4). Both types have the same overall structure. Light is detected and transduced in an *outer segment* that points toward the pigment epithelium. A narrow stalk, the *cilium*, connects the outer segment to a second expanded region called the *inner segment*, which contains mitochondria and produces the energy that maintains the cell. The cilium contains nine pairs of microtubules emanating from a basal body located in the inner segment. The nucleus and perikaryon of the cell are found in the outer nuclear layer, and, finally, the cell terminates in the outer plexiform layer in an expansion that makes synaptic contacts with neurons. This synaptic

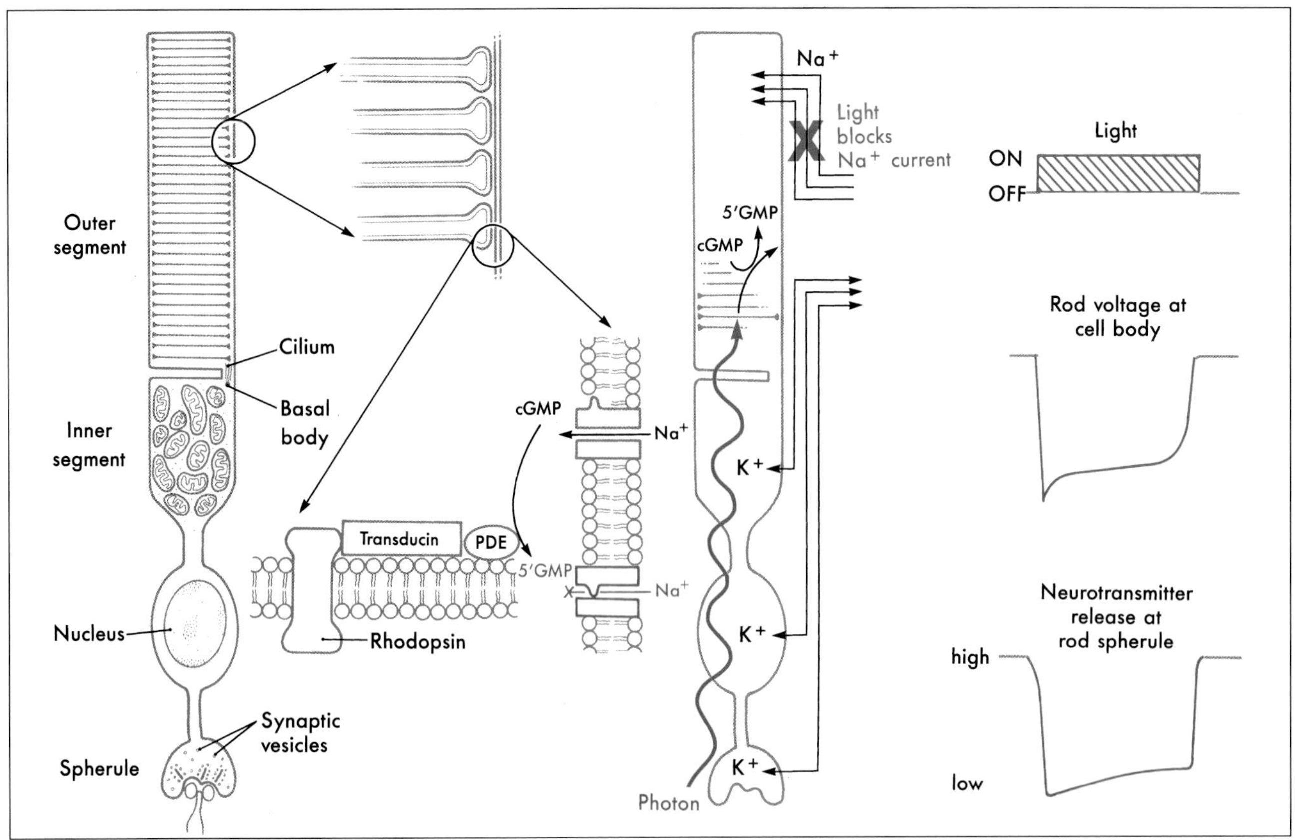

FIGURE 19-3 *The rod photoreceptor and the physiologic and chemical changes that occur in response to light. Events associated with light are shown in red.*

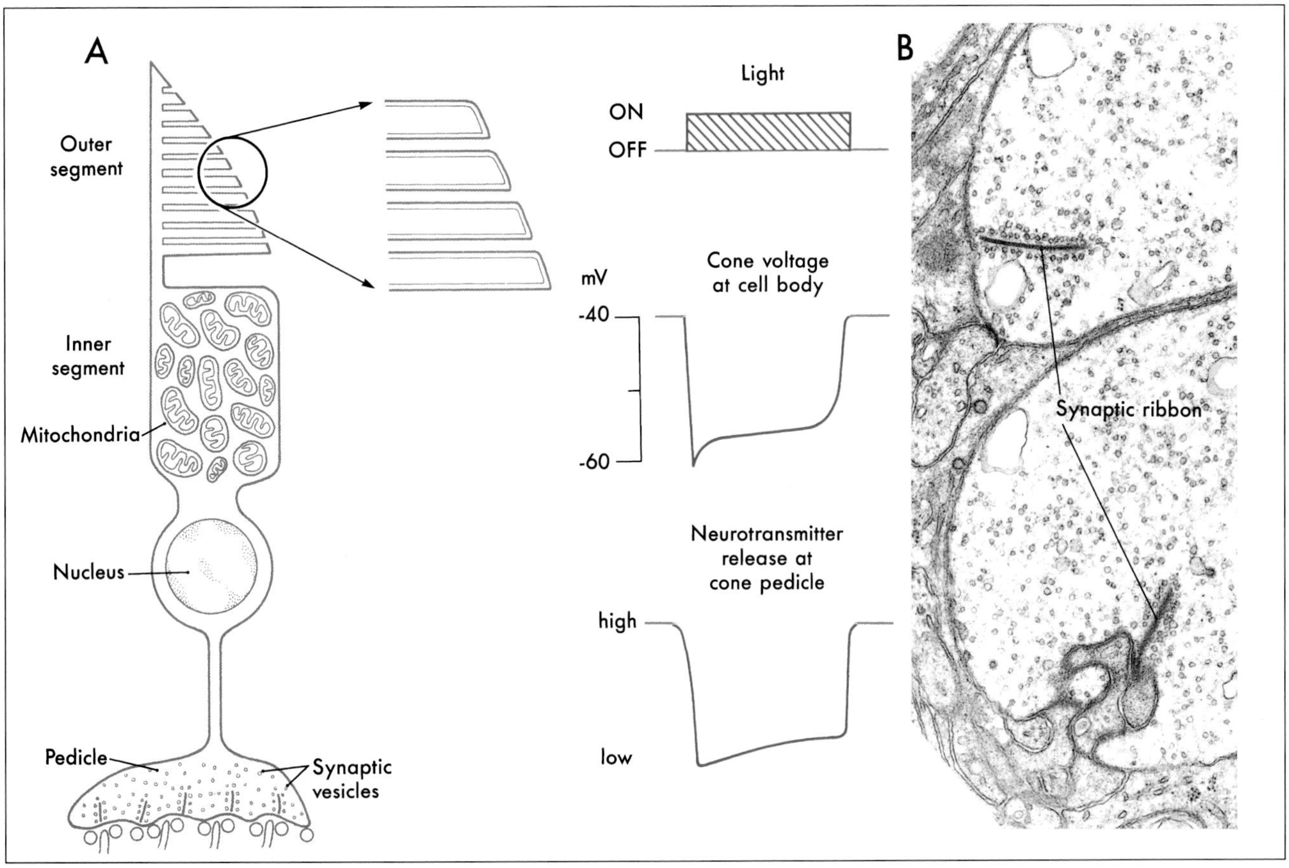

FIGURE 19-4 *The cone photoreceptor (A). Cones, like rods, reduce their levels of neurotransmitter release when stimulated by photons. Cones and rods are also distinguished by prominent electron-dense* synaptic ribbons *in their terminals (B).*

expansion is called the *spherule* in rod cells and the *pedicle* in cone cells. Both rod and cone synaptic terminals contain a characteristic dark sheet of protein called the *synaptic ribbon*. This structure may act as a "conveyor belt," organizing vesicular release of transmitter.

Rods Rod cells are named for the shape of their outer segment, which is a membrane-bound cylinder containing hundreds of tightly stacked membranous discs (Fig. 19-3).The rod outer segment is a site of *transduction* (Fig. 19-3). Photons travel through cells of the neural retina before striking the membranous discs of the rod outer segment. Molecules of *rhodopsin* within these membranes undergo a conformational change and along with tranducin and phosphodiesterase (PDE) induce biochemical changes in the rod outer segment, which reduce levels of cyclic GMP (cGMP). In the dark, cGMP levels in the rod outer segment are high. Cyclic GMP mediates a *standing sodium current*: sodium ions flow into the rod outer segment. This high resting level of sodium permeability results in a relatively high resting potential for rod cells, about –40 mV. These sodium channels of the outer segment membrane, which are normally open, close in response to increased calcium and/or a reduction in cGMP. This drives the membrane potential toward the potassium equilibrium potential, and the rod cell is *hyperpolarized* in response to a light stimulus (Fig. 19-3).

The hyperpolarization of the rod outer segment propagates passively (i.e., without firing an action potential) through the perikaryon to the rod spherule. In the absence of light, the photoreceptor terminals constantly release the transmitter *glutamate* at these synapses. The arrival of a light-induced wave of hyperpolarization causes a transient *reduction* in this tonic release of glutamate. As explained below, this event can *activate* some of the cells that receive synapses from photoreceptor terminals while *inhibiting* others.

Rhodopsin molecules are capable of a huge but finite number of photoisomerization events. Rather than replace individual rhodopsin molecules, the distal one-tenth of the outer segment is broken off and phagocytosed by the pigment epithelium. Through this process of *rod shedding*, the outer segment is constantly renewed. New discs are formed at the base of the outer segment and move outward so that the shed discs are replaced. In this way, the rod remains a constant length, and the outer segment is renewed about every 10 days.

Cones Like rod outer segments, *cone outer segments* also consist of a membranous stack (Fig. 19-4). Unlike rods, however, these stacks of cone membranes are of constantly decreasing diameter (from cilium to tip), giving the cell its characteristic shape. Also, they are not enclosed within a second membrane but are open to the extracellular space adjacent to the pigment epithelium (Fig. 19-4).

The process of transduction in cones is generally similar to that in rods. *Cone opsin* absorbs photons and undergoes a conformational change, resulting in a hyperpolarization of the cell membrane (Fig. 19-4). This hyperpolarization propagates passively to the cone's synaptic ending, the *cone pedicle*, in the outer plexiform layer. Cone pedicles are larger than rod spherules, but they also contain synaptic ribbons surrounded by vesicles (Fig. 19-4). Serial-section electron microscopy has shown that the synaptic ribbons are actually a single extensive sheet of protein. Like rods, cones release the neurotransmitter *glutamate* tonically in the dark and respond to light with a decrease in glutamate release.

There are three types of cones, each tuned to a different wavelength (Fig. 19-5). *L-cones* (red cones) are sensitive to long wavelengths, *M-cones* (green cones) to medium wavelengths, and *S-cones* (blue cones) to short wavelengths. Because any pure color represents a particular wavelength of light, each color will be represented by a unique combination of responses in the L-, M-, and S-cones.

If one of these cone types is absent because of a genetic defect in the corresponding opsin, the individual will confuse certain colors that look different to normal individuals and is said to be "*color blind*." Because the genes for the L-cone (red-absorbing) and M-cone (green-absorbing) opsins are located on the X chromosome, color blindness is more common in men. Alteration of the gene for the S-cone (blue-sensitive) pigment, which is located on an autosome, is much rarer. The inability to detect a pure red is known as *protanopia*, and inability to detect green is known as *deuteranopia*.

Macula and Fovea At the posterior pole of the eye is a yellowish spot, the *macula lutea*, the center of which is a

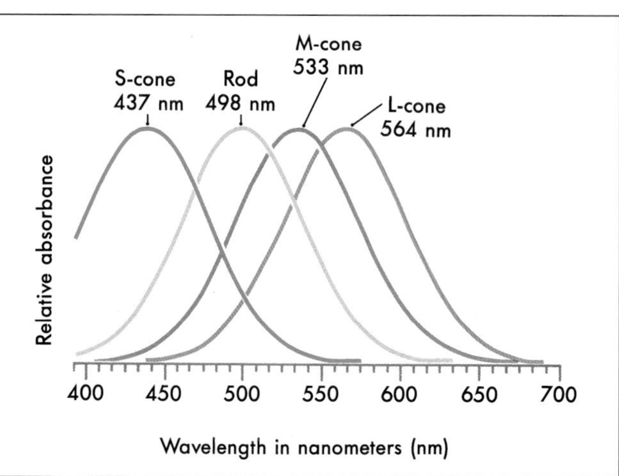

FIGURE 19-5 *Absorption spectra of rods and the three types of cones. Because the three cone spectra are different but overlapping, any wavelength of light in the visual spectrum (bottom scale) will elicit a set of response intensities in the three types of cones that is different from the set elicited by any other wavelength. Therefore, any color in the visual spectrum can be uniquely encoded. The rod spectrum is shown for comparison even though rod input is not used in color recognition. Dim red light can be used to adapt humans to maximum rod sensitivity because red light (620 to 700 nm) is not absorbed by rods to any significant extent.*

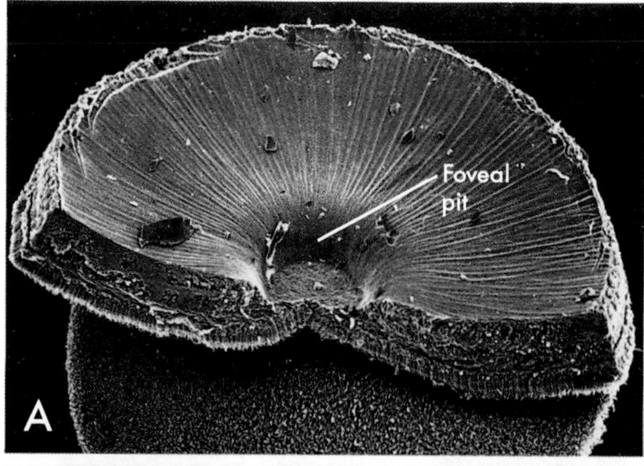

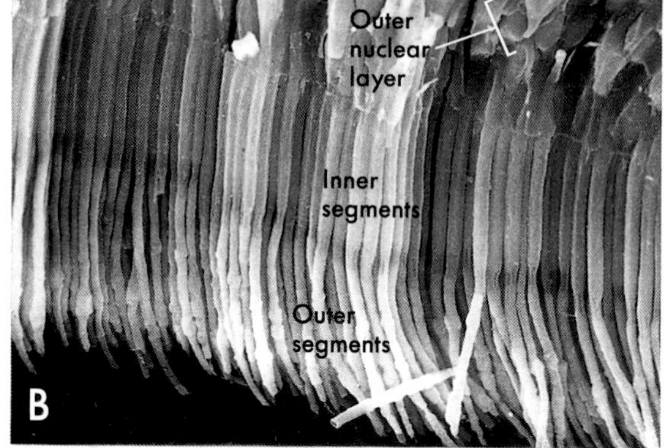

FIGURE 19-6 *Scanning electron micrographs of the primate fovea centralis (A) and of inner and outer segments of photoreceptors (B), mostly rods, in more peripheral areas of the retina. Only cones are present in the foveal pit. The surface striations (A) are ganglion cell axons en route to the optic nerve head. (Photographs courtesy of Dr. Bessie Borwein. From Borwien, 1983, with permission of Wiley-Liss, Inc.)*

depression called the *fovea centralis* (Fig. 19-6). Near the fovea, the inner retinal layers thin so that, at the bottom of the foveal pit, only the outer nuclear layer and photoreceptor outer segments remain. This allows a maximum amount of light to reach the photoreceptors with optimal fidelity.

Most of the visual input that reaches the brain comes from the fovea. Cones, which are responsible for color vision, are the only type of photoreceptor present in the fovea. In contrast, rods, which are most sensitive at low levels of illumination, are not present in the fovea but are the predominant type of photoreceptor in the periphery of the retina. The visual world is a composite formed from a succession of foveal images supplemented with information about motion contributed from the peripheral retina.

RECEPTIVE FIELDS

As with other sensory systems, the concept of a *receptive field* is key to an understanding of the visual system. Receptive fields in the visual system range from the very simple to quite complex.

The *receptive field* of any visually responsive cell is an exact location in the visual world (Fig. 19-7). Light originating in that location triggers a response in one or more retinal cells. This response can be either depolarization or hyperpolarization with a corresponding increase or decrease in the number of action potentials.

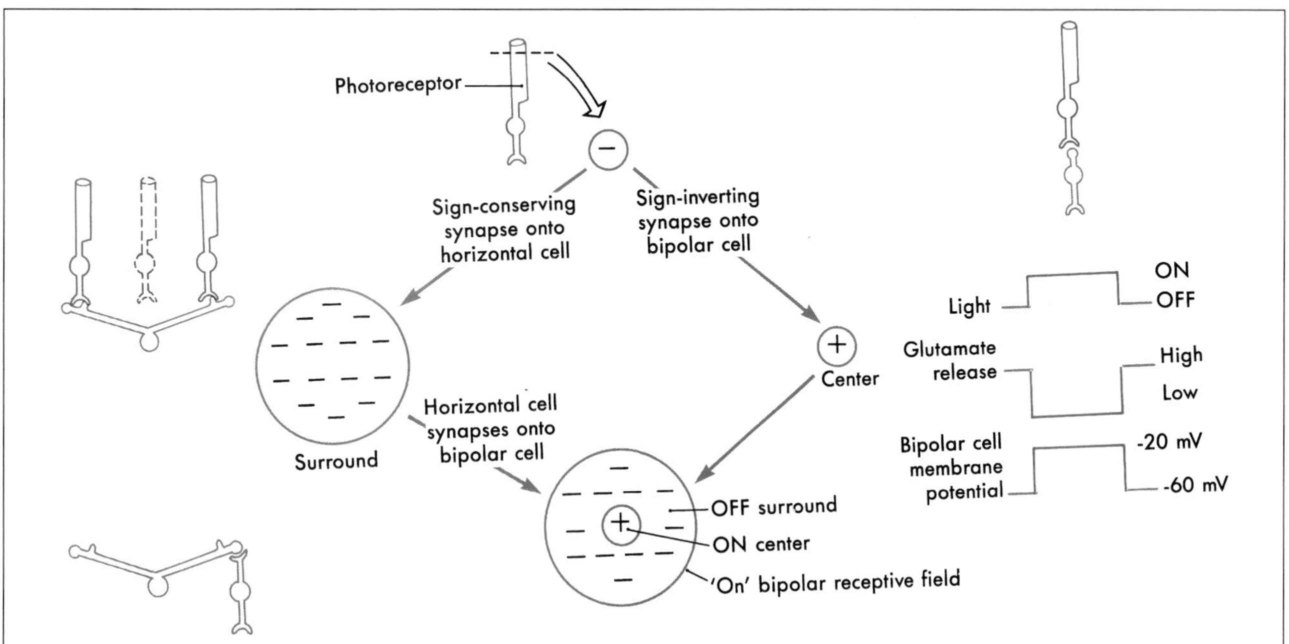

FIGURE 19-7 *How center-surround receptive fields are built in the visual system. Inputs from both receptors and horizontal cells contribute to the characteristic center-surround receptive fields of bipolar cells. A "sign-conserving" synapse is one in which depolarization in the postsynaptic cell promotes hyperpolarization in the postsynaptic cell. The example illustrated is of an "on" bipolar cell.*

In the early stages of visual information processing, receptive fields have a characteristic *concentric center-surround* organization. The receptive field is roughly circular (Fig. 19-7). Stimuli in the center of this circle tend to evoke one type of response (e.g., depolarization), whereas stimuli in the outer rim of the circle evoke the opposite response (e.g., hyperpolarization).

PROCESSING OF VISUAL INPUT IN THE RETINA

Only retinal ganglion cells have voltage-gated sodium channels on their axonal membranes. As a result, only ganglion cells use action potentials to carry information. So-called *calcium spikes* resulting from an increase in calcium permeability are seen in amacrine cells. All other retinal cells use only graded potentials to process information.

The receptive field properties of each retinal cell depend on the processing of information passing within and between the neurons insinuated between the photoreceptor and the retinal cell in question. For example, a bipolar cell's response is directly related to the activity of photoreceptors and horizontal cells (Figs. 19-7, 19-8). As with all sensory systems, the structural, electrical, and synaptic properties of the cell are reflected in receptive field properties.

Outer and Inner Plexiform Layers Synaptic contacts in the retina are concentrated into the *outer* and *inner plexiform layers* (Fig. 19-2). The outer plexiform layer contains synapses among and between retinal photoreceptors, horizontal cells, and bipolar cells. Contacts between a single *cone pedicle* or *rod spherule*, a centrally placed postsynaptic bipolar cell process, and two laterally placed horizontal cell processes form a *triad*.

The *inner plexiform layer* contains synaptic contacts among and between bipolar, amacrine, and ganglion cells. In this layer "off" and "on" bipolar cells terminate, making synaptic contact with the corresponding type ganglion cell. Amacrine cells also synapse with ganglion cells, other amacrine cells, and bipolar cells.

Horizontal Cells *Horizontal cells* consist of a cell body and its associated dendrites and an axon that courses parallel to the plane of the retina to nearby and distant photoreceptors (Fig. 19-2). These cells receive glutaminergic input from photoreceptors, and, in turn, form GABAergic synaptic contacts on adjacent rods and cones. This arrangement allows horizontal cells to sharpen the edge of a receptive field by inhibiting surrounding photoreceptor cells.

Bipolar Cells In their position between photoreceptor cells and ganglion cells, *bipolar cells* help to form a *straight-through* pathway for visual input (Fig. 19-2). *Cone bipolar cells* and *rod bipolar cells* are differentiated based on their principal synaptic inputs. Not surprisingly, cone bipolar cells predominate in the central retina, and rod bipolar cells are most common in the retinal periphery.

Bipolar cells are the comparators, or *edge-detectors*, of the retina. With the horizontal cells, they compare the activity in each region of the visual field with that in a nearby location. They are the first visual cells to exhibit the *center-surround receptive field* organization. In terms of physiologic response, there are two basic types of bipolar cells. "*On*," or *depolarizing*, cells respond to a light stimulus in the receptive field center by depolarizing, whereas "*off*," or *hyperpolarizing*, bipolar cells have the opposite center response (Figs. 19-7, 19-8).

In response to stimulation, photoreceptor cells hyperpolarize and decrease their neurotransmitter release. Consequently, *"on" bipolars* must then *hyperpolarize* in response to increased neurotransmitter release, whereas *"off" bipolars depolarize* when neurotransmitter binds to the cell's receptors. For example, as photons impinge on a cone, they cause a hyperpolarization of the cell (Fig. 19-7). Less glutamate is released onto the "off" bipolar cell. Because glutamate causes the cone bipolar to depolarize (e.g., by opening sodium channels), the *absence* of transmitter causes the opposite effect, hyperpolarization. The response of this particular bipolar cell type to stimulation in the receptive field center is hyperpolarization (the "off" response). The inverse process takes place at synapses with "on" bipolar cells. "On" and "off" bipolar cells have different types of glutamate receptor, each triggering the opening of a different ion-selective channel.

Amacrine Cells These cells have a small soma, no obvious axon, and dendrites that are few but highly branched (Fig. 19-2). Their cell bodies are usually found in the inner nuclear layer but may be displaced into the ganglion cell layer. Amacrine cells may contain two different types of transmitters, for example, GABA and acetylcholine or glycine and a neuropeptide.

Like horizontal cells, amacrine cells also have dendrites that extend over long distances, sampling and modifying bipolar cell output. Whereas horizontal cells *sense change*, amacrine cells *sense change in change*. For example, a rotating fan blade alters the activity of horizontal cells as the dark and light areas of the fan blade stimulate them, but the amacrine cell network will not change its activity. However, while the fan is speeding up or slowing down, the amacrine cell network is maximally stimulated.

Ganglion Cells *Ganglion cells* are the output cells of the retina (Fig. 19-2). Their somata form the ganglion cell layer and their axons converge on the *optic disc* and form the *optic nerve*. Ganglion cells are grouped in two ways: by size and by physiologic role. As described below, these classifications largely coincide. Like bipolar cells, ganglion cells have center-surround types of receptive fields.

The largest ganglion cells, called *alpha cells*, predominate in the peripheral retina and receive input mainly from rods. They have more extensive dendritic trees and thicker axons than other ganglion cell types. Physiologically, alpha cells correspond to the cell type called *Y* (or *M*). They participate little in color perception, in line with their largely rod input, and they show the "on" or "off" center-surround patterns of the bipolar cells with which they connect.

Medium-sized ganglion cells, *beta cells*, are found predominantly in the central retina and receive input mainly from cones. They correspond to the physiologic class *X* (or *P*). In keeping with their central location and small dendritic arbors, they have small receptive fields. They are responsive to color stimuli, and this fact gives the center-surround organization a new twist. The center will respond to one color, and the surround responds maximally to the color opposite it on a color wheel (Fig. 19-8C). For example, an X cell may have a yellow-responsive center and a blue-responsive surround. Still, the "*on-center*" and "*off-center*" categories remain.

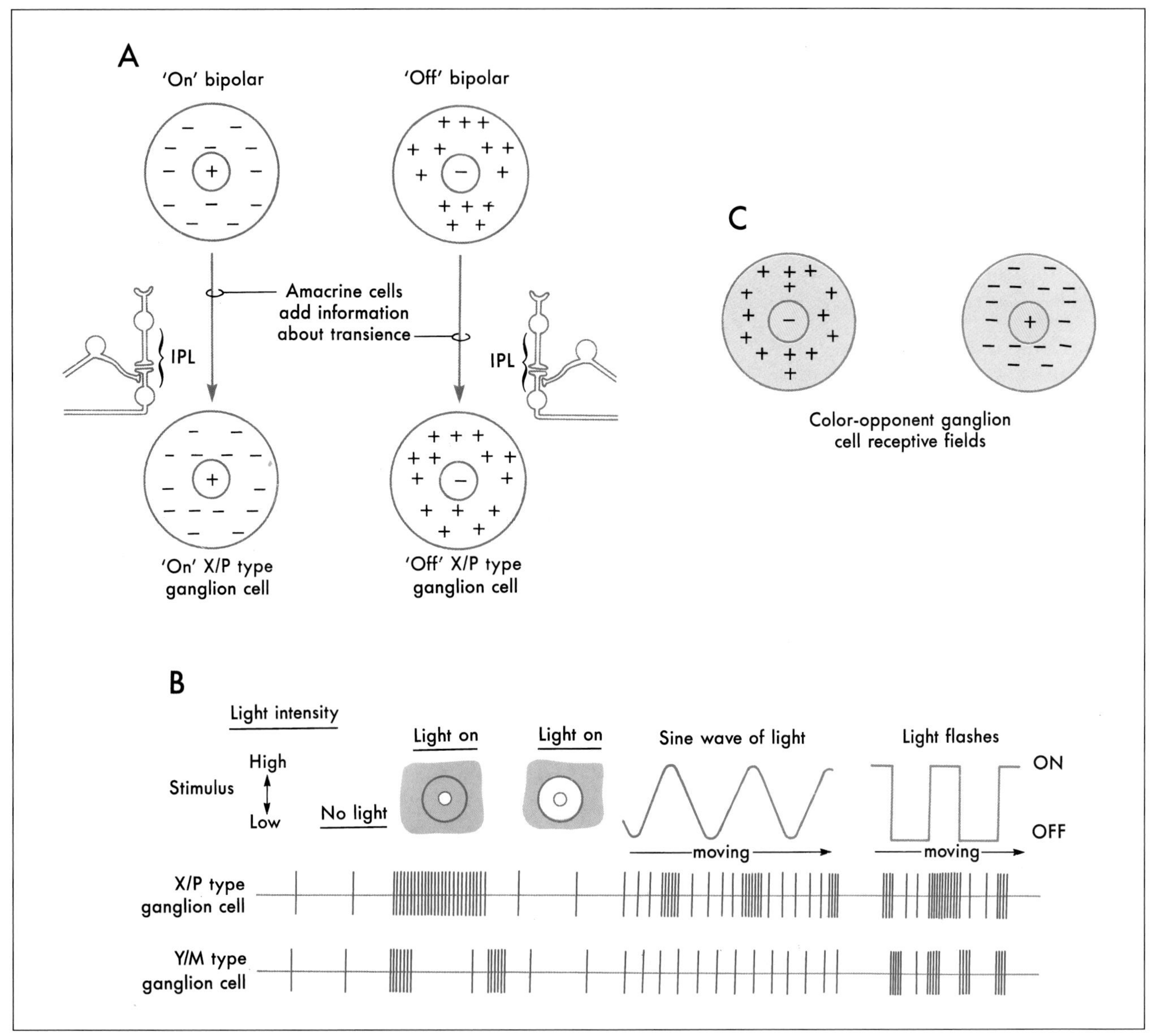

FIGURE 19-8 *How ganglion cell receptive fields are built in the visual system. Both" on"- and "off"- bipolar cells contribute to the formation of receptive fields (A) in ganglion cells. Amacrine cells add information about transience (i.e., how long since the light has changed from on to off). X- or P-type ganglion cells respond linearly to the sine wave grating; the frequency of action potentials rises and falls in synch with the sine wave of light intensity used to stimulate them (B). When light strikes both center and surround, there is no net change in ganglion cell activity in either cell type. On the other hand, Y- or M-type ganglion cells respond best to the changes between light onset and offset. X- or P-type ganglion cells are also responsive to color (C). Two examples are shown, called R^-G^+ (for red inhibitory center and green excitatory surround) and R^+G^-. There are also G^-R^+ and G^+R^-cells and blue-yellow combinations. IPL, inner plexiform layer.*

All ganglion cells left out of the preceding two categories are classified anatomically as *gamma*, *delta*, and *epsilon cells* and physiologically as *W cells*. By definition, these cells are a mixed bag; they tend to have smaller cell bodies and axons, and they show a variety of receptive field sizes and physiologic responses.

RETINAL PROJECTIONS

Retinal ganglion cells send axons to a variety of locations in the diencephalon and midbrain. Among the targets are the *suprachiasmatic nucleus*, a hypothalamic region that controls diurnal rhythms (see Chapter 29); the *accessory optic* and *olivary pretectal nuclei*, which subserve the pupillary light reflex (see Chapter 27); and the *superior colliculus*, which helps to control eye movements (see Chapter 27) and mediates so-called visual reflexes.The superior colliculus, in turn, projects to the *pulvinar*, the largest nucleus of the thalamus. The pulvinar receives input from the superior colliculus, pretectum, and visual cortex (see below) and sends information to *visual association areas*.

Retinogeniculate Projections Most retinal ganglion cells send their axons to the *lateral geniculate nucleus* by way of the optic nerve, chiasm, and tract. This connection is called the *retinogeniculate projection* (Fig. 19-9). In this pathway, an orderly map of visual space must be maintained. The receptive fields of photoreceptors (and of the ganglion cells to which they are connected) lie in a precise arrangement on the retinal surface. Adjacent points in the visual world are perceived by adjacent ganglion cells. This orderly representation of the visual world on the retina is called a *retinotopic map*.

The *visual field* is the part of the world seen by the patient with both eyes open and looking straight ahead (Fig. 19-9A). It consists of a *binocular zone*—a broad central region seen by both eyes—and right and left *monocular zones* (or *monocular crescents*) seen only by the corresponding eye. In the clinical setting, it is common to test the visual function of the two eyes separately by covering first one eye and then the other. Consequently, visual field deficits are commonly illustrated as losses from the visual field of each eye (see, for example, Fig. 19-12). Each *visual field* is divided into nasal

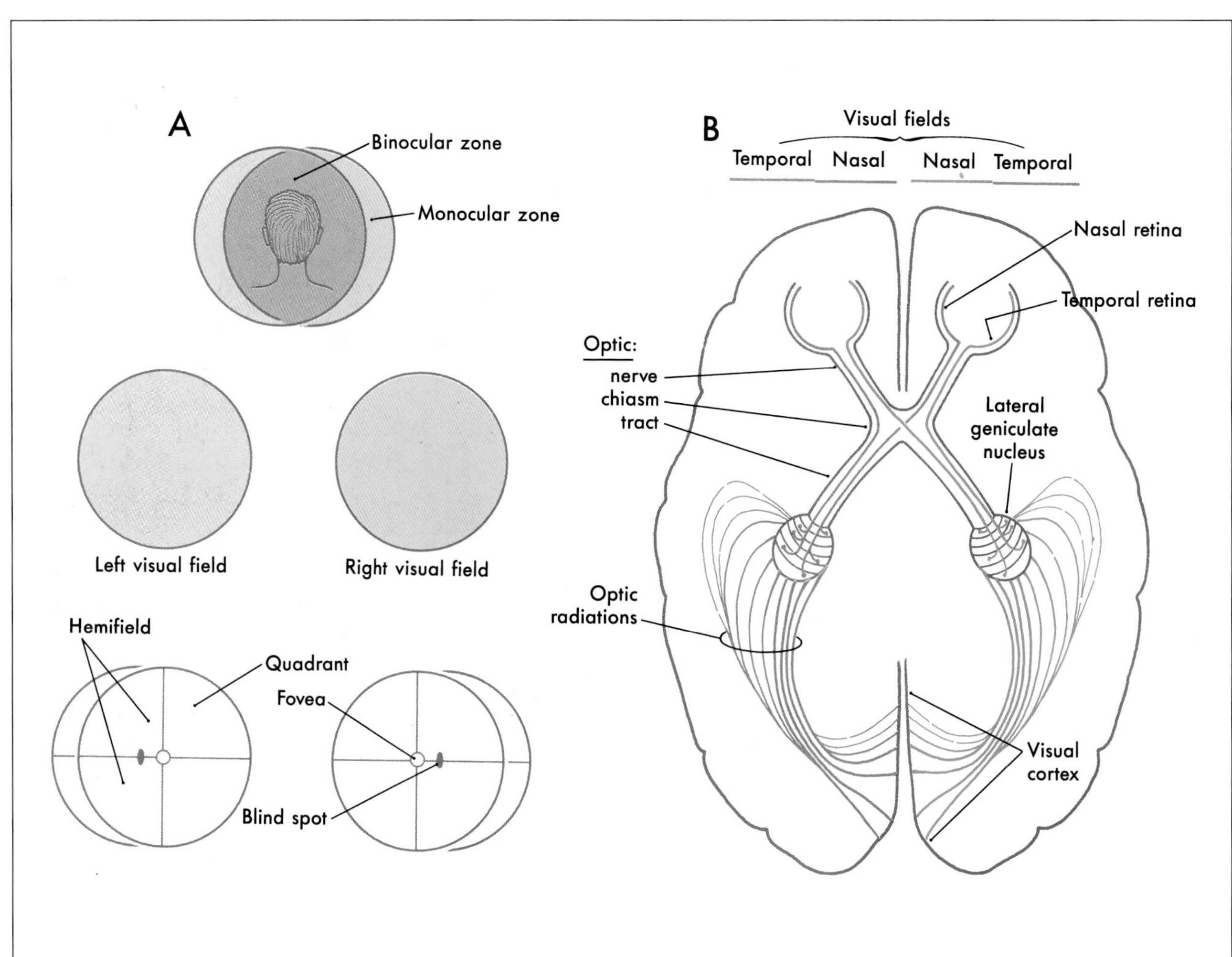

FIGURE 19-9 *Overview of the visual fields (A) and of the visual pathway as viewed from above (B).*

and temporal halves (*hemifields*), and each of these halves is divided into upper and lower parts (*quadrants*) (Fig. 19-9A). Consequently, each visual field is composed of four quadrants.

A stream of photons can be thought of as a ray of light that enters the eye. Stray light rays are blocked by the pigmented cells of the iris; only light passing through the pupil reaches the retina. The light ray is bent (*refracted*) by the cornea and lens so that the image is focused on the retina. Light from the inferior visual world strikes the superior retina. Light from the right visual world (in the binocular zone) strikes the temporal retina of the left eye and the nasal retina of the right eye (Fig. 19-9B). These patterns are essential to understanding normal vision and the defects in visual fields seen in patients with lesions in the visual pathways. The retinotopic map is maintained throughout the visual system.

Optic Nerve, Chiasm, and Tract Axons of retinal ganglion cells conveying input from all areas of the retina converge at the *optic disc* where they penetrate the choroid and sclera to form the *optic nerve*. Within the *nerve fiber layer* of the retina, ganglion cell axons are unmyelinated. However, as they pass through the sclera, they become ensheathed with myelin formed by oligodendrocytes. Because there are no photoreceptor cells in the optic disc (only ganglion cell axons), light striking this area is not perceived. Consequently, this part of the retina is commonly called the *blind spot* (Fig. 19-10A,B). Visual acuity is greatest at the fovea, but the peripheral retina has little form vision.

The *optic nerve* extends from the caudal aspect of the eye to the *optic chiasm* (Fig. 19-9B). This nerve is enclosed in a sleeve of dura and arachnoid mater that is continuous with the same layers around the brain. Thus,

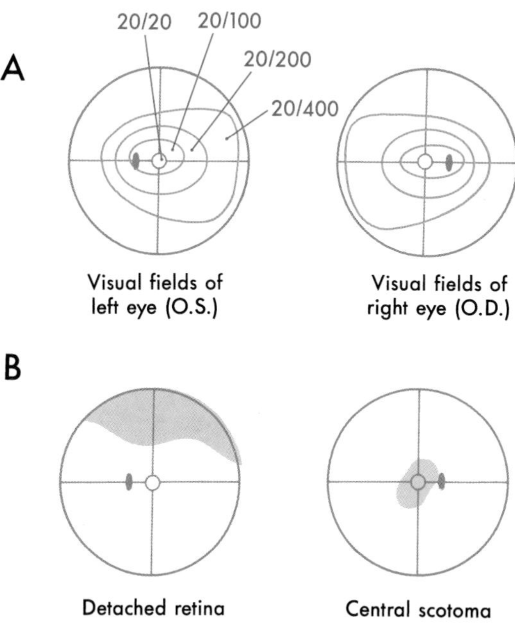

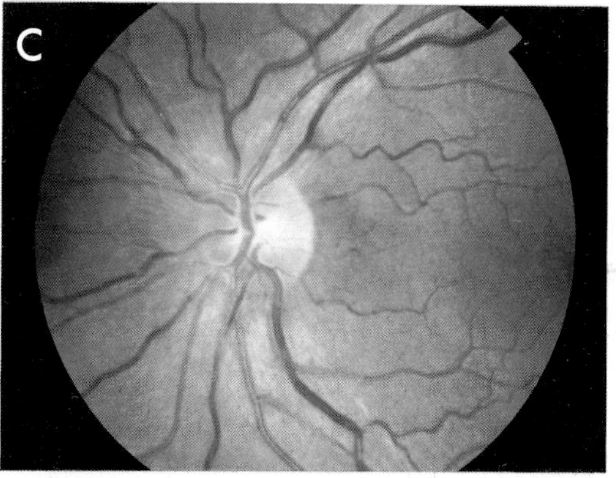

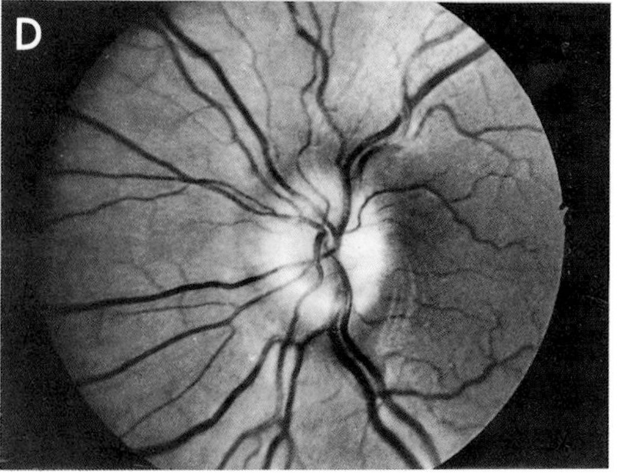

FIGURE 19-10 *Contour diagram of visual acuity on the retinal surface (A). The contour lines are isopters, lines of equal retinal sensitivity. Visual acuity is sharpest in the fovea (20/20) and drops precipitously in the outer parts of the retina (to 20/600). This drop correlates with a lower density of photoreceptors and ganglion cells in the peripheral regions of the retina. The standard abbreviations O.S. and O.D. stand for left eye (oculus sinister) and right eye (oculus dexter), respectively. Diagrams showing location of defects in the visual field (B). A detached retina in the lower part of the eye results in an irregular defect in the upper visual field (B, left), whereas an irregular lesion of the macula or compression of the optic nerve produces a central scotoma (area of reduced vision) in the center of the visual field (B, right). Ophthalmoscopic view of the fundus of a normal right eye (C). Increased intracranial pressure may produce a "choked disc" (papilledema), which is swelling of the optic nerve head visible through an ophthalmoscope (D). Blood vessels emerge from the optic disc, the light area in the center of the photograph.*

the subarachnoid space extends along the optic nerve, which is bathed in cerebrospinal fluid. For this reason, increases in intracranial pressure may be transmitted along the optic nerve(s) and can cause blockage of axoplasmic flow at the optic nerve head. This axoplasmic stasis results in swelling of the optic nerve head (*papilledema*) (Fig. 19-10D). Damage to an optic nerve results in loss of vision in that eye (Fig. 19-11A,C).

Terminal branches of the central retinal artery, a branch of the ophthalmic artery, issue from the optic disc and radiate over the retina. Examination of these vessels through an *ophthalmoscope* can help assess the health of the eye and the central nervous system (Fig. 19-10C, D). Changes in the configuration of the retinal vessels or of the size or shape of the optic disc may indicate diseases of the retina, the vascular system, or the central nervous system.

Just rostrolateral to the pituitary stalk, the optic nerves come together to form the *optic chiasm*, from which the *optic tracts* diverge as they pass caudally. In the chiasm, the fibers from the *nasal* half of each retina (corresponding to the temporal hemifields) cross to enter the contralateral optic tract, whereas the fibers from the *temporal* half of each retina (corresponding tot the nasal hemifields) remain on the same side and enter the ipsilateral optic tract. In this way, each half of the brain receives the fibers corresponding to the contralateral half of the visual world. In the most posterior portion of the optic nerve, as it joins the chiasm, some of these crossing fibers form a loop that arches through the caudal part of the contralateral optic nerve (as *Wilbrand's knee*) before joining the optic tract. This is an important morphologic point; it means that lesions of the optic nerve near the chiasm may result in visual loss in both eyes (Fig. 19-12A,C,D).

Although many clinical events can affect the optic chiasm (and consequently vision), this structure is especially susceptible to tumors of the pituitary gland. Enlarging pituitary tumors that damage the crossing fibers in the midline of the chiasm will interrupt visual input from the temporal halves of both visual fields, resulting in a *bitemporal hemianopia* (Fig. 19-12B,C,E). A lesion that damages the lateral part of the chiasm may interrupt only fibers conveying information from the nasal visual field on the same side, although in practice this situation is quite rare. This deficit is called an *ipsilateral* (either *right* or *left*) *nasal hemianopia*.

Extending caudolateral from the chiasm, the axons of retinal ganglion cells continue as a compact bundle, the *optic tract*. This structure courses over the surface of the crus cerebri at its junction with the hemisphere and ends in the lateral geniculate nucleus of the diencephalon (see Fig. 14-4). Because the optic tract contains fibers conveying visual input from the ipsilateral nasal hemifield and the contralateral temporal hemifield (Figs. 19-9, 19-13), lesions of the optic tract result in a *contralateral* (*right* or *left*) *homonymous hemianopia*.

The optic chiasm receives blood from the small *anteromedial branches* of the anterior communicating artery and A_1 segment of the anterior cerebral artery. As noted above, the optic nerve, nerve head, and retina are supplied by the central artery of the retina. The optic tract receives its main blood supply from the *anterior choroidal artery*, whereas the lateral geniculate nucleus is in the domain of the *thalamogeniculate artery*, a branch of the posterior cerebral artery (see Fig. 14-14).

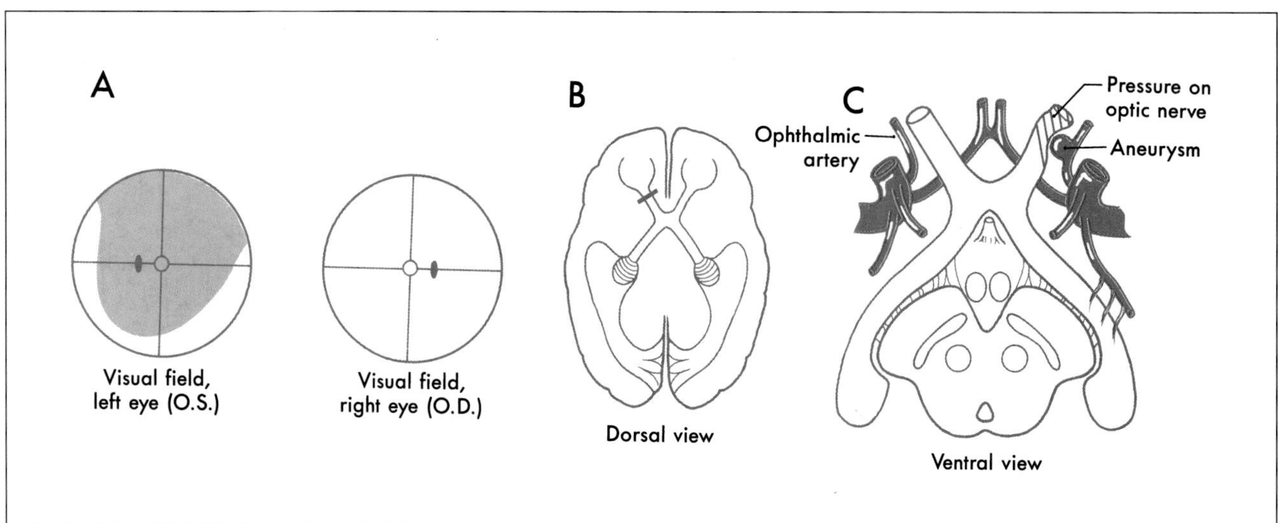

FIGURE 19-11 *A visual field deficit, such as blindness in the left eye (A), may result from a lesion of the left optic nerve (B, as seen from above). An aneurysm of the ophthalmic artery (C, seen from below) may cause damage to the optic nerve on that side.*

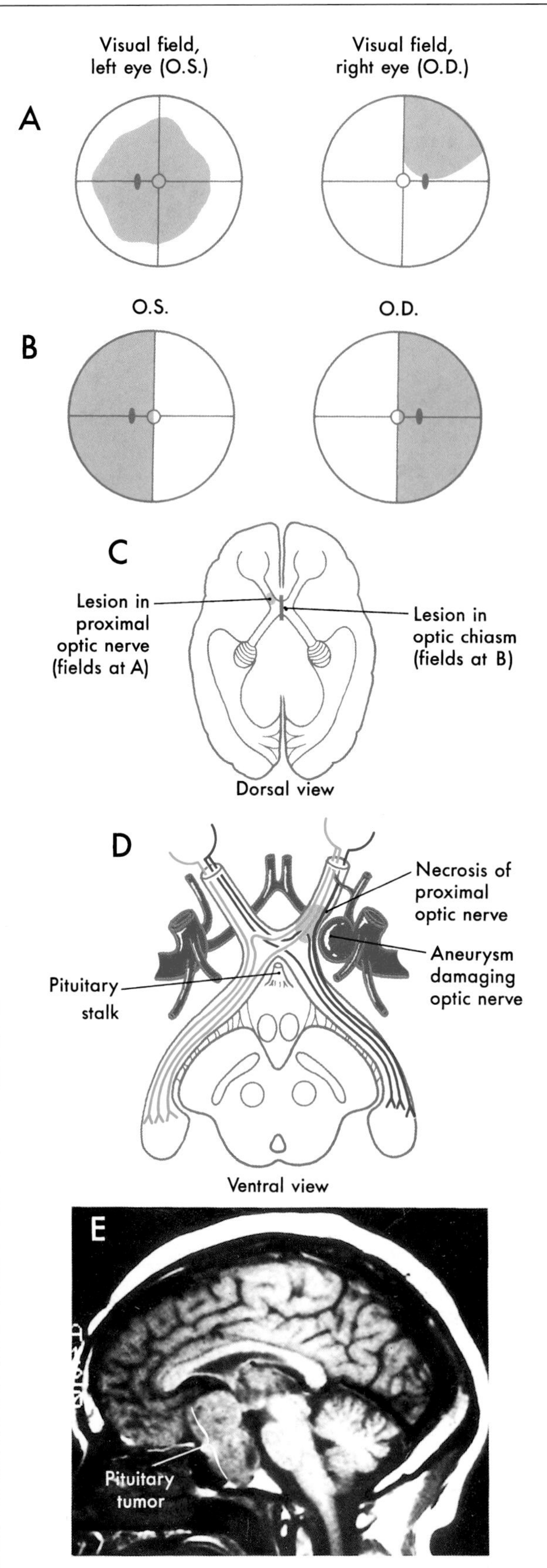

FIGURE 19-12 *Visual field deficits (A) resulting from a lesion of the left optic nerve (C, D) at its junction with the optic chiasm (a* junctional lesion*). This is caused by destruction of fibers of the left nerve plus the crossed fibers from the lower nasal hemiretina on the right (Wilbrand's knee) producing an upper temporal field deficit on the right. Visual field deficits (B, bitemporal hemianopia) resulting from damage to the crossing fibers in the optic chiasm (C). Pituitary tumors, such as the one shown in the magnetic resonance image (E), are a common cause of such deficits.*

LATERAL GENICULATE NUCLEUS

The *lateral geniculate nucleus* is located internal to an elevation on the caudoventral aspect of the diencephalon, the *lateral geniculate body* (Fig. 19-14A–C). The human lateral geniculate nucleus consists of six cellular layers with thin sheets of myelinated fibers sandwiched between them. The ventral base of this nucleus is formed by the incoming *optic tract* fibers, whereas its dorsal and lateral borders are formed by the outgoing *optic radiations*. The cell layers are numbered 1 through 6 from ventral to dorsal. As explained in the next two sections, the layers can be grouped by both the *type* of ganglion cell input they receive and the *side of the retina* from which the input originates.

Magnocellular and Parvocellular Layers Layers 1 and 2 of the lateral geniculate contain cells with large somata and are called the *magnocellular* layers. Layers 3 through 6 contain small cells and are, therefore, termed the *parvocellular* layers (Fig. 19-14C–E). The subdivision of the lateral geniculate into magnocellular and parvocellular layers correlates with the subdivision of the retinal ganglion cells into Y and X (M and P) classes. The Y (M) fibers terminate in the magnocellular layers (layers 1 and 2), whereas the X (P) fibers terminate in the parvocellular layers (layers 3 through 6)(Fig. 19-14E). (The M and P abbreviations are derived from the terms "magnocellular" and "parvocellular.") Recollect that the Y (M) ganglion cells draw their input mainly from rods and have larger receptive fields and thick, rapidly conducting axons. The X (P) ganglion cells receive input mainly from cones and have small receptive fields and slower-conducting axons; they arise mainly in the central retina and are responsible for high-acuity color vision. The ganglion cells of the remaining, mixed W class terminate on small cells scattered between the main layers.

Ipsilateral and Contralateral Layers The ganglion cell axons that arise in the *temporal* retina remain uncrossed as they pass through the chiasm and terminate in layers 2, 3, and 5 of the *ipsilateral* lateral geniculate nucleus. On

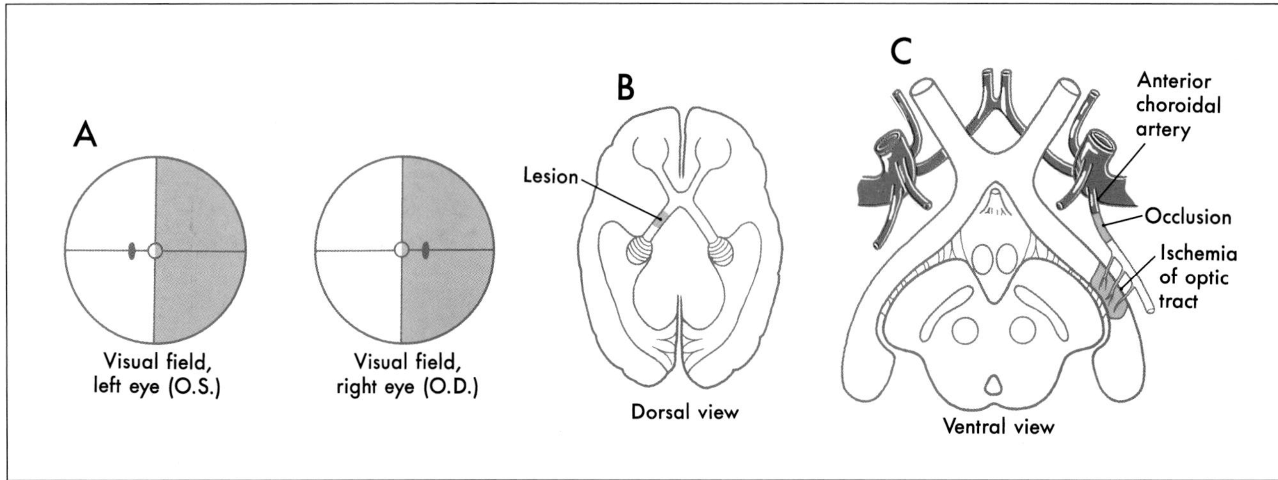

FIGURE 19-13 *Visual field deficits (A, right homonymous hemianopia) resulting from a lesion of the left optic tract (B, seen from above). Interruption in the blood supply to the optic tract (C, seen from below) may produce such deficits.*

the other hand, the axons that arise in the *nasal* retina cross in the chiasm and terminate in layers 1, 4, and 6 of the *contralateral* lateral geniculate (Fig. 19-9).

Ganglion cell axon terminals and relay cells on which they synapse are arranged so that the same point in visual space is represented six times, once for each layer of the lateral geniculate nucleus, and at the same medial-lateral point in each layer. The map progresses from the midline to the periphery in visual space as the layer runs from medial to lateral in the lateral geniculate nucleus. Layers also run rostral to caudal, representing the inferior-superior axis. Consider the position of the blind spot. About halfway through the lateral geniculate nucleus rostrocaudally is the position of the horizontal midline. The optic nerve disk lies about 15 degrees nasal to the fovea. Because there are no photoreceptors at that point on the retina, there are no relay cells in the lateral geniculate nucleus representing that location The optic nerve disc "representation" appears as a blank column extending through all six layers (Fig. 19-14C). This column lies halfway in a mediolateral direction, not one-sixth (15/90 degrees) of the way, because there are many more cells in the fovea and perifoveal region, and this part of the retina is expanded in the lateral geniculate nucleus. This enlargement of central visual areas is seen throughout the visual system.

OPTIC RADIATIONS

Relay cells forming the layers of the lateral geniculate nucleus receive input from ganglion cells (as *retinogeniculate fibers*) and send their axons to the ipsilateral *primary visual cortex* as a large bundle of myelinated fibers, the *optic radiations* (Figs. 19-9, 19-15,19-16). The primary visual cortex is located on the upper and lower banks of the *calcarine sulcus*. Consequently, the optic radiations are also called the *geniculostriate* (or *geniculocalcarine*) *pathway*.

The optic radiations can be divided into two main bundles, one serving the lower and one the upper quadrant of the contralateral hemifields (Figs. 19-15, 19-16). The fibers corresponding to the *lower* quadrant of the contralateral hemifields originate from the dorsomedial portion of the lateral geniculate nucleus, arch directly caudally to pass through the retrolenticular limb of the internal capsule, and synapse in the cortex of the superior bank of the calcarine sulcus, on the cuneus. Consequently, a lesion in the upper portion of the optic radiations results in a *contralateral* (*right* or *left*) *inferior quadrantanopia*.

The fibers corresponding to the *upper* quadrant of the contralateral hemifields originate from the ventrolateral portion of the lateral geniculate nucleus. These

FIGURE 19-14 *The lateral geniculate nucleus (LGN). The LGN has a characteristic layered structure in either nucleic acid-based stains or myelin stains (A, axial, B, coronal). Six layers can be distinguished. Note the absence of ipsilateral layers in the areas of the LGN representing the monocular crescents; only three contralateral layers are found here. A reconstruction of the human LGN is shown (C)(data from Hickey and Guillery, 1979). Note the "hole" in the LGN layers in the area representing the optic disc. Layers in the LGN are arranged so that ganglion cells receiving information from the same point in visual space innervate the LGN with axon terminals stacked one above the other (D). The LGN is not simply a relay; much information processing goes on there, as evidenced by the simplified wiring diagram (E) (data from Casagrande and Norton, 1990).*

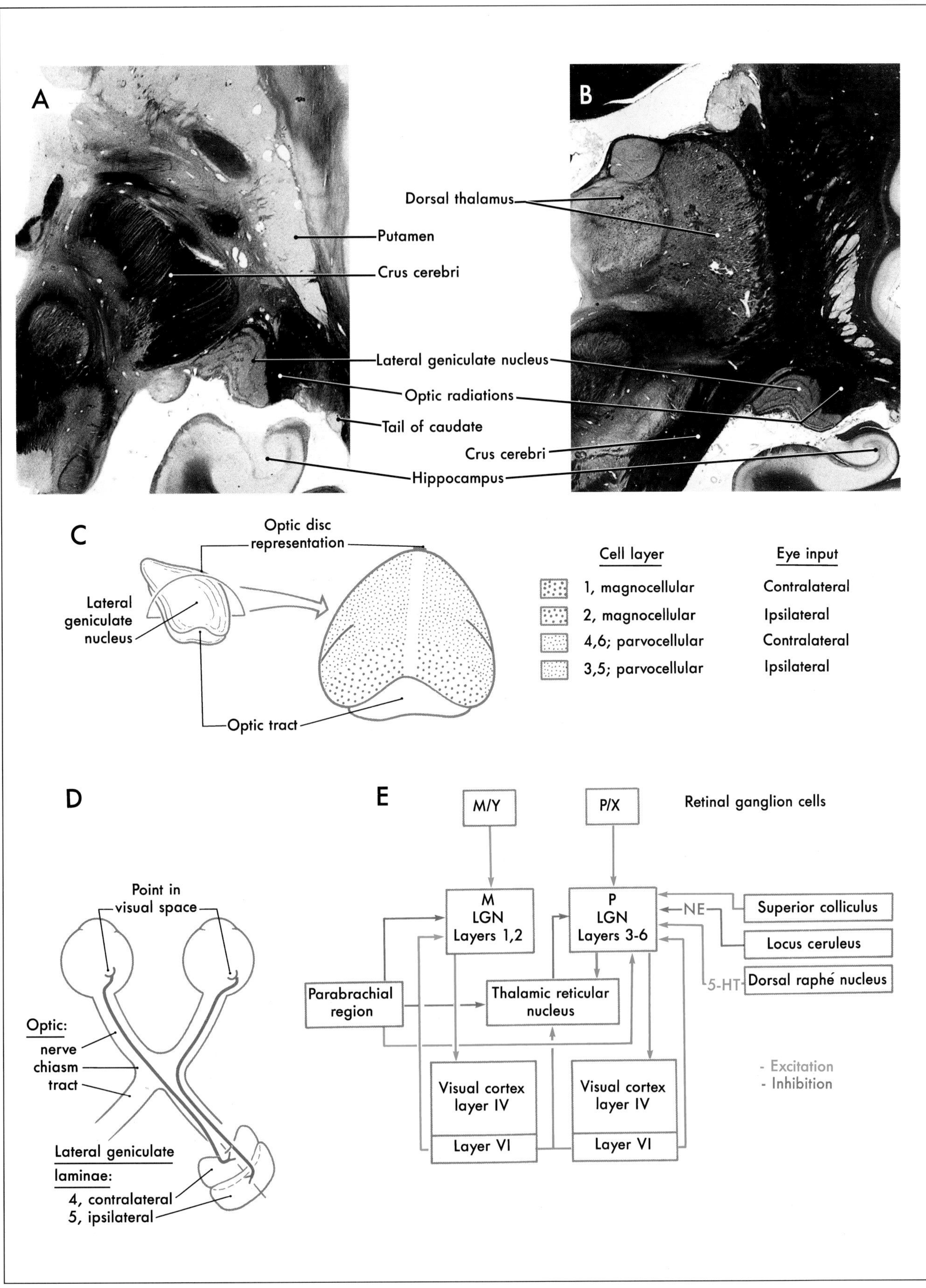
A
B
Dorsal thalamus
Putamen
Crus cerebri
Lateral geniculate nucleus
Optic radiations
Tail of caudate
Crus cerebri
Hippocampus
C
Optic disc representation
Lateral geniculate nucleus
Optic tract
Cell layer
Eye input
1, magnocellular
Contralateral
2, magnocellular
Ipsilateral
4,6; parvocellular
Contralateral
3,5; parvocellular
Ipsilateral
D
Point in visual space
Optic:
nerve
chiasm
tract
Lateral geniculate laminae:
4, contralateral
5, ipsilateral
E
M/Y
P/X
Retinal ganglion cells
M LGN Layers 1,2
P LGN Layers 3-6
Superior colliculus
NE
Locus ceruleus
5-HT
Dorsal raphé nucleus
Parabrachial region
Thalamic reticular nucleus
Visual cortex layer IV
Layer VI
Visual cortex layer IV
Layer VI
- Excitation
- Inhibition

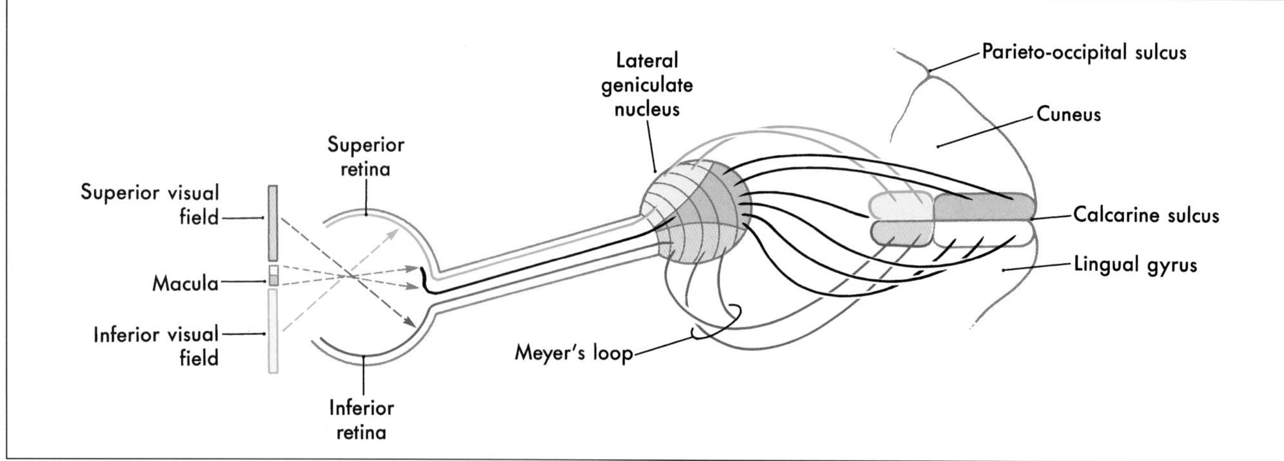

FIGURE 19-15 *The retinogeniculate and geniculostriate pathways in the sagittal plane. Input from the* superior visual field *is received by the inferior retina and is relayed to the lower bank of the calcarine cortex. Similarly, input from the* inferior visual field *reaches the upper bank of the calcarine sulcus. Note the disproportionately large representation of the macula; the central 10 degrees of visual field space occupies about one-half of the visual cortex.*

fibers do not pass directly caudal to the visual cortex. Instead they arch rostrally, passing ventral into the white matter of the temporal lobe, to form a broad U-turn (*Meyer's* or *Archambault's loop*) before passing caudally to synapse in the inferior bank of the calcarine sulcus, on the lingual gyrus (Figs. 19-15, 19-16). Damage to Meyer's loop in the temporal lobe, or to these fibers en route to the calcarine sulcus, results in a *contralateral* (*right* or *left*) *superior quadrantanopia* (Fig. 19-16). Geniculostriate fibers conveying information from the macula (and fovea) originate from central regions of the lateral geniculate nucleus and pass to caudal portions of the visual cortex.

Lesions of the optic radiations may be small and result in a *quadrantanopia*. Lesions in the optic tracts and optic radiations are largely thought of in terms of congruity and incongruity. The more anterior a lesion in the optic tract or radiations, the more likely it will be incongruous. A deficit is called *congruous* when the visual field loss of one eye is superimposable on that of the other eye.

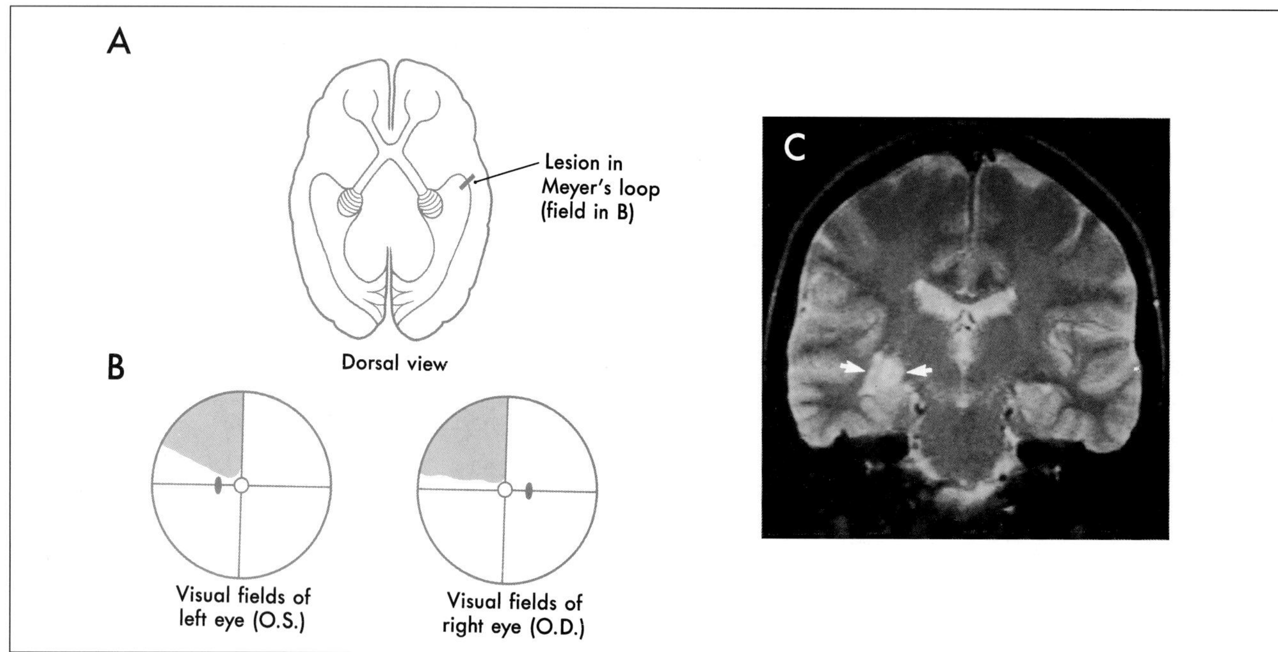

FIGURE 19-16 *Visual field deficits (B, right superior homonymous quadrantanopia) resulting from lesions in the lower part of the optic radiations (A). Such a defect can be produced by lesions in the left Meyer's loop, as shown in the magnetic resonance image (C).*

The blood supply to the optic radiations is via branches of the *middle* and *posterior cerebral arteries* that penetrate deep into the white matter. In general, the more laterally located fibers of the optic radiations and the fibers of Meyer's loop are served by branches of the middle cerebral artery. The more medially located fibers and the visual cortex receive their blood supply from the posterior cerebral artery.

PRIMARY VISUAL CORTEX

As mentioned above, the primary visual cortex, which receives most of the axons from the lateral geniculate nuclei, lies on either bank of the calcarine sulcus in the occipital lobe. It is also called *area 17*, the *striate cortex*, or *V1*. The superior bank of the calcarine sulcus, on the *cuneus*, receives input from the inferior part of the contralateral hemifields, whereas the inferior bank of the sulcus, on the *lingual gyrus*, receives input from the superior part of the hemifields (Fig. 19-15). Also, as mentioned above, the *central part* of the visual field (i.e., the macula and fovea) is represented in the portion of the primary visual cortex closest to the occipital pole, and more peripheral regions are represented more rostrally on the cuneus and lingual gyrus (Fig. 19-15). The central 10 degrees of the visual field occupies about one-half of the visual cortex.

The six-layered neocortex of area 17 is characterized by a wide layer IV. This layer contains an extra band of myelinated fibers, the *stria of Gennari* (Fig. 19-17). This structure accounts for the name *striate cortex* and is indicative of the large geniculocalcarine input to this layer. In addition, layer VI is prominent and is the source of a cortical feedback projection to the lateral geniculate nucleus. The visual cortex is organized into an elaborate array of *cortical columns* (see below), which extend perpendicularly from the pial surface to the white matter.

A large lesion of the visual cortex on one side (e.g., from occlusion of the calcarine artery) will result in a *contralateral* (*right* or *left*) *hemianopsia*. *Macular sparing* may result because caudal parts of the visual cortex can also be served by collateral branches of the middle cerebral artery.

Visual Cortical Columns In the visual cortex, the center-surround receptive field organization found at previous levels is transformed. Layer IV of the cortex receives input from the lateral geniculate nucleus. This layer contains cells that respond best to bars or edges of light rather than to spots or rings. These cells are termed simple cells (Fig. 19-18). In general, as one progresses toward the pial surface (i.e., toward layer I of cortex) or toward the white matter (i.e., toward layer VI of cortex), receptive field properties become even more complex. For this reason, some of the cells in these layers are termed complex cells. These cells respond vigorously to bars with a particular orientation. Unlike simple cells, the location within the receptive field is not important to complex cells (Fig. 19-18).

Cells lying directly above or below one another in visual cortex tend to respond to light stimuli in the same point in visual space. Thus, the *retinotopic order* seen at all levels of the visual system is preserved. However, there is an additional level of complexity: the simple cells that respond best to input from the right or left eye form narrow, parallel stripes called *ocular dominance columns* (Fig. 19-18C,D). If a special labeling method is used to mark the cells that respond best to input from only one eye, the result is a characteristic "zebra-stripe" pattern that can be seen from the external surface of the cortex (Fig. 19-18C).

Stripes called *orientation columns* cross the cortex at right angles to the ocular dominance columns (Fig. 19-18D). These orientation columns contain cells that respond best to bars or edges of light with a particular orientation.

ABNORMAL DEVELOPMENT OF VISUAL CORTEX

During development of the visual system, visually responsive cells *compete* for synaptic space on cortical cells. If both eyes receive the same visual information at the same time, this competition results in equal numbers of layer IV visual cortical cells being devoted primarily to inputs from the right or the left eye (Fig. 19-18D).

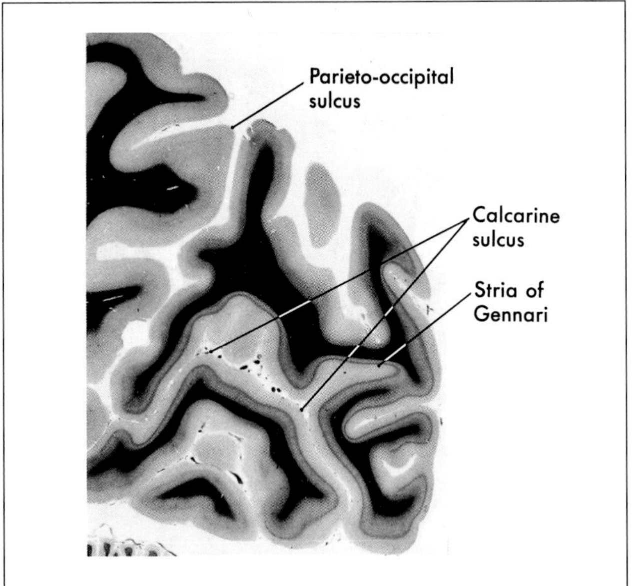

FIGURE 19-17 *The characteristic appearance of the stria of Gennari in the primary visual cortex bordering on the calcarine sulcus. This stria consists of the geniculocalcarine fibers that project to layer IV of the visual cortex.*

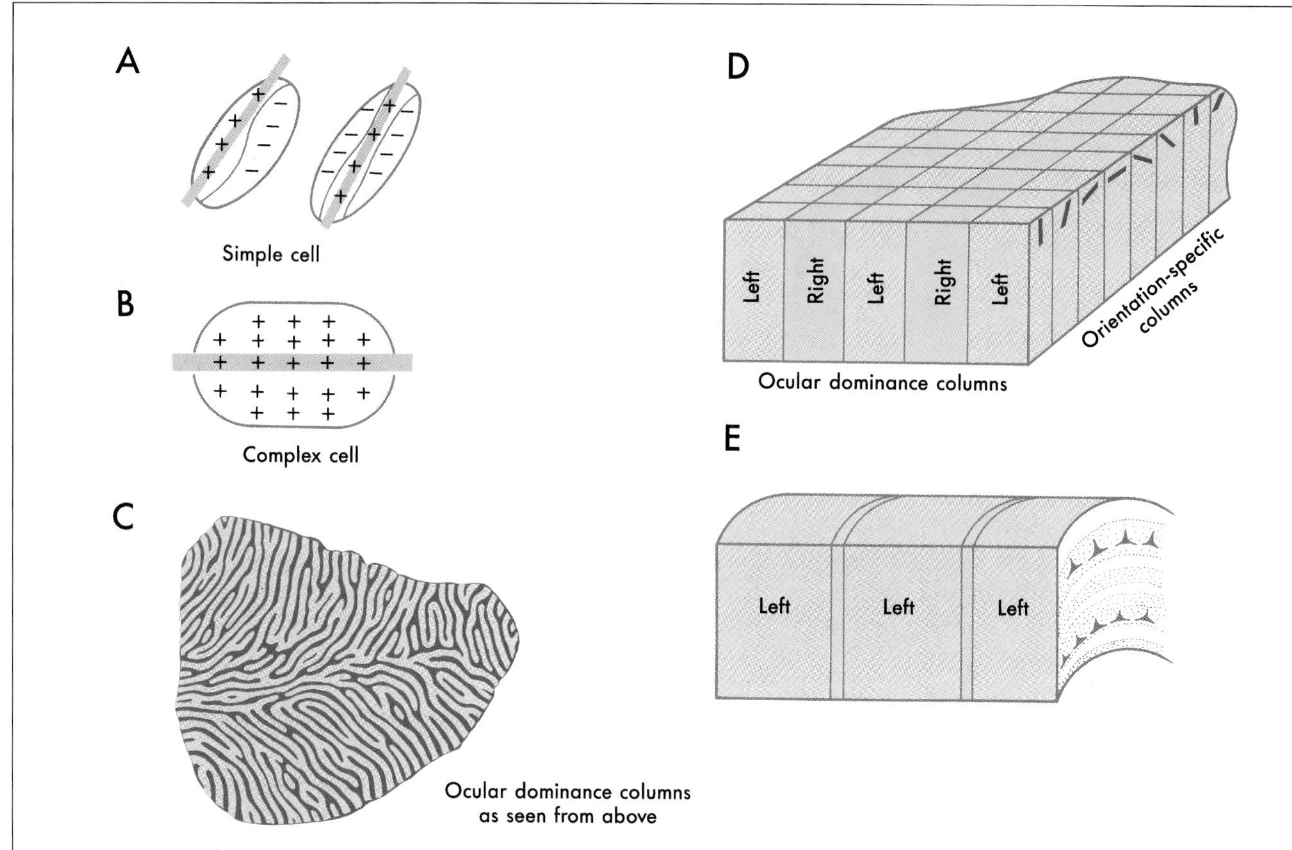

FIGURE 19-18 *The organization of visual cortex. Receptive fields of visual cortical cells differ from earlier levels of the visual system. The center-surround organization is supplanted by cells that respond best to light in a particular orientation. Two such examples, of* simple *(A) and* complex *(B) cells, are shown. Complex cells are estimated to comprise 70% of the visual cortical cells. Although cells in visual cortex respond to the contralateral visual field, they are driven more strongly by either the ipsilateral or contralateral eye. These eye-specific stripes were reconstructed by LeVay, as shown in C. Looking at a cross section of visual cortex (D), eye-specific stripes are actually columns that alternate with orientation-specific columns. In amblyopia (E), the territory occupied by one eye becomes much larger than that driven by the opposite eye, and a functional blindness occurs in the non-dominant eye. (C, Courtesy Simon LeVay. Modified from LeVay, 1975, with permission of Wiley-Liss, Inc.)*

Two problems arise if the competition for cortical territory is disrupted. First, accurate perception of depth relies on a comparison, made by visual cortical cells, between the information arising from *both* eyes for the same point in visual space. If only one eye is capable of driving cortical cells, almost all depth perception is lost. This problem occurs in about 3% of the population.

Second, there is a *critical period* for an effective competition. During the critical period, competition results in the formation or loss of synaptic contacts between axons of lateral geniculate neurons and visual cortical cells. Roughly speaking, the number and position of these synapses are then translated into the number of action potentials generated in response to a particular visual stimulus. At some point, this competition is declared closed and a victor named. The synaptic connections made during the competition phase become permanent, the lateral geniculate neurons that lost the competition are permanently shut out, and binocular vision cannot be regained. This condition is called *amblyopia*. Although the extent of the critical period in the human visual system is unknown, it probably does not extend beyond age 5 or 6 years.

Myopia (nearsightedness), *hyperopia* (farsightedness), *congenital cataracts*, or corneal abnormalities may all result in amblyopia if untreated. *Strabismus*, a deviation of one or both eyes, also may cause amblyopia; if the underlying deviation of the eyes cannot be treated, alternating a patch between both eyes prevents amblyopia.

OTHER VISUAL CORTICAL AREAS

We have seen how the visual world is broken down into elements (dots, stripes, and so forth) for efficient processing of the visual image. How the visual image is reconstructed from its component parts, so that a complete perception of visual space emerges, is unclear. Neuroscientists have jokingly proposed the existence of a "grandmother cell" or "Aunt Tillie cell" that is responsible for remembering how the face of one's grandmother appears. Others argue that such properties are the responsibility of small groups of cells, or even entire regions of the brain.

It is clear that a large part of the brain is devoted to the processing and perception of visual space (Fig. 19-19A,B). Areas 18 and 19, which surround area 17 in the occipital cortex, continue the general pattern of organization found in primary visual cortex. They receive inputs directly from area 17 and from the pulvinar.

Starting in area 18, the M and P pathways that originate in the retinal ganglion cells diverge. Up to this level, both of these pathways, or "streams," have been located in the same general region: M and P cells coexist in retina, LGN, and area 17 even though they process separate streams of information. This arrangement persists in the *V2 subregion* of area 18, but as the streams emerge from this subregion, they take different routes (Fig. 19-19C). The *M stream* proceeds to a subregion of area 18 called *V3*, then to the medial temporal area (*V5*), and finally goes to the posterior parietal area (area 7a) Remember that the information carried by this stream originates largely in rod cells and in the peripheral portions of the retinas and that the receptive fields involved are large. Appropriately, this information is used in determining *where relevant visual stimuli are located, and whether they are moving*.

The *P stream* proceeds from the V2 subregion to the *V4* subregion of Brodmann's area 19 and from there to the *inferior temporal cortex* (area 37). This stream, which originates mainly in cones and in the central area of the retina, codes for *form* and for *color* (Fig.19-19C). In fact, starting in the lateral geniculate nucleus, form and color information are carried by separate portions of the P stream. The portion that subserves form perception makes use of the small receptor fields and consequent high acuity of the P ganglion cells. The color-opponent receptive fields of these ganglion cells form the basis for color perception.

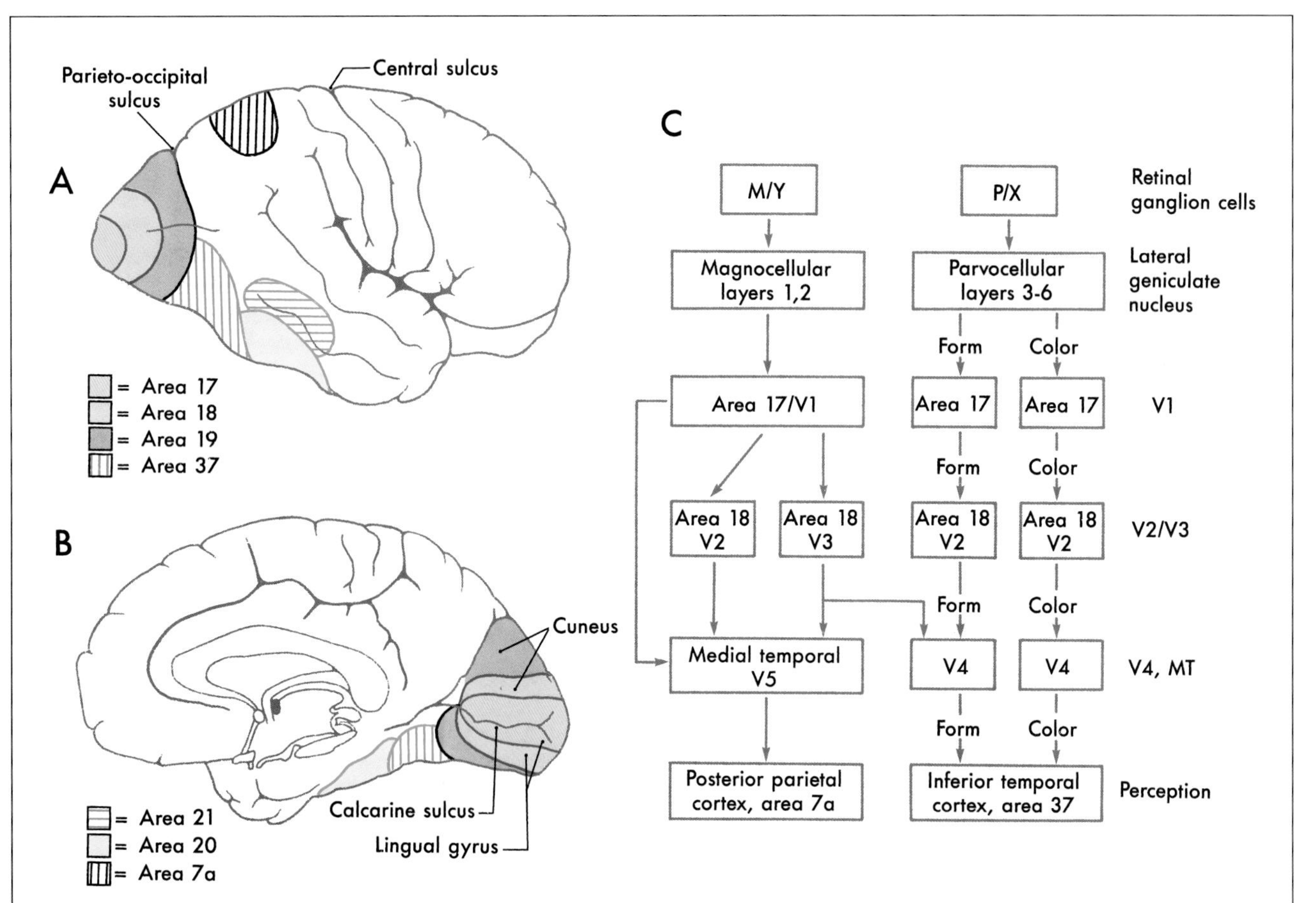

FIGURE 19-19 *Overview of the cortical processing of visual input. Lateral (A) and medial (B) views of the cerebral cortex showing the cortical areas (using Brodmann's numbers) involved in the processing of visual input. The paths through which these areas interact to create a perceived image (C). Area 18 is divided into V2 and V3 subregions on the basis of its cortical connections. The input from the magnocellular layers of the lateral geniculate nucleus, which originates in the X (P) retinal ganglion cells, is concerned mainly with detecting form and color. MT, medial temporal cortex.*

Severe deficits of *visual perception* arise from lesions of these areas. Although these patients are not blind, they are not able to correctly process information from visual space. *Agnosias* (conditions in which "percepts [are] stripped of their meaning" in the words of Lukas-Teuber) arise from lesions of areas 18, 20, and 21. In individuals with a dominant left hemisphere (the usual case), lesions in these regions on the left produce an *object agnosia*: the patient is unable to recognize real objects. Lesions in the analogous region on the nondominant (usually right) hemisphere produce agnosia for drawings of objects. Smaller lesions bilaterally can produce inability to recognize faces.

In general, inferior temporal lesions result in difficulties in perception of color or recognition of faces (*central achromatopsia* and/or *prosopagnosia*). Lesions of parietal visual processing areas result in *Balint's syndrome* (*optic ataxia*, *simultagnosia*, spasm of fixation, and visual disorientation) or *alexia* (inability to read) without *agraphia* (inability to write).

SOURCES AND ADDITIONAL READINGS

Borwein B: Scanning electron microscopy of monkey foveal photoreceptors. Anat Rec 205:363–373, 1983.

Curcio CA, Sloan KR, Kalina RE, Hendrickson AE: Human photoreceptor topography. J Comp Neurol 292:497–523, 1990.

Hubel DH: Eye, *Brain and Vision*. New York, Freeman, 1988.

Koretz JF, Handelman GH: How the human eye focuses. Sci Am 259:92–99, 1988.

LeVay S, Hubel DH, Wiesel TN: The pattern of ocular dominance columns in macaque visual cortex revealed by a reduced silver stain. J Comp Neurol 159:559– 576, 1975.

Masland RH: Functional architecture of the retina. Sci Am 255:102–111, 1986.

Stryker MP: Is grandmother an oscillation? Nature 338:297–298, 1989.

Werblin FS: The control of sensitivity in the retina. Sci Am 228:70–79, 1973.

Zeki S: *A Vision of the Brain*. Boston, Blackwell Scientific Publications, 1993.

CHAPTER TWENTY

The Auditory System

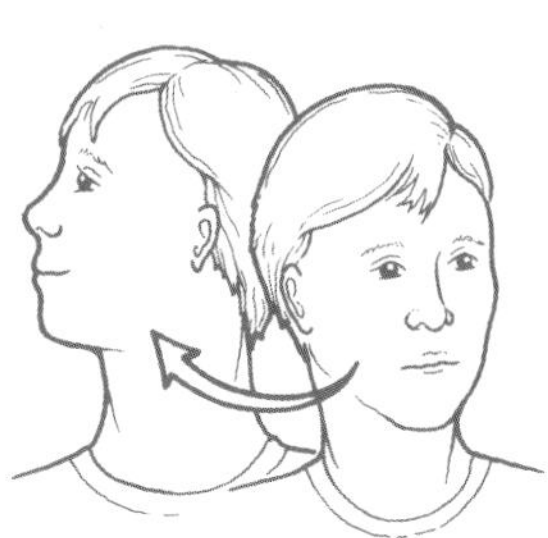

C. K. HENKEL

Hearing is one of the most important senses. In combination with vision and the ability to speak, it contributes, in a significant way, to the quality of life. In our daily routine, we unconsciously sort out meaningful sounds from background noise, localize the source of sounds, and react (many times in a reflex mode) to unexpected sounds. About 12% of the general population experiences a diminution or loss of hearing during their lifetime, which may represent a significant disability.

OVERVIEW

The auditory apparatus is adapted for receiving sound waves at the tympanic membrane and transmitting auditory signals to the central nervous system. Injury to elements of the peripheral apparatus, such as the ear ossicles, may result in *conductive deafness*. Alternatively, damage to the cochlea or the cochlear portion of the eighth cranial nerve may result in *sensorineural (nerve) deafness*. When central auditory pathways are injured, the apparent hearing dysfunction (*central deafness*) is usually combined with other signs and symptoms. Central lesions seldom result in complete deafness in one ear. To understand the neurophysiologic and audiologic methods used in assessing peripheral and central auditory disorders, it is essential to understand the structure and function of the cochlea and central auditory pathways.

PROPERTIES OF SOUND WAVES AND HEARING

Complex sounds are mixtures of pure tones that are either harmonically related, thus having *pitch*, or are randomly related and called *noise*. The cochlear apparatus is designed to analyze sounds by separating complex waveforms into their individual frequency components.

The *frequency* of audible sounds is measured in cycles per second, or hertz (Hz). A simple sine wave (Fig. 20-1) depicts the cyclic increase and decrease in the compression of air molecules that constitutes a pure tone. The time interval between two peaks is the *period*, the distance traveled is the *wavelength*, and the number of cycles per second is the *frequency*. The *intensity* is the peak to trough amplitude of force at the eardrum.

The normal frequency range for human hearing is 50 to 16,000 Hz. Most human speech takes place in the range of 100 to 8000 Hz, and the most sensitive part of the range is between 1000 and 3000 Hz. Exposure to loud noise can result in selective hearing loss for certain frequencies, and normal aging may reduce the range.

The hearing apparatus is exquisitely sensitive to sound *intensity* over an enormous *dynamic range*. Intensity of sound is related to the perception of loudness and is usually measured in units called *decibels* (dB). Intensity is also related to a measure of *sound pressure level* at the tympanic membrane. A sound that has 10 times the power of a just audible sound is said to have 20 dB sound pressure level. Normal conversational levels of sound are about 50 dB. Sounds above 120 to 130 dB elicit pain, and permanent damage to the hearing apparatus is probable for exposure to repeated sounds above 150 dB (e.g., jet engine).

The brain derives the location of a sound by computing differences in the shape, timing, and intensity of the waveforms that reach each ear. The path of the sound is affected by the distance to the ears and by obstacles such as the head (Fig. 20-1). Thus, *interaural time and intensity differences* are related to the angle between the direction in which the head is pointing and the direction of the sound source. *Interaural time differences are more important for localizing low-frequency sounds, whereas interaural intensity differences are more important for localizing high-frequency sounds.*

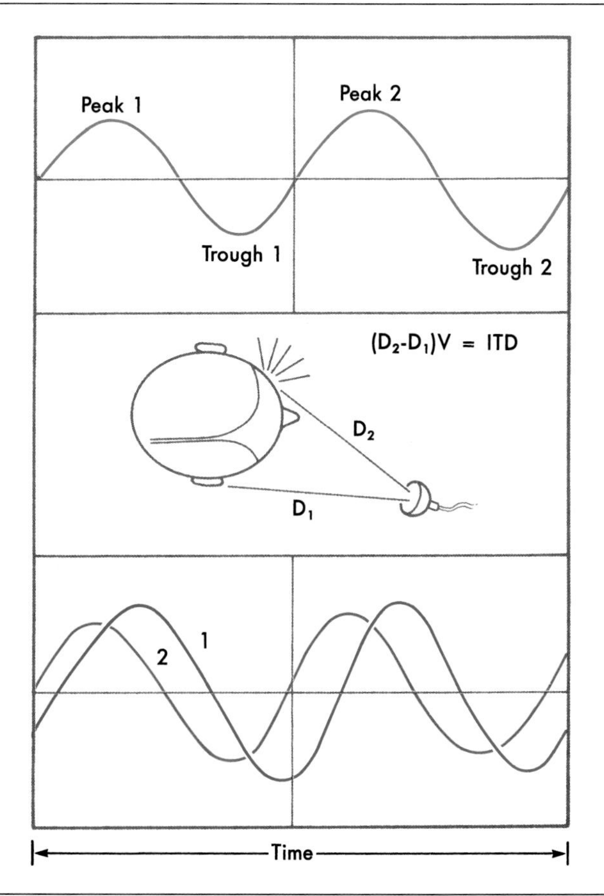

FIGURE 20-1 *Upper panel shows the cyclic changes in a simple, pure tone. Lower panel shows that the arrival of a tone at right (D_1) and left (D_2 ear is affected by the distance traveled and shadowing effect of the head (center panel) when the source of the sound is displaced from the midline. The interaural time difference (ITD) is calculated by the equation $(D_2 - D_1)v = ITD$, where v is the speed of sound.*

PROCESSING OF SOUND: THE EAR

External (Outer) Ear Sound waves are captured by the *external ear* (*pinna*) and channeled through the *external auditory meatus* to the *tympanic membrane* (Fig. 20-2A). Resonance features of the pinna and meatus enhance some frequencies more than others in a direction-dependent fashion. For example, sounds coming toward the back of the head are baffled compared to those coming toward the side of the head. *Monaural* (*single-ear*) *localization* (*one ear*) depends on such cues, and accuracy in localizing sound is impaired by damage to the pinna.

Middle Ear The *middle ear* or *tympanic cavity* is an air-filled spaced in the temporal bone that is interposed between the tympanic membrane and the inner ear structures (Fig. 20-2A) Sounds are transmitted across the space from the tympanic membrane to the fluid-filled inner ear by a chain of three bony *ossicles*: the *malleus*, *incus*, and *stapes*. On one end of this chain, the arm of the malleus is attached to the tympanic membrane, and at the other end the footplate of the stapes fits into the *oval window* of the membranous labyrinth of the inner ear. The three bones act as levers to reduce the magnitude of movements of the tympanic membrane while increasing their force at the oval window.

The mechanical stiffness of the ossicle chain acts to *compensate for the difference in impedance* between air and fluid environments (a function called *impedance matching*) so that there is optimal transfer of energy between the two media. Diseases such as *otosclerosis* and *otitis media* result in conductive hearing loss by affecting the efficiency of the ossicle movement. The stiffness of the ossicle chain can also be modified by two muscles of the middle ear, the tensor tympani and stapedius muscles (middle ear reflex).

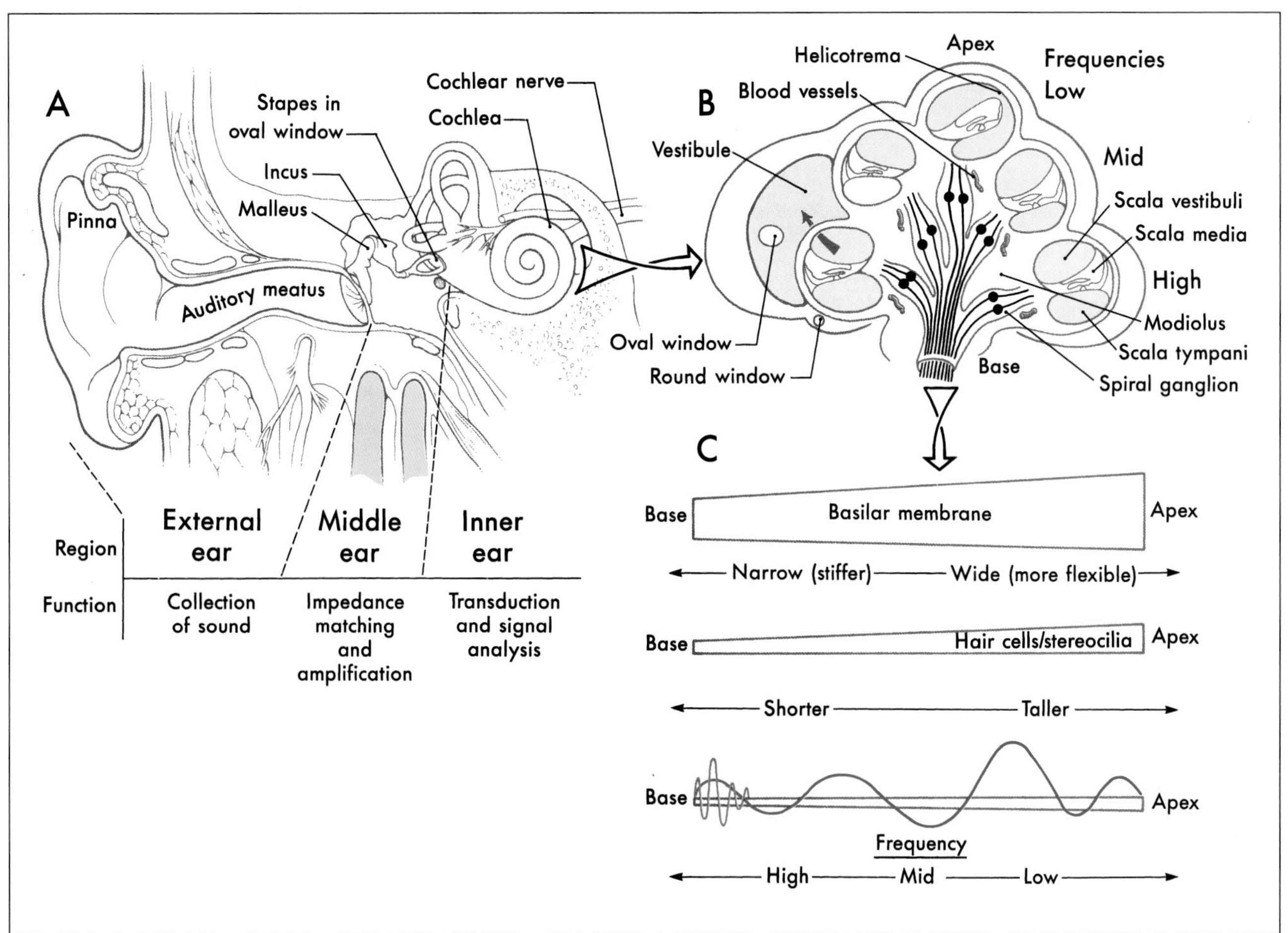

FIGURE 20-2 *The path of auditory signals from the external ear, through the middle ear, and into the inner ear (A). The cochlea (B) is shown in cross section from apex to base. The basilar membrane (C) functions to separate waves of different frequencies to sound. This membrane is narrow and stiff at its base and becomes wider and more flexible toward the apex, and the hair cell stereocilia increase correspondingly in height. These features "tune" the membrane so that each frequency of sound in the audible range will cause a wave in the basilar membrane that has its peak amplitude at a unique spot (near the base for high frequencies and near the apex for low frequencies). At this spot, the hair cells are excited most intensely, producing a peak in neural output.*

Inner Ear—Structure of the Cochlea The cochlea is named for its similarity to a conch shell (Fig. 20-2B). The *membranous cochlea*, the coiled portion of the inner ear, is encased in the osseous cochlea and consists of three spiraling chambers. The cochlea makes approximately two and two-thirds turns from base to apex. Uncoiled, it is about 34 mm long. The base of the cochlear spiral is connected to the saccule of the membranous labyrinth by the *ductus reuniens*.

The central chamber of the membranous cochlea is the *cochlear duct*, also called the *scala media* (Fig. 20-2B). Above it, the *scala vestibuli* is positioned to communicate with the vestibule, the portion of the membranous inner ear between the oval window and the cochlea. Below, the *scala tympani* ends at the *round window*, which separates this space from the middle ear cavity. In cross section, the scala media is bounded by the *basilar membrane* below, the *vestibular or Reissner's membrane* above, and the *stria vascularis* externally (Figs. 20-2B, 20-3). The screwlike bony core of the cochlea is the *modiolus*. A spiral osseous lamina extends outward from the modiolus to join the basilar membrane. The basilar membrane, in turn, is continuous laterally with the *spiral ligament*. The scala vestibuli is continuous with the scala tympani through an opening at the apex of the cochlea called the *helicotrema* (Fig. 20-2B). The scala vestibuli and tympani are filled with *perilymph*. The *endolymph*, which fills the cochlear duct, is elaborated by the cells and rich capillary bed of the *stria vascularis* (Fig. 20-3).

The *organ of Corti* is the specialized sensory epithelium resting on the basilar membrane (Fig. 20-3). It comprises inner and outer hair cells, supporting cells, and the tectorial membrane. The inner hair cells are separated from the outer hair cells by the *tunnel of Corti* (Fig. 20-3). This tunnel is formed by the filamentous arches of the *inner and outer pillar cells* and is filled with fluid.

Inner hair cells form a single line spiraling from base to apex, and the *outer hair cells* form three parallel lines that follow the same course (Fig. 20-4). In all, there are about 3500 inner and 12,000 outer hair cells in the cochlea. Projecting from the apical surface of each hair is a *hair bundle* consisting of 50 to 150 *stereocilia* arranged in curving rows (Fig. 20-4). Each hair bundle is polarized so that the longest stereocilia are on the outer border (Fig. 20-4), and the rows of stereocilia are linked by filamentous material at their tips.

The *tectorial membrane* is a gelatinous arm that extends outward over the sensory epithelium from the limbus of the osseous spiral lamina (Fig. 20-4). The taller stereocilia in each hair bundle are in contact with or embedded in the tectorial membrane. Consequently, movement of the basilar membrane and the organ of Corti will bend the stereocilia against the tectorial membrane and cause a graded depolarization of the hair cells.

The bony modiolus, around which the cochlear duct turns, houses the *spiral ganglion* (Figs. 20-2, 20-3). At the edge of the osseous spiral lamina, the peripheral processes of the bipolar cells of this ganglion lose their myelin and pass through perforations to the basilar

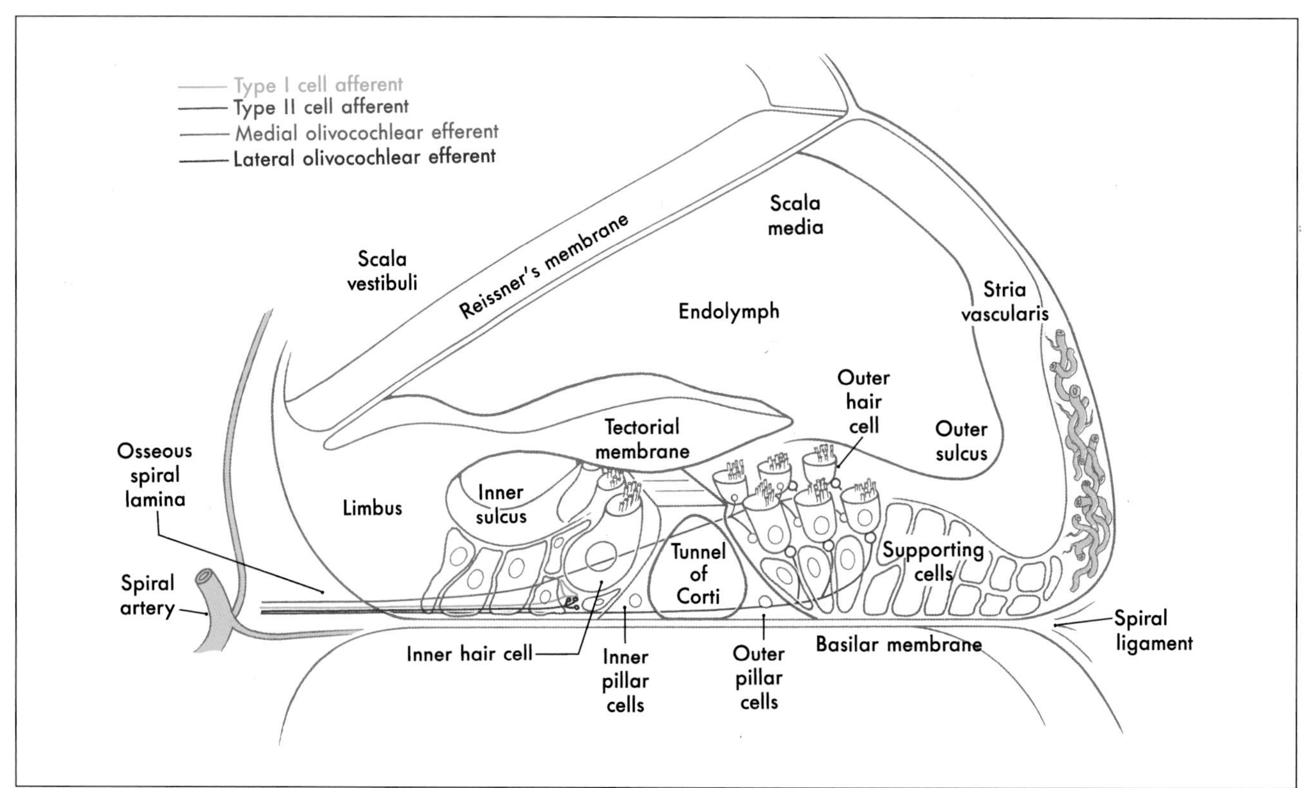

FIGURE 20-3 *Cross section through a typical turn of the membranous cochlea.*

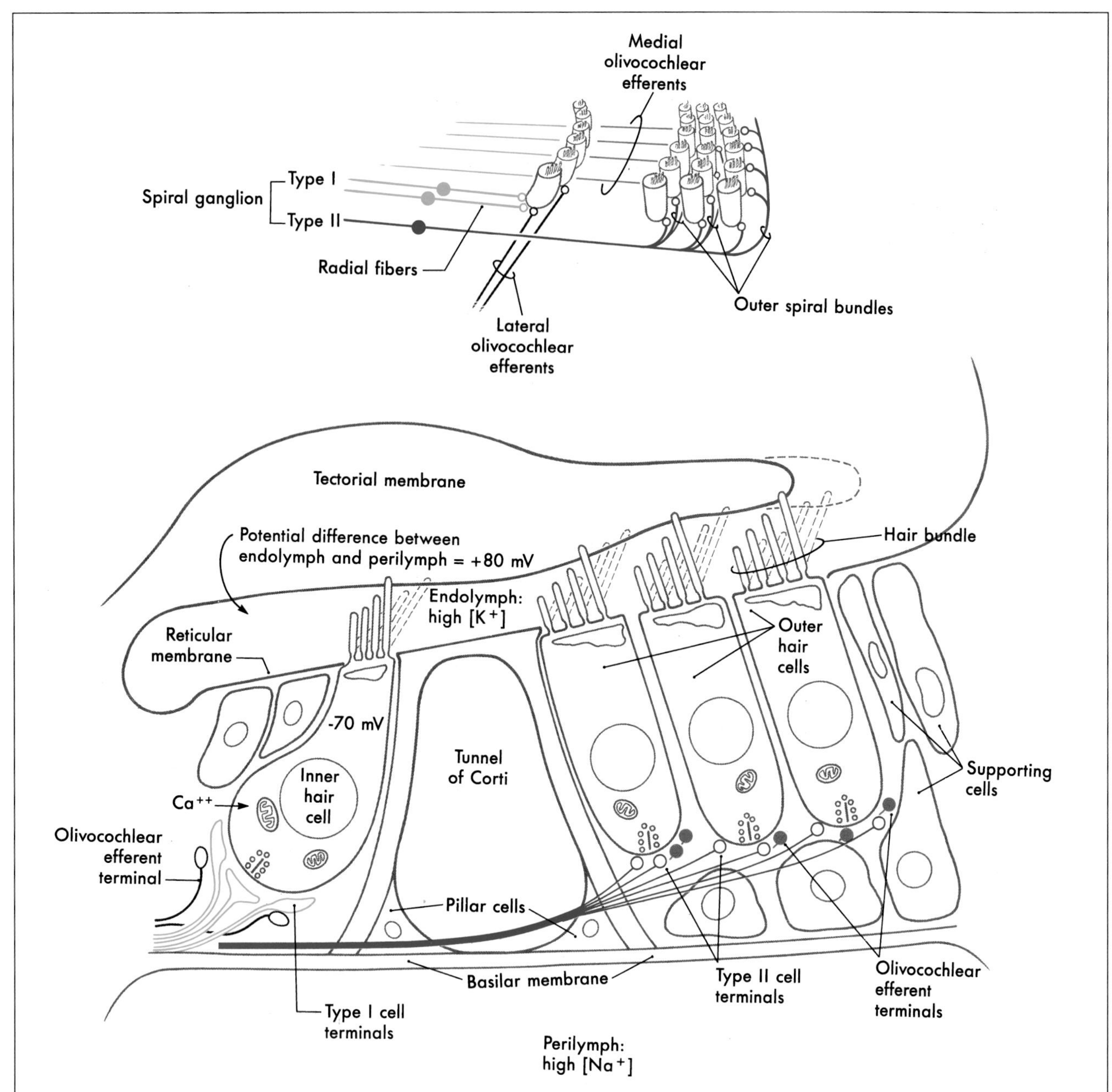

FIGURE 20-4 *The structure and function of the organ of Corti (lower) and the relation of type I and type II afferent fibers to the spiraling ranks of inner and outer hair cells (upper).*

membrane, where they synapse on the base of the inner and outer hair cells (Fig. 20-4). The central processes of the spiral ganglion cells form the *cochlear portion of the vestibulocochlear nerve* (*cranial nerve VIII*). *Efferent fibers* to the cochlea either spiral along the inner part of the basilar membrane to synapse on inner hair cells, or travel radially across the tunnel of Corti to contact outer hair cells (Fig. 20-4).

Mechanoelectrical Transduction Inner hair cells are extremely sensitive transducers that convert the mechanical force applied to the hair bundle into an electrical signal (Fig. 20-4). Endolymph, like extracellular fluid, has a high concentration of K^+. In contrast, perilymph, like cerebrospinal fluid, has a high concentration of Na^+. As indicated in Figure 20-4, the potential difference between the endolymph and the perilymph is +80 mV. This endolymphatic potential appears to be due to the selective secretion and absorption of ions by the stria vascularis. At the same time, ion pumps in the hair cell membrane produce a resting intracellular potential of about −70 mV (Fig. 20-4).

As the basilar membrane moves up in response to fluid movement in the scala tympani, the taller stereocilia are displaced against the tectorial membrane. This causes ion channels at the tips of the stereocilia to open, allowing K^+ flow along the electrical gradient to depolarize the cell

(Fig. 20-4). The large potential difference between the endolymph and the hair cell interior creates a force of 150 mV that drives K^+ into the cell and that increases the range of the cell's graded electrical response to mechanical displacement. Damage to the stria vascularis results in loss of the endolymphatic potential and failure of mechanoelectrical transduction.

When a hair cell depolarizes, voltage-gated Ca^{++} channels at the base of the cell open, and the resulting influx of Ca^{++} causes synaptic vesicles to fuse to the cell membrane and release a neurotransmitter into the synaptic cleft between the hair cell and the cochlear nerve fibers (Fig. 20-4). The transmitter causes depolarization of the afferent fiber, and an action potential is transmitted along the cochlear nerve fiber.

The stimulus-related changes in the electrical potential between the perilymph and the hair cells can be recorded anywhere in the cochlea. The potential varies synchronously with the sound stimulating the ear and is, therefore, referred to as the *cochlear microphonic*. This provides a clinically useful monitor of cochlear function.

Tuning of the Cochlea The cochlea acts as a frequency filter to separate and analyze individual frequencies from complex sounds. These tuning properties result from anatomic and physiologic characteristics of hair cells and the basilar membrane (Figs. 20-2, 20-3).

The plungerlike motion of the stapes in the oval window compresses the perilymph. In the fluid media of the cochlea, this pressure variation imparts motion to the basilar membrane, causing a wave to travel along it (Fig. 20-2C). The basilar membrane is stiffest at its base and becomes progressively more flexible toward its tip. Therefore, any given frequency of sound (pure tone) will cause a wave in the basilar membrane that has its maximum amplitude (maximum membrane displacement) at a unique point along the membrane. For high tones, this point is close to the base of the cochlea, and for lower frequencies it is more distal. The response of hair cells to the tone will be strongest at the point of greatest displacement. Therefore, the cochlea is organized to translate the frequency of sound into a place code—in other words, the *cochleotopic organization* is the basis for the *place theory of cochlear tuning*. The *tonotopic representation* along the cochlea is highly conserved through the auditory pathways.

In patients with profound sensorineural hearing loss, some audible sensation may be regained with *cochlear implants* having a number of fine wire electrodes. Each wire is tuned to a broad frequency band from an electric receiver, and the wires are implanted so that each stimulates nerve terminals at the appropriate tonotopic point along the cochlear spiral.

Primary Afferent Innervation and Function The spiral ganglion is made up of two types of bipolar sensory neurons. *Type I cells* make up 90–95% of the cells in the spiral ganglion and have radial branches that synapse with only one or two inner hair cells (Fig. 20-4). As many as 20 or more type I radial fibers converge on each inner hair cell. As a result, type I cochlear nerve fibers respond to a narrow frequency range. In contrast, *type II ganglion cells* have widely distributed peripheral processes, which traverse the tunnel of Corti and synapse with over 10 outer hair cells (Fig. 20-4). Thus, type II cochlear fibers are more sensitive to low-intensity sounds than are type I

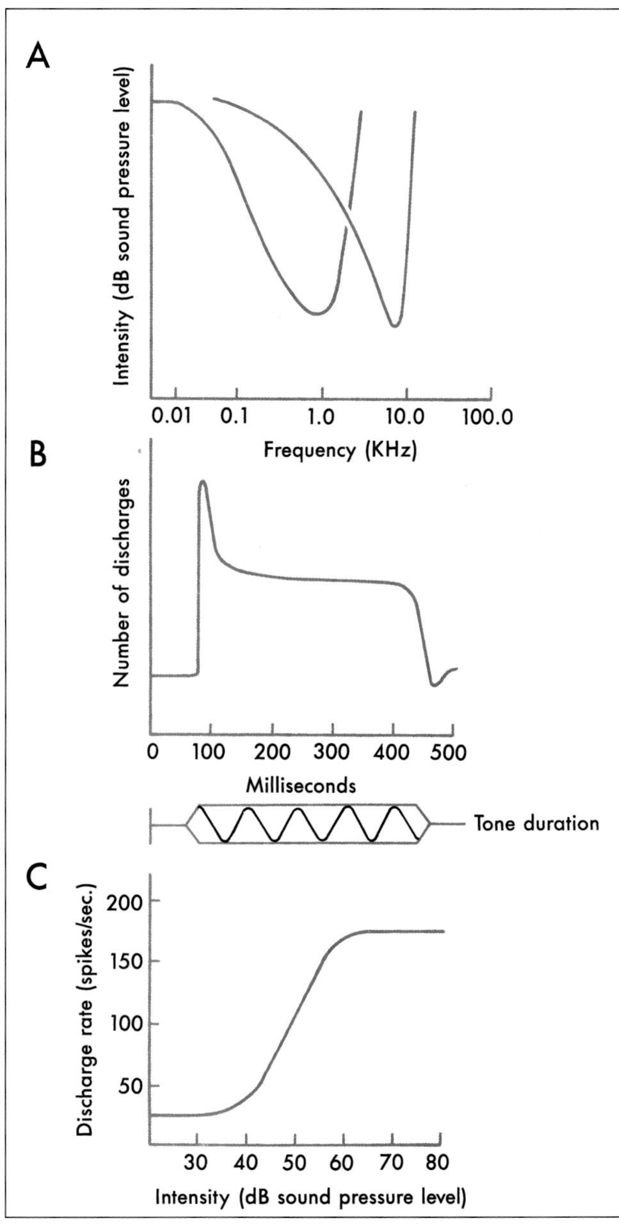

FIGURE 20-5 *Typical response characteristics of type I cochlear afferent fibers. Frequency tuning curves (A), post-stimulus time histogram of discharges through the duration of a tone burst at the characteristic frequency of a primary afferent fiber (B), and rate-intensity curve illustrating how a limited range of intensity can be coded in the response of primary afferent fibers to increasing intensity of a tone at its characteristic frequency (C).*

cells, but they may be less precisely tuned to frequency.

Frequency is coded in the cochlear nerve by the position of afferent fibers along the cochlear spiral. For loud sounds, each afferent fiber responds over a range of frequencies. As the intensity of the sound drops to near threshold, the frequency response range narrows. A *tuning curve* can be constructed that plots the threshold intensity for each frequency that will elicit a response (Fig. 20-5A). The *characteristic frequency* is the frequency at which the fiber has the lowest threshold. The discharge pattern of primary afferents over time to pure tone bursts is shown with poststimulus histograms of the number of action potentials summed over many presentations (Fig. 20-5B). Stimulus onset produces an initial high-frequency discharge followed by a lower sustained discharge level that is related to stimulus intensity. When the tone ends, the fiber drops back to a low, spontaneous discharge rate. For low-frequency fibers, the timing of each impulse is *phase-locked* with the stimulus cycle, so that the fiber output preserves the timing information of the signal.

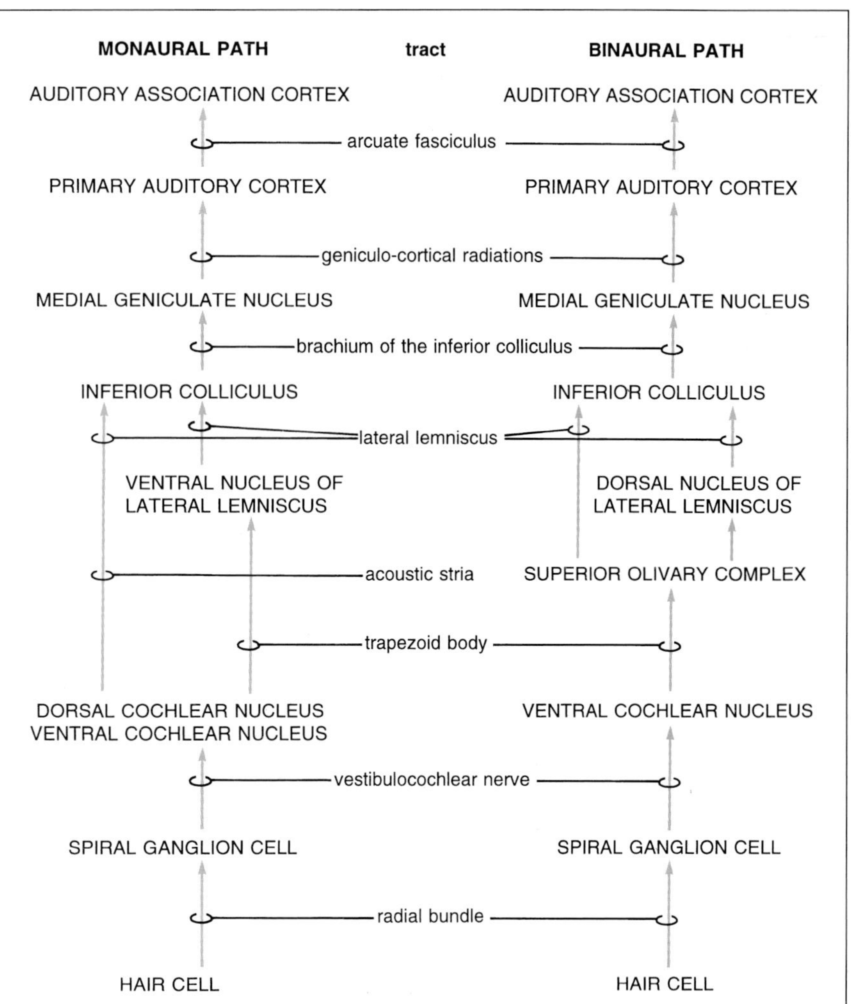

FIGURE 20-6 *Hierarchical order of central auditory pathways.*

Intensity is coded both by the discharge rate of cochlear nerve fibers and by recruitment of activity in additional afferents as stimulus intensity increases. The discharge rate for the cochlear nerve fibers increases proportionally with intensity over a range of about 40 dB sound pressure level and then plateaus (Fig. 20-5C). This limited response range cannot alone account for the large dynamic range in human hearing. At higher stimulus intensities, additional cochlear nerve fibers having sequentially greater thresholds are recruited. Therefore, stimulus intensity influences the response of individual fibers and the population of fibers responding.

AN OVERVIEW OF CENTRAL AUDITORY PATHWAYS

In the major ascending auditory connections from cochlea to cortex, *the place code of the cochlea is, as a rule, strictly maintained* (Fig. 20-6). Within this tonotopic framework, projections connect similar frequency regions of successive nuclei. Information processing is, therefore, hierarchical with increasing complexity of feature extraction.

All fibers in the cochlear nerve synapse in the *cochlear nuclei*. As cochlear information ascends to the auditory cortex, it passes through fiber bundles and is processed at synaptic sites in the brainstem and thalamus (Fig. 20-7; see also Fig. 20-10). The fiber bundles include the *trapezoid body*, *acoustic stria*, *lateral lemniscus*, and *brachium of the inferior colliculus*. Cell groups primarily involved in the relay of auditory signals are the *nuclei* of the *superior olivary complex* and *trapezoid body*, the *lateral lemniscus*, and the *inferior colliculus*. The *medial geniculate nucleus* receives auditory signals from the inferior colliculus and, in turn, projects to the *auditory cortex* via the sublenticular limb of the internal capsule.

Although fibers conveying auditory input decussate at several levels, this information is routed in one of two orderly ways: (1) *Monaural information* (information from each ear individually) *is routed to the contralateral side* and (2) *binaural information* (information from both ears at once) *is handled by central pathways that receive, compare, and transmit this input.* Binaural pathways perform the neural computation needed to localize brief sounds.

Unilateral damage to the cochlear nerve or cochlear nucleus results in monaural deafness. In contrast, unilat-

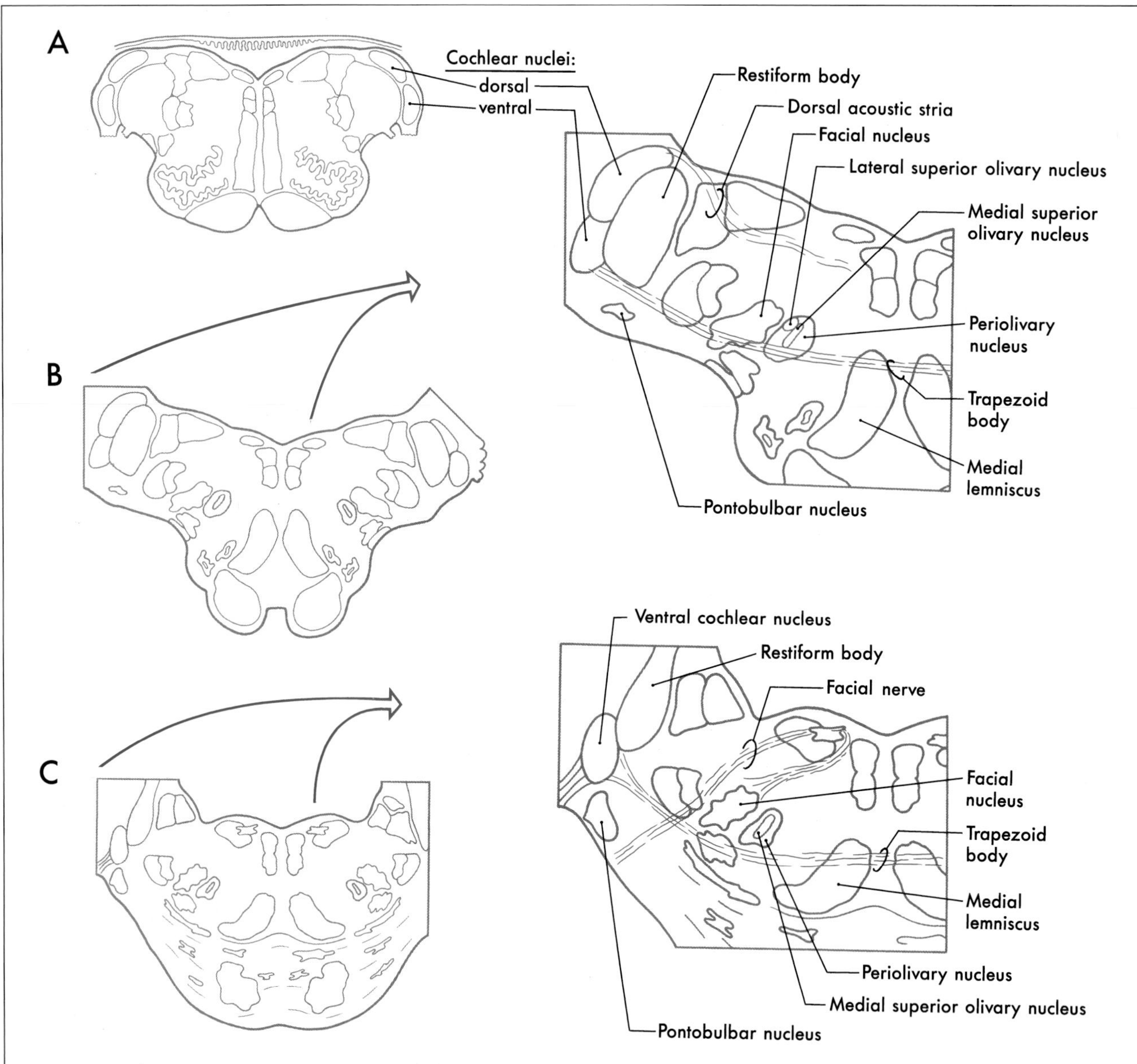

FIGURE 20-7 *Levels of the rostral medulla (A,B) and caudal pons (C) illustrating the relations of the cochlear nuclei and superior olivary complex.*

eral damage at or above the superior olivary complex leaves some binaural routes intact, so monaural deafness is not observed. The auditory decussations, particularly the trapezoid body, are functionally similar to the optic chiasm in the visual system and have been collectively referred to as a *functional acoustic chiasm*. Thus, central hearing dysfunction may result in inattention to stimuli on the contralateral side.

Vascular Supply of the Auditory Brainstem and Cortex The blood supply to the cochlea and the auditory nuclei of the pons and medulla originates from the *basilar artery*. The *internal auditory* (*labyrinthine*) *artery*, usually a branch of the *anterior inferior cerebellar artery* (AICA), supplies the inner ear and the cochlear nuclei. Occlusion of AICA will result in a monaural hearing loss. This lesion may also damage the emerging fibers of the facial nerve and the pontine gaze center, resulting in monaural deafness combined with ipsilateral facial paralysis and an inability to look toward the size of the lesion.

Vascular lesions higher in the ascending auditory system necessarily interrupt pathways conveying information from both ears. The superior olivary complex and lateral lemniscus are mainly supplied by *short circumferential branches of the basilar artery*. The *superior cerebellar* and *quadrigeminal arteries* supply the inferior colliculus, and the medial geniculate bodies lie in the vascular territory of the *thalamogeniculate arteries*. The blood supply to the primary auditory and association cortices is via branches of the M_2 segment of the *middle cerebral artery*.

BRAINSTEM AUDITORY NUCLEI AND PATHWAYS

Cochlear Nuclei The *dorsal* and *ventral cochlear nuclei* are located lateral and dorsal to the restiform body and are partially on the surface of the brainstem at the pontomedullary junction (Fig. 20-7A). The dorsal cochlear nucleus drapes over the restiform body just inferior to the pontomedullary junction. At this level the posterior part of the ventral cochlear nucleus is small in proportion to the dorsal cochlear nucleus (Fig. 20-7A,B). The ventral cochlear nucleus extends rostral to the dorsal cochlear nucleus (Fig. 20-7C), where it may be covered by the flocculus and by caudal fascicles of the middle cerebellar peduncle.

All cochlear nerve fibers end in the cochlear nuclei on the ipsilateral side (Fig. 20-8A–C). As these fibers enter the brainstem at the cerebellopontine angle, they divide into ascending and descending bundles. Fibers in the ascending bundle synapse in the anterior part of the ventral cochlear nucleus, whereas fibers in the descending bundle synapse in the posterior part of the ventral cochlear nucleus and in the dorsal cochlear nucleus.

In the cochlear nuclei, each afferent nerve fiber bifurcates into two branches, which make synaptic contacts in several portions of the nuclei. These contacts lie along a line that represents the point on the cochlear spiral from which the fiber originated (Fig. 20-8A,C). This point-to-line distribution produces tonotopic maps in the anterior and posterior parts of the ventral cochlear nucleus and in the dorsal cochlear nucleus. The frequency-related lines are organized so that low frequencies are represented laterally and high frequencies medially (Fig. 20-8C).

Ascending projections from individual cell types in the cochlear nuclei form multiple parallel channels for information processing. These projections are subdivided into pathways conveying monaural information to the inferior colliculus and those providing input to the superior olivary complex for binaural processing. Most fibers from the ventral cochlear nucleus course ventral to the inferior cerebellar peduncle in the *trapezoid body* (Fig. 20-7B,C). Projections from the dorsal cochlear nucleus and some from the ventral cochlear nucleus course dorsally over the restiform body as the *dorsal acoustic stria* and decussate in the pontine tegmentum before joining the lateral lemniscus.

The ventral cochlear nucleus is distinguished by the presence of anatomically and physiologically distinct output cell types (Figs. 20-8, 20-9). The anterior part of the ventral cochlear nucleus contains mainly *spherical* and *globular-shaped bushy cells*. There is almost a one-to-one relation between the synaptic endings of a cochlear nerve

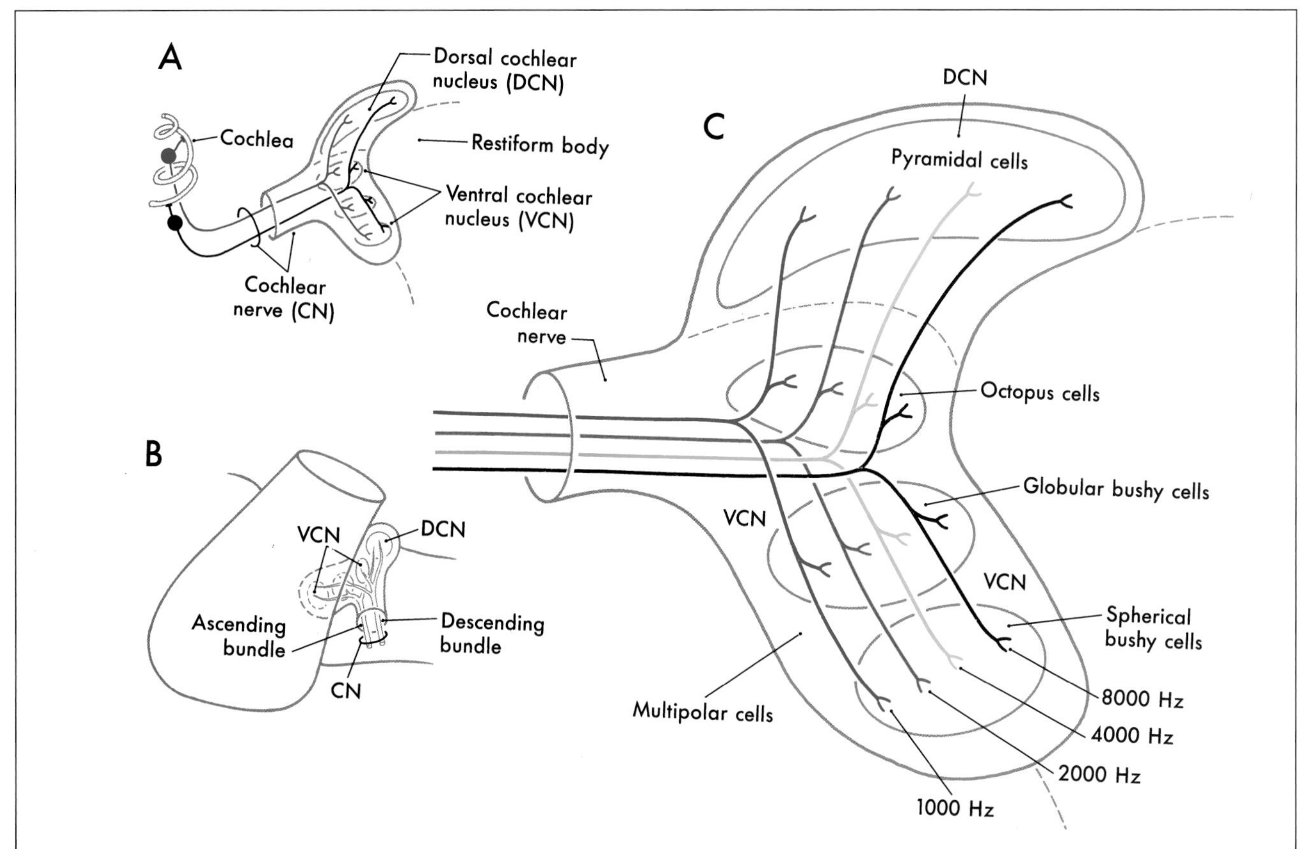

FIGURE 20-8 *The dorsal and ventral cochlear nuclei in cross section (A,C) and as viewed from the lateral aspect on the left (B). The ventral cochlear nucleus extends rostral to the dorsal nucleus (B). The course, tonotopic organization of cochlear afferent fibers, and the locations of main cell groups are also indicated (C).*

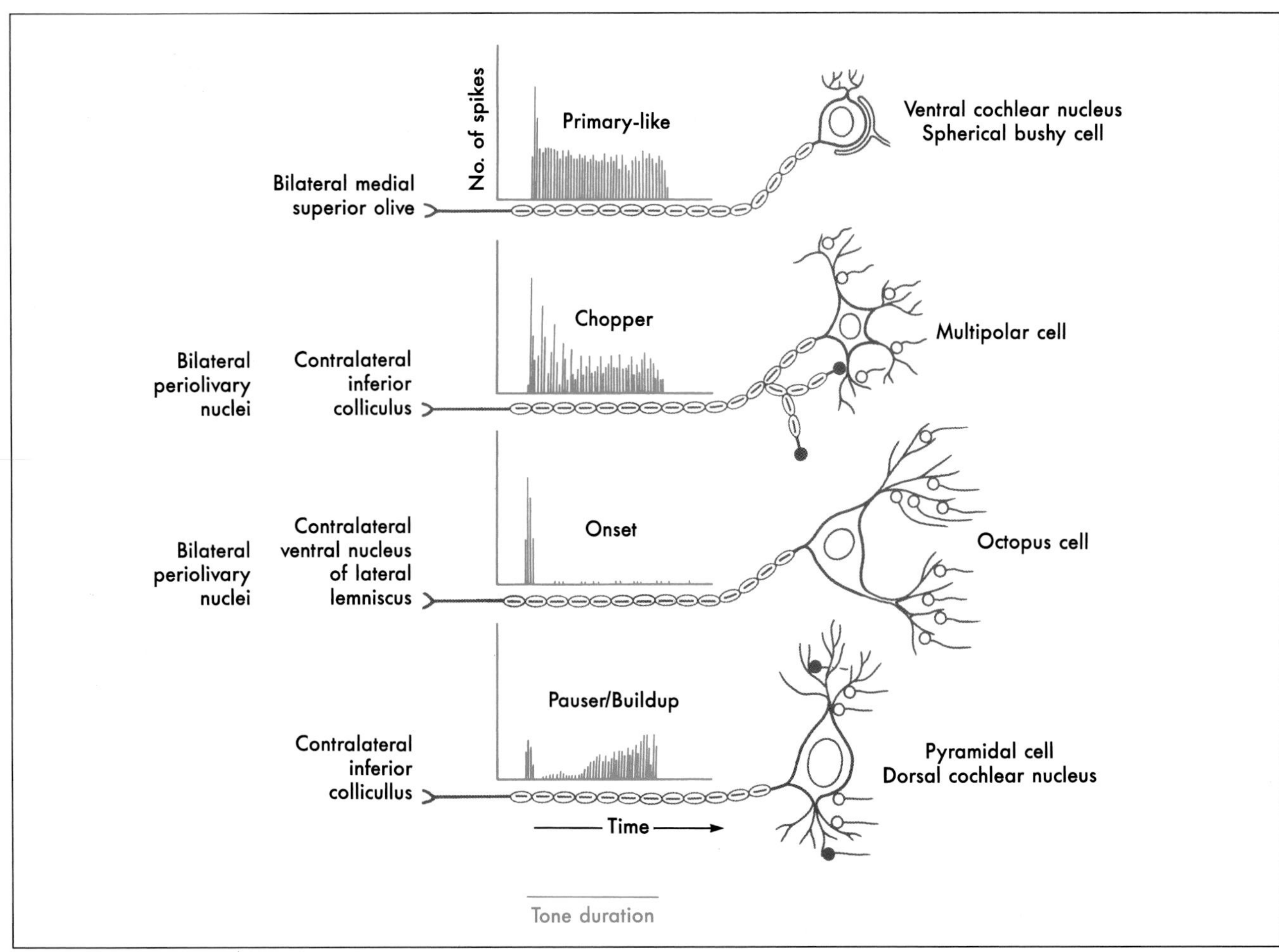

FIGURE 20-9 *Cell types in the cochlear nuclei, their typical response patterns and major ascending connections. The afferent synapses are shown as open profiles.*

fiber and each bushy cell. Consequently, the temporal pattern of activity in bushy cells is remarkably similar to that of the primary afferents, and these cells show sustained responses to short tones that convey information about the timing and phase of the tone (Fig. 20-9). Axons of bushy cells travel in the trapezoid body. They are the central origin of ascending channels that process *binaural information*, which is useful for sound localization (Figs. 20-6, 20-10).

Multipolar cells in the ventral cochlear nucleus receive input from a broader range of cochlear nerve fibers than do bushy cells. Multipolar cells often have bursting (chopper) discharges that are sensitive to changes in sound pressure level (Fig. 20-9). Accordingly, they convey information in a *direct monaural pathway*, primarily to the contralateral inferior colliculus, about the intensity of the sound.

Also in the posterior part of the ventral cochlear nucleus are *octopus cells*, which have long, relatively unbranched dendrites and integrate inputs from a wider array of cochlear afferents. Along with the relatively broad frequency tuning that results from their input, these cells show a characteristic onset response to stimuli (Fig. 20-9). Axons of octopus cells synapse mainly in the contralateral ventral nucleus of the lateral lemniscus, which, in turn, projects to the inferior colliculus. This *indirect monaural pathway* conveys information that is useful in analyzing brief components of speech sounds.

Many of the cells in the dorsal cochlear nucleus contribute to complex local circuits and are not easily correlated with distinct ascending channels. *Pyramidal cells* have fusiform cell bodies with apical and basal dendrites. The response pattern of these cells consists of a brief onset burst followed by a pause and then a slow buildup. A major output of the dorsal cochlear nucleus is via a direct pyramidal cell projection to the contralateral inferior colliculus (Fig. 20-9).

Superior Olivary Complex The superior olivary complex is located in the vicinity of the facial motor nucleus in the caudal pons (Figs. 20-7B,C, 20-10). It is the first site in the brainstem where information from both ears converges. This binaural processing is essential for accurate sound localization and the formation of a neural map of the contralateral auditory hemifield.

The *medial superior olivary* (*MSO*) nucleus, which forms a distinct vertical bar within a diffuse group of *periolivary nuclei*, is the principal nucleus in the human superior olivary complex (Figs. 20-7B,C, 20-10). The *lateral superior olivary nucleus, located lateral to the MSO nucleus*, is not distinct and contains fewer cells and is located lateral to the MSO nucleus.

The *trapezoid body* is a bundle of myelinated fibers passing ventral to the superior olivary complex and intermingling with fibers of the medial lemniscus as it crosses the midline (Figs. 20-7B,C, 20-10). Cells in the medial part of the trapezoid body constitute a rudimentary *medial nucleus of the trapezoid body*. Decussating fibers of the trapezoid body end in the contralateral

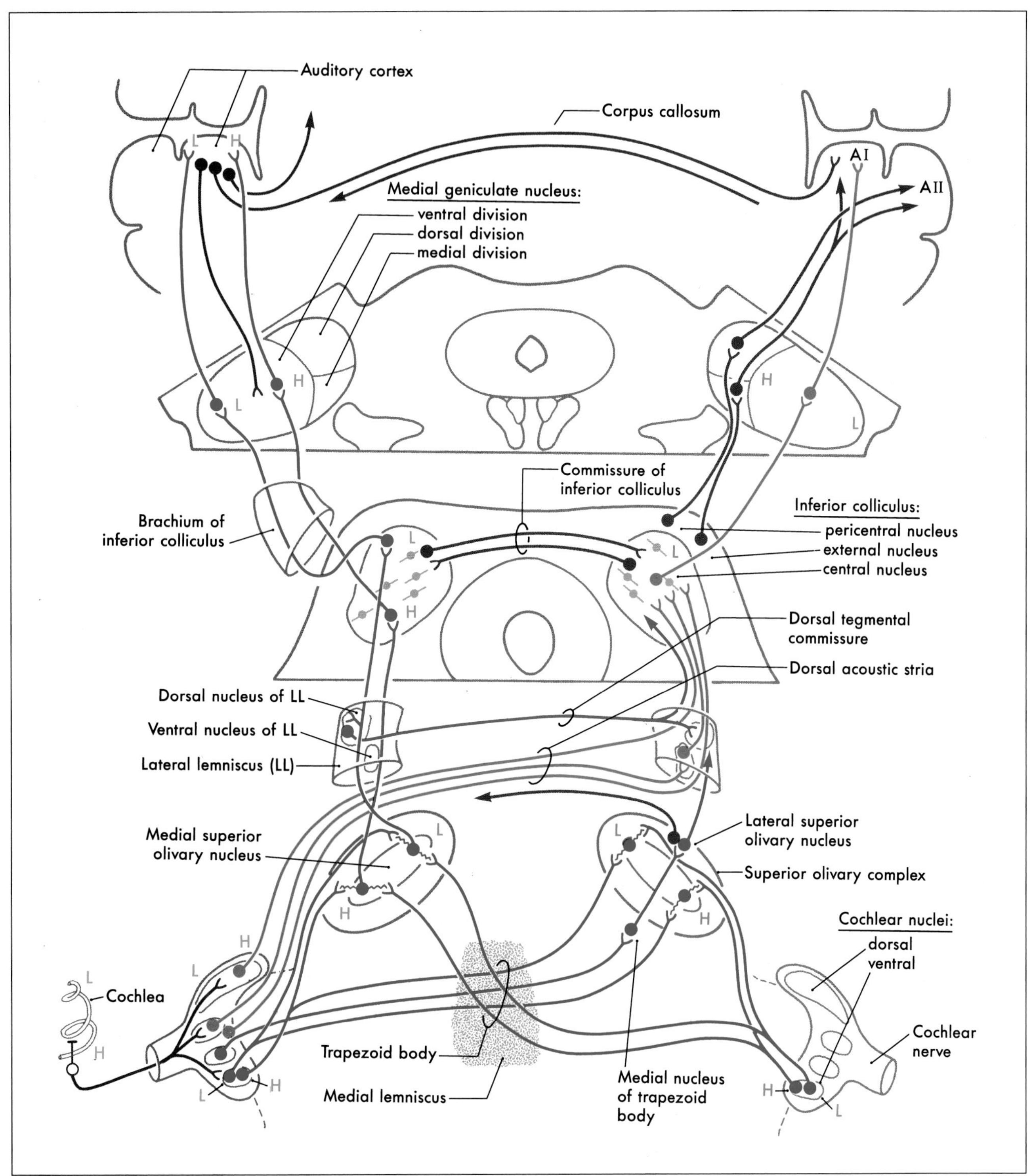

FIGURE 20-10 *Ascending central auditory pathways. Monaural pathways are shown in red, binaural pathways in blue, and other connections in black.*

superior olivary complex or ascend in the contralateral lateral lemniscus.

The topographic organization of the afferents to the MSO conserves the orderly representation of the cochlea (Fig. 20-10). Cells with low characteristic frequencies are found dorsally, and those with higher characteristic frequencies are found more ventrally. These are bipolar cells with medially and laterally directed dendrites (Fig. 20-11)}. Axons in the trapezoid body arising from ipsilateral spherical bushy cells make excitatory synapses with the laterally directed dendrites. Axons from the contralateral spherical bushy cells decussating in the trapezoid body make excitatory synapses with the medially directed dendrites. The excitatory neurotransmitter is probably *glutamate* or *aspartate*. Local inhibitory circuits in the superior olivary complex use *glycine* as a neurotransmitter.

Ascending projections from the MSO travel largely in the ipsilateral side in the lateral lemniscus and synapse in the *central nucleus* of the inferior colliculus (Figs. 20-10, 20-11). These fibers synapse in corresponding frequency regions (low to low, etc.) of the inferior colliculus (Fig. 20-10). The lateral superior olivary nucleus also contributes in a small way in humans to both of these pathways. Branches of these projections always end in the dorsal nucleus of the lateral lemniscus. This nucleus, in turn, projects to the contralateral inferior colliculus and constitutes an indirect binaural pathway from the superior olive to the inferior colliculus (Fig. 20-10).

Computation of Interaural Time and Intensity Differences in the Superior Olive as a Basis for Sound Localization The pathway from cochlear nucleus to the MSO nucleus is anatomically specialized so that the timing of low-frequency tones is preserved (Fig. 20-11). The delay between the production of impulses in the two ears and the arrival of this information at the MSO nucleus via the cochlear nucleus depends on the lengths of the ipsilateral and contralateral paths. If signals arrive close enough in time to summate, then the firing rate of MSO cells is increased. This is referred to as *coincidence detection*. For each cell there is a receptive field in the contralateral auditory hemifield where the stimulus has an interaural time delay that exactly compensates for the difference in neural delay between the ipsilateral and contralateral pathway to the superior olive. Thus, the MSO determines auditory stimulus location based on interaural time differences.

Detection of interaural intensity differences, which provide spatial cues for high frequency stimuli caused by shadowing of sounds by the path from the contralateral side of the head, is also accomplished by summation of excitatory and inhibitory inputs to superior olivary neurons. Humans are capable of detecting small sound differences between the two ears for high-frequency signals, which serve as cues to the source of the auditory signal.

Lateral Lemniscus and Its Nuclei The lateral lemniscus contains axons from second-order neurons in the cochlear nuclei, third-order neurons in the superior olive, and fourth-order neurons in the adjacent nuclei of the lateral lemniscus (Fig. 20-10). *It is precisely this heterogeneous collection that prevents a simple correlation of nuclei or tracts with specific wave components of the brainstem auditory evoked responses that are widely used to assess clinically the level of brainstem function.* The interposition of synap-

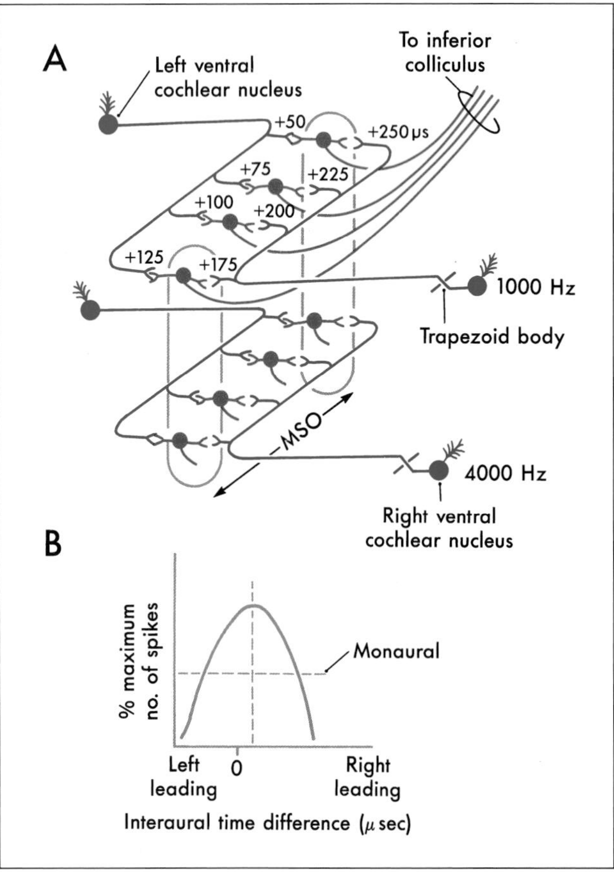

FIGURE 20-11 *The mechanism by which the brain calculates interaural time afferences. Diagrammatic representation of how time delay lines are established in connections between spherical bushy cells in VCN and two MSO isofrequency columns (A). Transmission times in micro seconds for the neural signals are shown in the delay line that maps interaural time delay along the rostral-caudal axis of MSO. When the interaural delay is precisely the reciprocal of this delay (that is, when the arrival at the right ear precedes the arrival at the left ear by the same increment as the signal on the left precedes the signal on the right), then summation occurs as shown in the delay-response curve (B). The dotted horizontal line indicates the response level to a monaural stimulus.*

tic delays in each of these components of the lateral lemniscus imparts temporal differences that contribute to at least the second, third, and fourth wave components of the evoked responses.

The *larger ventral nucleus of the lateral lemniscus* consists of cells scattered among the ascending fibers of the lateral lemniscus (Fig. 20-10). It extends from the rostral limit of the superior olive to just below the inferior colliculus. These cells project to the inferior colliculus, completing an *indirect monaural pathway* (Fig. 20-10).

The smaller *dorsal nucleus of the lateral lemniscus* is intercalated in the ascending fiber bundles of the lateral lemniscus just caudal to the inferior colliculus (Fig. 20-10). This nucleus receives input mainly from the superior olivary complex. Ascending projections from the dorsal nucleus decussate in the *dorsal tegmental commissure*. These fibers terminate in the contralateral inferior colliculus and, to a lesser degree, in the contralateral dorsal nucleus (Fig. 20-10). This pathway is largely inhibitory, using gamma-aminobutyric acid (GABA) as the neurotransmitter. It conveys binaural information and inhibits activity from the opposite hemifield.

Inferior Colliculus Virtually all ascending auditory pathways terminate in the inferior colliculus (Fig. 20-10). The egg-shaped core of the inferior colliculus, the prominent *central nucleus*, is nested in a base of afferent fibers formed by fibers of the lateral lemniscus. These fibers are the major source of input to the inferior colliculus. In a shell around the central nucleus, other cells form the smaller *paracentral nuclei* (Fig. 20-10). These are the *pericentral (dorsal) nucleus*, which lies dorsal and is traversed by fibers from the commissure of the inferior colliculus, and the external (lateral) nucleus, which lies lateral and is intersected by fibers that form the *brachium of the inferior colliculus*.

The central nucleus integrates information from multiple hindbrain auditory sources and, in turn, projects to the ventral division of the medial geniculate nucleus (Fig. 20-10). The central nucleus consists of parallel layers of cells with disc-shaped dendritic fields. Afferents from the lateral lemniscus course parallel to these dendritic fields, forming a series of *fibrodendritic laminae*. Ascending projections diverge and converge in a point-to-plane order in the central nucleus. As a result, each point along the cochlear spiral is represented in an *isofrequency lamina*. Functionally, cells in the central nucleus are narrowly tuned, with the lowest frequencies represented dorsolaterally and higher frequencies ventromedially (Fig. 20-10).

Many cells in the inferior colliculus respond to input from either ear. Among cells with low characteristic frequencies, many are sensitive to interaural time delays, and those with high characteristic frequencies are sensitive to interaural intensity differences. Thus, binaural responses of inferior collicular neurons resemble those of the superior olivary neurons, from which they receive a dominant binaural input. These responses are probably further modified by indirect binaural pathways from the dorsal nucleus of the lateral lemniscus and by intrinsic circuits in the fibrodendritic laminae. Other cells in the fibrodendritic laminae of the central nucleus are monaural and are mainly excited only by the contralateral ear. Their responses resemble those of cells in the contralateral cochlear nucleus.

Cells in the *paracentral nuclei* are broadly tuned to frequency, and they habituate rapidly to repetitive stimuli. They receive input from the central nucleus, the cerebral cortex, and nonauditory input from the spinal cord, dorsal column nuclei, and superior colliculus. These nuclei project to the medial geniculate nucleus (Fig. 20-10), superior colliculus, reticular formation, and precerebellar nuclei. Thus, the paracentral nuclei are probably involved in functions related to attention, multisensory integration, and auditory-motor reflexes (see Fig. 20-15).

Medial Geniculate Nucleus The medial geniculate nucleus forms a small protuberance on the lower caudal surface of the thalamus between the lateral geniculate body and the pulvinar (Fig 10-20; also see Fig. 14-7). The *ventral division* of the medial geniculate nucleus receives afferents from the central nucleus of the inferior colliculus and projects to the primary auditory cortex. Isofrequency contours in the ventral division are arranged so that low frequencies are represented laterally and higher frequencies medially (Fig. 20-10). As a result of collicular and thalamic integration, however, most cells are not reliably excited by simple tones and are probably involved in complex feature detection.

The *dorsal division* receives input from the pericentral nucleus of the inferior colliculus and projects to secondary auditory cortex (Fig. 20-10). These projections are also tonotopically arranged. More broadly tuned and sensitive to habituation, this pathway may convey information about moving or novel stimuli that direct auditory attention.

The *medial (magnocellular)* division receives afferents from the external nucleus of the inferior colliculus and projects to association areas of auditory cortex. It contains cells that are broadly tuned to auditory and other sensory stimuli, including vestibular and somesthetic inputs. The medial division projects to temporal and parietal association areas and to the amygdala, putamen, and pallidum. Given the multisensory convergence that occurs in this pathway, it may be a part of the reticular activating system.

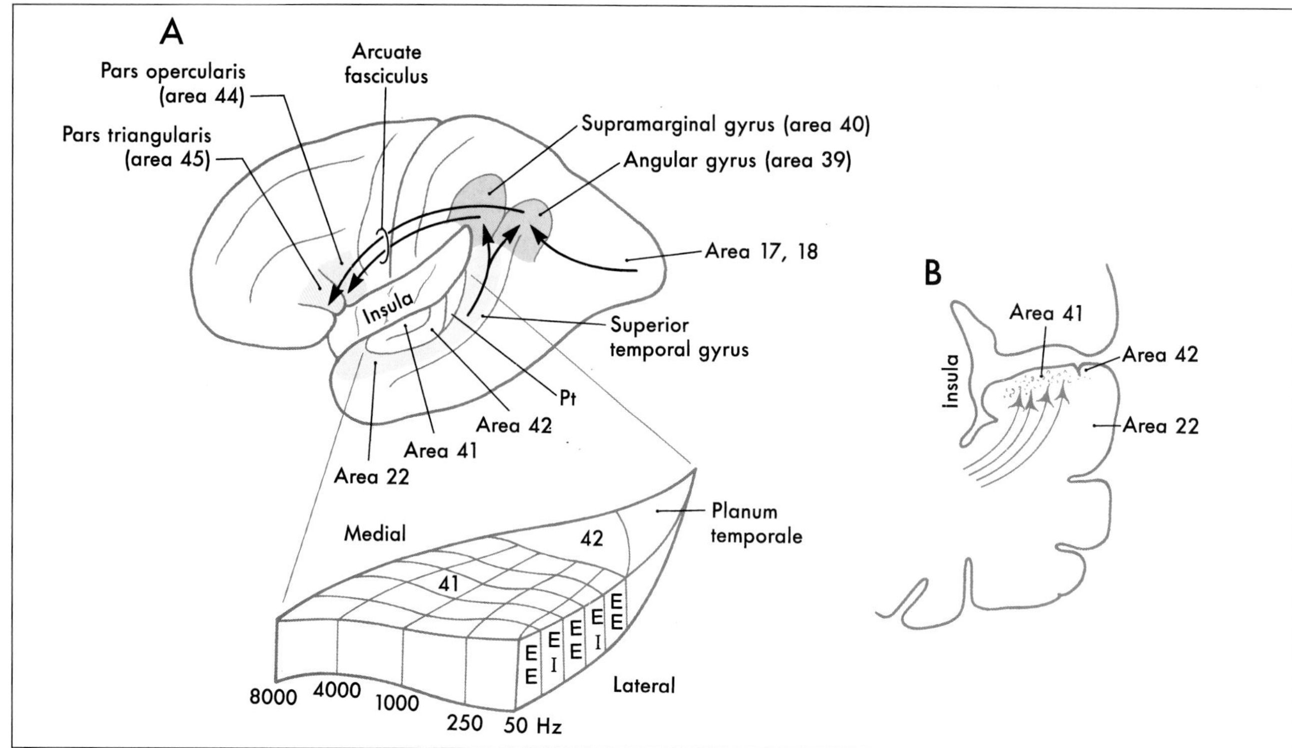

FIGURE 20-12 *The organization of auditory cortical areas. Location and interconnections of auditory cortical areas (A), of the granular cortex in area 41 (B), and the orthogonal isofrequency and binaural response columns in the primary auditory cortex (detail from A).*

AUDITORY AND RELATED ASSOCIATION CORTICES

The *primary auditory cortex (AI)* is located in the *transverse gyri of Heschl* (Figs. 20-10, 20-12). Two transverse temporal gyri are buried in the lateral sylvian sulcus, covered by parts of the frontal and parietal opercula, and continuous with the superior temporal gyrus. Caudal to the transverse temporal gyri is a smooth area, the *planum temporale*, which is usually larger on the left side than the right.

The primary auditory cortex (AI, Brodmann's area 41) is located in the first (anterior) transverse temporal gyrus but may extend into the second (posterior) gyrus (Fig. 20-12A shaded). Cytoarchitecturally, area 41 encompasses the *granular cortex*, with its well developed layer IV containing small granule cells and densely packed small pyramidal cells in layer VI (Fig. 20-12B). Adjacent to the granular cortex in the second transverse gyrus and planum temporale is area 42, which constitutes the *secondary (AII) auditory* cortex (Fig. 20-12A).

Area 41 is reciprocally connected with the ventral division, and area 42 with the dorsal division, of the medial geniculate body (Fig. 20-13). Through the corpus callosum, each auditory cortical area is connected with the reciprocal areas in the other cerebral hemisphere. The tonotopic organization of constituent cells of the cortical layers and incoming afferent fibers form a series of orderly isofrequency columns that extend through the primary auditory cortex as long stripes (Fig. 20-12). High frequencies are represented medially and low frequencies laterally. The series of stripes so formed have one subcomponent comprised of cells excited by stimulation of both ears (EE) alternating with a subcomponent comprised of cells excited by the contralateral ear and inhibited by the ipsilateral ear (EI).

The *auditory association cortex* surrounds the primary auditory area and is located mainly in the posterior portion of the superior temporal gyrus (Fig. 20-12A). It is connected to the primary auditory cortex by the *arcuate fasciculus* (Fig. 20-12A). Area 22 includes a part of the planum temporale and the posterior portion of the superior temporal gyrus. It receives connections from the primary auditory cortex, as well as visual and somesthetic information. This speech receptive area, known as *Wernicke's area*, may be as much as seven times larger on the left side than on the right. When this area is damaged by occlusion of branches of the middle cerebral artery, an *auditory aphasia (Wernicke's aphasia)* results. In such cases, comprehension of speech sounds is impaired, but discrimination of nonverbal sounds is largely unaffected.

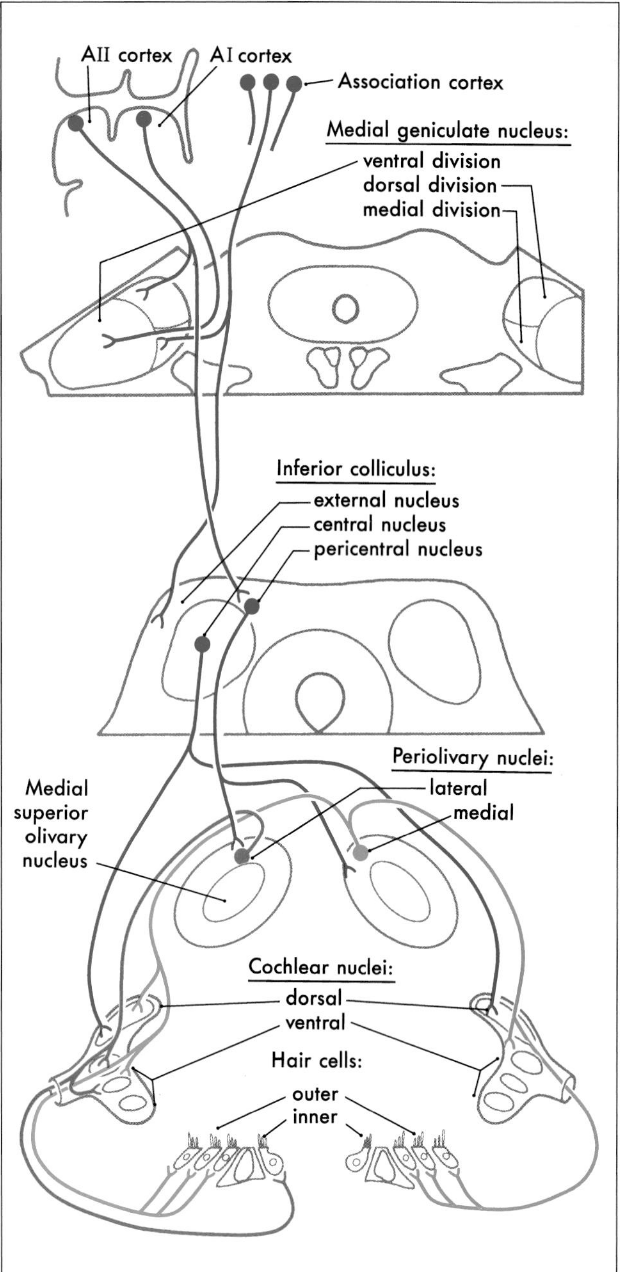

FIGURE 20-13 *Descending auditory pathways that modulate sensory processing at central and peripheral auditory sites. The lateral olivocochlear efferents are shown in red and the medial olivocochlear efferents in green.*

The higher association areas of auditory cortex also extend into the inferior parietal lobule (Fig 20-12A). This lobule is made up of the angular gyrus (area 39) and supramarginal gyrus (area 40). These two areas are important in aspects of language such as reading and writing and are sometimes included in Wernicke's area.

Brodmann areas 44 and 45 are known as *Broca's area* for expressive speech and language. They are located in the *pars opercularis* and *pars triangularis* of the *inferior frontal gyrus* (Fig. 20-12A). The major pathway connecting these areas with the primary and association auditory cortex is the *arcuate fasciculus* (Fig. 20-12). If areas 44 and 45 are damaged along with other motor cortices on the left side by a stroke involving branches of the middle cerebral artery, the result is *Broca's aphasia*. In this disorder, speech is nonfluent, but comprehension of verbal and nonverbal sounds is largely unimpaired.

DESCENDING AUDITORY PATHWAYS

Descending projections make reciprocal connections throughout the auditory pathway. They form feedback loops that provide circuits to modulate information processing from the peripheral level to the cortex (Fig. 20-13). For example, the auditory cortex projects to the medial geniculate nucleus and nuclei of the inferior colliculus. The inferior colliculus projects to the periolivary nuclei, which, in turn, send olivocochlear efferents to the cochlea. There are also descending projections from the periolivary nuclei to the cochlear nuclei.

The Olivocochlear Bundle The *olivocochlear efferent system* arises from groups of cells in the periolivary nuclei of the superior olivary complex (Fig. 20-13). These efferent systems travel as the *olivocochlear bundle* in the vestibular part of the vestibulocochlear nerve. *Lateral olivocochlear efferent* cells project to the ipsilateral inner hair cells, where they make axo-axonic synapses with type I spiral ganglion afferent fibers (Figs. 20-4, 20-13). *Medial olivocochlear efferent* cells have bilateral projections that terminate directly on outer hair cells (Figs. 20-4, 20-13).

Direct efferent feedback to outer hair cells, in particular, may influence cochlear mechanics and, consequently, the sensitivity and frequency selectivity of the cochlea. Efferent-induced changes in outer hair cell membrane potentials result in changes in the height of the cells and the stiffness of their stereocilia. These changes modulate basilar membrane motion and thereby influence cochlear function.

MIDDLE EAR REFLEX

The small striated muscles of the middle ear affect the mechanical impedance of the ossicular chain. These muscles are activated by the *middle ear reflex* (Fig. 20-14).

The *stapedius muscle* is innervated by *facial motor neuron*, and the *tensor tympani muscle* by *trigeminal motor neurons*. These motor neurons are intimately associated with

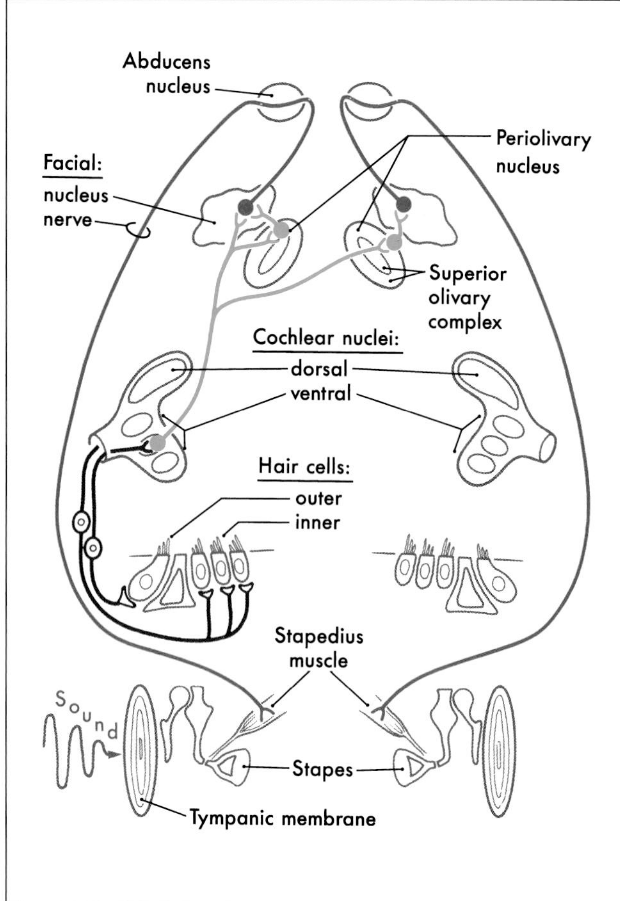

FIGURE 20-14 *The pathway of the middle ear reflex arc. For simplicity, only the stapedius reflex is shown.*

the caudal end of the superior olivary complex, in the case of the stapedius muscle, and the rostral end of the superior olivary complex, in the case of the tensor tympani muscle. In these positions, auditory input via axons of neurons in the cochlear nuclei or the superior olivary complex provides the sensory limb of the reflex. The sensory pathways are bilateral, so that stimuli may be presented by earphones to one ear while the device to measure impedance is placed in the ear canal on the other side.

ACOUSTIC STARTLE REFLEX, ORIENTATION, AND ATTENTION

Reflexive and learned responses to sound require sensory-motor integration. In addition to corticocortical interconnections for the dissemination of auditory information, there is also integration of auditory sensory input with motor pathways in the brainstem. Reticulospinal neurons in the region of the lateral lemniscus have dendrites that sample lemniscal activity and are involved in rapid *acoustic startle reflex* pathways. In addition, the *deep layers of the superior colliculus* receive auditory information from the inferior colliculus and auditory cortical areas (Fig. 20-15). The deep layers of the superior colliculus integrate auditory, visual, and somesthetic information and project to brainstem and cervical spinal cord nuclei via tectobulbospinal fibers, which are involved in controlling orientation of the head, eyes, and body to sound (Fig. 20-15).

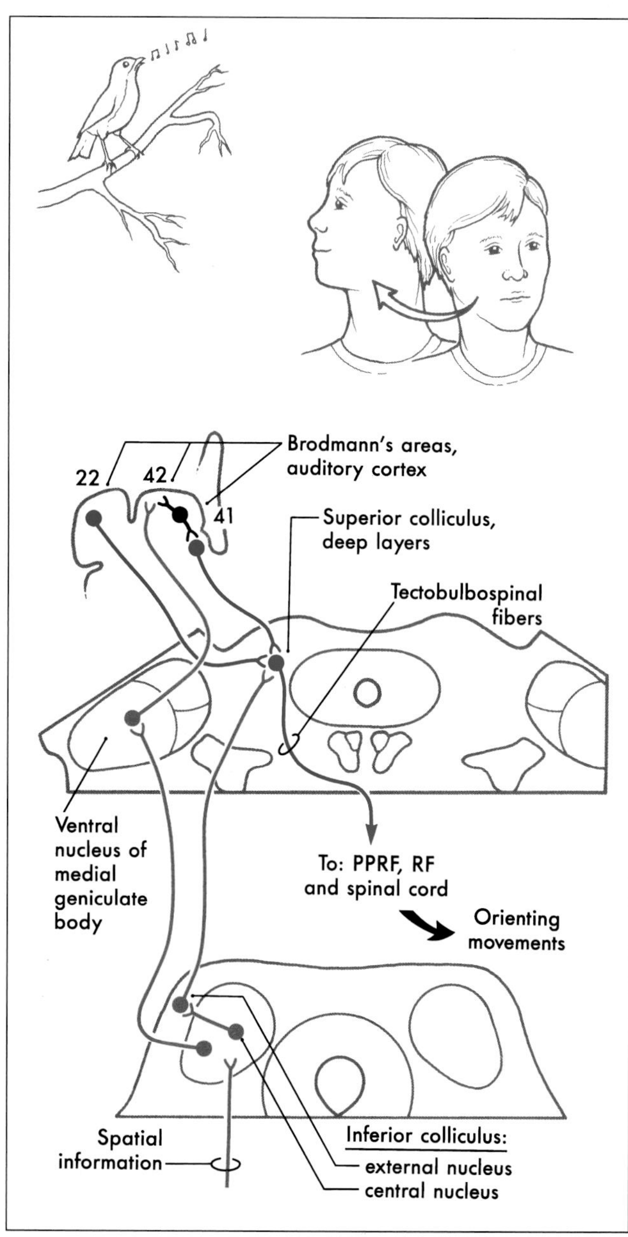

FIGURE 20-15 *The pathways that subserve auditory-motor integration involved in simple orientation to a novel auditory stimulus. (RF, reticular formation; PPRF, paramedian pontine reticular formation)*

SOURCES AND ADDITIONAL READINGS

Altschuler RA, Bobbin RP, Hoffman DW: *Neurobiology of Hearing: The Cochlea*. Raven Press, New York, 1986.

Altschuler RA, Bobbin RP, Clopton BM, Hoffman DW: *Neurobiology of Hearing: The Central Auditory System*. Raven Press, New York, 1991.

Gelfand SA: Hearing: *An Introduction to Psychological and Physiological Acoustics*. Marcel Dekker, New York, 1990.

Pickles JO: *An Introduction to the Physiology of Hearing*, 2nd Ed. Academic Press, London, 1988.

Webster D, Fay RR, Popper AN: *Springer Handbook of Auditory Research. Vol I. The Auditory Pathway: Neuroanatomy*. Springer-Verlag, New York, 1992.

Yost WA: *Fundamentals of Hearing: An Introduction*. Academic Press, San Diego, 1994.

CHAPTER TWENTY-ONE

The Vestibular System

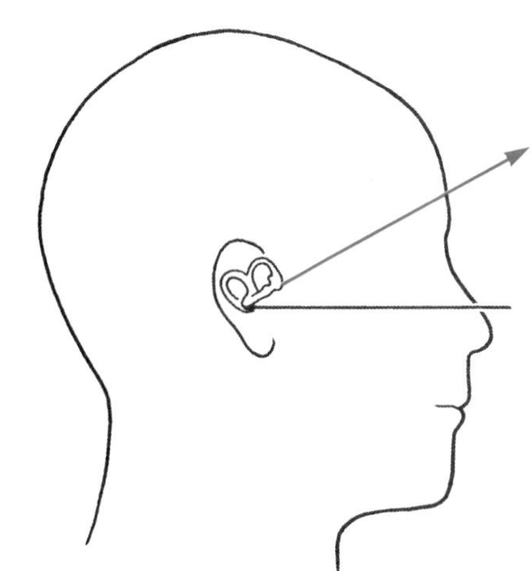

J. D. DICKMAN

Humans have the ability to control posture and movements of the body and eyes relative to the external environment. The *vestibular system* mediates these motor activities through a network of receptors and neural elements. This system integrates peripheral sensory information from vestibular, somatosensory, and visual receptors, as well as motor information from the cerebellum and cerebral cortex. Central processing of these inputs occurs rapidly, with the output of the vestibular system providing an appropriate signal to coordinate relevant muscle movements. Although the vestibular system is considered as a special sense, most vestibular activity is conducted at a subconscious level. However, in situations producing unusual or novel vestibular stimulation, such as rough air in a plane flight or wave motion on ships, vestibular perception becomes acute, with dizziness, vertigo, or nausea possibly resulting.

OVERVIEW

The vestibular system is an essential component in the production of motor responses that are critical for daily function and survival. Throughout evolution, the highly conserved nature of the vestibular system is revealed through striking similarities in the anatomic organization of receptors and neuronal connections in fish, reptiles, birds, and mammals.

For the present discussion, the vestibular system can be divided into five components: (1) The *peripheral receptor apparatus* resides in the inner ear and is responsible for transducing head motion and position into neural information. (2) The *central vestibular nuclei* comprise a set of neurons in the brainstem that are responsible for receiving, integrating, and distributing information that controls motor activities such as eye movements, head movements, and postural reflexes. (3) The *vestibulo-ocular network* arises from the vestibular nuclei and is involved in the control of eye movements. (4) The *vestibulospinal network* coordinates head movements, axial musculature, and postural reflexes. (5) The *vestibulo-thalamo-cortical network* is responsible for the conscious perception of movement and spatial orientation.

PERIPHERAL VESTIBULAR LABYRINTH

The vestibular labyrinth contains specialized sensory receptors and is located lateral and posterior to the cochlea in the inner ear (Fig. 21-1). The vestibular labyrinth consists of five separate receptor structures, *three semicircular canals* and *two otolith organs*, which are contained in the petrous portion of the temporal bone. The labyrinth is actually composed of two distinct components. The *bony labyrinth* is a surrounding shell that contains and protects the sensitive underlying vestibular sensory structures (Fig. 21-1). In humans, the bony

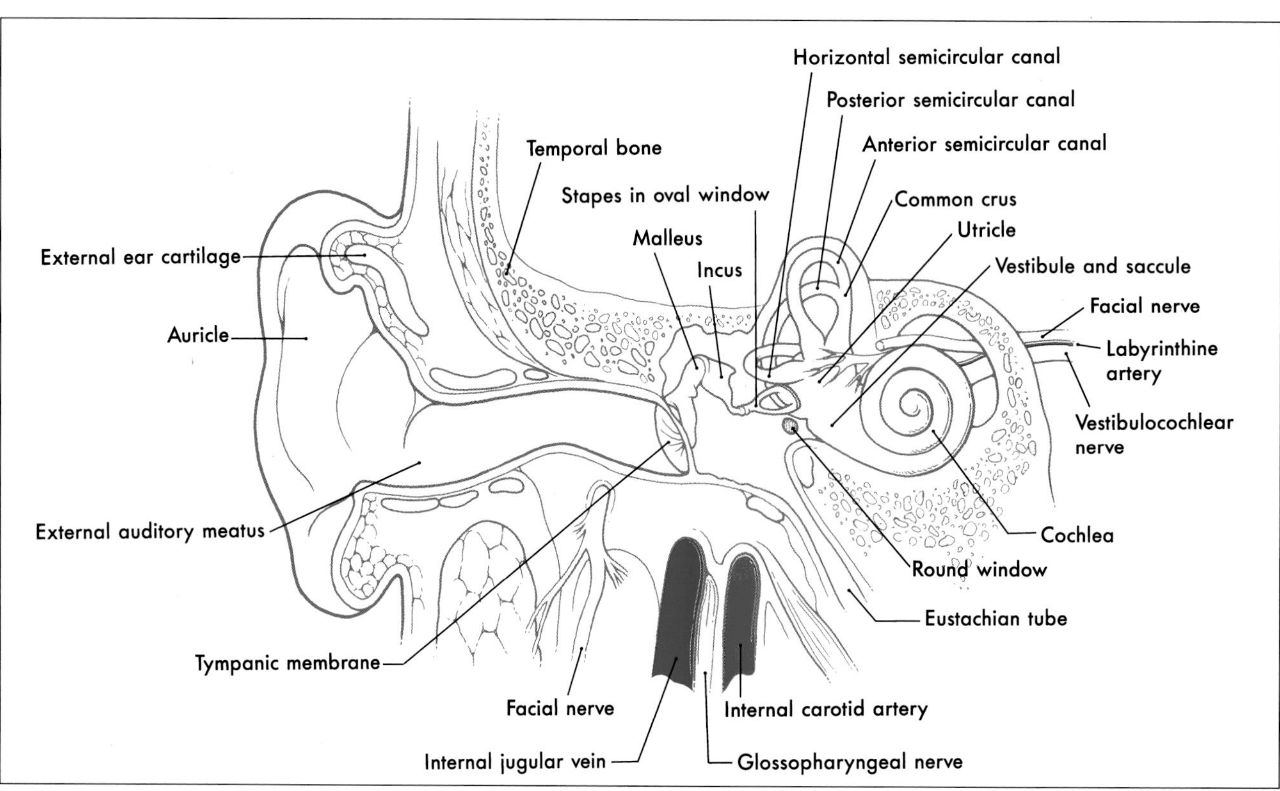

FIGURE 21-1 *A cross section of the outer, middle, and inner ear.*

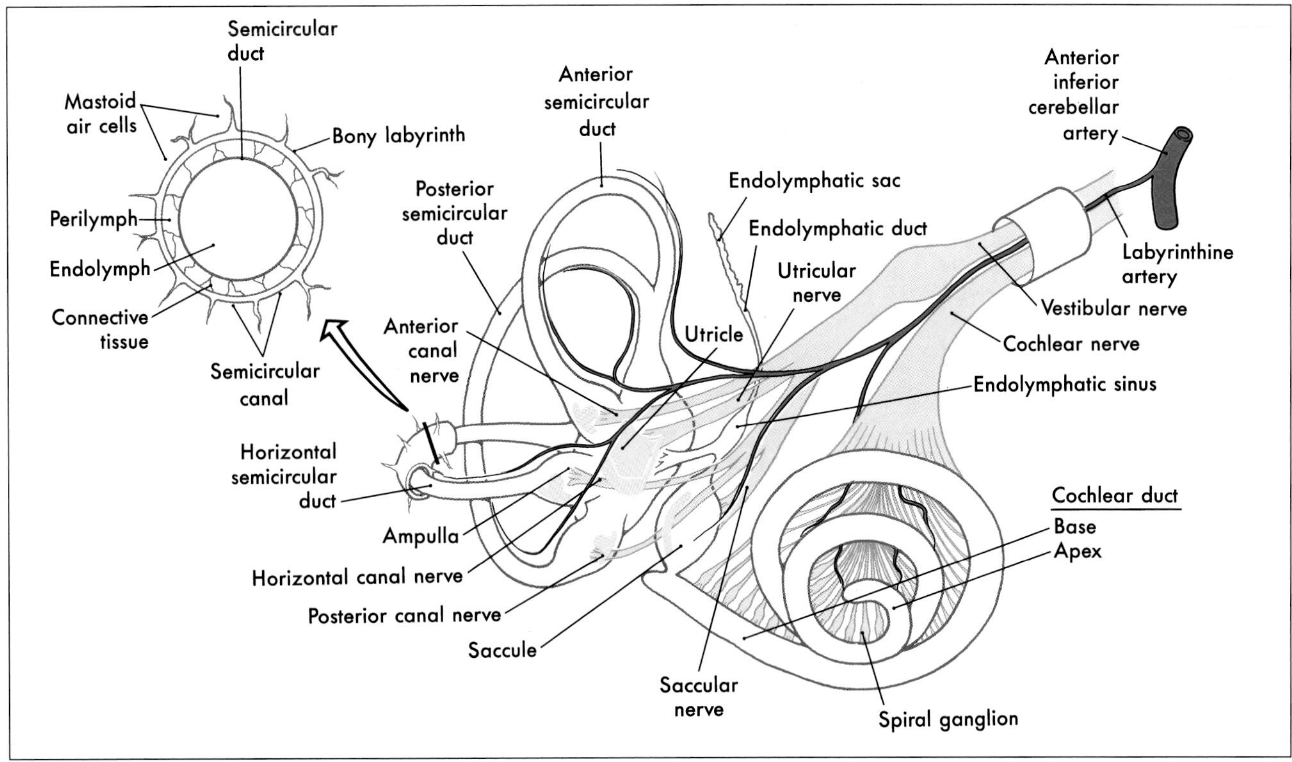

FIGURE 21-2 *The membranous labyrinth and associated vessels and nerves. The approximate configuration of the receptor sites in the ampullae, utricle, and saccule are shown in green. The detail shows the relationship between bony and membranous labyrinths.*

labyrinth can be visualized only on excision of the mastoid process. Inside the bony labyrinth is a closed, fluid-filled system, the *membranous labyrinth*, which consists of connecting tubes and prominences (Fig. 21-2). Vestibular receptors are located in specialized regions of the membranous labyrinth.

Between the membranous and bony labyrinths is a space containing a fluid called *perilymph*, which is similar to cerebrospinal fluid. Perilymph has a high sodium content (150 mM) and a low potassium content (7 mM), and it bathes the vestibular portion of the eighth cranial nerve.

The membranous labyrinth is filled with a different type of fluid, called *endolymph*, which covers the specialized sensory receptors of both the vestibular and the auditory systems. Endolymph has a high concentration of potassium (150 mM) and a low concentration of sodium (16 mM). It is important to note the differences in these two fluids because both are involved in the normal functioning of the vestibular system. Disturbances in the distribution or ionic content of endolymph often lead to vestibular pathology.

Vestibular Receptor Organs The five vestibular receptor organs in the inner ear complement each other in function. The semicircular canals (horizontal, anterior, and posterior) transduce rotational head movements (angular accelerations). The otolith organs (utricle and saccule) respond to translational head movements (linear accelerations) or to the orientation of the head relative to gravity. Each semicircular canal and otolith organ is spatially aligned so as to be most sensitive to movements in specific planes in three-dimensional space.

In humans, the horizontal semicircular canal and the utricle both lie in a plane that is slightly tilted anterodorsally relative to the naso-occipital plane (Fig. 21-3). When a person walks or runs, the head is normally declined (pitched downward) by approximately 30 degrees, so that the line of sight is directed a few meters in front of the feet. This orientation causes the plane of the horizontal canal and utricle to be parallel with the earth horizontal and perpendicular to gravity. The anterior and posterior semicircular canals and the saccule are arranged vertically in the head, orthogonal to the horizontal semicircular canal and utricle (Fig. 21-3). The two vertical canals on each side are positioned orthogonal to each other, whereas the plane of the anterior canal on one side of the head is colinear with the plane of the contralateral posterior canal (Fig. 21-3).

The receptor cells in each vestibular organ are innervated by primary afferent fibers that join with those from the cochlea to comprise the *vestibulocochlear (eighth) cranial nerve*. The cell bodies of these bipolar vestibular afferent neurons are in the vestibular ganglion (Scarpa's ganglion), which lies in the internal acoustic meatus (Fig.

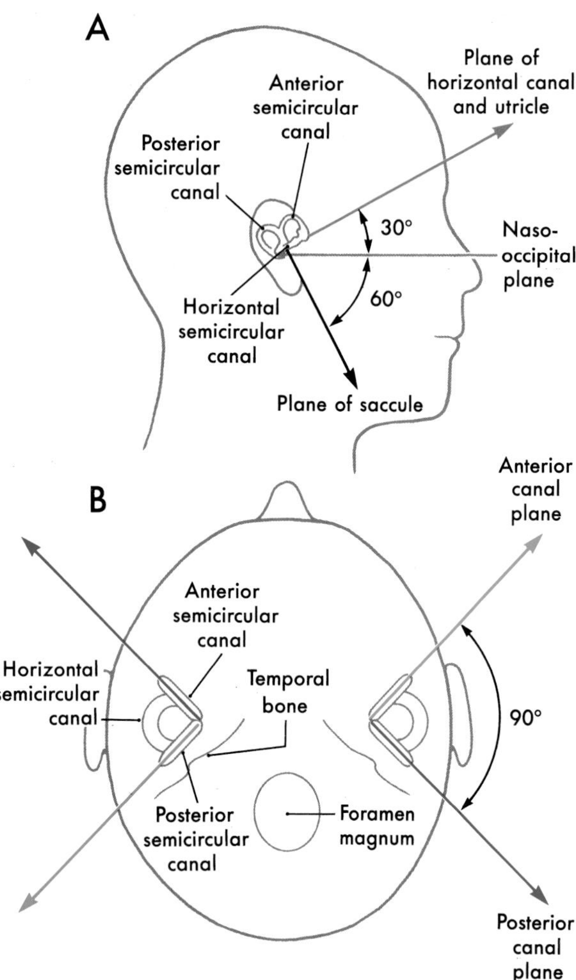

FIGURE 21-3 *Orientation of the vestibular receptors. In lateral view (A), the horizontal semicircular canal and the utricle lie in a plane that is tilted relative to the naso-occipital plane. In the axial view (B), the vertical semicircular canals lie at right angles to each other.*

21-4). The central processes of these bipolar cells enter the brainstem and terminate in the ipsilateral vestibular nuclei and cerebellum.

The blood supply to the labyrinth is primarily via the *labyrinthine artery*, a branch of the anterior inferior cerebellar artery. This vessel enters the temporal bone through the internal auditory meatus. Although not as important as the labyrinthine artery, the *stylomastoid artery* also provides branches to the labyrinth, mainly to the semicircular canals. An interruption of blood supply to the labyrinth will compromise vestibular (and cochlear) function, resulting in labyrinth-associated symptoms, such as dizziness and unstable gait.

Membranous Labyrinth The membranous labyrinth is supported inside the bony labyrinth by connective tissue. The three *ducts of the semicircular canals* connect to the utricle, and each duct ends with a single prominent enlargement, the *ampulla* (Fig. 21-2). Sensory receptors for the semicircular canals reside in a neuroepithelium at the base of each ampulla. The receptors in the utricle are oriented longitudinally along its base, and in the saccule they are oriented vertically along the medial wall (Fig. 21-2). Endolymph in the labyrinth is drained into the endolymphatic sinus via small ducts. In turn this sinus communicates through the *endolymphatic duct* with the *endolymphatic sac*, which is located adjacent to the dura mater (Fig. 21-2). The saccule is also connected to the cochlea by the *ductus reuniens*.

The balance between the ionic contents of endolymph and perilymph is maintained by specialized secretory cells in the membranous labyrinth and the endolymphatic sac. In cases of advanced *Ménière's disease*, there is disruption of normal endolymph volume resulting in *endolymphatic hydrops* (an abnormal distention of the membranous labyrinth). Symptoms of Ménière's disease include severe vertigo, positional nystagmus, and nausea. These patients often suffer unpredictable attacks of auditory and vestibular symptoms, including vomiting, tinnitus (ringing in the ears), and a complete inability to make head movements or even stand passively. For patients with

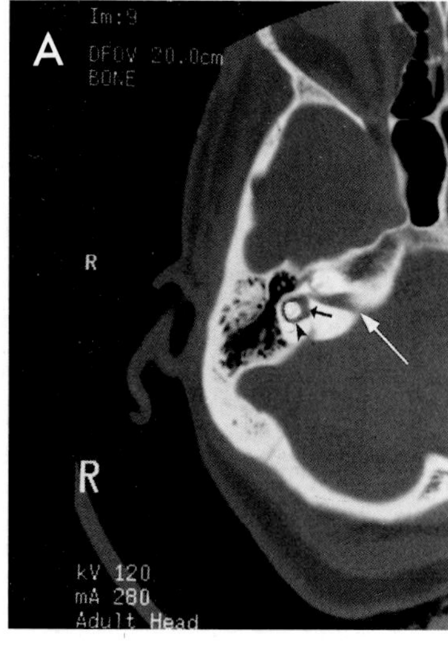

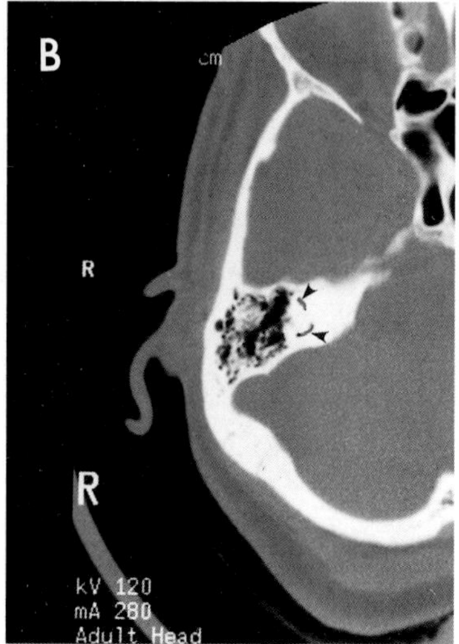

FIGURE 21-4 *Computed tomography (CT) scans of the human temporal bone. The horizontal (A, arrowhead) and anterior and posterior (B, arrowheads) semicircular canals, utricle (A, small arrow), and internal acoustic canal (A, large arrow) are visible.*

frequent debilitating attacks, one course of treatment involves the implantation of a small tube or shunt into the abnormally swollen endolymphatic sac.

VESTIBULAR SENSORY RECEPTORS

Hair Cell Morphology The sensory receptor cells in the vestibular system, like those in the auditory system, are called *hair cells* owing to the *stereocilia* that project from the apical surface of the cell (Fig. 21-5A). Each hair cell contains 60 to 100 hexagonally arranged stereocilia and a single longer *kinocilium*. The stereocilia are oriented in rows of ascending height, with the tallest lying next to the lone kinocilium. The stereocilia arise from a region of dense actin, the *cuticular plate*, located at the apical end of the hair cell. The cuticular plate acts as an elastic spring to return the stereocilia to the normal upright position after bending. Each stereocilium is connected to its neighbor by small filaments.

There are two types of hair cells, and they differ in their pattern of innervation by fibers of the eighth cranial nerve (Fig. 21-5A). Type I hair cells are chalice-shaped and typically are surrounded by an afferent terminal that forms a *nerve calyx*. Type II hair cells are cylindrical and are innervated by simple synaptic boutons. Excitatory amino acids such as aspartate and glutamate are the neurotransmitters at the receptor cell-afferent fiber synapses. Both types of hair cells, or their afferents, receive synapses from *vestibular efferent fibers* that control the sensitivity of the receptor. These efferent fibers contain acetylcholine and arise in the brainstem just rostral to the vestibular nuclei. They are activated by behaviorally arousing stimuli or by trigeminal stimulation.

Within each ampulla, the hair cells and their supporting cells lie embedded in a saddle-shaped neuroepithelial ridge, the *crista*, which extends across the base of the ampulla (Fig. 21-5B). Type I hair cells are concentrated in central regions of the crista, and type II hair cells are more numerous in peripheral areas. Arising from the crista and completely enveloping the stereocilia of the hair cells is a gelatinous structure, the *cupula*. The cupula attaches to the roof and walls of the ampulla, forming a fluid-tight partition that has the same specific density as endolymph. Rotational head movements produce angular accelerations that cause the endolymph in the membranous ducts to be displaced, so that the cupula is pushed to one side or the other like the skin of a drum. These cupular movements displace the stereocilia (and kinocilium) of the hair cells in the same direction.

For the otolith organs, a structure analogous to the crista, the *macula*, contains the receptor hair cells (Fig. 21-5C). The hair cell stereocilia of otolith organs extend into a gelatinous coating called the *otolith membrane*, which is covered by calcium carbonate crystals called *otoconia* (from Greek, "ear stones"). Otoconia are about three times as dense as the surrounding endolymph, and they are not displaced by normal endolymph movements. Instead, changes in head position relative to gravity, or linear accelerations (forward/backward, upward/downward) produce displacements of the otoconia, resulting in bending of the underlying hair cell stereocilia.

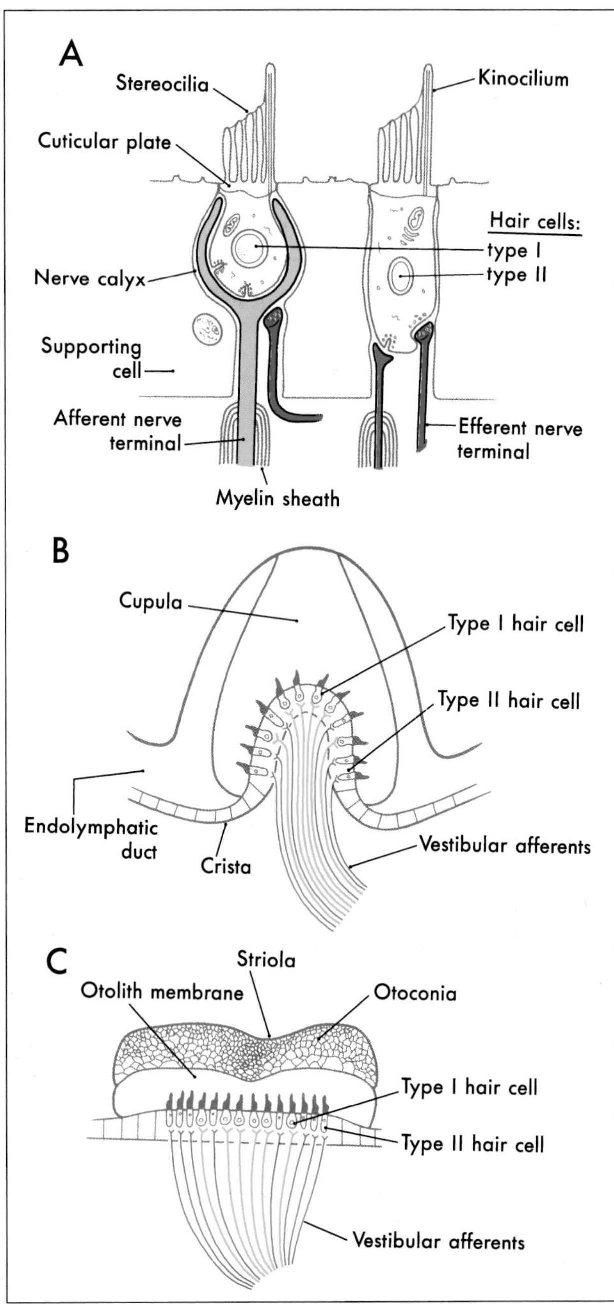

FIGURE 21-5 *The receptor cells (A, types I and II hair cells) of the vestibular system. The relation of these cells to the crista and cupula (B) in the ampullae and to the macula and otolith membrane (C) of the otolith organs.*

Hair Cell Transduction The response of hair cells to deflection of their stereocilia is highly polarized (Figs. 21-6, 21-7A). Movements of the stereocilia *toward the kinocilium* cause the hair cell membranes to *depolarize*, which results in an increased rate of firing in the vestibular afferent fibers. If the stereocilia are *deflected away from the kinocilium*, however, the hair cell is *hyperpolarized* and the afferent firing rate decreases.

The mechanisms underlying the depolarization and hyperpolarization of vestibular hair cells depend, respectively, on the potassium-rich character of endolymph and the potassium-poor character of the perilymph that bathes the basal and lateral portions of the hair cells. Deflection of the stereocilia *toward* the kinocilium causes K^+ channels in the apical portions of the stereocilia and kinocilium to open. K^+ flows into the cell from the endolymph, depolarizing the cell membrane (Fig. 21-6). This depolarization in turn causes voltage-gated calcium channels at the base of the hair cells to open, allowing Ca^{++} to enter the cell. The influx of Ca^{++} causes synaptic vesicles to release their transmitter (aspartate or glutamate) into the synaptic clefts, and the afferent fibers respond by undergoing depolarization and increasing their rate of firing. When the stimulus subsides, the stereocilia and kinocilium return to their resting position, allowing most Ca^{++} channels to close and voltage-gated K^+ channels at the base of the cell to open. K^+ efflux returns the hair cell membrane to its resting potential (Fig. 21-6).

Deflection of the stereocilia *away* from the kinocilium causes K^+ channels in the basolateral portions of the hair cell to open, allowing K^+ to flow out from the cell into the interstitial space. The resulting hyperpolarization of the cell membrane decreases the rate at which the neurotransmitter is released by the hair cells and, consequently, decreases the firing rate of afferent fibers.

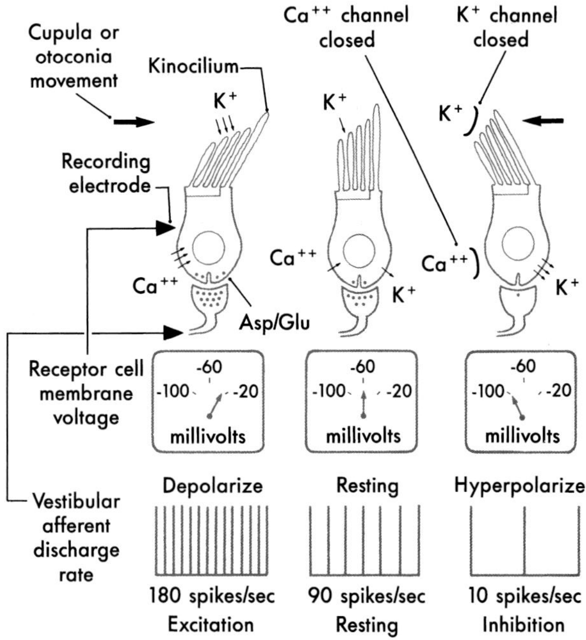

FIGURE 21-6 *Physiologic responses of vestibular hair cells and their vestibular afferent fibers.*

Almost all vestibular primary afferent fibers have a moderate spontaneous firing rate at rest (approximately 90 spikes per second). Therefore, it is likely that some hair cell Ca^{++} channels are open at all times, causing a slow, constant release of neurotransmitter. The ototoxic effects of some aminoglycoside antibiotics (e.g., streptomycin and gentamicin) may be due to direct reduction of the transduction currents of hair cells.

Morphologic Polarization of Hair Cells Given that deflection of the stereocilia toward and away from the kinocilium cause opposed physiologic responses, it is clear that the directional orientation of the hair cells on the vestibular organs will play an essential role in signaling the direction of movements. On the cristae of the horizontal semicircular canal, the hair cells are all arranged with their kinocilium on the side closest to the utricle (Fig. 21-7B). Thus, *movement of endolymph toward the ampulla in the horizontal canals* causes the stereocilia to be deflected toward the kinocilium, resulting in depolarization of the hair cell. In the vertical semicircular canals, the hair cells are arranged with their kinocilia on the side farthest from the utricle (closest to the endolymphatic duct). Thus the *hair cells of the vertical canals are hyperpolarized by movement of endolymph toward the ampulla* (ampullopetal movement) and are depolarized by movement away from the ampulla (ampullofugal movement).

In both the utricle and the saccule, the otolith membrane overlying the hair cells contains a small, curving depression, the *striola*, that roughly bisects the underlying macula (Fig. 21-7C). Hair cells on the utricular macula are polarized so that the kinocilium is always on the side toward the striola (Figs. 21-5C, 21-7C), which effectively splits the receptors into two morphologically opposed groups. In contrast, the kinocilia of saccular hair cells are oriented on the side away from the striola. Because the striola curves through the macula, otolith hair cells are polarized in *many different directions* (Fig. 21-7C). In this way, utricular and saccular hair cells are directionally sensitive to a wide variety of head positions and linear movements.

SEMICIRCULAR CANALS AND OTOLITH ORGANS

As stated above, the vestibular receptors transduce *movement and position* stimuli into neural signals that are sent to the brain. The *semicircular canals are responsive to rotational acceleration* resulting from turns of the head and/or body. The *otolith organs are responsive to linear accelerations*. The most prominent linear acceleration on earth is the force of gravity, which typically remains con-

FIGURE 21-7 *Morphologic polarization of vestibular receptor cells showing polarity of stereocilia and kinocilia (A), and the orientation of receptors in the ampullae (B) and maculae (C).*

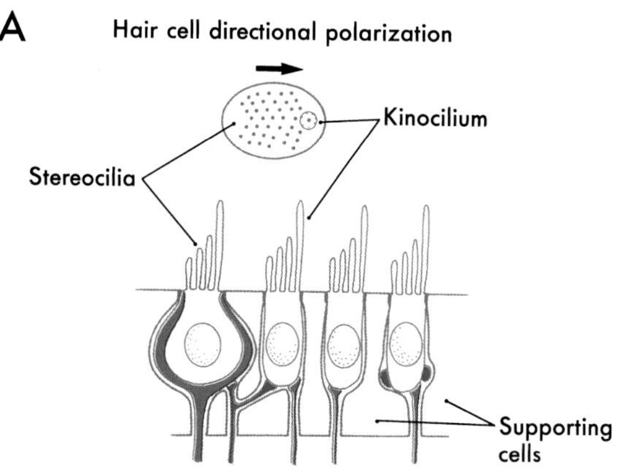

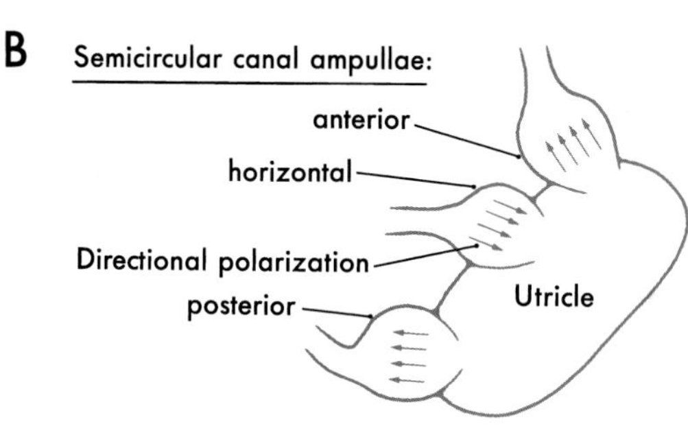

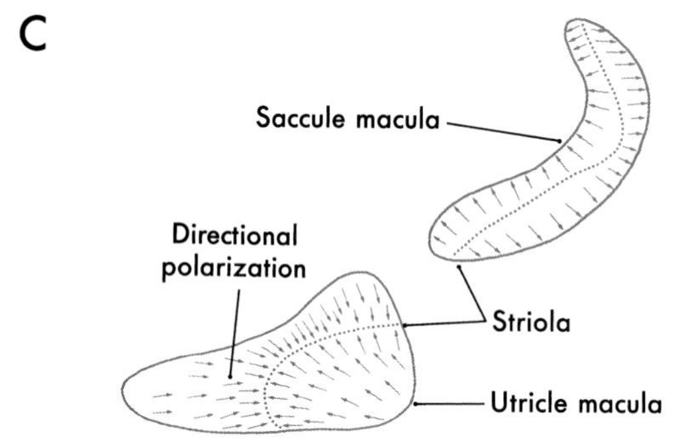

stant. The effect of gravity can be increased or decreased in situations such as swinging on a swing or flying in an airplane through turbulence and even eliminated during space flight. Linear accelerations also occur in situations such as halting from a run or accelerating a car, and they accompany tilting of the head. (Forward and backward tilting is called *pitch*; side-to-side tilting is called *roll.*)

Function of Semicircular Canals The membranous semicircular duct can be thought of as a fluid-filled tube with a partition (the cupula) in the middle (Fig. 21-5B). Because the utricles are located medially on each side of the head, the hair cells of the complementary left and right semicircular canals are *oppositely polarized.* An example is seen in rotational head movements made in the horizontal plane (Fig. 21-8) When the head is stationary (no angular acceleration), the endolymph and the cupula remain still, and the afferents from the two horizontal semicircular canals fire at the same (resting) rate (Fig. 21-8A). When the head turns to the right or left, however, the horizontal semicircular ducts turn with it, but the endolymph lags owing to inertial forces and the viscous drag between the fluid and the duct wall. The lagging endolymph deflects the cupula, which in turn deflects the stereocilia of the hair cells. As Figure 21-8B shows, a leftward turn of the head causes the stereocilia in the left horizontal canal ampulla to be deflected toward their kinocilia, resulting in an increase in the discharge rate of the eighth nerve afferents on the left side. Simultaneously, the hair cells in the right horizontal canal ampulla are hyperpolarized, so their afferents show a decreased rate of firing. A rightward head turn produces the opposite pair of responses (Fig. 21-8C).

It is important to realize that the left and right semicircular canals of each functional pair (such as the left and right horizontal canals) *always* respond oppositely to any head movement that affects them. This fact leads to the "push-pull" concept of vestibular function, which states *that directional sensitivity to head movement is coded by opposing receptor signals.* Because of commissural connections, neurons in the vestibular nuclei receive information from receptors on both sides of the head. These neurons act as *comparator units* that interpret head rotation on the basis of the relative discharge rates of left and right canal afferents. This pattern of connections also increases the sensitivity of the system, so that even small differences in the discharge rates of afferents from corresponding canal pairs (as in slow head movements) can be perceived. During a leftward head turn, the comparator units receive impulses at a higher frequency from the left horizontal canal compared to the right horizontal canal; this is interpreted as a left head turn. Similar conditions exist when the head is pitched or rolled so that the vertical semicircular canals are stimulated by rotational accelerations in their respective planes. However, in the case of the vertical canals, the opposing push-pull responses occur between the anterior semicircular canal in one ear and the posterior semicircular canal of the opposite ear (Fig. 21-3).

Head trauma or disease can change the normal resting activity in eighth nerve afferent fibers. This may be interpreted by the brain as turning, even though the head is stationary. For example, a lesion of the eighth nerve, such as produced by a glomus tumor or acoustic neuroma (Fig. 21-9), may reduce the frequency of impulses in the ipsilateral afferent fibers or block their impulse transmission entirely. The comparator units of the vestibular nuclei will then consistently receive a

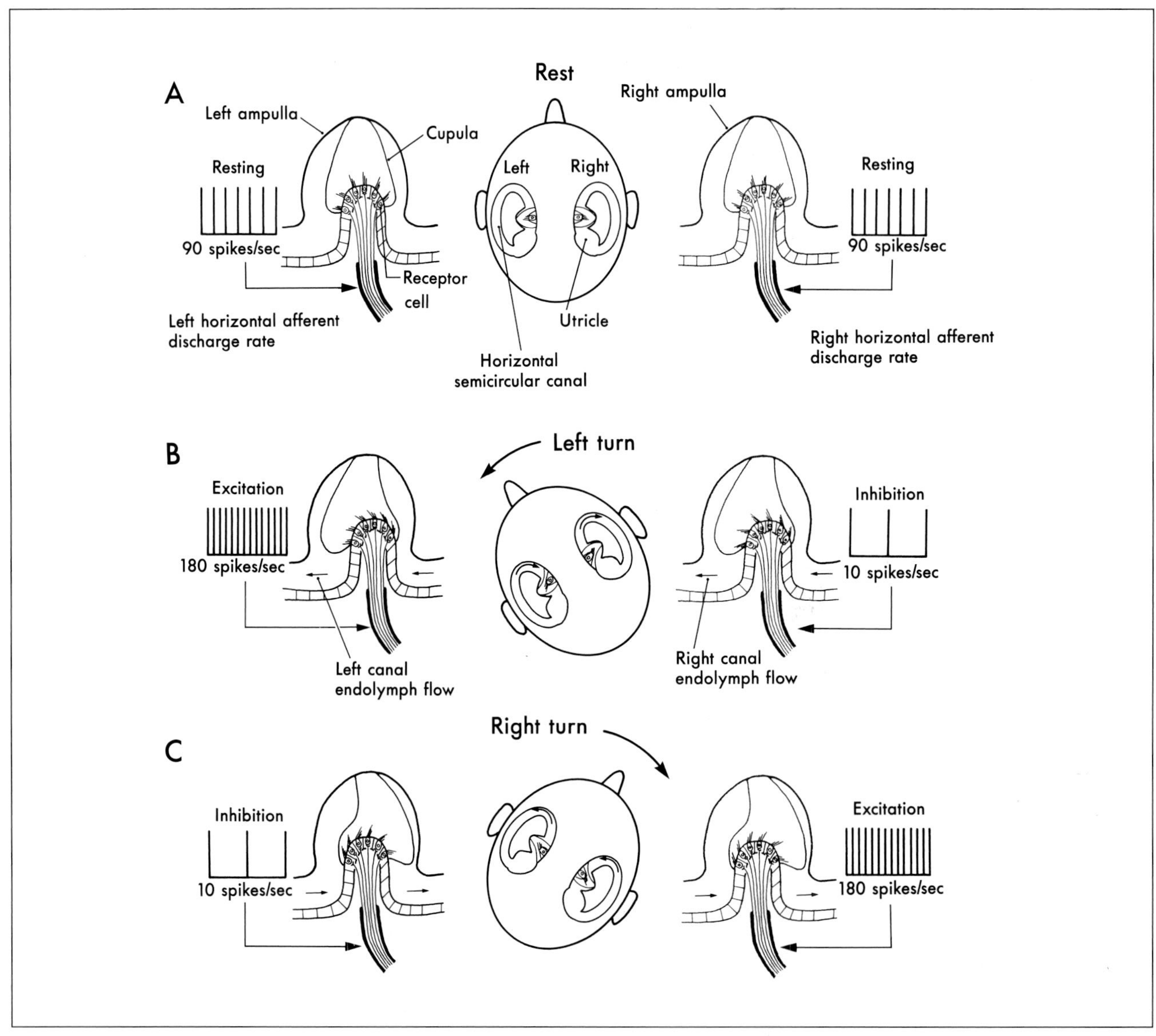

FIGURE 21-8 *Response of the horizontal semicircular canals to head rotations in the horizontal plane. At rest (A) the firing rate of horizontal canal afferents is equivalent on both sides. With a leftward head turn (B) or a rightward head turn (C), there is receptor depolarization and afferent fiber excitation toward the side of the turn and corresponding inhibition on the opposite side.*

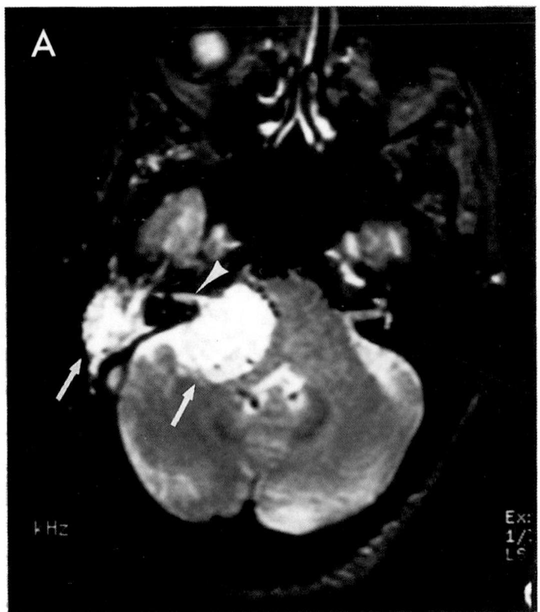

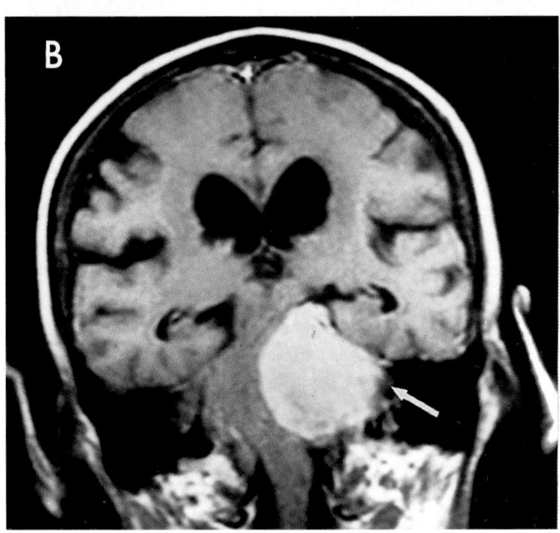

FIGURE 21-9 *Magnetic resonance imaging (MRI) scan of a glomus tumor (A) and an acoustic neuroma (B) involving the vestibular nerve. Both patients complained of dizziness, nausea, and spatial disorientation.*

higher impulse frequency from the intact side, which will be interpreted as a head turn away from the lesioned side.

Function of Otolith Organs The receptor hair cells in the maculae do not respond to head rotation but are sensitive to linear acceleration and tilt of the head (Fig. 21-10). When the head is moved with respect to gravity (*rolled* or *pitched*), the otoconia crystals are displaced because of their density with respect to the surrounding endolymph. This shifts the underlying gelatinous coating on the maculae and produces stereocilia deflection in the hair cells. Similar to the responses of semicircular canal hair cells, otolith organ hair cells are either depolarized or hyperpolarized with stereocilia deflection toward, or away from, the kinocilium, respectively. However, hair cells on the maculae are oriented according to their position relative to the striola (Fig. 21-7). Hair cells on one side of the striola will be depolarized, and hair cells on the other side of the striola will be hyperpolarized (Fig. 21-10). Because the striola is curved, only certain groups of cells will be affected by a specific direction of head tilt or linear acceleration. Thus, movement is encoded by a macular map of directional space. The eighth nerve fibers maintain the directional signal because each afferent only innervates hair cells from a small region on the macular neuroepithelium.

VESTIBULAR NUCLEI

Neural information carried on vestibular afferent fibers is transmitted to the four vestibular nuclei, which lie in the rostral medulla and caudal pons (Fig. 21-11). The *superior vestibular nucleus* lies dorsally in the central pons and is bordered by the restiform body and the fourth ventricle (Fig. 21-11B). The *medial vestibular nucleus* lies in the lateral floor of the fourth ventricle throughout most of its rostrocaudal extent (Fig. 21-11B–E). The *lateral vestibular nucleus* lies lateral to the medial vestibular nucleus (Fig. 21-11B,C) and contains some large neurons known as Deiter's cells. Located lateral to the medial vestibular nucleus, the *inferior* (or *descending*) vestibular nucleus extends through much of the medulla (Fig. 21-11D–F).

The processing of positional and movement information for control of visual and postural reflexes largely takes place in the vestibular nuclei. Consequently, the major targets for efferents of the vestibular nuclei include the oculomotor nuclei, the vestibulocerebellum, the contralateral vestibular nuclei, the spinal cord, the reticular formation, and the thalamus. Each vestibular nucleus differs in its cytoarchitecture and its afferent and efferent connections.

Vestibular Afferent Inputs Vestibular primary afferent fibers enter the brainstem at the pontomedullary junction. These fibers traverse the restiform body, then bifurcate into ascending and descending branches. Afferent fibers from the semicircular canals project primarily to the superior and medial vestibular nuclei, although lesser inputs also reach the lateral and inferior vestibular nuclei (Fig. 21-12). The otolith organs project primarily to the lateral, medial, and inferior vestibular nuclei. Saccular afferents also project to cell group Y which, in turn, excites neurons in the contralateral oculomotor nucleus and influences vertical eye movements.

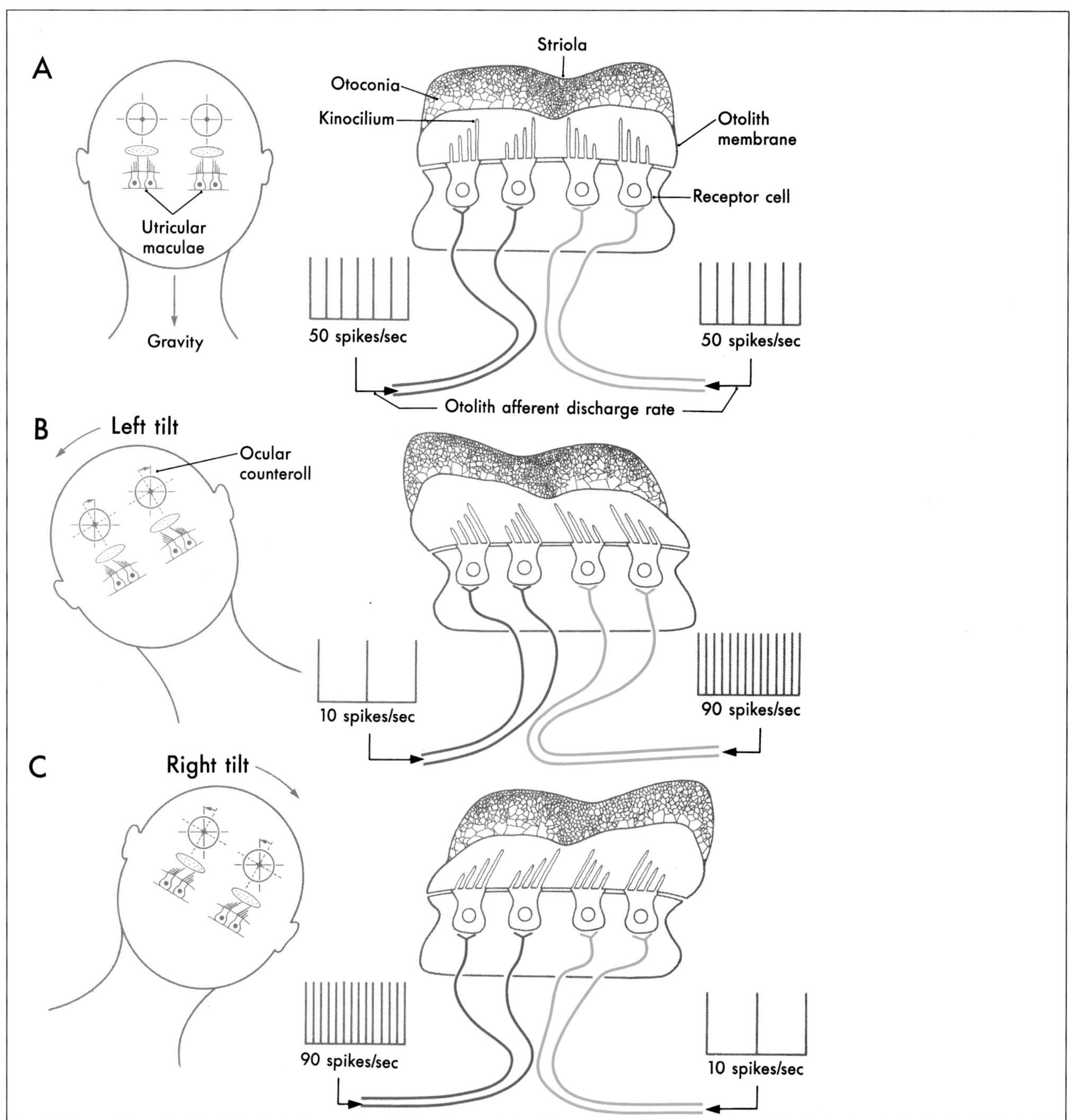

FIGURE 21-10 *Responses of the utricular maculae to tilts of the head. When the head is upright (A), the afferent fibers have equivalent firing rates on both sides of the striola (red and blue lines). With leftward tilt (B) or rightward tilt (C), hair cells and their innervating afferents are either excited or inhibited, depending on their position relative to the striola; the weight of the otoconia causes the stereocilia to be deflected. Hair cells on the "upslope" side of the striola increase their firing rate, and those on the "downslope" side decrease their firing rate.*

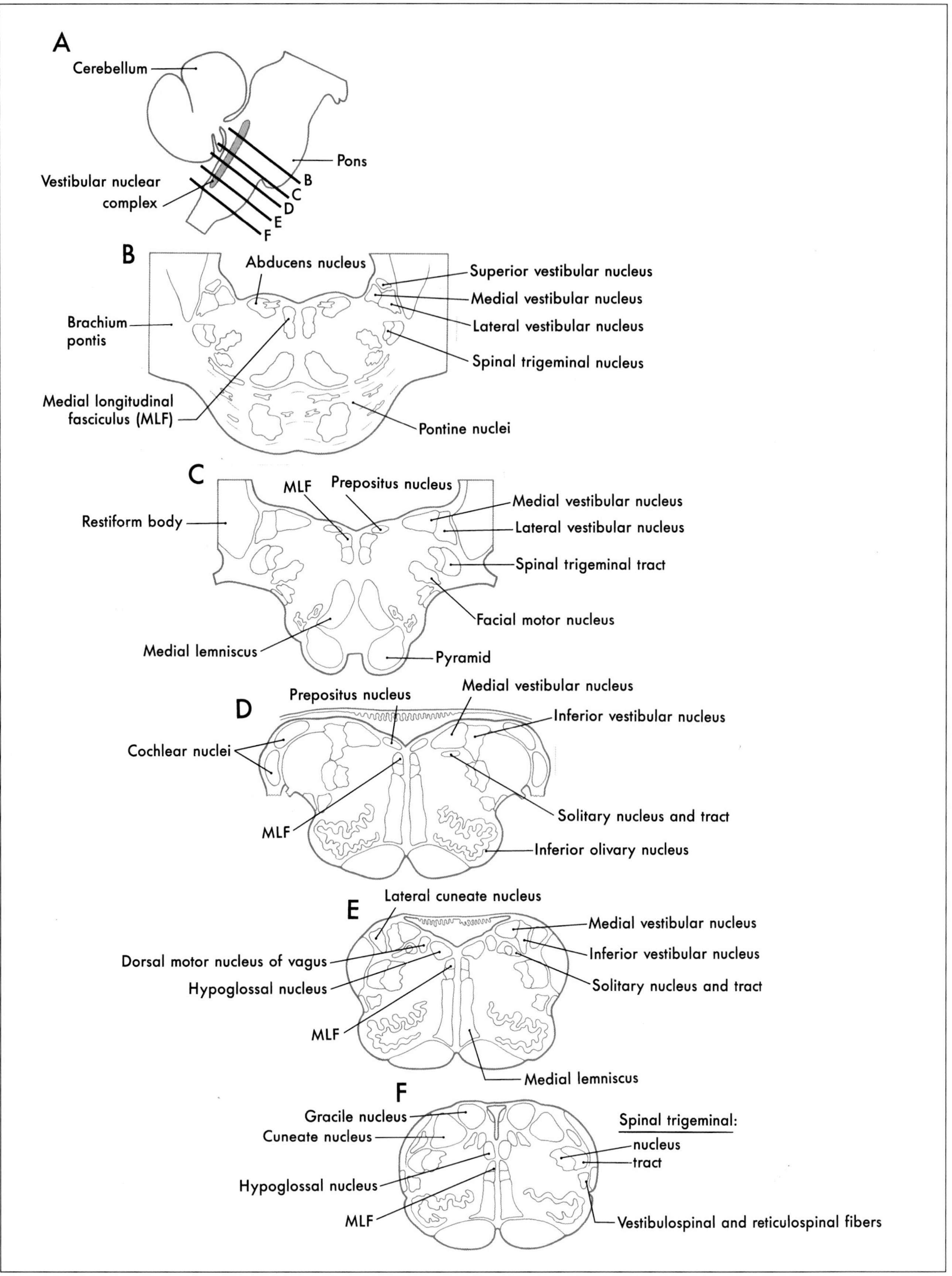

FIGURE 21-11 *Location of the vestibular nuclei in the brainstem in sagittal (A) and representative cross-sectional planes (B–F from levels indicated in A).*

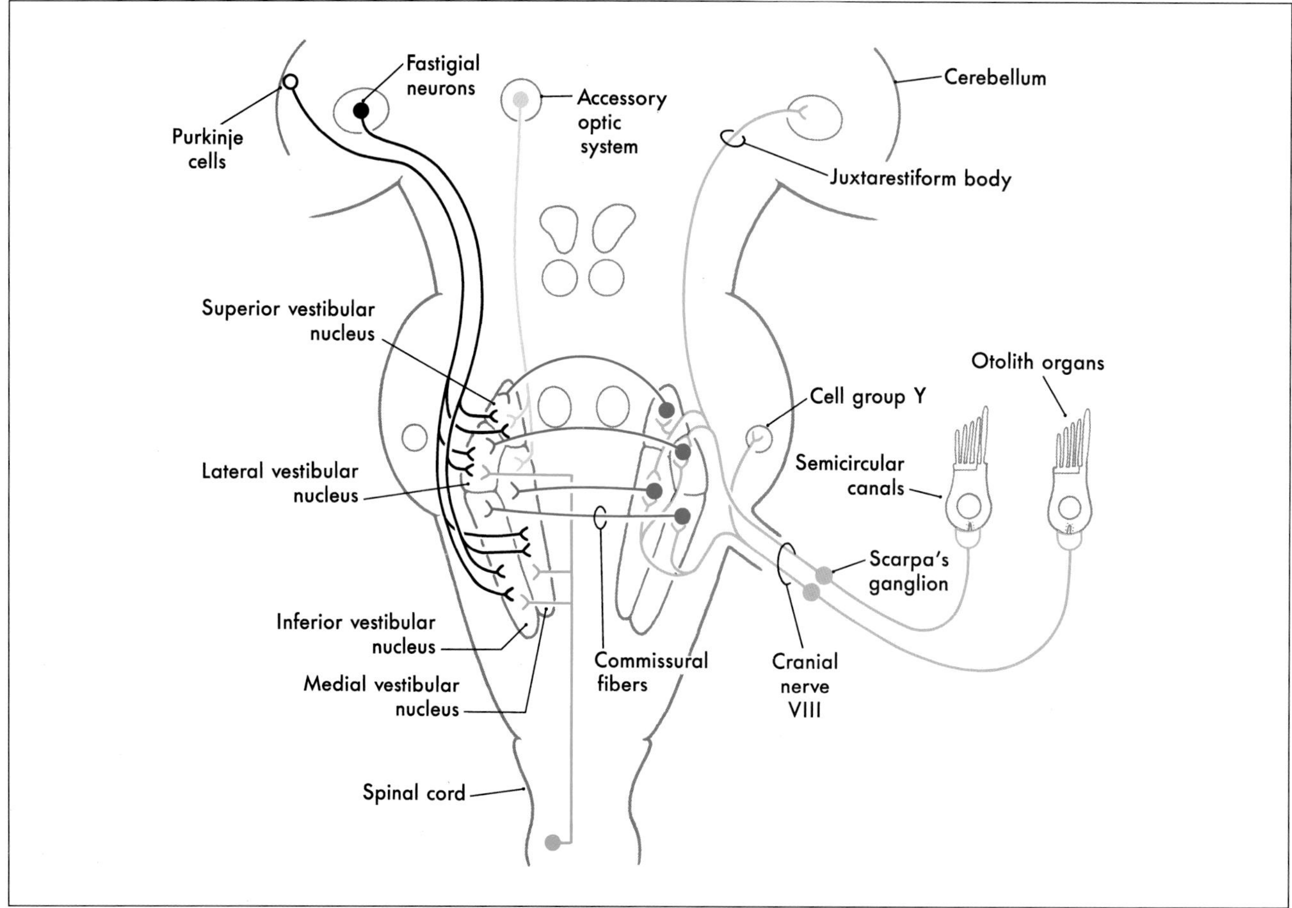

FIGURE 21-12 *Afferents to the vestibular nuclei. Open cell bodies represent inhibitory projections.*

The termination of vestibular afferent fibers on neurons of the vestibular nuclei is highly ordered. Individual central neurons in the superior and medial vestibular nuclei appear to receive information from otolith receptors and from one semicircular canal pair (either horizontal or vertical). Vestibular neurons in the lateral and inferior nuclei mostly receive information from several canal pairs and otolith receptors. As a result of their inputs, neurons in the vestibular nuclei show directional selectivity for particular head movements and can encode both the angular and linear components of head movements. These cells distribute information about both the direction and speed of the head movement, as well as the position of the head with respect to gravity, to many different regions of the brain.

Cerebellar Connections The vestibular labyrinth is the only sensory organ in the body that sends direct primary afferent projections to the cerebellar cortex and nuclei (Fig. 21-12). These *primary vestibulocerebellar* fibers course through the *juxtarestiform body*, the smaller medial part of the inferior cerebellar peduncle. Primary vestibulocerebellar fibers send collaterals to the dentate nucleus and terminate as mossy fibers in the nodulus, the uvula, and perhaps the flocculus. Neurons in all four vestibular nuclei also send axons to the cerebellum as *secondary vestibulocerebellar* projections. These axons end in the nodulus, uvula, flocculus, fastigial nucleus, and dentate nucleus.

The cerebellum forms reciprocal connections with the vestibular nuclei. This cerebellovestibular projection includes Purkinje cell axons (*cerebellar corticovestibular fibers*) from the nodulus, uvula, flocculus, and other areas of the cerebellar vermis. In addition, projections from the fastigial nucleus (*fastigiovestibular fibers*) also innervate the vestibular nuclei. Purkinje cells are GABAergic, and therefore inhibitory, whereas the fastigiovestibular fibers use glutamate or aspartate and are excitatory. These *vestibulocerebellar* and *cerebellovestibular* fibers all pass through the juxtarestiform body. The reciprocal connections between the cerebellum and the vestibular nuclei comprise important regulatory mechanisms for the control of eye movements, head movements, and posture.

Commissural Connections Commissural *vestibulovestibular fibers* arise from all vestibular nuclei, but they appear to be most prominent from the superior and medial nuclei.

Many of these fibers form reciprocal connections with the analogous contralateral nucleus. Most vestibulovestibular cells contain the inhibitory neurotransmitters GABA or glycine, although some may use the excitatory amino acids. These commissural fibers provide the pathways by which information from pairs of corresponding semicircular canals and otolith organs can be compared. Commissural fibers also play a major role in *vestibular compensation*, a process by which reflexes and postural control that are impaired as a result of unilateral loss of vestibular receptor function (through trauma or disease) are restored gradually by means of central adjustment.

Other Afferent Connections *Spinovestibular fibers* arise from all levels of the spinal cord and provide proprioceptive input primarily to the medial and lateral vestibular nuclei. Information concerning the movement of the head through the visual world also reaches vestibular nuclei neurons through the *accessory optic system* (see Chapter 27). Finally, vestibular nuclear neurons receive input from the reticular formation, primarily from cells relaying information regarding proprioception.

VESTIBULO-OCULAR NETWORK

It is often necessary to keep the gaze fixed on an object of interest while the head is moving—as in reading a sign on a building while walking down the street. The vestibular system provides this capability by eliciting compensatory eye movements through a network of neural connections. These elicited stabilizing eye movements, collectively known as the *vestibulo-ocular reflex*, are said to be *compensatory* because they are equal in magnitude and opposite in direction to the head movement perceived by the vestibular system. The vestibulo-ocular reflex occurs for any direction or speed of head movement, whether the movement is rotational, linear, or a combination of both. The reflex can also be suppressed at will if, for example, one wishes to focus on a moving target while turning the head in the same direction (as when watching tennis volleys from a position next to the net).

Rotational Vestibulo-ocular Reflex There are three types of rotationally induced eye movements: *horizontal*, *vertical*, and *torsional*. Each of the six pairs of eye muscles must be controlled in unison to produce the appropriate response. Thus, the vertical semicircular canals and the saccule are responsible for controlling vertical eye movements, whereas the horizontal canals and utricle control horizontal eye movements. Torsional eye movements are controlled by the vertical semicircular canals and the utricle.

For purposes of example, the horizontal vestibulo-ocular reflex is described (Fig. 21-13). Primary afferents from the horizontal semicircular canals project to specific neurons in the medial and lateral vestibular nuclei. Most of these cells send excitatory signals through the *medial longitudinal fasciculus* to the *contralateral abducens nucleus*. Abducens motor neurons send impulses via the sixth cranial nerve to excite the *ipsilateral lateral rectus muscle*. At the same time, abducens interneurons send excitatory signals to motor neurons in the *contralateral oculomotor nucleus*, which innervates the *medial rectus muscle*. A second population of vestibular neurons sends excitatory signals to the medial rectus subdivision of the ipsilateral oculomotor nucleus. A third group of vestibular neurons carry inhibitory signals to the ipsilateral abducens nucleus.

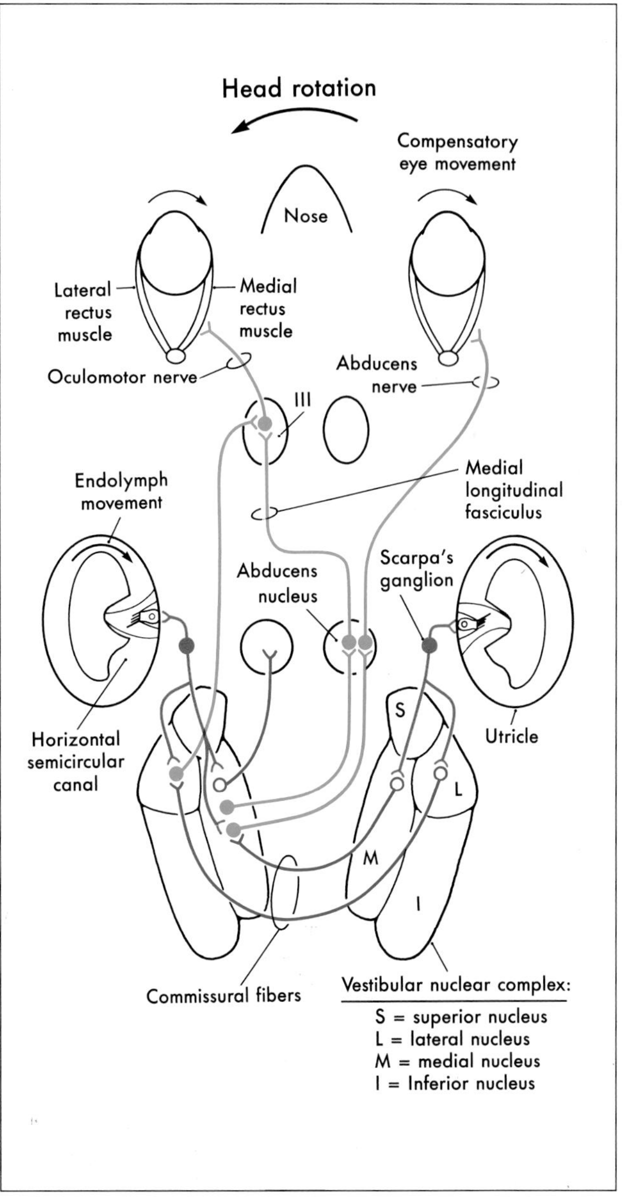

FIGURE 21-13 *The connections subserving the horizontal vestibulo-ocular reflex. Open cell bodies represent inhibitory projections. III, oculomotor nucleus.*

During a *leftward head turn*, excitatory signals from the left horizontal semicircular canal afferents increase the firing rate of neurons in the left vestibular nuclei neurons (Fig. 21-13). At the same time, inhibitory signals from the right vestibular nuclei are decreased via commissural neurons. Neurons in the left vestibular nuclei then excite contralateral abducens motor neurons and interneurons, which, in turn, produce contraction in the right lateral rectus and the left medial rectus muscles (Fig. 21-13). The resulting *rightward eye movement* keeps the object of interest on the fovea. Through matching bilateral connections, the left lateral rectus and right medial rectus eye muscles are inhibited.

A similar pattern of connections links the vertical semicircular canals with the motor neurons in the trochlear and oculomotor nuclei to control vertical and torsional responses (see also Chapter 27). The vertical vestibulo-ocular reflex originates primarily from neurons in the superior vestibular nucleus, although some medial vestibular nucleus neurons also participate.

Linear Vestibulo-ocular Reflex During linear movements that do not involve head rotation, an appropriate vestibulo-ocular reflex also occurs. These reflexes depend on input from the otolith organ receptors and involve connections to the extraocular motor neuron pools that are similar to those described above for the rotational vestibulo-ocular reflex. For example, side-to-side head movements result in a horizontal eye movement in a direction opposite to the head movement. Vertical displacements of the body, such as occur during walking or running, elicit oppositely directed vertical eye movements to stabilize the gaze. During roll tilts of the head, the compensatory eye movement is termed *counter-roll* and is actually a torsional eye movement (Fig. 21-10).

Nystagmus With large head rotations, such as a 360-degree body turn, compensatory eye movements take another form (Fig. 21-14). Initially, the vestibulo-ocular reflex directs the eyes slowly in the direction opposite to the head motion. This is called the *slow phase*. When the eye reaches the limit of how far it can turn in the orbit, it springs back rapidly to a central position, moving in the same direction as the head. This is the *fast phase*. Another slow phase then begins. This combination of slow compensatory phases punctuated by fast return phases is called *nystagmus*. Nystagmus movements are named for the direction of the fast return phase—for example, as leftward-beating nystagmus or downward-beating nystagmus. Nystagmus takes many forms and is often observed clinically (see also Chapter 27). In cases of head injury with acute temporal bone fracture, the semicircular canals can be affected, producing rapid spontaneous nystagmus that can persist for hours or days.

Nystagmus can be used as a diagnostic indicator of vestibular system integrity. Typically, in patients complaining of dizziness or vertigo, the function of the vestibular labyrinth is assessed by administering a *caloric test*. Either warm (40°C) or cold (30°C) water is introduced into the external auditory canal. In normal individuals, warm water induces nystagmus that beats toward the ear into which the water has been introduced, whereas cold water induces nystagmus that beats away from the ear into which the water has been introduced. (This relationship is encapsulated in the mnemonic COWS: *C*old water produces nystagmus beating to the *o*pposite side; *w*arm water produces nystagmus beating to the *s*ame side.) In normal individuals, the two ears give equal responses. If there is a unilateral lesion in the vestibular pathway, however, nystagmus will be reduced or absent on the side of the lesion.

VESTIBULOSPINAL NETWORK

The vestibular system influences muscle tone and produces reflex postural adjustments of the head and body through two major descending pathways to the spinal cord, the *lateral vestibulospinal tract* and the medial vestibulospinal tract (Fig. 21-15). There is also a reticulospinal pathway that receives input from the vestibular system.

Lateral Vestibulospinal Tract The lateral vestibulospinal tract (LVST) arises primarily from neurons in the lateral and inferior vestibular nuclei and projects to all

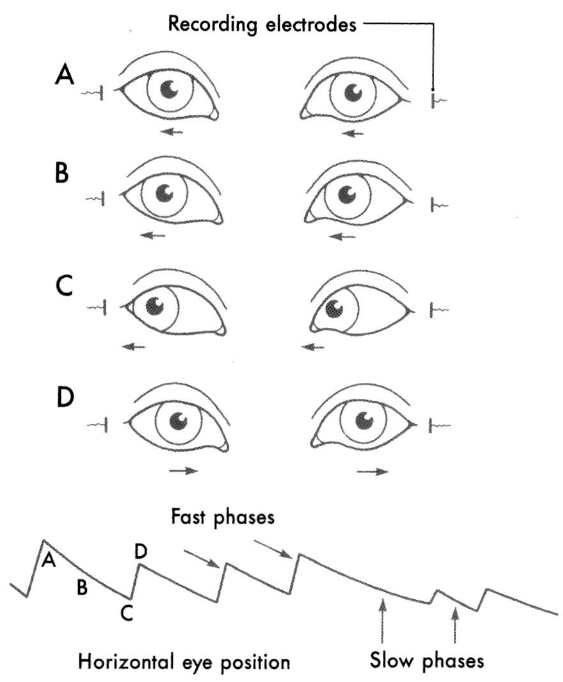

FIGURE 21-14 *Vestibular nystagmus in the horizontal plane showing slow phase (A–C) and fast phase (D) eye movements.*

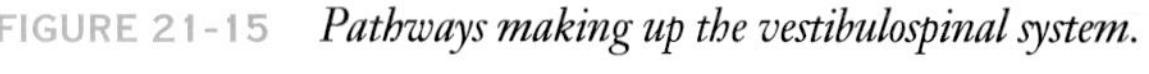

FIGURE 21-15 *Pathways making up the vestibulospinal system.*

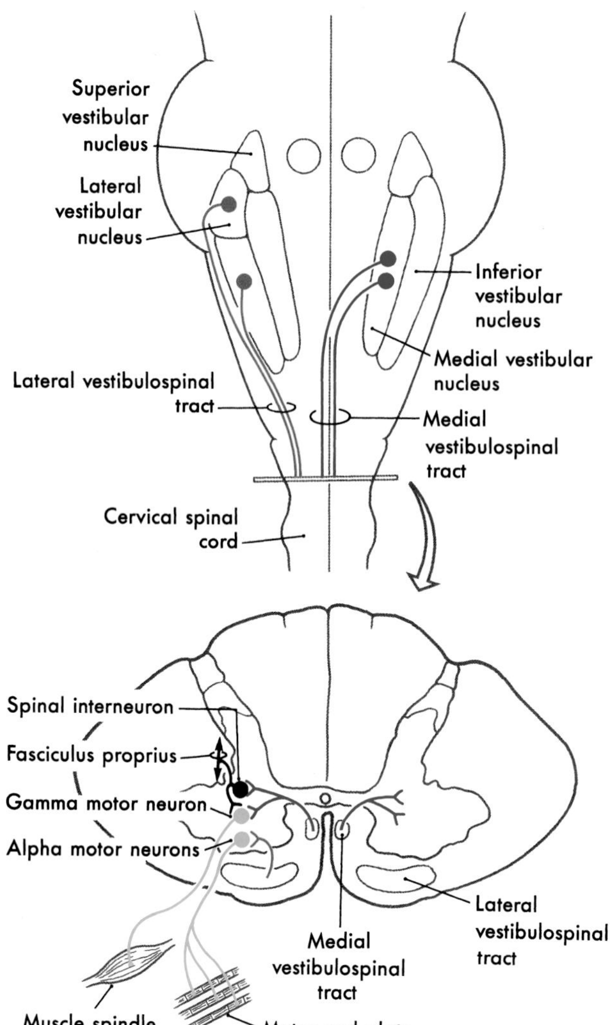

levels of the ipsilateral spinal cord. This projection is topographically organized. Cells in rostroventral areas of the lateral nucleus project to the cervical cord, while cells in dorsocaudal regions project to the lumbosacral cord. These vestibulospinal neurons receive substantial input from orthogonal semicircular canal pairs, from the otolith organs, the vestibulocerebellum, and the fastigial nucleus, as well as proprioceptive inputs from the spinal cord.

Fibers of the LVST course through the lateral medulla dorsal to the inferior olivary complex (Fig. 21-11), then through the ventral funiculus of the cord (Fig. 21-15) to terminate directly on alpha and gamma motor neurons and on interneurons in lamina VII to IX. Axons of many LVST neurons give off collaterals in different segments of the cord, thus ensuring that different muscle groups will be coordinated during postural control. The LVST neurons contain either acetylcholine or glutamate as a neurotransmitter and exert an excitatory influence on extensor muscle motor neurons. The coordinated actions of neurons that make up the LVST and provide postural stabilization are not completely understood. However, if a person begins tilting to the right, ipsilateral LVST fibers elicit extension of the left axial and limb musculature. Concurrently, right extensor muscles are inhibited. These actions stabilize the body's center of gravity and preserve upright posture.

Medial Vestibulospinal Tract The action of vestibular stimulation on neck muscles arises primarily through neurons in the medial vestibulospinal tract (MVST). These fibers originate primarily from the medial vestibular nucleus, although lesser projections arise from the inferior and lateral vestibular nuclei. Similar to LVST neurons, cells of the MVST receive input from vestibular receptors and the cerebellum, as well as somatosensory information from the spinal cord. Fibers of the MVST descend bilaterally through the medial longitudinal fasciculus to terminate in laminae VII to IX of the cervical spinal cord (Fig. 21-15). These MVST fibers carry both excitatory and inhibitory signals, and they terminate on neck flexor and extensor motor neurons, as well as propriospinal neurons.

The effects of vestibular-induced responses can be seen in the *vestibulocollic* reflex, which is actually a series of responses that stabilize the head in space. If, for example, one falls forward, MVST neurons will receive signals on downward linear acceleration from the saccule, signals on the changing head position relative to gravity from both the utricle and saccule, and signals on forward rotational acceleration from the vertical semicircular canals. The MVST neurons process this information and transmit excitatory signals to the dorsal neck flexor muscles (splenius, biventer cervicus, and complexus muscles). At the same time, inhibitory signals are sent to the ventral neck extensor muscles. The result is a neck movement upward, opposite to the falling motion, to protect the head from impact.

VESTIBULO-THALAMO-CORTICAL NETWORK

Vestibular Thalamus The cognitive perceptions of motion and spatial orientation arise through the convergence of information from the vestibular, visual, and somatosensory systems at the thalamocortical level. Neurons in the superior, lateral, and inferior vestibular nuclei project bilaterally to two thalamic areas (Fig. 21-16). The first is located in the *ventral posterolateral (VPL) nucleus and includes adjacent cells in the ventral posteroinferior (VPI) nucleus*. The second is the *posterior nuclear group*, located near the medial geniculate body. In humans, electrical stimulation of these areas produces sensations of movement and dizziness. Thalamic

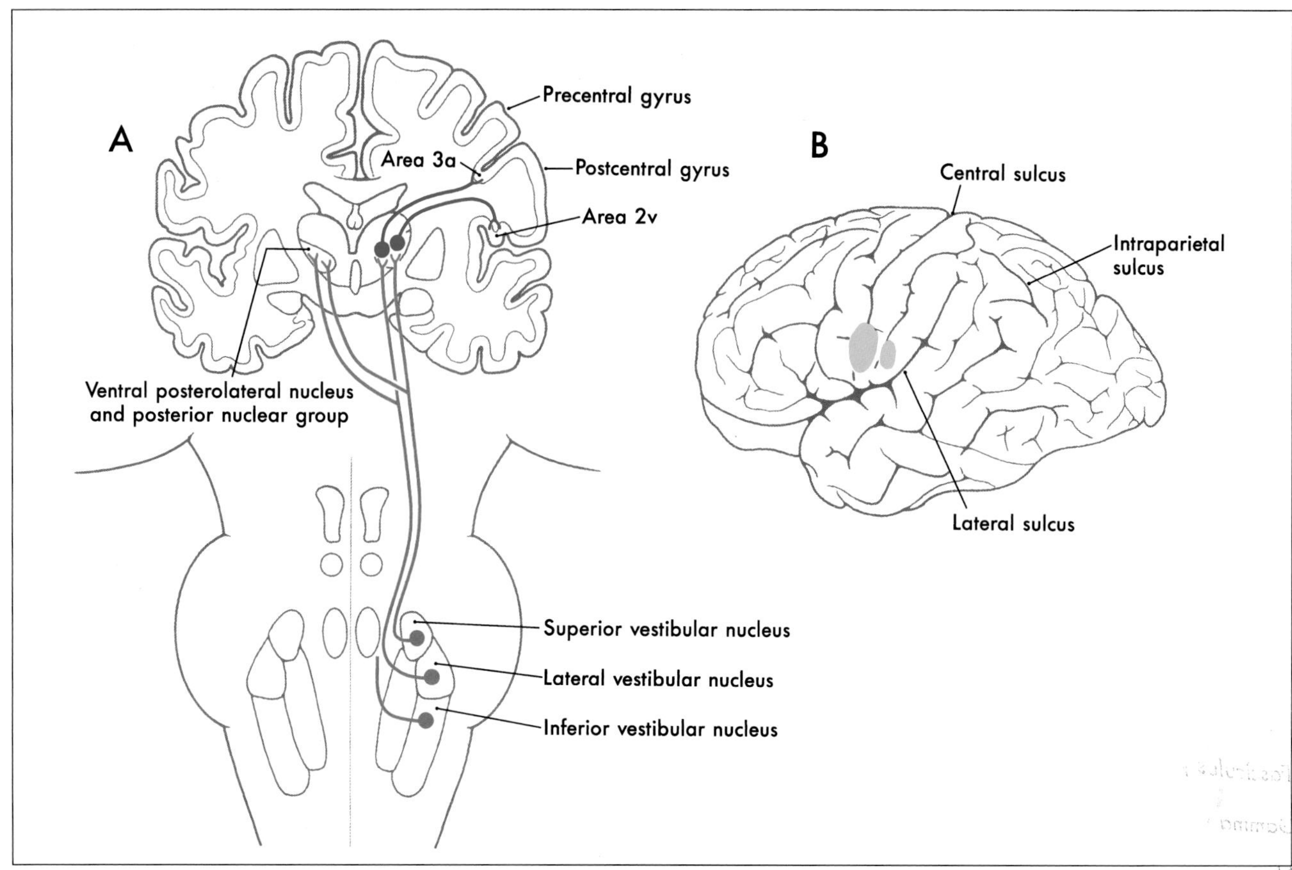

FIGURE 21-16 *The vestibulo-thalamo-cortical pathway. Vestibular input arises from the vestibular nuclei as vestibulothalamic fibers and is relayed to the cortex as thalamocortical fibers (A). Areas 3a and 2v (B) are the main cortical regions that receive this input.*

VPL and posterior nucleus neurons constitute separate, parallel pathways transmitting vestibular information from the brainstem to the cortex, because their connections with cortical areas are distinct.

Vestibular Cortex Two cortical areas respond to vestibular stimulation (Fig. 21-16). One region, *area 2v*, lies at the base of the intraparietal sulcus just posterior to the hand and mouth representations in the postcentral gyrus. Electrical stimulation of this area in humans produces sensations of moving, spinning, or dizziness. Area 2v neurons respond to head movements and receive projections from the posterior thalamic nucleus. These 2v cells also receive visual and proprioceptive inputs. Area 2v is probably involved with motion perception and spatial orientation, because it has reciprocal connections with other parietal regions involved in similar functions (e.g., areas 5 and 7). Lesions of parietal cortical areas result in confusions in spatial awareness.

The second cortical area responding to vestibular stimulation, *area 3a*, lies at the base of the central sulcus, adjacent to the motor cortex (Fig. 21-16) and receives input from VPL or VPI thalamic nuclear neurons. In addition, area 3a cells receive inputs from the somatosensory system. Because these cells project to area 4 of the motor cortex, it is believed that one of their functions is to integrate motor control of the head and body.

DIZZINESS AND VERTIGO

Dizziness is a nonspecific term that generally means a spatial disorientation that may or may not involve feelings of movement. Dizziness may be accompanied by nausea and/or postural instability. A large number of factors may produce a dizzy sensation, and many are not exclusively vestibular in origin.

Vertigo is a specific perception of body motion, often spinning or turning, when no real motion is taking place. As children, we all learn to produce vertigo by whirling in place as fast as possible and then abruptly stopping. For a minute the world seems to be spinning in the opposite direction. Examination of the eyes during this phase will reveal a nystagmus that beats in the direction opposite to the original direction of rotation. Vertigo can also be elicited optokinetically if the visual surroundings are revolved while the body remains stationary. Many modern amusement games take advantage of this phenomenon to produce the sensation of motion.

One of the most common vestibular disorders observed clinically is *benign positional vertigo*. This condition is characterized by brief episodes of vertigo that coincide with particular changes in body position. Typically, episodes may be triggered by turning over in bed, getting up in the morning, bending over, or rising from a bent position. The pathophysiology of benign positional vertigo is not clearly understood, but posterior canal abnormalities are implicated. One possible explanation is that otoconial crystals from the utricle separate from the otolith membrane and become lodged in the cupula of the posterior canal (a condition called *cupulolithiasis*). The resulting increased density of the cupula produces abnormal cupula deflections when the head changes position relative to gravity.

SOURCES AND ADDITIONAL READINGS

Baloh RW, Honrubia V: *Clinical Neurophysiology of the Vestibular System*, FA Davis, Philadelphia, 1990.

Buttner U, Lang W: The vestibulocortical pathway: Neurophysiological and anatomical studies in the monkey. In Grant R, Pompeiano O (eds): *Progress in Brain Research*, Vol 50. *Reflex Control of Posture and Movement*. Elsevier, Amsterdam, 1979.

Goldberg JM: The vestibular end organs: morphological and physiological diversity of afferents. Curr Opin Neurobiol 1:229–235, 1991.

Highstein SM, McCrea RA: The anatomy of the vestibular nuclei. In Buttner-Ennever JA (ed): *Reviews of Oculomotor Research*, Vol 2. Neuroanatomy of the Oculomotor System. Elsevier, Amsterdam, 1988.

Hudspeth AJ: How the ear's works work. Nature 341:397–404, 1989.

Wilson VJ, Melvill Jones G: *Mammalian Vestibular Physiology*. Plenum Press, New York, 1979.

Wilson VJ, Peterson BW: Vestibular and reticular projections to the neck. In Peterson BW, Richmond FJ (eds): *Control of Head Movement*. Oxford University Press, Oxford, 1988.

CHAPTER TWENTY-TWO

Olfaction and Taste

R. D. SWEAZEY

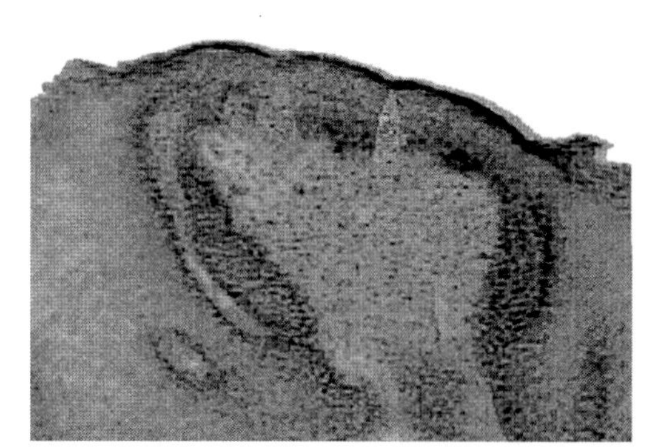

The olfactory and taste systems sample the rich chemical environment that surrounds us. Information provided by these systems is intimately associated with the enjoyment of foods and beverages. When we refer to the taste of food, what we mean is a complex sensory experience correctly called *flavor*. Flavor perception results from a combination of the olfactory, taste, and tactile cues present in foods and beverages. *Olfaction* is the sensation of odors that results from the detection of odorous substances aerosolized in the environment. In contrast, taste (*gustation*) is the sensation evoked by stimulation of taste receptors located in the oropharyngeal cavity.

OVERVIEW

For many mammals, smell is the principal means by which information about the environment is received. *Macrosmatic* animals have a well-developed sense of smell on which they rely for recognizing food, detecting predators and prey, and locating potential mates. In animals, such as humans, who are less dependent on smell (*microsmatic* animals), the olfactory system is less well developed. However, humans are still able to distinguish thousands of odors, many at extremely low concentrations. Through connections with cortical and limbic structures, the olfactory system plays a role in the pleasures associated with eating and with the many scents that make up our world.

In contrast to olfaction, the taste system exhibits a limited range of sensations. Traditionally, taste sensations are divided into sweet, salty, sour, and bitter. Combinations of these four qualities account for much of our taste experience. Taste input, which originates from receptors in the oropharyngeal cavity, is important for determining the acceptance or rejection of foods. This information is relayed by neural pathways that underlie various ingestive and digestive functions.

Disorders of olfaction and/or taste may adversely affect a patient's quality of life. The intimate association between the chemical senses and ingestion means that chemosensory disorders impair the patient's ability to enjoy eating. In addition, these disorders can render the patient unable to detect hazards such as gas leaks or spoiled foods.

OLFACTORY RECEPTORS

The receptors responsible for transduction of odor molecules are found in the *olfactory mucosa*. This portion of nasal mucosa is about 1 to 2 cm^2 and is located in the roof of the nasal cavity on the ventral surface of the cribriform plate and along the nasal septum and medial wall of the superior turbinate (Fig. 22-1). The olfactory mucosa is composed of a superficial acellular layer of *mucus* that covers the *olfactory epithelium* and underlying *lamina propria*. The olfactory epithelium is differentiated from the adjacent pinkish respiratory epithelium by its faint yellowish color and greater thickness. In humans the transition between olfactory and respiratory epithelia is gradual.

The olfactory epithelium is pseudostratified and contains three main cell types: *olfactory receptor neurons*, *supporting cells* (*sustentacular cells*), and *basal cells* (Fig. 22-2A,B). The small (5 μm), bipolar somata of olfactory receptor neurons are found in the basal two-thirds of the epithelium. Each has a single thin apical dendrite and a basally located unmyelinated axon. The apical dendrite extends to the surface of the epithelium, where it terminates in a knoblike *olfactory vesicle* from which 10 to 30 nonmotile *cilia* arise and protrude into the overlying mucus layer (Fig. 22-2C,D). These olfactory cilia contain receptors for odorant molecules.

The unmyelinated axon of an olfactory receptor neuron is about 0.2 μm in diameter, making it one of the smallest in the nervous system. These axons pass through the lamina propria and group together into bundles called *olfactory filia*, which collectively make up the *olfactory nerve* (cranial nerve I) (Fig. 22-2A). The olfactory filia pass through the cribriform plate to terminate in the olfactory bulb.

Olfactory receptor cells are true neurons because they originate embryologically from the central nervous system. However, they are unique in being the only mammalian neurons that undergo continuous turnover. Each has a life span of between 30 and 60 days, and they are replaced by receptors arising from undifferentiated basal cells by mitotic division (Fig. 22-2A,C). Thus, basal cells are stem cells that give rise to the receptor cells.

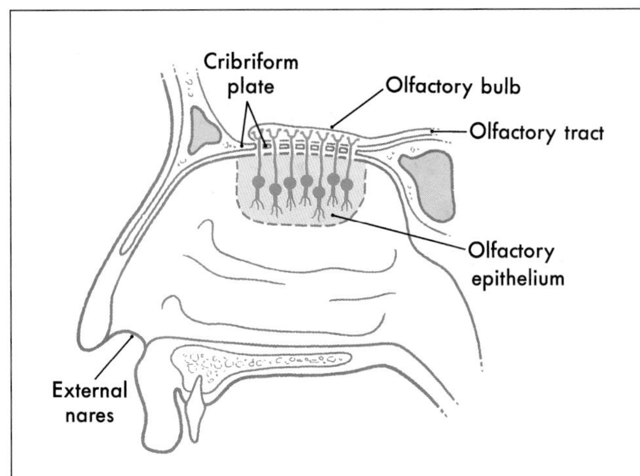

FIGURE 22-1 *Sagittal section through the human nasal cavity showing the relationship of the olfactory epithelium and bulb to the cribriform plate.*

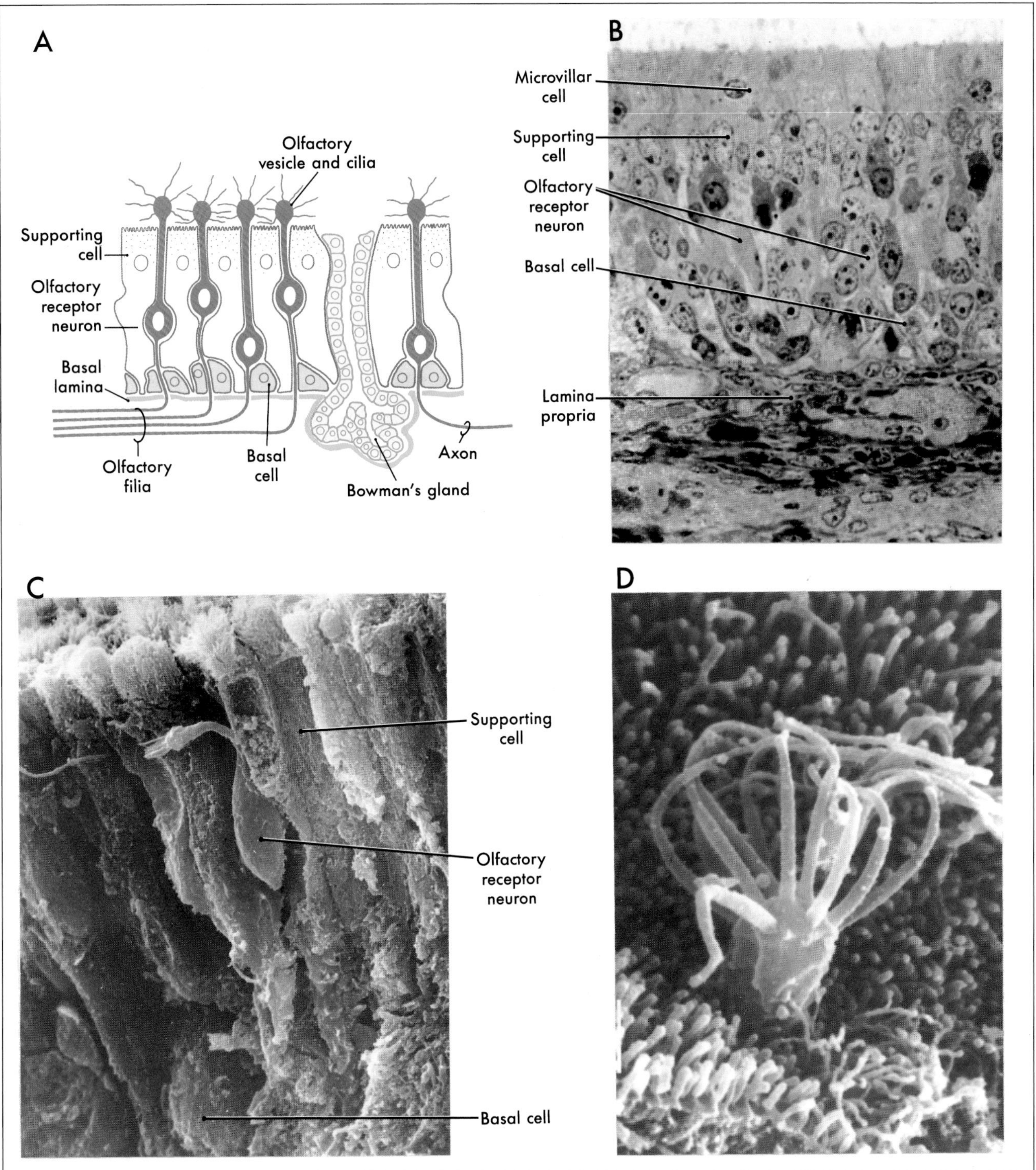

FIGURE 22-2 *Schematic drawing of the olfactory epithelium (A). Light micrograph of the human olfactory epithelium and the underlying lamina propria (B). Scanning electron micrographs of the human olfactory epithelium showing its characteristic cell types (C) and the dendritic knob and cilia of a receptor neuron (D). (Photomicrographs courtesy of Dr. Richard M. Costanzo, Virginia Commonwealth University.)*

The supporting cells are columnar and extend from the lamina propria to the surface of the epithelium, where they end in short microvilli that extend into the overlying mucus (Fig. 22-2A,C,D). Nuclei of the sustentacular cells are found near the surface of the epithelium. These cells provide mechanical support for the olfactory receptor cells (Fig. 22-2B). In addition, they contribute secretions to the overlying mucus that may play a role in the binding or inactivation of odorant molecules.

A fourth and minor cell type, the *microvillar cell*, is found in the human olfactory epithelium (Fig. 22-2B). These cells have an apical process that projects into the mucus and a basal process that extends to the lamina propria. Although their function is unknown, they may be a second type of receptor neuron.

The lamina propria contains bundles of olfactory axons, blood vessels, fibrous tissue, and numerous *Bowman's glands* (Fig. 22-2A). The serous secretions of the Bowman's glands, combined with the secretions of the sustentacular cells, provide the mucus covering of the olfactory mucosa.

Disorders of smell are normally classified according to the type of loss experienced by the patient. The loss of smell (*anosmia*) or decreased sensitivity to odorants (*hyposmia*) is frequently associated with upper respiratory infections, sinus disease, and head trauma. Nasal and paranasal diseases (*rhinitis*, *sinusitis*) block the access of odorants to the olfactory epithelium. In addition, the viruses associated with such infections may permanently damage the olfactory epithelium. Head trauma can produce olfactory deficits by damaging central olfactory pathways or olfactory receptor axons as they pass through the cribriform plate. In the latter case, shearing movements of the olfactory bulb relative to the cribriform plate transect these thin axons. For example, anosmia or hyposmia is common in boxers.

OLFACTORY TRANSDUCTION

Olfactory perception begins when volatile odor molecules are inhaled and contact the mucus layer that bathes the olfactory epithelium. This mucus is an aqueous solution of proteins and electrolytes. Odorants, particularly hydrophobic ones such as musk, cross the mucus by interacting with small, water-soluble proteins called *odorant binding proteins*. These proteins are ubiquitous in the mucus layer.

After crossing the mucus, odor molecules bind to odorant receptors on the cilia of the olfactory receptor neurons, where transduction occurs (Fig. 22-3). The *odorant receptors* are membrane proteins belonging to a superfamily of G-protein coupled receptors. Binding of the odorant to the receptor leads to activation of a second-messenger pathway involving an *olfactory-specific G-protein*, which, in turn, activates adenylyl cyclase to produce cyclic AMP (cAMP). The transient rise in ciliary cAMP opens a cyclic-nucleotide gated cation channel in the ciliary membrane, allowing cations to flow into the cell (Fig. 22-3). The flow of cations into the cell results in a gradual depolarization (*generator potential*) that travels down the dendrite to the soma of the olfactory receptor neuron. A sufficiently large depolarization initiates an action potential that travels along the axon to the olfactory bulb.

There is also evidence for another intracellular second messenger pathway in olfactory transduction. This pathway, involving inositol 1,4,5-trisphosphate (IP_3), is thought to act either separately or with the cAMP pathway. In this pathway, binding of odorant to the receptor activates a G-protein that, in turn, activates phospholipase C to produce IP_3. The IP_3 opens a channel in the ciliary membrane that permits Ca^{++} to enter the cell (Fig. 22-3).

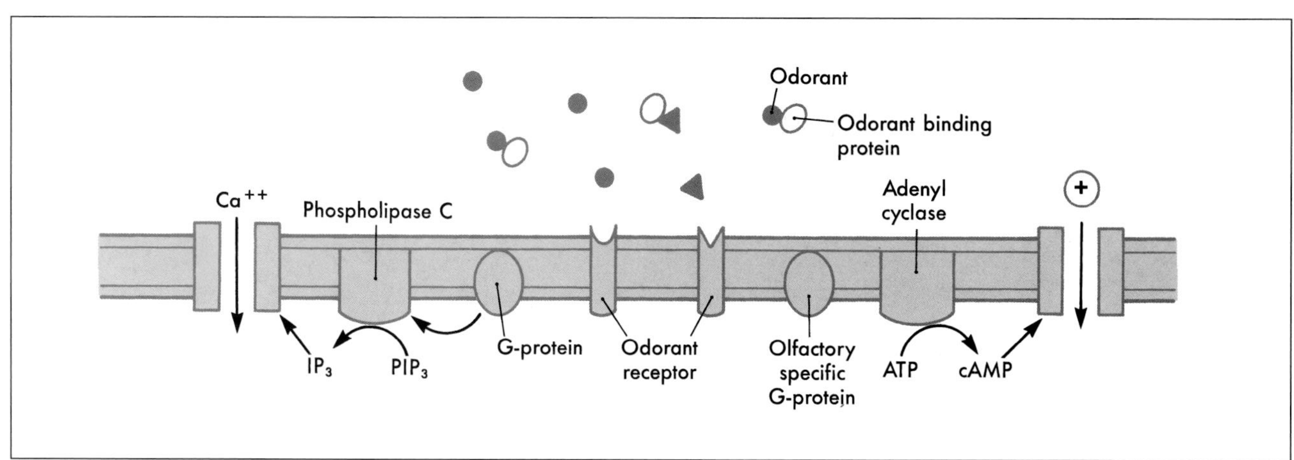

FIGURE 22-3 *Pathways of olfaction transduction. Odorants are transported through the mucus by odorant binding proteins. Binding of odorants to receptors on the olfactory cilia activates a second-messenger pathway involving either cyclic AMP (cAMP) or inositol 1,4,5-trisphosphate (IP_3). Both pathways lead to the opening of membrane cation channels and depolarization of the olfactory receptor neuron. PIP_3, phosphatidylinositol 4,5-bisphosphate.*

CENTRAL OLFACTORY PATHWAYS

Olfactory Bulb The olfactory bulb, a forebrain structure, is located on the ventral surface of the frontal lobe in the olfactory sulcus and is attached to the rest of the brain by the *olfactory tract*. The olfactory tract is an inclusive structure that contains fibers of the *lateral olfactory tract*, cells of the *anterior olfactory nucleus*, and fibers of the *anterior limb of the anterior commissure*. The latter part of the olfactory tract is the route through which many centrifugal fibers reach the olfactory bulb (see Fig. 22-6).

The olfactory bulb consists of five well-defined layers of cells and fibers, which give it a laminated appearance. From superficial to deep these are the *olfactory nerve layer*, *glomerular layer*, *external plexiform layer*, *mitral cell layer*, and *granule cell layer* (Fig. 22-4).

The afferent projections from the olfactory epithelium form the *olfactory nerve layer* on the surface of the olfactory bulb. These axons terminate exclusively in structures called *olfactory glomeruli*, which are found in the glomerular layer of the bulb (Figs. 22-4, 22-5). Their terminations, in general, preserve the topographic arrangements of the receptors.

Glomeruli are the most prominent feature of the olfactory bulb. The core of an olfactory glomerulus is made up of the axons of olfactory receptor neurons, which branch and synapse on the bushy endings of the *primary dendrites* (apical dendrites) of *mitral* and *tufted* cells (Fig. 22-4). These two cells are functionally similar and together constitute the efferent neurons of the olfactory bulb. Adjacent to the glomerulus are small interneurons (juxtaglomerular cells), of which *periglomerular cells* are the principal type. This cell has short bushy dendrites that arborize extensively within a glomerulus and a short axon that distributes within a radius of about five glomeruli.

There is significant neural convergence at the level of the olfactory glomerulus; thousands of olfactory receptor neurons form axodendritic synapses on mitral, tufted, and periglomerular cells (Fig. 22-5). These synapses are excitatory, although the transmitter is not known. The other major synaptic connections within the glomerulus are reciprocal and serial dendrodendritic synapses between mitral or tufted cells and periglomerular cells. It appears that the synapses of mitral and tufted cells onto periglomerular cells are excitatory (glutaminergic), whereas those of periglomerular cells onto mitral and tufted cells are inhibitory (GABAergic).

The glomerular layer also receives input from other central nervous system areas via *centrifugal afferents* (Figs. 22-4, 22-5). *Noradrenergic* centrifugal afferents from the *locus ceruleus* and *serotonergic* fibers from the *raphe nuclei* of the midbrain and rostral pons terminate in the glomeruli. Centrifugal fibers from the ipsilateral *anterior olfactory nucleus* and the *diagonal band* terminate in the periglomerular spaces, primarily on periglomerular cells. Excitatory amino acids, such as *glutamate*, are present in centrifugal fibers that arise in cortical structures.

The *external plexiform layer* is composed of the somata of tufted cells, along with the *primary* and *secondary dendrites* (basal dendrites) of tufted and mitral cells and the apical dendrites of *granule cells* (Fig. 22-4). Within this layer, the apical dendrites of granule cells form reciprocal dendrodendritic *GABAergic* synapses with the secondary dendrites of tufted and mitral cells. These synapses modulate tufted and mitral cell output through lateral and feedback inhibition. Mitral and tufted cells, in turn, have excitatory (*glutaminergic*) synapses on granule cell dendrites (Fig. 22-5).

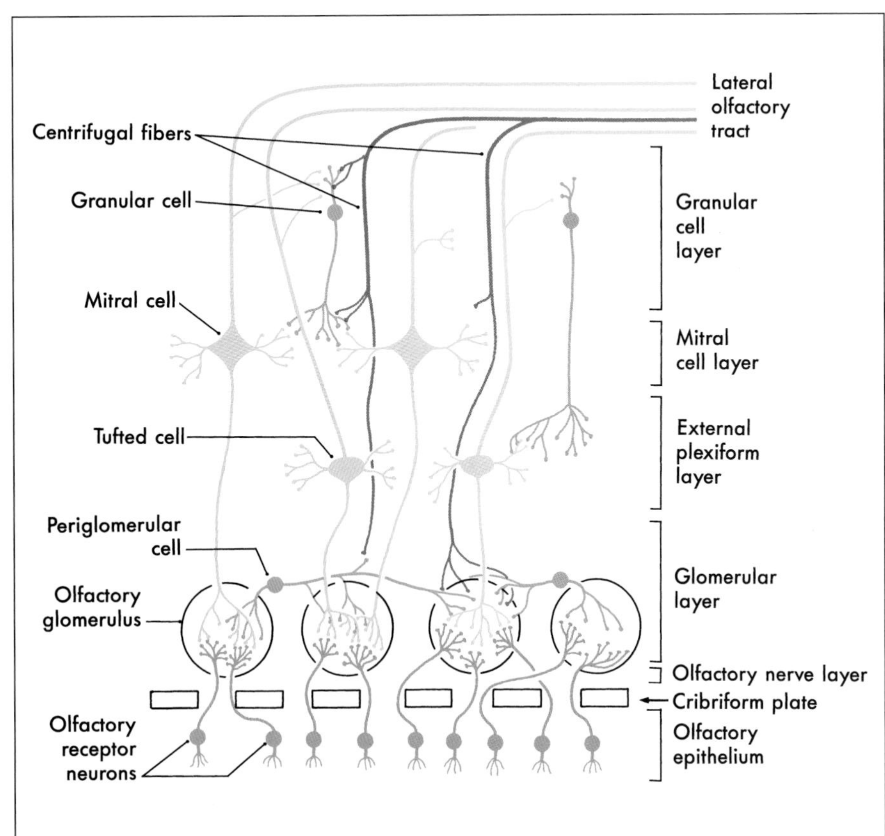

FIGURE 22-4 *Schematic drawing of the olfactory bulb, showing the laminar organization, the major cell types, and the basic neuronal circuitry. Receptor neurons are shown in blue, interneurons in red, the efferent neurons of the bulb in green, and centrifugal fibers in black.*

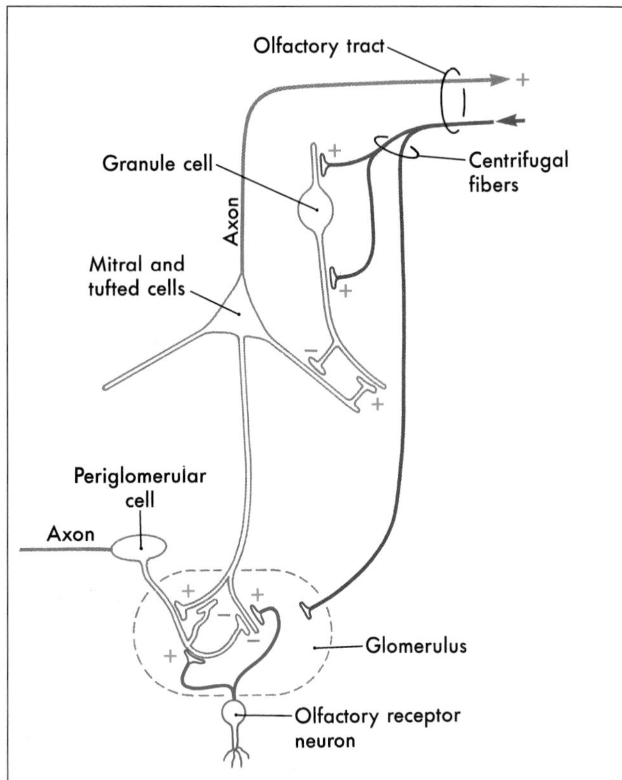

FIGURE 22-5 *Synaptic interaction between the principal cell types of the olfactory epithelium and bulb. Excitatory synapses (+) are shown in green and inhibitory ones (–) in red. The action of centrifugal axons in the glomerulus is not clearly established.*

The *mitral cell layer* is a thin layer containing the large somata of mitral cells. In addition, the axons of tufted cells, granule cell processes, and centrifugal fibers traverse this layer (Fig. 22-4).

Internal to the mitral cell layer, the *granular cell layer* contains the cell bodies of *granule cells*, the principal interneuron of the olfactory bulb. This layer also contains primary and collateral axons of mitral and tufted cells and centrifugal afferents from the anterior olfactory nucleus, olfactory cortex, cells of the diagonal band, locus ceruleus, and raphe nucleus (Fig. 22-4). Granule cells lack axons, their only output being via dendrodendritic GABAergic synapses with mitral and tufted cells. In addition, granule cells receive numerous synaptic inputs from both mitral and tufted cell axon collaterals and centrifugal afferent fibers. Granule cells presumably modulate olfactory bulb activity via an inhibitory feedback loop that shuts down the activity of the mitral and tufted neurons.

Olfactory Bulb Projections Axons of mitral and tufted cells emerge from the caudal portion of the olfactory bulb to form the *lateral olfactory tract*. Although *glutamate* is the major neurotransmitter of these efferent fibers, a few tufted cells may use dopamine. These fibers course caudally to terminate in areas on the ventral surface of the telencephalon, which are broadly defined as the *olfactory cortex* (Fig. 22-6). The principal areas making up the olfactory cortex are the *anterior olfactory nucleus, olfactory tubercle, piriform cortex, anterior cortical amygdaloid nucleus, periamygdaloid cortex*, and *lateral entorhinal cortex*. The olfactory cortex is an example of *paleocortex*, a phylogenetically older type of cortex that is less complex than the neocortex. Throughout most of its extent, olfactory cortex has three cell layers, as opposed to the six layers characteristic of neocortex. A unique aspect of the olfactory system is that the olfactory bulb projects directly to the cortex. In other sensory systems, information reaches the cortex after a relay in the thalamus.

Lateral olfactory tract axons send collaterals to the anterior olfactory nucleus, to other areas of olfactory cortex, and to subcortical limbic structures. The major targets of the anterior olfactory nucleus are the olfactory bulbs bilaterally and the contralateral anterior olfactory nucleus (Figs. 22-6, 22-7). The large numbers of interbulbar connections, via the anterior olfactory nucleus, suggest that interhemispheric processing of odors plays an important role in olfactory functions.

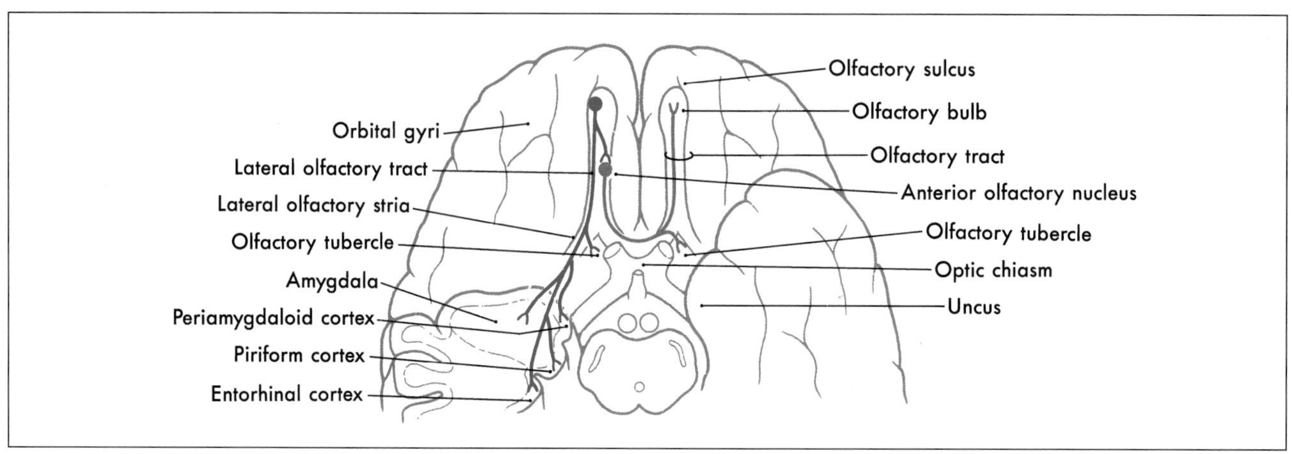

FIGURE 22-6 *Major efferent projections of the olfactory bulb.*

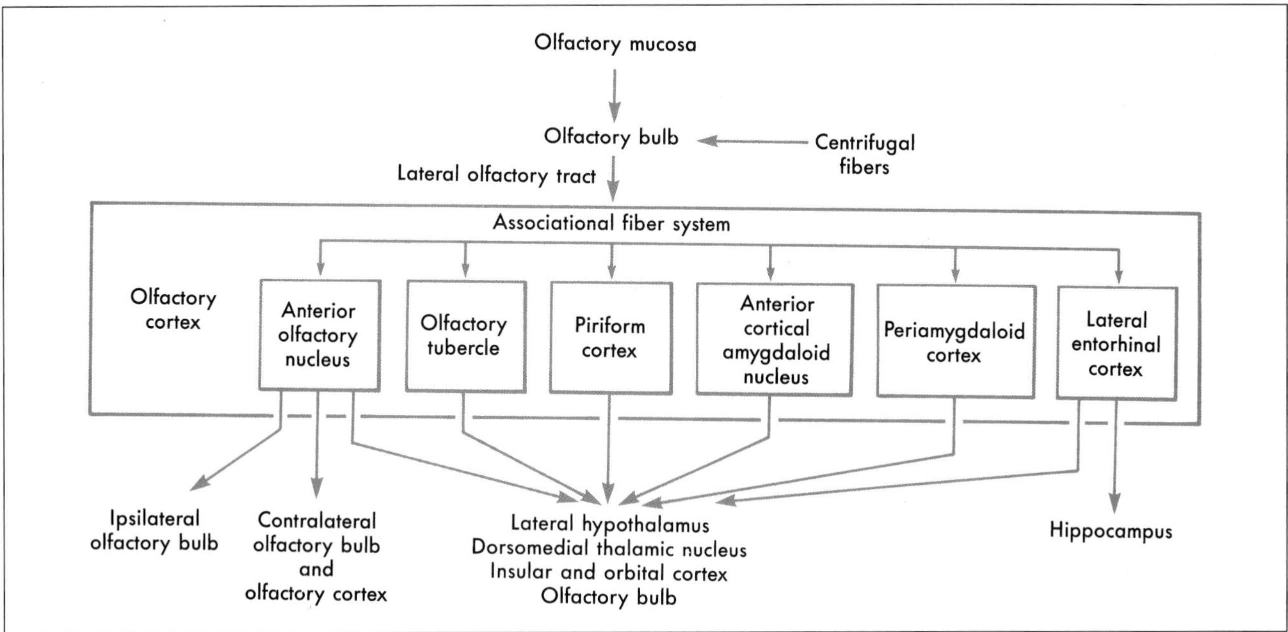

FIGURE 22-7 *Major projections of the olfactory cortex.*

Axons of the lateral olfactory tract course caudally in the form of the *lateral olfactory stria* to terminate in the *olfactory tubercle* and the *piriform cortex* (Fig. 22-6). The piriform cortex is a major component of the olfactory cortex. Fibers of the lateral olfactory tract also continue posteriorly to terminate in the *anterior cortical amygdaloid nucleus*, the *periamygdaloid cortex* (a part of the piriform cortex overlying the amygdala), and the *lateral entorhinal cortex*.

There is little evidence of a topographic projection from the olfactory bulb onto the various structures constituting the olfactory cortex. Mitral cells project to all areas of the olfactory cortex, whereas tufted cells terminate primarily in its anterior parts. However, each region of olfactory cortex is regarded as receiving input from all areas of the olfactory bulb.

Olfactory Cortex Projections Cells of the olfactory cortex have reciprocal connections with other regions of the olfactory cortex (*intrinsic* or *associational connections*) and connections with regions outside the olfactory cortex (*extrinsic connections*) (Fig. 22-7). Most intrinsic connections arise from the anterior olfactory nucleus, piriform cortex, and lateral entorhinal cortex. As a group, these associational fibers distribute to all areas of the olfactory cortex (Fig. 22-7).

Extrinsic connections include extensive projections back to the olfactory bulb. These centrifugal fibers originate from most parts of the olfactory cortex, with the exception of the olfactory tubercle. As in other sensory systems, olfactory information is also relayed to the neocortex. This connection occurs through a direct projection from olfactory cortex to *orbitofrontal* and *ventral agranular insular cortices* or via a relay in the thalamus. The latter pathway originates from cells in the olfactory cortex that project to the *dorsomedial nucleus of the thalamus* (Fig. 22-7). This neocortical representation of olfaction is important for discrimination of odors. Another noteworthy point is that the insular and orbitofrontal cortex also receives taste input. Although there is little evidence of overlap between these two modalities, it is possible that these neocortical areas may play a role in the integration of taste and olfaction that produces the experience of flavor.

In addition to neocortical projections, the olfactory cortex also sends fibers directly to the lateral hypothalamus and hippocampus. Those to the lateral hypothalamus arise primarily from the piriform cortex and anterior olfactory nucleus and are probably important for feeding behavior. The projection to the hippocampus arises from the entorhinal cortex and links olfactory input to centers concerned with learning and behavior.

Olfactory disorders are especially common among individuals over the age of 65. The chief complaint of most patients with chemosensory disturbances is the loss or alteration of taste. However, clinical studies reveal that in all but a small number of patients, the dysfunction actually resides in the olfactory system. The reason for this discrepancy is that most individuals confuse taste with flavor.

Olfactory dysfunction is also encountered in neurodegenerative diseases such as Alzheimer's and Parkinson's disease, or in Huntington's chorea. These neurodegenerative diseases involve central olfactory pathways and produce marked reductions in a patient's olfactory capabilities. Most notably, these olfactory deficits appear very early in the course of the disease and may be among its first manifestations.

Disorders of olfaction are also associated with epilepsy and various depressive and psychiatric disorders. These patients frequently experience *parosmia* (dysosmia), a distortion in a smell experience or the perception of a smell when no odor is present.

TASTE RECEPTORS

The taste experiences of sweet, salty, sour, and bitter result from an interaction between gustatory stimuli and receptor cells located in sensory organs called *taste buds*. Although found throughout the oropharyngeal cavity, taste buds are most obvious on the tongue, where they are ovoid structures with a constriction at their apical end.

Each taste bud contains 40 to 60 *taste receptor cells* that extend from a basal lamina to the surface of the epithelium. The apical ends of these receptor cells are covered with microvilli of variable lengths that extend into a *taste pore*. The pore forms a pocket to permit contact between the microvilli of the taste receptor cells and the external milieu (Figs. 22-8, 22-9). Numerous junctional complexes located between the apices of receptor cells restrict access of stimuli to the microvilli where taste transduction occurs. The taste pore is filled with a protein-rich substance through which substances must pass to reach the receptor cell microvilli. Taste receptor cells undergo a continuous process of turnover, having a life span of 10 to 14 days. New taste cells are thought to arise from polygonal *basal cells* located in basolateral areas of the taste bud. These cells are not involved in taste transduction.

Afferent fibers penetrate the basement membrane and then branch within the base of the taste bud (Fig. 22-8). Each taste bud is typically innervated by more than one afferent fiber, and an individual fiber may innervate multiple taste buds. The taste afferent fibers form the postsynaptic element of a chemical synapse near the base of the taste receptor cell.

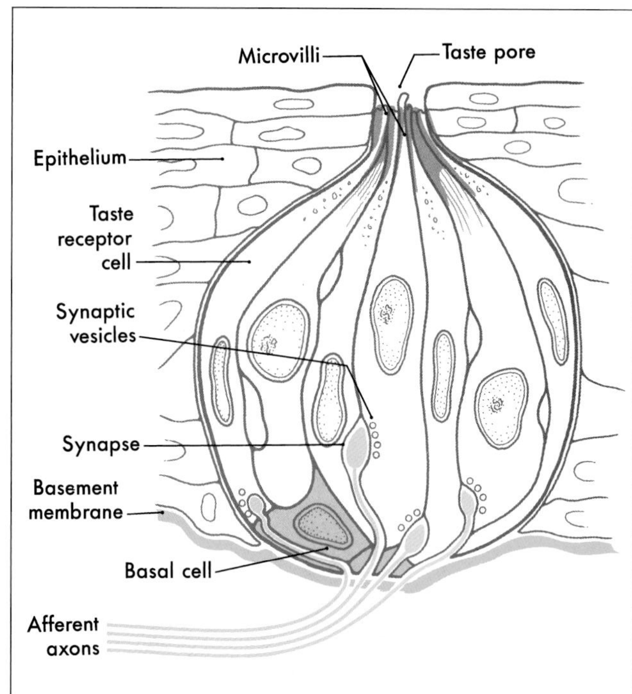

FIGURE 22-8 *The mammalian taste bud and associated structures. (Adapted from Mistretta, 1989, with permission.)*

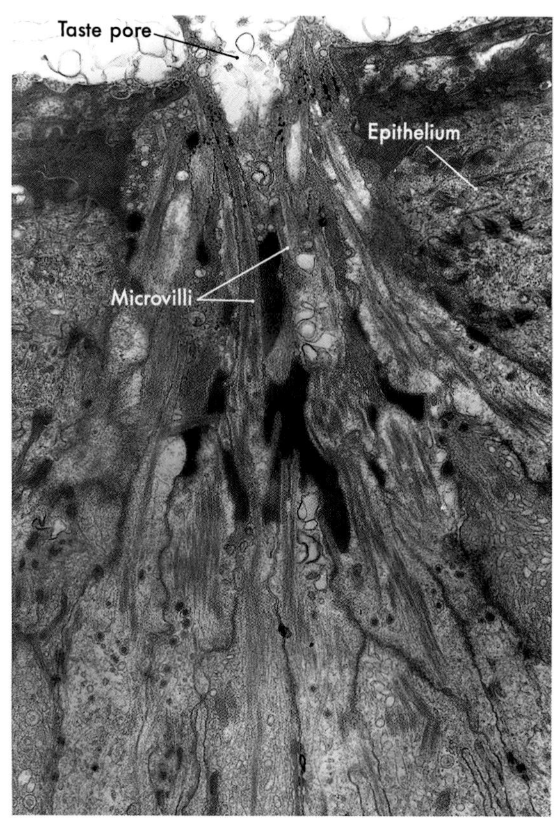

FIGURE 22-9 *Electron micrograph of the pore region of a mouse taste bud in the circumvallate papilla. (Courtesy of Drs. J. Kinnamon and H. Linnen, University of Denver.)*

DISTRIBUTION OF TASTE RECEPTORS

Lingual Taste Buds Taste buds are found, in variable numbers, on the human tongue, palate, pharynx, and larynx. On the tongue, taste buds are located exclusively in specialized structures called *papillae*, of which there are three types (Fig. 22-10A). Taste buds on the anterior two-thirds of the tongue reside in mushroom-shaped *fungiform papillae* (Fig. 22-10B). These are scattered among the more numerous nongustatory *filiform papillae* distributed over the surface of the tongue. The size, shape, and number of fungiform papillae vary widely, and usually 2 to 4 taste buds are found in the dorsal epithelium of each. The *circumvallate papillae* are located on the dorsal surface of the tongue at the junction of the oral and pharyngeal cavities (Fig. 22-10A). There are from 8 to 12 circumvallate papillae, each of which is composed of a central papilla surrounded by a cleft containing taste buds in its epithelium (Fig. 22-10C). A single *foliate papilla* on each side of the tongue appears as a series of clefts along the lateral margin of the tongue (Fig. 22-10A). Each is composed of 2 to 9 clefts, with 5 being the most common number. Taste buds in foliate papillae are also located in the epithelium that lines the clefts (Fig. 22-10D).

Associated with both the circumvallate and foliate papillae are the *von Ebner's* lingual salivary glands. These glands drain into the base of the clefts and influence their microenvironment. Taste stimulation of circumvallate and foliate papillae influences the secretions of von Ebner's glands via circuits located in the brainstem.

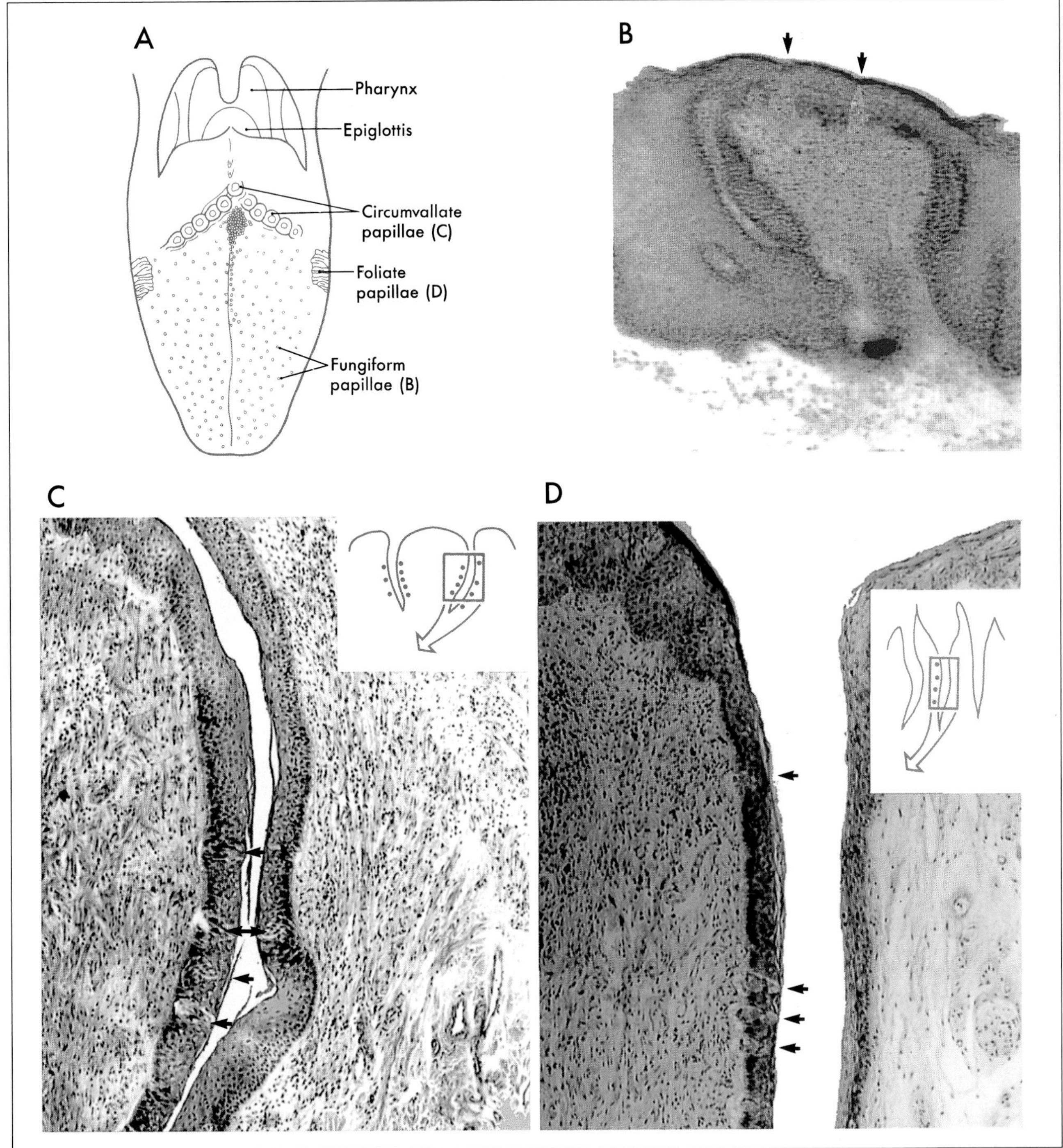

FIGURE 22-10 *Distribution of papillae (A), and their associated taste buds, on the human tongue. Light micrographs of transverse sections through human fungiform (B), circumvallate (C) and foliate (D) papilla. Examples of taste buds are shown at the arrows. (Photomicrographs courtesy of Dr. I.J. Miller, Bowman Gray School of Medicine.)*

Extralingual Taste Buds Additional taste buds are located on the human soft palate, oral and laryngeal pharynx, larynx, and upper esophagus. Extralingual taste buds are not located in papillae, but rather are situated in the epithelium. Palatal taste buds are located at the juncture of the hard and soft palate and on the soft palate. Laryngeal taste buds are found on the laryngeal surface of the epiglottis and adjacent aryepiglottal folds. The number of extralingual taste buds is substantial, and they may contribute to the taste experience. Stimulation of some extralingual taste buds, particularly those near the larynx, elicits brainstem-mediated reflexes that prevent accidental aspiration of ingested materials.

TASTE TRANSDUCTION MECHANISMS

In general, taste transduction is initiated when soluble chemicals diffuse through the contents of the taste pore and interact with the exposed apical microvilli of the receptor cells. The interaction of the chemical stimuli with the taste cell microvilli results in either a depolarization or a hyperpolarization of the receptor cell (*receptor potential*). Depolarizing receptor potentials produce an increase in intracellular Ca^{++}, either by the release of Ca^{++} from internal stores or by the activation of voltage-gated calcium channels located in the basolateral membrane of the taste receptor cells (Fig. 22-11). This Ca^{++} release results in a release of chemical transmitters at the afferent synapse, which, in turn, leads to an action potential in the afferent fiber. The transmitter at the synapses between taste receptor cells and primary afferent fibers is unknown.

Transduction of stimuli leading to salty and perhaps some sour and bitter tastes appears to be the result of a direct interaction of these tastants with specific ion channels located in the apical membrane of the taste receptor cells. Transduction of sodium salts such as NaCl involves movement of Na^+ into the taste receptor cell through apically located amiloride-sensitive cation channels. Similar mechanisms have been proposed for K^+ salts (Fig. 22-11). One pathway responsible for transduction of some sour and bitter stimuli is blockage of apical voltage-sensitive K^+ channels. At the resting potential, there is a small outward K^+ current through the apical membranes of taste receptor cells. Protons provided by sour stimuli such as HCl block this outward current, causing the cell to depolarize.

At least some sweet and bitter tasting compounds are transduced by receptors that activate intracellular G-protein mediated second-messenger pathways (Fig. 22-11). Binding of sweet-tasting compounds such as sucrose to apically located receptors stimulates an adenylyl cyclase-cAMP second-messenger pathway that closes basolateral K^+ channels, leading to depolarization of the taste receptor cell. Additional mechanisms for sweet transduction have been proposed, and there is good evidence for multiple sweet receptor types. Some bitter compounds are thought to act through a different second-messenger pathway (IP_3) that releases Ca^{++} from intracellular stores. Taste stimuli such as amino acids may initiate a response by binding receptors that are coupled directly to cation channels having properties similar to those of the nicotinic acetylcholine receptor.

PERIPHERAL TASTE PATHWAYS

The afferent fibers of first-order taste neurons (special visceral afferent, SVA) innervating oropharyngeal taste buds travel in the *facial* (VII), *glossopharyngeal* (IX), and *vagus* (X) nerves (Fig. 22-12). The *chorda tympani* branch of the facial nerve innervates taste buds in the fungiform papillae on the anterior two-thirds of the tongue and in the most anterior clefts of the foliate papillae. The *greater superficial petrosal nerve*, also a branch of the facial nerve, innervates taste buds on the soft palate. The cell bodies of facial nerve fibers subserving taste are located in the *geniculate ganglion*, and their central processes enter the brainstem at the pontomedullary junction in the *intermediate nerve*, which is actually a part of the facial nerve. These primary afferent taste fibers enter the *solitary tract*, travel caudally, and terminate on cells of the surrounding *solitary nucleus* (Fig. 22-12).

The chorda tympani branch leaves the facial nerve just distal to the geniculate ganglion. Consequently, lesions of the root of the seventh cranial nerve, or tumors in the internal auditory meatus (such as an acoustic neuroma), will result in loss of taste perception from the anterior two-thirds of the tongue on the ipsilateral side. Accompanying this deficit are paralysis of the ipsilateral facial muscles, *hyperacusis* (paralysis of the stapedius muscle), and impaired secretion of the nasal and lacrimal glands and of submandibular and sublingual salivary glands. Damage just distal to the geniculate ganglion may or may not result in taste loss, depending on the origin of chorda tympani branches, but an ipsilateral facial paralysis will be seen.

Taste buds located in the circumvallate papillae and posterior clefts of the foliate papillae are innervated by the *lingual-tonsillar branch* of the glossopharyngeal nerve (cranial nerve IX). Those located on the epiglottis and esophagus are innervated by the *superior laryngeal nerve*, a branch of the vagus nerve (cranial nerve X) (Fig. 22-12). Taste fibers in cranial nerves IX and X have their cell bodies of origin in the *inferior ganglia* (*petrosal* and *nodose*, respectively) of these cranial nerves (Fig. 22-12).

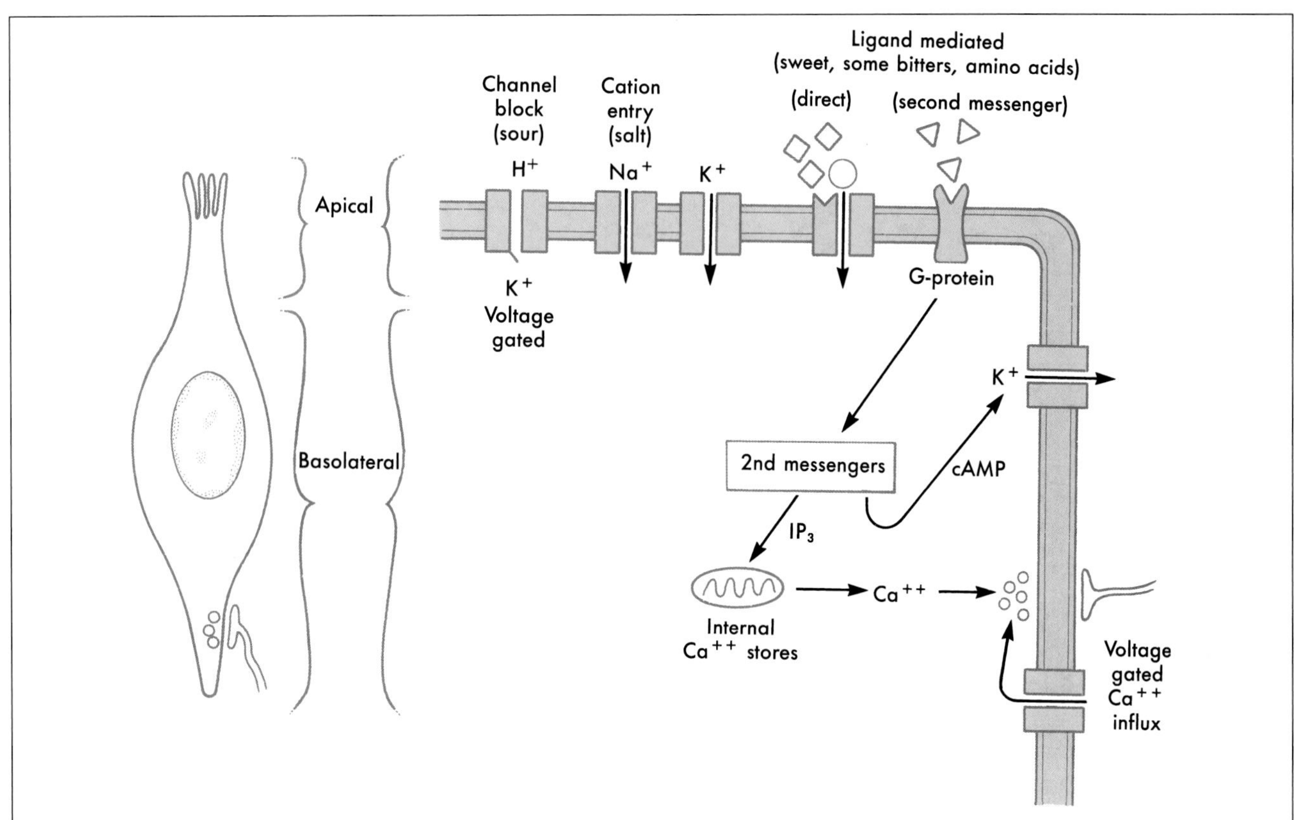

FIGURE 22-11 *Pathways of transduction in taste receptors. Some sour and bitter substances are transduced by the closing of apical voltage-sensitive K^+ channels. The transduction of salts such as NaCl involves the movement of ions (such as Na^+ and K^+) through amiloride-sensitive cation channels in the apical membrane. Sweet and some bitter compounds are thought to activate intracellular second-messenger pathways (cAMP, IP_3), which leads to activation of membrane channels in the basolateral membrane of the taste cell. (Adapted from Kinnamon, 1988, with permission.)*

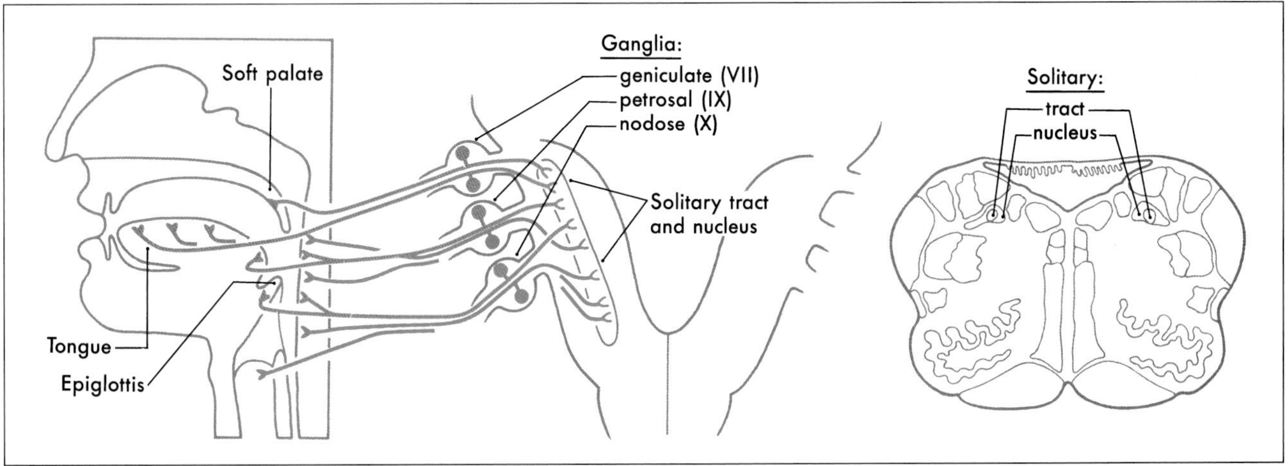

FIGURE 22-12 *Summary of peripheral taste pathways. Special visceral afferent fibers for taste (in red) terminate in the rostral (gustatory) areas of the solitary nucleus, whereas general visceral afferent fibers (in blue) terminate in the caudal portion of the nucleus.*

The central processes of these fibers, like those of the facial nerve, enter the medulla, descend in the solitary tract, and terminate on neurons in the adjacent solitary nucleus (Fig. 22-12).

CENTRAL TASTE PATHWAYS

The solitary nucleus is the principal visceral afferent nucleus of the brainstem. Based on functional characteristics, it is divided into a *rostral (gustatory) nucleus* and a *caudal (visceral* or *cardiorespiratory) nucleus*. Taste fibers traveling in cranial nerves VII, IX, and X terminate primarily in the rostral portions of the solitary nucleus because this region contains most second-order neurons in the taste pathway (Fig. 22-12). There is considerable overlap in the distribution of the terminals of the primary taste cranial afferents from these three nerves in the gustatory nucleus. General visceral afferent fibers of the vagus, and those that travel in the glossopharyngeal nerve, terminate in the caudal part of the solitary nucleus (Fig. 22-12). These visceral fibers are involved in the central control of respiration, cardiac function, and certain aspects of swallowing.

Axons arising from second-order taste neurons in the gustatory nucleus ascend in association with the ipsilateral *central tegmental tract* and terminate in the *parvicellular division of the ventral posteromedial nucleus* of the thalamus (*VPMpc*), medial to the head representation (Fig. 22-13). Axons from these neurons in the VPMpc travel through the ipsilateral posterior limb of the internal capsule to terminate in the inner portion of the *frontal operculum* and *anterior insular cortex*, and in the rostral extension of *Brodmann's area 3b* on the lateral convexity of the postcentral gyrus (Fig. 22-13). This pathway (solitary nucleus → VPMpc → cortex) is responsible for the discriminative aspects of taste and, in contrast to other sensory pathways, is exclusively ipsilateral.

Physiologic studies in primates indicate there is an additional region of the cortex that processes taste information. The *lateral posterior orbitofrontal cortex* receives inputs from primary taste cortex and acts as a site of integration for taste, olfactory, and visual cues associated with the ingestion of foods. Taste-responsive cells have also been found in the primate amygdala and hypothalamus. These cells do not respond exclusively to taste, and their connecting pathways or role in taste-mediated behavior is not fully understood. Taste information is also relayed from cells in the solitary nucleus into medullary reflex connections that influence salivary secretion, mimetic responses, and swallowing.

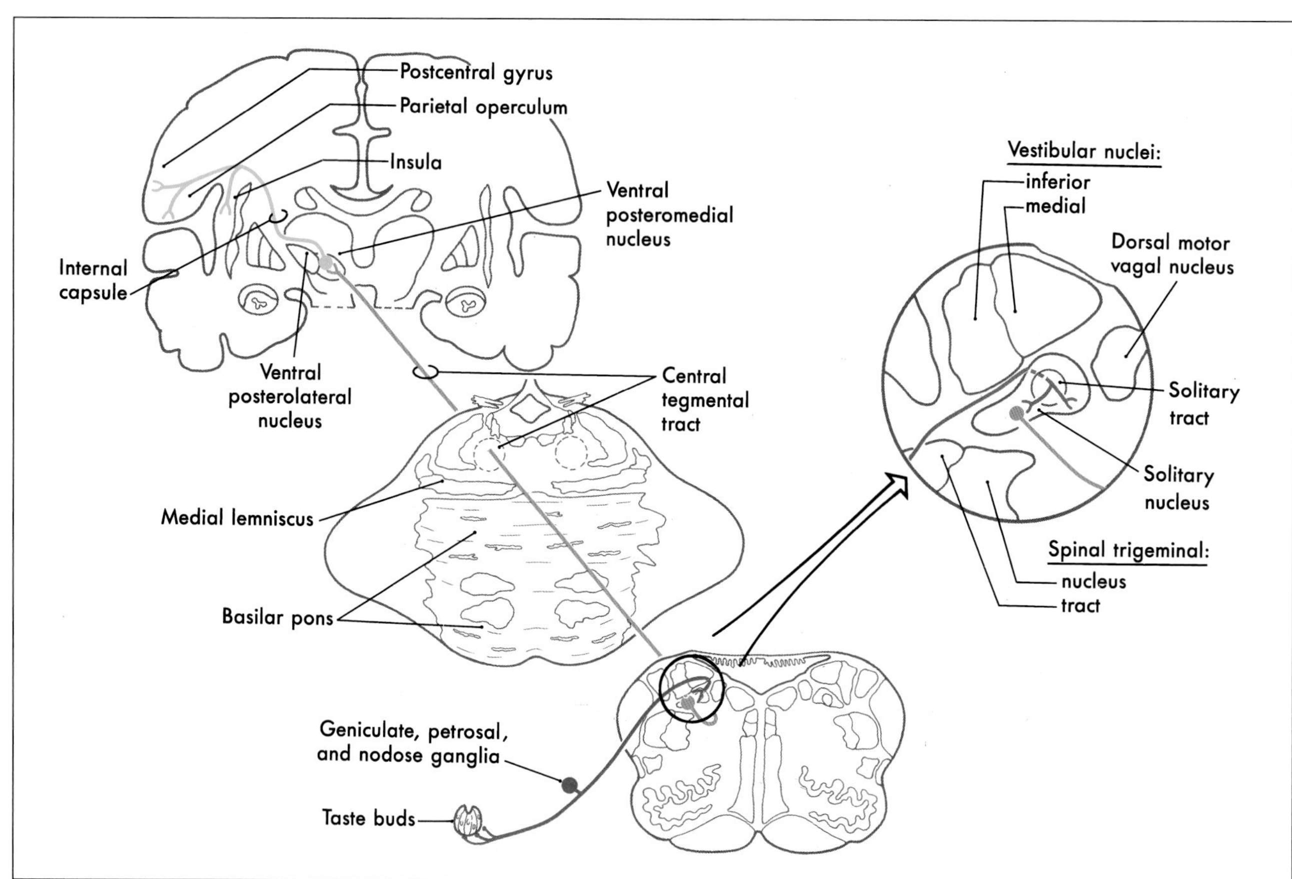

FIGURE 22-13 *The ascending taste pathway to thalamus and cortex.*

The complete loss of taste (*ageusia*) is rarely encountered, in part because of the large numbers of nerves that relay taste information to the central nervous system. Rarely does a patient suffer from bilateral injury to all the nerves innervating the oropharyngeal region. More frequently a patient suffers from *hypogeusia*, decreased taste sensitivity, or *parageusia* (*dysgeusia*), distortions in the perception of a taste. Like olfactory disorders, taste disorders are associated with head trauma, viral infections, and various psychiatric disorders. Taste changes are one of the most frequent complaints of cancer patients undergoing radiation and chemotherapy treatments, and there is a progressive taste loss in diabetic patients.

SOURCES AND ADDITIONAL READINGS

Anholt RRH: Molecular neurobiology of olfaction. Crit Rev Neurobiol 7:1–22, 1993.

Getchell TV, Doty RL, Bartoshuk LM, Snow JB (eds): *Smell and Taste in Health and Disease*. Raven Press, New York, 1991.

Kinnamon SC: Taste transduction: a diversity of mechanisms. Trends Neurosci 11:491–496, 1988.

Mistretta CM: Anatomy and neurophysiology of the taste system in aged animals. In Murphy C, Cain WS, Hegsted DM (eds): Nutrition and the Chemical Senses in Aging: Recent Advances and Current Research Needs. Ann NY Acad Sci 561:277–290, 1989.

Mori K: Membrane and synaptic properties of identified neurons in the olfactory bulb. Prog Neurobiol 29:275–320, 1987.

Morrison EE, Costanzo RM: Morphology of the human olfactory epithelium. J Comp Neurol 297:1–13, 1990.

Norgren R: Gustatory system. In Paxinos G (ed): *The Human Nervous System*. Academic Press, San Diego, 1990, pp 845–861.

Scott JW, Wellis DP, Riggott MJ, Buonviso N: Functional organization of the main olfactory bulb. Microsc Res Tech 24:142–156, 1993.

CHAPTER TWENTY-THREE

Motor System I: Peripheral Sensory, Brainstem, and Spinal Influence on Ventral Horn Neurons

G. A. MIHAILOFF • D. E. HAINES

Spinal ventral horn motor neurons whose axons innervate skeletal muscles are called *lower motor neurons*. These cells stimulate muscles to produce characteristic movements of a body part. The activity of these motor neurons is influenced from two sources. First, *peripheral sensory input* arrives via dorsal roots and is transmitted to ventral horn neurons. Second, extensive descending projections from the cerebral cortex and brainstem, called *supraspinal systems*, terminate at all levels of the spinal cord and are responsible for a mixture of excitatory and inhibitory influence on ventral horn motor neurons. This chapter focuses on the peripheral sensory and brainstem systems that influence ventral horn neurons.

OVERVIEW

The *lower motor neurons* of the spinal cord ventral horn are topographically arranged according to the muscle groups they innervate. This is particularly evident in the cervical and lumbosacral enlargements, the levels of the spinal cord that innervate the musculature of the upper and lower limbs, respectively. Motor neurons that supply flexor muscles generally are located more dorsal in the ventral horn than are extensor motor neurons. In addition, motor neurons that innervate paravertebral and proximal limb muscles are most medial, whereas those that innervate distal musculature are most lateral (Fig. 23-1). The ventral horn motor neurons receive sensory feedback from the muscles they control, as well as from synergist and antagonist muscles. The linkage of peripheral sensory input and ventral horn neurons forms the substrate for a number of spinal reflexes (see Figs. 9-8 to 9-10).

In addition to sensory feedback, the activity of lower motor neurons in the spinal cord is greatly influenced by descending projections from cells in the brainstem and cerebral cortex. These brainstem and cortical neurons are referred to as *upper motor neurons*, and, unlike lower motor neurons, they have no direct synaptic link with muscles. Because of their origin, these descending projections are also called *supraspinal systems*.

Ventral horn motor neurons represent the only direct link (the *final common path*) between the nervous system and skeletal muscle. As such, these neurons play a central role in the production of movement. The regulation of motor neuron activity by peripheral sensory input and descending brainstem influences is critical to the performance of normal movement.

VENTRAL HORN MOTOR NEURONS

Types and Distribution There are two varieties of ventral horn motor neurons, alpha and gamma, which are mingled in the ventral horn. *Alpha motor neurons* innervate the ordinary, working fibers of skeletal muscles called *extrafusal fibers*, and *gamma motor neurons* innervate a special type of striated muscle fiber, the *intrafusal fibers*, which are found only within *muscle spindles*. Recall that the ventral horn also contains small *interneurons* whose axons distribute locally within the spinal gray. Interneurons are numerous in the intermediate zone and ventral horn and are functionally quite essential in the regulation of alpha and gamma motor neurons. Their action on motor neurons may either be excitatory or inhibitory.

The axons of ventral horn motor neurons exit the spinal cord via the ventral roots and course distally in peripheral nerves. These fibers represent the *final common path* that links the nervous system and skeletal muscles. As the motor axon reaches the muscle it innervates, it loses its myelin sheath and forms a series of flattened boutons that indent the surface of a group of muscle fibers. This specialized type of synapse is called a *neuromuscular junction* or *motor end-plate* (Fig. 23-2).

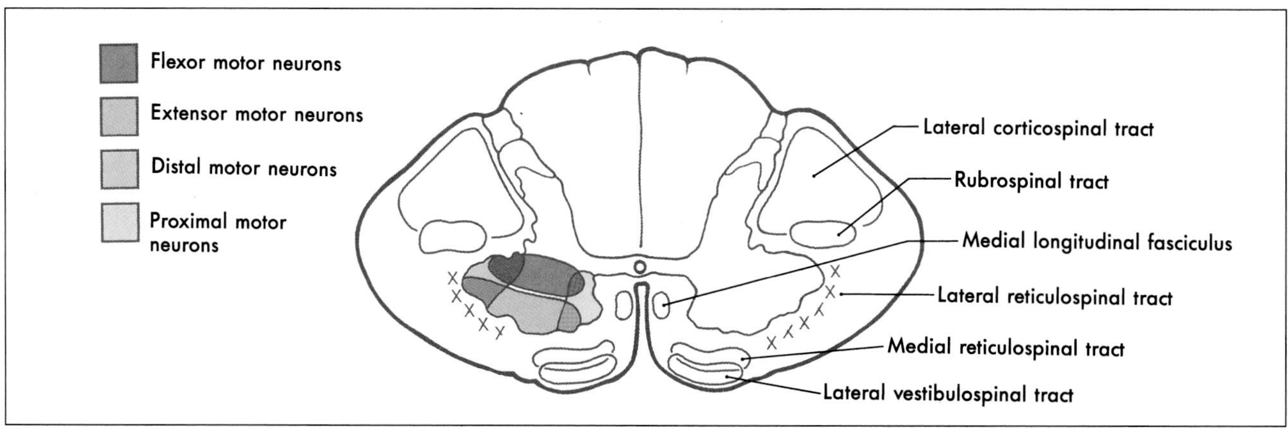

FIGURE 23-1 *The locations of vestibulospinal and reticulospinal tracts at a representative cervical level of the spinal cord. Medial vestibulospinal fibers are located in the medial longitudinal fasciculus. The general positions of motor neuron pools are shown on the left.*

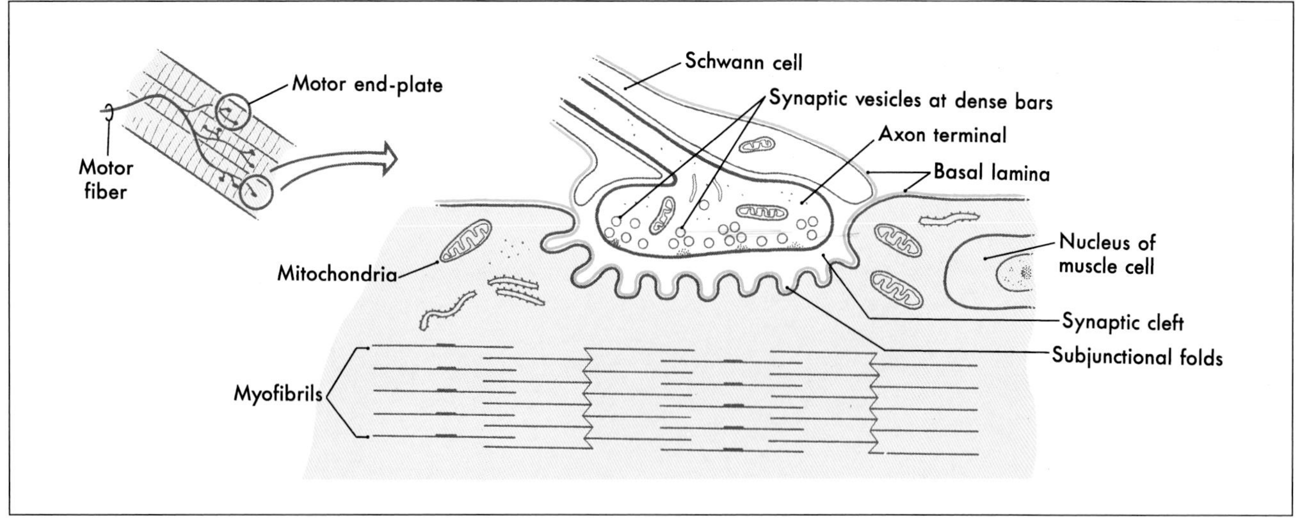

FIGURE 23-2 *The structural elements characteristic of a motor end-plate (neuromuscular junction).*

Neuromuscular Junction Like synapses in the central nervous system, the junction between a motor axon and striated muscle fibers consists of presynaptic and postsynaptic components (Fig. 23-2). The *presynaptic element*, the axon terminal, contains round, clear synaptic vesicles (filled with the neurotransmitter *acetylcholine*), mitochondria, and small patches of dense material around which the vesicles aggregate at the active site. The presynaptic element is separated from the postsynaptic element by the space of the *synaptic cleft*. The *postsynaptic membrane*, the specialized portion of the muscle cell plasma membrane subjacent to the axon terminal, exhibits a large number of folds that effectively increase the surface area of the muscle cell in contact with the axon terminal (Fig. 23-2). These irregularities, called *subjunctional folds*, contain *nicotinic acetylcholine receptors* near their summit at the synaptic cleft. The receptors are linked to ligand-gated ion channels in the postsynaptic membrane. These channels mediate the ion flux that underlies the transmission of electrical impulses from nerve to muscle. Surrounding the entire muscle fiber and extending into the synaptic cleft is a *basal lamina* whose composition is similar to basement membranes present in other tissues.

When an action potential depolarizes the presynaptic element, there is an influx of calcium through voltage-gated membrane channels. Synaptic vesicles fuse with the presynaptic membrane at the active sites (which are marked by structures called *dense bars*) and release *acetylcholine* into the synaptic cleft. The transmitter binds to receptors on the postsynaptic membrane, ion flux occurs, and a depolarizing potential called an *end-plate-potential* spreads over the surface of the muscle fiber. This potential triggers the release of Ca^{++} (from the sarcoplasmic reticulum), which elicits the movement of actin and myosin filaments, resulting in muscle contraction. Synaptic transmission is terminated by an enzyme called *acetylcholinesterase*, which is located in the matrix of the basal lamina in the depths of the postjunctional folds. This enzyme inactivates acetylcholine by hydrolyzing it to acetate and choline.

Motor Units Each muscle fiber receives only one motor end-plate, but the number of muscle fibers innervated by a single alpha motor neuron axon varies from a few to many. The aggregate of a motor neuron axon and *all* the muscle fibers it innervates is called a *motor unit* (Fig. 23-3). In general, as the need for fine control of a muscle increases, the size or innervation ratio of its motor unit decreases. That is, the number of muscle fibers innervated by a single axon decreases. The size of a motor unit is also related to the mass of the muscle and its speed of contraction. Small muscles that generate low levels of force and contract rapidly typically have *small motor units* (10 to 100 muscle fibers

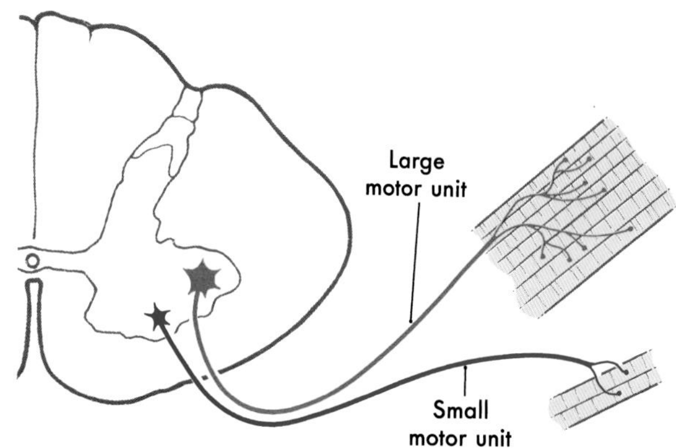

FIGURE 23-3 *Large and small motor units.*

per motor axon). In contrast, large, powerful muscles that generate high levels of force are usually innervated by *large motor units* (600 to 1,000 muscle fibers per motor neuron axon).

Motor units can be divided into three categories on the basis of the physiologic properties of their muscle and nerve components. *Type S* units have a slow contraction time and generate relatively low tension but are fatigue resistant. They are innervated by relatively small and slow-conducting alpha fibers. Muscles in which S units predominate are dark red and are called *slow twitch muscles. Type FF* units generate large contractile force and have a fast conduction time, but they are easily fatigued. They are innervated by large alpha neurons with heavily myelinated, fast-conducting axons. Muscles in which these units predominate are pale-colored and are called *fast twitch muscles. Type FR* units have a slightly slower contraction time than FF units and generate nearly as much force but are almost as fatigue-resistant as S units.

Muscles generally have a mixture of motor units, and the proportions vary according to the demands placed on the muscle. For example, the soleus muscle is a slow-twitch muscle containing mainly S-type units. The relatively slow conduction time of the narrow alpha axons serving these motor units are adequate for the demands of this muscle. By contrast, the muscles controlling the fingers need to contract very rapidly. Therefore, these are fast-twitch muscles, containing mainly FF units and innervated mainly by large-diameter, fast-conducting axons.

Size Principle The nervous system uses the size and functional properties of motor units as a means of grading the force of muscle contraction. When a group of motor neurons in the ventral horn is activated, those with the *smallest somata* are recruited first because they have the *lowest threshold* for synaptic activation and respond to the weakest input. They typically form small motor units and produce relatively weak movements. More intense input recruits successively *larger neurons* (and larger motor units), producing more powerful movements. This is called the *size principle*. Increased muscle force output can also be achieved by increasing the firing rate of motor units. Muscle contraction generally remains smooth because the component motor units are activated asynchronously (owing to the size principle), and this tends to "even out" the differences in motor unit properties.

PERIPHERAL SENSORY INPUT TO THE VENTRAL HORN

Muscle Spindles Signals that transmit information from skeletal muscles into the nervous system enter the spinal cord via the dorsal roots. For the most part, these signals are generated in specialized structures in muscles called *neuromuscular spindles* (*muscle spindles*). The output of the muscle spindle signals a change in muscle length and the rate of change in muscle length.

A muscle spindle (Fig. 23-4) is a long, thin encapsulated structure that typically contains about seven striated *intrafusal muscle fibers*. Spindles range in length from 4 to 10 mm. The capsule of the spindle is attached to, and oriented in parallel with, the *extrafusal fibers* that comprise the bulk of the muscle.

There are two basic types of intrafusal fibers: *nuclear bag fibers* and *nuclear chain fibers* (Fig. 23-4). Like other striated muscle cells, intrafusal fibers are multinucleated, and the arrangement of the nuclei is the most obvious feature distinguishing the two types. In both types, the nuclei occupy the central (*equatorial*) region of the cell. In *nuclear bag fibers*, the nuclei are clustered, and the equatorial region is somewhat swollen. In *nuclear chain fibers*, the nuclei are arranged in a single row and the equatorial region is not obviously expanded. The contractile elements of both types of cell are located almost entirely in the distal (polar) regions of the cell. Because the ends of the cell are anchored, contraction of the intrafusal fibers causes the equatorial region to be stretched between the two polar regions.

The different types of intrafusal fiber perform different sensory functions. There are actually two subtypes

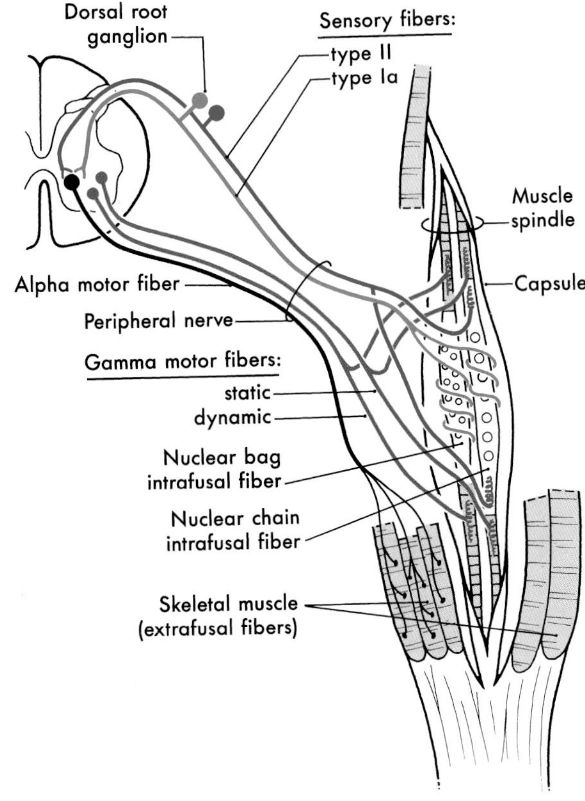

FIGURE 23-4 *Structure of a muscle spindle and the relation of afferent and efferent nerve fibers to intrafusal and extrafusal muscle fibers.*

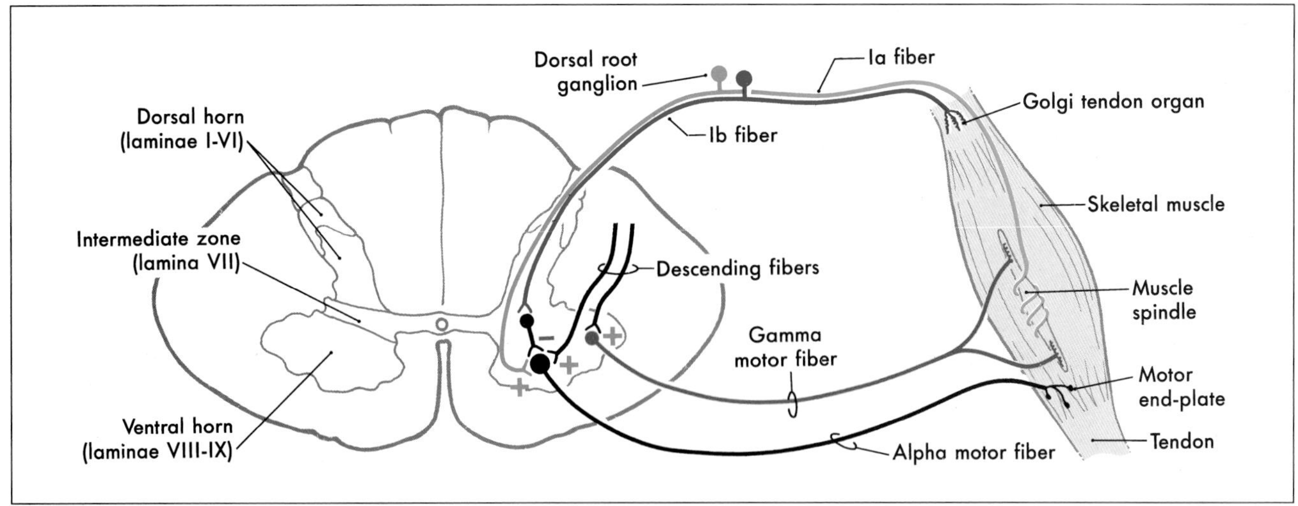

FIGURE 23-5 *Circuits related to input and output mediated by Golgi tendon organs and to alpha-gamma coactivation.*

of nuclear bag fiber, which have different elastic properties and, correspondingly, different functions. One type, the *dynamic bag fiber*, is sensitive mainly to the rate of change in muscle length. The other, the *static bag fiber*, signals only a change in muscle length. Nuclear chain fibers, like static bag fibers, are mainly sensitive to changes in muscle length.

Intrafusal muscle fibers are associated with two types of sensory fibers, both of which are concentrated at the equatorial (noncontractile) region of infrafusal fibers. The *type Ia* fiber is heavily myelinated, has a conduction velocity of 80 to 120 m/sec, and is typically associated with nuclear bag fibers. The distal end of this fiber is wrapped around the central (noncontractile) region of an intrafusal muscle fiber. Because of this relationship, the type *Ia* afferent terminations are called *annulospiral endings*. These endings are, in effect, mechanoreceptors. Stretching of the central region of the intrafusal fiber will also stretch the sensory fiber and mechanically open ion channels in its membrane. If the induced ion flux raises the membrane potential above threshold, an action potential is initiated. The firing frequency is directly proportional to the degree to which the spindle is stretched.

The other type of muscle spindle sensory fiber, the *type II fiber*, is principally associated with nuclear chain fibers. Its connection with the equatorial region of the target infrafusal fiber has the form of a cluster of thin, radiating branches and is called a *secondary ending* or *flower-spray ending*. This sensory fiber is also activated by mechanical stretch, but it only *codes the event*, *not the rate* of the stretch.

Each type of intrafusal fiber is also innervated by a gamma motor neuron. Dynamic nuclear bag fibers are associated with *dynamic gamma motor neurons*, whereas static nuclear bag fibers and nuclear chain fibers are innervated by *static gamma motor neurons*. When the gamma motor neuron is active, contractile elements at either pole of the intrafusal muscle fiber contract, resulting in increased stretch on its central region. This increases the frequency of action potentials that travel over the *Ia* sensory fibers. As explained below, dynamic and static gamma motor neurons function to maintain spindle sensitivity and length, respectively.

The Gamma Loop Muscle spindles play an essential role in movement and in the maintenance of muscle tone. Consider two situations: one in which a muscle, for example the biceps brachii, is passively stretched, and another in which it contracts and shortens actively against a load.

A passive stretching of the biceps muscle, produced, for example, by tapping on its tendon, will elongate the muscle spindles. The stretching of the equatorial region of the nuclear bag fibers results in an increase in the firing rate of the Ia fibers (see Fig. 23-6A). These fibers enter the cervical spinal cord and form monosynaptic excitatory synapses with alpha motor neurons that innervate the biceps brachii (Figs. 23-5, 23-6). This is the circuit that forms the basis of the monosynaptic tendon reflex explained in Chapter 9 (see Fig. 9-8).

The connection between the Ia fibers and the alpha motor neurons of a muscle also functions in a more complex mechanism called the *gamma loop*, which is critical to the maintenance of stretch reflexes and muscle tone. In this mechanism, muscle contraction is produced by supraspinal activation of *gamma* rather than alpha motor neurons (Fig. 23-5). Like alpha neurons, gamma motor neurons receive supraspinal input from the cerebral cortex and brainstem. In the gamma loop, this supraspinal input activates the gamma neurons so that the infrafusal muscle fibers contract. Because the contraction of an infrafusal fiber has the effect of stretching the equatorial region between the two shortened polar regions, it

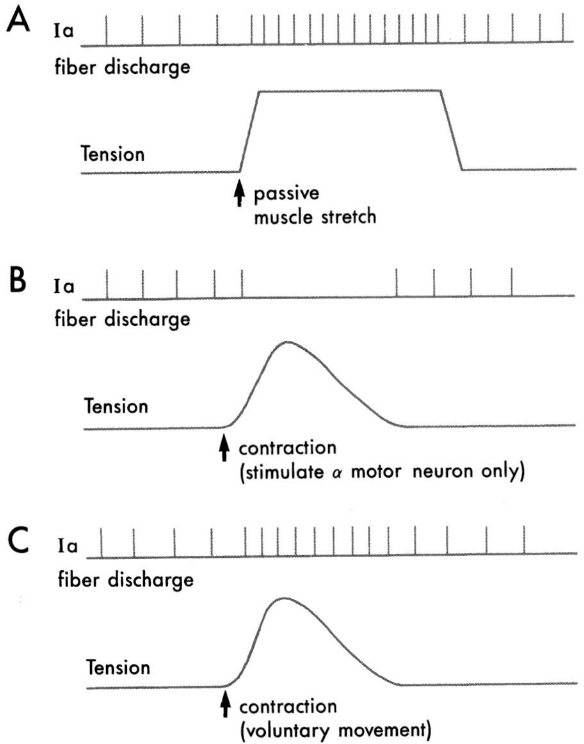

FIGURE 23-6 *Responses shown by Ia fibers in three situations: passive stretching of the muscle (A), experimentally induced contraction of the muscle in which only the alpha motor neurons are stimulated (B), and a voluntary contraction of the muscle in which gamma and alpha motor neurons are co-activated (C). When only alpha neurons are activated (B), the infrafusal fibers stay relaxed while the surrounding muscle contracts, and Ia discharge ceases. In a normal contraction (C), the shortening of the spindle caused by extrafusal fiber contraction is offset by contraction of infrafusal fibers, and Ia discharge persists.*

results in an increase in Ia fiber activity. In the spinal cord, this increase in Ia fiber discharge activates the alpha motor neurons of the muscle, resulting in muscle contraction (Fig. 23-5). This circuit involving gamma motor neurons, Ia primary afferent fibers, and alpha motor neurons is called the *gamma loop*.

Now, consider the situation in which a muscle is contracting actively against a load. Because a muscle spindle is attached parallel to the adjacent extrafusal fibers, its overall length is determined, in part, by the length of the surrounding muscle; when the muscle contracts, the spindle shortens. If the infrafusal fibers remained passive during muscle contraction, the shortening of the spindle would relax the equatorial regions of the infrafusal fibers, and the Ia fibers would cease firing (Fig. 23-6B). This slack, inactive bundle would be useless for reporting muscle dynamics. In reality, the Ia fibers continue firing during voluntary muscle contraction, and the spindle retains its normal sensitivity (Fig. 23-6C) because when the brain signals the alpha motor neurons to initiate muscle contraction, it sends parallel impulses to the gamma neurons to cause the infrafusal fibers to contract. Therefore, when the muscle shortens, the infrafusal fibers also shorten. The relaxation of the equatorial regions that would result form extrafusal muscle fiber contraction is offset by equatorial stretching because of infrafusal fiber contraction, and the equatorial regions remain under nearly constant tension. This phenomenon is called *alpha-gamma coactivation* (Fig. 23-5).

Golgi Tendon Organ Sensory feedback to the spinal ventral horn is also derived from the Golgi tendon organ. These structures are located in tendons near their junctions with muscle fibers and consist of networks of thin nerve fibers intertwined with the collagen fibers of the tendon (Fig. 23-5). These nerve fibers, like the sensory fibers of muscle spindles, are mechanoreceptors. When force is applied to the tendon, the sensory fibers are stretched, which opens ion channels in the nerve fiber membrane. The fibers that lead from the tendon organs to the spinal cord are *type Ib* fibers. These fibers are large in diameter and heavily myelinated, with a conduction velocity of 70 to 110 m/sec. After entering the spinal cord, the type Ib fibers traverse the intermediate zone to reach the ventral horn, where they form *excitatory* synapses with interneurons. There interneurons in turn *inhibit* alpha motor neurons that innervate the muscle associated with the activated Golgi tendon organ. This action of the Golgi tendon organ is exactly opposite that of the muscle spindle; activation of the latter leads to excitation of alpha motor neurons innervating the muscle associated with the activated spindle.

Reflex Circuits Afferent fibers from muscle spindles and Golgi tendon organs take part in a variety of reflex circuits that directly or indirectly influence the activity of ventral horn motor neurons. Several of the more prominent of these circuits were described in Chapter 9 (see Figs. 9-8 to 9-10); they are summarized only briefly here. As mentioned earlier, many type *Ia* spindle afferents form monosynaptic excitatory connections with alpha motor neurons that innervate the muscle from which the afferents originated. This is the basis for the *tendon* or *stretch reflex* (see Fig. 9-8). At the same time, these *Ia* fibers activate *Ia interneurons* that inhibit motor neurons innervating *antagonist* muscles; this is *reciprocal inhibition* (see Fig. 9-8). Incoming muscle afferents can also activate interneurons that project to the contralateral side of the spinal cord, as well as to propriospinal neurons that link the spinal segment of entry to more rostral or caudal spinal cord levels. Circuits of the first type, which convey cutaneous somatic inputs, form the basis for the *crossed extensor reflex* (see Fig. 9-10).

Generally the various spinal reflex pathways primarily target alpha motor neurons, although some may influence gamma motor neurons as well. For the most part, the activity of these basic spinal reflexes occurs in the background and is not under direct volitional control. However, other long loop reflexes involve ascending pathways that reach the cerebral cortex by way of a thalamic relay. These reflexes are subject to voluntary regulation and their spinal effects can be enhanced or diminished on a volitional basis through various supraspinal systems.

BRAINSTEM AND SPINAL SYSTEMS: ANATOMY AND FUNCTION

Of the several pathways that project to the spinal cord from the brainstem or cerebral cortex, four are particularly relevant to voluntary movement. All four originate from cell groups in the brainstem. Two of them, the *vestibulospinal* and *reticulospinal systems*, travel in the ventral funiculus of the spinal cord. The other two, the *rubrospinal* and *lateral corticospinal tracts*, travel in the lateral funiculus. The following sections focus on vestibulo-, reticulo-, and rubrospinal tracts.

Vestibulospinal Tracts The vestibulospinal system comprises medial and lateral vestibulospinal tracts (Figs. 23-1, 23-7). The *medial vestibulospinal tract* is made up of axons that originate in the medial and inferior vestibular nuclei and *descend bilaterally* into the spinal cord as part of the medial longitudinal fasciculus. The *lateral vestibulospinal tract* is formed by axons that originate in cells of the lateral vestibular nucleus and *descend ipsilaterally* through the ventral portion of the brainstem to course in the ventral funiculus of the spinal cord.

The medial vestibulospinal tract projects only as far as cervical or upper thoracic spinal cord levels and influences motor neurons controlling neck musculature. The lateral vestibulospinal tract, in contrast, extends throughout the length of the cord. Cells in rostral portions of the lateral vestibular nucleus project to the cervical cord, cells in the midportion project to thoracic cord, and cells in the caudal part terminate in lumbosacral levels. The fibers of this tract terminate in the medial portions of laminae VII and VIII and *excite motor neurons that innervate paravertebral extensors and proximal limb extensors* (Fig. 23-7). These muscles function to counteract the force of gravity and, therefore, are commonly called *antigravity muscles*. Through their effects on these extensor muscles, lateral vestibulospinal fibers function in the control of posture and balance. Evidence from experimental studies suggests that some vestibulospinal axons synapse directly on alpha motor neurons but that most exert their influence through spinal interneurons.

Activity in the lateral vestibulospinal tract is driven primarily by three ipsilateral inputs, two excitatory and one inhibitory (Fig. 23-8). The two sources of excitatory input are the vestibular sensory apparatus and the cerebellar nuclei, mainly the fastigial nucleus. The inhibitory input consists of Purkinje cell axons from the cerebellar cortex.

The lateral vestibulospinal tract is the path by which input from the vestibular sensory apparatus is used to coordinate orientation of the head and body in space. Maintenance of body and limb posture is also influenced by extensive cerebellovestibular projections, which can be either excitatory or inhibitory. The cerebral cortex essentially has no direct projections to the vestibular nuclei and, consequently, the vestibulospinal tract is not influenced by cortical mechanisms.

Reticulospinal Tracts Cells at many levels of the reticular formation contribute to the reticulospinal system, and these fibers can be found in the lateral and ventral funiculi throughout the spinal cord. Reticulospinal fibers participate in a wide variety of functions ranging from pain modulation to visceromotor activity. Most of the fibers involved in somatomotor function originate either from the oral and caudal pontine nuclei or from the gigantocellular reticular nucleus (Fig. 23-9). The fibers from the *oral* and *caudal pontine reticular nuclei* descend bilaterally, but with an ipsilateral predominance, in the ventral funiculus. They constitute the *medial reticulospinal* (or *pontine reticulospinal*) *tract*, which runs the full length of the spinal cord. The fibers from the *gigantocellular reticular nucleus* originate at medullary levels. Most of these *medullary reticulospinal fibers* remain ipsilateral and descend in the ventral funiculus, although a few decussate (Fig. 23-9). Most take up a new position somewhat lateral and ventral to the ventral horn, where they are often called the *lateral reticulospinal tract*.

Like the vestibulospinal fibers, reticulospinal fibers terminate in ventromedial portions of laminae VII and VIII, where they influence *motor neurons supplying paravertebral and limb extensor musculature*. However, in contrast to the vestibulospinal tract, individual reticulospinal fibers commonly terminate at multiple spinal levels by means of collateral branches, and there is little evidence for monosynaptic contact with alpha motor neurons.

The reticulospinal system is activated by ipsilateral descending cortical projections (*corticoreticular fibers*) as well as ascending somatosensory systems (*spinoreticular fibers*), mainly those conveying nociceptive signals. Through influences on gamma motor neurons, the reticulospinal system is involved in the maintenance of posture and in the modulation of muscle tone. Pontine reticulospinal fibers tend to mediate excitatory effects, and medullary reticulospinal fibers usually produce inhibitory effects.

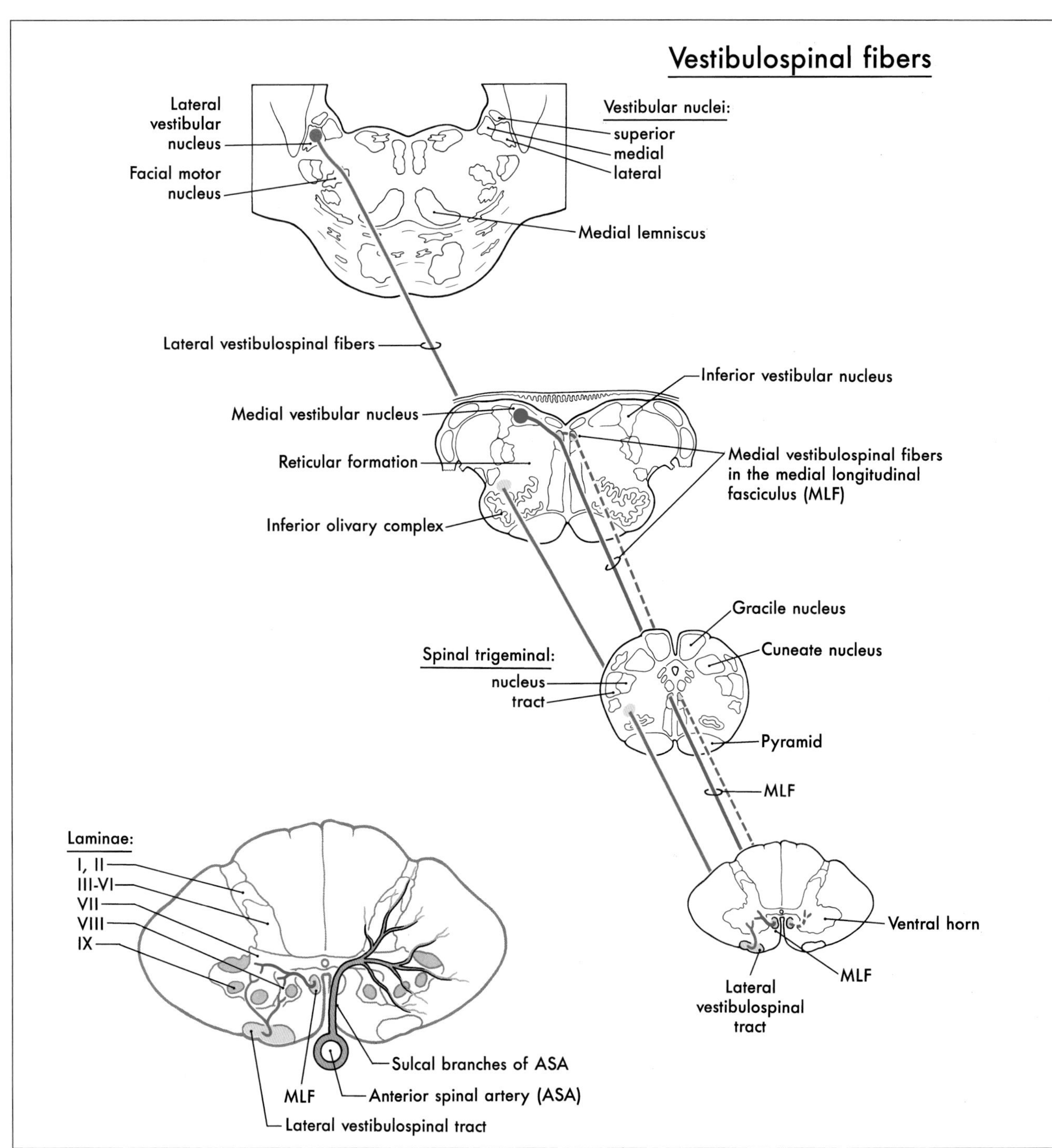

FIGURE 23-7 *Medial and lateral vestibulospinal tracts.*

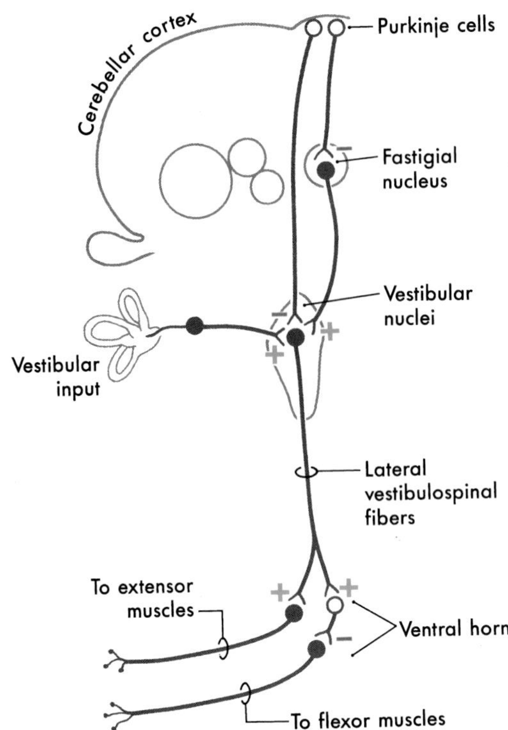

FIGURE 23-8 *Summary of vestibular and cerebellar inputs to vestibular nuclei and the subsequent action of lateral vestibulospinal fibers on spinal motor neurons. Inhibitory neurons have open cell bodies.*

Rubrospinal Tract In the midbrain, neurons in the red nucleus give rise to axons that cross the midline in the *ventral tegmental decussation* (Fig. 23-9). These fibers descend through the brainstem contralateral to their origin and enter the spinal cord ventrally adjacent to the lateral corticospinal tract. The red nucleus consists of magnocellular and parvicellular subdivisions. In mammals that have been investigated, and probably also in humans, the magnocellular part gives rise to *rubrospinal fibers, and the parvicellular part gives rise to rubro-olivary fibers*. Each rubrospinal fiber terminates in a restricted area of the spinal cord; they do not innervate multiple cord levels by means of collaterals as do reticulospinal fibers. In the spinal gray, rubrospinal fibers terminate in laminae V, VI, and VII. For the most part, they provide *excitatory influence to motor neurons innervating proximal limb flexors* (Fig. 23-9).

The magnocellular portion of the red nucleus is relatively smaller in humans than in other mammals, and the rubrospinal tract is correspondingly small. In addition, relatively few rubrospinal axons extend caudal to the cervical enlargement, suggesting that this system is primarily involved with the upper extremity. Clinical findings in patients are consistent with this conclusion, indicating that the rubrospinal system exerts its control mainly over the upper extremity and has little influence over the lower extremity.

The rubrospinal system is influenced by the cerebral cortex and the cerebellar nuclei via *corticorubral* and *cerebellorubral* fibers, respectively. Precentral and premotor cortices project to the ipsilateral red nucleus, and the supplementary area contributes contralateral input. The latter pathways provide a route through which the cortex might influence flexor motor neurons and thus serve as a supplement to the corticospinal system. Connections that link the cerebellar nuclei, inferior olive, red nucleus, and rubrospinal tract may represent circuitry important for modifying motor performance or acquiring new motor skills.

FUNCTIONAL ROLE OF BRAINSTEM AND SPINAL INTERACTIONS

Insight into the functional role of the brainstem and spinal systems has come from animal studies in which lesions have been created in specific locations of the brainstem. The resulting deficits mimic those of humans known or suspected to have damage in the same structures.

Decerebration In the basic experiment, under deep anesthesia, the brainstem was completely transected between the superior and inferior colliculi (Fig. 23-10A). This procedure results in deficits that closely resemble those seen in humans after traumatic insults, vascular disease, or tumors in the midbrain. This lesion results in hyperactivity of extensor musculature in all four extremities, a condition called *decerebrate rigidity*.

In this situation, all descending cortical systems are interrupted. This includes the corticospinal tract, as well as the corticorubral and corticoreticular projections. In addition, the rubrospinal tract is transected, but the excitatory and inhibitory components of the reticular formation are intact. Also unaffected is the ascending somatosensory input to the reticular formation, most of which is directed to the excitatory elements of the reticulospinal system.

Dorsal Root Section An important question that arose in relation to decerebration experiments was whether the extensor hypertonus was due to excessive activation of alpha or of gamma extensor motor neurons. To answer this question, the dorsal root input from one extremity was interrupted in a decerebrate animal (Fig. 23-10B). Immediately, the extensor hypertonus in that limb collapsed. What does this result indicate? Remember that supraspinal input can produce muscle contraction by two routes: by activation of alpha motor neurons or via the *gamma loop*. In the latter, the supraspinal input stimulates

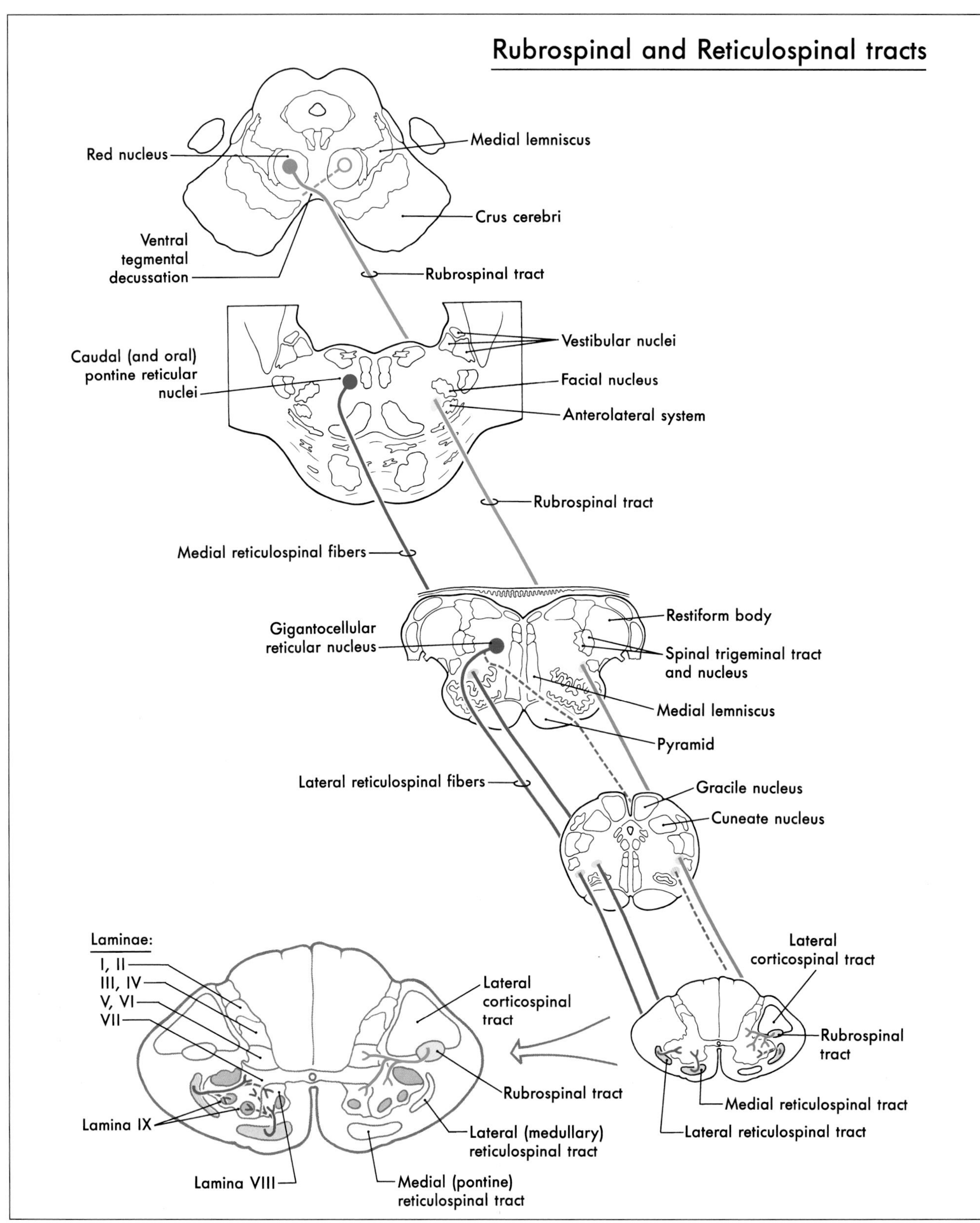

FIGURE 23-9 *Rubrospinal and reticulospinal tracts.*

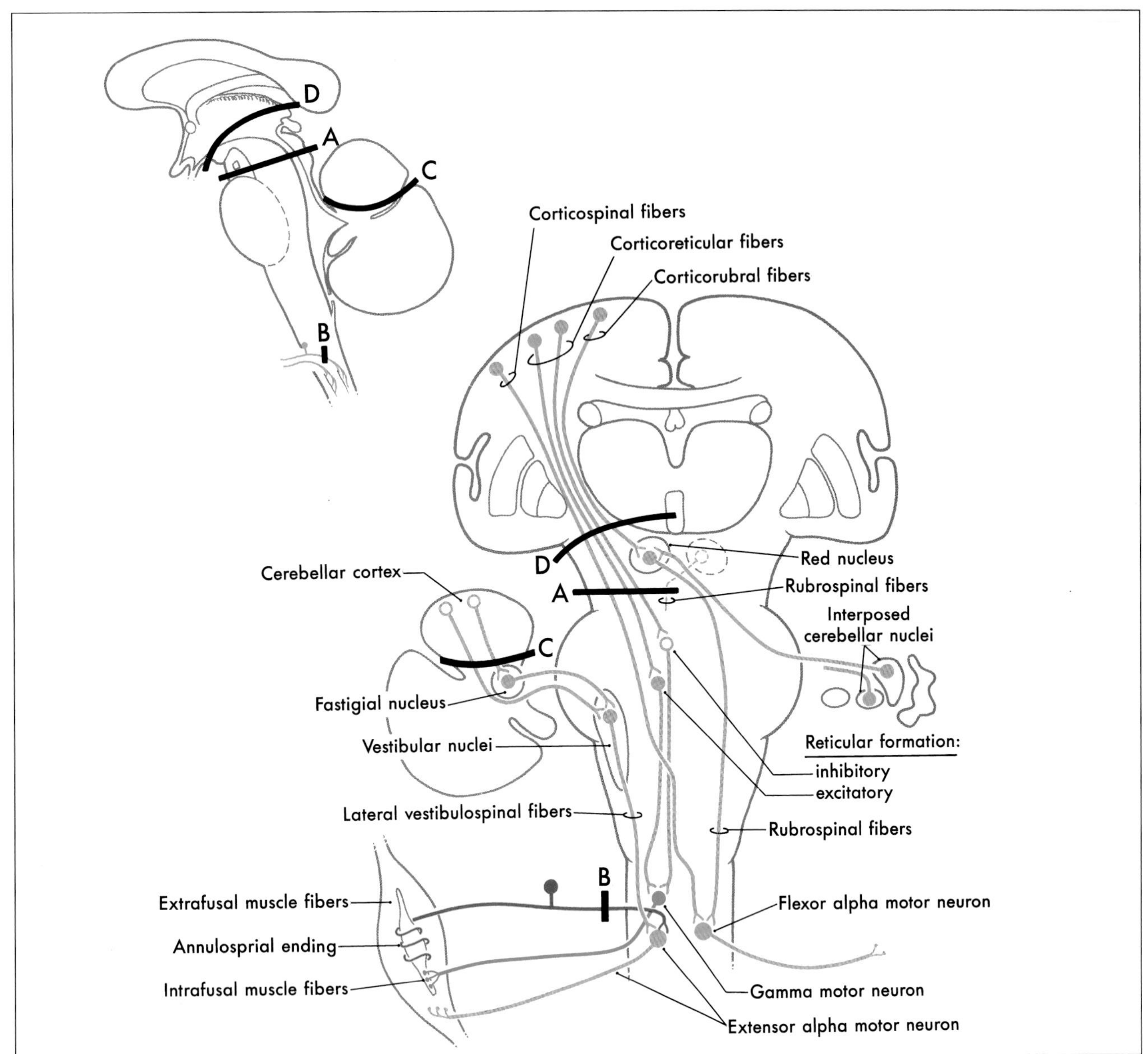

FIGURE 23-10 *Locations of lesions and circuits involved in decerebrate rigidity (A), dorsal root section in a decerebrate or decerebrate preparation (B), decerebellate rigidity (C), and decorticate posturing (D). Inhibitory neurons have open cell bodies.*

gamma motor neurons, leading to contraction of infrafusal fibers, and the resulting increase in Ia sensory input activates alpha motor neurons. Section of a dorsal root eliminates the Ia input to the spinal cord and thus breaks the gamma loop. The collapse of extensor hypertonus caused by dorsal root section therefore indicates that excitation of alpha motor neurons alone (in the absence of the gamma loop) is not sufficient to produce extensor hypertonus. Thus, decerebrate rigidity is due to excessive excitatory drive on extensor *gamma* motor neurons, coupled with diminished activation of all flexor motor neurons as a result of interruption of the corticospinal and cortico-rubro-spinal system.

Having reached this conclusion, one may ask what produces the excessive activation of gamma motor neurons in decerebrate animals. Two supraspinal systems are intact and could be the source of this excessive drive: the vestibulospinal and reticulospinal systems. The vestibular nuclei do not normally receive significant input from the cerebral cortex, however, and the major inputs they do receive—from the vestibular apparatus and from the cerebellar cortex and nuclei (see Fig. 23-8)—are not disrupted by decerebration. Therefore, the *vestibulospinal system* is not likely to be the culprit. By contrast, the *reticulospinal system* does receive important cortical input. The inhibitory portion of the reticulospinal system is driven largely from the cortex and becomes nonfunctional after decerebration. Although the excitatory elements have lost their cortical input, they can still be activated by ascending sensory inputs not affected by the lesion. Thus, decerebration upsets the balance between excitatory and inhibitory reticulospinal influences on spinal neurons, creating a bias toward excitation. The conclusion is, therefore, that *extensor hypertonus in the decerebrate patient results from excessive excitatory input to extensor gamma motor neurons via reticulospinal fibers*. For this reason, decerebrate rigidity is sometimes called *gamma rigidity*.

Cerebellar Anterior Lobe Section If extensor gamma motor neurons receive preferential input from the reticulospinal system, it is reasonable to ask if extensor alpha motor neurons might be preferentially driven by vestibulospinal fibers. To investigate this point, the cerebellar anterior lobe was removed in an animal also rendered decerebrate (midcollicular transection) (Fig. 23-10C). Under these conditions the extensor hypertonus actually proved to be enhanced compared to decerebration alone and the condition was called *decerebellate rigidity*. Section of the dorsal roots from one extremity in such an animal produced only a slight decline in extensor rigidity of the limb.

Removal of the cerebellar anterior lobe cortex has two effects (Fig. 23-10C). First, *it eliminates direct Purkinje cell inhibition of the vestibular nuclei*, resulting in enhanced output over the vestibulospinal tract. Second, Purkinje cell *inhibition of fastigial neurons is eliminated*. The resulting increase in fastigial excitatory output to the vestibular nuclei further augments the activity in the vestibulospinal tract. Therefore, the effect of cerebellar cortex removal is to substantially increase activity in the vestibulospinal system. Is this excessive drive delivered to the extensor alpha motor neurons, as hypothesized? Apparently it is, as is evident from the failure of the dorsal root section to abolish the extensor hypertonus in this animal. The persistent hypertonus therefore cannot be due to activity in the gamma loop. Instead, the increased vestibulospinal output must reach *alpha* motor neurons directly. Consequently, *decerebellate extensor rigidity* is referred to as *alpha rigidity*.

Decortication An extension of these experiments was undertaken to explain the neural substrate for the phenomenon called *decorticate posturing* or *decorticate rigidity* observed in humans. In this situation, the patient presents with flexion of the upper extremities at the elbow combined with extensor hypertonus in the lower extremities. In experimental animals, this condition can be mimicked by transecting the brainstem at a level *rostral to the superior colliculus* (Fig. 23-10D). This lesion leaves the rubrospinal tract intact while eliminating the cortical input to the red nucleus. The rubrospinal system is still functional because excitatory projections to the red nucleus from the cerebellar nuclei are unaffected. The rubrospinal tract influences primarily *flexor muscles*, and most of this activity, particularly in humans, is limited to the upper extremity. Therefore, under these conditions, the lower extremities exhibit extensor hypertonus for the same reasons as in decerebration. The upper extremities do not exhibit extensor hypertonus but instead show an increase in flexor tone. This characteristic type of posturing is called *decorticate rigidity*. These two conditions, *decerebration* and *decortication*, are frequently seen in patients, and knowledge of these symptoms and the underlying brain pathology is important in the diagnosis and clinical management of these patients. In some cases, the patient may be comatose and exhibit decorticate posturing that converts to decerebrate posturing. This is an ominous sign, as it suggests that the lesion has involved more caudal portions of the brainstem. The patient's cardiovascular and respiratory control centers in the medulla may soon be compromised, necessitating prompt intervention.

SOURCES AND ADDITIONAL READINGS

Boyd IA: The isolated mammalian muscle spindle. Trends Neurosci 3:258–265, 1980.

Brodal A: *Neurological Anatomy in Relation to Clinical Medicine*, 3rd Ed. Oxford University Press, New York, 1981, pp 148–211, 264–282.

Brooks VB (ed): *Handbook of Physiology*, Section 1: *The Nervous System*, Vol II. *Motor Control, Part 1*. American Physiological Society, Bethesda, 1981.

Desmedt JE (ed): *Spinal and Supraspinal Mechanisms of Voluntary Motor Control and Locomotion*, Vol 8. *Progress in Clinical Neurophysiology*. Karger, Basel, 1980.

Mathews PBC: Evolving views on the internal operation and functional role of the muscle spindle. J Physiol (Lond) 320:1–30, 1981.

Shepherd GM: *Neurobiology*. Oxford University Press, New York, 1994.

Sherrington CS: Decerebrate rigidity and reflex coordination of movements. J Physiol (Lond) 22:319–332, 1898.

Taylor A, Prochazka A (eds): *Muscle Receptors and Movement*. Macmillan, London, 1981.

CHAPTER TWENTY-FOUR

Motor System II: Corticofugal Systems and the Control of Movement

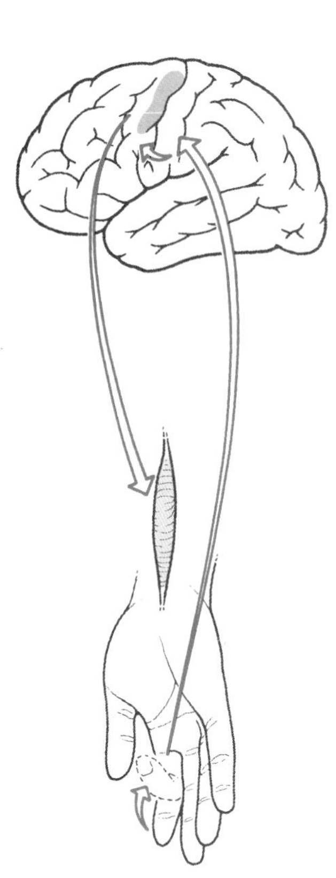

G. A. MIHAILOFF • D. E. HAINES

Brushing your teeth seems like a simple voluntary movement. However, understanding the neural basis for this action is a complex problem. For example, muscles in the upper limb are used cooperatively with jaw muscles while neck and back muscles provide postural support. Sensory feedback from the teeth and gums is linked to muscle afferents conveying tension and proprioceptive signals from the arm and hand. Even other, less obvious aspects, such as visual system input and memory of prior experience, are involved. This chapter focuses on elements of voluntary movement that are regulated by the cerebral cortex.

OVERVIEW

The control of voluntary movement is a complex, multifaceted process that involves many areas of the brain. One of the principal control sites is the cerebral cortex, specifically the *primary motor*, *premotor*, and *supplementary motor cortices* in the frontal lobe, along with portions of the parietal lobe. Although the latter two cortical regions have direct projections to the spinal cord, they also work cooperatively through the *primary motor cortex* (*upper motor neurons*) and the *corticospinal* and *corticobulbar systems* to influence the activity of ventral horn and cranial nerve motor neurons (*lower motor neurons*). The latter cells and their axons represent the *final common path* that links the central nervous system with skeletal muscles. Lesions that damage the descending cortical systems or lower motor neurons produce signs and symptoms of *upper* or *lower motor neuron disease*, respectively. These signs are among the most useful in diagnosing neurologic deficits related to the control of movement.

GENERAL FEATURES OF MOTOR DEFICITS

Lower Motor Neuron Signs *Lower motor neurons* are those cells whose axons synapse directly on skeletal muscle. When these neurons or their axons are damaged, the innervated muscles will show some combination of the following signs: (1) *flaccid paralysis* followed eventually by atrophy, (2) *fibrillations* or *fasciculations* (involuntary contractions of one or a group of motor units), and (3) *hypotonia* (decreased muscle tone) and weakening or absence of tendon (stretch) reflexes (*hyporeflexia, areflexia*).

Upper Motor Neuron Signs The term *upper motor neuron* is commonly used in reference to corticospinal cell bodies and their axons. Other neurons, such as rubrospinal or reticulospinal neurons, can also be included under the strict definition of this term. Corticospinal neurons are also called *pyramidal neurons* because their axons pass through a medullary pyramid. Therefore, *upper motor neuron signs* and *pyramidal tract signs* are often used synonymously. However, as described later in this chapter, these characteristic signs of "pyramidal tract damage" are, in fact, the result of injury to other descending motor systems in *combination* with damage to the pyramidal tract. Damage to upper motor neurons results in muscles that (1) are *initially weak* and flaccid but (2) eventually become *spastic* and exhibit increased tone (*hypertonia*) and deep tendon reflexes (*hyperreflexia*). Upper motor neuron lesions usually *affect groups of muscles*, and certain pathologic reflexes and signs often appear. One of the most common is the *inverted plantar reflex*, also known as the *Babinski sign*. This is a dorsiflexion of the great toe in response to firm stroking of the lateral aspect of the sole of the foot with a blunt instrument. The response in the normal patient is plantar flexion of the great toe.

Spasticity Muscles that are no longer under the influence of upper motor neurons exhibit spasticity. That is, when tested by the examiner, the affected muscles offer an *increased resistance to passive movement or manipulation*. These effects are most pronounced in the antigravity muscles, which in humans include the proximal flexors in the upper extremity and extensors in the lower extremity. Also, the increased resistance to passive movement is velocity-dependent: *the more rapidly the examiner moves the affected extremity, the greater the resistance*. However, after a relatively brief period of applied force, the increased resistance totally collapses; this is known as the *clasp-knife effect*.

Several hypotheses have been advanced to explain spasticity. One suggests that spasticity, and its associated hypertonia and hyperreflexia, is the result of excessive activation (or release from inhibitory control) of *dynamic gamma motor neurons*. This would lead to increased activity on type Ia muscle spindle afferents, resulting in increased excitatory tonic drive on the associated alpha motor neurons. Another suggestion is that spasticity may represent a *generalized failure of the descending cortical activation of spinal cord inhibitory interneurons*. For example, supraspinal fibers activate *type Ia* inhibitory interneurons that contact extensor motor neurons (Fig. 24-1A). If the upper motor neuron input were removed, the extensor would be *released from inhibitory control*, and the result would be hypertonia and spasticity.

Descending cortical fibers also activate a type of interneuron called a *Renshaw cell* (Fig. 24-1B). The Renshaw cell receives excitatory input from a lower motor neuron via an axon collateral, and in turn it inhibits the same lower motor neuron. It is also known, for example, that cortical fibers activating ankle flexors also contact Renshaw cells (as well as type Ia inhibitory interneurons) that inhibit ankle extensors (Fig. 24-1B). This serves to prevent reflex stimulation of the extensors when flexors are active. Therefore, when the cortical fibers are lost (upper motor neuron lesion), the inhibition of antagonists is absent. This can result in repetitive, alternating

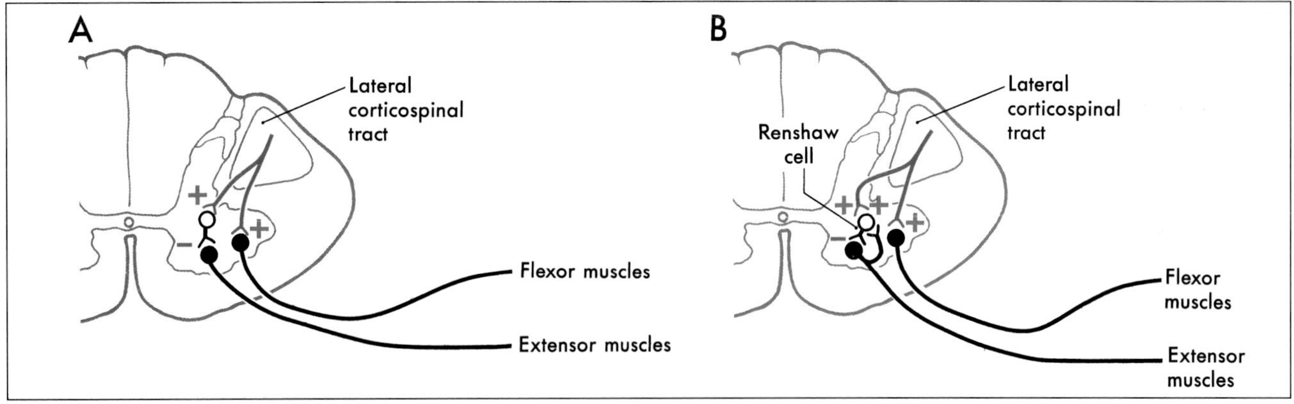

FIGURE 24-1 *Connections by which descending fibers influence lower motor neurons. Direct influence and influence via an inhibitory interneuron (A), and influence via a Renshaw cell, which participates in a recurrent inhibition circuit (B).*

contraction of ankle flexors and extensors. Such a phenomenon is referred to as *clonus* and is often present in combination with spasticity and hyperreflexia.

THE CORTICOSPINAL SYSTEM

Before considering the specifics of corticospinal projections, a general point about laterality should be made. The fibers that form the core of the corticospinal system cross the midline in the pyramidal decussation. Corticospinal fibers arising in the left motor cortex therefore influence muscles on the right side of the body, and vice versa. Consequently, *lesions of corticospinal fibers rostral to the pyramidal decussation result in contralateral motor deficits while lesions in the spinal cord result in ipsilateral deficits.* This concept of laterality is essential in the diagnosis of the neurologically impaired patient.

Origin Neurons that give rise to *corticospinal* axons are located in deep portions of layer V of the cerebral cortex (Fig. 24-2A). A small number of these pyramidal neurons are especially large with somata that may reach 100 µm or more in diameter. These are called *Betz cells*, and at one time it was believed that they were the sole source of corticospinal axons. Now it is known that they account for only 1–2% of this fiber bundle.

Corticospinal neurons are found primarily in six cortical locations (Fig. 24-2B). The single largest concentration is in *Brodmann's area 4*, which occupies the posterior portion of the *precentral gyrus* bordering on and extending into the depth of the *central sulcus*. This region is also called *MI*, the *primary motor cortex*. The *premotor* and *supplementary motor cortices*, which are located in *area 6*, also give rise to corticospinal axons. About two-thirds of all

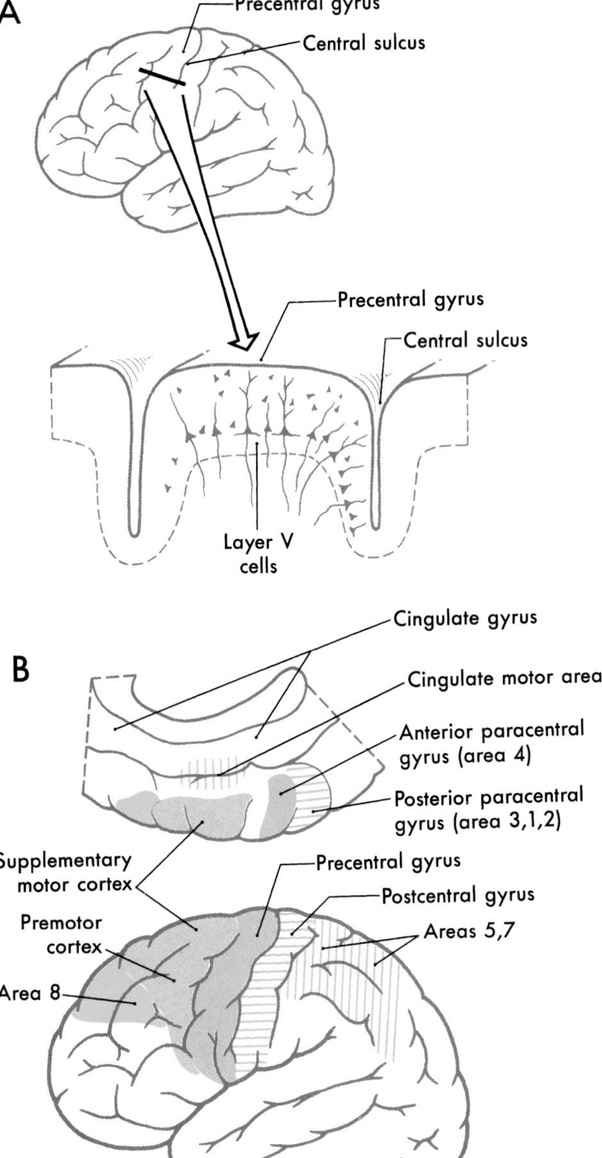

FIGURE 24-2 *The motor-related areas of the cortex. A cross section through the precentral gyrus, showing pyramidal cells in layer V (A). The main areas of the cortex that give rise to corticospinal axons (B).*

corticospinal axons arise from neurons in the *frontal lobe*, with more than one-half of this group coming from the *precentral* and *anterior paracentral gyri* (Fig. 24-2B). The remaining one-third arise from the *parietal lobe* and a few other regions. Included are cells in the *postcentral gyrus* (areas 3, 1, and 2), the *superior parietal lobule* (areas 5 and 7), and portions of the *cingulate gyrus*.

Within *MI*, corticospinal neurons are somatotopically organized in patterns that reflect their influence over specific muscles. The caricature thus created is called the *motor homunculus* (Fig. 24-3A). Neurons in medial *MI*, the anterior paracentral gyrus, project to lumbosacral cord levels to influence motor neurons which innervate muscles of the foot, leg, and thigh. Thoracic and cervical cord levels, which contain motor neurons innervating the trunk and upper extremity, receive input from neurons in the medial two-thirds of the precentral gyrus. The musculature of the head, face and oral cavity is influenced by neurons in the ventrolateral part of the precentral gyrus (Fig. 24-3A). These cells contribute to the *corticobulbar tract* which projects to cranial nerve motor nuclei. The disproportion in body part size in the *homunculus* reflects the density, and distribution of corticospinal neurons devoted to the control of musculature in each particular region of the body (Fig. 24-3A). Complete but less precise body representations are also found in other motor cortical regions. Thus, any single muscle may be influenced from multiple locations in the cerebral cortex.

The blood supply to MI arises from branches of the anterior and middle cerebral arteries (Fig. 24-3B). The lower extremity area of MI is served by terminal branches of the A_2 segment of the anterior cerebral artery. Specifically, these branches arise from the *callosomarginal artery*. The trunk, upper extremity, and head areas of the motor cortex are supplied by branches of the M_2 segment of the middle cerebral artery, mainly its *Rolandic* and *pre-Rolandic branches*.

Lesions that involve only areas of motor cortex outside the MI usually do not result in paralysis, and the effects may dissipate over time. For example, vascular infarcts of the premotor cortex may produce an *apraxia*. This involves difficulty in using the affected part of the body to perform voluntary actions, such as grasping a pencil, even though there is no obvious spasticity, paralysis, or altered tone in the muscles. For example, a premotor lesion might make a person unable to perform voluntary actions with the contralateral hand, although the strength and tone of the hand muscles would be normal. Similarly, unilateral lesions in the supplementary motor cortex affect the ability to coordinate actions on the two sides of the body. The muscles, again, are normal. In contrast, lesions that affect both the primary motor cortex in combination with another motor cortical region usually result in spastic paralysis and hyperreflexia, signs characteristic of upper motor neuron lesions. Very small lesions, which affect only MI, can lead to flaccid paralysis, but this situation is rare.

Course The largest axons in the corticospinal tract are myelinated, range from 12 to 15 μm in diameter, and have conduction velocities of 70 m/sec. Surprisingly, however, these large axons make up less than 10% of the total population. The remainder are less than 5 μm in diameter, and many are lightly myelinated or unmyelinated.

Corticospinal fibers pass through the corona radiata and converge to enter the *posterior limb of the internal capsule* (Fig. 24-4A). In this region, these fibers are somatotopically organized such that the axons that will terminate at the highest cord levels are located most anteriorly and axons that will terminate progressively lower are located more posteriorly (Fig. 24-4B).

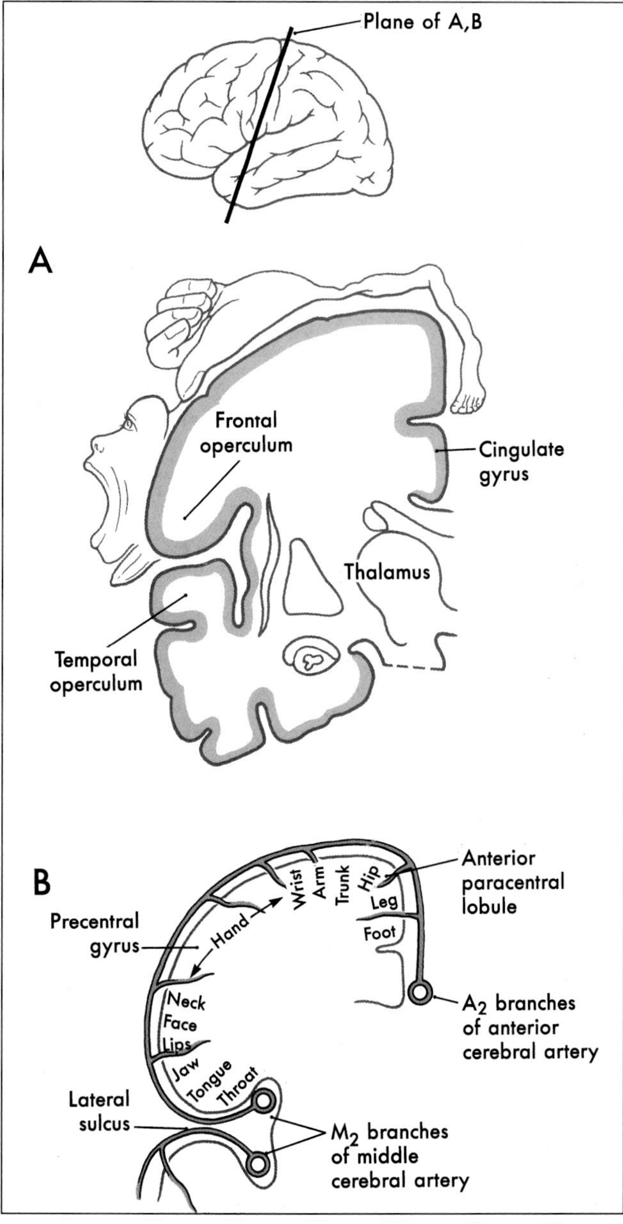

FIGURE 24-3 *Coronal views of the cerebral hemisphere showing the somatotopy of the primary motor cortex (A, B) and the blood supply of the anterior paracentral lobule and the precentral gyrus (B). (A is adapted from Penfield and Rasmussen, 1968, with permission.)*

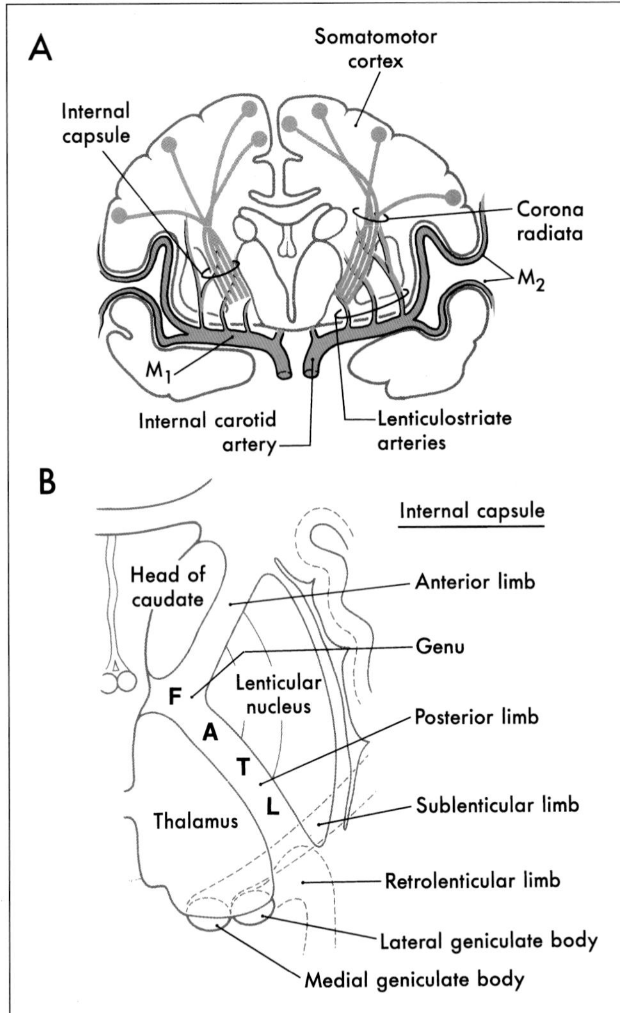

FIGURE 24-4 *Descending fibers of the corticospinal and corticobulbar systems in the internal capsule in coronal (A) and axial (B) planes. The positions of fibers from face (F), arm (A), trunk (T), and leg (L) areas are shown in the internal capsule in the axial plane (B).*

Unlike lesions of the cortical gray matter, interruption of axons in the posterior limb of the internal capsule often results in catastrophic motor deficits. A common cause of lesions in this area is hemorrhage from *lenticulostriate branches* of the M_1 segment of the middle cerebral artery (Fig. 24-4A). Motor symptoms of capsular infarcts appear in the contralateral upper and lower extremities and consist of weakness and eventual spasticity (upper motor neuron damage). These symptoms appear because not only corticospinal fibers but also many other types of cortical axons are interrupted. Included are axons projecting to the striatum, thalamus, and brainstem, as well as thalamocortical axons involved in somatic sensation and vision. Damage to the latter fibers explains why *hemisensory loss or homonymous hemianopia may accompany the motor deficit.* It is important to note that deficits such as spasticity, hypertonia, and hyperreflexia, while commonly associated with pyramidal tract lesions, are in fact due to damage of other descending systems in *combination* with corticospinal fibers.

As they pass caudally from the internal capsule, corticospinal fibers traverse the various divisions of the brainstem. In the midbrain they coalesce to form the middle third of the crus cerebri (Figs. 24-5A, 24-6). Within this part of the crus, fibers from arm/hand areas of MI are located medially, while those from leg/foot areas are lateral.

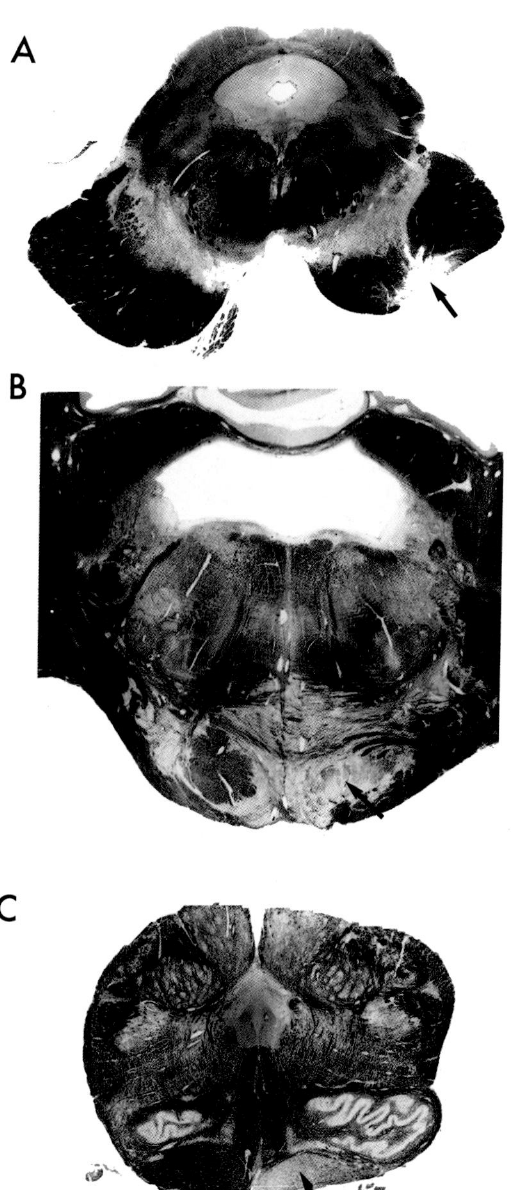

FIGURE 24-5 *Degeneration of corticospinal fibers caused by an infarction in the posterior limb of the internal capsule. The degeneration serves as a marker to show the position of these fibers (at arrows) in the middle third of the crus cerebri (A), the basilar pons (B), and the pyramid of the medulla (C).*

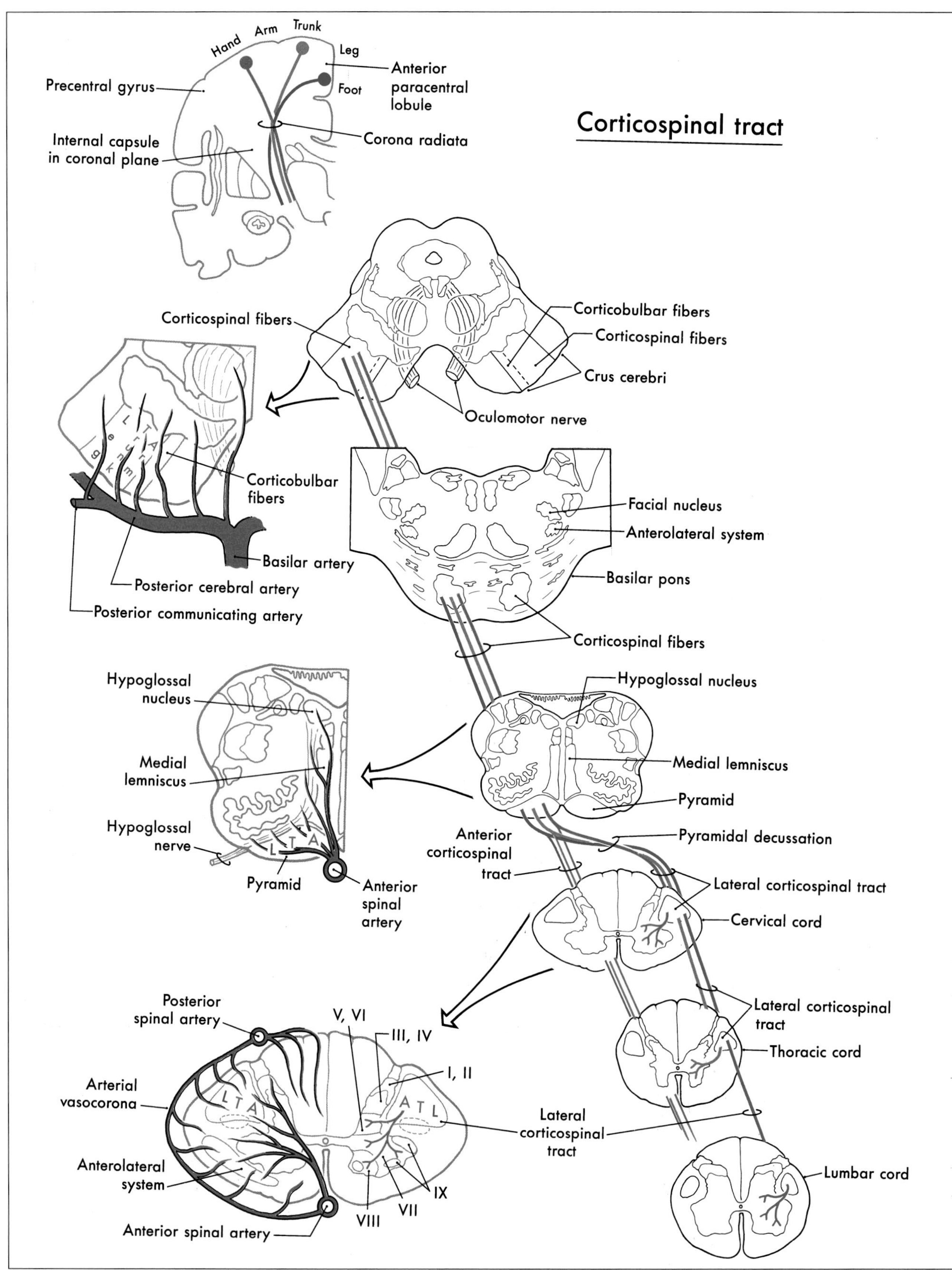

FIGURE 24-6 *The corticospinal system with details showing the blood supply to these fibers in the midbrain, medulla, and spinal cord.*

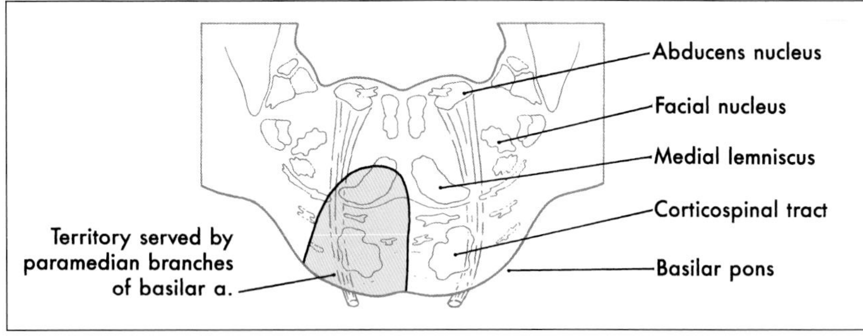

FIGURE 24-7 *The blood supply to corticospinal fibers, the abducens fibers, and parts of the medial lemniscus in the caudal pons. The motor deficits seen in this lesion constitute a middle alternating hemiplegia.*

Fibers in the medial two-thirds of the crus cerebri (frontopontine, corticobulbar, and corticospinal) and the exiting rootlets of the oculomotor nerve fibers are served by *paramedian branches of P_1* and branches from the adjacent *posterior communicating artery* (Fig. 24-6). Hemorrhage of these vessels will damage these groups of fibers, resulting in (1) *contralateral hemiparesis* of the arm and leg with spasticity and (2) *deviation of the ipsilateral eye* down and laterally, owing to the unopposed action of the superior oblique and lateral rectus muscles. *Direct* and *consensual light reflexes* and accommodation may also be lost in the eye on the side of the lesion. This condition is known as *superior alternating hemiplegia* since cranial nerve signs are seen on one side and corticospinal signs on the "alternate" side.

From the midbrain, corticospinal fibers continue into the basilar pons, where they make their way between the masses of neurons forming the basilar pontine nuclei (Figs. 24-5B, 24-6). As corticospinal axons pass through the pontine gray, they give rise to collaterals which synapse on these neurons.

Corticospinal fibers in the basilar pons and the exiting fibers of the abducens nerve in the caudal pons are in the domain of the *paramedian branches of the basilar artery*. Occlusion or rupture of these vessels results in *hemiplegia* and *upper motor neuron* signs in the contralateral extremities. If the lesion extends caudally, it may involve *intra-axial* abducens fibers, resulting in lower motor neuron paralysis of the ipsilateral lateral rectus muscle (Fig. 24-7). This combination of deficits is called *middle alternating hemiplegia*. The paramedian branches of the basilar artery may penetrate deep into the pons and also serve the medial lemniscus (Fig. 24-7). In such cases, damage to these vessels can produce not only the motor deficits described above but also *contralateral loss of vibratory sense and two-point tactile discrimination*.

In the medulla, corticospinal fibers aggregate on the ventral surface of the brainstem to enter the medullary pyramids (Figs. 24-5C, 24-6). Within the pyramid, fibers that will terminate at cervical levels tend to be located medially, while those projecting to lumbar and sacral levels are more lateral. Collaterals of these axons innervate the inferior olivary complex, dorsal column nuclei, and various medullary reticular nuclei.

The pyramid, the laterally adjacent exiting fibers of the hypoglossal nerve, and the medial lemniscus receive their blood supply through penetrating branches of the *anterior spinal artery* (Fig. 24-6). Occlusion of these branches results in a *contralateral hemiparesis of the extremities* (with spasticity) and an *ipsilateral flaccid paralysis of the tongue*. When protruded, the tongue deviates toward the side of the lesion. This combination of symptoms is called *inferior alternating hemiplegia*. Since branches of the anterior spinal artery also serve the medial lemniscus, an inferior alternating hemiplegia may be accompanied by a *contralateral loss of two-point discrimination and vibration sense*.

At the medullospinal junction, some 85–90% of the corticospinal fibers cross the midline as the *pyramidal decussation* (Fig. 24-6). Fibers that cross in the rostral and caudal portions of the pyramidal decussation originate, respectively, from the arm and leg areas of MI cortex. This explains why small vascular lesions in the decussation (which is also served by branches of the *anterior spinal artery*) may result in bilateral hemiplegia of either the arms or the legs. The decussating fibers extend into the lateral funiculus to form the *lateral corticospinal tract*. The corticospinal axons that do not cross in the decussation continue into the ipsilateral ventral funiculus of the spinal cord as the *ventral (anterior) corticospinal tract* (Fig. 24-6). Damage to this tract is of little clinical significance.

Termination Fibers of the lateral corticospinal tract are topographically organized (Fig. 24-6). Axons terminating in cervical cord levels are most medial in this tract, whereas those distributing to lumbosacral levels are most lateral. This pattern means that, as the medially located fibers enter and terminate in the spinal gray, the adjacent, more lateral fascicles shift medially.

Corticospinal fibers that arise from the frontal lobe terminate primarily in the intermediate zone and ventral horn (laminae VII to IX), whereas fibers arising from the parietal lobe terminate in the base of the dorsal horn (laminae IV to VI). As might be expected, most fibers terminate in the spinal cord enlargements that serve the extremities; about 55% terminate in the cervical enlargement, and about 25% in the lumbosacral enlargement. The remaining fibers terminate in thoracic levels. As mentioned earlier, some corticospinal fibers terminate via collaterals at multiple levels. However, the influence

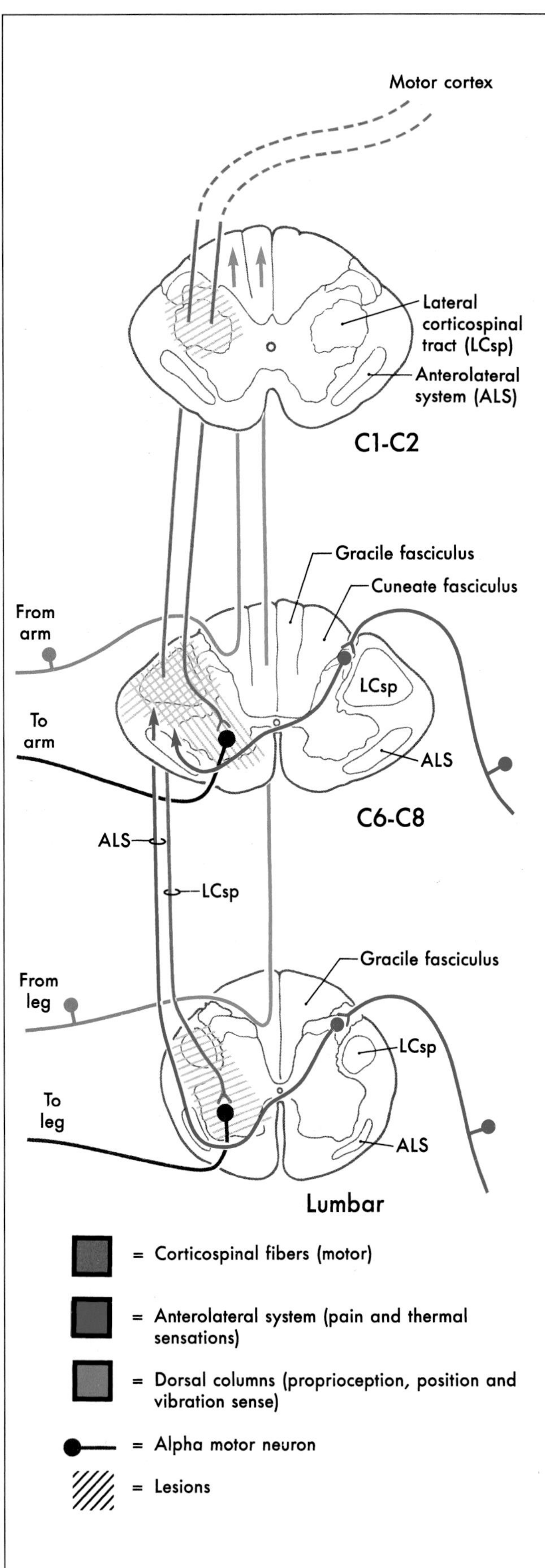

FIGURE 24-8 *Examples of spinal cord lesions at C1 to C2 (involving only corticospinal fibers), at C6 to C8 (involving corticospinal fibers only, or these plus the ventral horn), and at lumbar levels (involving corticospinal fibers and the ventral horn).*

exerted by any single axon or collaterals depends on the *number* of synapses it forms and the locus of the synaptic contacts on the postsynaptic neuron. Thus, a given corticospinal axon may have a powerful action on some spinal cord neurons and only a weak influence on others.

At their level of termination, particularly in the cord enlargements, corticospinal fibers synapse primarily on *interneurons* in laminae V to VII. In animals capable of dexterous finger movements, such as monkeys and humans, some corticospinal fibers terminate among clusters of lamina IX alpha motor neurons, which innervate distal flexor muscles. However, most corticospinal fibers, at least in nonhuman primates, synapse with excitatory and inhibitory interneurons, which, in turn, influence flexor and extensor motor neurons, respectively.

Interruption of lateral corticospinal axons in the *upper cervical cord* (C1, C2) results in spastic hemiplegia involving the ipsilateral upper and lower extremities (Fig. 24-8). Common upper motor neuron signs such as *hypertonia*, *hyperreflexia*, and the *Babinski sign* will be present ipsilateral to the lesion. If the lesion is sufficiently large, the innervation of the diaphragm (from C3 to C5 via the phrenic nerve) may be disrupted, necessitating the use of a respirator.

A lesion of the *cervical enlargement* results in a different pattern of motor deficits (Fig. 24-8). If the damage involves only the lateral funiculus, then the ipsilateral upper and lower extremities will exhibit typical upper motor neuron signs. If, however, the C6 to C8 ventral horn and the lateral funiculus are both included in the lesion, then *lower motor neuron signs will appear in the upper extremity ipsilaterally, whereas upper motor neuron signs are seen in the ipsilateral lower extremity*. When ventral horn motor neurons or their axons are damaged, the affected muscles exhibit lower motor neuron signs, despite the fact that supraspinal systems providing input to these cells may also have been interrupted.

At *lumbosacral levels*, injury to the spinal cord frequently affects motor neurons in the ventral horn as well as descending supraspinal fibers (Fig. 24-8). Characteristically, such patients exhibit lower motor neuron signs in the ipsilateral lower extremity.

The blood supply to the lateral corticospinal tract is derived from *penetrating branches of the arterial vasocorona* and *sulcal branches of the anterior spinal artery* (Fig. 24-6). The former serve fibers in lateral parts of the tract, and the latter serve the medially located fibers. Hyperextension of the neck may result in injury to the cord or in occlusion of the sulcal arteries (*central cord syndrome*); either can result in bilateral hemiparesis of the arms because of vascular infarcts involving medial regions of both lateral corticospinal tracts. In addition, these patients exhibit both urinary retention and a bilateral, patchy loss of pain and temperature sensations below the lesion.

A functional hemisection of the spinal cord, such as may be caused by a tumor or by trauma, results in a characteristic set of deficits known as *Brown-Séquard syndrome* (Fig. 24-9; see also Fig. 17-8). These deficits begin about two levels below the lesion and consist of (1) *ipsilateral* loss of two-point discrimination and vibration (from damage to the dorsal columns), (2) *contralateral* loss of pain and thermal sensation (from damage to the anterolateral system), and (3) an *ipsilateral* paresis or paralysis (from damage to the corticospinal tract). The latter involves the arm and leg or only the leg, depending on the level of the injury. In addition, damage to primary afferent fibers entering the cord may result in a narrow band of anesthesia on the ipsilateral side at the level of the lesion.

THE CORTICOBULBAR SYSTEM

Origin Organized in parallel with the corticospinal system is the *corticobulbar system* (Fig. 24-10). As defined here, the corticobulbar system consists of the cortical neurons that influence the movements of striated muscles innervated by the motor nuclei of cranial nerves V, VII, and XII, by the nucleus ambiguus, and by the accessory nucleus. (The term is sometimes used in a wider sense to include all cortical systems that project to the brainstem. The suffix "bulbar" comes from "bulb," an obsolete term for the medulla oblongata.) Some corticobulbar axons project directly to cranial motor neurons, but most terminate on reticular formation interneurons immediately adjacent to these nuclei. The corticobulbar system originates for the most part from the face and head area of the precentral gyrus (Fig. 24-10). Because most of the musculature innervated by cranial nerves is located in the facial region, this area of MI is typically called *face motor cortex*.

The oculomotor, trochlear, and abducens nuclei do not receive direct input from the face motor cortex. Instead, voluntary control of eye movement is mediated via cortical projections from *frontal* and *parietal motor eye fields* to eye movement (gaze) control centers in the midbrain and pons. These centers in the *midbrain reticular formation* and the *paramedian pontine reticular formation*, in turn, relay input from the cortical eye fields to somatic motor neurons in the nuclei of cranial nerves III, IV and VI (see Chapter 27 for details). Although the cortex of each hemisphere influences these nuclei bilaterally, the resulting eye movements are conjugate and are toward the side contralateral to the cortex from which the input originated. Because these cortical axons are often several synapses removed from the actual cranial nerve motor neurons, they are not considered part of the corticobulbar system as defined in this text.

Course Corticobulbar axons that originate from cells in layer V of the face motor cortex funnel into the *genu of the internal capsule*, where they are positioned just anterior to corticospinal fibers (Fig. 24-4). Corticobulbar fibers continue into the crus cerebri where they are located medial to corticospinal fibers traveling to cervical cord levels (Figs. 24-6, 24-10). From here, these axons descend into the pons and medulla in association with corticospinal fibers. Corticobulbar fibers in the genu and in the crus cerebri receive their blood supply from lenticulostriate arteries and from the paramedian branches of the basilar bifurcation, respectively.

Termination As corticobulbar fibers pass through the basilar pons, branches arch dorsally into the pontine tegmentum to terminate in the area of the trigeminal and facial motor nuclei (Fig. 24-10). The fibers to the trigemi-

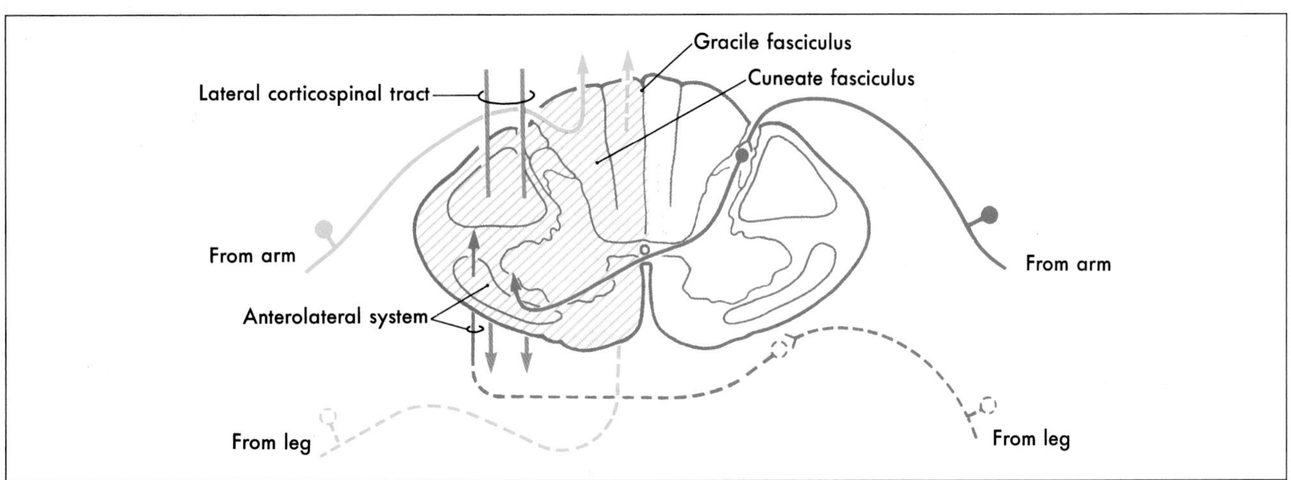

FIGURE 24-9 *Hemisection of the spinal cord (Brown-Séquard syndrome).*

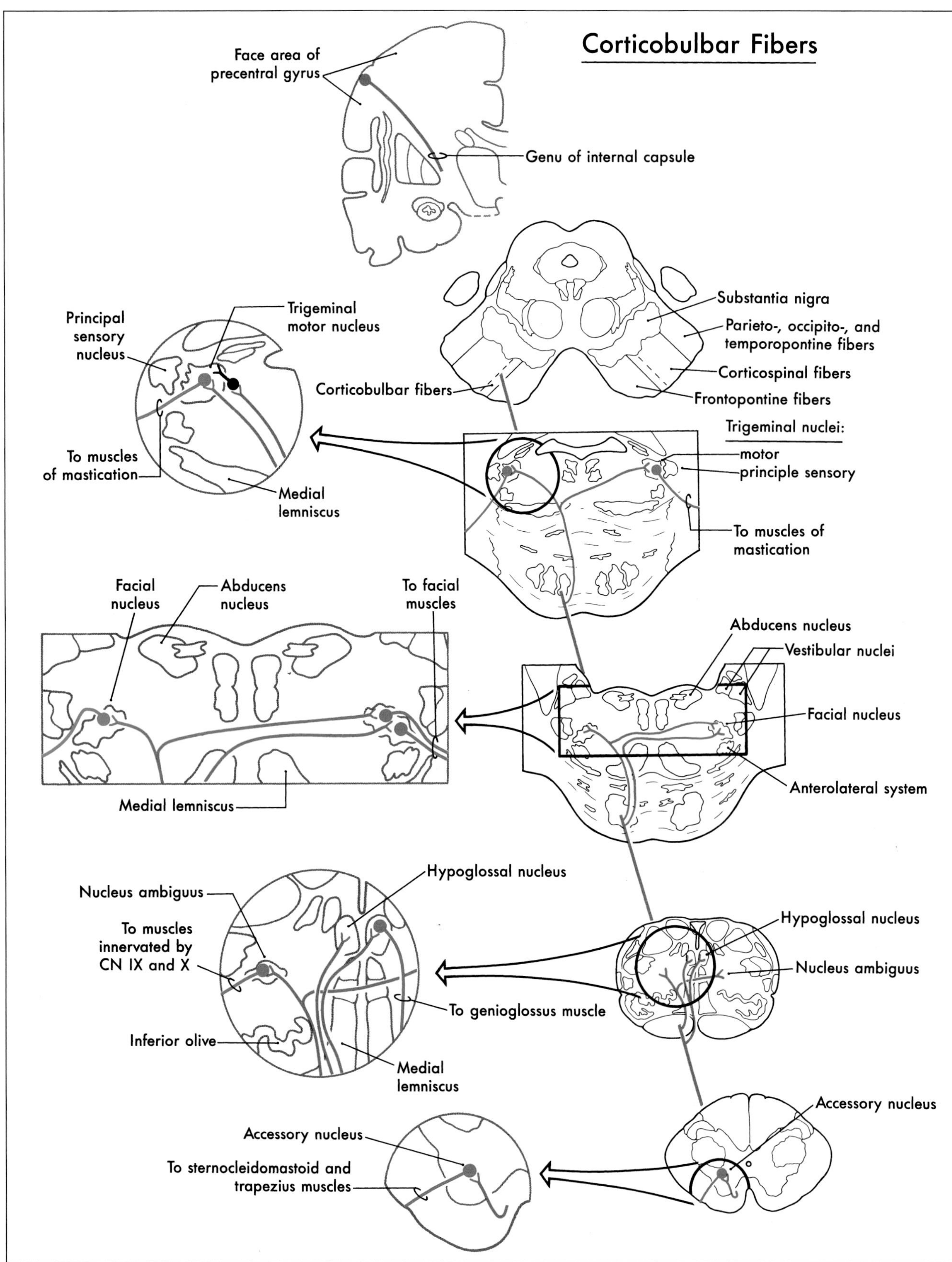

FIGURE 24-10 *The corticobulbar system with details shown at the levels of the trigeminal motor, facial motor, hypoglossal and ambiguus, and accessory nuclei.*

nal motor nuclei terminate on interneurons adjacent to the nuclei. The corticobulbar system sends nearly equal numbers of fibers to the left and right trigeminal motor nuclei. Likewise, nearly equal numbers of fibers are sent to the left and right facial motor nuclei. Whereas the muscles of facial expression in the upper half of the face are controlled about equally from both hemispheres, *muscles in the lower half of the face are influenced primarily from the contralateral hemisphere*. Consequently, a lesion of corticobulbar fibers rostral to the facial motor nucleus results in drooping of muscles at the corner of the mouth and on the lower portion of the face *on the side opposite the lesion* (Fig. 24-11A). This is called *central facial paralysis* (*central seven*). In contrast, a lesion of the root of the facial nerve will result in a flaccid paralysis of facial muscles of *upper and lower portions of the face on the ipsilateral side* (Fig. 24-11B); this is *Bell's (facial) palsy*.

At midmedullary levels corticobulbar fibers pass dorsally to reach the ambiguus and hypoglossal nuclei (Fig. 24-10). Projections to nucleus ambiguus motor neurons are generally bilateral. However, the motor neurons that innervate muscular parts of the soft palate and uvula receive mainly a contralateral input. Consequently, a lesion of corticobulbar fibers to the nucleus ambiguus may produce a weakness in the affected muscles and *result in failure of the soft palate to elevate on the contralateral side and in deviation of the uvula toward the side of the lesion* (Fig. 24-12).

In the case of the hypoglossal nuclei, those *motor neurons that innervate the genioglossus muscles receive primarily contralateral corticobulbar input*. Each genioglossus muscle pulls its half of the tongue anteriorly and slightly medially. When they function together and symmetrically, the tongue protrudes straight out of the mouth. However, a lesion of corticobulbar fibers to the hypoglossal nucleus will cause the tongue to *deviate toward the weak side, when protruded, because of the unopposed pull of the intact muscle* (Fig. 24-12).

The final contingent of corticobulbar fibers innervates the spinal portion of cranial nerve XI (*accessory nucleus*) (Fig. 24-10). These fibers continue into the upper cervical spinal cord along with corticospinal fibers. Clinical observations from patients with cortical or internal capsule lesions reveal that the sternocleidomastoid and trapezius muscles (targets of accessory motor neurons) are mainly affected on the side of the lesion. The patient is unable to shrug that shoulder or turn the head away from the side of the lesion. This suggests that *corticobulbar fibers distribute primarily to the ipsilateral accessory nucleus*.

Because the corticospinal and corticobulbar systems course adjacent to each other, it is common for lesions of the internal capsule or midbrain to affect both fiber bundles. For example, lenticulostriate arteries serve the genu and most of the posterior limb of the internal capsule (Fig. 24-4A). Consequently, hemorrhage of these vessels

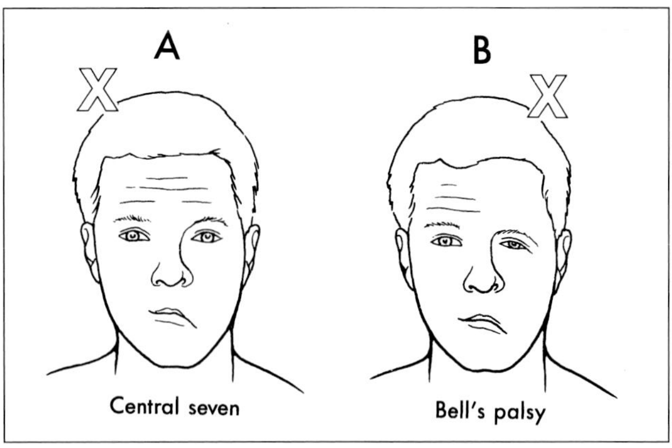

FIGURE 24-11 *Appearance of the face following a lesion of corticobulbar fibers to the facial nucleus (A) versus a lesion of the root of the facial nerve (B). The large X indicates the side of the lesion.*

on the *left side* results in (1) a *right spastic hemiparesis* of the extremities (corticospinal damage), (2) a *central facial paralysis* on the right, (3) a *deviation of the uvula to the left* on phonation, and (4) a *deviation of the tongue to the right* when protruded. The latter three deficits are the result of damage to corticobulbar fibers. Effects on the trapezius and sternocleidomastoid muscles are variable and are absent in some patients. Lesions at lower brainstem levels may produce corticospinal and corticobulbar signs in combination with symptoms of cranial nerve root injury. For example, a lesion in the territory of the paramedian branches of the basilar artery will damage corticospinal fibers, corticobulbar fibers to relatively caudal targets (nucleus ambiguus, hypoglossal nucleus), and the exiting rootlets of the abducens nerve (Fig. 24-7). A patient with such a lesion presents with (1) a *contralateral spastic hemiparesis* of the extremities, (2) *ipsilateral deviation of the*

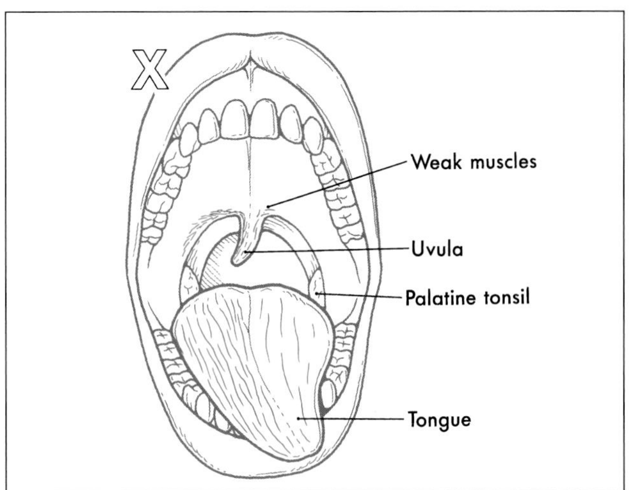

FIGURE 24-12 *Deviation of the uvula and of the tongue following a lesion of corticobulbar fibers to the nucleus ambiguus and hypoglossal nucleus, respectively. The large X indicates the side of the lesion.*

uvula on phonation, (3) *deviation of the tongue* (on protrusion) *to the contralateral side* and (4) an *inward rotation of the ipsilateral eye* caused by a flaccid paralysis of the lateral rectus muscle and the unopposed action of the medial rectus. Comparison of corticospinal with corticobulbar deficits is essential to localizing in neurologically compromised patients.

OTHER CORTICOFUGAL SYSTEMS

The Corticorubral System Cortical projections to the red nucleus arise primarily from areas 4 and 6, and to a lesser extent from areas 5 and 7. Both large and small neurons in the red nucleus receive ipsilateral corticorubral input. Although pyramidal tract neurons provide some collaterals to the red nucleus, many corticorubral axons are not collaterals of pyramidal tract fibers. In general, the corticorubral-rubrospinal projection is topographically organized. For example, the arm region of the MI cortex projects to cells of the red nucleus that, in turn, send their axons to contralateral cervical levels of the spinal cord (see Fig. 23-10). Because the rubrospinal system primarily influences flexor musculature, this pathway may supplement the function of the corticospinal tract. It is known from experimental studies that section of corticospinal fibers in the medullary pyramid leaves the animal still able to walk, climb, and pick up food but unable to perform fine, dexterous movements with its digits. This suggests that the *corticorubrospinal system may partially compensate for the loss of the corticospinal tract.*

The red nucleus also receives input from the contralateral interposed nuclei of the cerebellum (see Chapter 26). Consequently, a fairly small population of brainstem upper motor neurons are in the unusual position of integrating signals from motor-related areas of the cerebral cortex and from the cerebellum. Input from the interposed nuclei is excitatory, and this projection may be part of a circuit specialized for rapid control or adjustment of movements based on sensory processing by the cerebellum.

The Corticoreticular System The pontine and medullary nuclei that give rise to the reticulospinal tracts receive cortical input from the premotor cortex and, to a lesser extent, from the supplementary motor cortex. Because reticulospinal systems influence extensor muscles, including the paravertebral extensors as well as those of the limbs, the *corticoreticulospinal system* may provide the cortex with the means to influence extensor musculature in parallel with its regulation of flexors (see Fig. 23-9). It should also be noted that the cerebellar nuclei project to the motor-related areas of the reticular formation, thus providing for a cerebellar influence on extensor musculature.

The Corticopontine System Axons from nearly all regions of the cerebral cortex contribute to the corticopontine projection, and this pathway is particularly well developed in the human brain. Although most of these fibers originate from motor-related areas, nonmotor regions in the frontal lobe and in parietal, temporal, and occipital association cortices also contribute fibers. Corticopontine axons descend through the internal capsule (see Fig. 15-11) and continue into medial and lateral parts of the crus cerebri. Frontopontine fibers are located medially, and parieto-, occipito-, and temporopontine fibers are laterally placed (see Fig. 13-8). These corticopontine projections synapse in the ipsilateral basilar pontine nuclei. Although most neurons in the pontine nuclei send their axons into the contralateral cerebellum via the middle cerebellar peduncle, there is a notable ipsilateral projection.

Although little is known about the function of this vast system, it certainly must be involved in some aspects of motor control. However, recent studies in humans have shown that the cerebellum is also active during mental problem solving and internal (nonvocal) language functions. This implies that the corticopontine system is an important route of communication between the cerebral cortex and the cerebellum, structures that have no direct connections in the mature brain.

MOTOR CORTEX AND THE CONTROL OF MOVEMENT

The classic view of voluntary movement control is that the various motor-related areas of the cerebral cortex act hierarchically. At one time, the primary motor cortex was thought to be at the apex of this hierarchy with the output of the other cortical areas being funneled through it. Recent finding suggest that the motor-related cortical areas outside MI and their respective descending projections carry out the tasks involved in planning and executing a movement *in parallel* with MI and its projections.

Primary Motor Cortex Recall that the primary motor cortex (MI) is organized into a detailed somatotopic map of the body (Fig. 24-2) and that many corticospinal fibers originate from the pyramidal neurons in its layer V. What is this area's contribution to the control of movement?

Like other cortical areas, such as the striate and primary somatosensory cortices, MI is organized into a series of modules or *vertical columns*. Microstimulation in MI can result in discrete movements of individual muscles. For example, stimulation in a vertical column in the hand area of *MI* may evoke flexion of a digit (Fig. 24-13). Neurons in the same columnar array will receive somatosensory feedback from the patch of skin on the volar (glabrous) surface of the digit, which is the area

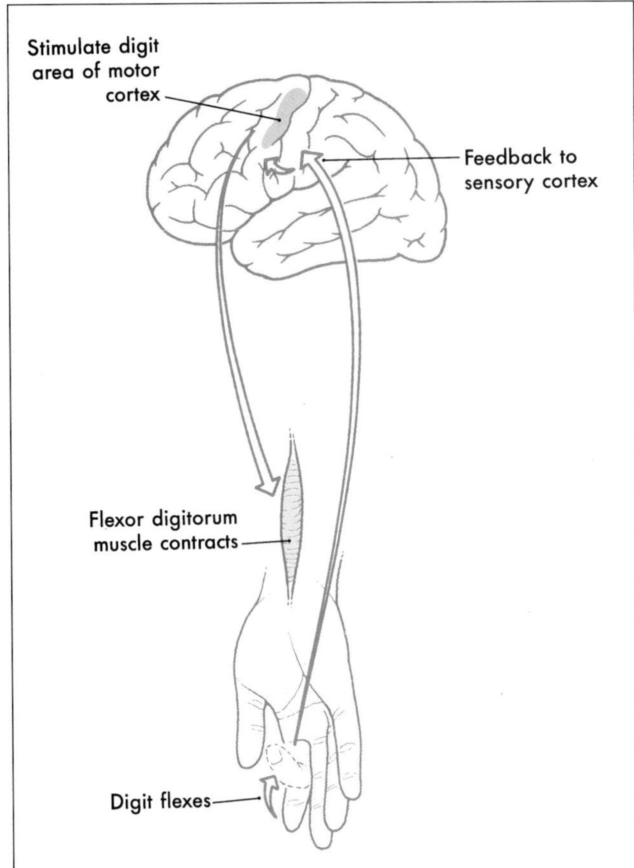

FIGURE 24-13 *Schematic representation of a long-latency reflex.*

that would come in contact with a surface when the digit is flexed, as to grasp an object. These connections are part of *long-latency reflex* circuits. The sensory information reaches MI from the ascending sensory systems indirectly via synapses in the thalamus and the primary sensory cortex (SI). The inference here is that *motor cortical neurons are informed of the result of their output.*

Recent studies have shown that certain muscles, particularly distal muscles of the upper extremity, are regulated from more than one cortical location. Conversely, microstimulation at one cortical locus can activate more than one muscle. Thus, the classical view that a discrete, somatotopically organized projection emanating from area 4 is primarily responsible for the control of individual muscles may be an oversimplification.

The activity of corticospinal neurons in MI can also be modified during a movement. In experiments involving nonhuman primates, microelectrodes placed in layer V identified single corticospinal neurons by their response to *antidromic stimulation* of a medullary pyramid. These monkeys were trained to make wrist flexion or extension movements in a situation where the movement was either assisted or impeded by a weight attached to the wrist by a pulley system. Under these conditions, corticospinal neurons were found to be *active slightly in advance of the movement*, and they did not simply code for flexion or extension but rather for *the amount of force required to make the movement*. For example, when the weight was arranged to oppose the movement, increases in the amount of weight were matched by increases in the activity of corticospinal neurons.

Small populations of MI cortical neurons encode *movement direction*. When a monkey is trained to move a handle toward one of several targets arranged concentrically around a central starting location, the activity of individual cortical neurons varies with the direction of the required movement. That is, some neurons fire rapidly for a movement in one direction but are silent for a movement in the opposite direction. This directional tuning is rather broad however, with most neurons firing with movement in a preferred direction but exhibiting less vigorous activity in relation to movements in other directions.

Supplementary Motor Cortex The supplementary motor cortex occupies the portion of Brodmann's area 6 that lies anterior to MI near the convexity and extends onto the medial wall of the hemisphere near the paracentral gyri (Fig. 24-2). It contains a map of the body musculature that is complete, although less precisely organized than that of MI. It receives input from the parietal lobe and projects to MI and directly to the reticular formation and spinal cord. Stimulation of the supplementary cortex can evoke movements. In contrast to the single-muscle movements evoked by MI stimulation, however, these movements involve sequences or groups of muscles and orient the body or limbs in space. In addition, higher stimulus intensities are required than for MI, and bilateral movements of the hands or upper extremities are often produced.

Experiments performed in humans illustrate the functional role of the supplementary motor cortex. By using a device similar to a CT scanner and injecting small amounts of a solution containing radioactive xenon into the vascular system, it is possible to measure small increases in local blood flow (as enhanced radioactivity) in brain regions where neuronal activity is increased. When a volunteer made a series of random finger flexion movements, an increase in neural activity was observed only over the hand region of MI (Fig. 24-14A). The subject was then asked to make flexion movements with several fingers of the same hand, but in a specific sequence. Activity was increased both in the *supplementary motor cortex* and in the hand region of MI (Fig. 24-14B). Finally, when the subject was asked to mentally rehearse the sequence of finger movements without actually moving the fingers, increased activity was restricted to the supplementary cortex and the hand region of *MI* was silent (Fig. 24-14C). These findings indicate that the supplementary motor cortex is involved in *organizing* or *planning* the sequence of

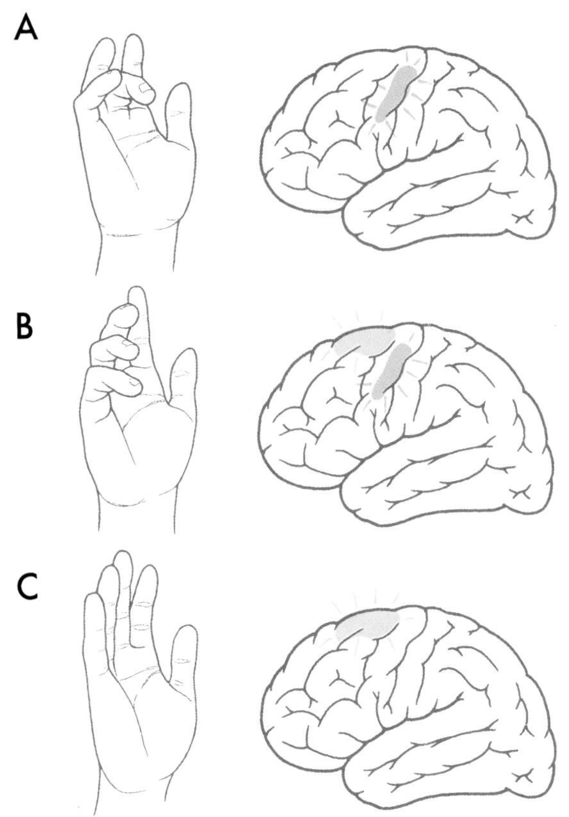

FIGURE 24-14 *Random movements made without prior planning and in no particular order (A) result in increased activity in only the hand area of motor cortex. When the movement is planned and executed in a specific sequence (B) both motor and supplementary cortices are active. When the movement is mentally planned and rehearsed but never executed (C) only the supplementary cortex is activated.*

muscle activation required to make a movement, whereas the primary motor cortex mainly functions to *execute* the movement.

Premotor Cortex The premotor cortex occupies the portion of area 6 lying just anterior to the ventrolateral part of MI (Fig. 24-2). Like the supplementary motor cortex, this region contains a somatotopic representation of the body musculature that is complete, although less precisely organized than that of MI. The premotor cortex receives considerable input from sensory areas of the parietal cortex and projects to *MI*, the spinal cord, and the reticular formation. The latter gives rise to reticulospinal fibers, which, in turn, influence spinal motor neurons that innervate paravertebral and proximal limb musculature.

On the basis of these connections, it was suggested that premotor cortex, like the supplementary cortex, is involved in the *preparation to move*. That is, it organizes those postural adjustments that are required to make a movement. To test this concept, monkeys were trained to move one hand to a specific target location that differed from trial to trial. The monkey was first given a cue signaling which target to reach for and then a "go" signal to actually make the movement. Recordings of *cell activity* revealed that premotor neurons were active only *during the interval between presentation of the cue and the "go" signal*. The premotor cortex is most active in directing the control of proximal limb muscles that are used to position the arm for movement tasks or more generally to orient the body for movement.

Posterior Parietal Cortex The motor regions of the posterior parietal cortex comprise Brodmann's areas 5 and 7, which largely occupy the superior parietal lobule (Fig. 24-2). These areas carry out some of the "background computations" necessary for making movements in space. To organize such a movement, it is necessary to collate input from a variety of sensory systems to create a map of space and to compute a trajectory by which a body part can reach its target. Area 5 receives extensive projections from somatosensory cortex and input from the vestibular system, whereas area 7 processes visual information related to the location of objects in space. Both areas project primarily to supplementary and premotor cortices and have few spinal or brainstem targets.

Experiments in monkeys provide the best insight into the function of areas 5 and 7. In area 5, *arm projection neurons* are active only when the monkey reaches for a specific object of interest. They are not active when the same arm movement is made but the object is not present. In area 7, many different types of neurons are present. One of these, the *eye-hand coordination neurons* are vigorously active only when the eyes fixate a target and the hand reaches for that target.

Cingulate Motor Cortex Two aggregates of corticospinal neurons are associated with the cingulate gyrus (Fig. 24-2). One occupies the ventral bank of the cingulate sulcus, and the other, located slightly more posteriorly, occupies the dorsal and ventral banks of the cingulate sulcus. Each is topographically organized with respect to its spinal cord projections, and each also projects to primary motor cortex. Little is known about the functional role of these areas, other than that stimulation in either area produces motor effects. Because of their proximity to limbic cortex, these motor neurons may be involved in movements that have an intense motivational or emotional component.

Cerebellar and Pallidal Influences The basal ganglia and cerebellum play essential roles in the control of movement through their interaction with motor-related areas of the cerebral cortex. Although these pathways are detailed in Chapters 25 and 26, their general relationships are summarized here. The cerebellar nuclei and the globus pallidus project primarily to spatially segregated regions in the ventral anterior, ventral lateral, and oral part of the ventral posterolateral nuclei of the dorsal thalamus. These so-called motor areas of the thalamus give rise to thalamocortical projections to MI and the supplementary motor area. The thalamic areas that receive input from the globus pallidus project mainly to the supplementary motor cortex, and the thalamic areas that receive input from the cerebellum project to MI. The degree to which these two lines of communication are separate is not clear. Therefore, it is important to realize that signals transmitted through the corticospinal and corticobulbar systems can be modified by outputs that reach the cortex from the cerebellum, basal ganglia, and thalamus.

HIERARCHICAL ORGANIZATION VERSUS PARALLEL DISTRIBUTED PROCESSING IN THE MOTOR SYSTEM

Until recently, it was generally thought that the control of voluntary movement could be satisfactorily described by the hierarchical scheme shown in Figure 24-15. In this plan, lower motor neurons and their associated interneurons are influenced by (1) segmental peripheral sensory feedback circuits, (2) descending brainstem-spinal systems modulated by the cerebral cortex, and (3) the corticospinal system.

The motor areas in the cortex are organized such that the output of the corticospinal system is regulated and modulated by higher-order motor cortices. When a voluntary movement is desired, a plan for the movement is organized by the combined efforts of the higher-order motor areas and then transmitted to MI. The primary motor cortex then executes the plan by communicating with the spinal motor apparatus either directly or indirectly via brainstem-spinal systems (Fig. 24-15). However, a reexamination of the origins of the corticospinal tract indicates that more of these fibers than previously thought originate from cells located *outside MI*. In addition, circuits linking the basal ganglia, thalamus, and motor cortical regions are segregated to some extent from pathways linking the cerebellum, thalamus, and motor cortical areas.

These observations have led to an evolving hypothesis that motor system control is achieved by a series of *parallel systems* formed by somatotopically organized, descending cortical projections that link the various motor-related areas of cortex more directly with spinal motor circuits. Included here are spinal projections originating from the so-called higher-order motor regions such as the premotor and supplementary motor areas. The concept is that each descending cortical pathway contributes its own element or series of elements to movement control. This idea is supported by the fact that stimulation in different cortical locations can elicit movements involving the same muscles or muscle groups, but the characteristics of the movement differ somewhat for each site of stimulation. However, additional information is needed before this hypothesis can be validated.

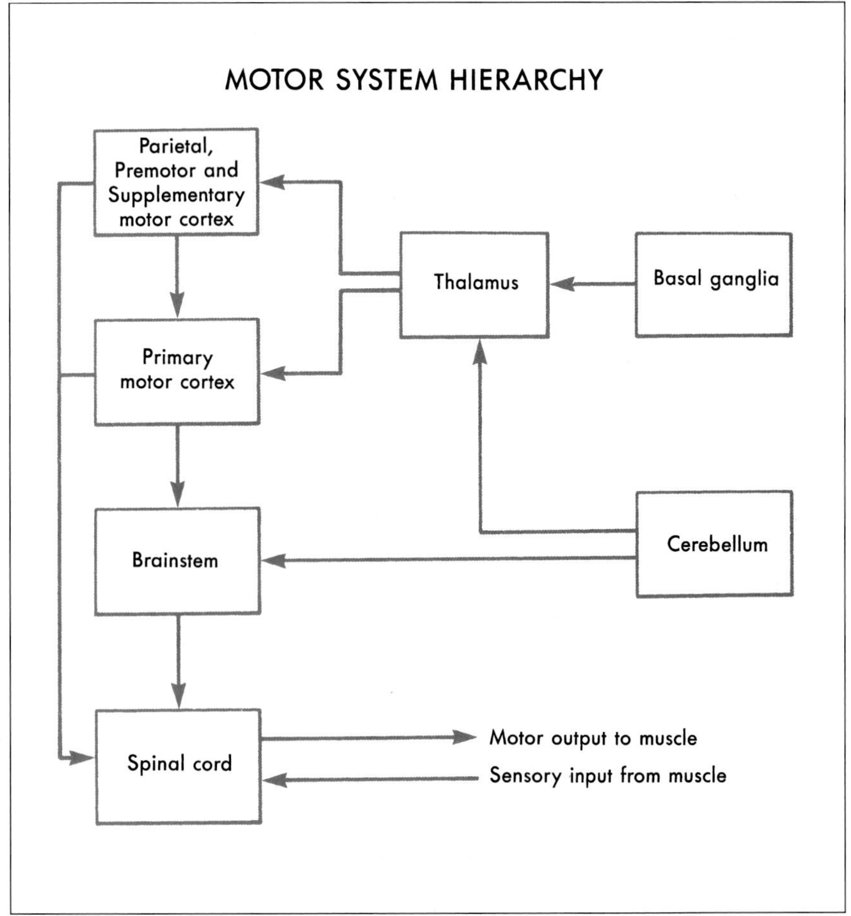

FIGURE 24-15 *The hierarchical organization of the motor system. Recent evidence suggests that the direct links from the various brainstem and cortical centers to the spinal cord, which operate as a series of systems, may be more important than previously thought, and that hierarchical organization may be less important.*

SOURCES AND ADDITIONAL READINGS

Asanuma H, Rosen I: Topographical organization of cortical efferent zones projecting to distal forelimb muscles in the monkey. Exp Brain Res 14:243–256, 1972.

Asanuma H: *The Motor Cortex*. Raven Press, New York, 1988.

Brodal A: *Neurological Anatomy in Relation to Clinical Medicine*, 3rd Ed. Oxford University Press, New York, 1981.

Evarts EV: Relation of pyramidal tract activity to force exerted during voluntary movement. J Neurophysiol 31:14–27, 1968.

Georgopoulos AP: Higher order motor control. Annu Rev Neurosci 14:361–377, 1991.

Humphrey DR: On the cortical control of visually directed reaching: contributions by nonprecentral motor areas. In Talbot RE, Humphrey DR (eds): *Posture and Movement*. Raven Press, New York, 1979, pp 51–112.

Humphrey DR: Representation of movements and muscles within the primate precentral motor cortex: historical and current perspectives. Fed Proc 45:2687–2699, 1986.

Penfield W, Rasmussen T: *The Cerebral Cortex of Man: A Clinical Study of Localization of Function*. Hafner Publishing, New York, 1968 (Facsimile of 1950 ed).

Roland PE, Larsen B, Lassen NA, Skinholf E: Supplementary motor area and other cortical areas in organization of voluntary movements in man. J Neurophysiol 43:118–136, 1980.

CHAPTER TWENTY-FIVE

The Basal Ganglia

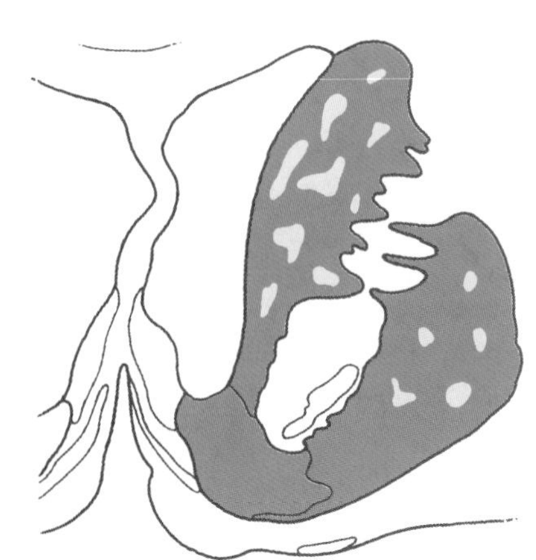

T. P. MA

Voluntary movement is essential to the well-being of living animals. Such behaviors are accomplished by signals that direct the action of individual muscles. Although these signals originate in the cerebral cortex, they are modulated by a variety of subcortical structures. One such group of structures is the basal ganglia (nuclei) and their functionally associated cell groups. Classically, motor systems have been divided into "pyramidal" and "extrapyramidal" on the basis of whether the pathway is mediated by corticofugal neurons (pyramidal), or by the basal ganglia, cerebellum, or descending brainstem pathways (extrapyramidal). However, recent observations have shown this distinction to be overly simplistic, if not inaccurate. Consequently, it is not used here. The basal ganglia are involved in a wide variety of motor and affective behaviors, in sensorimotor integration, and in cognitive functions.

OVERVIEW

The nuclei considered in this chapter are traditionally called the *basal ganglia* rather than the *basal nuclei*, even though "ganglia" is usually reserved for groups of nerve cell bodies in the peripheral nervous system. In recognition of the fact that the official term is *basal nuclei*, the two designations will be used interchangeably.

The *basal ganglia* consist of cell groups embedded in the cerebral hemisphere. Although not classified as nuclei of the basal ganglia in a strict sense, the *subthalamic nucleus*, *substantia nigra*, and *pedunculopontine tegmental nucleus* are integral parts of the pathways passing through these forebrain cell groups. Collectively, the basal ganglia and their associated nuclei function primarily as components in a series of parallel circuits from the cerebral cortex through the basal ganglia to the thalamus and then back to the cerebral cortex.

Four fundamental concepts are critical in understanding the basal ganglia. First, damage to or disorders of the basal ganglia result in disruption of movements, and may also cause significant deficits in other neural functions such as cognition, perception, and mentation. Second, the basal nuclei are anatomically and functionally segregated into parallel circuits that process different types of behaviorally significant information. Third, the basal ganglia function primarily through *disinhibition* (release from inhibition). Fourth, diseases of the basal ganglia can be described as disruptions of the neurochemical interactions between elements of the basal nuclei. These neurochemical relationships rely not simply on the neurotransmitters involved, but also on the characteristics of the transmitter receptors, on the locations of the synapses, and on other inputs received by these cells.

In summary, the basal ganglia integrate and modulate cortical information along multiple parallel channels. These channels affect behavior indirectly by feedback to the cerebral cortex and directly by providing information to subcortical centers that influence movements. Disruption of these channels by stroke or disease results in dysfunctions characteristic of the circuits damaged.

COMPONENTS OF THE BASAL GANGLIA

The basal ganglia are typically divided into dorsal and ventral divisions. The dorsal basal ganglia include the *caudate* and *putamen* (together constituting the *neostriatum*) and the *globus pallidus* (constituting the *paleostriatum*) (Fig. 25-1A). Associated with the dorsal basal ganglia are the *substantia nigra*, the *subthalamic nucleus*, and the *parabrachial pontine reticular formation* (containing the *pedunculopontine tegmental nucleus*). The ventral basal ganglia are located inferior to the anterior commissure and include the *substantia innominata*, *nucleus basalis of Meynert*, *nucleus accumbens*, and *olfactory tubercle*. This ventral region is intimately associated with the amygdala and ventral tegmental area. For the purposes of this chapter, the basal ganglia are regarded as making up two complexes: the striatal complex and the pallidal complex (Fig. 25-1B).

The telencephalic regions of the basal nuclei are supplied by the *medial striate artery*, *lenticulostriate branches* of the M_1 segment of the middle cerebral artery, and the *anterior choroidal artery* (Fig. 25-2A,B). The diencephalic and mesencephalic regions are supplied by the posteromedial branches of the P_1 segment of the posterior cerebral artery and branches of the posterior communicating artery. Diseases of these vessels may result in various behavioral/motor deficits, depending on which vessel and region is affected.

The Striatal Complex The *striatal complex* is a functional unit composed of the *neostriatum* and *ventral striatum* (Fig. 25-3A,B). The neostriatum consists of the *caudate nucleus* and *putamen*. These two nuclei have the same origin and similar connections. Although fused rostroventrally, they are separated throughout most of their extent by fibers of the internal capsule. The ventral striatum is composed of the *nucleus accumbens* and the *olfactory tubercle* (Figs. 25-1A,B, 25-3A). The nucleus accumbens is located rostroventrally in the hemisphere, subjacent to where the putamen is continuous with the head of the caudate. It is internal to part of the anterior perforated substance. Portions of the olfactory tubercle are considered part of the ventral striatum because of functional, cytoarchitectural, and chemoarchitectural similarities. The olfactory tubercle is unique in that it has striatal characteristics yet receives primary olfactory information.

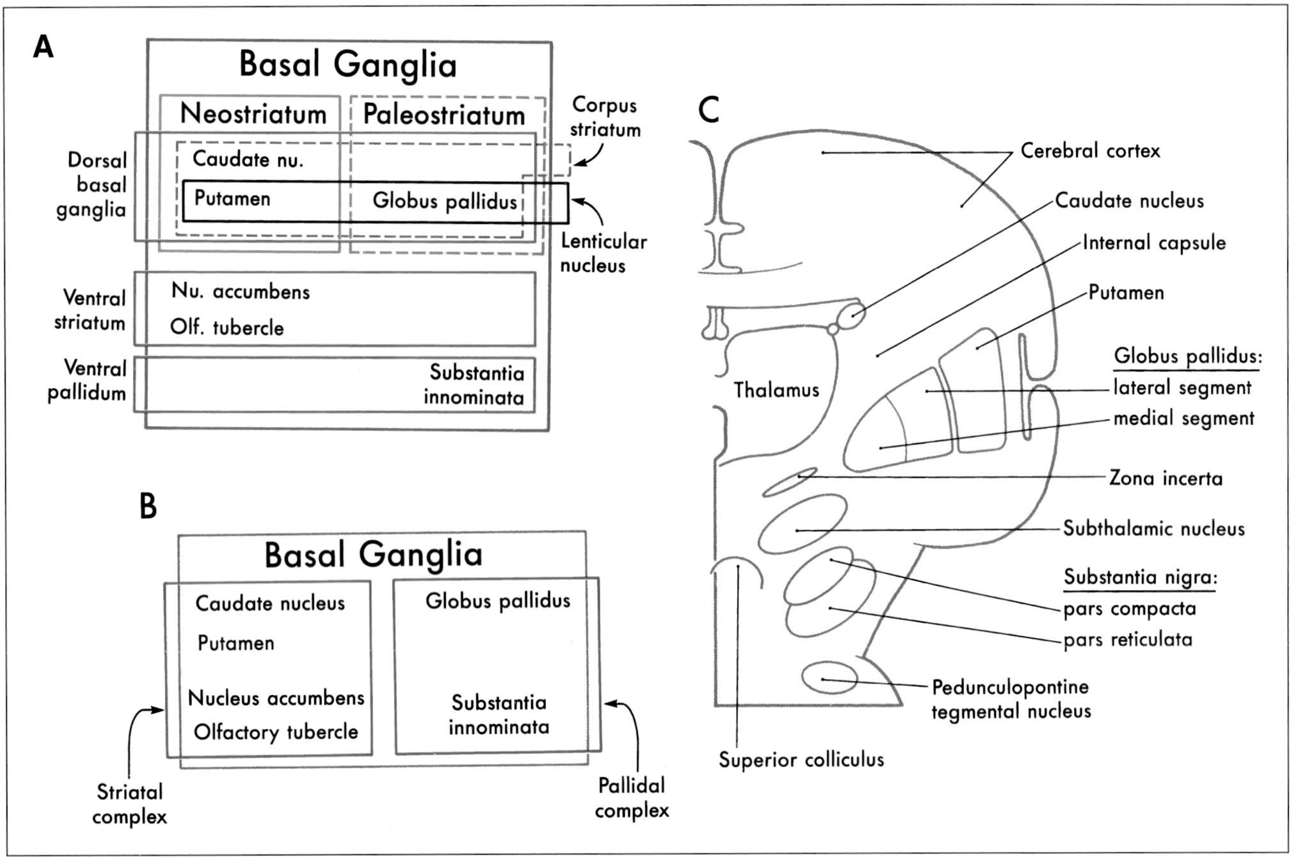

FIGURE 25-1 *A series of stacked boxes (A) illustrating which nuclei form the various parts of the basal ganglia and how these groups are used in this chapter (B). A standard drawing of the basal ganglia (C) used throughout this chapter.*

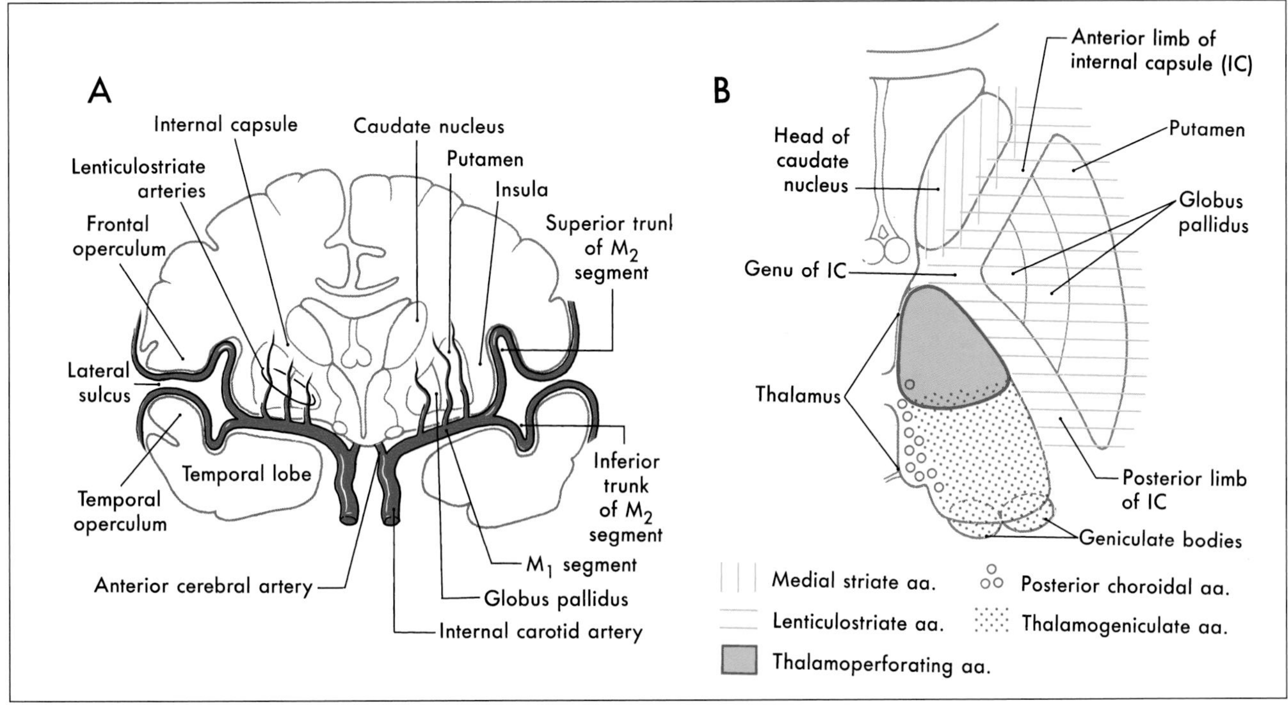

FIGURE 25-2 *Blood supply to the basal ganglia in coronal (A) and axial (B) planes.*

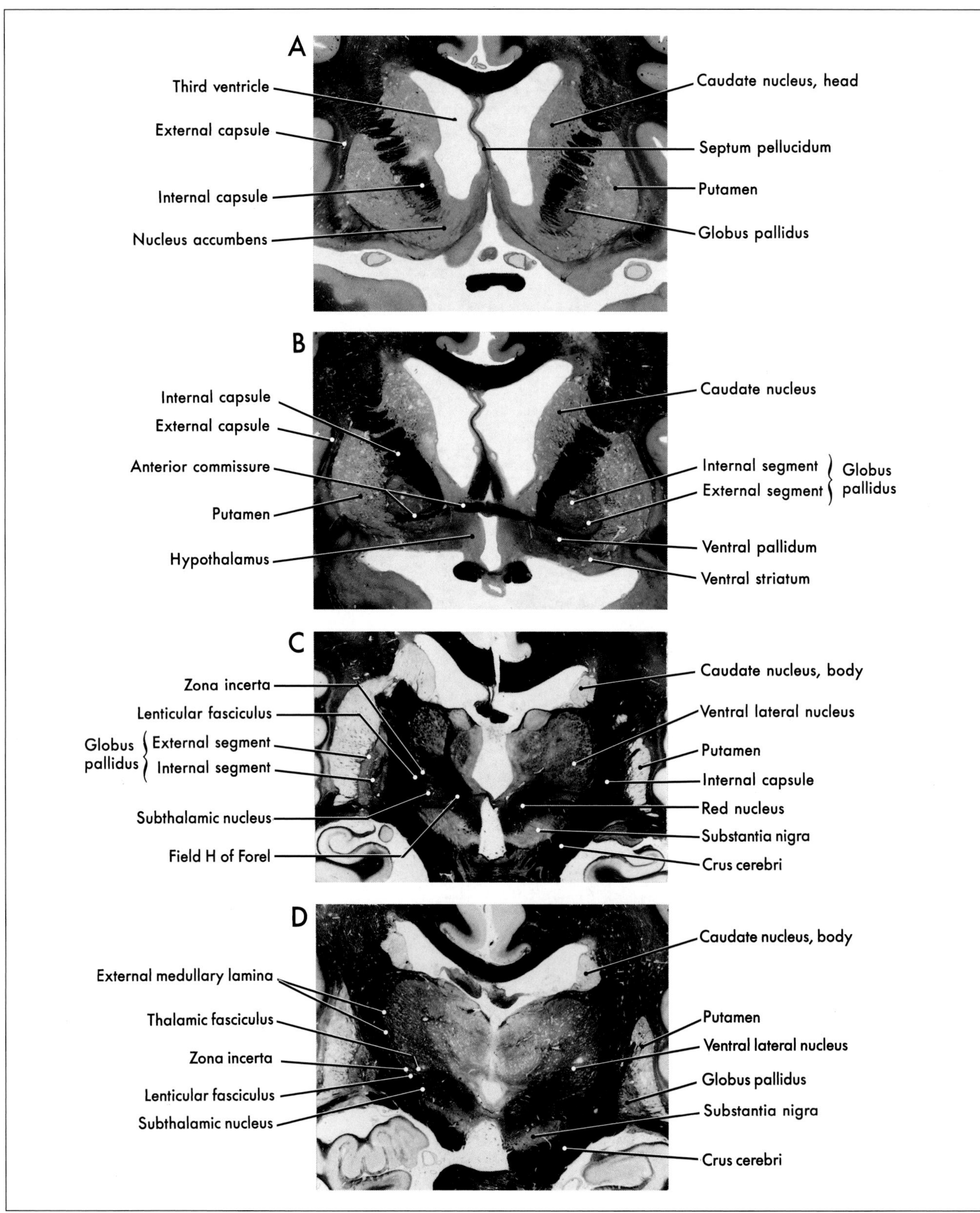

FIGURE 25-3 *Cross sections of the human brain from rostral (A) to caudal (D) showing the basal ganglia and related nuclei. Myelin stain.*

A defining characteristic of the striatal complex, *striosomes* (also called *patches*) are particularly prominent in the head of the caudate (Fig. 25-4). Striosomes are acetylcholinesterase-poor regions within the striatal complex. They contain large amounts of one or more neuropeptides and one or more types of opiate receptors. The remainder of the striatal complex, called the *matrix*, contains high concentrations of acetylcholinesterase and, therefore, stains darkly when tissue is histochemically reacted to reveal this enzyme.

The largest afferent projections to the striatum are from the cerebral cortex (*corticostriatal fibers*) (Fig. 25-5A). Other afferents are from the thalamus (*thalamostriatal fibers*), substantia nigra (*nigrostriatal fibers*), and parabrachial pontine reticular formation (*pedunculopontostriatal fibers*) (Fig. 25-5A). The efferent projections of the striatum reach primarily the pallidum (*striatopallidal fibers*) and the nigral complex (striatonigral fibers), and to a small degree the subthalamic nucleus.

Most of the neurons in the neostriatum are called *medium spiny neurons*, so named because of their medium-sized cell bodies and the large numbers of spines on their dendrites (Fig. 25-6). Most medium spiny cells have dendritic fields that are restricted to the region in which the cell bodies are located. That is, medium spiny neurons in striosomes have dendrites that usually ramify within striosomes; medium spiny neurons within the matrix primarily ramify within the matrix.

Medium spiny neurons fire few action potentials spontaneously and thus require activation by their afferent fibers. These cells use the inhibitory neurotransmitter GABA and may also contain neuroactive peptides such as substance P and enkephalin. Thus, when medium spiny neurons are activated, they subserve both direct inhibitory and neuromodulatory functions at their targets. These cells are the only efferent neurons of the neostriatum.

Also found in the neostriatum are large, acetylcholine-containing local circuit neurons that modulate local activity within the neostriatum. Huntington's disease is characterized by progressive loss of medium spiny neurons and acetylcholine-containing neurons throughout the striatal complex.

The Pallidal Complex The pallidal complex is composed of the globus pallidus and the ventral pallidum. The latter is largely synonymous with the substantia innominata (Fig. 25-3B,C). The pallidal complex contains primarily GABAergic neurons with high rates of spontaneous activity. Consequently, these cells tonically inhibit their targets.

The globus pallidus is divided into *medial* (*internal*) and *lateral* (*external*) *segments* by a dorsoventrally oriented sheet of white matter (the medullary lamina) (Fig. 25-3C). The substantia innominata is located ventral to the anterior commissure and internal to the anterior perforated substance. One important cell group in the substantia innominata is the basal nucleus of Meynert. This nucleus has large acetylcholine-containing neurons, which are lost in Alzheimer's disease. However, this disease is not considered a basal ganglia disorder because acetylcholine-containing cells in the cerebral cortex, hippocampus, and septum are also lost in Alzheimer patients. This disease is further characterized by other biochemical and pathologic features such as senile plaques.

The two divisions of the globus pallidus are reciprocally connected (*pallidopallidal fibers*) (Fig. 25-5B) but subserve different functions. The main afferent input to the pallidum is from the striatal complex. Medium spiny neurons from the striatum that project to the medial segment and substantia nigra use GABA and substance P; those that project to the lateral segment use GABA and enkephalin (Fig. 25-5B).

The medial division is composed of the medial segment of the globus pallidus. It subserves the *direct basal ganglia pathway* (described below) and projects primarily to the thalamus (*pallidothalamic fibers*) (Fig. 25-5B). These fibers exit the globus pallidus as two bundles: the *ansa lenticularis* and the *lenticular fasciculus* (Fig. 25-7). The *ansa lenticularis* originates from lateral portions of the medial segment and loops around the posterior limb of the internal capsule to enter field H of Forel. The *lenticular fasciculus*, on the other hand, originates in the dorsomedial portion of the medial segment. These fibers traverse the internal capsule as small groups of axons, merge to form the lenticular fasciculus between the zona incerta and subthalamic nuclei, and then enter field H of Forel. In Forel's field, the ansa lenticularis and lenticular fasciculus join the *thalamic fasciculus*, which courses dorsal to the zona incerta (Fig. 25-7). These fibers ultimately terminate in ventral anterior, ventral lateral, and centromedian nuclei of the thalamus. The medial division of the pallidal complex is a principal efferent nucleus of the basal ganglia.

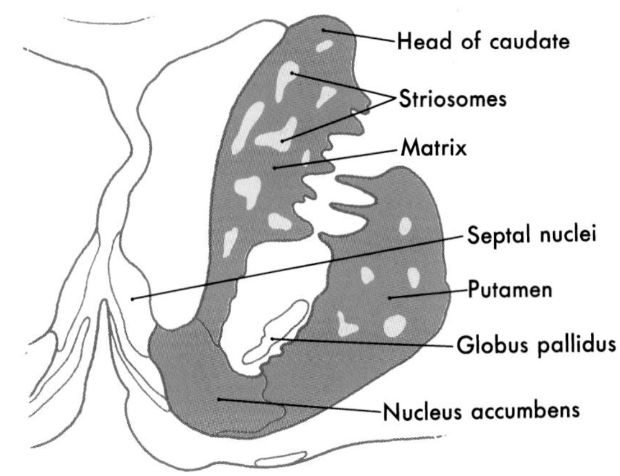

FIGURE 25-4 *Schematic representation of striosomes and matrix in the striatal complex.*

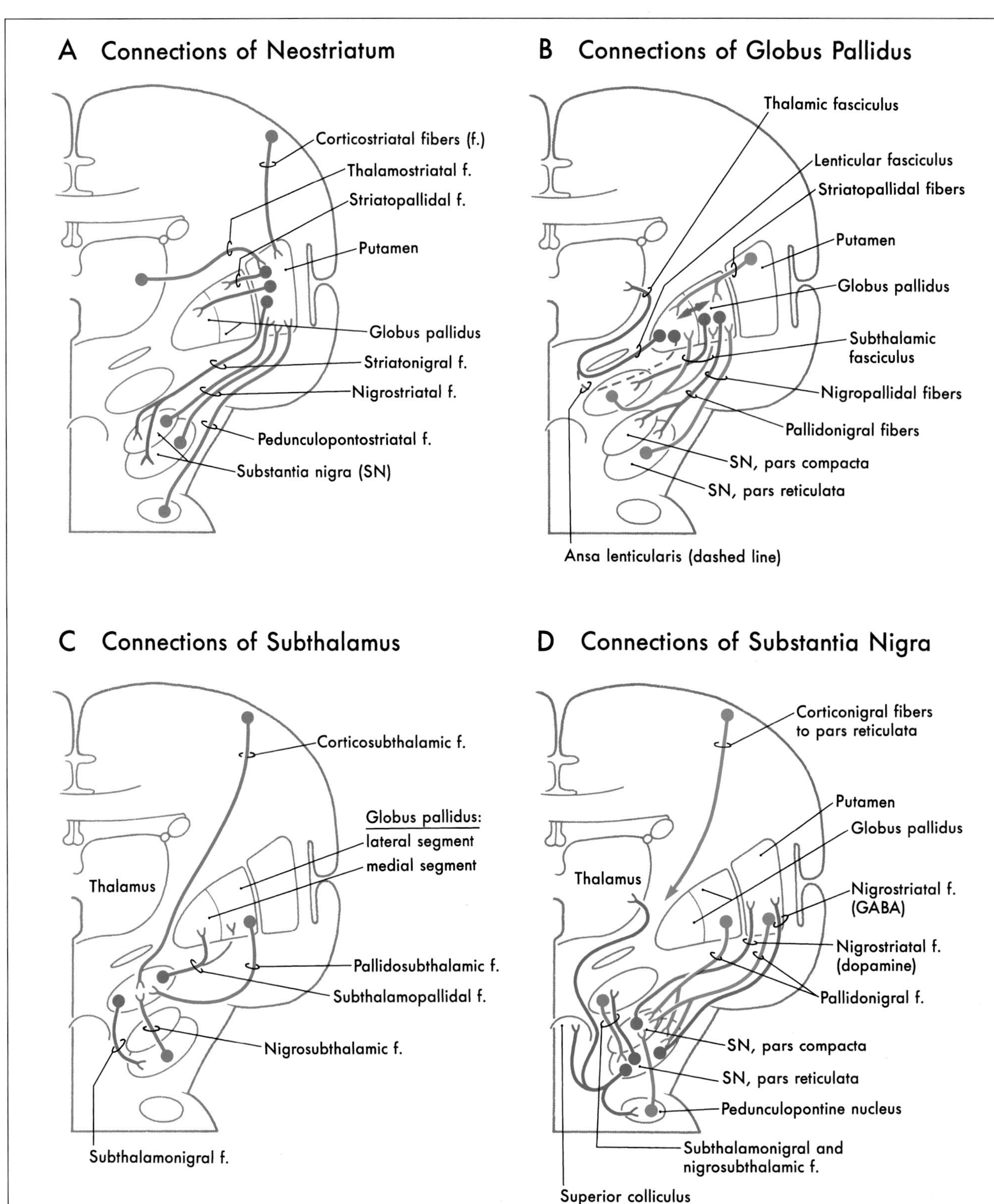

FIGURE 25-5 *Schematic representations of the afferent (in green) and efferent (in red) connections of the neostriatum (A) and subthalamus (C) and of the afferent (in green) and efferent (in red and blue) connections of the globus pallidus (B) and substantia nigra (D). The double-headed arrow in B represents pallidopallidal fibers.*

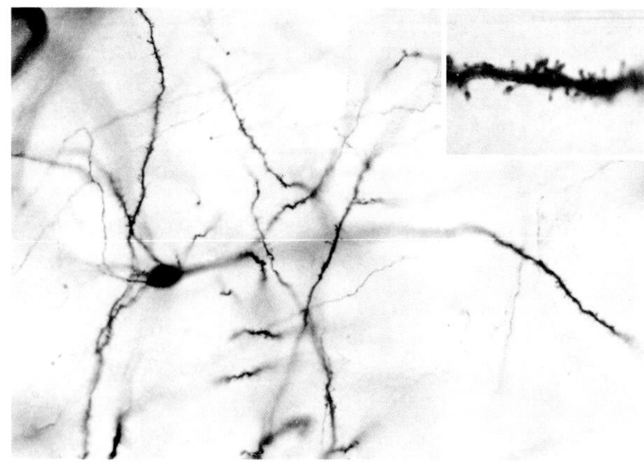

FIGURE 25-6 *Medium spiny neuron from the primate neostriatum. The detail shows the characteristic appearance of dendritic spines on these cells. (Photos courtesy of Dr. José Rafols.)*

The lateral division is composed of the external (or lateral) segment of the globus pallidus and the ventral pallidum. This division subserves the *indirect basal ganglia pathway* (see below). These nuclei receive a large input from the striatal complex (*striatopallidal fibers*) and small projections from the subthalamic nucleus (*subthalamopallidal fibers*) and the substantia nigra pars reticulata (*nigropallidal fibers*). They project strongly to the subthalamic nucleus (*pallidosubthalamic fibers*) and are also connected with the substantia nigra (*pallidonigral fibers*) (Fig. 25-5B).

The Subthalamic Nucleus The subthalamic nucleus is a lens-shaped cell group that makes up the largest part of the ventral thalamus. It is immediately ventral to the zona incerta and rostral to the substantia nigra (Fig. 25-3C,D). It receives projections from the lateral pallidal division (*pallidosubthalamic fibers*), cerebral cortex (*corticosubthalamic fibers*), nigral complex (*nigrosubthalamic fibers*), and parabrachial pontine reticular formation. The subthalamic nucleus projects to both pallidal divisions (*subthalamopallidal fibers*) and to the substantia nigra (*subthalamonigral fibers*) (Fig. 25-5C). These connections, especially the subthalamopallidal projections to the medial globus pallidus, are an essential part of the indirect pathway underlying basal ganglia function.

Subthalamic neurons use the excitatory neurotransmitter glutamate. Under normal conditions, subthalamic cells are inactive because of the constant inhibition by cells of the external pallidal segment. However, if these afferents are removed, subthalamic neurons have a high level of activity. This activity is mediated, in part, by a large corticosubthalamic projection.

The Nigral Complex The nigral complex is composed of the *substantia nigra* and the *ventral tegmental area* (Fig. 25-3C,D). The substantia nigra is divided into a cell-dense portion (*pars compacta*) and a reticulated portion (referred to here as the *pars reticulata*, although it can be divided into a *pars reticulata* and a *pars lateralis*). The pars reticulata is located at and within the medial edge of the descending corticofugal fibers that form the crus cerebri. The pars compacta and the adjacent ventral tegmental area appear to subserve similar functions and to have a similar chemoarchitectural organization. The major afferents to the nigral complex are from the striatal and pallidal complexes. The nigral complex also receives cortical (*corticonigral*), subthalamic (*subthalamonigral*), and pendunculopontine fibers (Fig. 25-5D).

The pars compacta includes a large number of neuromelanin-containing cells, whose dark color gives the nucleus its name (*substantia nigra*, dark substance). Neurons in the pars compacta use the neurotransmitter dopamine and project primarily to the neostriatum as *nigrostriatal fibers*. The dopamine released by these cells may excite or inhibit striatal neurons, depending on the *type of receptor* on the postsynaptic membrane.

The pars reticulata is formed by loose aggregations of medium- to large-sized GABAergic neurons that are indistinguishable from those of the medial pallidum. Neurons in the pars reticulata have axons with an extensive system of collaterals, and, consequently, they may project to and inhibit one or more target structures. These targets include the neostriatum (*nigrostriatal fibers*), thalamus (*nigrothalamic fibers*), superior colliculus

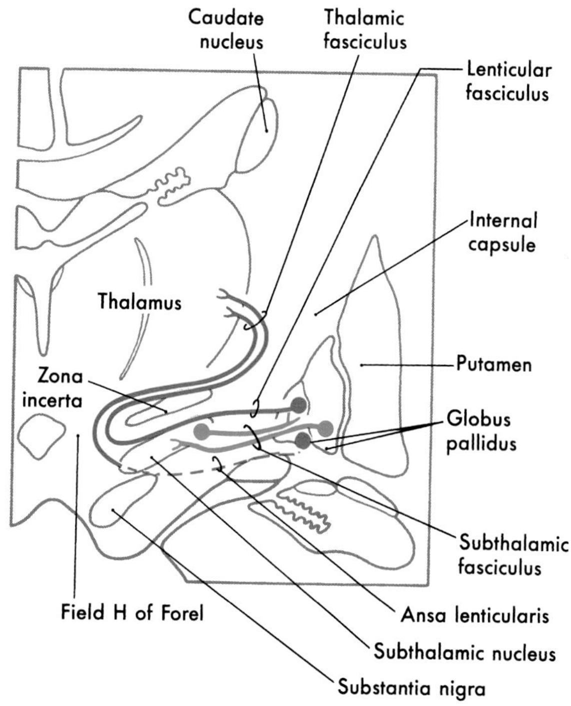

FIGURE 25-7 *Schematic representation of pallidal connections to the thalamus and with the subthalamic nucleus.*

(*nigrotectal fibers*), and the parabrachial pontine reticular formation. These cells have a high rate of discharge and tonically inhibit their targets. Projections of pars reticulata neurons represent an important pathway by which the basal ganglia influence other motor centers.

The pars compacta and pars reticulata are interconnected. Dendrites of dopaminergic pars compacta neurons extend into the pars reticulata, where they release free dopamine (by a nonvesicular mechanism). The level of dopamine modulates the resting membrane potential of pars reticulata cells, making them either more or less likely to discharge, depending on the subtype of dopamine receptor they possess. In turn, pars reticulata neurons have axon collaterals that ramify extensively in the pars compacta and form GABAergic synapses. Collectively, these interactions form modulatory loops between neurons of the pars compacta and pars reticulata.

The ventral tegmental area is located medial to the substantia nigra. It contains large numbers of dopaminergic neurons and forms connections with the ventral striatum, the amygdala, and other limbic system structures. Cells of the ventral tegmental area project to, and terminate on, striatal neurons that have postsynaptic D2 (dopamine) receptors. In schizophrenia there is an increase in number and in sensitivity of these receptors. Neuroleptic drugs help to control schizophrenia by blocking (down-regulating) these D2 receptors.

The Parabrachial Pontine Reticular Formation Nuclei in the region of the parabrachial pontine reticular formation, primarily the pedunculopontine tegmental nucleus, are intimately connected with all portions of the basal ganglia and their associated nuclei. For example, GABAergic substantia nigra pars reticulata neurons project onto cells of the pedunculopontine tegmental nucleus. In return, acetylcholine-containing pedunculopontine tegmental neurons project to the substantia nigra pars compacta. In addition, the parabrachial pontine nuclei are connected with motor centers in the brainstem, which project to the spinal cord via the descending spinal pathways. Thus, these nuclei serve as an efferent pathway for the basal ganglia. It has been suggested that damage to the connections between the pedunculopontine nucleus and the basal ganglia may partially account for motor deficits, such as tremor or chorea, in some patients.

DIRECT AND INDIRECT PATHWAYS OF BASAL GANGLIA ACTIVITY

Pathways though the basal ganglia consists of parallel circuits that share certain features. The basic circuit is divided into *direct* and *indirect* pathways that have opposing actions on target nuclei of the basal ganglia (Fig. 25-8). As a general concept, the *direct pathway facilitates* a flow of information through the thalamus, and the *indirect pathway inhibits* this flow. These pathways create a balance in the inhibitory outflow of the basal ganglia and function by modulating the extent of this inhibition on target nuclei.

The *direct pathway* (Fig. 25-9A) begins as an excitatory, glutaminergic projection from the cerebral cortex to the striatal complex. Striatal neurons inhibit cells in the internal (or medial) segment of the globus pallidus and in the substantia nigra pars reticulata. These *striatopallidal* and *striatonigral fibers* use GABA and substance P. Cells of the internal segment of the globus pallidus (as *pallidothalamic fibers*) and of the substantia nigra pars reticulata (as *nigrothalamic fibers*) project to thalamic neurons. These fibers have a high rate of spontaneous activity and thus tonically inhibit target thalamic neurons. Inhibition of these pallidal and nigral projections by striatal cells decreases the inhibitory inputs to thalamocortical neurons (*thalamic disinhibition*). The *net effect of the direct pathway is to increase the activity of the thalamus and the consequent excitation of the cerebral cortex* (Fig. 25-9B).

The *indirect pathway* includes a loop through the globus pallidus and subthalamic nucleus (Fig. 25-9C). *Striatopallidal* neurons involved in this pathway contain GABA and enkephalin. They project into the lateral pallidal segment, which, in turn, sends *pallidosubthalamic fibers* into the subthalamic nucleus. These pallidosubthalamic fibers are GABAergic, have high spontaneous

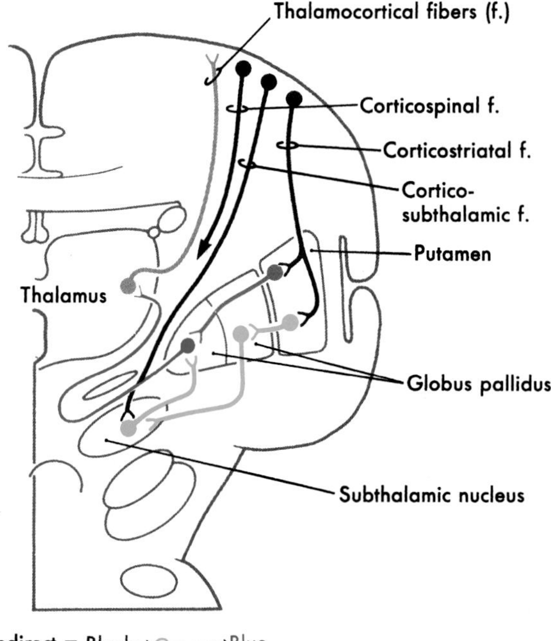

FIGURE 25-8 *Schematic representation of direct and indirect pathways through the basal ganglia.*

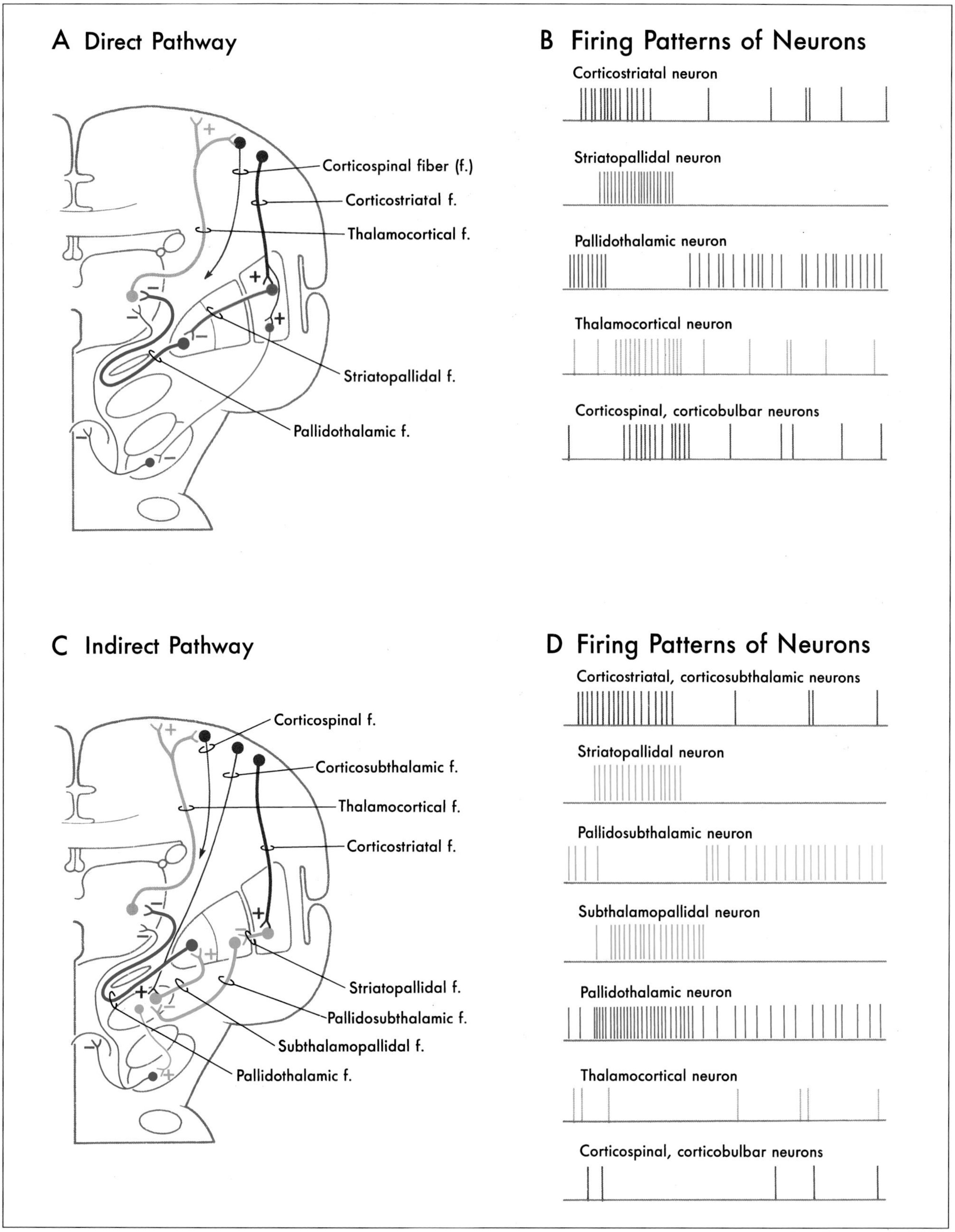

A Direct Pathway

Corticospinal fiber (f.)
Corticostriatal f.
Thalamocortical f.
Striatopallidal f.
Pallidothalamic f.

B Firing Patterns of Neurons

Corticostriatal neuron
Striatopallidal neuron
Pallidothalamic neuron
Thalamocortical neuron
Corticospinal, corticobulbar neurons

C Indirect Pathway

Corticospinal f.
Corticosubthalamic f.
Thalamocortical f.
Corticostriatal f.
Striatopallidal f.
Pallidosubthalamic f.
Subthalamopallidal f.
Pallidothalamic f.

D Firing Patterns of Neurons

Corticostriatal, corticosubthalamic neurons
Striatopallidal neuron
Pallidosubthalamic neuron
Subthalamopallidal neuron
Pallidothalamic neuron
Thalamocortical neuron
Corticospinal, corticobulbar neurons

FIGURE 25-9 *The direct and indirect pathways (A,C) and the corresponding firing patterns of their neurons (B, D). The color of each fiber correlates with the color of the firing pattern (action potentials) of that neuron (B or D). The bold fibers in A and C represent the primary pathway in each example. +, excitatory synapse; –, inhibitory synapse.*

firing rates, and tonically inhibit subthalamic cells. Inhibition of these fibers by the striatum releases these subthalamic cells from their tonically inhibited state (*subthalamic disinhibition*). These subthalamic neurons have spontaneous activity and are also influenced by an excitatory *corticosubthalamic* projection. Together, these inputs increase the firing rates of glutaminergic *subthalamopallidal fibers* to the medial pallidal segment. As a consequence, the firing rates of inhibitory *pallidothalamic fibers* is increased, with a resultant decrease in the activity of thalamocortical neurons. The *net effect of the indirect pathway is to decrease activity of the thalamus and, consequently, decrease activity of the cerebral cortex* (Fig. 25-9D).

Disinhibition as the Primary Mode of Basal Ganglia Function The function of the direct pathway is to release the thalamus from its pallidal inhibition. This is accomplished by striatopallidal inhibition of pallidothalamic neurons. In the indirect pathway, the subthalamic nucleus is released from inhibition by the lateral pallidal segment so that it can excite the inhibitory pallidothalamic cells. This mechanism that *releases cells from inhibition* is called *disinhibition*. Balance between *thalamic disinhibition* by the direct pathway and *subthalamic disinhibition* by the indirect pathway results in normal basal ganglia function. Behavioral deficits that accompany basal ganglia disorders can ultimately be traced to imbalances between the direct and indirect pathways.

PARALLEL CIRCUITS OF INFORMATION FLOW THROUGH THE BASAL GANGLIA

Information flow through the basal ganglia is separated into five distinct parallel circuits. They are the *motor loop*, *oculomotor loop*, *dorsolateral prefrontal loop*, *lateral orbitofrontal loop*, and *limbic loop* (Fig. 25-10). The *motor loop* is involved in somatosensory and somatomotor control. The *oculomotor loop* is primarily related to the control of orientation and gaze. The *dorsolateral prefrontal* and *lateral orbitofrontal loops* are related to cognitive processes. The *limbic loop* is concerned with emotional and visceral functions. All of these circuits have both the direct and indirect pathways described above.

The five circuits originate from functionally distinct regions of the cerebral cortex, pass through distinct regions of each basal ganglia component, modulate different areas of the thalamus, and return to functionally distinct cortical regions (Fig. 25-10). That is, each circuit can be considered an independent channel, which processes information from one functional type of cortex by way of its own areas of the basal ganglia and thalamus, and returns to the appropriate functionally related part of cortex.

Anatomic studies in nonhuman primates have demonstrated that each loop projects to a restricted portion of each nucleus. Therefore, there is an anatomic, as well as a functional, separation of the basal ganglia circuits. This separation is called the *closed component* of the circuits. However, at every stage of each circuit, the information is modulated and integrated with input from other centers by intrinsic basal ganglia connections. Thus, although the circuits are anatomically and functionally distinct, their activities are modulated by the other functions of the basal ganglia. This integration is called the *open component* of basal ganglia circuits.

The Motor Loop Because the most obvious basal ganglia symptoms are those associated with the motor system, it is important to understand the motor loop. Although this loop contains both direct and indirect pathways, which are associated with specific parts of each nucleus, only the direct pathway of the motor loop is described here (Fig. 25-10).

The motor loop originates mainly in supplementary motor (SMA), primary motor (MC), and premotor (PMC) cortices (Fig. 25-10). These corticostriatal projections terminate in the putamen (Put), which also receives projections from somatosensory cortex (SC). Efferents from the putamen terminate in the ventral portion of the internal segment of the globus pallidus (GPi-v) and dorsolateral substantia nigra pars reticulata (SNr-dl). These two regions project to the oral part of the ventral lateral nucleus (VLo), the ventral anterior nucleus (VA), and the centromedian nucleus (CM) of the thalamus. In turn, the VLo and VA nuclei project to the supplementary motor cortex, the VA nucleus to premotor cortex, and the VLo and CM nuclei to motor cortex. Although the primary projections of the globus pallidus and substantia nigra are to the thalamus, they also project to the superior colliculus and brainstem reticular formation. In this way, this basal ganglia circuit affects both cortical motor efferents and brainstem motor centers.

Comparable circuits, as described above for somatomotor function, also exist in each of the other four loops passing through the basal ganglia. The specific synaptic interactions for each of these loops is shown in Figure 25-10.

BEHAVIORAL FUNCTIONS OF THE BASAL GANGLIA

The best understood functions of the basal ganglia are associated with the motor systems, in particular, the somatomotor (motor loop) and visuomotor (oculomotor loop) systems. The role of the dorsolateral prefrontal, lateral orbitofrontal, and limbic loops in causing the cognitive and associative disturbances associated with basal ganglia syndromes is less clear. Lesions in the basal ganglia resulting from stroke or other disease processes lead to significant changes in the motor behavior of the patient. Examination of the pathways damaged in these

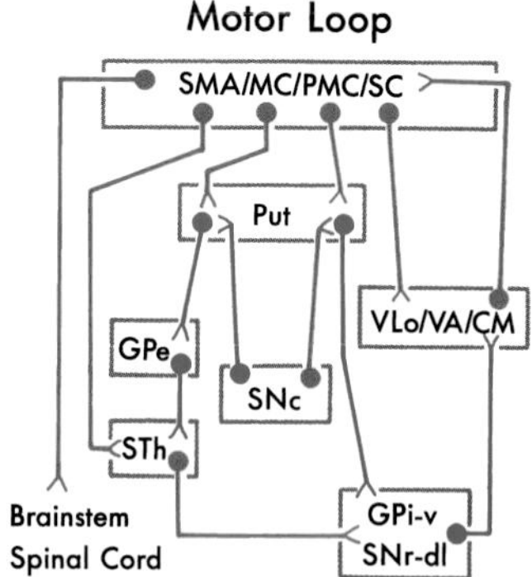

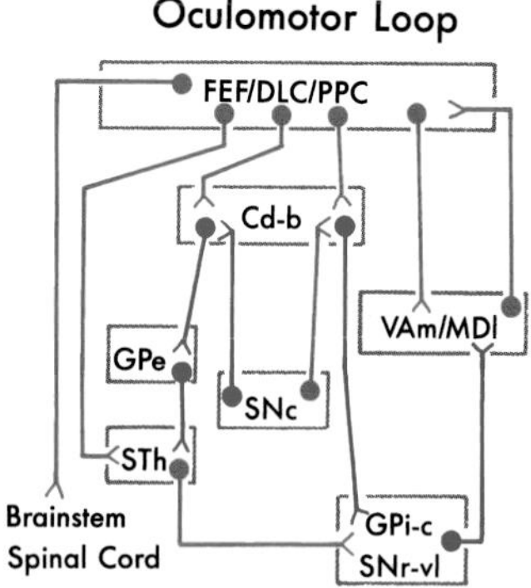

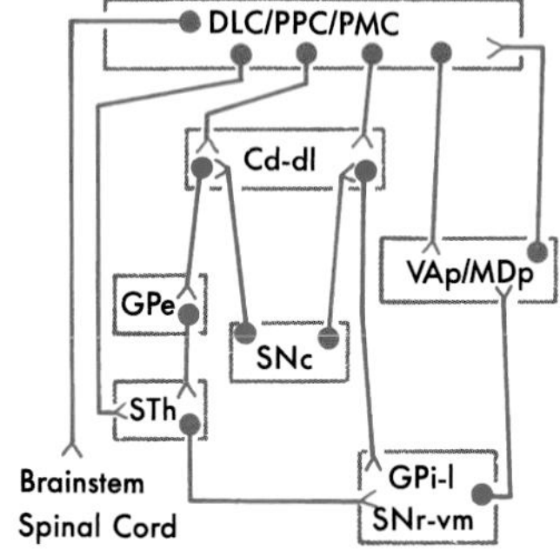

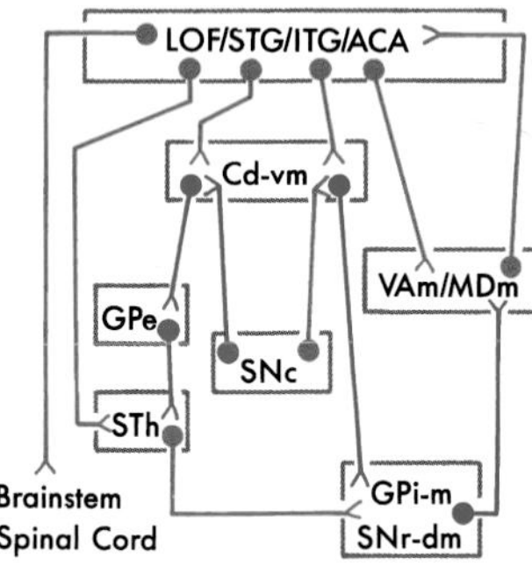

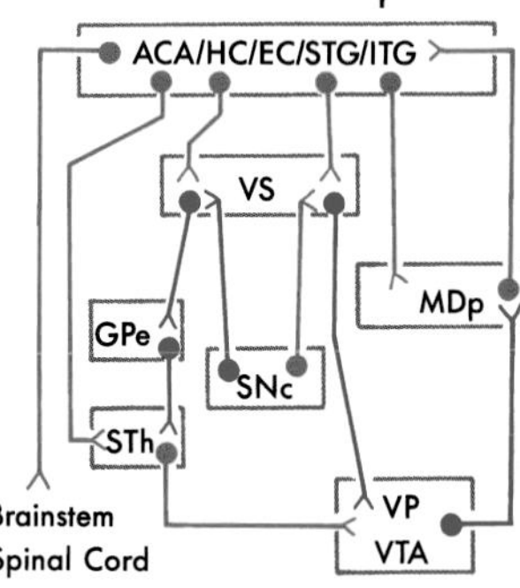

ACA; anterior cingulate area
CB; body of caudate nucleus
CH-dl; head of caudate nucleus, dorsolateral part
CH-vm; head of caudate nucleus, ventromedial part
CM; centromedian nucleus
DLC; dorsolateral prefrontal cortex
EC; entorhinal cortex
FEF; frontal eye fields
GPe; external segment of globus pallidus
GPi-c; internal segment of globus pallidus, central portion
GPi-l; internal segment of globus pallidus, lateral portion
GPi-m; internal segment of globus pallidus, medial portion
GPi-v; internal segment of globus pallidus, ventral portion
HC; hippocampal cortex
ITG; inferior temporal gyrus
LOF; lateral orbitofrontal cortex
MC; motor cortex
MDl; medial dorsal nucleus, paralamellar part
MDm; medial dorsal nucleus, magnocellular part
MDp; medial dorsal nucleus, parvocellular part
PMC; premotor area PPC, posterior parietal cortex
Put; putamen
SC; somatosensory cortex
SMA; supplementary motor area
SNc; substantia nigra pars compacta
SNr-dl; substantia nigra pars reticulata, dorsolateral portion
SNr-dm; substantia nigra pars reticulata, dorsomedial portion
SNr-vl; substantia nigra pars reticulata, ventrolateral portion
SNr-vm; substantia nigra pars reticulata, ventromedial portion
STG; superior temporal gyrus
STh; subthalamic nucleus
VA; ventral anterior nucleus
VAm; ventral anterior nucleus, magnocellular part
VAp; ventral anterior nucleus, parvocellular part
VLo; ventral lateral nucleus, oral part
VP; ventral pallidum VS, ventral striatum
VTA; ventral tegmental area

FIGURE 25-10 *Flow diagrams of the direct and indirect pathways of the five parallel circuits through the basal ganglia. The motor loop is discussed in the text as an example.*

cases reveals how disruption of the basal ganglia could lead to seemingly opposite effects in different patients. For example, movements can be either reduced (*hypokinetic disturbances*) or increased (*hyperkinetic disturbances*).

Hypokinetic Disturbances The two major types of hypokinetic disturbances seen in patients with basal ganglia disorders are *akinesia* and *bradykinesia*. Akinesia is an impairment in the initiation of movement; bradykinesia is a reduction of velocity and amplitude of movement. Both are characteristic in patients with Parkinson's disease.

Akinesia, the impaired ability to initiate voluntary movements, may be due to disruption of the ability to plan a movement or to guide a movement to some desired position. Experimental studies in nonhuman primates reveal that many basal ganglia neurons are most active during the *planning phase of a movement* or when the subject is making an *internally guided movement*. The latter is a movement to a location at which no particular stimulus is present. Thus, *patients with akinesia have a generalized disruption of the role of the basal ganglia in planning and generating programmed movements*.

Bradykinesia, the reduction in velocity and amplitude of movements, is due to disruption of the balance between the outflows of the direct and indirect pathways to the thalamus. The result is an increase in the activation of antagonist muscles. Thus, the observed symptoms are due to an inappropriate activation of the antagonistic muscles, and not necessarily to an overall decrease in muscular activity.

Hypokinetic disorders can be considered as lesions of the neostriatum (Fig. 25-11A,B). These lesions result in the loss of inhibitory connections between the neostriatum and internal segment of the globus pallidus. Thus, tonically active pallidothalamic neurons continuously inhibit their thalamic targets. The thalamus is not disinhibited, so there is a decreased flow of information through the thalamus to the cerebral cortex. This, in turn, causes a decrease in the activity of the appropriate corticospinal and other corticofugal neurons. In addition, most of the connections that subserve the indirect pathway remain intact. Therefore, when those connections are activated, for example by a larger than normal burst in cortical neurons, subthalamopallidal cells excite pallidothalamic neurons, which results in increased inhibition of the thalamus. The combination of a *lack of disinhibition of the thalamus* by the direct pathway and an *increased inhibition of the thalamus* by the indirect pathway significantly *decreases the level of appropriate activity in the cerebral cortex and increases the level of inappropriate cortical activity*. Therefore, the patient becomes less able to execute the appropriate movements. Experiments in nonhuman primates have shown that bradykinetic animals have levels of neuronal activity that are consistent with this view.

Parkinson's disease is correlated pathophysiologically with the loss of the melanin-containing dopaminergic neurons of the nigral complex that project to the striatum. Dopamine has opposite effects on the direct and indirect pathways: It is excitatory to striatal cells feeding into the direct pathway and inhibitory to those of the indirect pathway (Fig. 25-12A,B). The loss of these dopaminergic neurons causes a significant change in the chemistry of the striatal neuropil. This change favors an increase in the activity of striatal projections to the lateral pallidum and a decrease in the activity of striatal projections to the medial pallidum and substantia nigra. The result of these changes is similar to that of a neostriatal lesion (Fig. 25-12C,D).

Hyperkinetic Disturbances Hyperkinetic disturbances take the form of dyskinesias. The three most common forms are *ballismus*, *choreiform movements*, and *athetoid movements*. *Ballismus* is most typically seen as *hemiballismus* because it usually occurs on one side. It consists of uncontrolled flinging (ballistic) movements of an upper or lower extremity. This motor disorder is most commonly seen in patients with vascular lesions localized to the contralateral subthalamic nucleus (Fig. 25-11C,D). *Choreiform movements*, which are present in Huntington's disease and sometimes in treated Parkinson's disease, are generalized irregular dancelike movements of the limbs. Similar movements may occur in oral and facial musculature. In Huntington's disease, there is an initial selective loss of the medium spiny cells in the striatum, which project to the lateral pallidum, and of acetylcholine-containing neurons in the striatal complex. It is thus likely that the neurons specifically associated with the indirect pathway are lost. Finally, *athetoid movements* are a continuous writhing of distal portions of the extremity.

These hyperkinetic disturbances can most easily be explained by the disruption of the indirect pathway through the motor loop, resulting from the loss of excitatory subthalamopallidal neurons (Fig. 25-11C,D). The balance between excitation of pallidothalamic neurons (by the direct pathway) and their inhibition (by the indirect pathway) is skewed. The result is a decrease in the net amount of inhibition of thalamic cells, which results in more activity in the cerebral cortex.

ETIOLOGY OF BASAL GANGLIA–RELATED DISORDERS

The hallmark of basal ganglia disorders involves changes in the neurochemical environment within the striatal complex. Careful examination of patients with basal ganglia syndromes reveals that the classic motor symptoms are only one characteristic of these disorders and that associative memory and limbic dysfunctions also occur. Recent clinical and basic science studies of the chemistry of basal ganglia disorders have provided significant insight into their etiology and treatment.

Huntington's Disease As originally described, Huntington's disease is a progressive, untreatable disorder in which patients lose their ability to function and show increasing dementia; death occurs 10 to 15 years after onset. About 10,000 cases have been diagnosed in the United States. In the early stages, this disease is characterized by absent-mindedness, irritability, depression, clumsiness, and sudden falls. Gradually, choreiform movements increase until the patient is bedridden. Cognitive functions and speech progressively deteriorate. The later stages of this disease are characterized by severe dementia. Postmortem examination reveals decrease in the size of the striatal complex caused by loss of about 90% of all striatal neurons. In particular,

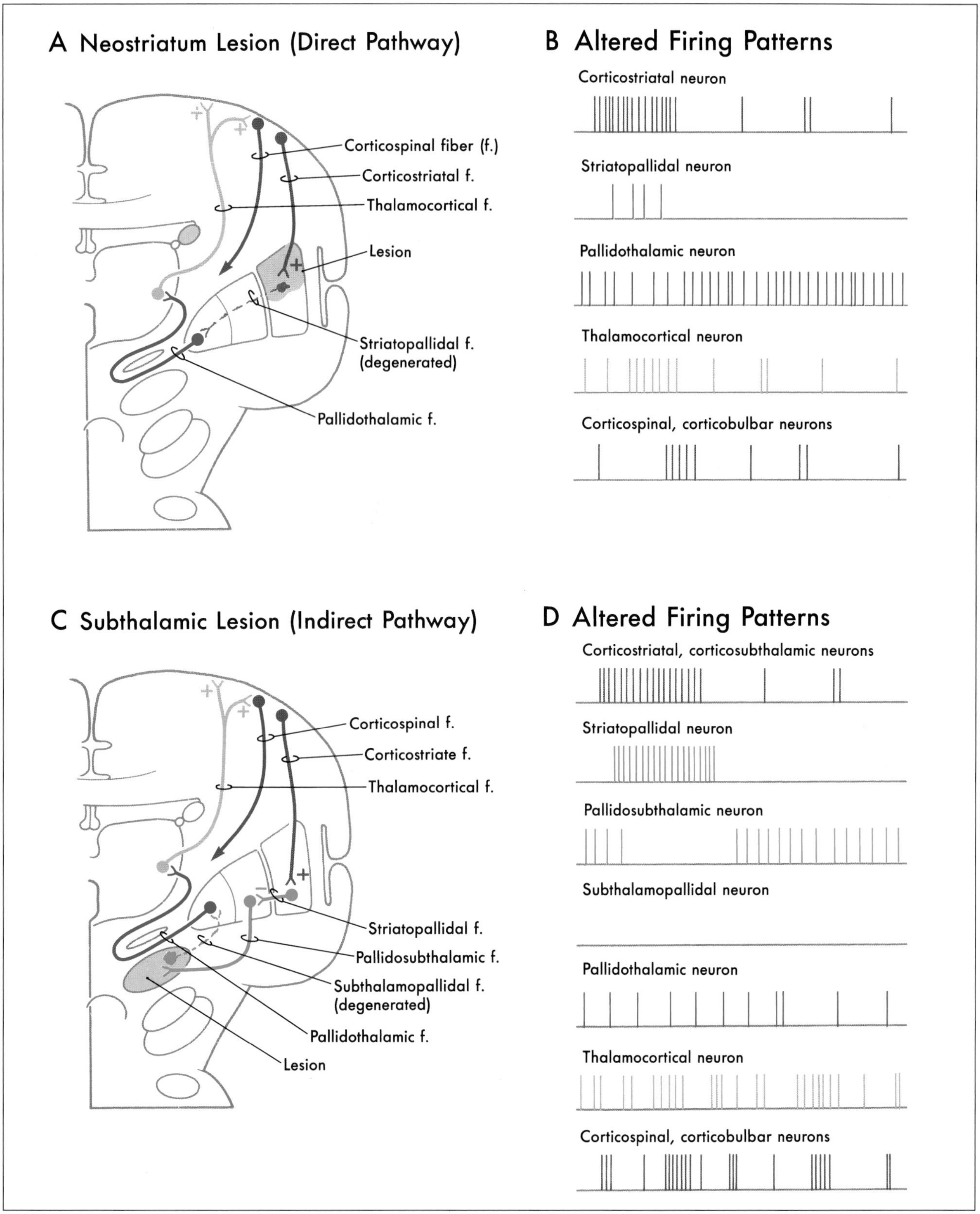

A Neostriatum Lesion (Direct Pathway)

Corticospinal fiber (f.)
Corticostriatal f.
Thalamocortical f.
Lesion
Striatopallidal f. (degenerated)
Pallidothalamic f.

B Altered Firing Patterns

Corticostriatal neuron
Striatopallidal neuron
Pallidothalamic neuron
Thalamocortical neuron
Corticospinal, corticobulbar neurons

C Subthalamic Lesion (Indirect Pathway)

Corticospinal f.
Corticostriate f.
Thalamocortical f.
Striatopallidal f.
Pallidosubthalamic f.
Subthalamopallidal f. (degenerated)
Pallidothalamic f.
Lesion

D Altered Firing Patterns

Corticostriatal, corticosubthalamic neurons
Striatopallidal neuron
Pallidosubthalamic neuron
Subthalamopallidal neuron
Pallidothalamic neuron
Thalamocortical neuron
Corticospinal, corticobulbar neurons

FIGURE 25-11 *Schematic representation of a lesion in the neostriatum (A) and a lesion of the subthalamic nucleus (C), and the corresponding alterations of neuronal firing patterns of all fibers involved in each pathway (B,D). The color of each fiber in A and C correlates with the colors of the altered firing patterns in B and D. +, excitatory synapses; –, inhibitory synapses.*

Direct and Indirect Pathways (Including the Substantia Nigra)

A Connections

Thalamocortical fiber (f.)
Corticostriatal f.
Striatopallidal f.
Striatopallidal f.
+
−
Striatonigral f.
Pallidosubthalamic f.
Nigrostriatal f.
Subthalamopallidal f.
Pallidothalamic f.

B Firing Patterns of Neurons

Corticostriatal/corticosubthalamic neurons
Striatopallidal (direct)/striatonigral neurons
Striatopallidal (indirect) neuron
Nigrostriatal dopaminergic neuron
Pallidosubthalamic neuron
Subthalamopallidal neuron
Pallidothalamic neuron
Thalamocortical neuron

Loss of Dopaminergic Nigral Connections

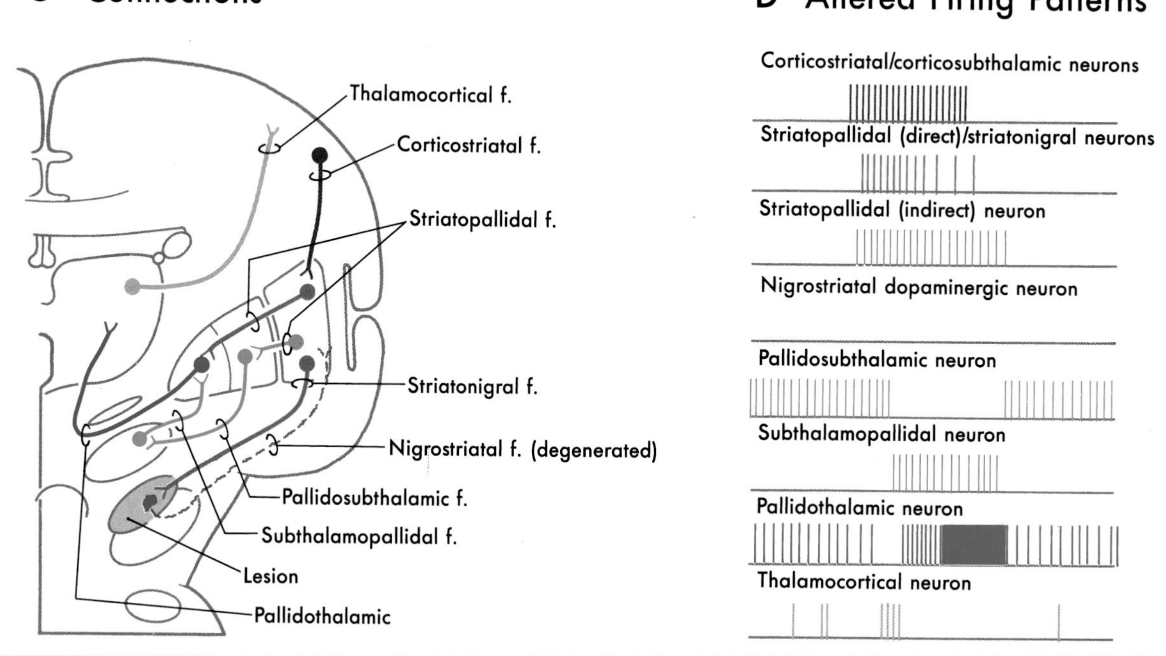

FIGURE 25-12 *The role of the substantia nigra pars compacta in the direct and indirect pathways. Fibers and their corresponding firing patterns for the intact pathway are shown in A and B. Lesions of the pars compacta (C), as in Parkinson's disease, is characterized by the loss of dopaminergic neurons and results in the altered firing patterns of neurons in the entire pathway (D). The colors of the fibers correlate with the colors of the action potentials. +, excitatory synapses;–, inhibitory synapses.*

medium spiny cells, which project to the lateral pallidum, and the large acetylcholine-containing local circuit cells are lost. This disease can be demonstrated on magnetic resonance imaging (MRI) as a flattening of the head of the caudate nucleus (Fig. 25-13).

Huntington's disease is an autosomal dominant genetic disorder, and any person who inherits the gene will develop the disease. It is known to be due to a mutation on the short arm of chromosome 4. Thus, there are now reliable markers for the disease.

Much effort has been devoted to understanding the mechanism underlying the death of striatal neurons in this disease. *Glutamate excitotoxicity* is thought to be primarily responsible for this process. Normally, cortical axons release glutamate as their neurotransmitter in the caudate and putamen. Glutamate binds with its receptor on the medium spiny neurons and opens the receptor channel to an influx of ions. This depolarizes the membrane, resulting in an excitatory postsynaptic potential. As glutamate dissociates from the receptor, it is cleared from the extracellular space by uptake by astrocytes. In Huntington's disease, an unknown mechanism causes glutamate to persist at one type of receptor, the *N*-methyl-D-aspartate (NMDA) receptor, which opens calcium ion channels. The resulting excessive influx of calcium causes an increase in intracellular calcium, which triggers a cascade that leads to cell death. Glutamate excitotoxicity is also thought to be the primary cause of localized neuronal death following acute brain injury, such as stroke.

Parkinson's Disease Parkinson's disease also is a progressive, debilitating disorder. It affects about 500,000 Americans, and is the third most common neurologic disorder. Parkinson's disease usually affects persons over 55 years of age. Some familial groups with Parkinson's disease have been described.

This disorder is characterized by a progressive onset of movement and affective disturbances. The movement disorders include *tremor at rest*, *cogwheel (gamma) rigidity* (increased muscular tone), *akinesia*, *bradykinesia*, *disturbances of eye movements*, and *loss of postural reflexes*. A classic picture of a patient with Parkinson's disease is a person sitting or standing with *pill-rolling tremor*, a *blank stare* (reptilian or decreased blink), a *flexed posture*, and a *paucity of movements*. When the patient starts to move there is a shuffling start, as if the feet were stuck in place (also called a *festinating gait*), followed by nearly normal gait.

At autopsy, the nigral complex is found to be devoid of dopaminergic neurons. Although there is degeneration of serotoninergic and noradrenergic pathways, it is this specific loss of dopamine that results in the observed symptoms. The standard treatment aims at replacing the lost dopamine. Because dopamine itself will not cross the blood-brain barrier, patients are given L-3,4-hydroxyphenylalanine (L-DOPA), which will. This agent is now combined with a second drug, carbidopa, which does not cross the blood-brain barrier but has the effect of inhibiting peripheral uptake of L-DOPA and thus increasing the amount of L-DOPA available to brain tissue. Individuals treated with this combination therapy show significant improvements in their symptoms, although progression of the disease is not arrested. Why L-DOPA works is not clear. The method by which L-DOPA is converted to dopamine in the brain of Parkinson's patients is not known, as these persons have very little tyrosine hydroxylase, the enzyme that is necessary for this catabolism. Moreover, the dopamine is not localized to specific nerve terminals nor to basal ganglia. Thus, it appears that dopamine simply needs to be in the neural environment of the striatal complex to reduce the symptoms of the patients.

Insight into one possible mechanism of Parkinson's disease was inadvertently realized by a group of illicit drug manufacturers whose heroin was contaminated by a

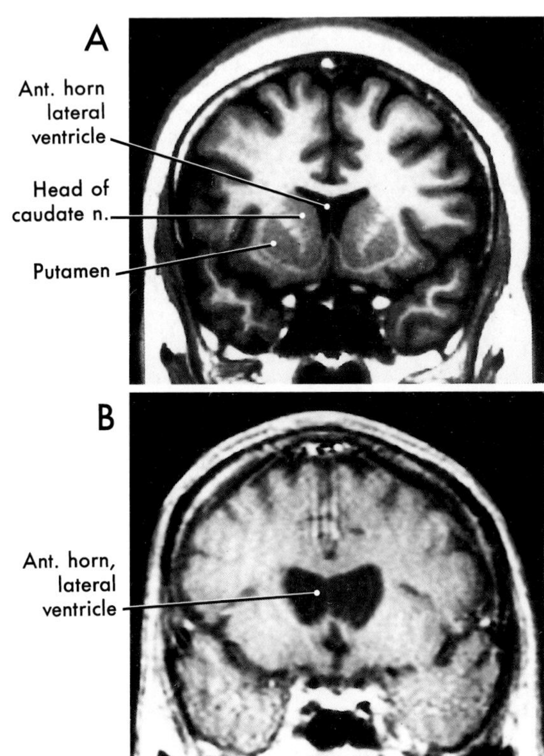

FIGURE 25-13 *Coronal MRI through the frontal lobe, and head of the caudate nucleus, of a normal individual (A) and of a patient with Huntington's disease (B). The head of the caudate normally forms a prominent bulge into the anterior horn of the lateral ventricle (A, inversion recovery image). Profound cell loss in the neostriatum in Huntington's disease greatly diminishes the size of the caudate and renders the lateral wall of the ventricle flat (B, T_1-weighted image). The slightly wavy appearance of the MRI in B is the result of movement (tremor) while the scan was being done.*

compound called 1-methyl-4-phenyl-1,2,3,6-tetrahydropyridine (MPTP). Individuals who used this drug developed symptoms exactly like those of patients with Parkinson's disease, but at an age (early twenties) when this disease is normally never manifest. Autopsy revealed a profound loss of dopaminergic neurons in the substantia nigra pars compacta.

Studies on the mechanism of action of MPTP in animal models showed that MPTP is converted into an active form (MPP+) before it causes a loss of dopaminergic cells. This metabolic pathway requires monoamine oxidase (MAO). Thus, it was hypothesized that the use of MAO inhibitors might affect the progression of Parkinson's disease. In fact, clinical trials have now shown that the drug L-deprenyl, a MAO-B inhibitor, slows the progression of Parkinson's disease and increases the levels of dopamine in the brain. The increase in dopamine may result both from protection of neurons against toxicity and from blocking of the degradation pathway for dopamine, which requires the MAO enzyme.

A controversial method of treatment for Parkinson's disease patients involves the use of human embryonic or autologous transplants. Tissues that produce dopamine, such as substantia nigra (embryonic) and the adrenal cortex (autologous), are obtained and separated into cell suspensions. These are then injected into the lateral ventricles of the patient. The idea is that these cells would adhere to the walls of the ventricle and produce dopamine. The dopamine would then diffuse into the nearby cerebral cortex and basal ganglia. This procedure has now been performed in humans as well as in animals. In the clinical trials to date, only a small number of patients have benefited from this approach.

Tardive Dyskinesia ***Tardive dyskinesia*** is a basal ganglia disorder that is iatrogenic in nature, that is, caused by medical intervention for another disease. This condition is caused by chronic treatment with neuroleptic medications such as the phenothiazines (e.g., chlorpromazine, thioridazine) and butyrophenones (e.g., haloperidol). The manifestation of this condition is uncontrolled involuntary movements, particularly of the face and tongue, and cogwheel rigidity. These symptoms may be temporary or permanent. The action of these neuroleptic drugs is to block dopaminergic transmission throughout the brain. The primary target cells are those in the ventral tegmental area that form the mesolimbic dopaminergic pathway. Prolonged treatment with neuroleptic drugs leads to a hypersensitivity at the D3 dopamine receptor, which causes an imbalance in the nigrostriatal influence on the basal ganglia motor loop and ultimately results in movement disorders.

SOURCES AND ADDITIONAL READINGS

Alexander GE, DeLong MR, Strick PL: Parallel organization of functionally segregated circuits linking basal ganglia and cortex. Annu Rev Neurosci 9:357–381, 1986.

Gerfen CR: The neostriatal mosaic: multiple levels of compartmental organization in the basal ganglia. Annu Rev Neurosci 15:285–320, 1992.

Goldman-Rakic PS (ed): Basal ganglia research (Special Issue). Trends Neurosci 13:241–308, 1990.

Graybiel AM, Aosaki T, Flaherty AM, Kimura M: The basal ganglia and adaptive motor control. Science 265:1826–1831, 1994.

Middleton FA, Strick PL: Anatomical evidence for cerebellar and basal ganglia involvement in higher cognitive function. Science 266:458–461, 1994.

Parent A, Hazrati L-N: Functional anatomy of the basal ganglia. I. The cortico-basal ganglia-thalamo-cortical loop. Brain Res Rev 20:91–127, 1995.

Parent A, Hazrati L-N: Functional anatomy of the basal ganglia. II. The place of the subthalamic nucleus and external pallidum in basal ganglia circuitry. Brain Res Rev 20:128–154, 1995.

Pedro BM, Pilowsky LS, Costa DC, Hemsley DR, Ell PJ, Verhoeff NP, Kerwin RW, Gray NS: Stereotypy, schizophrenia and dopamine D2 receptor binding in the basal ganglia. Psychol Med 24:423–429, 1994.

CHAPTER TWENTY-SIX

The Cerebellum

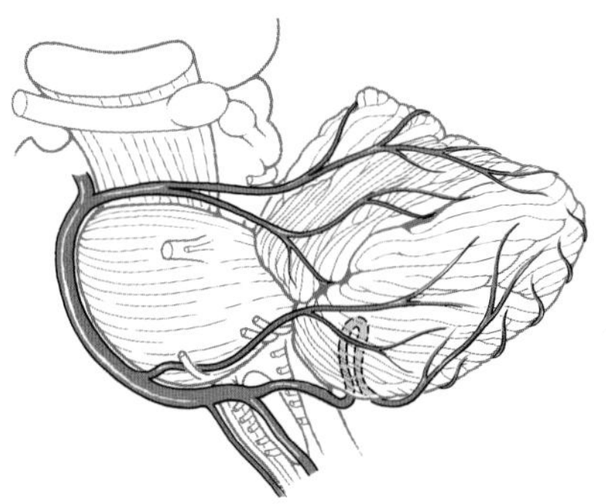

D. E. HAINES • G. A. MIHAILOFF • J. R. BLOEDEL

As indicated by its relative size (about 10% of the weight of the central nervous system), the cerebellum is important in brain function. However, it executes these responsibilities in unique ways. First, it receives extensive sensory input, but it is not involved in sensory discrimination. Second, although it profoundly influences motor function, resection of relatively large portions of the cerebellar cortex does not result in lasting paralysis. Third, the cerebellum is not critical for most cognitive functions, but it may play a role in motor learning and higher mental function.

OVERVIEW

The cerebellum is composed of a highly convoluted *cerebellar cortex* and a core of white matter containing the *cerebellar nuclei*. This structure is anchored to the brainstem via the *cerebellar peduncles*. The cerebellum is located dorsal to the brainstem, inferior to the tentorium cerebelli, and internal to the occipital bone. The cerebellum has a *superior surface* apposed to the tentorium and a convex *inferior surface* that abuts the inner surface of the occipital bone.

The cerebellum receives input from many areas of the neuraxis and influences motor performance through connections with the dorsal thalamus and, ultimately, the motor cortices. Lesions of these pathways result in characteristic motor dysfunctions, which may involve either proximal (axial) or distal musculature. These deficits are actually the result of altered activity in the motor cortex and its descending brainstem and spinal projections, which influence lower motor neurons of the spinal cord.

BASIC STRUCTURAL FEATURES

Cerebellar Peduncles The cerebellum is connected to the brainstem by three pairs of *cerebellar peduncles* (Fig. 26-1A,B). The *inferior cerebellar peduncle* is composed of a *restiform body* and a *juxtarestiform body*. The former is the large ridge on the dorsolateral aspect of the medulla rostral to the level of the obex. This bundle contains mainly fibers that arise in the spinal cord or medulla. The juxtarestiform body is located in the wall of the fourth ventricle. Its fibers form reciprocal connections between the cerebellum and vestibular structures.

The exiting fibers of the trigeminal nerve represent the boundary between the basilar pons and the *middle cerebellar peduncle* (*brachium pontis*) (Fig. 26-1A,B). This large peduncle conveys fibers from the basilar pons into the cerebellum.

The *superior cerebellar peduncle* (*brachium conjunctivum*) sweeps rostrally out of the cerebellum and courses into the midbrain just caudal to the exit of the trochlear nerve (Fig. 26-1A,B). This bundle predominately contains *cerebellar efferent fibers* that originate from neurons of the cerebellar nuclei and distribute to the diencephalon and brainstem.

Cerebellar Lobes, Lobules, and Zones At the most general level, it is common to divide the cerebellum into a narrow midline *vermis* and expansive lateral *hemispheres* (Figs. 26-2, 26-3). The cerebellum is further divided into *anterior*, *posterior*, and *flocculonodular lobes* by the *primary* and *posterolateral fissures*, respectively. The anterior and posterior lobes are composed of yet smaller divisions called *lobules* (Fig. 26-2). The anterior lobe comprises *lobules I to V*, and the posterior lobe, *lob-*

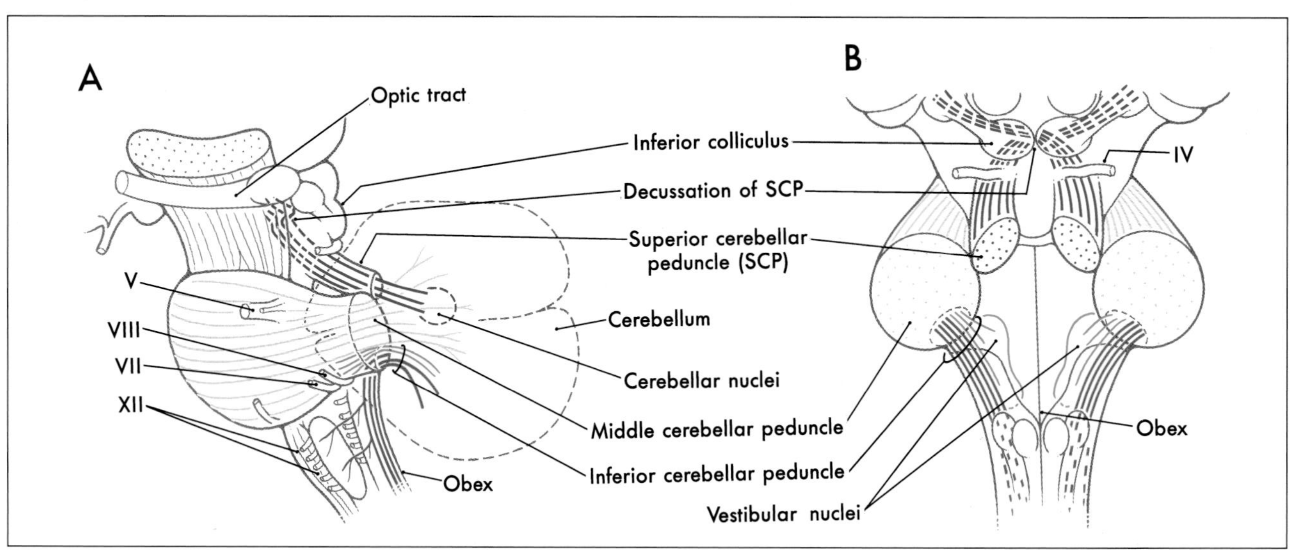

FIGURE 26-1 *Lateral (A) and dorsal (B) views of the brainstem showing the inferior (dark green + red), middle (light green), and superior (blue) cerebellar peduncles. The inferior peduncle is composed of juxtarestiform (dark green) and restiform (red) bodies. Cranial nerves are identified by Roman numerals.*

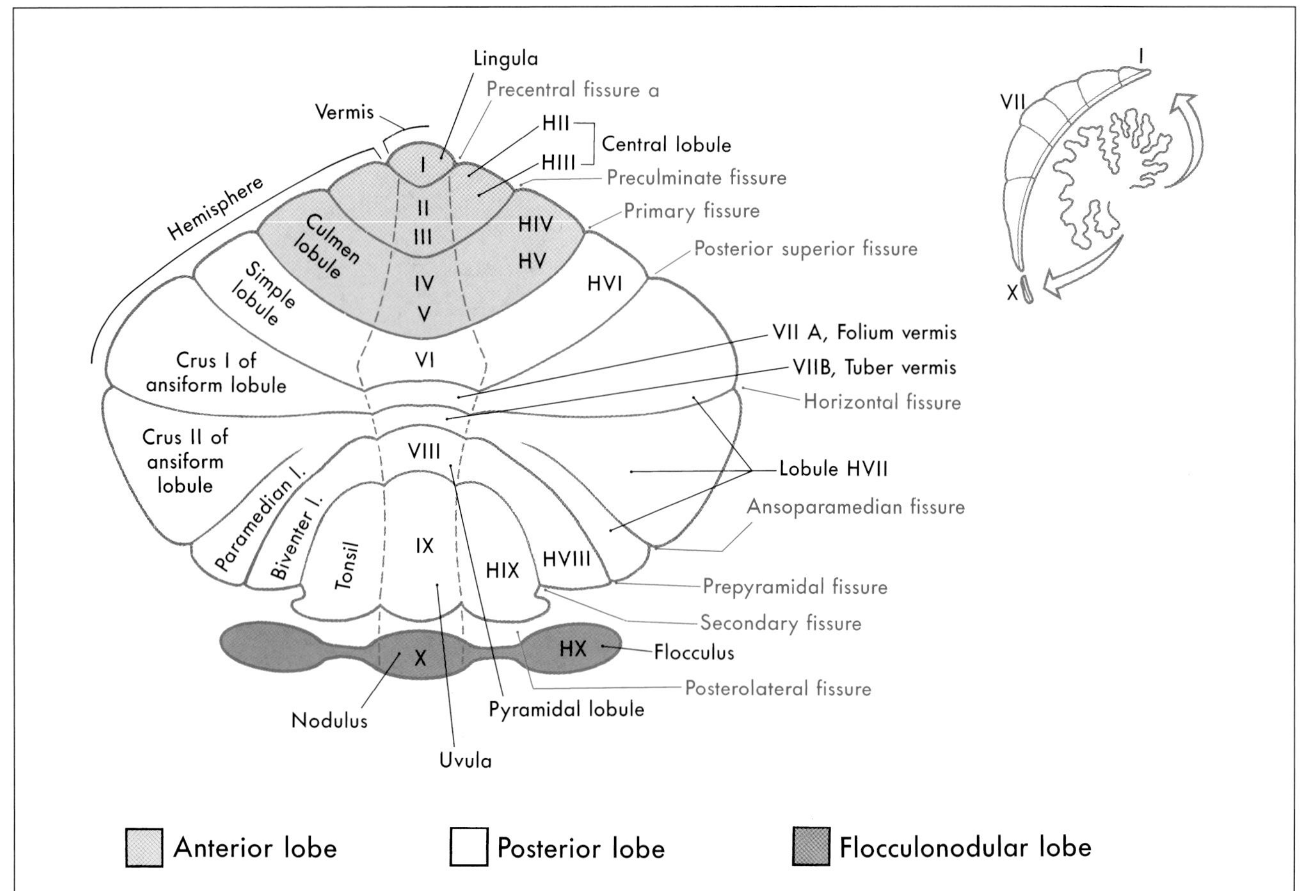

FIGURE 26-2 *Unfolded view (see upper right) of the cerebellar cortex showing lobes, lobules (by name and number), and main fissures (printed in blue). The lobules of the hemisphere are designated by the prefix "H." This specifies which lobule of the hemisphere (H) is continuous with its corresponding (by the numeral) vermal lobule. l., lobule.*

ules VI to IX. The flocculus and nodulus collectively form the flocculonodular lobe (*lobule X*). In turn, each lobule consists of a series of individual ridges of cortex called *folia* (singular, *folium*).

The individual folia are continuous from one hemisphere to the other, across the midline (Fig. 26-3A). This pattern, obvious on the superior cerebellar surface, is disrupted on the inferior surface by the enlargement of the lateral parts of the cerebellum and consequent infolding of the midline area (Fig. 26-3B).

Superimposed on the lobes and lobules of the cerebellum are rostrocaudally oriented cortical zones that are defined on the basis of their connections. There are three principal zones on each side: the medial (vermal), intermediate (paravermal), and lateral (hemisphere) zones (Fig. 26-4). These can be further subdivided into seven smaller zones, two (zones A and B) in the medial zone, three (C1, C2, and C3) in the intermediate zone, and two (D1 and D2) in the lateral zone. The clinical deficits that result from a cerebellar lesion depend mainly on which of the three principal zones is involved; consequently, the latter terminology is used in this chapter.

The *medial* (*vermal*) *zone* is a narrow strip of cortex adjacent to the midline that extends throughout anterior and posterior lobes and includes the nodulus (Figs. 26-3A,B, 26-4). This zone is widest in lobule VI and tapers rostrally and caudally. The *intermediate* (*paravermal*) *zone* lies adjacent to the medial zone and extends throughout anterior and posterior lobes but has little representation in the flocculonodular lobe (Fig. 26-4). The *lateral* (*hemisphere*) *zone* occupies by far the largest part of the cerebellar cortex. It includes large portions of anterior and posterior lobes and the flocculus (Figs. 26-2, 26-4).

Cerebellar Nuclei The four pairs of cerebellar nuclei are located within the white matter core of the cerebellum and are accessed easily by fibers traveling to and from the overlying cortex (Figs. 26-4, 26-5). The *fastigial* (*medial cerebellar*) *nucleus* lies immediately adjacent to the midline and is functionally related to the overlying medial zone of the cerebellar cortex. Lateral to the fastigial nucleus are the two interposed nuclei: the *globose* (*posterior interposed*) *nucleus* and the *emboliform* (*anterior interposed*) *nucleus*. These nuclei are functionally related to the intermediate zone of the cortex. Lateral to the

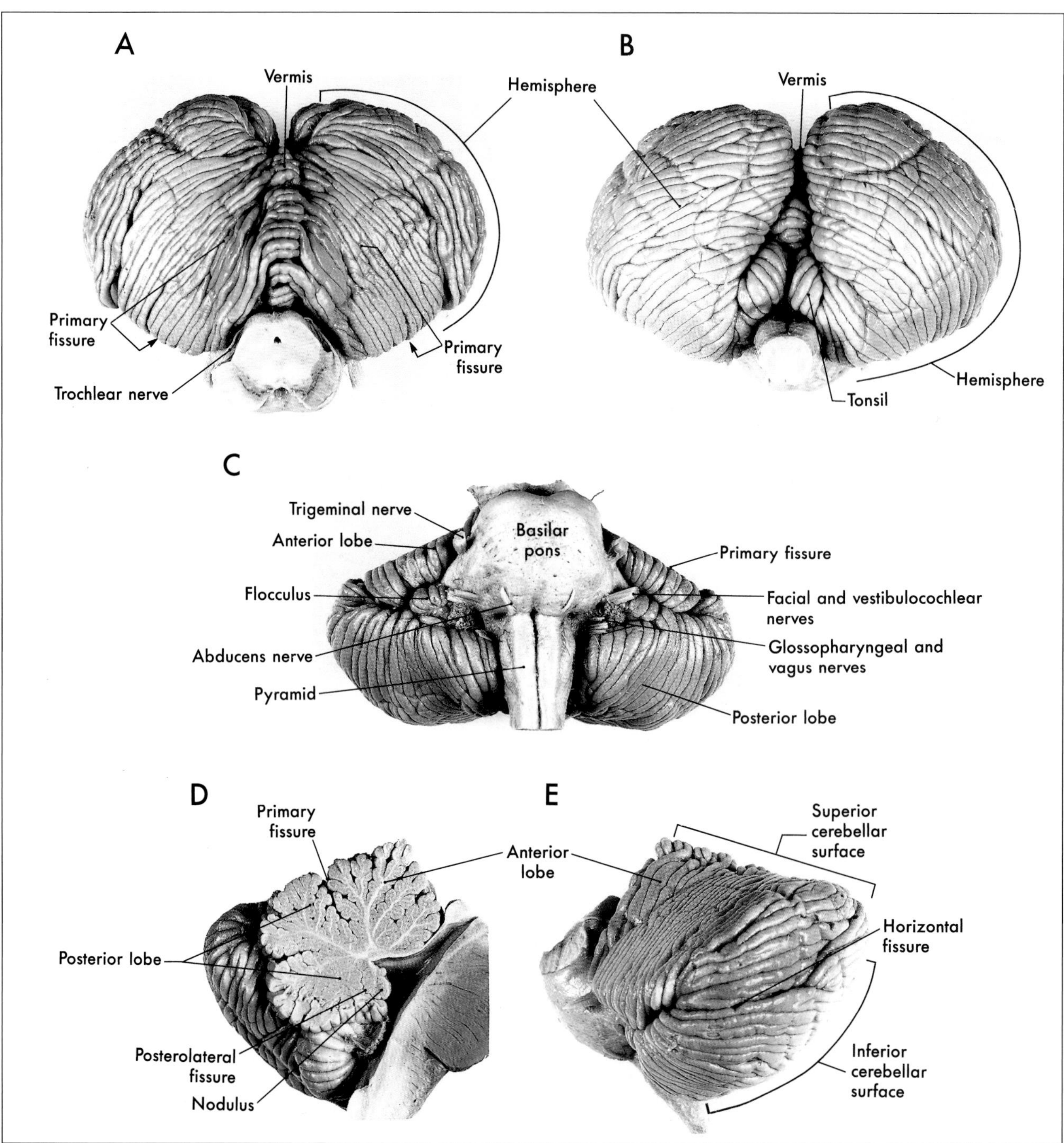

FIGURE 26-3 *Anterior (A, superior surface), posterior (B, inferior surface), ventral (C, with brainstem intact), midsagittal (D), and lateral (E) views of the cerebellum. Only the major structures and relationships are indicated.*

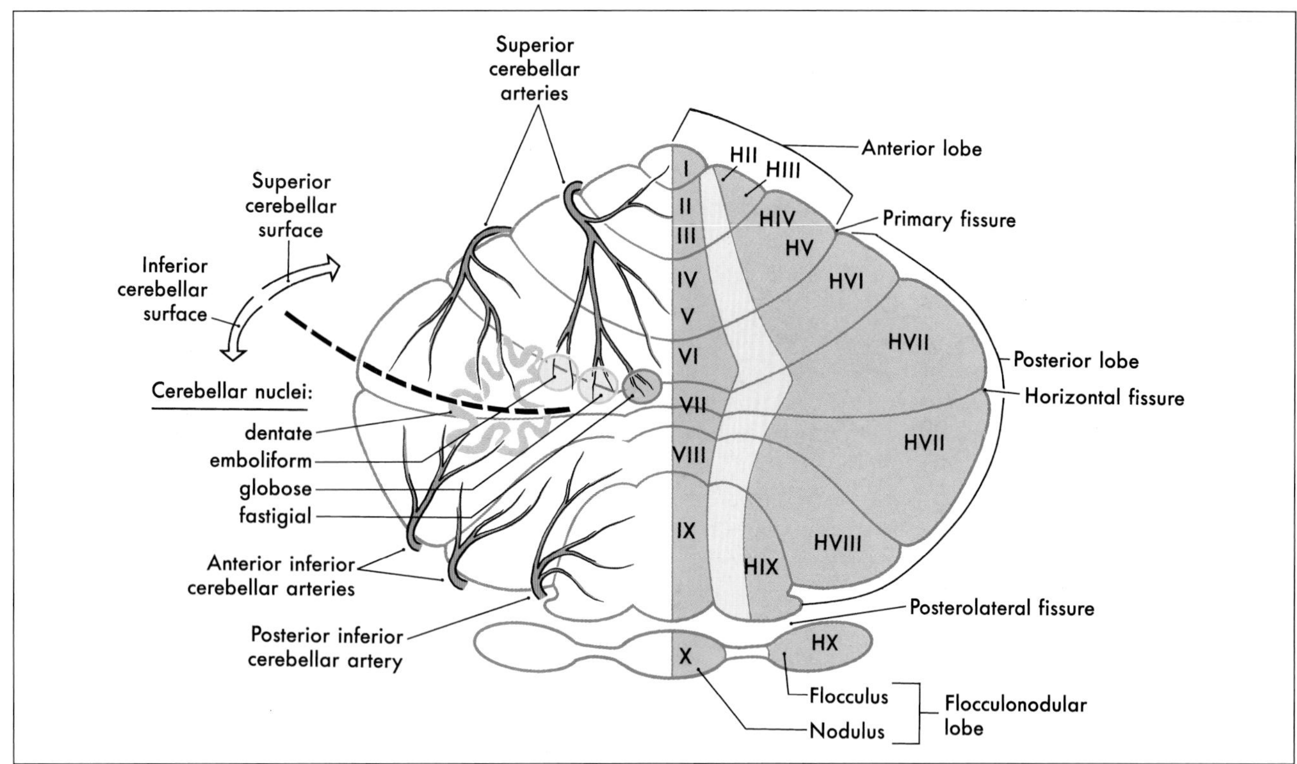

FIGURE 26-4 *An unfolded view of the cerebellum showing medial (gray), intermediate (green), and lateral (blue) zones on the right and the nuclei with which each zone is functionally associated on the left in the corresponding color. The general areas of cortex and nuclei served by the cerebellar arteries are also indicated on the left.*

emboliform nucleus is the *dentate* (*lateral cerebellar*) *nucleus*, which appears as a large, undulating sheet of cells shaped like a partially crumpled paper bag. Its opening, the *hilus*, is directed rostromedially (Fig. 26-5). This nucleus is functionally related to the lateral zone of the cortex; its large size correlates with the large size of this cortical zone.

Most of the signals that leave the cerebellum do so via axons that arise in the cerebellar nuclei; the remainder travel on fibers that originate in the cerebellar cortex. Collectively, the former constitute *cerebellar efferent projections* (or *fibers*). These axons originate from cells in the cerebellar nuclei and use the excitatory neurotransmitters *glutamate* or *aspartate* and thus function to activate their targets. The fastigial nuclei generally project bilaterally to the brainstem through the juxtarestiform bodies. Fibers originating in the dentate, emboliform, and globose nuclei exit the cerebellum via the superior cerebellar peduncle and cross in its decussation.

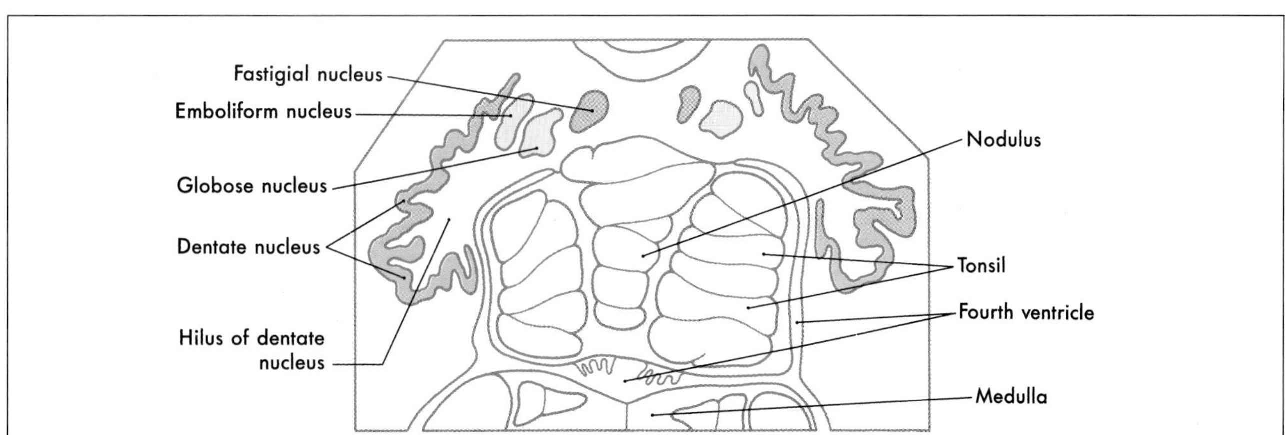

FIGURE 26-5 *The cerebellar nuclei in cross section, drawn from a slide. The color coding of each nucleus corresponds to its appropriate zone in Figure 26-4.*

Some neurons in each cerebellar nucleus send axons or axon collaterals into the overlying cortex, where they terminate in the granular layer as mossy fibers. These axons are called *nucleocortical fibers*, and they exert an excitatory influence on the cerebellar cortex.

Blood Supply to Cerebellar Structures The blood supply to the cerebellar cortex, nuclei, and peduncles is via the *posterior inferior* (PICA), *anterior inferior* (AICA), and *superior cerebellar arteries* (Figs. 26-4, 26-6). The PICA originates from the vertebral artery and supplies the dorsolateral medulla (including the restiform body), the choroid plexus of the fourth ventricle, and caudomedial regions of the inferior cerebellar surface (including the vermis) (Figs. 26-4, 26-6). Caudal parts of the middle cerebellar peduncle and caudolateral portions of the inferior cerebellar surface are served by the AICA. This vessel may also supply caudal parts of the dentate nucleus (Fig. 26-4). The entire superior surface of the cerebellum, most of the cerebellar nuclei, the anterior parts of the middle cerebellar peduncle, and the superior cerebellar peduncle are served by the superior cerebellar artery (Fig. 26-6).

CEREBELLAR CORTEX

Histologically, each folium of the cerebellum has a superficial cellular layer, the *cerebellar cortex*, and a core of myelinated fibers traveling to (*afferent*) or from (*efferent*) the overlying cortex. The cortex consists of a *Purkinje cell layer* insinuated between a cell-dense inner region immediately adjacent to the white matter core, the *granular cell layer*, and an outer pale and relatively cell-sparse portion, the *molecular layer* (Fig. 26-7).

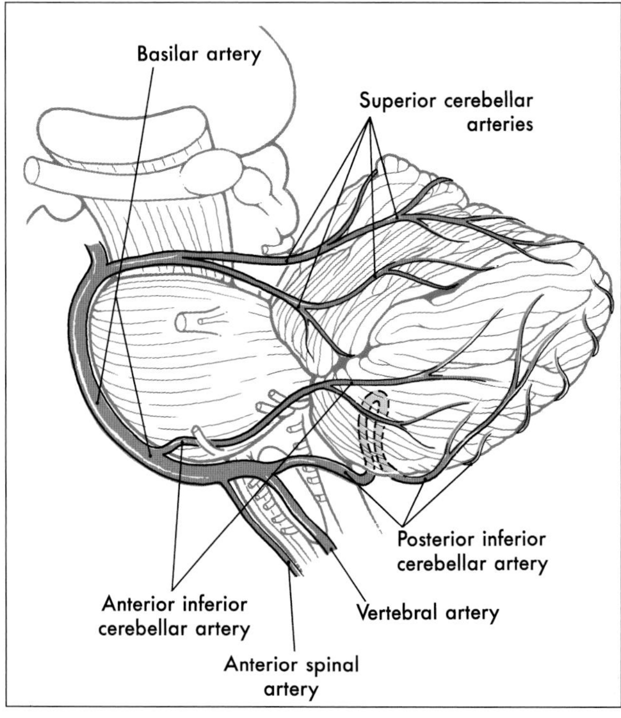

FIGURE 26-6 *Origin and course of arteries serving the cerebellum as seen from the lateral aspect.*

Purkinje Cell Layer The large (40 to 65 μm) somata of Purkinje cells form a single layer at the interface of the granular and molecular layers. Each Purkinje cell gives rise to an elaborate dendritic tree that radiates into the molecular layer. This dendritic tree is shaped like a fan with its wide flattened side oriented perpendicular to the long axis of the folium (Figs. 26-7, 26-8). The "trunk" of the tree is a single primary dendrite, which gives rise to several secondary dendrites, which in turn branch into many tertiary dendrites. Synaptic contacts are formed mainly on short *terminal branchlets* that emerge primarily from the secondary and tertiary dendrites (Figs. 26-7, 26-8). There are two types of branchlets. *Smooth branchlets* emerge from secondary and tertiary dendrites, whereas *spiny branchlets* (*gemmules*) arise mainly from tertiary dendrites.

Purkinje cells are the only efferent neurons of the cerebellar cortex. Axons of Purkinje cells arise from the basal aspect of the cell body and may give rise to recurrent collaterals. The former processes traverse the granular layer and the subcortical white matter to eventually terminate in either the cerebellar or vestibular nuclei. Purkinje cells projecting into the cerebellar nuclei (as *cerebellar corticonuclear fibers*) arise from all areas of the cortex, whereas those projecting into the vestibular nuclei (as *cerebellar corticovestibular fibers*) originate only from parts of the vermis and the flocculonodular lobe. Purkinje cells release gamma aminobutyric acid (GABA) at their synaptic terminals and inhibit target neurons in the cerebellar and vestibular nuclei.

Granular Layer The most numerous neuron of the granular layer is the small (5 to 8 μm) *granule cell* (Figs. 26-7, 26-8C). The dendrites of these cells form clawlike endings (*dendritic digits*) that ramify in the vicinity of the cell body. Their axons ascend into the molecular layer, where they bifurcate to form *parallel fibers*. As indicated by their name, *parallel fibers run parallel to the long axis of the folium*. Consequently, these fibers pass through the fanlike dendritic trees of the Purkinje cell and, when doing so, make synaptic contacts with spiny branchlets (Figs. 26-7, 26-8B,F). They also synapse with the cells intrinsic to the molecular layer (discussed below). The distance spanned by the parallel fibers of a granule cell ranges from about 0.3 to 5.0 mm, and the number of Purkinje and other cells contacted varies accordingly. Granule cells use *glutamate* (or perhaps *aspartate*) as their neurotransmitter and thus have an excitatory effect on their target cells. In fact, the *granule cells are the only excitatory neurons of the cerebellar cortex*. All the others, as we shall see, are inhibitory.

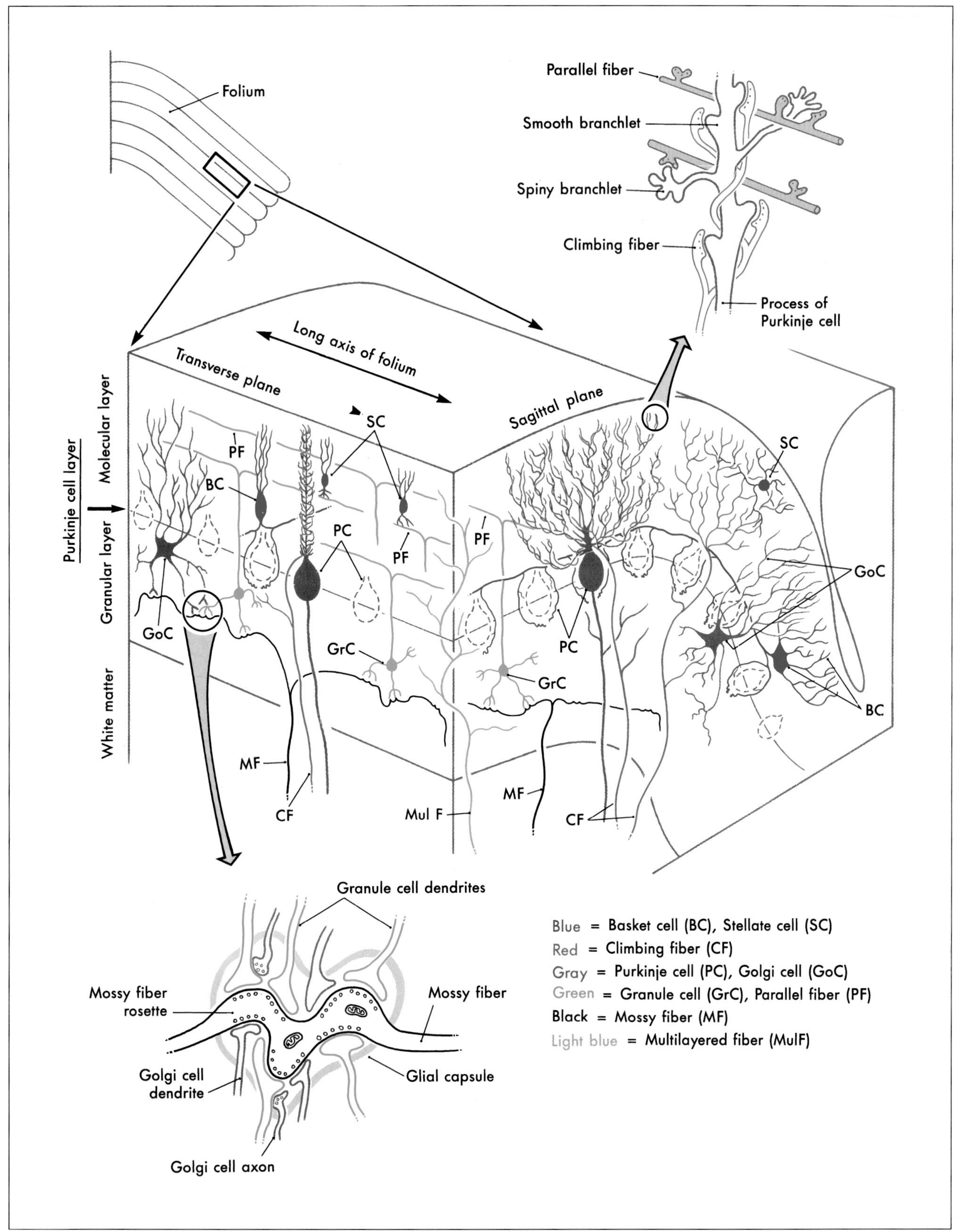

FIGURE 26-7 *Cell types and synaptic relations in the cerebellar cortex in transverse and sagittal planes. Note the structure of the cerebellar glomerulus (lower left) and the interaction of parallel and climbing fibers (upper right) with the dendritic processes of Purkinje cells. Compare with Figure 26-8.*

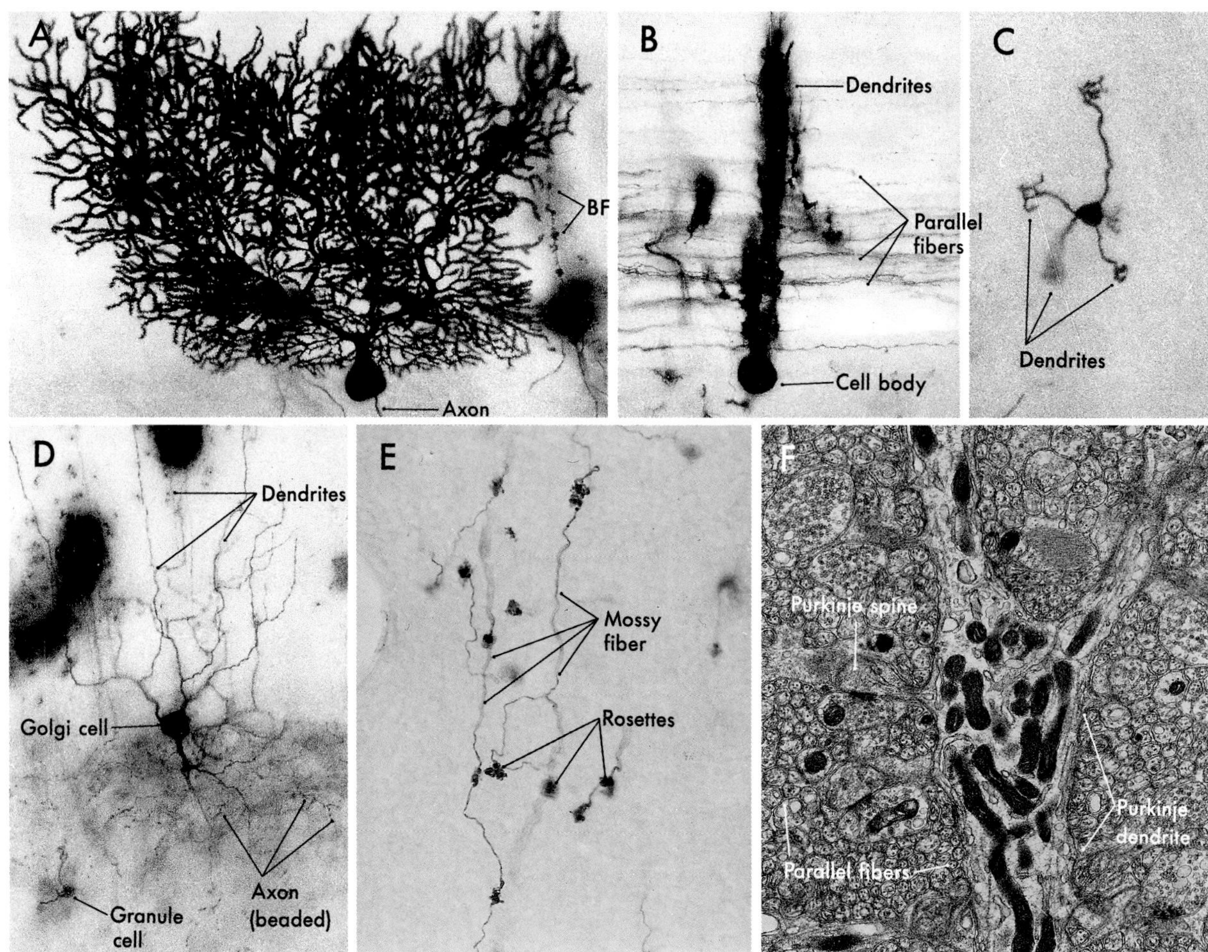

FIGURE 26-8 *Examples of cells of the cerebellar cortex (cf. Fig. 26-7). The Purkinje cells are shown in sagittal (A, note the beaded appearance of the dendrites) and transverse (B, note the many parallel fibers) planes. Granule cell (C,D) dendrites end as a cluster of short clawlike processes. Dendrites of Golgi cells (D) branch into molecular and granular layers while their axons (D, beaded structures) ramify in the granular layer. Mossy fibers (E) branch profusely and have many rosettes; synaptic contacts between mossy fibers and granule cell dendrites take place at the rosette in the cerebellar glomerulus (see Fig. 26-7). At the ultrastructural level, the Purkinje cell dendrite is surrounded by the numerous small profiles of parallel fibers. BF, Bergmann fiber, a type of glial cell process. (A–D, Courtesy of Dr. José Rafols, Wayne State University.)*

The second cell type of the granular layer is the *Golgi cell.* The soma of this neuron is larger (at 18 to 25 μm) than that of the granule cell and is usually found in the granular layer adjacent to the Purkinje cells (Figs. 26-7, 26-8D). Dendrites of Golgi cells branch in the granular layer but extend primarily into the molecular layer without regard to plane of orientation. Axons of Golgi neurons branch in the granular layer and form synaptic contacts on granule cell dendrites (Fig. 26-7). The Golgi cell utilizes GABA as a neurotransmitter and is an inhibitory interneuron in the cerebellar cortex.

The *cerebellar glomerulus* is a synaptic complex found in the neuropil of the granular layer. The glomerulus includes granule and Golgi cell dendrites and Golgi cell axons, and a specialized synaptic segment of a *mossy fiber*, one type of cerebellar afferent axon (Figs. 26-7, 26-8C-E). The mossy fiber terminal is centrally located and forms synapses with several granule cell dendrites. Golgi cell axons contact granule cell dendrites in the glomerulus and the entire complex is encapsulated by glial cell processes.

Molecular Layer The molecular layer has considerably fewer cell bodies than the granular layer but has proportionately more cell processes. These processes include *parallel fibers*, *Purkinje cell dendrites*, *Golgi cell dendrites*, *climbing fibers* (see below), and the processes of cells intrinsic to the molecular layer (Fig. 26-7).

The intrinsic cell types of the molecular layer are *stellate cells* and *basket cells* (Fig. 26-7). Stellate cells are usually found in outer regions of the molecular layer and are frequently referred to as *superficial* or *outer stellate cells*. The somata of basket cells are located immediately above the Purkinje cell layer. The basket cell axon travels in the sagittal plane and gives rise to descending branches that form elaborate "baskets" around the Purkinje cell body. This cell derives its name from this characteristic feature.

In general, the dendritic and axonal plexuses of basket and stellate cells are *oriented primarily in the sagittal plane*, much like that of Purkinje cell dendrites (Fig. 26-7). Although basket and stellate cells are similar in general shape, the *extent* of the dendritic and axonal fields is much larger in basket cells than in stellate cells. Consequently, basket cells may influence a large number of Purkinje cells, mainly in the sagittal plane, whereas stellate cells influence a much smaller population, but also in the sagittal plane. Stellate and basket cells receive excitatory inputs from parallel fibers.

Basket and stellate cells are GABAergic and inhibit their target neurons. Although these cells influence several targets in the molecular layer, for our purposes the Purkinje cell is the most important. Similarly, Purkinje and Golgi cells are also inhibitory, thus making *the granule cell the only neuron in the cerebellar cortex whose output evokes an excitatory response*.

Cerebellar Afferent Fibers The afferent fibers to the cerebellar cortex are grouped into three types on the basis of the *morphology and connections of their terminals in the cortex*. These three types of cortical terminals arise from *mossy fibers*, *climbing fibers*, and *multilayered* (monoaminergic) *fibers*.

The axons that end as *mossy fibers* originate from cell bodies in the cerebellar nuclei (*nucleocortical fibers*) and from a variety of other nuclei in the spinal cord, medulla, and pons (Table 26-1). En route to the cerebellar cortex, many of these cerebellar afferent fibers send collaterals into a cerebellar nucleus. In the granular layer, *mossy fibers* branch profusely, and their large terminals contact other cells at irregular intervals (the *mossy fiber rosette*). The rosette, which is the central element of the cerebellar glomerulus, gives the fiber a mossy appearance (Figs. 26-7, 26-8E). Single mossy fibers may form up to 50 rosettes, and each rosette may participate in synaptic contacts with up to 10 to 15 granule cells in a cerebellar glomerulus. In addition, a mossy fiber may branch and distribute to more than one folium. Mossy fibers utilize *glutamate* as their neurotransmitter and are excitatory to granule cell and Golgi cell dendrites in the cerebellar glomerulus and to cerebellar nuclear neurons on which their collaterals terminate.

The inferior olivary nuclei are the only source of axons that end as *climbing fibers* in the cerebellar cortex (Table 26-1). Olivocerebellar fibers send collaterals to the appropriate cerebellar nucleus. The climbing fibers then terminate in the molecular layer by entwining, ivy-like, up the dendritic trees of Purkinje dendrites (Fig. 26-7). Each Purkinje cell is innervated by a single climbing fiber, but olivocerebellar axons may branch to serve several Purkinje cells. Climbing fibers use the neurotransmitter *aspartate*, and they excite Purkinje cells and cerebellar nuclear neurons.

Multilayered fibers (monoaminergic or peptide-containing) originate from cells of the locus ceruleus (*noradrenergic*), the raphe nuclei (*serotoninergic*), the hypothalamus (some are *histaminergic*), and other select locations. These fibers enter the cerebellum via the cerebellar peduncles and, in the case of some hypothalamocerebellar fibers, by passing through the periventricular gray and then into the cerebellum. In the cortex, these axons branch diffusely and terminate in molecular and granular layers, where they may influence all major cell types. In general, these fibers modulate the output of the cerebellar cortex through two mechanisms. First, they decrease the spontaneous discharge rates of Purkinje cells. Second, both directly and via interneurons, multilayered fibers alter the responsiveness of Purkinje cells to excitation by climbing fibers and by the mossy fiber-granule cell projection.

Topographic Localization The cerebellum receives input from a wide range of sources. Some input that originates in the periphery is conveyed via spinocerebellar and vestibulocerebellar pathways that project *directly* to the cerebellum. Other afferent information is *indirect*, having passed through multiple central pathways before entering the cerebellum. For example, responses can be recorded in selected regions of the cerebellar cortex following stimuli that activate visual, auditory, or sensorimotor areas of the cerebral cortex in primates (Fig. 26-9A). These pathways involve a cerebropontine-pontocerebellar connection. Visual and auditory cortices project to cells of the basilar pons, which, in turn, provide a mossy fiber input to areas of the cerebellar cortex where the eye and ear are represented (Fig. 26-9A). Similarly, the sensorimotor cortex, also via projections through the basilar pons, influences cerebellar cortical regions containing representations of the body.

At a finer level of resolution, experimental studies in mammals have shown that body parts are not represented continuously over a large area of cerebellar cortex but instead are broken into smaller, discontinuous patches. In this pattern, a small area of cortex that receives sensory input from the arm (via mossy fiber-granule cell connections) may be located adjacent to an area that receives input from a noncontiguous region of the same upper extremity (Fig. 26-9B). In addition, each body part is represented in several locations. This pattern of spatial representation is referred to as *fractured somatotopy*.

TABLE 26-1

SYNOPSIS OF SELECTED AFFERENT AND EFFERENT FIBERS OF THE CEREBELLUM AS CONTAINED IN, OR ASSOCIATED WITH, THE CEREBELLAR PEDUNCLES

STRUCTURE/FIBERS	LATERALITY	TERMINATE AS, OR IN
INFERIOR CEREBELLAR PEDUNCLE		
Restiform Body		
Dorsal spinocerebellar f.	–	mossy fibers
Cuneocerebellar f.	–	mossy fibers
Olivocerebellar f.	X	climbing fibers
Reticulocerebellar f. (from reticulotegmental pontine nucleus)	X,–	mossy fibers
Reticulocerebellar f. (from lateral reticular nucleus)	–(X)	mossy fibers
Reticulocerebellar f. (from paramedian reticular nucleus)	–,(X)	mossy fibers
Trigeminocerebellar f.	–	mossy fibers
Raphecerebellar f.	–,(X)	multilayered fibers
Juxtarestiform Body		
Vestibulocerebellar f. (primary and secondary)	–	mossy fibers
Cerebellar corticovestibular f.	–	vestibular nuclei
Fastigiovestibular f.	–,X	vestibular nuclei
Fastigioreticular f.	–,X	reticular nuclei
Fastigio-olivary f.	X	caudal parts of accessory olivary nuclei
Fastigiospinal	X	spinal cord
MIDDLE CEREBELLAR PEDUNCLE		
Pontocerebellar f.	X,(–)	mossy fibers
Raphecerebellar f.	–,(X)	multilayered fibers
SUPERIOR CEREBELLAR PEDUNCLE		
Ventral spinocerebellar f.	X,–	mossy fibers
Rostral spinocerebellar f.	X,–	mossy fibers
Ceruleocerebellar f.	–,X	multilayered fibers
Hypothalamocerebellar f.	–,X	multilayered fibers
Raphecerebellar f.	–	multilayered fibers
Cerebellar efferent f.		
Dentatothalamic	X	dorsal thalamus
Dentatorubral	X	red nucleus
Dentatoreticular	X	reticular nuclei
Dentatopontine	X	pontine nuclei
Dentato-olivary	X	principal olivary nucleus
Dentatohypothalamic	X	hypothalamus
Interpositothalamic	X	dorsal thalamus
Interpositorubral	X	red nucleus
Interpositoreticular	X	reticular nuclei
Interposito-olivary	X	rostral parts of accessory olivary nuclei
Interpositohypothalamic	X	hypothalamus
Interpositospinal	X	spinal cord

– = uncrossed; X = crossed; (–) = some uncrossed; (X) = some crossed

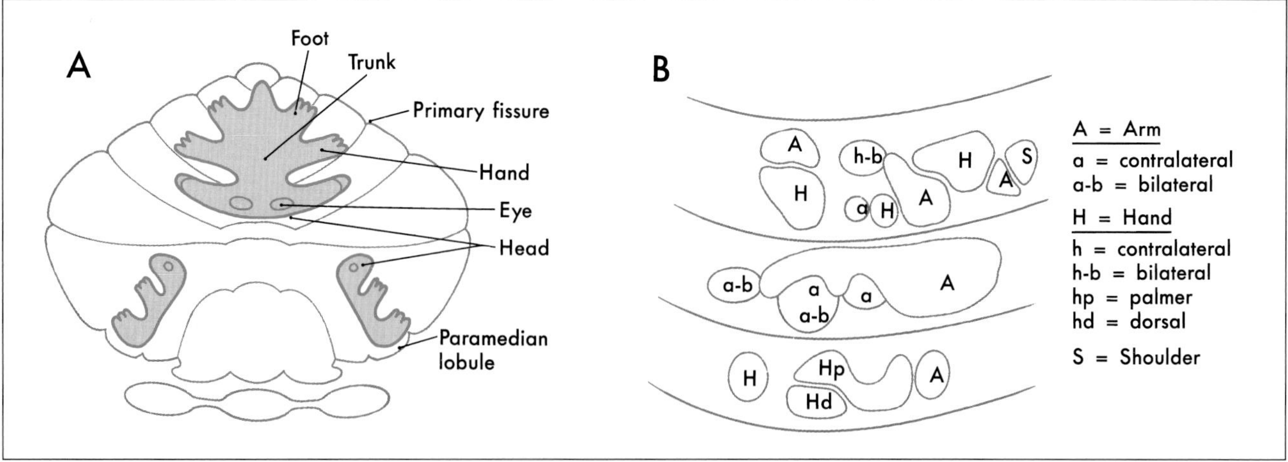

FIGURE 26-9 *Somatopy in the cerebellar cortex (A) and a summary representation of fractured somatopy in the paramedian lobule (B) of a primate. In the somatotopic map, body areas were thought to be continuous (A), but recent studies suggest that discontinuous body parts (or areas) may be represented in immediately adjacent cortical regions (B). (B is adapted from Welker et al, 1988, with permission.)*

Synaptic Interactions in the Cerebellar Cortex In general, cerebellar function involves an ongoing series of *excitatory inputs to the cerebellar nuclei*, the effects of which are modulated by the *inhibitory action of Purkinje cell axons* (*corticonuclear fibers*) descending from the overlying cortex (Fig. 26-10). These synaptic interactions continuously modify the efferent signals generated by cerebellar nuclear neurons.

Climbing fibers synapse directly on Purkinje cells, whereas mossy fibers act through granule cells. Because a *single climbing fiber* makes numerous synaptic contacts on a *single Purkinje cell*, its influence over that cell is substantial. Consequently, the Purkinje cell response to input from one climbing fiber is represented by a complicated waveform called a complex spike (Fig. 26-11). These spikes are unique and are the result of the combined action of multiple excitatory climbing fiber synapses formed throughout the Purkinje cell dendritic tree. In contrast, each Purkinje cell receives excitatory input from *many granule cells* via their parallel fibers. Summation, both spatial and temporal, of parallel fiber input is responsible for the generation of a single Purkinje cell response, the *simple spike* (Fig. 26-11). At any given moment in time, there is an ongoing background level of simple spike activity in the cerebellum. This level can be modulated by phasic increases or decreases in afferent inputs to the mossy fiber–granule cell–parallel fiber system. Collectively, mossy fibers have a powerful influence over Purkinje cells, and this influence may be modulated by climbing fibers through mechanisms not fully understood.

Let us review the mossy fiber–granule cell connection and its effects (Figs. 26-7, 26-10). In the cerebellar glomerulus, mossy fibers excite granule and Golgi cell dendrites. The Golgi cell axon, in turn, synapses on and inhibits granule cell dendrites in the glomerulus. Thus, the Golgi cell provides feedback inhibition to granule cells previously excited by mossy fiber activity. The granule cell axon enters the molecular layer, branches into parallel fibers, and excites Purkinje, stellate, basket, and

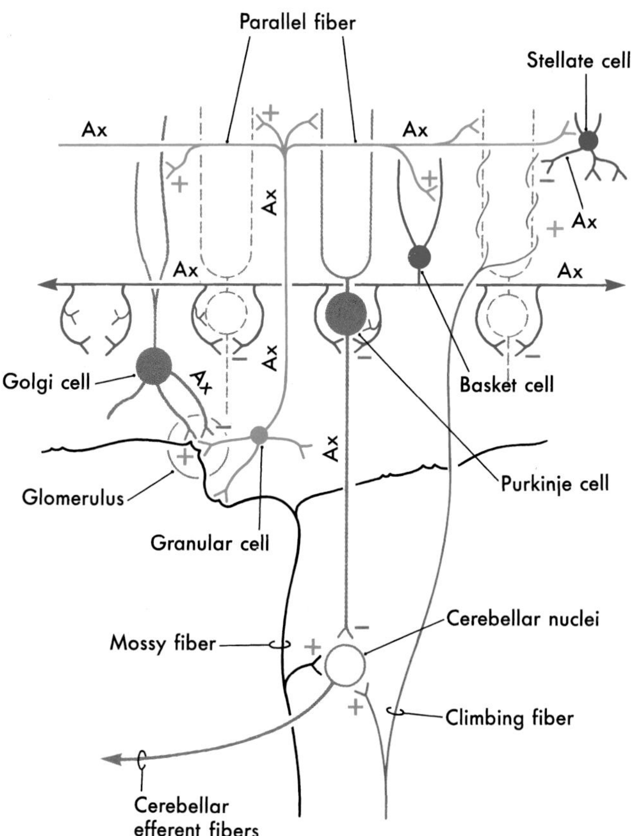

FIGURE 26-10 *A diagrammatic representation of synaptic interactions in the cerebellar cortex. +, excitatory synapses; –, inhibitory synapses.*

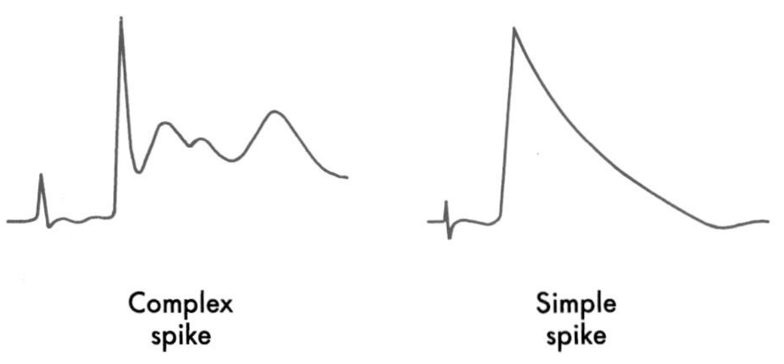

FIGURE 26-11 *Complex and simple spikes of Purkinje cells, as recorded intracellularly, following excitation, by climbing and mossy fibers, respectively.*

Golgi cells (Fig. 26-10). At a basic level, mossy fiber input leads to excitation of Purkinje cells via parallel fibers, and the GABAergic Purkinje cells respond by inhibiting the cerebellar nuclei.

The synaptic interactions within the cerebellar cortex are described in the following simplified model, which considers the cytoarchitectural and electrophysiologic properties of cerebellar cortical neurons. The inhibitory Purkinje cell outflow is modulated, in part, by the feed-forward inhibition resulting from stellate and basket cell activation (Figs. 26-10, 26-12). Parallel fibers excite a specific population of Purkinje cells, as well as stellate and basket cells located within their domain (Fig. 26-12). The latter two GABAergic interneurons, in turn, inhibit Purkinje cells located adjacent to (via stellate synapses), or at some distance from (basket synapses), the row of activated parallel fibers. When a narrow bundle of parallel fibers is activated, under certain experimental conditions, basket and stellate cells can define a central row, or beam, of excited Purkinje cells. Within the beam, Purkinje cells are activated by parallel fibers and, in turn, inhibit cells in the cerebellar or vestibular nuclei (Fig. 26-12). Purkinje cells on either side of the activated row are inhibited by stellate and basket axons and, consequently, do not inhibit their target neurons in the cerebellar (or vestibular) nuclei (Fig. 26-12). These target cells are removed from their normal (background) inhibitory influence; that is, they are disinhibited.

Synaptic interactions between cells of the cerebellar cortex contribute to the activity of cerebellar nuclear neurons. The unique structural/functional properties of the cortex provide circuits for *the temporal and spatial processing of information* that contributes substantially to the cerebellar capacity to coordinate movement. The exact nature of these interactions has not been fully determined. However, the cytoarchitecture and synaptology within the cerebellum suggest one hypothetical model. This model suggests that *the excitatory outflow from each cerebellar nucleus varies dynamically* in response to the combined effects of (1) the excitatory input from cerebellar afferent collaterals and (2) the inhibitory influence mediated by Purkinje cells.

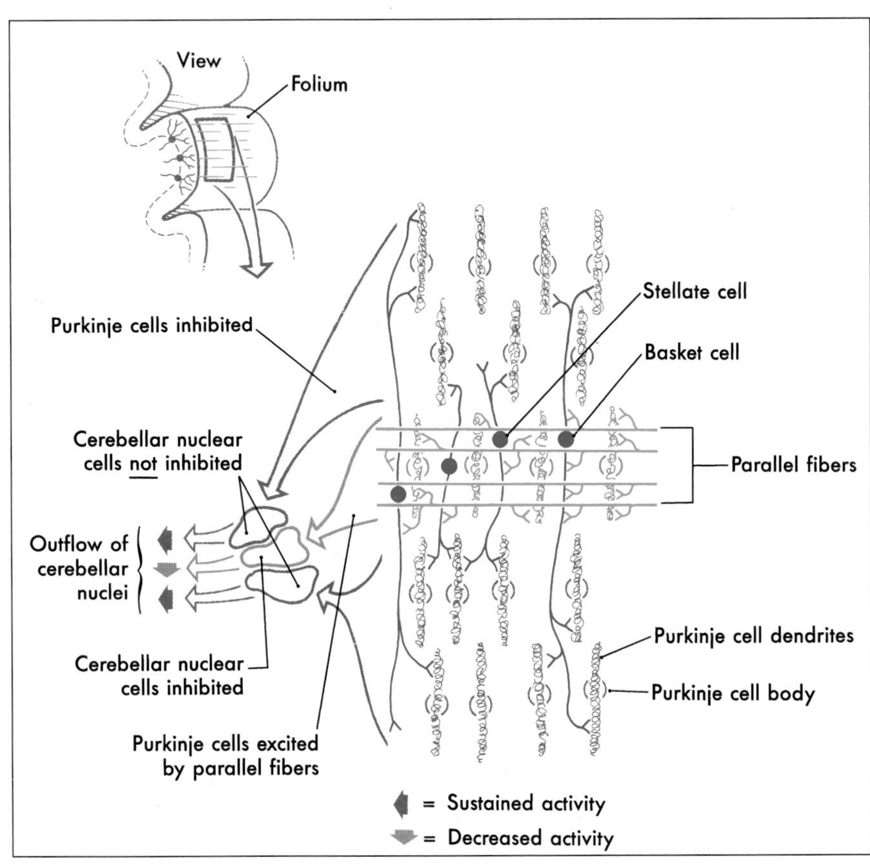

FIGURE 26-12 *A diagrammatic representation of a portion of a cerebellar folium as viewed from the surface. Activation of a bundle of parallel fibers (green) will lead to activation of a row of Purkinje cells (red) located within their domain. Simultaneously, the Purkinje fibers in the flanking zones (gray) will be inhibited by the action of basket and stellate cells, which are also activated by parallel fibers. These populations of activated and inhibited Purkinje cells will cause, respectively, inhibition (red) and disinhibition (gray) of cerebellar nuclear cells via the corticonuclear pathway.*

FUNCTIONAL CEREBELLAR MODULES

It is convenient to think of the cerebellum as being arranged into *compartments* or *modules.* Each module consists of (1) an area of cortex (usually a cortical zone), (2) a white matter core that contains afferent and efferent fibers to and from that cortical area, and (3) a nucleus (or nuclei) that is functionally related to the overlying cortical area. A cerebellar cortical zone and its corresponding nucleus (nuclei) constitute a module.

Vestibulocerebellar Module The flocculonodular lobe and adjacent portions of vermal lobule IX receive afferents from the ipsilateral vestibular ganglion (*primary vestibulocerebellar fibers*) and vestibular nuclei (*secondary vestibulocerebellar fibers*). Therefore, these cortical areas are commonly called the *vestibulocerebellum*. Along with the *fastigial nucleus*, they form the *vestibulocerebellar* module (Fig. 26-13). Because this is phylogenetically the oldest part of the cerebellum, it is also called the *archicerebellum*.

Vestibulocerebellar fibers access the flocculonodular cortex and fastigial nucleus via the juxtarestiform body and convey information concerning the position of the head and body in space, as well as information useful in orienting the eyes during movements. This information is supplemented by inputs carried on *olivocerebellar fibers* from the contralateral olivary nuclei and *pontocerebellar fibers* (only to the flocculus) from the contralateral basilar pons. The latter pathways convey *indirect* inputs from nuclei of the diencephalon and brainstem, which are concerned with a broad spectrum of information regarding visual processing and eye movements (see also Chapter 27).

The outflow of the vestibulocerebellar module consists of *cerebellar corticovestibular fibers* from the flocculonodular lobe, *cerebellar corticonuclear fibers* from the

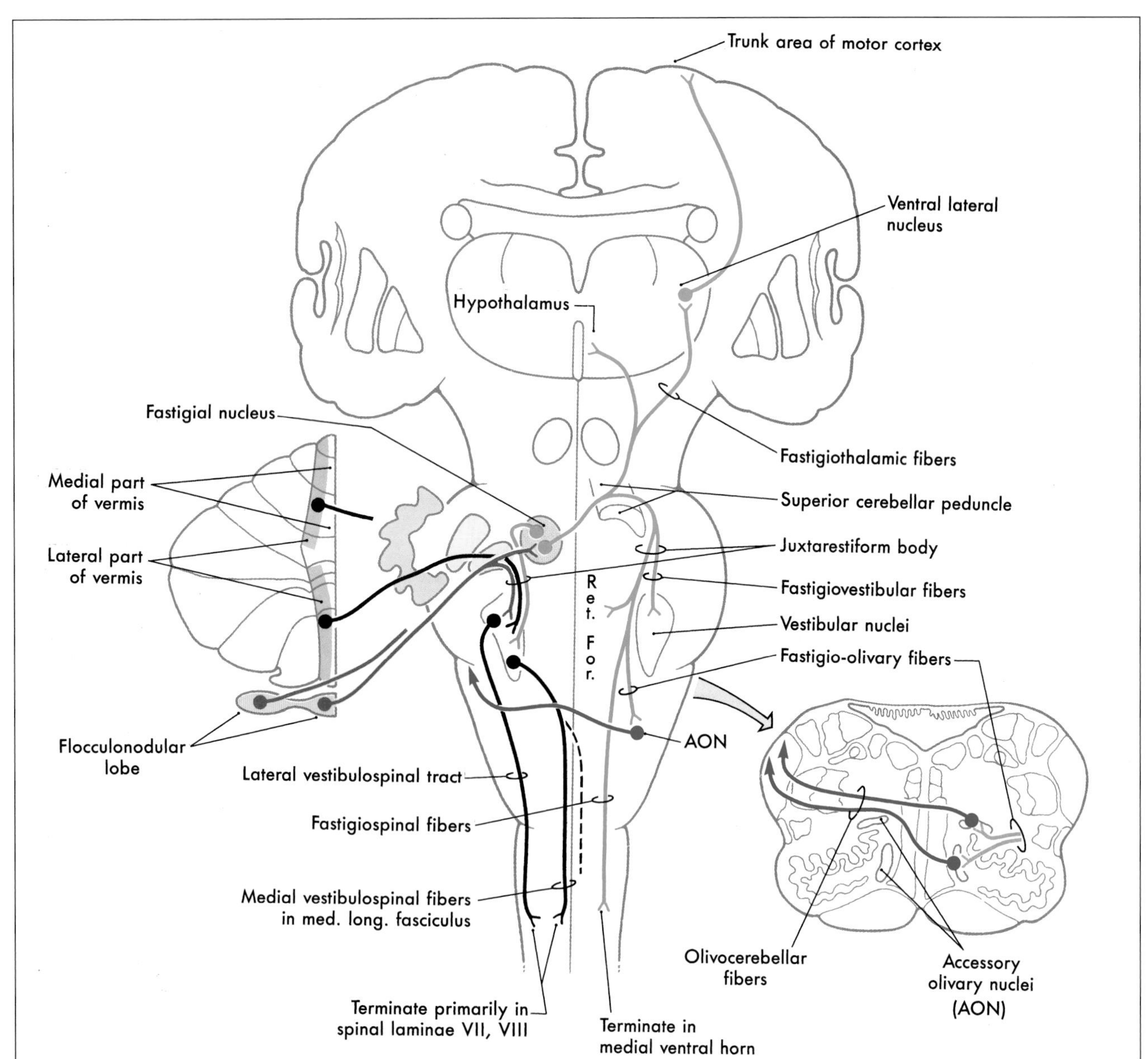

FIGURE 26-13 *Projections of the vestibulocerebellum and of the lateral part of the medial zone through the fastigial and vestibular nuclei. Ret. For., reticular formation.*

nodulus to the fastigial nucleus, and efferent fibers arising in the *fastigial nucleus* (Fig. 26-13). Purkinje cells in the flocculonodular cortex project, via the juxtarestiform body, directly to the ipsilateral vestibular nuclei (*cerebellar corticovestibular fibers*). Other Purkinje cells in the nodulus project into caudal regions of the fastigial nucleus. Both of these projections are inhibitory (GABAergic) pathways. Fastigial neurons provide bilateral excitatory inputs to the vestibular and reticular nuclei (Fig. 26-13). On the ipsilateral side, these axons pass directly through the juxtarestiform body. Fibers passing to the contralateral side cross in the cerebellar white matter and form the *uncinate fasciculus* as they loop over the superior cerebellar peduncle. These fibers enter the vestibular complex via the juxtarestiform bundle.

Vestibulocerebellar Dysfunction The vestibulocerebellum influences posture, balance, and equilibrium through vestibulospinal and reticulospinal projections to extensor motor neurons that control axial and proximal limb muscles. The vestibular nuclei also bilaterally innervate the motor nuclei of cranial nerves III, IV, and VI through fibers that ascend in the medial longitudinal fasciculus.

Damage to the flocculonodular lobe, or to midline structures such as the nodulus and fastigial nucleus, will result in an unsteady lurching gait (*truncal ataxia*) that resembles drunkenness. This instability is also manifest as exaggerated movements of the legs and a *tendency to fall* to the side, forward, or backward. The patient may stand with feet further apart than usual (*wide-based stance*) in an effort to maintain balance. These individuals are *unable to walk in tandem* (heel-to-toe) or to walk on their heels or toes. Midline lesions may also result in a tremor of the axial body and/or head called *titubation*. This tremor can range in amplitude from barely noticeable to so powerful that the patient is unable to sit or stand unsupported. *Nystagmus* is frequently seen, and deficits in pursuit eye movements are also common. In addition, the patient's head may *tilt* or turn to one side, the direction being unrelated to the laterality of the lesion.

Vestibular Connections of the Vermis In addition to the nodulus (of the flocculonodular lobe), most lobules of the vermal zone also have vestibular connections (Fig. 26-13). For example, lateral portions of the vermal cortex receive secondary vestibulocerebellar fibers and project to the ipsilateral vestibular nuclei. Like the nodular cortex of the vestibulocerebellar module, the medial portions of the vermal cortex send cerebellar corticonuclear fibers into the ipsilateral fastigial nucleus (Fig. 26-14). Consequently, the *fastigial nucleus* links vestibulocerebellar cortex *and* portions of the vermal cortex with the vestibular and reticular nuclei of the brainstem. In this respect, the vermal cortex and fastigial nucleus share the task of influencing axial musculature along with vestibulocerebellar and spinocerebellar modules.

Spinocerebellar Module The vermal and intermediate zones receive input mainly via the *dorsal* and *ventral spinocerebellar tracts* and, from the upper extremity, through *cuneocerebellar fibers*. Owing to this predominant input, these zones are collectively called the *spinocerebellum* (sometimes the *paleocerebellum*) (Figs. 26-14, 26-15). Dorsal spinocerebellar and cuneocerebellar fibers enter the cerebellum via the restiform body, whereas ventral spinocerebellar fibers course into the cerebellum in association with the superior cerebellar peduncle. Fibers that enter the vermal zone send collaterals into the *fastigial nucleus*, and those passing into the intermediate zone branch into the *emboliform* and *globose nuclei*.

The output of the spinocerebellum is focused primarily on the control of axial musculature through vermal cortex and fastigial efferents, and on the control of limb musculature through efferents of the globose and emboliform nuclei. Dorsal spinocerebellar and cuneocerebellar fibers inform the cerebellum of limb position and movement. This information is processed in the cerebellum and, through connections with the motor cortex via the thalamus, influences movements of the extremities and muscle tone. Cells in the spinal cord that give rise to ventral spinocerebellar fibers receive primary sensory inputs and are also under the influence of descending reticulospinal and corticospinal fibers. In this respect, ventral spinocerebellar fibers provide afferent signals *and* feedback to the cerebellum regarding motor circuits in the spinal cord.

There are additional inputs to the spinocerebellar cortex. These arise in the contralateral accessory olivary nuclei (*olivocerebellar fibers*), the vestibular nuclei (*secondary vestibulocerebellar fibers*), the contralateral pontine nuclei (*pontocerebellar fibers*), and the reticular nuclei (*reticulocerebellar fibers*). These afferent axons also send collaterals into the fastigial and interposed nuclei.

The outflow of the spinocerebellar module consists of *cerebellar corticonuclear fibers* from vermal and intermediate cortex to *fastigial*, *emboliform*, and *globose nuclei* and of *cerebellar efferent axons* arising in these nuclei (Fig. 26-14, 26-15). Corticonuclear fibers project in a topographic sequence into their respective nuclei on the ipsilateral side. For example, fibers from anterior parts of the vermis enter rostral portions of the fastigial nucleus, whereas those of the posterior vermis project caudally in the same nucleus. In general this pattern is repeated between the intermediate zone and the emboliform and globose nuclei.

As indicated above, the *fastigial nucleus* projects bilaterally to vestibular and reticular nuclei, which, through their spinal projections, influence axial muscles. The fastigial nucleus also projects to (1) the contralateral *medial accessory olivary* nucleus, from which it receives input; (2) the medial areas of the ventral horn in upper levels of the

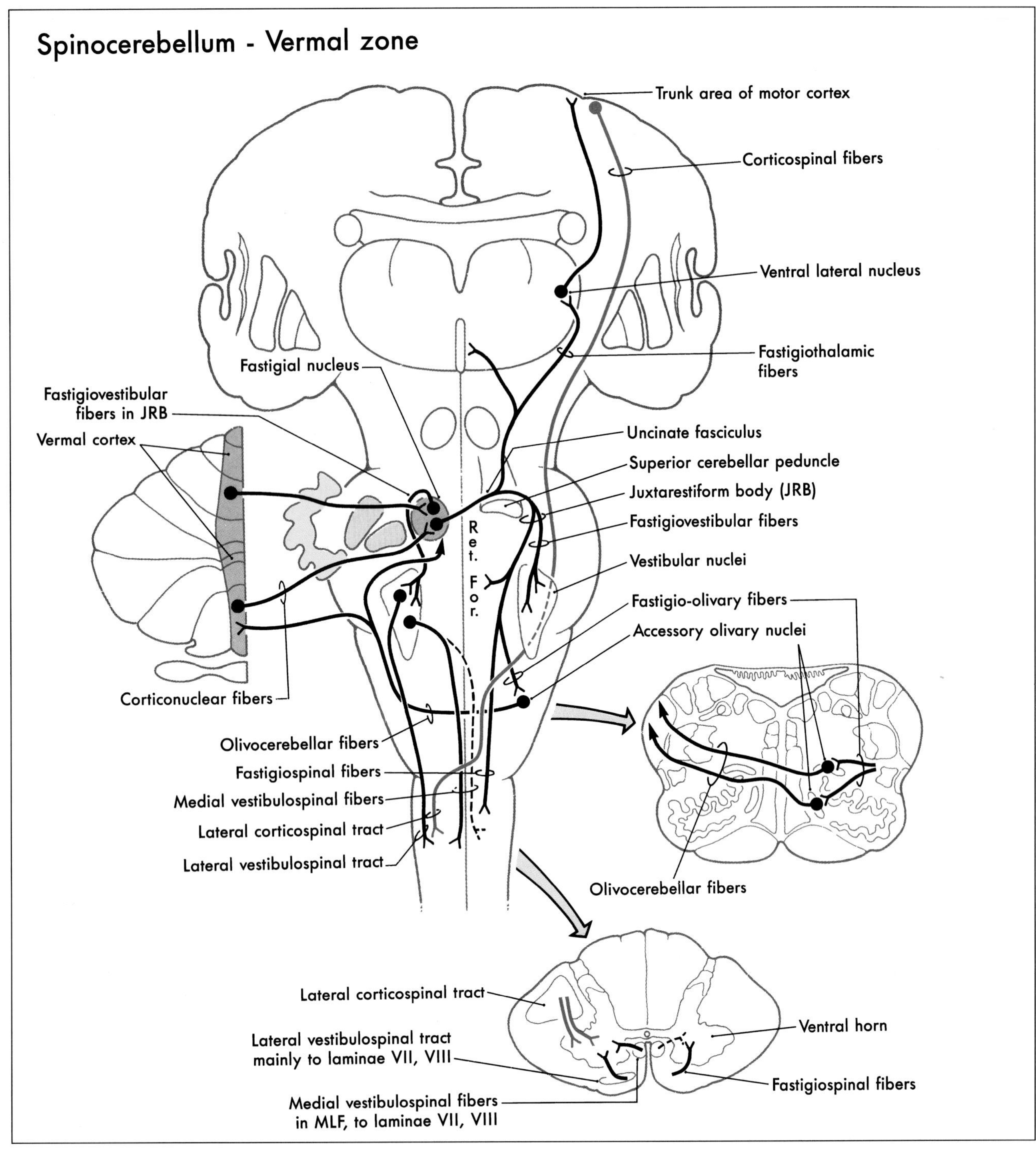

FIGURE 26-14 *Projections of the spinocerebellum (vermal zone) through the fastigial and vestibular nuclei. Ret. For., reticular formation.*

spinal cord as *fastigiospinal fibers*; and (3) the ventral lateral nucleus of the thalamus, which, in turn, projects to trunk regions of the motor cortex (Fig. 26-14).

Axons from the *globose* and *emboliform nuclei* exit the cerebellum via the superior cerebellar peduncle and cross in its decussation (Fig. 26-15). From this point, some of these cerebellar efferent fibers course rostrally to terminate in the magnocellular part of the red nucleus (*cerebellorubral fibers*) and in the ventral lateral nucleus of the thalamus (*cerebellothalamic fibers*). These particular thalamic neurons project mainly to areas of the primary motor cortex. The red nucleus, via *rubrospinal fibers*, and the motor cortex, through *corticospinal fibers*, influence motor neurons in the contralateral spinal cord that control distal limb musculature (Fig. 26-15). Other globose and emboliform efferents travel caudally to terminate in the reticular formation (*cere-*

belloreticular fibers) and in the inferior olivary complex (*cerebello-olivary fibers*). Reticular cells influence spinal motor neurons and project back to the spinocerebellum as *reticulocerebellar fibers*. The globose and emboliform nuclei also receive *olivocerebellar fibers* from the accessory olivary nuclei to which they project (Fig. 26-15).

Damage to spinocerebellar structures is frequently the result of extensions from more medially or laterally located lesions. Consequently, the clinical picture is dominated by deficits characteristic of these medial or lateral regions. The latter is more frequently the case (see below).

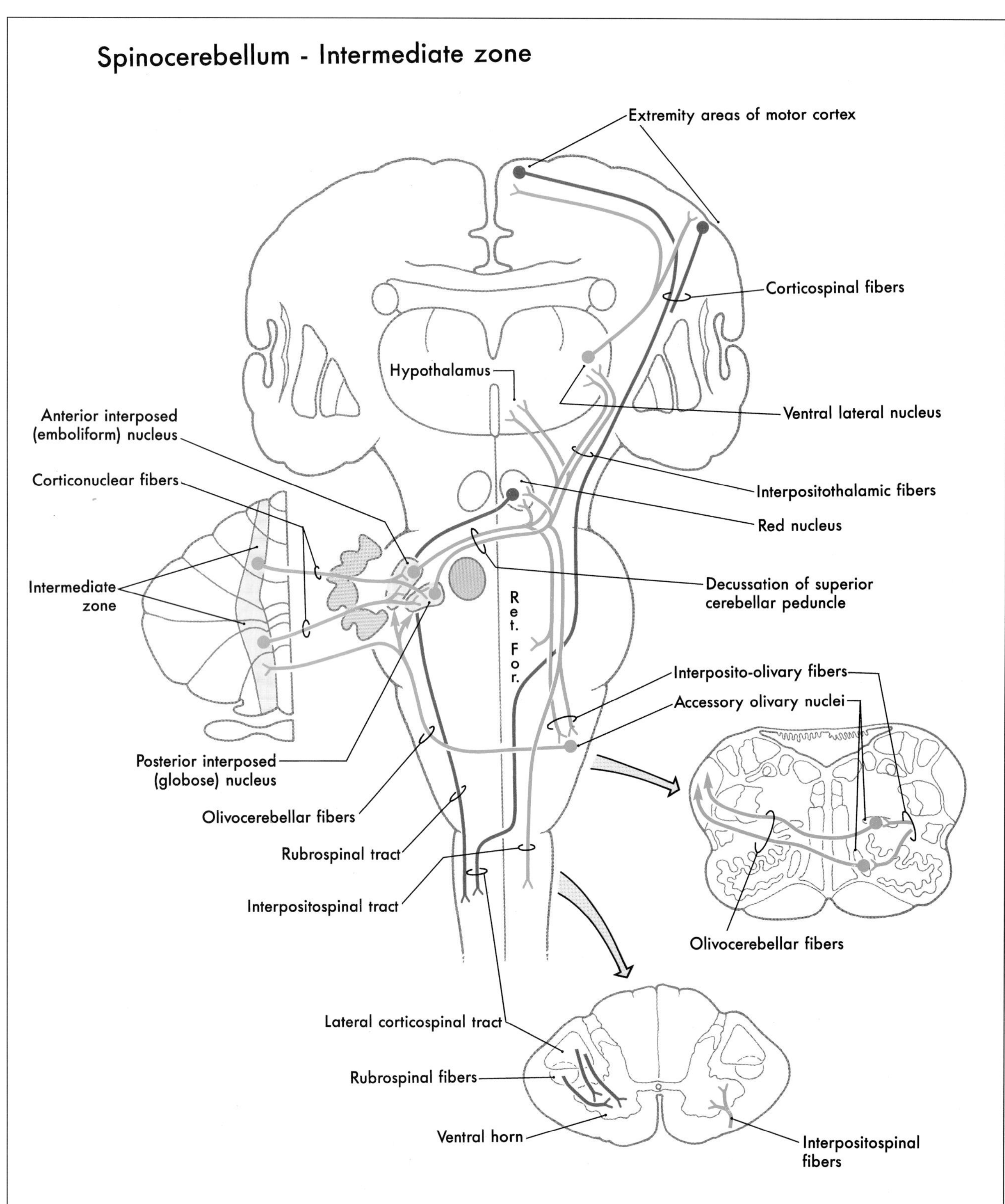

FIGURE 26-15 *Projections of the spinocerebellum (intermediate zone) through the emboliform and globose nuclei. Ret. For., reticular formation.*

Pontocerebellar Module The large lateral zone receives significant input from the *basal pontine nuclei* via a primarily crossed *pontocerebellar fiber* projection. These fibers enter the cerebellum via the middle cerebellar peduncle. Because of this predominant source of afferent fibers, the lateral zone is called the *pontocerebellum* (Fig. 26-16). Since the pontine nuclei receive a major projection from the ipsilateral cerebral cortex (as *corticopontine fibers*), this lateral zone may also be called the *cerebrocerebellum* (or sometimes the *neocerebellum*). Afferent fibers that enter the lateral zone send collaterals into the *dentate nucleus* (Fig. 26-16).

The pontocerebellum functions in the planning and control of precise dexterous movements of the extremities, particularly in the arm, forearm and hand, and in the timing of these movements. Through its connections to motor cortical areas, the dentate nucleus is capable of modulating activity in cortical neurons that project to the contralateral spinal cord.

Another important source of afferents to the pontocerebellum is from the *principal inferior olivary nucleus* (Fig. 26-16). These *olivocerebellar fibers* are exclusively crossed. They enter the cerebellum via the restiform body, send collaterals to the dentate nucleus, and end in the molecular layer as climbing fibers.

The outflow of the pontocerebellar module consists of *cerebellar corticonuclear fibers* from the lateral zone to the dentate nucleus and *cerebellar efferent fibers* originating in the dentate nucleus (Fig. 26-16). As for other cerebellar regions, corticonuclear fibers of the lateral zone are topographically organized; rostral and caudal areas of the zone project to the corresponding portions of the dentate nucleus.

Neurons of the *dentate nucleus* send their axons out of the cerebellum via the superior cerebellar peduncle and through its decussation (Fig. 26-16). The fibers that pass rostrally project mainly to the parvicellular part of the red nucleus (*dentatorubral fibers*) and to the intralaminar and ventral lateral nuclei of the thalamus (*dentatothalamic fibers*). Some neurons of the red nucleus project, as part of the central tegmental tract, to the ipsilateral inferior olivary complex (*rubro-olivary fibers*). At the same time, cells of the ventral lateral nucleus project to wide areas of the motor and premotor cortices. The motor cortex, in turn, projects to the contralateral spinal cord (as *corticospinal fibers*) to influence motor neurons that innervate distal limb musculature (Fig. 26-16). Descending crossed projections from the dentate nucleus pass mainly to the principal olivary nucleus (*dentato-olivary fibers*) and, in limited numbers, to the reticular and pontine nuclei. *Olivocerebellar fibers* arising in the principal nucleus cross the midline and distribute to the cortex of the lateral zone and to the dentate nucleus. There is also feedback to the dentate nucleus via pontocerebellar and reticulocerebellar fibers.

This relationship between the dentate nucleus and movement has been explored in experiments in monkeys. A cooling probe implanted in the cerebellar white matter adjacent to the dentate nucleus halts most of the electrical activity in dentate neurons. This temporarily disconnects the dentate nucleus from its targets without permanently destroying the nucleus. Reversing the cooling results in a resumption of normal neuronal activity. After dentate cooling, the electrical activity of neurons in the primary motor cortex (MI) that signals the initiation of a movement (in response to a visual stimulus) was delayed, as was the execution of the movement itself. This observation indicates that dentate projections, which reach the cortex via the ventral lateral thalamic nucleus, are essential for the initial activation of MI corticospinal neurons at the beginning of a movement.

As a result of the delay of excitatory output from the motor cortex, there are corresponding delays in muscle contraction. For example, the *initial activation* of an agonist muscle (biceps brachii) to a load is slowed and its overall contraction time is longer. Similarly, the activation of the antagonist muscle (triceps brachii) that occurs when the load is removed is also delayed. This observation indicates that the reciprocal pattern of activation in agonists and antagonists that accompanies some movements is dramatically disrupted. Thus, cerebellar output is involved in *timing* muscle activation (and inactivation), as well as influencing the duration of muscle contraction.

Pontocerebellar Dysfunction Before discussing the consequence of lesions affecting the pontocerebellar module, two important points merit emphasis. First, damage that involves *only* the cerebellar cortex rarely results in permanent motor deficits. However, damage to the cortex plus nuclei or to only the nuclei results in a wide range of motor problems. Second, *lesions of the cerebellar hemisphere result in motor deficits on the ipsilateral side of the body because the motor expression of cerebellar injury is mediated primarily through corticospinal and rubrospinal pathways.* In brief, the right lateral and interposed nuclei influence the left motor cortex and red nucleus, which, in turn, project to the right side of the spinal cord. Thus, a lesion in the cerebellum on the right results in deficits on the right side of the body. The exception is a midline lesion, which produces bilateral deficits restricted to axial/truncal parts of the body. Lesions that involve the cerebellar hemisphere frequently affect portions of the lateral and intermediate modules. It is common for these disorders to be categorized as disorders of the *lateral* (or *hemisphere*) *zone* or as *neocerebellar disorders.*

In general, lesions of the lateral cerebellum result in a deterioration of coordinated movement, which is sometimes referred to as a *decomposition of movement* (or *dyssynergia*). This deficit consists of the breakdown of a movement into its individual component parts. There may also be a decrease in muscle tone (*hypotonia*) and in deep tendon reflexes. *Ataxia* involving the extremities generally is also seen in patients with lateral cerebellar lesions. Because of ataxia of the lower extremity, these patients may also have an *unsteady gait* and a tendency to lean or fall to the side of the lesion.

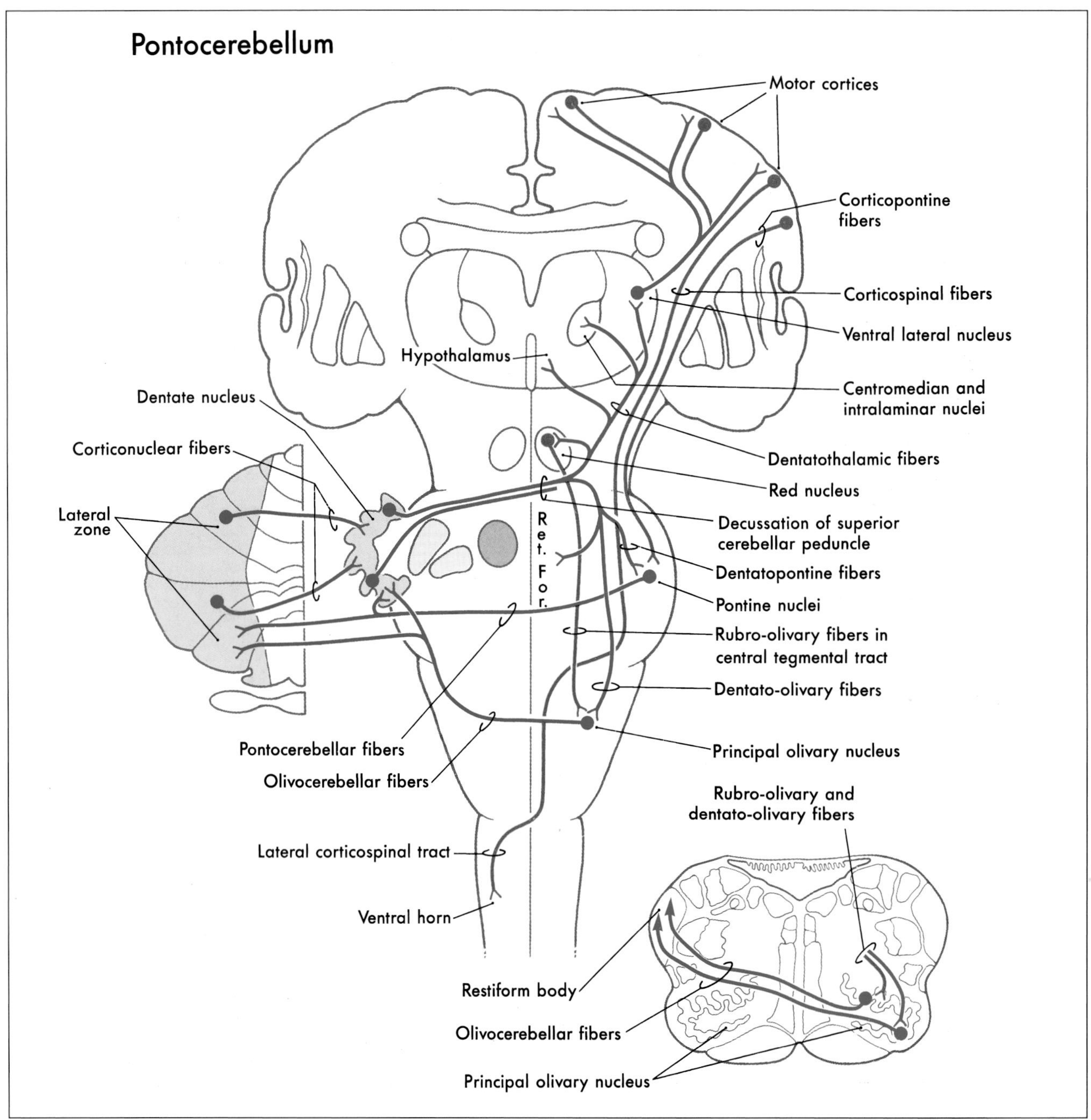

FIGURE 26-16 *Projections of the pontocerebellum (lateral zone) through the dentate nucleus. Ret. For., reticular formation.*

Dysmetria (also called *past-pointing*) is apparent in patients when they attempt to point accurately and/or rapidly to moving or stationary targets. The patient may reach past the target (*hypermetria*) or fall short of the target (*hypometria*).

Tremor is a consistent finding in patients with lateral cerebellar lesions. A *kinetic tremor*, commonly called an *intention tremor*, is evident when the patient performs a voluntary movement and is most obvious as the endpoint, or target, is approached. This deficit is commonly seen when the patient extends the arm and then attempts to touch the index finger to the nose. At rest there is little or no tremor, but as the finger approaches the nose the tremor is markedly accentuated. This finding is opposite to that seen in patients with Parkinson's disease, whose tremor is evident at rest (*resting tremor*) but largely diminishes during a voluntary movement. Patients with cerebellar lesions may also demonstrate a *static tremor*. This tremor is manifest when the patient stands with the arms extended (muscles contracted against gravity). There is rhythmic movement of the shoulders that also involves the upper extremities.

Awkward performance of rapid alternating movements, such as supinating and pronating the hand against the thigh, is called *dysdiadochokinesia*. The patient may also be unable to perform rhythmic movements. This deficit is demonstrated by asking the patient to rapidly tap the table three times with the index finger, pause two seconds, tap three times, etc. For patients with lateral cerebellar lesions, this task will be difficult or impossible.

Other lateral zone deficits include rebound phenomena, dysarthria, and ocular motor dysfunction. The *rebound phenomenon* (or *impaired check*) is an inability of agonist and antagonistic muscles to adapt to rapid changes in load. For example, if the patient is asked to push against the physician's hand and the hand is then unexpectedly removed, the patient's arm will overshoot beyond the point where it would normally halt. Tremors or oscillations may also be evident as the arm returns to its starting point. Patients with *dysarthria* have slurred, garbled speech that may also be alternatively slow and/or staccato in nature (*scanning speech*). This is a motor problem (not an aphasia) because the patient is still able to use words and grammar correctly. Characteristic ocular motor dysfunctions seen in patients with lateral cerebellar lesions are *nystagmus* and abnormalities of target-directed eye movements. Nystagmus most commonly presents as abnormal horizontal eye movements that consist of a slow conjugate movement away from the side of the lesion and a fast movement toward the side of the lesion. This abnormality is opposite to that seen in lesions of the vestibular receptors, primary sensory fibers, and nuclei. In other cases, the velocity of these conjugate movements may be the same in both directions (*pendular nystagmus*). Abnormal target-directed eye movements may present as an inability to follow a slowly moving target (disturbed pursuit movements) or difficulty in maintaining fixation on a stationary target.

CEREBELLAR INFLUENCE ON VISCEROMOTOR FUNCTIONS

The cerebellum receives input from the *solitary* and *dorsal motor vagal nuclei* and from a number of nuclei in the *hypothalamus*. These areas are directly involved in the control and modulation of a variety of visceromotor functions. The hypothalamus, in addition to having direct connections with the cerebellar cortex and nuclei, also projects to brainstem and spinal nuclei that are involved in the regulation of visceral functions.

Neurons in several hypothalamic areas and nuclei project to the cerebellar cortex and nuclei (*hypothalamocerebellar fibers*). The cerebellar nuclei, in turn, send a primarily crossed projection to the hypothalamus (*cerebellohypothalamic fibers*) via the superior cerebellar peduncle (Figs. 26-14, 26-16). Through these reciprocal connections the cerebellum may receive visceral input and influence neurons that control visceral functions.

Visceral deficits related to cerebellar lesions are rarely reported for two reasons. First, the somatomotor deficits seen in such cases are overwhelmingly diagnostic and there is no need to look further. Second, cerebellar lesions may cause an increase in intracranial pressure with resultant pressure on the medulla. Consequently, it is difficult to separate visceral deficits that relate to the cerebellar lesion from those that relate to pressure on the medulla.

In some situations, however, visceral deficits can be related directly to the cerebellar lesion. For example, deficits of this type occurred in a patient who had an occlusion of branches of the left superior cerebellar artery with resultant damage to the cerebellar nuclei on that side. There was no indication of increased intracranial pressure. This patient had characteristic somatomotor tremors in the left arm and leg *during attempted movement* on that side and showed two distinct visceral responses *that occurred concurrent with the somatomotor tremor*, but not before or after. First, the patient's pupils dilated *during* the tremor. Second, the skin on the patient's face flushed and felt warm to the touch *during* the tremor. Immediately upon cessation of the attempted movement (and resulting tremor) of the left hand, the patient fanned his face with his right hand and complained that "it's hot in here." In other cases, small lesions restricted to the fastigial nucleus result in decreases in heart rate and blood pressure. These observations suggest that the cerebellum is actively involved in the regulation of visceral function concurrent with its classically recognized influence in the somatomotor sphere.

THE CEREBELLUM AND MOTOR LEARNING

The cerebellum is unequivocally involved in the learning of a variety of relatively simple reflexive motor behaviors. It is extremely difficult to produce specific modifications in certain reflexes without this structure. Such modifications include the adaptation of the vestibulo-ocular reflex and the classic (Pavlovian) conditioning of reflexes evoked by aversive stimuli, such as the eyeblink and withdrawal reflexes.

Although the cerebellum is believed to be involved in the acquisition of voluntary, complex motor skills, its role in this process is not well understood. Initial studies demonstrated that normal subjects performed better after practicing a new task than did patients with cerebellar damage. This finding was initially ascribed to the presence of a learning deficit in these patients. However, recent experiments that examined the *rate* at which a new motor skill is acquired revealed that cerebellar patients can learn to perform new movements even though the *quality* of the performance is affected by their deficit. These studies, together with animal experiments showing that volitional limb movements can be acquired during inactivation of the cerebellar nuclei, confirm that new (and complex) motor tasks can be learned despite impairment of cerebellar function.

Under normal conditions, the cerebellum plays an important role in the *acquisition* component of motor learning. For example, in animals, complex movements learned during the inactivation of the cerebellar nuclei are more variable. In other studies, imaging of the living brain has revealed that specific regions in the cerebellum are activated during the learning of novel movements. In addition, the acquisition of complex behaviors in animals is associated with increased modulation of cells in the cerebellar nuclei when the animal first learns to perform the task correctly and consistently on successive trails. Together, these data support the view that the cerebellum is actively involved in the acquisition of novel movements.

The participation of the cerebellum in the storage of memory engrams required for the recall of previously learned movements remains controversial. It has been proposed that the climbing fiber input to the dendrites of Purkinje cells can induce plastic changes that modify the responsiveness of these neurons to specific inputs mediated by parallel fibers. However, the relevance of this mechanism to normal physiologic conditions continues to be debated.

There is evidence that memory storage sites also exist outside the cerebellum. For example, it is known that modification of the vestibulo-ocular reflex (see Chapter 21) involves persistent changes in synaptic transmission in the vestibular nuclei. However, some scientists contend that such modifications also occur within the cerebellum when this reflex is changed.

Studies that examined the basis for the classic conditioning of the eyeblink and withdrawal reflexes have implicated the cerebellum as the critical storage site for the plastic changes established during this process. This argument is not yet supported by a satisfactory body of evidence. Some researchers contend that sites outside the cerebellum are involved in this storage process. However, it is unclear which locations are most important, because the cerebellar nuclei are so critical for the *performance* of these reflexes once they are learned. As a consequence, learning may appear to be deficient only because of the difficulty in performing the desired movement.

The issue of memory storage for volitional movements is clearer. The cerebellum is not essential for the retention of previously learned complex motor behaviors. The patterns of movement required to perform complex volitional motor tasks can be recalled during the inactivation of the ipsilateral dentate and interposed nuclei.

In summary, certain reflexes cannot be modified in the absence of the cerebellum, and lesions of this structure impair the *performance* of certain previously learned behaviors. This structure plays an important role in the *acquisition of several motor behaviors including skilled volitional movements*, although the nature of its contribution has not been characterized. The cerebellum's participation in memory storage for certain reflex behaviors is unclear, although it is not considered to be an essential storage site for engrams related to complex volitional movements.

SOURCES AND ADDITIONAL READINGS

Bloedel JR, Dichgans J, Precht W: *Cerebellar Functions*. Springer Verlag, New York, 1985.

Brooks VB, Thach WT: Cerebellar control of posture. In Brooks VB (ed): *Handbook of Physiology*, Section 1. *The Nervous System*, Vol II. *Motor Control*, Part 2. American Physiological Society, Bethesda, Maryland, 1981.

Dietrichs E, Haines DE, Roste GK, Roste LS: Hypothalamocerebellar and cerebellohypothalamic projections—circuits for regulating nonsomatic cerebellar activity. Histol Histopathol 9:603–614, 1994.

Gilman S, Bloedel JR, Lechtenberg R: *Disorders of the Cerebellum, Contemporary Neurology Series*, Vol 21. FA Davis, Philadelphia, 1981.

Haines DE, Patrick GW, Satrulee P: Organization of cerebellar corticonuclear fiber systems. Exp Brain Res Suppl 6:320–371, 1982.

Hore J, Flament D: Evidence that a disordered servo-like mechanism contributes to tremor in movements during cerebellar dysfunction. J Neurophysiol 56:123–136, 1986.

Ito M: *The Cerebellum and Neural Control*. Raven Press, New York, 1984.

Mihailoff GA: Identification of pontocerebellar axon collateral synaptic boutons in the rat cerebellar nuclei. Brain Res 648:313–318, 1994.

Ojakangas CL, Ebner TJ: Purkinje cell complex and simple spike changes during a voluntary arm movement learning task in the monkey. J Neurophysiol 68:2222–2236, 1992.

Robertson L: Organization of climbing fiber representation in the anterior lobe. In King JS (ed): *New Concepts in Cerebellar Neurobiology*. Alan R Liss, New York, 1987, pp 281–320.

Thach WT, Goodkin HP, Keating JG: The cerebellum and the adaptive coordination of movement. Annu Rev Neurosci 15:403–442, 1992.

Welker W, Blair C, Shambes GM: Somatosensory projections to cerebellar granule cell layer of giant bush-baby, Galago crassicaudatus. Brain Behav Evol 31:150–160, 1988.

CHAPTER TWENTY-SEVEN

Visual Motor Systems

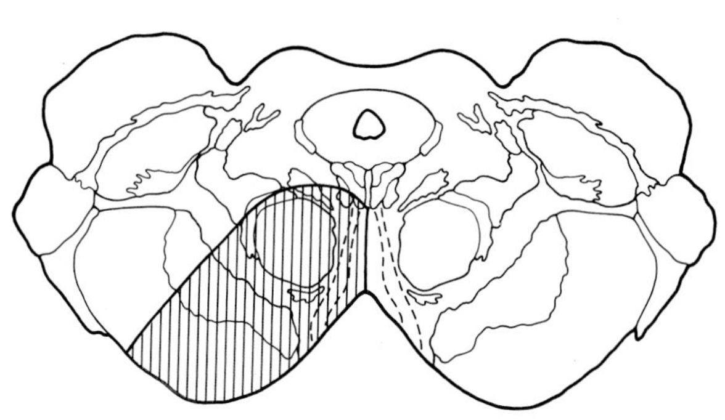

P. J. MAY • J. J. CORBETT

All animals use their sensory organs to scan the environment in search of information. Often these organs, for example the eyes, are oriented toward relevant targets to allow closer inspection. This *orienting behavior* is exhibited by creatures from honey bees to human beings. However, human eyes have a fovea, a small portion of the central retina that has exquisite visual sensitivity. Accurately directing the fovea to targets of interest represents a crucial orienting behavior in humans. Extraocular muscles orient our highly mobile eyes, and the *oculomotor* (or ocular motor) system controls these muscles. It is one of several *visual motor* systems that support the function of visual sensation.

OVERVIEW

The oculomotor system includes *gaze systems* that redirect the eyes to each new target. There are three basic types of "targeting" movements: (1) *saccades*, rapid movements that direct the eyes to each new target; (2) *smooth pursuit*, slower movements that allow the eyes to follow moving targets; and (3) *vergence movements*, that adjust for target distance by changing the angle between the eyes. Vergence is coupled with changes in the curvature of the *lens* and the size of the *pupil* that focus the target image on the fovea. Saccades and smooth pursuit are *conjugate* movements in which the eyes move in the same direction, often with accompanying movements of the head and body. Vergence movements are *disconjugate*.

Visual motor systems also mediate a set of reflex actions. *Compensatory reflexes* keep the eyes on target despite body movements. Sensory inputs from the vestibular and visual systems tell the brain that the body is in motion. During movement, the *vestibulo-ocular reflex* compensates for acceleration, which is sensed by the vestibular labyrinth, whereas the *optokinetic reflex* compensates for velocity, which is indicated by movement of the whole visual field. Visual motor systems also compensate for the amount of light falling on the retina. The *pupillary light reflex* maintains the level of retinal illumination within the working range of the photopigments in the photoreceptor cells (rods or cones). Finally, the *blink reflex* protects the eye.

Disturbances of the visual motor systems are common and often produce the first symptoms recognized by a patient. Understanding these ocular signs provides for timely and effective diagnosis. For instance, *strabismus* is a defect in which the eyes are misaligned. Left untreated, the brain reacts to the constant diplopia (double vision) by ignoring the input from one eye and failing to focus it (*amblyopia*) and eventually, failing even to orient it. However, amblyopia can be avoided through early treatment of strabismus.

PERIPHERAL STRUCTURES

Extraocular Muscles The eye is moved in the orbit by six extraocular muscles (Fig. 27-1). These muscles produce movements in the horizontal plane (left and right) around a vertical axis, movements in the vertical plane (up and down) around a horizontal axis, and torsional movements (clockwise and counterclockwise) around an axis running through the center of the pupil to the fovea. There are two pair of recti muscles, with the members of each antagonistic pair arranged opposite one another on the globe, and a single pair of oblique muscles, which also act as antagonists. For horizontal eye movements, the *medial rectus muscle* rotates the eye toward the nose (*adduction*) and the *lateral rectus muscle* rotates the eye toward the temple (*abduction*). For vertical eye movements, the primary action of the *superior rectus* is to rotate the eye upward (*elevation*), and the primary action of the *inferior rectus* is to rotate the eye downward (*depression*). The direction of pull of the *superior oblique muscle* is modified because its tendon passes through a loop of connective tissue, the *trochlea*, on the medial wall of the bony orbit. Its insertion on the globe is caudal to that of the superior rectus. As a consequence, the actions of the superior oblique are *intorsion*, depression, and abduction. On the other hand, the *inferior oblique muscle actions* are *extorsion* and elevation, as well as abduction.

The extraocular muscles are supplied by three cranial nerves. The *abducens nerve* (cranial nerve VI) supplies the lateral rectus muscle, the *trochlear nerve* (cranial nerve IV) supplies the superior oblique muscle, and the rest are innervated by the *oculomotor nerve* (cranial nerve III). The extraocular muscles are striated and contain fibers adapted to produce extremely high velocities and nearly constant tension. The trigeminal nerve carries sensory information from the extraocular muscles. This proprioceptive signal appears to be critical for normal development of *stereoscopic vision* (perception of three-dimensional space), but visual feedback is the primary source of information on eye movement accuracy.

Intraocular Muscles The eyes contain three intrinsic smooth muscles (Fig. 27-2). The *ciliary muscle* changes the curvature of the lens in order to bring visual targets into focus on the retina. The *sphincter* (or *constrictor*) *pupillae muscle* and *dilator pupillae muscle* control the size of the pupils in an antagonistic fashion, to regulate the amount of light entering the eyes and the depth of field.

The ciliary muscle, found in the *ciliary body*, is connected to the lens by the *suspensory ligaments*, or *zonule of Zinn*. These fine connective tissue threads resemble the spokes of a bicycle wheel. The action of the ciliary muscle changes the shape of the lens (via the zonule) to adjust its refractive state and focus the image on the retina (Fig. 27-2E,F). These changes are termed *lens*

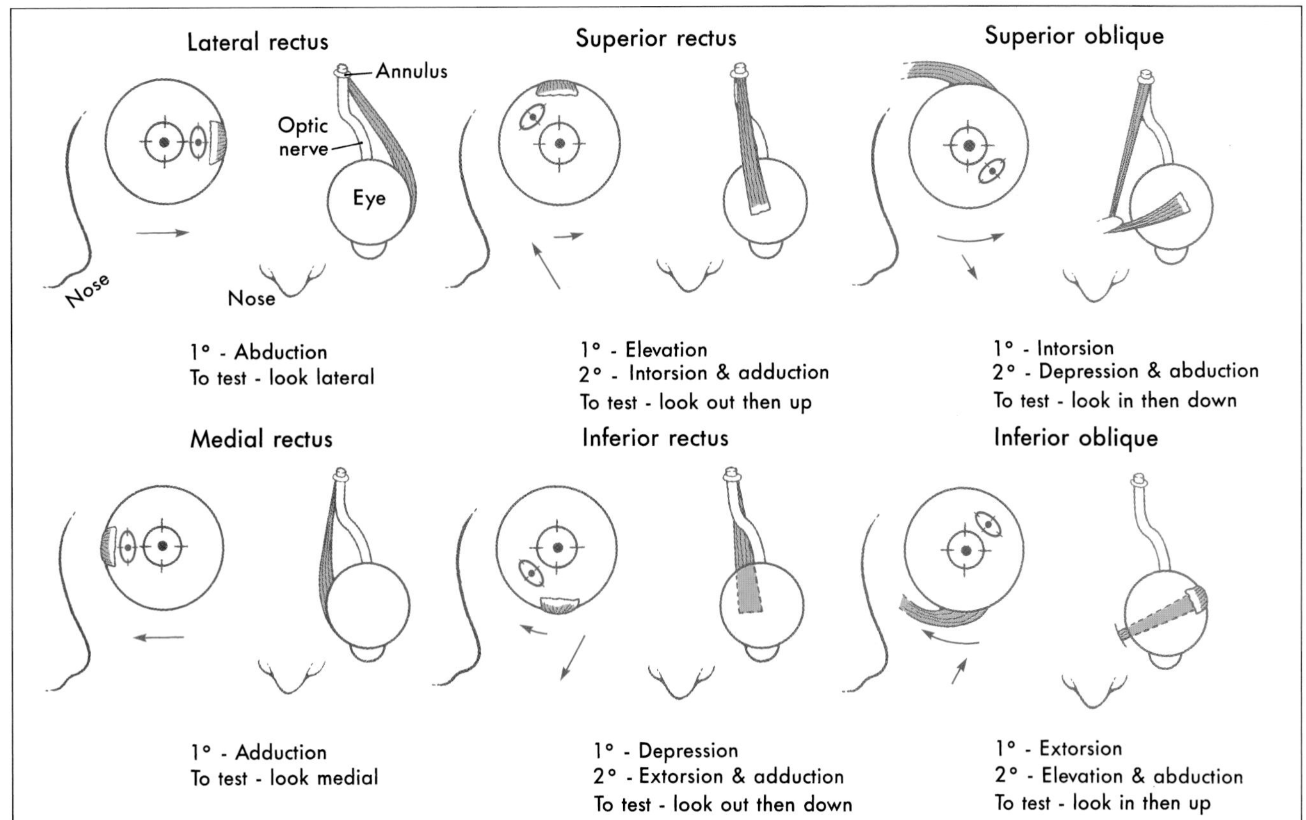

FIGURE 27-1 *The six extraocular muscles of the left eye in frontal (left) and dorsal (right) views. The primary (long arrow) and secondary (short arrow) actions of each muscle and the position of the pupil after the movement (red oval) are indicated on the frontal view. These muscles originate at the common tendon (annulus of Zinn), with the exception of the inferior rectus, which originates from the orbit's nasal wall. Recti muscles insert in front of the equator of the globe, and the oblique muscles insert behind it. Secondary actions are important for clinical evaluation of muscle function. In these cases, the eye is first rotated to align its axis with that of the muscle; for example, for the inferior oblique, the eye is first adducted, then elevated.*

accommodation. Constriction of the ciliary muscle is produced by activation of cholinergic parasympathetic fibers from the ciliary ganglion. With age, the lens grows less elastic, so that the actions of the ciliary muscle have less effect on refraction. This loss of accommodation produces a blurring of near vision termed *presbyopia*. *Myopia*, a loss of distant acuity, generally appears at an early age and may be due to genetic or environmental factors.

The sphincter pupillae is a ring-shaped muscle that lies along the pupillary margin (Fig. 27-2). It contracts in response to activation of cholinergic parasympathetic fibers from the ciliary ganglion, to constrict the pupil (*miosis*). The dilator muscle is radially arranged, so that its action folds the iris and draws open the pupil (*mydriasis*). The dilator is activated by adrenergic sympathetic postganglionic fibers from the superior cervical ganglion.

Eyelid The eyelid is controlled by the *levator palpebrae superioris*, the *orbicularis oculi*, and the tarsal or *Müller's muscles* (Fig. 27-3). The levator palpebrae is supplied by cranial nerve III. It originates with and travels parallel to the superior rectus but continues forward to insert into the upper lid. The levator holds the eyelid up when the eyes are open, and it functions in concert with the superior rectus, increasing the elevation of the lids when the eyes look up. The orbicularis oculi, supplied by cranial nerve VII, closes the eyes by depressing the upper lid and elevating the lower lid. The tarsal muscles are small smooth muscles at the edge of the bony orbit. They are supplied by postganglionic sympathetic fibers and help keep the lids open.

CENTRAL STRUCTURES

Oculomotor Nucleus The oculomotor nucleus (Fig. 27-4) lies near the midline in the ventral periaqueductal gray of the rostral midbrain. Beneath it lie the fibers of the *medial longitudinal fasciculus*, many of which synapse within the oculomotor nucleus. Axons from motor neurons in the oculomotor nucleus pass either medial to or through the red nucleus and exit the midbrain just

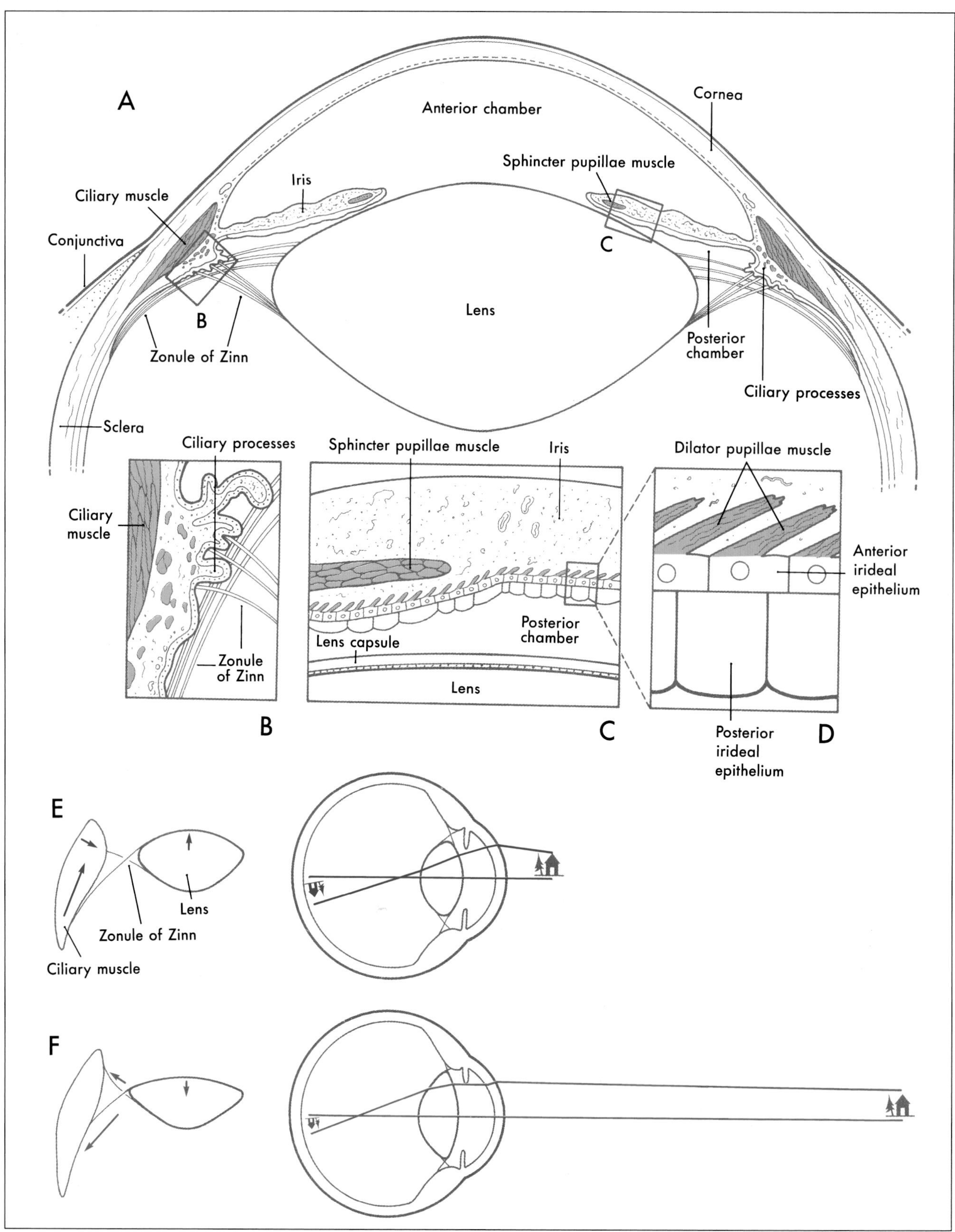

FIGURE 27-2 *The anterior segment of the eye showing the intrinsic eye muscles and optical components (A). The zonule of Zinn (suspensory ligament) extends from the lens capsule and inserts into the epithelium covering the ciliary muscle (B). The pupillary sphincter and dilator muscles are separate muscles in the iris (C). The latter is actually made up of the myoid processes of the anterior irideal epithelium, as shown in D. The mechanism (left) and effects (right) of accommodation (E,F) are shown. When looking at a nearby target (E), the ciliary muscle contracts (arrows), releasing tension in the zonule and allowing the anterior surface of the lens to round up (arrow) owing to its own elasticity. When looking at a distant target (F), the ciliary muscle relaxes, and the unopposed tension in the zonule (arrows) flattens the lens (arrow).*

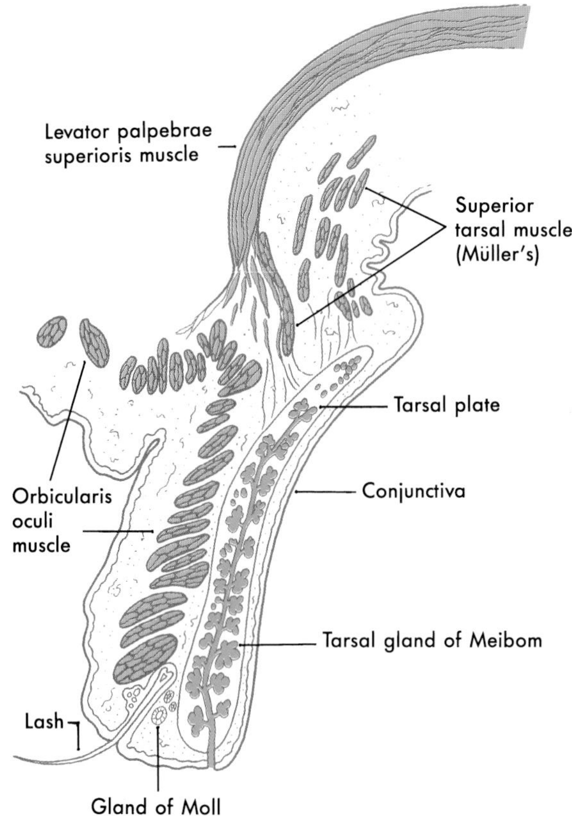

FIGURE 27-3 *The structure of the upper eyelid.*

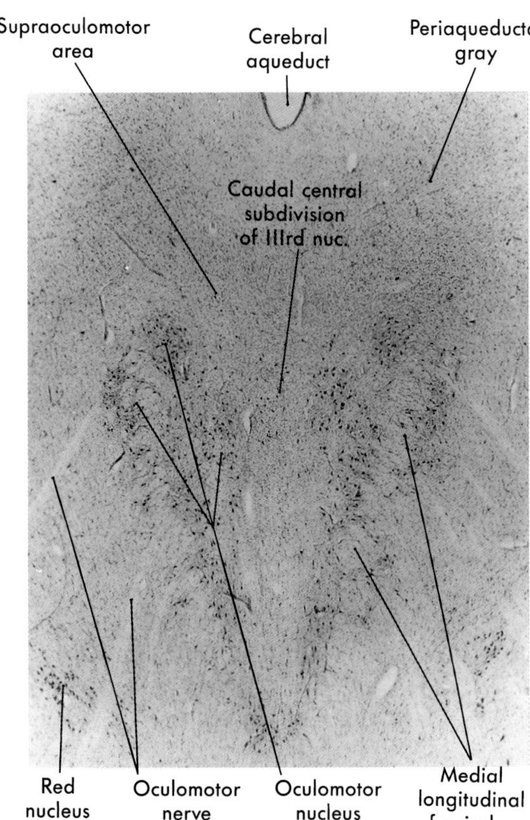

FIGURE 27-4 *The human oculomotor nucleus and adjacent structures in a Nissl-stained section.*

medial to the crus cerebri. These structures are supplied by paramedian branches of the basilar artery and the proximal part of the posterior cerebral artery (P_1 segment). Consequently, vascular lesions in this region produce oculomotor deficits in combination with other symptoms (Table 27-1).

The third cranial nerve passes along the wall of the cavernous sinus and enters the ipsilateral orbit by way of the superior orbital fissure. Branches supply the superior rectus and levator palpebrae muscles, the inferior rectus and inferior oblique muscles, and the medial rectus muscle. The motor neurons supplying each of these individual muscles form rostrocaudally oriented columns within the nucleus (Fig. 27-5). The motor neurons supplying the levator palpebrae superioris muscle form a separate dorsal midline subnucleus called the *caudal central subdivision*.

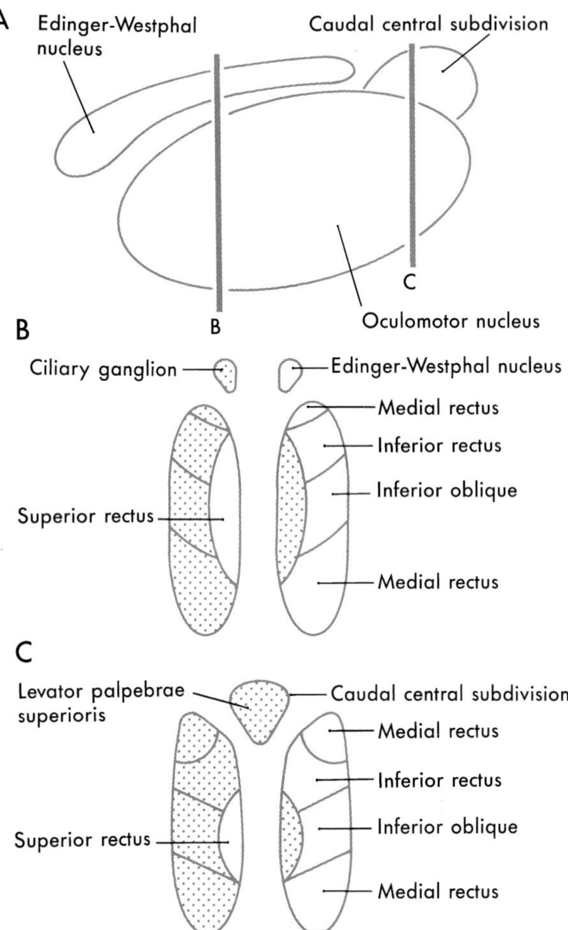

FIGURE 27-5 *Lateral view of the oculomotor and related nuclei (A). The motor neuron pools in the nucleus are shown in frontal sections through anterior (B) and posterior (C) parts of the nucleus. The targets of each subdivision are indicated in B and C. As indicated by the stippled and clear areas, these motor neurons project ipsilaterally with the exception of the contralaterally projecting superior rectus motor neurons and the bilaterally distributed levator motor neurons. The axons of contralateral motor neurons cross immediately to join the ipsilateral oculomotor nerve. Thus, the oculomotor nerve projects entirely to ipsilateral muscles.*

TABLE 27-1

SUMMARY OF LESIONS OF THE EXITING OCULOMOTOR NERVE

STRUCTURES INVOLVED	DEFICITS
Weber's syndrome:	
Oculomotor nerve	Ipsilateral oculomotor palsy and muscle atrophy, ptosis, mydriasis
Corticospinal and corticobulbar fibers in the crus cerebri	Contralateral hemiparesis of arm and leg, contralateral paralysis of lower face and tongue
Claude's syndrome:	
The structures involved in Weber's syndrome plus red nucleus and cerebellothalamic fibers	The deficits of Weber's syndrome, plus contralateral tremor and cerebellar ataxia

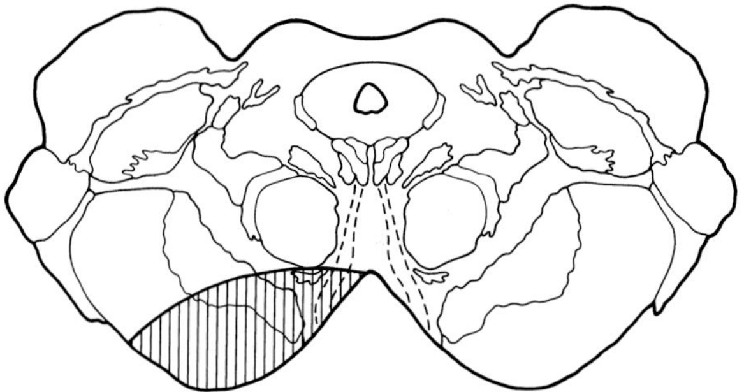

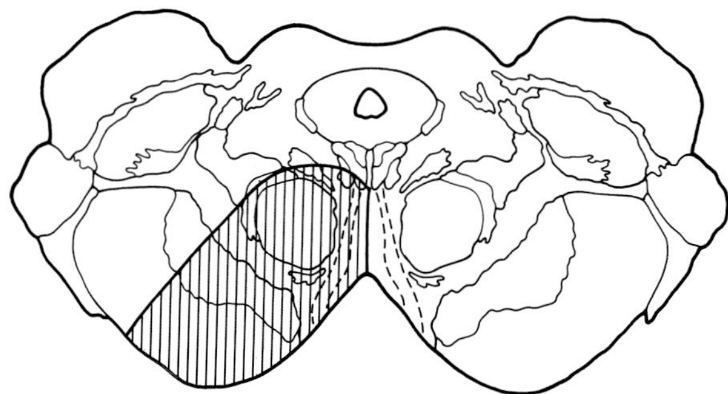

Edinger-Westphal Nucleus The nucleus of *Edinger-Westphal* contains the cholinergic, preganglionic parasympathetic motor neurons that control lens accommodation and pupillary constriction. In humans, the Edinger-Westphal nuclei form a pair of cell columns that lie near the midline, dorsal to the oculomotor nucleus (Fig. 27-5). The preganglionic fibers travel with the ipsilateral oculomotor nerve and synapse in the *ciliary ganglion*. Cholinergic, postganglionic motor neurons send their axons to the globe via the *short ciliary nerves*. These supply the ciliary muscle and pupillary constrictor, with the great majority supplying the former. Thus, in Weber's syndrome (Table 27-1) or other lesions of the oculomotor nerve, loss of the preganglionic fibers may result in ipsilateral *mydriasis* (dilation of the pupil) and paralysis of accommodation.

Trochlear Nucleus The trochlear nucleus is a small, ovoid group of neurons nestled in the medial longitudinal fasciculus of the caudal midbrain (Fig. 27-6). These motor neurons supply the contralateral superior oblique muscle. The axons forming the trochlear nerve arch dorsally and caudally around the periaqueductal gray, cross the midline in the anterior medullary vellum, and exit the dorsal surface of the brainstem at the base of the inferior colliculus. This nerve then courses ventrally, hugging the surface of the midbrain. It passes rostrally along the wall of the cavernous sinus, before entering the orbit via the superior orbital fissure, to innervate the superior oblique muscle.

Abducens Nucleus The abducens nucleus is a spherical cell group located in the facial colliculus, adjacent to the internal genu of the facial nerve in the caudal pons (Fig. 27-7; see also Fig. 12-10). Abducens fibers pass caudoventrally to exit near the midline at the pontomedullary junction. En route, they pass adjacent to the medial lemniscus and corticospinal tract. Owing to this relationship, loss of paramedian branches of the basilar artery compromises both the corticospinal tract and the exiting abducens fibers, resulting in a *contralateral hemiplegia* and paralysis of abduction in the *ipsilateral* eye (*Foville's syndrome*). This pairing of motor symptoms is

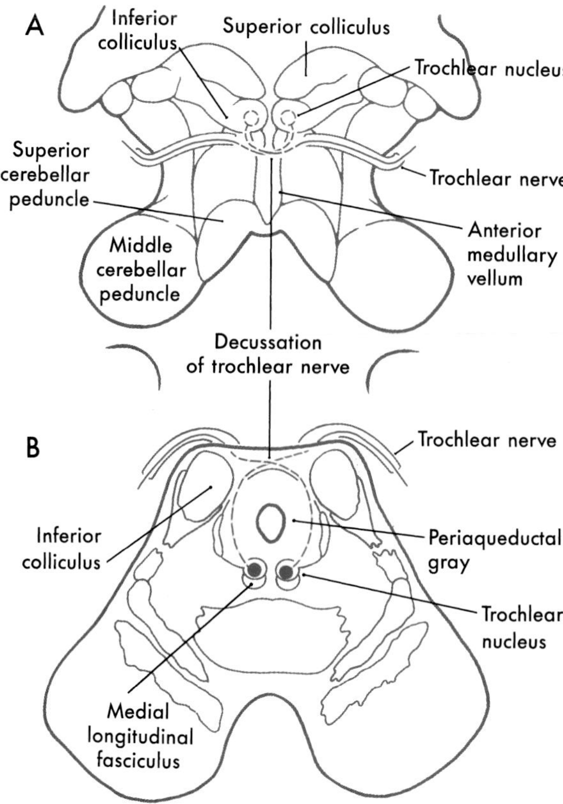

FIGURE 27-6 *The trochlear nucleus and nerve shown in a dorsal view of the midbrain (A) and in a coronal section through the nucleus (B).*

called an *alternating hemiplegia* and can also occur in association with cranial nerves III and XII. After exiting the brainstem, the abducens nerve (cranial nerve VI) enters the dura and ascends the clivus. It then passes beneath the petroclinoid ligament (Dorello's canal) to enter and pass through the cavernous sinus. Cranial nerves III, IV, VI, and branches of V may be damaged in this sinus, singly or in combination, by pituitary tumors or carotid aneurysms. Cranial nerve VI supplies the lateral rectus muscle, so damage to this nerve in isolation will paralyze abduction ipsilaterally (Fig. 27-8).

Abducens Internuclear Neurons Conjugate eye movement is achieved in the horizontal plane by a pathway from the abducens nucleus to the contralateral medial rectus subdivision of the oculomotor nucleus (see Fig. 27-12). Inputs to the abducens nucleus not only supply motor neurons, but also drive *abducens internuclear neurons* found in this nucleus. Abducens internuclear neurons transmit signals to the medial rectus motor neurons on the contralateral side via the medial longitudinal fasciculus (MLF). In this way the lateral rectus of one eye works in tandem with the medial rectus of the other eye. Lesions in the MLF *between the abducens and oculomotor nuclei* damage the abducens internuclear neuron axons, producing *internuclear ophthalmoplegia*. In this situation, the lateral rectus contracts appropriately during horizontal eye movements, but the medial rectus in the opposite eye does not. However, because the oculomotor nerve and nucleus are intact, no deficits are present upon convergence. The presence of internuclear neurons also explains why the symptoms of abducens nerve lesions differ from those of abducens nucleus lesions. In the latter, the action of the contralateral medial rectus muscle is impaired during conjugate horizontal movements, in addition to the expected paralysis of the ipsilateral lateral rectus muscle.

Sympathetic Supply to the Orbit The sympathetic innervation of the orbital contents is from the ipsilateral *superior cervical ganglion* (Fig. 27-9). These adrenergic postganglionic fibers enter the cranium on the internal carotid artery. In the cavernous sinus, they run briefly with cranial nerve VI and then join cranial nerves III and V. Sympathetic fibers traveling on cranial nerve III branch to the levator palpebrae muscle and supply the superior tarsal muscle. Fibers traveling with the trigeminal nasociliary nerve exit as the *long ciliary nerves* to supply the eye, including the dilator pupillae muscle.

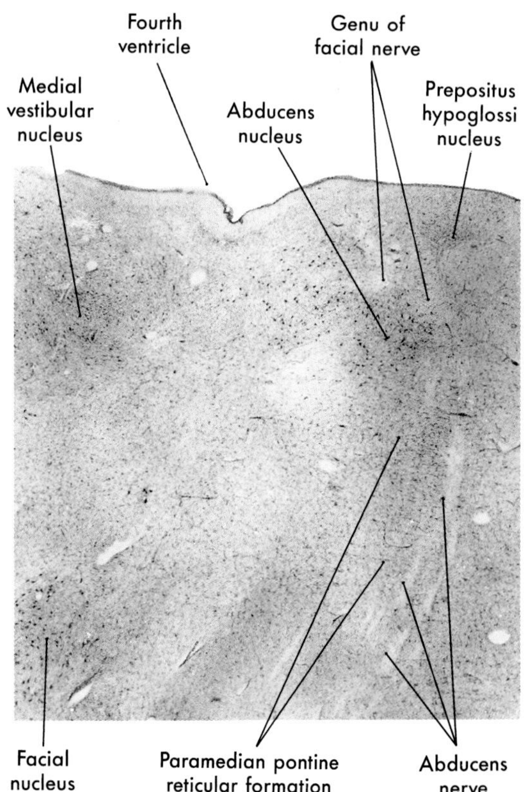

FIGURE 27-7 *The human abducens nucleus and adjacent structures, in a Nissl-stained section.*

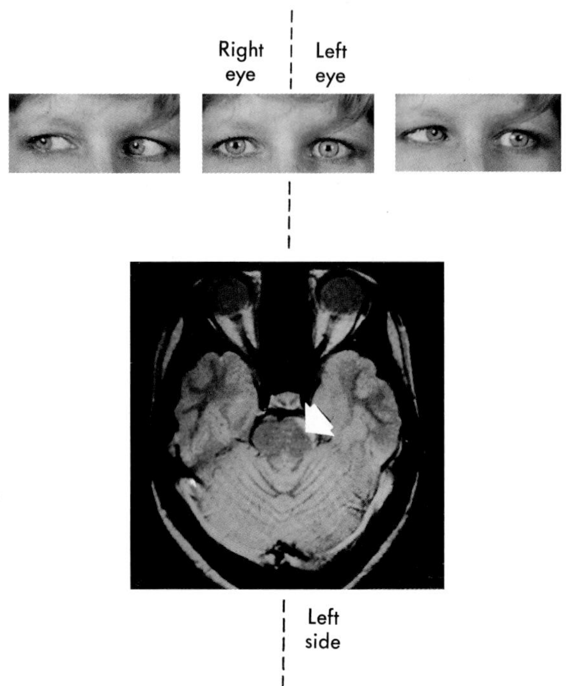

FIGURE 27-8 *Paralysis of abduction in the patient's left eye when looking to the left. The computed tomography scan shows demyelination of the pons at the level of the exiting left abducens nerve (arrow).*

The cholinergic preganglionic motor neurons that supply the superior cervical ganglion are located at spinal cord levels T1 to T3. Their axons enter and ascend in the sympathetic trunk, to terminate in the ipsilateral superior cervical ganglion. Damage along this long path produces *Horner's syndrome*. The cardinal symptoms of this loss of sympathetic input to the head are partial *ptosis* (a drooping lid resulting from relaxation of the superior tarsal muscle), *miosis* (constriction of the pupil because the action of the sphincter is no longer opposed by the action of the dilator), and *anhidrosis* (loss of facial sweating). Horner's syndrome may also result from interruption of pathways linking the hypothalamus and brainstem to preganglionic motor neurons in the thoracic cord.

TARGETING MOVEMENTS

Saccades To obtain detailed information about the visual world, the eyes make a series of very rapid movements (200 to 700 degrees/sec) from one point to another, stopping briefly at each point to allow detailed foveal inspection. These rapid conjugate eye movements are *saccades*, and the places where the detailed visual inspection occurs are *fixation points*. During the saccade, the visual system suppresses incoming visual input. Consequently, one is not aware of these movements. Other processes in visual association cortex provide *visual constancy* by weaving together the information obtained at each fixation into a seamless view of the visual world.

Extraocular Muscle Motor Neurons Extraocular muscle motor neurons have a characteristic *burst-tonic* firing pattern for saccadic eye movements. For example, before a leftward horizontal movement (Fig. 27-10A,B), left lateral and right medial rectus motor neurons show an initial burst of action potentials, which is then reduced to a sustained tonic level of activity. At the new eye position, the firing rate is slower than during the initial burst but higher than the rate for the previous eye position (Fig. 27-10B,C).

These two parts of the motor neuron and target muscle response are referred to as a *pulse* and *step* of activity. The pulse, the initial burst of motor neuron activity, directs the phasic portion of the movement, producing the muscle contraction necessary to overcome the viscosity of the orbit and send the globe towards the

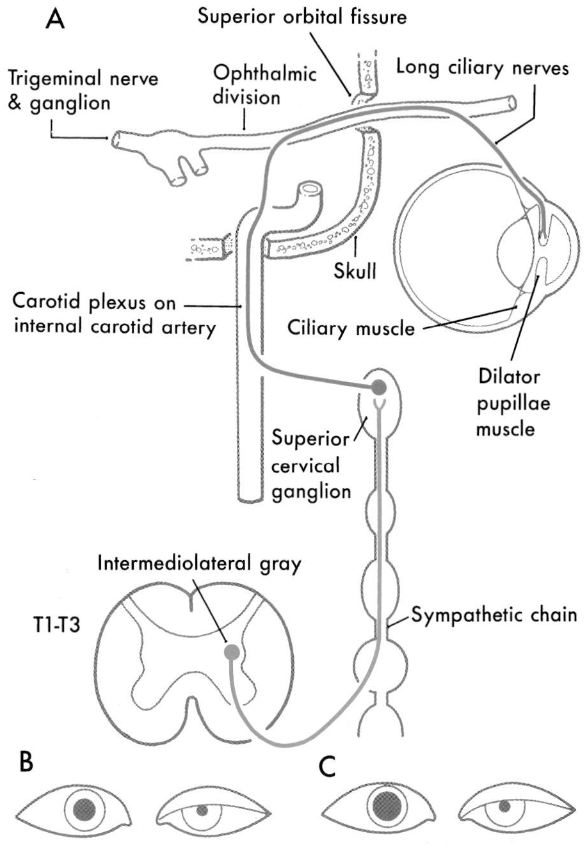

FIGURE 27-9 *The autonomic pathway to the dilator pupillae muscle (A). A patient with a presumed Horner's syndrome (disruption of this pathway) presents with a partial* ptosis *and* miosis *(B). Application of cocaine drops to both eyes establishes that the* anisocoria *is not due to natural or pharmacologic causes (C). The normal eye dilates in response to blockade of norepinephrine reuptake, but the deafferented eye shows no change.*

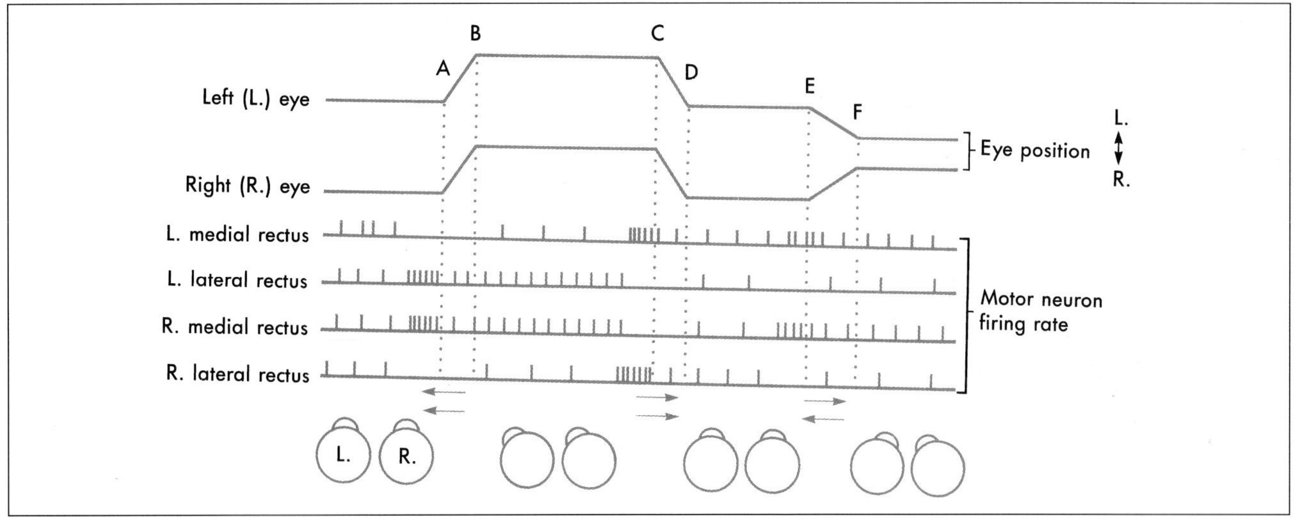

FIGURE 27-10 *Relationship between eye movements and motor neuron firing patterns for horizontal saccades and vergence movements. The top two traces illustrate the changes in eye position. The bottom four traces show idealized firing patterns for motor neurons during horizontal movements. The cartoons at the bottom indicate that the eyes make a saccade to the left (A,B), then the right (C,D), and finally a convergent movement (E,F).*

target. The step in the activity supports the tonic action of the muscle, which is required to maintain the eye at its new position. The activity of the antagonists is silenced for the saccade and then resumes at a lower rate. However, when the eyes look to the right (Fig. 27-10C,D), motor neurons for the left medial rectus and right lateral rectus muscles are activated. The brainstem distribution of activated motor neurons defines which muscles are activated and hence the direction of movement. The number of cells activated and their firing rate define the speed and distance (metrics) of the movement.

Horizontal and Vertical Gaze Centers The brainstem circuitry that controls saccades is subdivided into systems controlling horizontal and vertical eye movement. The pontine reticular formation near the midline, which contains neurons that project to the extraocular motor nuclei, is sometimes called the *paramedian pontine reticular formation* (PPRF) or the *horizontal gaze center* (Figs. 27-11, 27-12). The PPRF occupies portions of the *oral* and *caudal pontine reticular nuclei*. Cells of this premotor region show activity related to horizontal saccades, and lesions in this region produce *horizontal gaze palsies*. PPRF cells that project to extraocular motor neurons include *excitatory burst neurons* (EBNs), found rostral to the sixth nucleus, and *inhibitory burst neurons* (IBNs), found caudal to the sixth nucleus. Both of these cell types have phasic activity patterns; that is, they produce a burst of action potentials that slightly precedes the activity of the motor neurons (Fig. 27-12). When gaze shifts to the right, the EBNs activate the abducens nucleus neurons on the right side, as the IBNs suppress the abducens nucleus neurons on the left side. If this inhibition of antagonists does not occur, eye movements are slowed and undershoot.

The pattern of activity in PPRF neurons is a product of the signal (coded in the cell firing pattern) sent from supranuclear structures, including the superior colliculus and the frontal eye field, and processed through interneurons within the PPRF. In addition, the burst of action potentials produced by EBNs and IBNs is gated by inhibition from cells found in the midline of the pontine tegmentum (Figs. 27-11, 27-12). These inhibitory neurons are called *omnipause cells* because they fire spontaneously during fixation but are silent during a saccadic eye movement in any direction. Thus, the burst of action potentials in PPRF neurons is due, in part, to release from inhibition. PPRF neurons directly contribute to the pulse of motor neuron activity that produces a saccade but not to the step change in activity that maintains eye position. The "where the eye is going" signal produced by the PPRF is transformed (integrated) into a "keep the eye in that position" signal by another brainstem structure. The likely source of this tonic position signal is the *nucleus prepositus hypoglossi* (Figs. 27-11, 27-12).

Vertical gaze palsies are often encountered with lesions of the midbrain-diencephal junction. The *vertical gaze center* is located in the *rostral interstitial nucleus* of the *medial longitudinal fasciculus* (*riMLF*), found at the rostral end of the MLF (Fig. 27-11). This region receives supranuclear input from the superior colliculus and frontal eye field, as well as input from omnipause cells. It also contains burst neurons. These provide the phasic signal for the saccade-related pulse of activity present in vertical gaze motor neurons. The tonic signal

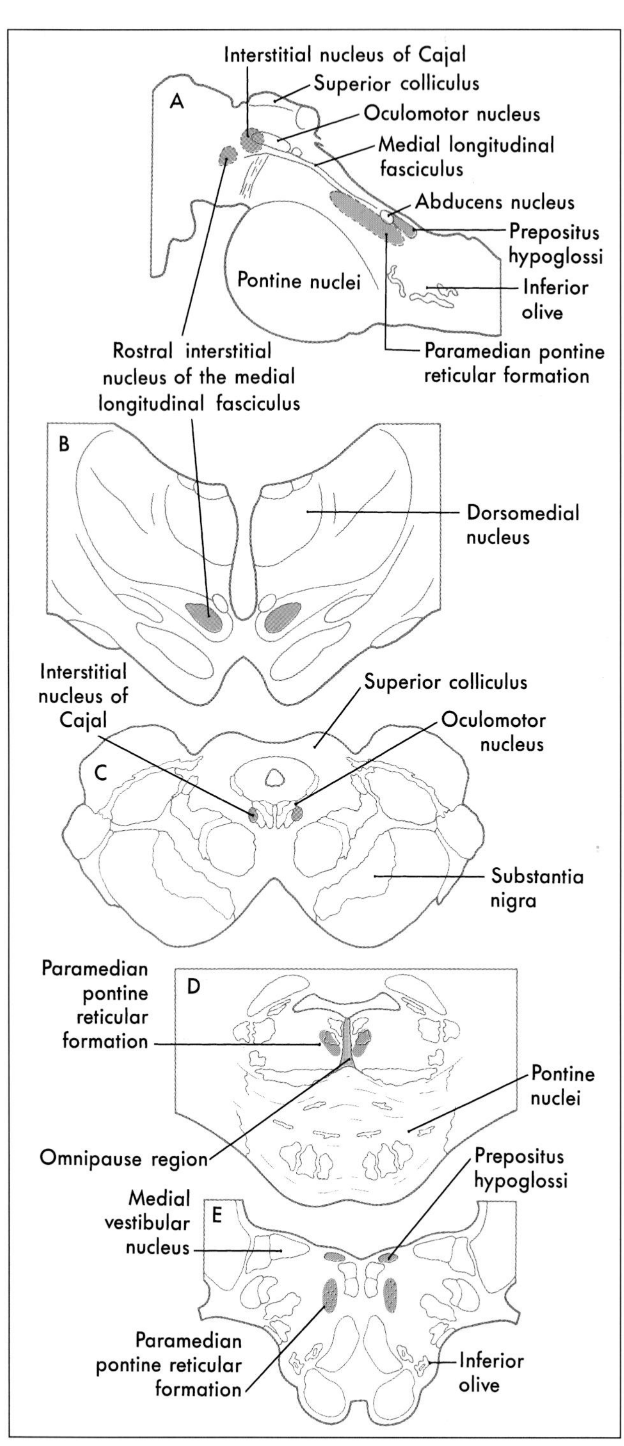

FIGURE 27-11 *The positions of the horizontal and vertical gaze centers in the sagittal (A) and frontal (B–E) planes.*

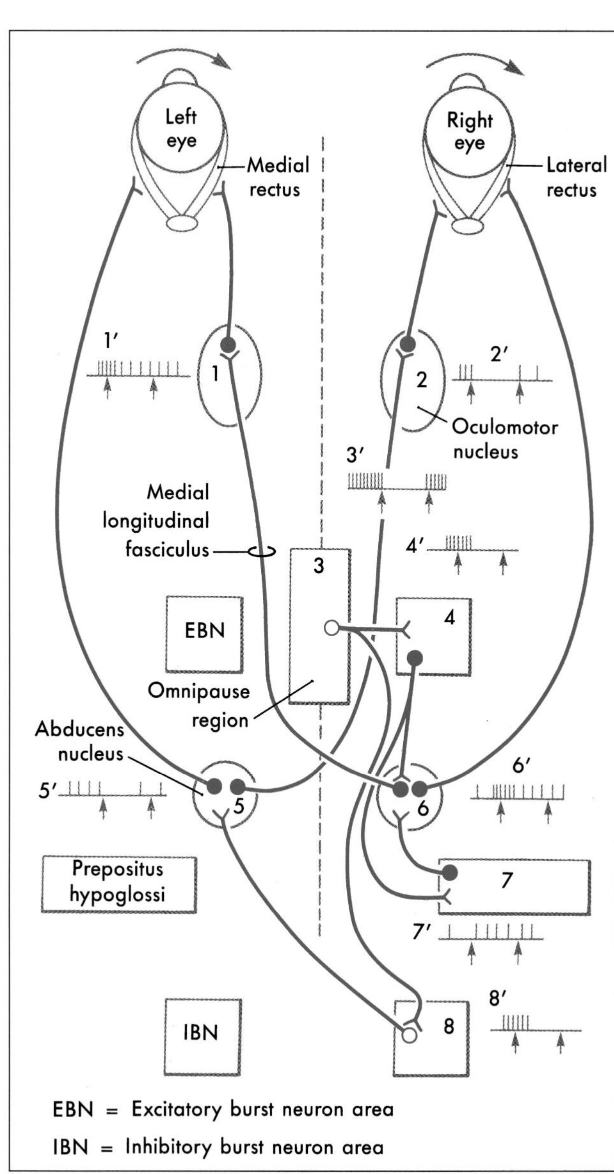

FIGURE 27-12 *Pathway for horizontal saccades. Inhibitory circuits are indicated by open circles. Idealized firing patterns (1′, etc) for a* saccade to the right *are indicated for each nucleus (numbers). The beginning and end of the saccade are indicated by arrows on the firing patterns.*

for the step in vertical gaze motoneuron activity is probably provided by neurons in the *interstitial nucleus of Cajal* (Fig. 27-11). Because the superior and inferior recti act in pairs during vertical eye movements, the projections of the riMLF and interstitial nucleus of Cajal to the extraocular motor neurons are generally bilateral. Oblique saccades are produced by the vertical and horizontal gaze centers working in concert.

Saccadic eye movements are often accompanied by orienting movements of the head. Areas lateral to the PPRF, riMLF, and interstitial nucleus of Cajal are responsible for combined head and eye movements. These areas receive collicular and cortical projections, and they project to the extraocular motor neurons and to the cervical spinal cord as *reticulospinal* and *interstitiospinal* fibers. The superior colliculus also projects to the cervical spinal cord, but the tectospinal portion of the tectobulbospinal system is very small.

Supranuclear Control For each saccade, the central nervous system must determine the position of the next target of interest and transform this position, which is coded in a sensory map, into the appropriate pattern of motor neuron activity. The areas of the brain that direct saccadic eye movements include the *cortical eye fields* and the *superior colliculus*. In each area, stimulation will initiate contralaterally directed saccades, and single cell recordings reveal neuronal activity before saccades occur. Specifically, stimulation of a location where cells are active before a 20-degree saccade to the left will produce a 20-degree leftward saccade.

The *frontal eye field* (area 8 of Brodmann) is located rostral to motor cortex (Figs. 27-13 and 27-14). It is apprised of the location of targets via input from visual association cortex and from basal ganglia and thalamic relays (paralamellar dorsomedial nucleus) (Fig. 27-14). The frontal eye field influences eye movements through projections to the vertical and horizontal gaze centers and the superior colliculus. Additional cortical regions influencing saccades include the *supplemental eye field* and the *parietal eye field* in the lateral intraparietal cortex (area 7 of Brodmann) (Fig. 27-13). They have features similar to those of the frontal eye field but are less directly connected to the brainstem saccade circuits. These three cortical eye fields are reciprocally connected, and all three project to the superior colliculus. Perhaps for this reason, loss of any one of these four structures produces few visual motor symptoms.

The *superior colliculus* (*optic tectum*) is a layered structure found in the roof of the midbrain (Fig. 27-15). This area of the midbrain tectum receives its blood supply from the *quadrigeminal artery*, a branch of the posterior cerebral artery. The superficial layer of the superior colliculus is visual sensory. It is a target of retinal axons with Y- and W-type physiologic characteristics and projects to the dorsal lateral geniculate and pulvinar nuclei. In con-

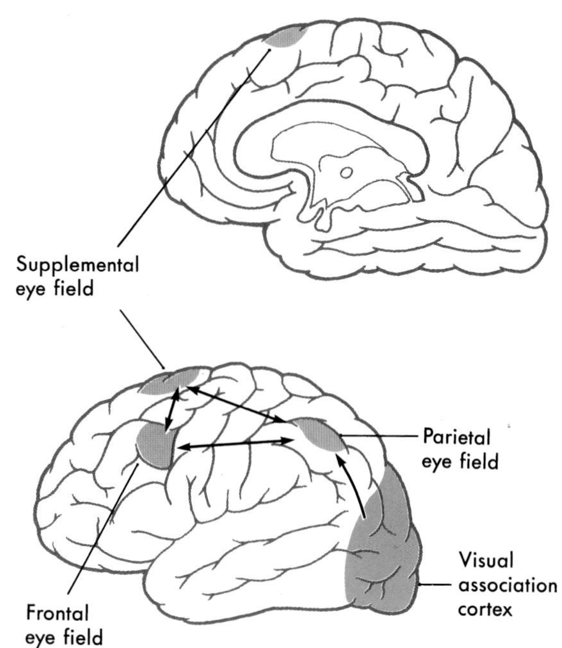

FIGURE 27-13 *Locations and interconnections of the cortical eye fields.*

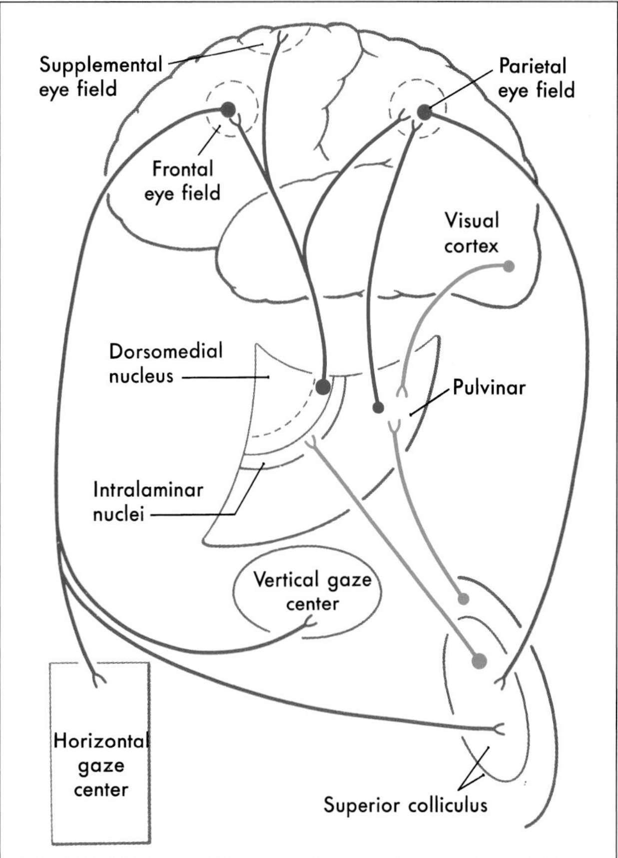

FIGURE 27-14 *Inputs to and outputs of the frontal and parietal eye fields. The dashed line in the dorsomedial nucleus indicates the paralamellar subdivision. The pathway to the horizontal gaze center is crossed.*

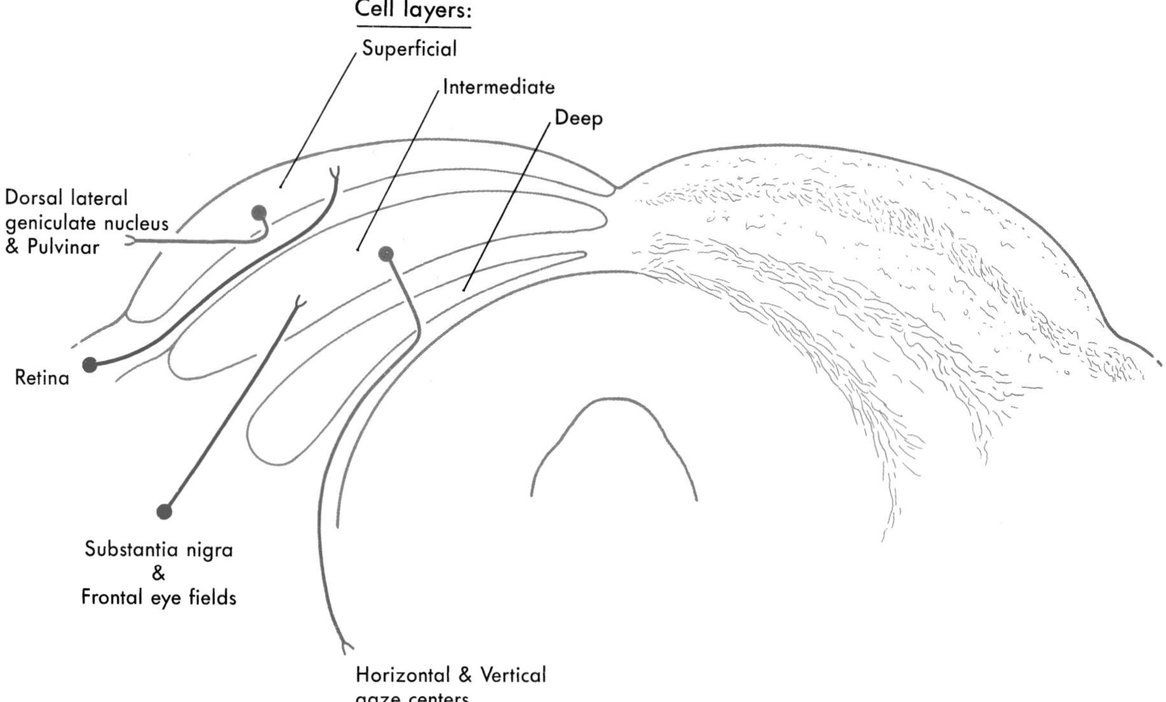

FIGURE 27-15 *The superior colliculus in frontal section. The right side shows the layering seen in myelin stains, and the major inputs and outputs are indicated on the left.*

trast, the intermediate layer is visual motor. It is the source of the *crossed tectobulbospinal system* (predorsal bundle) that runs ventral to the MLF and terminates in the vertical and horizontal gaze centers (Fig. 27-16).

Like the frontal eye field, the major inputs to the intermediate layers of the superior colliculus arise from parietal association cortex and basal ganglia circuits (Fig. 27-16). GABAergic nigrotectal cells in the substantia nigra pars lateralis and reticulata are spontaneously active but cease firing before saccadic eye movements. Tectal saccade-related activity is partly the result of release from this nigral inhibition. Basal ganglia diseases produce eye movement disorders; for example, patients with parkinsonism have a dearth of spontaneous eye movements because of unmodulated activity in the nigrotectal pathway.

The superior colliculus and the frontal eye field differ in the types of saccades they control. The frontal eye field is important for voluntary and memory-guided eye movements, and the superior colliculus directs reflexive orienting movements. Loss of either structure produces few deficits after recovery because the remaining structure compensates for the loss, but loss of both produces profound visuomotor impairment.

Smooth Pursuit The eyes also make conjugate movements that allow the foveas to follow a moving target

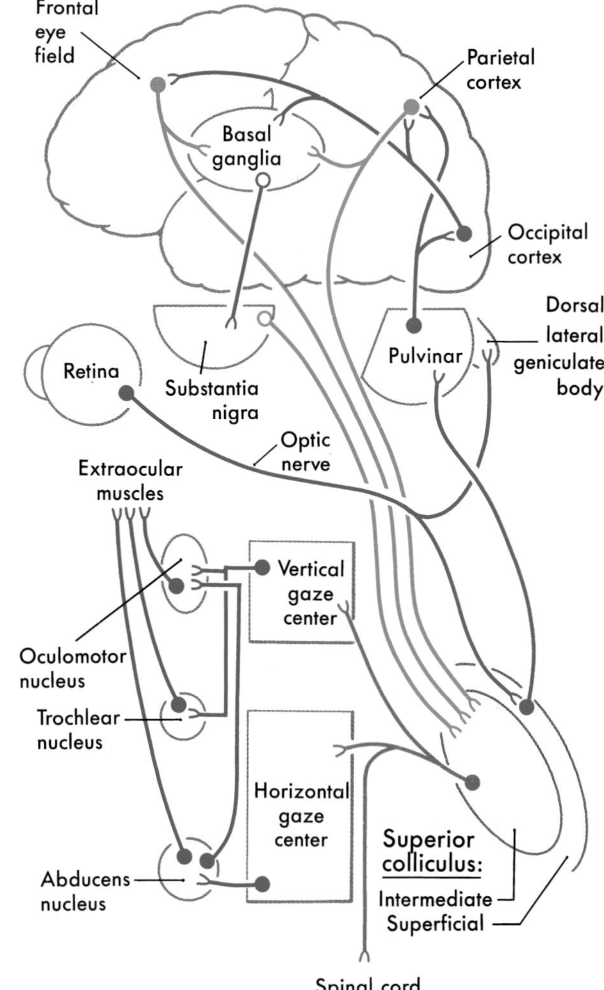

FIGURE 27-16 *Pathways for the superior colliculus. Inhibitory circuits are indicated by open circles. The pathway to the horizontal gaze center and spinal cord (*tectobulbospinal system*) is crossed.*

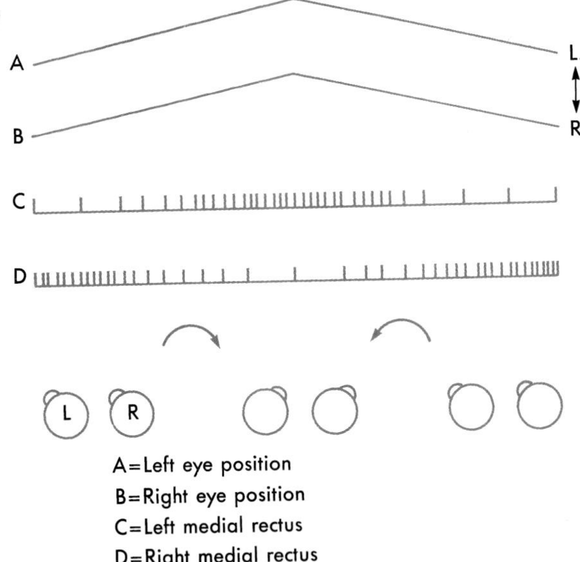

FIGURE 27-17 *The relationship between eye movements and motor neuron firing for smooth pursuit. The eyes change position as they follow a slow moving target from the left to the right and back to the left (A,B). Idealized and graded firing patterns in left (C) and right (B) medial rectus motor neurons.*

(Fig. 27-17). Although usually used to follow slow-moving, predictable targets (30 degree/sec or less), *smooth pursuit* eye movements are capable of following targets at speeds up to 100 degree/sec. The lateral parietal and midtemporal cortices contain neurons that are sensitive to the speed and direction of a target moving across the retina (Fig. 27-18). This input determines the speed and direction of the pursuit eye movements needed to keep the foveas on target. Neurons that display pursuit-related motor activity are found in a portion of the frontal eye field. These three cortical regions project to the *flocculus* and *paraflocculus* of the cerebellum, by way of the *dorsolateral pons*. This portion of the cerebellum, in turn, provides input to the *vestibular nuclei*. Although located in a "sensory" nucleus, vestibular smooth pursuit cells fire with respect to the position and velocity of the eyes, not the visual sensory input. They are, in fact, *premotor neurons*, which project to the third, fourth, and sixth cranial nerve nuclei.

Smooth pursuit premotor neurons fire in a graded manner, depending on the degree and rate of eye excursion. This firing pattern is similar to that displayed by their motor neuron targets during smooth pursuit movements (Fig. 27-17). They do not have the pulse-and-step form seen with saccades. Presumably, the cerebellum plays a role in precisely determining the rate of movement and predicting target trajectory. Floccular lesions do, in fact, produce deficits in smooth pursuit movements.

Vergence Movements and the Near Triad Foveation involves directing the eyes toward targets in three-dimensional space. To look from a distant target to a closer one, three changes are made in the eyes. First, the eyes converge by the simultaneous activation of both medial rectus muscles to point both foveas at the closer target (Fig. 27-10E,F). This is a disconjugate movement because the eyes move in opposite directions. Second, the curvature of the lens is increased, producing an increase in refractive power, to focus the closer target on the fovea (Fig. 27-2). Third, the pupil is constricted, thereby increasing the depth of field of the eye. These three actions are termed the *near triad* or *near response*. The opposite effects (*divergence*, flattening of the lens, and pupillary dilation) occur when the gaze is shifted from a closer target to one farther away.

As the term "near triad" suggests, the three actions are generally yoked. Although vergence and accommodation can be dissociated under special conditions (closing one eye and bringing the target straight at the open eye), under normal conditions the vergence angle is used by the brain to adjust the accommodation of the lens. Other cues used to direct the near response include *retinal disparity* and focus. Cells in visual cortex with binocular visual fields are activated when the retinal images

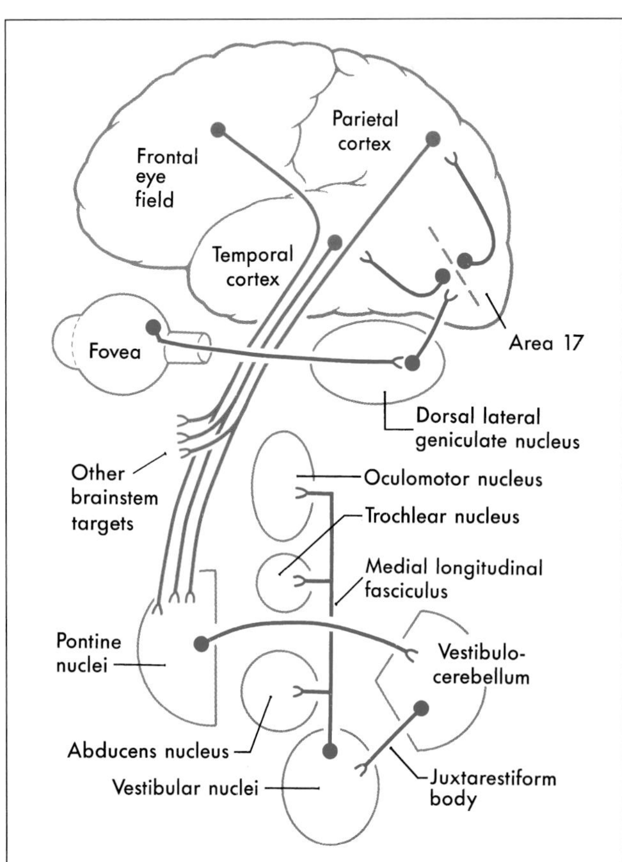

FIGURE 27-18 *Pathways for smooth-pursuit eye movements. The retinogeniculostriate pathway is the source of relevant visual sensory input to parietal and temporal association cortex.*

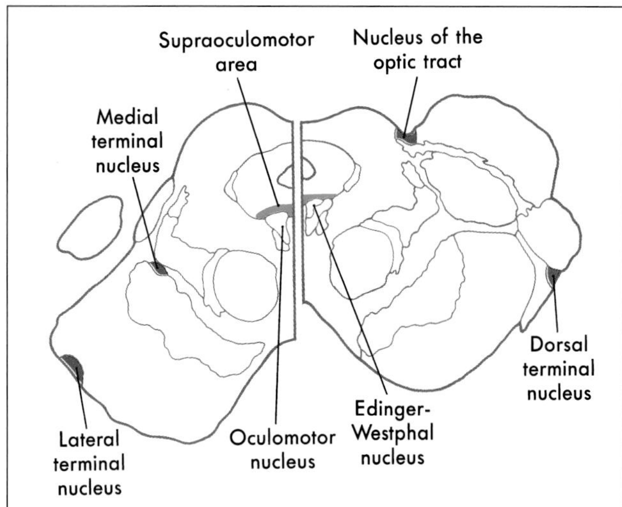

FIGURE 27-19 *Nuclei of the accessory optic system and the supraoculomotor area in caudal (left) and rostral (right) coronal sections through the rostral midbrain.*

have a specific degree of disparity (the difference in the points on each retina where an image falls). This is an appropriate signal for the control of vergence. Lack of focus, or blurring of the image, supplies a feedback signal for adjusting the lens until *blur* is minimized.

Vergence movements are generally slower than saccades, and the burst of motor neuron activity related to the initial pulse in muscle activity is less evident (Fig. 27-10E,F). Premotor neurons whose activity correlates with the near response are found in a *midbrain near response region* located in the *supraoculomotor area* (SOA) (Fig. 27-19). Cells in the SOA project to the medial and lateral rectus motor neurons and also to the preganglionic motor neurons in the Edinger-Westphal nucleus that control the lens and pupil. Although the near response pathways to the SOA remain undetermined, it is likely that the cerebellum influences vergence and accommodation through projections to this region.

Although combined vergence and saccade movements occur and have a similar time course, the saccadic and vergence systems can act independently. For instance, following a lesion in the paramedian pontine reticular formation, horizontal saccades are disrupted but vergence movements in the horizontal plane are unaltered.

REFLEX MOVEMENTS

The body is equipped with compensatory systems that keep the eyes directed at a target despite external perturbations of the body or head. The vestibular system specializes in sensing the acceleration that usually occurs at the beginning of a movement, and it also senses gravity. Activation of the vestibular labyrinths elicits a series of compensatory eye movements commonly termed the *vestibulo-ocular reflexes* (VOR). The pathways subserving these reflexes are discussed in Chapter 21. In contrast, the *optokinetic system* compensates for continuous-velocity movements. This section covers the optokinetic system as well as the pupillary and blink reflexes.

Optokinetic Eye Movements As we move in the world, or move our heads, the entire visual scene moves across the retina. This whole-field movement of the visual scene is called *retinal slip*. Under these conditions, the eyes automatically move in a compensatory fashion to stabilize the image on the retina. For example, if the body is rotating to the left, the visual world will seem to move to the right, and rightward compensatory eye movements match the apparent movement of the visual scene (Fig. 27-20). These *optokinetic movements* are generally slow and match the *velocity* of retinal slip. As with smooth pursuit, they are produced by graded increases and decreases in the tonic firing rate of the appropriate motor neurons (Fig. 27-20). When the eyes approach the limit of their rotation, a quick saccade brings them back to their primary position and another slow following movement begins. This set of alternating *slow* and *fast* (saccadic) *phases* of movement is called *optokinetic nystagmus* (OKN, named for fast phase) (Fig. 27-20). A series of stripes can be moved in front of a subject to elicit nystagmus and test the *optokinetic reflex*.

The afferent limb of this reflex begins with stimulation of wide-field retinal ganglion cells that are sensitive to slow movements of the whole receptive field (Fig. 27-

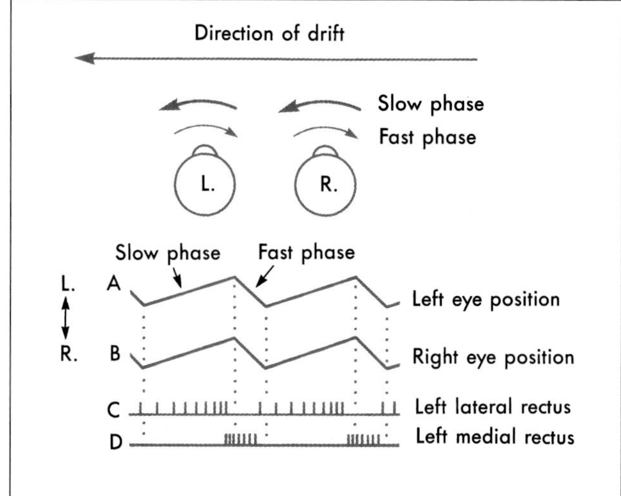

FIGURE 27-20 *The relationship between eye movements and motor neuron firing during optokinetic nystagmus. The long arrow indicates the drift of the visual scene. The thick arrow shows the slow optokinetic movement of the eye as it follows the visual scene, and the thin arrow indicates the fast saccadic resetting movement. These changes in eye position are illustrated in A and B. Idealized motor neuron firing patterns are shown in C and D.*

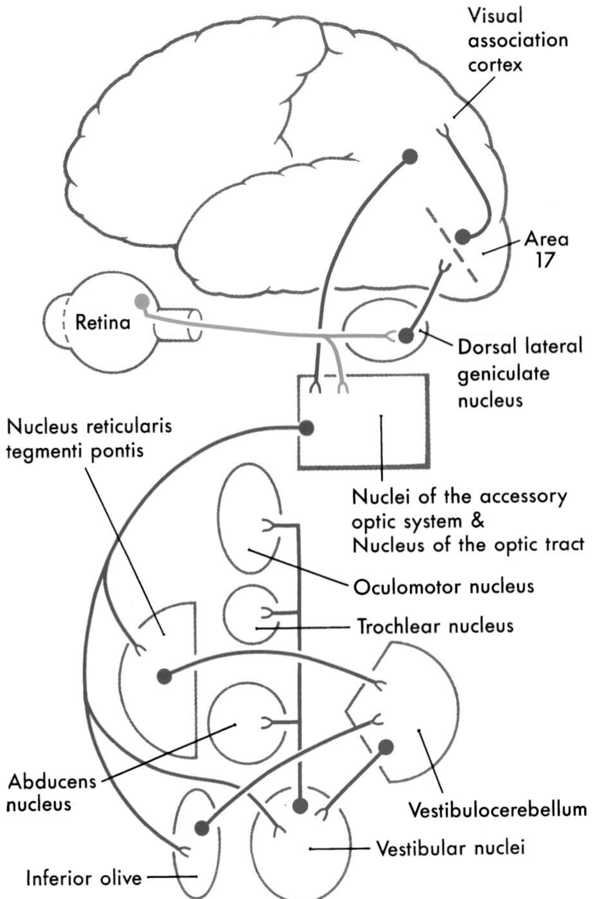

FIGURE 27-21 *Pathway for the optokinetic system.*

21). The receptive fields of these retinal cells are tuned to directions of movement that are comparable to the spatial planes in which the vestibular labyrinths are oriented. Axons of these retinal ganglion cells terminate in a series of small nuclei along the incoming optic tract, termed the *accessory optic system* (AOS) (Fig. 27-19). These consist of the *nucleus of the optic tract* and the *medial*, *lateral*, and *dorsal accessory nuclei*. Each of these four nuclei contains cells that are activated by retinal slip in specific directions; for example, nucleus of the optic tract cells are sensitive to temporal-to-nasal movements. The AOS nuclei also receive input from visual sensory association cortex, presumably from neurons of the smooth pursuit system (Fig. 27-21). In humans, the optokinetic reflex only comes into play when the broad field movement cells in the accessory optic system indicate retinal slip and the pursuit system subserved by the geniculocortical pathways notes equivalent foveal target movement. Otherwise the pursuit system overrides the optokinetic system.

The AOS nuclei project to the portions of the *nucleus reticularis tegmenti pontis* and the *inferior olive* that supply the vestibulocerebellum, and to the vestibular nuclei (Fig. 27-21). In addition to vestibulo-ocular premotor neurons, the latter nuclei contain optokinetic neurons that influence extraocular motor neurons. In fact, lesions in the vestibular nuclei produce severe deficits in both reflexes. The pathways through the cerebellum are involved in adapting the gains of the optokinetic and vestibulo-ocular reflexes so that they combine to produce appropriate eye movements.

Pupillary Light Reflex In addition to the changes in pupillary size that result from the near response, the pupil also responds to the amount of ambient light. The actions of the iris in each eye are yoked, so that light directed into one eye results in pupillary constriction in both the illuminated eye (*direct response*) and the opposite eye (*consensual response*). The pupil changes diameter to maintain luminance of the retina in the optimal range of the receptor photopigments.

The pupillary light reflex is a four-neuron arc (Fig. 27-22). Retinal ganglion cells with broad receptive fields that respond in a linear fashion to luminance levels project via the optic nerve and tract to the midbrain. The decussation of approximately half of these fibers in the chiasm is one of the structural features responsible for the consensual response. The retinal axons terminate in the pretectum within the *olivary pretectal nucleus*, which, in turn, projects bilaterally to the Edinger-Westphal nucleus. Parasympathetic, preganglionic fibers from the Edinger-Westphal nucleus exit with the oculomotor nerve and terminate in the ciliary ganglion. The cholinergic postganglionic fibers reach the iris, where they excite the pupillary constrictor and inhibit the pupillary dilator muscles. The pupillary light reflex is a useful diagnostic tool for testing brainstem and cranial nerve function (Fig. 27-22). Lesions may result in loss of the direct or consensual pupillary responses or in uneven pupil size (*anisocoria*). A dilated, unresponsive (fixed) pupil (or pupils) in the unconscious victim of head trauma is a grave sign. For example, it may indicate that a space-occupying lesion has forced the parahippocampal gyrus and/or uncus over the edge of the tentorium (*uncal herniation*), compressing the third nerve. The pupillary fibers are superficially located in the oculomotor nerve and are particularly sensitive to pressure. Their loss may indicate that compression of the brainstem is imminent.

The effects of lesions in the sympathetic pathways (Horner's syndrome) have already been discussed. Another syndrome, called *Argyll Robertson pupil*, is found in cases of *tabes dorsalis* (central nervous system syphilis). These patients show small pupils with very weak or absent pupillary light reflexes bilaterally, but there is no loss of visual acuity, and the pupils do constrict in the near response. This sparing indicates that the afferent and efferent limbs of the pupillary light reflex must be intact. Consequently, it is assumed that bilateral degeneration in either the olivary pretectal nuclei or the pathways connecting them to the Edinger-Westphal nucleus must be the source of the pupillary dysfunction.

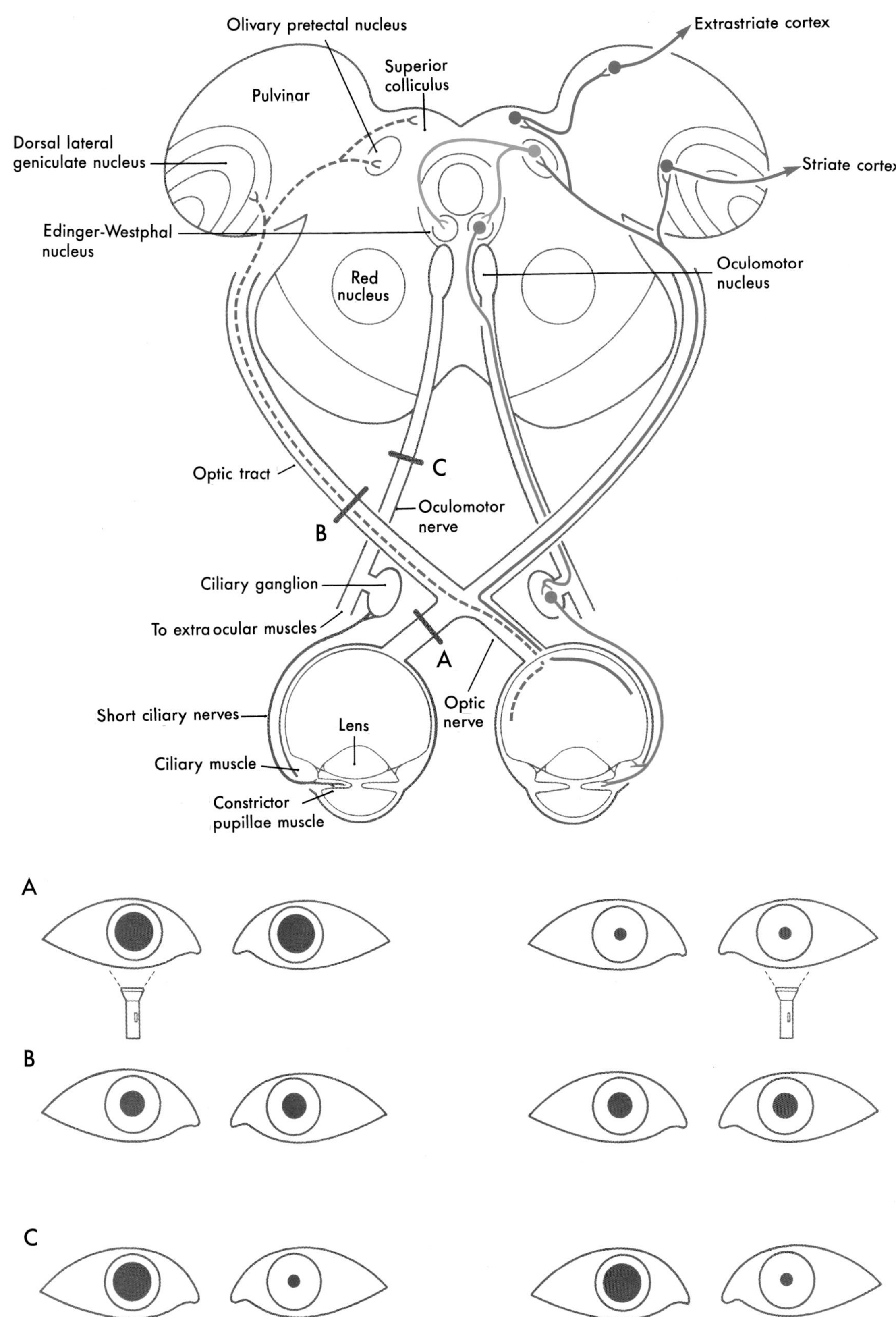

FIGURE 27-22 *Pupillary light reflex pathways. If the optic nerve is partially damaged (A), shining a light into that eye will produce a diminished direct and consensual response (left); but both will be present when the undamaged side is illuminated (right). This is termed a* relative afferent pupillary defect. *A total lesion at A would produce a blind eye, which would induce neither a direct nor a consensual response when illuminated. If the lesion occurs in the optic tract (B) or pretectum, neither response is lost. Although the reflexes may be weaker, this is* not *easily discerned clinically. If the lesion occurs in the oculomotor nucleus or nerve (C), both direct and consensual responses will be lost in the eye on the lesion side, but they will be present in the other eye.*

Greater levels of excitement, including desire, result in dilation of the pupil via sympathetic activation. This fact was known to Elizabethan women, who used tinctures of belladonna to dilate their eyes for cosmetic purposes. Today, cholinergic blockers are used to dilate the pupils for an ophthalmologic examination.

Blinking and Other Lid Movements The delicate structures of the eye are protected by the eyelids. Blinks, some of which occur in response to somatosensory stimulation, ensure protection for the eye. The *blink reflex* is used to assess trigeminal sensory and facial nerve function, as well as the integrity of the lid pathways through the lateral pons. Trigeminal nerve fibers with free nerve endings in the cornea or lid have central processes terminating in the spinal portion of the trigeminal sensory nucleus (Fig. 27-23). Second-order trigeminal neurons project directly and indirectly to the facial nucleus, where they excite orbicularis oculi motor neurons, which produce lid closure. In addition, an inhibitory pathway suppresses the activity of antagonist levator palpebrae motor neurons in the oculomotor nucleus.

Blinks also occur at regular intervals (averaging 12 blinks/minute), as automatically triggered movements that spread the tear film over the cornea. Although the precise origin of these rhythmic blinks is unknown, the constant dispersal of the tear film prevents corneal lesions and scarring. *Blepharospasm* is a disorder in this rhythmic behavior that results in bouts of high frequency blinking, while Parkinsonism produces a decreased blink rate. To keep the lids out of the line of vision, they also move with the eyes during vertical eye movements. These movements are produced by actions of the levator palpebrae muscle, which works in concert with the superior rectus. The orbicularis oculi muscle is not involved. Consequently, *Bell's palsy*, in which the facial nerve is damaged along its peripheral course, results in a loss of the blink reflex on the affected side, but no ptosis or loss of vertical gaze-related lid movements. Lesions of the oculomotor nerve produce just the opposite results.

The tarsal muscles help to keep the lids open, as indicated by the partial ptosis present in Horner's syndrome. Their sympathetic innervation suggests that they regulate lid position with respect to emotional state. For example, high sympathetic tone produces widely opened eyes. Relaxation of the tarsal muscles leads to the feeling of "heavy lids," which signals the general tone of the autonomic system, as the brain prepares to rest.

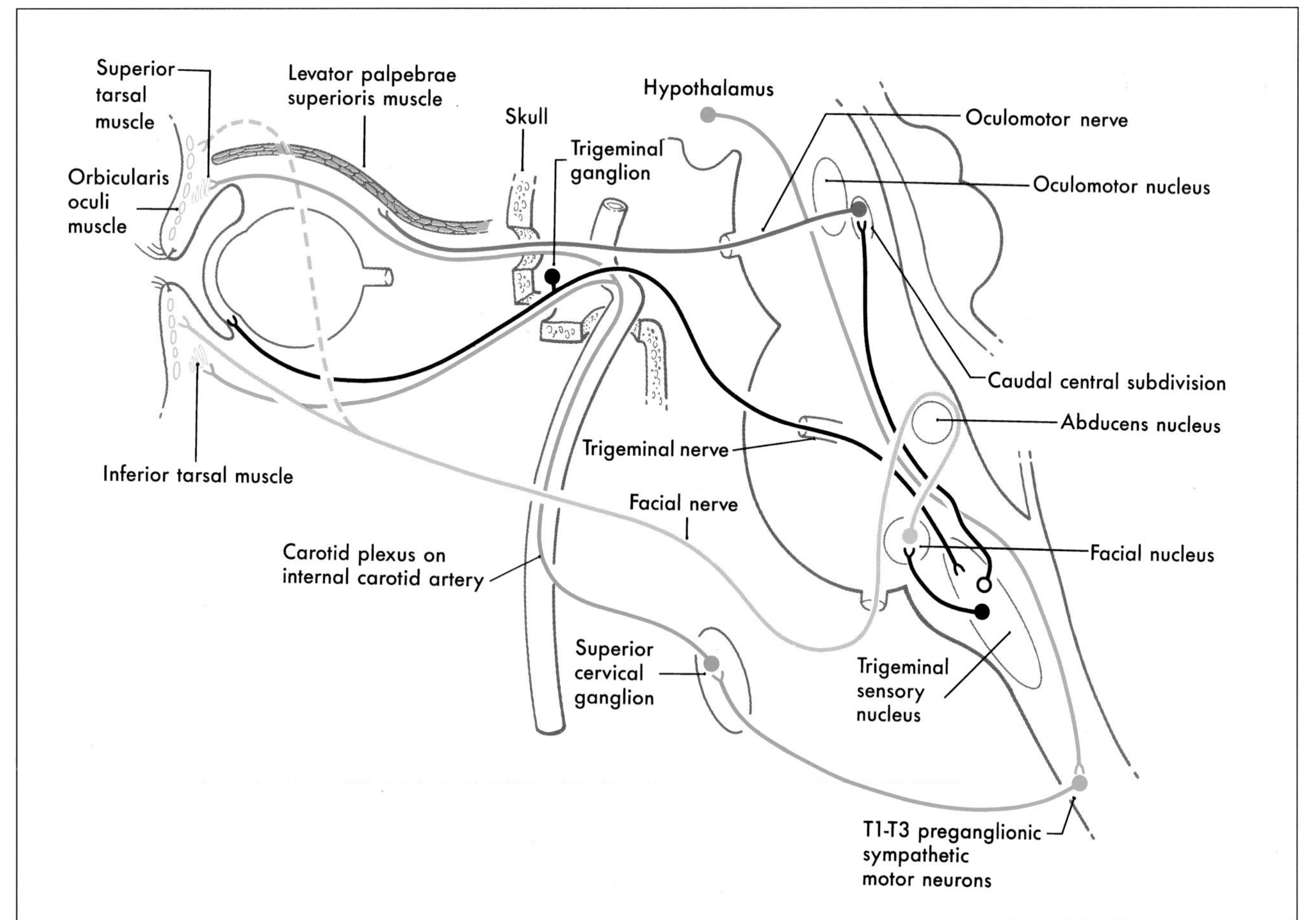

FIGURE 27-23 *Pathways for control of eyelid movements. The motor nerves and their muscle targets are coded in the same color. Inhibitory projections are indicated by open circles.*

SOURCES AND ADDITIONAL READINGS

Buttner-Ennever JA (ed): *Neuroanatomy of the Oculomotor System*. Vol. 2 in the Series *Reviews of Oculomotor Research*. Elsevier, Amsterdam, 1988.

Hall WC, May PJ: The anatomical basis for sensorimotor transformations in the superior colliculus. In Neff WD (ed): *Contributions to Sensory Physiology*, Vol. 8. Academic Press, San Diego, 1984, pp 1–40.

Huerta MF, Halting JK: The mammalian superior colliculus: studies of its morphology and connections. In Vanegas H (ed): *Comparative Neurology of the Optic Tectum*. Plenum Press, New York, 1984, pp 678–773.

Keller EL, Heinen SJ: Generation of smoot-pursuit eye movements: neuronal mechanism and pathways. Neurosci Res 11:79–107, 1991.

Leigh RJ, Zee DS: *The Neurology of Eye Movements*. Vol. 23 in the Series *Contemporary Neurology*. FA Davis, Philadelphia, 1983.

Loewenfeld IE: *The Pupil. Anatomy, Physiology and Clinical Applications*. Iowa State University Press, Ames, 1993.

Mays LE: Neural control of vergence eye movements: convergence and divergence neurons in midbrain. J Neurophysiol 51:1091–1108, 1984.

Schiller PH, True SC, Conway JL: Deficits in eye movements following frontal eye-field and superior colliculus ablations. J Neurophysiol 44:1175–1189, 1980.

Schlag J, Schlag-Rey M: Evidence for a supplementary eye field. J Neurophysiol 57:179–200, 1987.

Scudder CA, Fuchs AF, Langer TP: Characteristics and functional identification of inhibitory burst neurons in the trained monkey. J Neurophysiol 59:1430–1454, 1988.

Sparks DL: Functional properties of neurons in the monkey superior colliculus: coupling of neural activity with saccade onset. Brain Res 156:1–16, 1978.

Wurtz RH, Goldberg ME (eds): *The Neurobiology of Saccadic Eye Movements*. Vol. 3 in the Series *Reviews of Oculomotor Research*. Elsevier, Amsterdam, 1989.

CHAPTER TWENTY-EIGHT

Visceral Motor Pathways

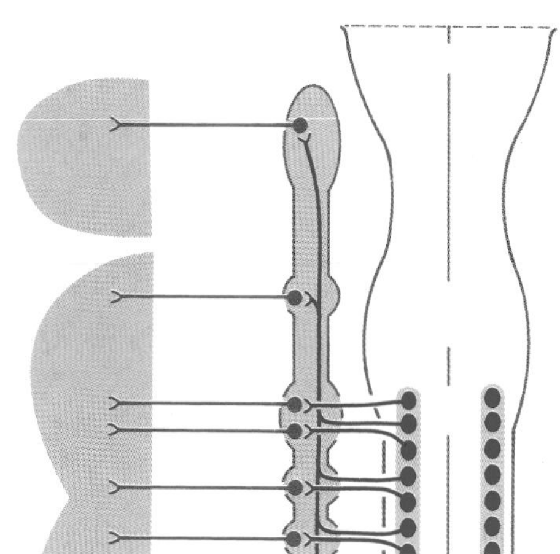

J. P. NAFTEL • S. G. P. HARDY

The primary function of the visceral motor system is the regulation of cardiovascular, respiratory, digestive, urinary, and reproductive organs. These are the main effectors of *homeostasis*, the maintenance of a stable internal environment against perturbing influences, both external and internal. In general, visceral motor neurons innervate smooth and cardiac muscles and glandular epithelium, or structures made up of combinations of these tissues.

OVERVIEW

The *visceral motor* (*autonomic*) *system* ensures that tissues of the body receive appropriate nutrients, electrolytes, and oxygen, and that functions such as osmolarity and temperature are properly regulated. The nervous system contributes significantly to the control and coordination of homeostatic mechanisms in response to continually changing requirements. Two overlapping control systems influence visceral effectors. One is *humoral* (*endocrine*). Hormonal responses tend to develop slowly, but the effects are prolonged. The other system is *neural* (*autonomic*). Visceral motor responses tend to be immediate, but their effects are short term.

The endocrine and autonomic systems are interdependent. They are both under the control of widely distributed central nervous system (CNS) structures, which generate commands after integrating inputs from a wide variety of sources. Thus, visceral motor output is influenced by emotional status, as well as by sensory signals reporting conditions inside and outside the body.

The visceral motor system has two major subdivisions, *sympathetic* and *parasympathetic*. In addition, neurons located in the wall of the alimentary canal form a somewhat autonomous component, called the *enteric nervous system*. This is sometimes regarded as a third subdivision of the autonomic system. Within each of these components are populations of chemically coded, target-specific neurons.

TABLE 28-1

COMPARISON OF EFFECTS OF SYMPATHETIC AND PARASYMPATHETIC ACTIVITY ON SOME VISCERAL FUNCTIONS

PHYSIOLOGIC PROCESS	SYMPATHETIC STIMULATION	PARASYMPATHETIC STIMULATION
Eye		
Pupil diameter	+	–
Lens refraction	0	+
Palpebral fissure width	+	0
Tear flow	0	+
Salivary gland flow	–	+
Skin		
Piloerection	+	0
Sweating	+	0
Blood flow	–	0
Skeletal muscle blood flow	±	0
Cardiovascular system		
Cardiac output	+	–
Total peripheral resistance	+	0
Bronchial diameter	+	–
Gut		
Peristalsis	–	+
Secretion	–	+
Sphincter tone	+	–
Blood flow	–	+
Liver glycogenolysis	+	0
Pancreatic insulin secretion	–	+
Pancreatic glucagon secretion	+	+
Urinary bladder detrusor tone	–	+
Urethra sphincter tone	+	±
Penile or clitoral erection	0	+
Ejaculation	+	0

+ = positive effect, – = negative effect, 0 = no effect, ± = variable effect

ORGANIZATION OF THE VISCERAL MOTOR SYSTEM

Targets of Visceral Motor Outflow The autonomic system provides neural control of *smooth muscle*, *cardiac muscle*, and *glandular secretory cells*. The *sympathetic* and *parasympathetic divisions* have overlapping and generally antagonistic influences on those viscera located in body cavities and on some structures of the head, such as the iris (Table 28-1). There are also visceral targets in the body wall and limbs. These are found in skeletal muscle (blood vessels) and in the skin (blood vessels, sweat glands, and arrector pili muscles). Visceral structures of the body wall and extremities are generally regulated by the sympathetic division alone. The sympathetic outflow thus has a global distribution in that it innervates visceral structures in all parts of the body, whereas the parasympathetic outflow serves only targets in the head and body cavities (Table 28-1).

General Features of Peripheral Visceral Motor Outflow There are similarities and differences between the neural control of skeletal muscle and visceral effector such as smooth muscle (Fig 28-1). As described in Chapter 23,

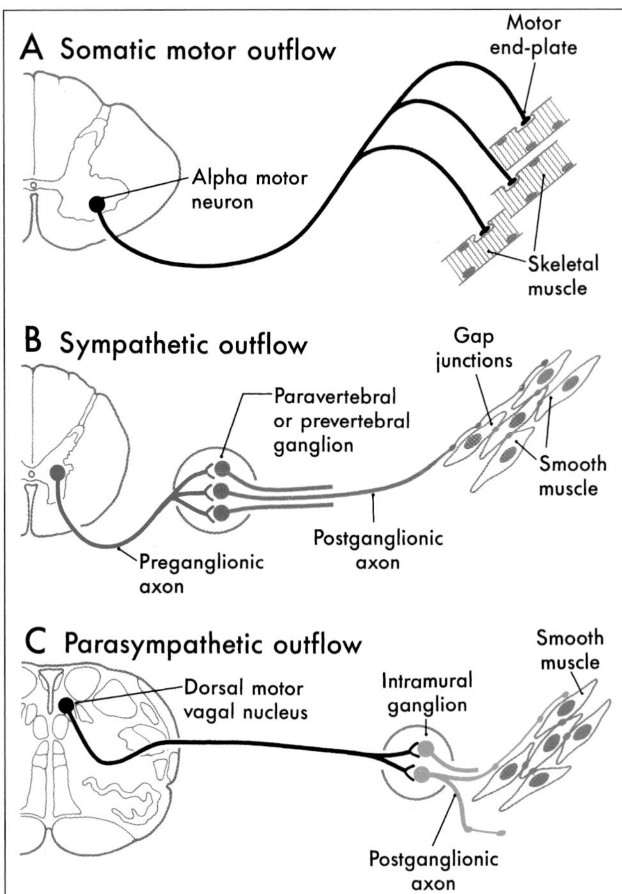

FIGURE 28-1 *Comparison of somatic motor outflow (A) with sympathetic (B) and parasympathetic (C) outflow.*

lower motor neurons (alpha motor neurons) function as the final common pathway linking the CNS to skeletal muscle fibers (Fig. 28-1A). Similarly, sympathetic and parasympathetic outflows serve as the final, but often dual, common neural pathway from the CNS to visceral effectors. However, unlike the somatic motor system, the peripheral visceral motor pathway consists of two neurons (Fig. 28-1B,C). The first, the *preganglionic neuron*, has its cell body in either the brainstem or spinal cord. Its axon projects as a thinly myelinated *preganglionic fiber* to an autonomic ganglion. The second, the *postganglionic neuron*, has its cell body in the ganglion and sends an unmyelinated axon (*postganglionic fiber*) to visceral effector cells such as smooth muscle. In general, parasympathetic ganglia are close to the effector tissue, and sympathetic ganglia are close to the CNS. Consequently, *parasympathetic pathways typically have long preganglionic fibers and short postganglionic fibers, whereas sympathetic pathways more often have short preganglionic fibers and long postganglionic fibers.*

Visceral motor neurons and their targets are not organized into discrete motor units like those of the somatic motor system. Recall that an alpha motor neuron makes synaptic contacts with a definite group of skeletal muscle fibers over which it has exclusive control. In contrast, the terminal branches of a postganglionic visceral motor axon typically have a series of swellings containing neurotransmitter vesicles along their length, giving them a beaded (varicose) appearance (Fig. 28-1B,C). The neurotransmitters released from these terminals may act on effector cells at a distance of up to 100 µm. Moreover, unlike skeletal muscle fibers, cardiac muscle fibers and the smooth muscle cells of some organs are electrically coupled by *gap junctions*. Because of this, neurochemical signaling to a few cells is sufficient to regulate a large group of cells that act as a unit. The main features of sympathetic and parasympathetic divisions are summarized in Table 28-2.

DEVELOPMENT

Preganglionic Visceral Motor Neurons Cell bodies of these neurons are located in nuclei or cell columns embryologically derived from the *general visceral efferent* cell column. This column arises from neuroblasts in the dorsal part of the basal (motor) plate of the brainstem and spinal cord portions of the neural tube.

Postganglionic Visceral Motor Neurons Cell bodies of these multipolar neurons are located in autonomic ganglia, which may be either well-defined, encapsulated structures, such as the superior cervical ganglion, or clusters of somata found in nerve plexuses or in the walls and capsules of visceral organs. Like most primary sensory neurons, autonomic ganglion cells are derived from *neural crest cells* that migrate to appropriate locations during development. *Congenital megacolon* or *Hirschsprung's syndrome* results from failure of enteric neuronal precursor cells to migrate into the wall of the lower gut. As a result, the affected segment of the colon is paralyzed in a constricted state, with consequent distention of the proximal, normally innervated intestine.

Neurotrophins are a family of proteins, each of which regulates development of specific populations of neurons. The existence of these proteins was established when *nerve growth factor* (NGF) was identified as a target-tissue-derived messenger molecule essential for the survival and development of sympathetic postganglionic neurons (Fig. 28-2). The pathologic changes in animals deprived of NGF during development are similar to those associated with familial dysautonomia (Riley-Day syndrome), an autosomal recessive disease. Patients with Riley-Day syndrome have a range of neurologic deficits. Those related to the visceral motor system include inadequate tearing, dilations of the intestines and esophagus, and poor vasomotor control resulting in cold feet and hands. This pattern suggests that the defect may involve a gene coding for NGF or its receptor.

TABLE 28-2

COMPARISON OF THE SYMPATHETIC AND PARASYMPATHETIC DIVISIONS OF AUTONOMIC OUTFLOW

FEATURE	SYMPATHETIC (THORACOLUMBAR)	PARASYMPATHETIC (CRANIOSACRAL)
Location of preganglionic cell bodies	Spinal segments T1 to L2, mainly intermediolateral cell column	Spinal segments S2 to S4, intermediate gray; GVE nuclei of cranial nerves III, VII, IX, X
Location of preganglionic fibers	White rami T1 to L2, sympathetic trunks, splanchnic nerves	Pelvic nerves, cranial nerves III, VII, IX, X
Location of postganglionic cell bodies	Paravertebral ganglia, prevertebral ganglia (celiac, aorticorenal, superior mesenteric, inferior mesenteric)	Ganglion cell clusters in walls of viscera, cranial nerve autonomic ganglia (ciliary—III; pterygopalatine and submandibular—VII; otic—IX)
Location of postganglionic nerve fibers	Fibers to structures of body wall and limbs in gray rami and spinal nerves, plexuses associated with arteries supplying visceral structures of the head and body cavities	Within the viscera of body cavities; short nerves or plexuses extending from cranial ganglia to target organs; often accompany trigeminal nerve branches in head
Target effectors	Smooth muscle, cardiac muscle, and secretory cells throughout body	Mostly viscera of the head and the thoracic, abdominal, and pelvic cavities
Primary neurotransmitter of preganglionic neurons	Acetylcholine	Acetylcholine
Primary neurotransmitters of postganglionic neurons	Norepinephrine; cells supplying sweat glands use acetylcholine	Acetylcholine
Neuropeptides of postganglionic neurons	Neuropeptide Y and others	Vasoactive intestinal polypeptide and others
General physiologic effects	Mobilization of resources for intensive activity	Promotion of restorative processes

SYMPATHETIC DIVISION

Sympathetic Preganglionic Neurons Although the sympathetic outflow influences visceral targets throughout the body, sympathetic preganglionic neurons are found only in spinal cord segments T1 through L2 (plus sometimes in C8 and L3). These cell bodies are located in Rexed's lamina VII, primarily in the *intermediolateral nucleus* (*cell column*) of the lateral horn. The axons of these preganglionic cells exit the spinal cord in the *ventral root* and enter the *sympathetic trunk* via the *white communicating ramus*. Once in the sympathetic chain, a preganglionic fiber may (1) synapse at that level, with the postganglionic fiber joining that spinal nerve; (2) ascend or descend in the sympathetic chain to synapse on postganglionic neurons whose axons join higher or lower spinal nerves; or (3) pass through the chain ganglion as a preganglionic fiber to form part of a *splanchnic nerve* (Fig. 28-3A–C).

The sympathetic outflow for the entire body originates from thoracic and upper lumbar spinal cord segments. Although the segmental pattern of innervation is not straightforward, there is a general viscerotopic organization (Figs. 28-4, 28-5). Neurons of the *superior*, *middle*, and *inferior cervical ganglia* receive input, via the sympathetic trunk, from preganglionic neurons of the upper thoracic spinal segments. Lower lumbar and sacral ganglia are supplied by neurons of the lower thoracic and upper lumbar spinal segments. The ganglia between these regions are supplied by their corresponding spinal levels. Thus, visceral targets in the head, neck, and upper extremity, as well as the viscera of the thoracic cavity, are served by preganglionic sympathetic neurons in the upper thoracic segments. The main abdominal viscera and other targets in the trunk are served by the central and lower thoracic spinal cord segments, whereas the pelvic viscera, the lower trunk, and the lower extremity are served by the lower thoracic and upper lumbar spinal cord segments.

Preganglionic sympathetic neurons also target the adrenal gland (Fig. 28-5). Chromaffin cells of the

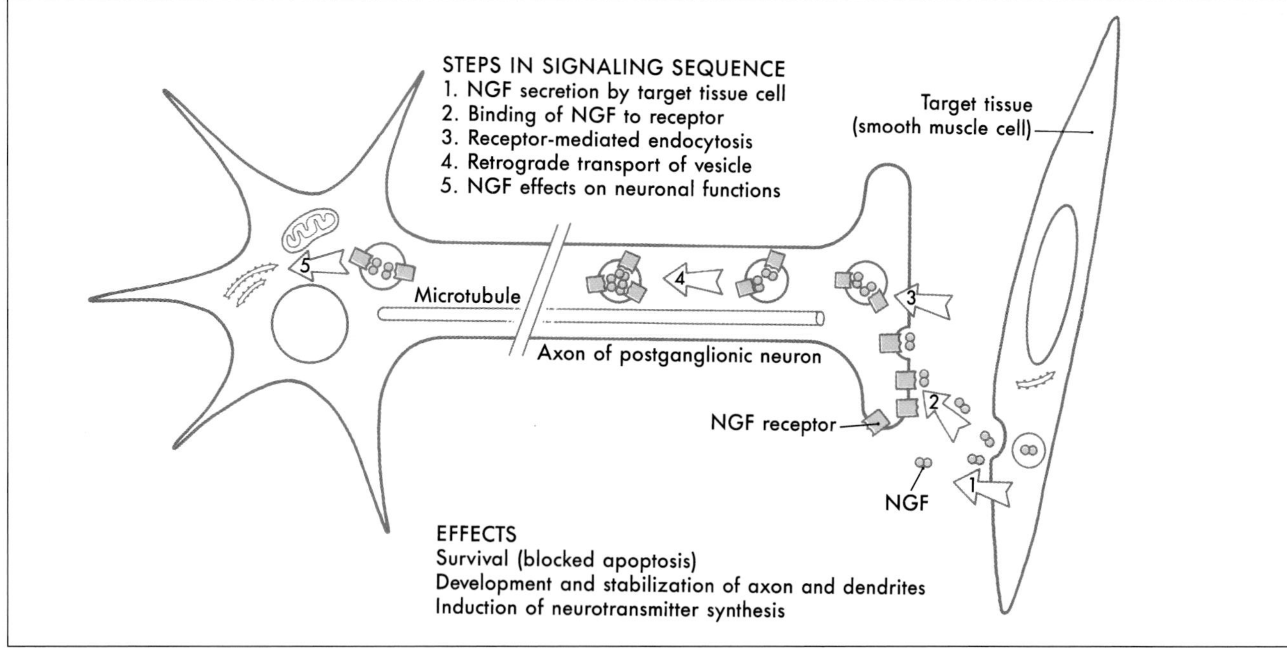

FIGURE 28-2 *Mechanism by which the neurotrophin, nerve growth factor (NGF), regulates the development of sympathetic ganglion neurons.*

adrenal medulla are related to sympathetic ganglion neurons in both derivation (neural crest) and function. These cells secrete catecholamines (mostly epinephrine) into the bloodstream in response to signals from preganglionic neurons. Thus, the sympathetic system, through this endocrine pathway, regulates functions of cells that are not directly contacted by nerve terminals.

Sympathetic Ganglia Cell bodies of sympathetic postganglionic neurons are generally grouped into discrete ganglia that are located at some distance from the target tissue. Most of them make up the *sympathetic chain (paravertebral) ganglia* and the *prevertebral ganglia* associated with the abdominal aorta and/or its large branches (the celiac, aorticorenal, superior mesenteric, and inferior mesenteric ganglia; Fig. 28-5). In addition, small clusters of cell bodies are also scattered among nerve fibers in communicating rami, the sympathetic trunk, and in peripheral plexuses.

The *sympathetic chain* extends along the full length of the vertebral column, but the number of ganglia does not exactly match the number of spinal nerves. Generally, there are 3 cervical, 10 to 11 thoracic, 3 to 5 lumbar, and 3 to 5 sacral sympathetic ganglia on each side, and a single *coccygeal ganglion* (*ganglion impar*) where the two chains meet caudally (Figs. 28-4, 28-5). Typically, the inferior cervical ganglion and the first thoracic ganglion fuse to form the *stellate ganglion*.

The sympathetic chain ganglia are connected to spinal nerves via *white* and *gray communicating rami*. The former contains *preganglionic fibers*, and the latter is composed of *postganglionic fibers* (Fig. 28-3). Consequently, only spinal nerves T1 to L2 have white rami (and contain preganglionic axons), whereas every spinal nerve is connected to the sympathetic trunk by a gray ramus that conveys postganglionic axons (Fig. 28-4).

Sympathetic postganglionic neurons in paravertebral ganglia send their axons in two general directions. First, some postganglionic fibers rejoin the spinal nerve via a gray ramus. These fibers distribute to blood vessels, sweat glands, and arrector pili muscles in the body wall and extremities (Figs. 28-3A,B; 28-4). These structures receive little or no parasympathetic innervation so they are exceptions to the dual arrangement of visceral innervation. Second, some of the postganglionic fibers that arise from cervical and upper thoracic sympathetic chain ganglia form *cervical* and *thoracic cardiac nerves* and *pulmonary nerves* that emerge directly from the ganglia (Figs. 28-3C, 28-5). These fibers innervate the vascular smooth muscle of the esophagus, heart, and lung; the glandular epithelium of respiratory structures; the smooth muscles of the esophagus; and cardiac muscle. Axons of these sympathetic neurons mingle with parasympathetic fibers of the vagus nerve to form the autonomic plexuses of the thorax.

The largest of the paravertebral (sympathetic chain) ganglia is the *superior cervical ganglion*. Postganglionic fibers from these cells innervate blood vessels and cutaneous targets of the face and scalp and neck of the territories supplied by the first four cervical nerves (Figs. 28-4, 28-5). The superior cervical ganglion also innervates the salivary glands, nasal glands, lacrimal gland, and structures of the eye such as the pupillary dilator muscle and the superior and inferior tarsal muscles (Fig. 28-5). Accordingly, a constellation of symptoms results from interruption (central or peripheral) of the sympathetic

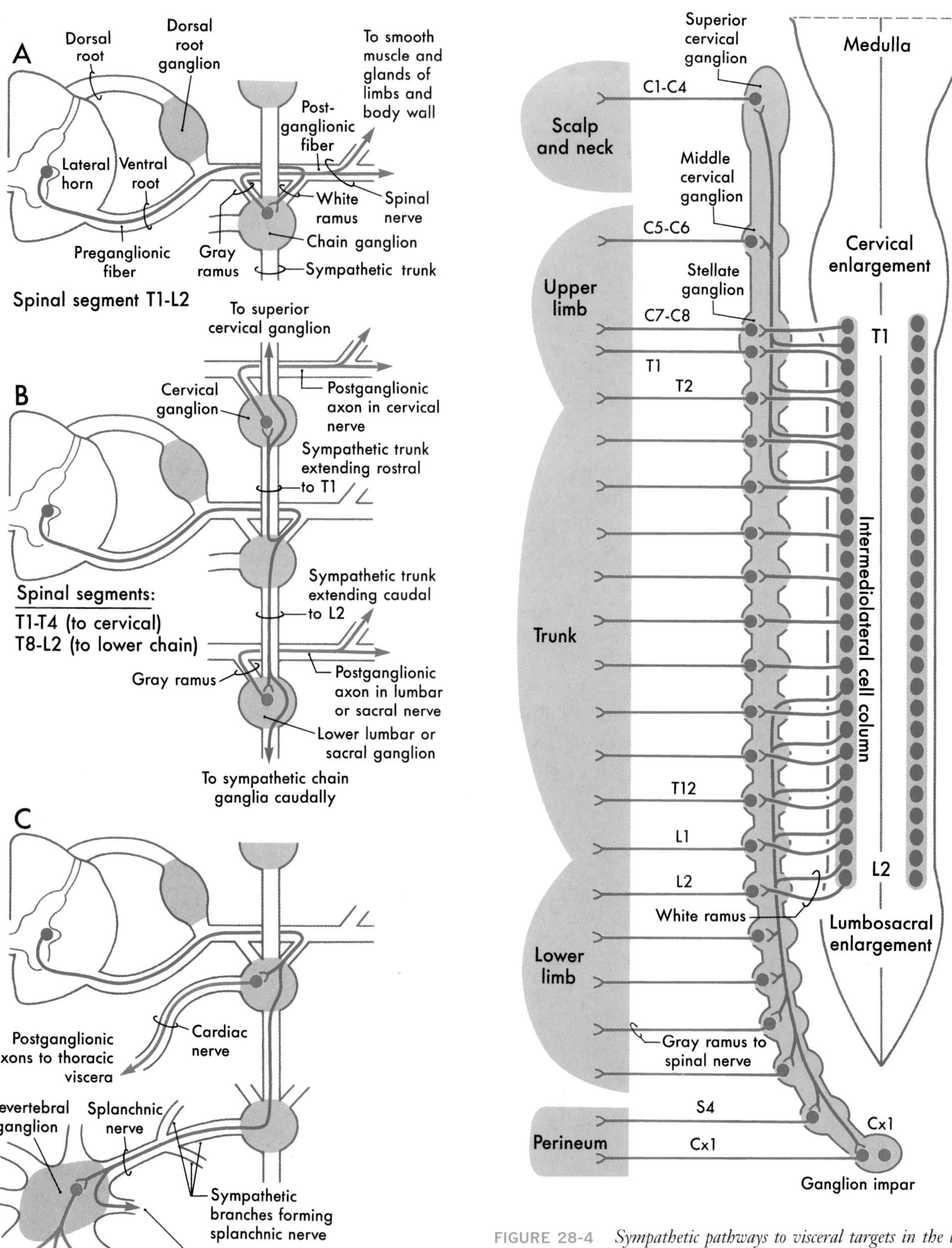

FIGURE 28-4 *Sympathetic pathways to visceral targets in the body wall, limbs, and scalp and neck. Postganglionic sympathetic fibers that distribute to targets of the head are shown in Figure 28-5.*

FIGURE 28-3 *The types of routes that can be taken by peripheral sympathetic pathways. A preganglionic fiber may either (A) terminate in the chain ganglion at its level of origin, (B) ascend or descend in the chain to terminate in different ganglia, or (C) traverse the ganglion to enter a splanchnic nerve and terminate in a prevertebral ganglion. Postganglionic neurons that originate in a sympathetic chain ganglion may exit via either the gray ramus and spinal nerve (A and B) or directly via a nerve such as the cardiac nerve shown in C.*

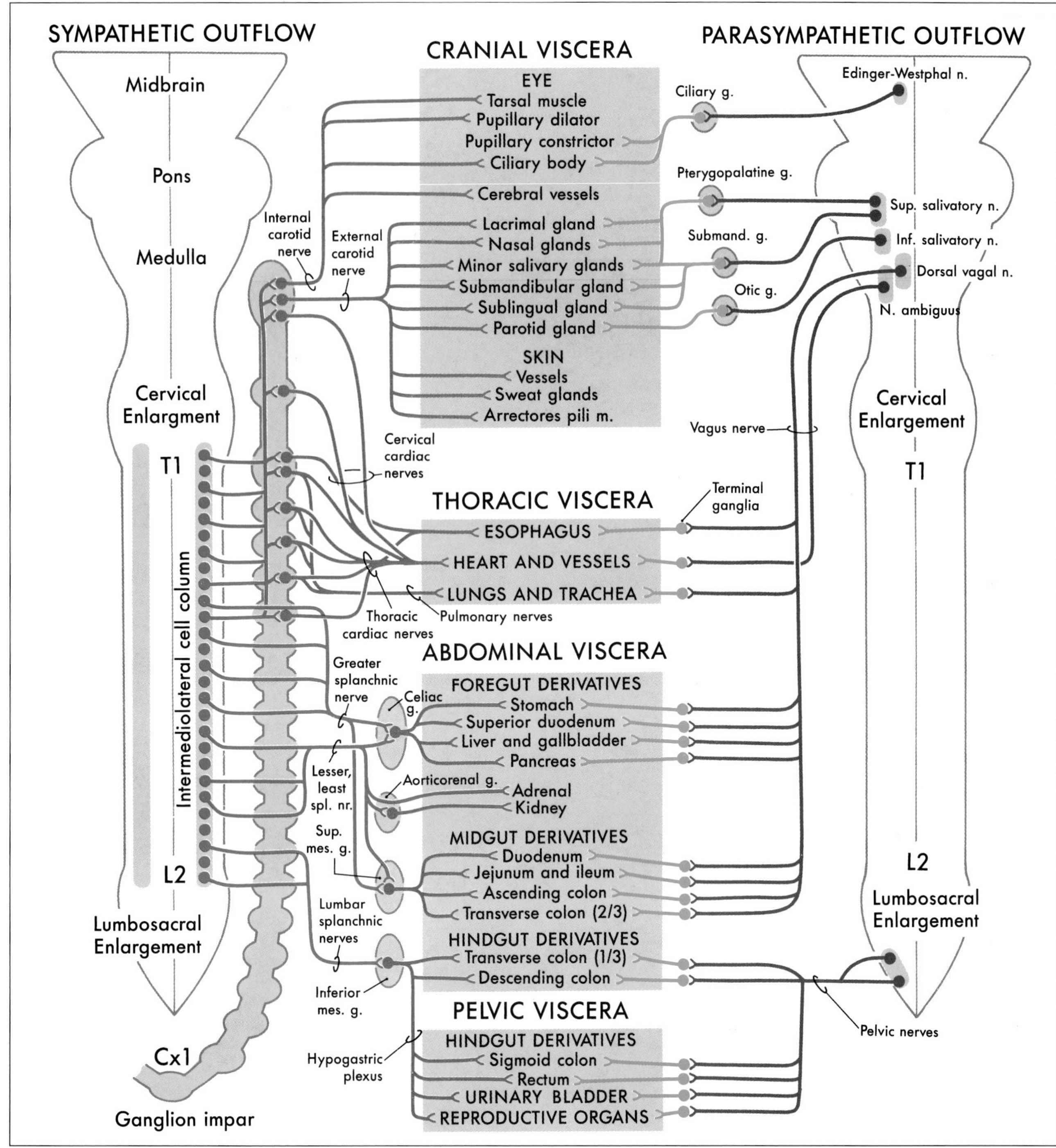

FIGURE 28-5 *Sympathetic and parasympathetic pathways to visceral targets in the head and body cavities. g., ganglion; inf., inferior; m., muscle; mes., mesenteric; n., nucleus; nr., nerve; submand., submandibular; sup., superior; spl., splanchnic.*

pathway through the superior cervical ganglion. These include constriction of the pupil (*miosis*) caused by the unopposed action of the parasympathetically innervated pupillary constrictor, drooping of the upper eyelid (*ptosis*) resulting from paralysis of the superior tarsal (Müller's) muscle, *flushing of the face* from loss of sympathetically mediated vascular tone, and diminished or absent sweating (*anhidrosis*) on the face. Collectively, these symptoms are known as *Horner's syndrome*.

The *prevertebral sympathetic ganglia* are found in visceral motor plexuses associated with the abdominal aorta and its major branches, and they receive input via the splanchnic nerves (Figs. 28-3C, 28-5). In general, the postganglionic fibers arising from each of these ganglia supply the same visceral targets as the corresponding branch of the aorta. Thus, the *celiac ganglion* is located at the origin of the celiac artery and supplies postganglionic fibers to the spleen and to viscera derived from the embryonic foregut. The *aorticorenal ganglion*, associated with the renal arteries, contains cell bodies of neurons that innervate the blood vessels of the kidneys. The *superior mesenteric ganglion* provides postganglionic fibers that supply the territory of the superior mesenteric artery (derivatives of the midgut). *Inferior mesenteric ganglion* neurons project to hindgut derivatives and to the urinary bladder, urethra, and reproductive organs. All abdominal aortic plexuses contain sympathetic fibers mixed with parasympathetic preganglionic fibers of either vagal or sacral origin.

Internal Organization of Sympathetic Ganglia Axons of preganglionic sympathetic neurons branch in the periphery and synapse on many postganglionic neurons; the output of these cells is thus widely *divergent* (Figs. 28-1, 28-6). The number of postganglionic neurons exceeds preganglionic neurons by over a hundred-fold. Each postganglionic neuron, however, receives synaptic input from a number of preganglionic neurons, so there is also considerable *convergence* within sympathetic ganglia.

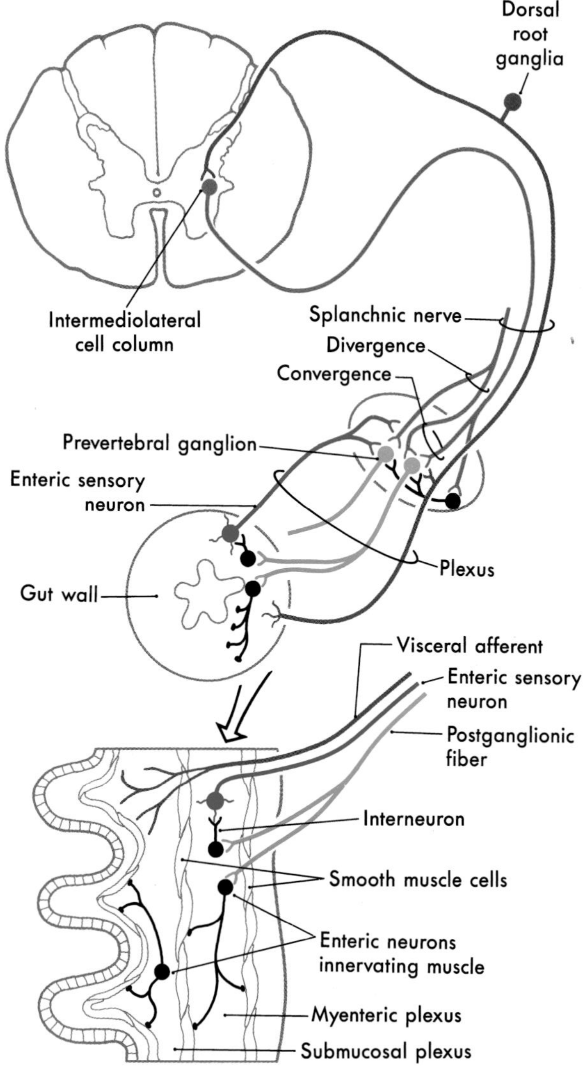

FIGURE 28-6 *Sources of synaptic input to postganglionic neurons in prevertebral ganglia, and the organization of the enteric nervous system. The lower drawing is a detail of the gut wall.*

Sympathetic ganglia are commonly referred to as "relay ganglia," implying they are simply sites of signal transduction between pre- and postganglionic neurons. However, it is clear that more complicated signal processing takes place in prevertebral ganglia. In addition to receiving diverging and converging input from functionally coded preganglionic neurons, postganglionic neurons are influenced by a variety of other sources. These include synaptic inputs from collaterals of general visceral afferent fibers and from local neurons (Fig. 28-6). These connections indicate a high degree of integration in prevertebral ganglia.

Functional and Chemical Coding Conditions of extreme excitement or exertion bring about a comprehensive ("en masse") activation of sympathetic outflow, with widespread effects. These include increases in heart rate, blood pressure, blood flow to skeletal muscles, blood glucose level, sweating, and pupil diameter. Concurrently, there are decreases in gut motility, digestive gland secretion, and blood flow to abdominal viscera and skin (Table 28-1). This constellation of effects has led to the concept that the sympathetic system acts in a global, nonselective manner. However, in less extreme conditions, there is ongoing, selective control of function-specific and target-specific subpopulations of pre- and postganglionic neurons. Such selectivity can be found, for example, among cell groups that control vascular tone. For instance, sympathetic pathways to blood vessels in the skin are primarily influenced by temperature, whereas sympathetic outflow to vessels in skeletal muscle responds mainly to changes in blood pressure signaled by baroreceptors. Also, stabilization of blood flow to the head during movement from a reclining to a standing position is a function of the sympathetic division. This requires rapid changes in vascular tone that must vary according to the region of the body.

Just as components of the sympathetic system can be regulated independently, there is evidence that distinct populations of preganglionic and postganglionic neurons exist. For example, even though preganglionic neurons are all cholinergic, some also express one or more neuropeptides, such as substance P and enkephalins. In addition, distinct populations of preganglionic neurons have been identified on the basis of (1) location of the cell body in the visceral motor cell groups of the spinal cord, (2) morphology of the dendritic tree, and (3) specific target cell type among ganglion cells.

More is known about the functional significance of chemical coding in postganglionic sympathetic neurons. While most of these cells use *norepinephrine* as a transmitter, some are *cholinergic*. The latter neurons provide secretomotor innervation to sweat glands and vasodilator innervation to blood vessels in skeletal muscle. In addition, postganglionic cells express a variety of neuropeptides, some of which have been linked to specific functional populations of cells (Fig. 28-7, Table 28-2). Most prevalent among sympathetic peptides is *neuropeptide Y*, which is released along with norepinephrine by vasoconstrictor postganglionic fibers. This peptide has multiple effects at the adrenergic ending. These include stimulation of vascular smooth muscle contraction, potentiation of epinephrine effects, and, paradoxically, inhibition of norepinephrine release.

Receptor Types in Sympathetic Targets The effect of a neurotransmitter on a target cell is determined by the nature of the target cell receptor and the particular signal transduction mechanism to which it is linked. Thus, the effects of norepinephrine, the main neurotransmitter of most postganglionic neurons, and epinephrine, the main hormone of the adrenal medulla, vary among different target cells according to the type or types of adrenergic receptor they express (subclasses of α- and β-adrenergic receptors). For example, α_2 receptors on vascular smooth muscle cells mediate vasoconstriction, whereas β_2 receptors on this cell type mediate relaxation. Increases in heart rate and cardiac output are mediated by β_1 receptors on cardiac muscle. Epinephrine is a more potent ligand than norepinephrine at both α- and β-adrenergic receptors. Consequently, epinephrine or a related agonist is administered clinically to elicit a sympathetic response, for example to induce bronchiolar dilation during an asthma attack. As mentioned, some tissues receive cholinergic sympathetic innervation (Fig. 28-7). For example, stimulation of sweat gland secretion is mediated by muscarinic acetylcholine receptors, as is the relaxation of vascular smooth muscle in skeletal muscles.

PARASYMPATHETIC DIVISION

Preganglionic and Postganglionic Neurons Compared to the sympathetic division, the parasympathetic division is more restricted in its distribution. The cell bodies of parasympathetic preganglionic neurons are located in either sacral segments S2 to S4 or in the nuclei that provide the *general visceral efferent* (GVE) fibers that travel in cranial nerves III, VII, IX, and X (Table 28-3, Fig. 28-5). The parasympathetic division of the visceral motor system is accordingly called the *craniosacral* system, as distinct from the sympathetic division, which is *thoracolumbar*.

The cell bodies of postganglionic parasympathetic neurons supplying cranial structures are located in discrete ganglia, and, in general, their axons travel distally with branches of the trigeminal nerve. Cell bodies of parasympathetic postganglionic neurons supplying viscera of the body cavities are not grouped into macroscopic ganglia. Rather, these cells are scattered within nerve plexuses of the target organ or in the wall of the gut (*terminal* or *intermural ganglia*), where they intermingle with neurons of the *enteric nervous system*.

Parasympathetic Outflow Pathways The visceral motor component of the oculomotor nerve arises from the *Edinger-Westphal nucleus*. These preganglionic fibers terminate in the ciliary ganglion. Axons of postganglionic cells of the ciliary ganglion innervate the sphincter muscle of the iris (for pupillary constriction) and the ciliary muscle (for near-vision accommodation) (Table 28-3, Fig. 28-5).

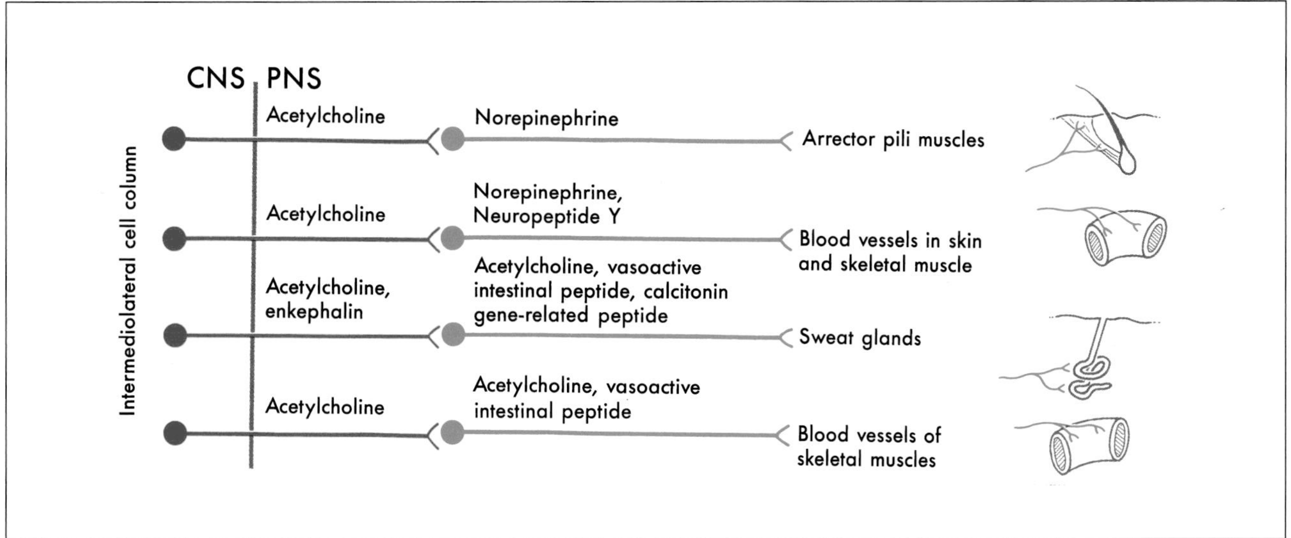

FIGURE 28-7 *Chemical coding of sympathetic pre- and postganglionic neurons.*

TABLE 28-3

PERIPHERAL PATHWAYS OF PARASYMPATHETIC OUTFLOW

NERVE	LOCATION OF PREGANGLIONIC CELL BODIES	COURSE OF PREGANGLIONIC FIBERS	LOCATION OF POSTGANGLIONIC CELL BODIES	TARGET TISSUES	EFFECT ON TARGET
Oculomotor	Midbrain: Edinger-Westphal n.	With nerve III	Ciliary ganglion	Ciliary body, pupillary constrictor	Ciliary muscle contraction; contraction of pupillary sphincter
Facial	Pons: Superior salivatory n.	Nervus intermedius, greater petrosal nr. to pterygopalatine g. or chorda tympani to submandibular g.	Pterygopalatine ganglion and submandibular ganglion	Lacrimal gland, nasal glands, submandibular and sublingual glands	↑ secretion
Glossopharyngeal	Medulla: Inferior salivatory nucleus	Tympanic branch of IX, tympanic plexus, lesser petrosal nr.	Otic ganglion	Parotid gland	↑ secretion
Vagus	Medulla: Dorsal motor vagal nucleus and nucleus ambiguus*	Various branches of X	Terminal ganglia in or on wall of target organ	Heart and great vessels, respiratory system, esophagus, foregut and midgut derivatives	↓ heart rate; bronchial constriction; ↑ blood flow to gut; ↑ peristalsis and secretion
Sacral splanchnic	S2 to S4 of spinal cord: intermediate gray matter	Pelvic nr.	Terminal ganglia in or on wall of target organ	Hindgut derivatives, reproductive organs, urinary bladder	

* Some GVE preganglionic parasympathetic cells that innervate the heart are found in this nucleus, although its main function is to provide SVE fibers that distribute on cranial nerves IX and X.
↑ = increase, ↓ = decrease

The GVE preganglionic parasympathetic fibers of the *facial nerve* originate in the *superior salivatory nucleus*. Some distinguish a separate lacrimal nucleus that targets the lacrimal gland. Preganglionic GVE axons from the superior salivatory nucleus exit the brainstem in the *intermediate nerve*, which is classically considered a part of the facial nerve. Some of these fibers course via the greater petrosal nerve to terminate in the *pterygopalatine ganglion*, which supplies the lacrimal gland and nasal and palatal mucous glands. Other preganglionic fibers travel via the chorda tympani to the *submandibular ganglion*, which innervates the submandibular and sublingual salivary glands (Table 28-3, Fig. 28-5).

The *glossopharyngeal nerve* contains preganglionic parasympathetic fibers that originate in the *inferior salivatory nucleus*. These fibers take a tortuous course, via the tympanic nerve and plexus, to form the lesser petrosal nerve, which ends in the *otic ganglion*. Postganglionic fibers from the otic ganglion join the auriculotemporal nerve to reach the parotid gland (Table 28-3, Fig. 28-5).

The visceral motor component of the vagus nerve provides parasympathetic innervation to organs of the thoracic and abdominal cavities. Preganglionic GVE fibers of the vagus nerve originate in the *dorsal motor vagal nucleus*. In addition, a part of the *nucleus ambiguus* contains GVE preganglionic cells, whose axons travel with the vagus to innervate the heart. It should be emphasized, however, that the main outflow from the nucleus ambiguus consists of SVE fibers to the glossopharyngeal and vagus nerves. Preganglionic fibers of the vagus terminate on postganglionic neurons located in the walls of viscera of the thorax and abdomen (Table 28-3, Fig. 28-5). Thus, unlike cranial nerves III, VII, and IX, postganglionic neurons of the vagus nerve are not aggregated into discrete ganglia.

The *sacral component of the parasympathetic division* innervates the lower digestive tract (beginning at about the left colic flexure) and the urinary bladder, urethra, and reproductive organs. Preganglionic neurons of the *sacral parasympathetic nucleus* occupy a position at sacral levels S2 to S4 comparable to the intermediolateral cell column in thoracic levels (see Fig. 28-9). Preganglionic fibers exit the spinal cord via ventral roots and form the *pelvic nerves* (*nervi erigentes*). These nerves mingle with sympathetic fibers of the inferior hypogastric plexuses to form the pelvic visceral plexus lateral to the rectum, bladder, and uterus (Table 28-3).

Functional and Chemical Coding Preganglionic parasympathetic neurons, like preganglionic sympathetic neurons, use *acetylcholine* as their main neurotransmitter. Postganglionic parasympathetic neurons are also cholinergic. Both pre- and postganglionic parasympathetic neurons release molecules in addition to the principal transmitter at their terminals. These include neuropeptides, most prominently vasoactive intestinal peptide, which act as modulators of the postsynaptic response to the main transmitter.

Receptor Types in Parasympathetic Targets Nicotinic receptors are limited to skeletal muscle and cholinergic synapses in autonomic ganglia and the CNS. Muscarinic cholinergic receptors appear to be the only class involved in the response of smooth muscle, cardiac muscle, and glandular cells to acetylcholine. The nature of the response depends on which type of muscarinic receptor (M_1, M_2, etc.) is expressed. For example, parasympathetic stimulation of gastric acid secretion is mediated by M_1 muscarinic receptors, whereas M_2 receptors mediate parasympathetic depression of heart rate and contraction of cardiac muscle.

The cholinergic receptor blocker atropine has effects that are clinically useful in certain circumstances. These include pupillary dilation, relaxation of bronchiolar muscle, and reduction of peristalsis and secretion in the stomach.

ENTERIC NERVOUS SYSTEM

It has been estimated that the digestive tract contains about as many nerve cells as the spinal cord. This enormous population of enteric neurons is concentrated mostly in the *myenteric* and *submucosal plexuses* of the gut wall. Neurons of the gut exhibit a wide range of structural, chemical, and physiologic properties, and they form an elaborate system of neural connections (Fig. 28-6).

The *enteric nervous system* is influenced by inputs of the sympathetic and parasympathetic divisions, and some of these enteric cells may be regarded as postganglionic parasympathetic neurons. However, the enteric nervous system is largely self-sufficient in its regulation of digestive tract activities. This regulation is a result of intrinsic neuronal circuits involving both sensory and motor neurons, as well as interneurons within the myenteric and submucosal plexuses (Fig. 28-6). For example, although smooth muscle cells undergo spontaneous contraction, there are elaborate neural circuits that coordinate waves of relaxation and contraction (peristalsis). These intrinsic neuronal circuits include mechanoreceptors, which signal stretch, and burst-type oscillator cells that spontaneously generate a chain of action potentials. Noradrenergic cells excited by these signals inhibit smooth muscle contraction. In addition to intrinsic circuits that control peristalsis, other circuits regulate blood flow and secretion in response to the contents of the lumen.

Enteric neurons express a large variety of neuropeptides, such as vasoactive intestinal peptide, neuropeptide Y, and cholecystokinin. Many of these peptides were first identified in the gut but have since been found in other central and peripheral neurons.

REGULATION OF VISCERAL MOTOR OUTFLOW

Sensory input occurs at every level of the visceral motor pathway, including prevertebral postganglionic neurons, preganglionic neurons, and a wide variety of CNS structures that project, either directly or indirectly, to preganglionic neurons. Many different kinds of sensory information are integrated by a series of CNS structures collectively termed the *central autonomic network* (CAN), which generates coordinated signals to the visceral motor, endocrine, and somatic motor outflow pathways. Although visceral motor activities are generally beyond conscious control, emotional status and mental activity clearly influence visceral structures. Accordingly, the CAN integrates input from higher CNS centers involved in cognition and complex behavioral functions.

Major CNS Components The preganglionic neurons of the autonomic motor pathways are influenced by cells in various brainstem and forebrain areas. The *hypothalamus* is the highest integrator of autonomic and endocrine functions. It directly regulates the secretory activity of the anterior and posterior pituitary and has reciprocal connections with the solitary nucleus and other components of the CAN in the forebrain and brainstem. Some hypothalamic nuclei project directly to preganglionic visceral motor neurons in the dorsal vagal nucleus, nucleus ambiguus, and intermediolateral cell column. The organization and functions of the hypothalamus are considered in the next chapter.

Because of its diverse connections, the solitary nucleus is the most important brainstem structure coordinating autonomic functions. It receives general and special visceral sensory input, and it projects to vagal motor neurons, to salivatory and reticular nuclei, and to populations of brainstem neurons, which in turn project to sympathetic preganglionic neurons. The solitary nucleus also has reciprocal connections with other components of the CAN.

Other cell groups that are important in autonomic regulation reside in the reticular formation of the brainstem. These neurons are not always restricted to specific nuclei, and they are therefore sometimes designated by their relative positions. For example, cells in the *rostral ventrolateral medulla* project to the intermediolateral cell column, particularly to preganglionic neurons involved in cardiovascular regulation. This area is called the *vasopressor center* because stimulation results in increased peripheral vascular resistance and increased cardiac output. Areas such as this have been designated *centers*, for example, the respiration center, micturition center, and vomiting center. Although this terminology is convenient, it should be understood that these are not well-defined anatomic entities, but are components of widely distributed neural networks.

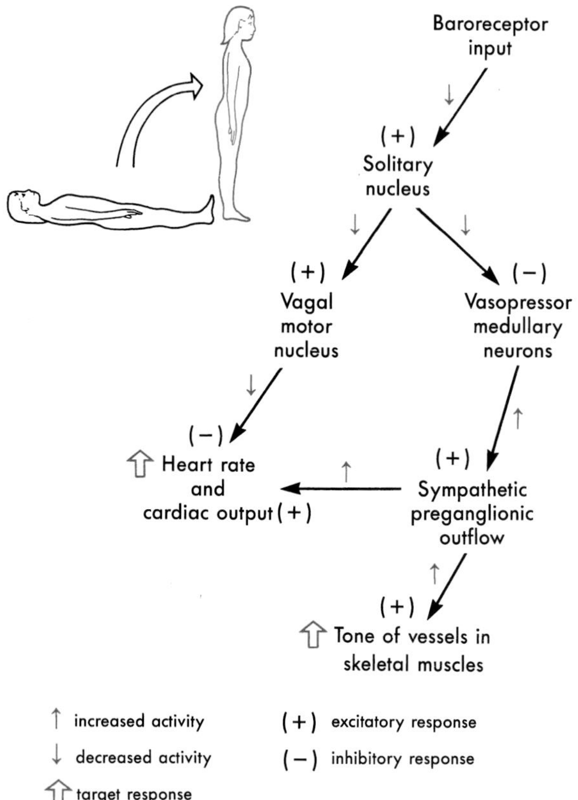

FIGURE 28-8 *Pathways of the baroreceptor reflex. A sudden movement to an upright position results in modification of neuronal activity (red) in these pathways to produce the compensatory changes in the cardiovascular system.*

Cardiovascular System The function of the cardiovascular system is influenced by mental activity, emotional state, posture, muscular exertion, visceral activity, body temperature, and concentrations of blood gases and electrolytes. In addition to mechanisms that regulate blood pressure, there is precise neural control of blood flow to specific organs and regions of the body.

The *baroreceptor reflex* (Fig. 28-8) functions to buffer blood pressure against a sudden change in posture. Failure of this reflex results in *orthostatic hypotension*, a severe drop in blood pressure when the patient assumes an upright position. Primary visceral sensory neurons of the glossopharyngeal and vagus nerves convey signals from mechanoreceptors in the carotid and aortic sinuses centrally, where they terminate in the solitary nucleus (see Fig. 18-8). Projections of solitary neurons influence the tonic activity of parasympathetic (vagal) output to the heart and sympathetic output to the heart and peripheral vessels.

When a reclining individual stands, there is a rapid reduction in baroreceptor discharge that results in a *decrease* in excitatory signals from the solitary nucleus to two brainstem targets (Fig. 28-8). One of these targets is vagal preganglionic parasympathetic neurons that suppress heart rate and cardiac output. Thus, a reduction of baroreceptor discharge results in a release of the heart from this inhibitory parasympathetic drive. The second target of solitary cells consists of *vasopressor neurons* in the *rostral ventrolateral medulla*. The neurons in this region have intrinsic pacemaker features and receive inhibitory inputs from the solitary nucleus. When these rostral neurons are released from the inhibitory drive of solitary neurons, the result is increased sympathetic outflow (Fig. 28-8). This input is mediated via a major descending projection from these rostral medullary cells to the sympathetic preganglionic neurons in the intermediolateral cell column. This mediates increased cardiac output and increased resistance in vascular beds of skeletal muscle and abdominal visceral organs, but not of the skin, heart, and brain. Some of this solitary input to vasopressor neurons may be relayed through vasodepressor cell groups in the caudal ventrolateral medulla. Thus, when *an individual moves from a reclining to standing posture, the resultant pooling of blood in the lower half of the body is quickly countered by increased vascular tone and increased cardiac output*. Without this reflex, movement to a standing position results in dizziness or fainting because of decreased blood flow to the brain. This is a serious consequence of many forms of autonomic dysfunction.

The *chemoreceptor reflex* maintains homeostasis of blood gas composition by adjusting respiration, cardiac output, and peripheral blood flow. Decreased pO_2 and increased pCO_2, detected by receptors in the carotid and aortic bodies, are signaled by glossopharyngeal and vagal afferents that terminate in the solitary nucleus. Within the medulla, the reflex pathway for cardiovascular effects parallels that for the baroreceptor reflex. A decrease in blood pO_2 activates this reflex and promotes increased heart rate and vascular tone. This results in a decreased blood flow to skeletal muscles and viscera, whereas blood flow to the brain is maintained. Thus, proportionately more oxygenated blood is available to the brain than to skeletal muscle and viscera. The resulting conservation of oxygen preserves vital functioning of the CNS.

The cardiovascular component of the chemoreceptor reflex is closely coordinated with respiration, a somatic motor function coordinated by other neurons of the brainstem reticular formation. For example, if breathing is suspended (as in diving), heart rate is slowed (*bradycardia*) rather than accelerated (*tachycardia*).

Urinary Bladder and Micturition Emptying of the urinary bladder, *micturition*, is brought about by contraction of smooth muscle of the bladder wall (*detrusor muscle*) and relaxation of skeletal muscle of the *external urethral sphincter* (Fig. 28-9). Contraction of the detrusor is mediated by parasympathetic outflow. Preganglionic neurons from the sacral cord innervate postganglionic neurons in the bladder wall (Fig. 28-9). The bladder wall also has a sympathetic innervation. Its influence is mainly inhibitory on both the detrusor muscle and the parasympathetic postganglionic neurons in the bladder wall (Fig. 28-9). The external urethral sphincter, which is subject to both reflex and voluntary control, is supplied by alpha motor neurons in segments S3 to S4.

During periods of urine storage, activity of bladder afferent neurons is low. This results in (1) low activity of parasympathetic excitatory innervation to the detrusor, (2) tonic activity of sympathetic neurons that inhibit both the parasympathetic ganglion cells in the bladder wall and the detrusor muscle directly, and (3) tonic activity of sacral somatic motor neurons mediating constriction of the external sphincter. As urine accumulates, pressure on the bladder wall activates tension receptors until bladder afferent activity rises to a threshold level. This induces micturition by way of both spinal and brainstem reflexes that result in inhibition of sympathetic outflow, activation of parasympathetic outflow, and inhibition of somatic motor neurons supplying external sphincter muscle.

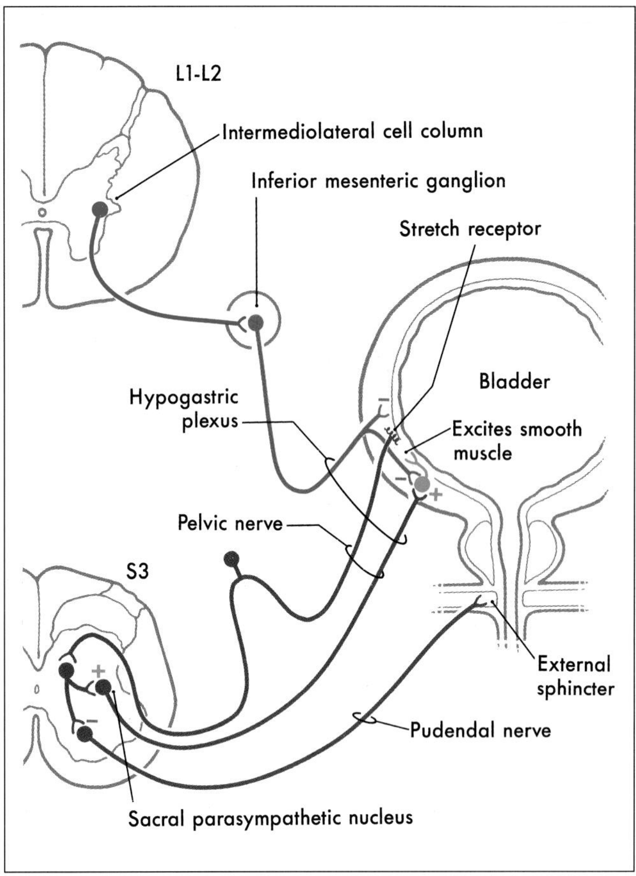

FIGURE 28-9 *Neural pathways mediating control of the urinary bladder.*

SOURCES AND ADDITIONAL READINGS

Appenzeller O: *The Autonomic Nervous System*, 4th Ed. Elsevier, Amsterdam, 1990.

Bannister R, Mathias CJ (eds): *Autonomic Failure. A Textbook of Clinical Disorders of the Autonomic Nervous System*, 3rd Ed. Oxford University Press, Oxford, 1992.

Benarroch EE: Neuropeptides in the sympathetic system: presence, plasticity, modulation, and implications. Ann Neurol 36:6–13, 1994.

Brodal A: *Neurological Anatomy*, 3rd Ed. Oxford University Press, New York, 1981.

Gabella G: *Structure of the Autonomic Nervous System*. Chapman and Hall, London, John Wiley & Sons, New York, 1976.

Jänig W, Schmidt RF (eds): *Reflex Sympathetic Dystrophy*. VHC, Weinheim (Federal Republic of Germany), VHC Publishers, New York, 1990.

Loewy AD, Spyer KM (eds): *Central Regulation of Autonomic Functions*. Oxford University Press, New York, Oxford, 1990.

Low PA (eds): *Clinical Autonomic Disorders. Evaluation and Management*. Little, Brown, Boston, 1992.

Pick J: *The Autonomic Nervous System. Morphological, Comparative, Clinical and Surgical Aspects*. JB Lippincott, Philadelphia, 1970.

Shephard GM: *Neurobiology*, 3rd Ed. Oxford University Press, New York, 1994.

CHAPTER TWENTY-NINE

The Hypothalamus

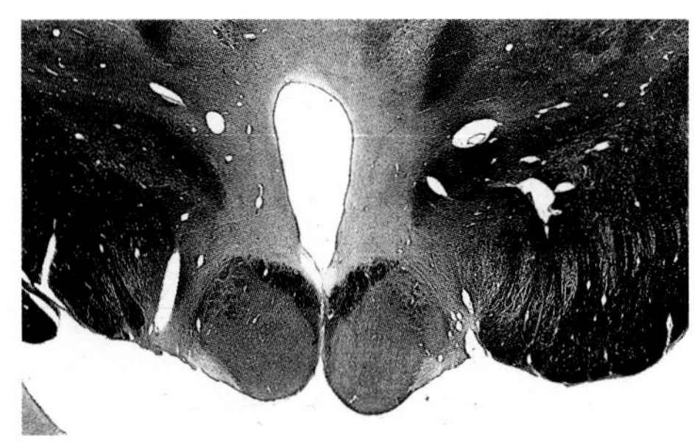

S. G. P. HARDY • R. B. CHRONISTER

One of the most rostral cell groups to influence visceral function, and the one that has direct input to all other visceral nuclei in the neuraxis, is the hypothalamus. In addition to its role in regulating visceromotor functions, the hypothalamus also influences neural circuits that modify behavior.

OVERVIEW

The hypothalamus is the part of the diencephalon concerned with the central control of visceral functions (through the *visceromotor* and *endocrine systems*) and with affective or emotional behavior (via the *limbic system*) (Fig. 29-1). It is primarily directed at maintaining *homeostasis*. Among the many functions partially regulated by the hypothalamus are regulation of water and electrolyte balance, food intake, temperature, blood pressure, possibly the sleep-waking mechanism, circadian rhythmicity, and general body metabolism. The hypothalamus (at about 4 g) is dwarfed in size by the rest of the brain (weighing approximately 1,400 g). However, it is perhaps the most important 4 g in the entire body. In short, the hypothalamus influences our responses to both the internal and external environments (Fig. 29-1) and is necessary for life.

BOUNDARIES OF THE HYPOTHALAMUS

The anterior boundary of the hypothalamus is the *lamina terminalis*, a thin membrane that extends ventrally from the anterior commissure to the rostral edge of the optic chiasm (Fig. 29-2A). The lamina terminalis separates the hypothalamus from the more anterior septal nuclei. Dorsally, the hypothalamus is bounded by the *hypothalamic sulcus*, a shallow groove that separates the hypothalamus from the dorsal thalamus (Fig. 29-2A,B). The lateral boundary of the hypothalamus is formed anteriorly by the substantia innominata and posteriorly by the medial edge of the posterior limb of the *internal capsule* (Fig. 29-2B,C; see also Fig. 14-6). Medially, the hypothalamus is bordered by the inferior portion of the *third ventricle*. Posteriorly, the hypothalamus is not sharply demarcated, merging instead into the *midbrain tegmentum* and the *periaqueductal gray*.

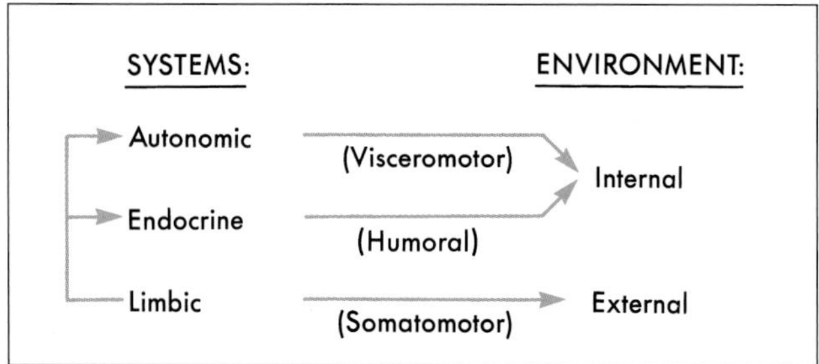

FIGURE 29-1 *The interrelationships among the autonomic, endocrine, and limbic systems. All three systems are under the control of the hypothalamus.*

DIVISIONS OF THE HYPOTHALAMUS

The hypothalamus can be divided into the *preoptic area* and the *lateral*, *medial*, and *periventricular zones* (Fig. 29-3). The preoptic area is a transition region that extends rostrally, by passing laterally to the lamina terminalis, to form a continuation with structures in the basal forebrain. Three zones are located posterior to the preoptic area. The thin periventricular zone is the most medial and is subjacent to the ependymal cells that line the third ventricle. The medial zone is located lateral to the periventricular zone, and a line drawn from the postcommissural fornix to the mammillo-thalamic tract separates it from the lateral zone (Fig. 29-3).

Preoptic Area The preoptic area, although functionally a part of the hypothalamus (and diencephalon) is embryologically derived from the telencephalon. This area is composed primarily of the medial and lateral preoptic nuclei (Fig. 29-3). The *medial preoptic nucleus* contains neurons that manufacture luteinizing hormone-releasing hormone (LHRH) or gonadotropin-releasing hormone. LHRH is transported along the *tuberoinfundibular tract* to capillaries of the hypothalamohypophyseal portal system and thence to the anterior lobe of the pituitary gland (Fig. 29-4), where it causes the release of gonadotropins (luteinizing hormone and follicle stimulating hormone). Because gonadotropin release is continuous in males and cyclical in females, the medial preoptic nucleus of males tends to be more active and consequently larger than that of females. Accordingly, the medial preoptic nucleus is often referred to as the *sexually dimorphic* nucleus of the preoptic area. The medial preoptic nucleus also influences behaviors that are related to eating, reproductive activities, and locomotion. The *lateral preoptic nucleus* is located immediately rostral to the lateral hypothalamic zone (see Fig. 29-3). The function of this nucleus is not fully established. However, through its connections with the ventral pallidum, it may function in part in locomotor regulation. Some consider the nuclei of the preoptic area to be part of the supraoptic region of the medial hypothalamic zone.

Lateral Zone The *lateral zone* (Fig. 29-3) contains a large bundle of axons collectively called the *medial forebrain bundle* (Fig. 29-3; see also Fig. 29-6). This diffuse bundle of fibers traverses the lateral hypothalamic zone and interconnects the hypothalamus with rostral areas such as the septal nuclei and with caudal regions such as the brainstem reticular formation.

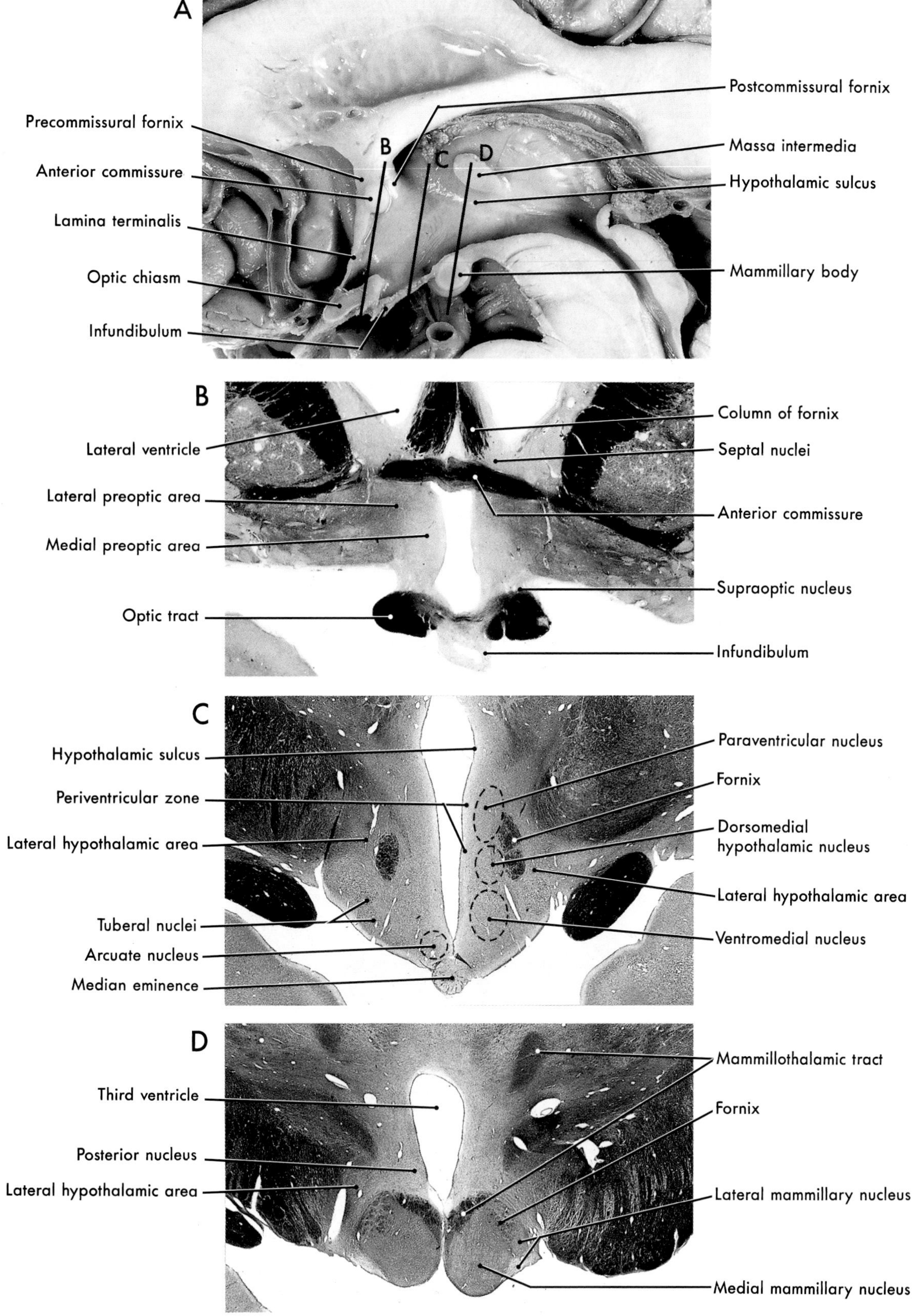

FIGURE 29-2 *Midsagittal view (A) of the brain emphasizing hypothalamic structures. Cross sections of the hypothalamus through supraoptic (B), tuberal (C), and mammillary (D) regions. The myelin-stained sections in B, C, and D correspond to the comparably labeled lines in A.*

Hypothalamic sulcus
Postcommissural fornix
Mammillary nuclei
Lamina terminalis
Lamina terminalis
Medial preoptic nucleus
Lateral preoptic nucleus
Preoptic area
Postcommissural fornix
Supraoptic region
Lateral zone
Periventricular zone
Medial forebrain bundle
Tuberal region
Mammillothalamic tract
Third ventricle
Lateral zone
Mammillary region
Medial zone

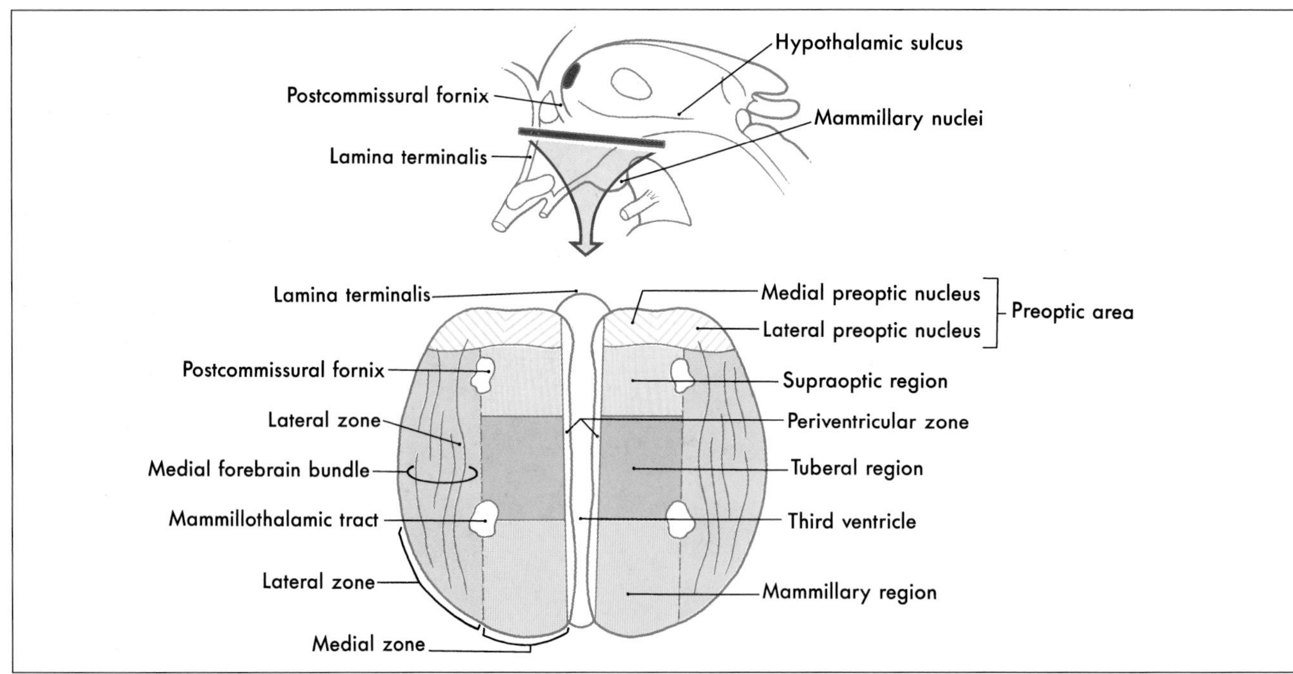

FIGURE 29-3 *Diagrammatic representation of the hypothalamus in the axial (horizontal) plane showing the zones and regions.*

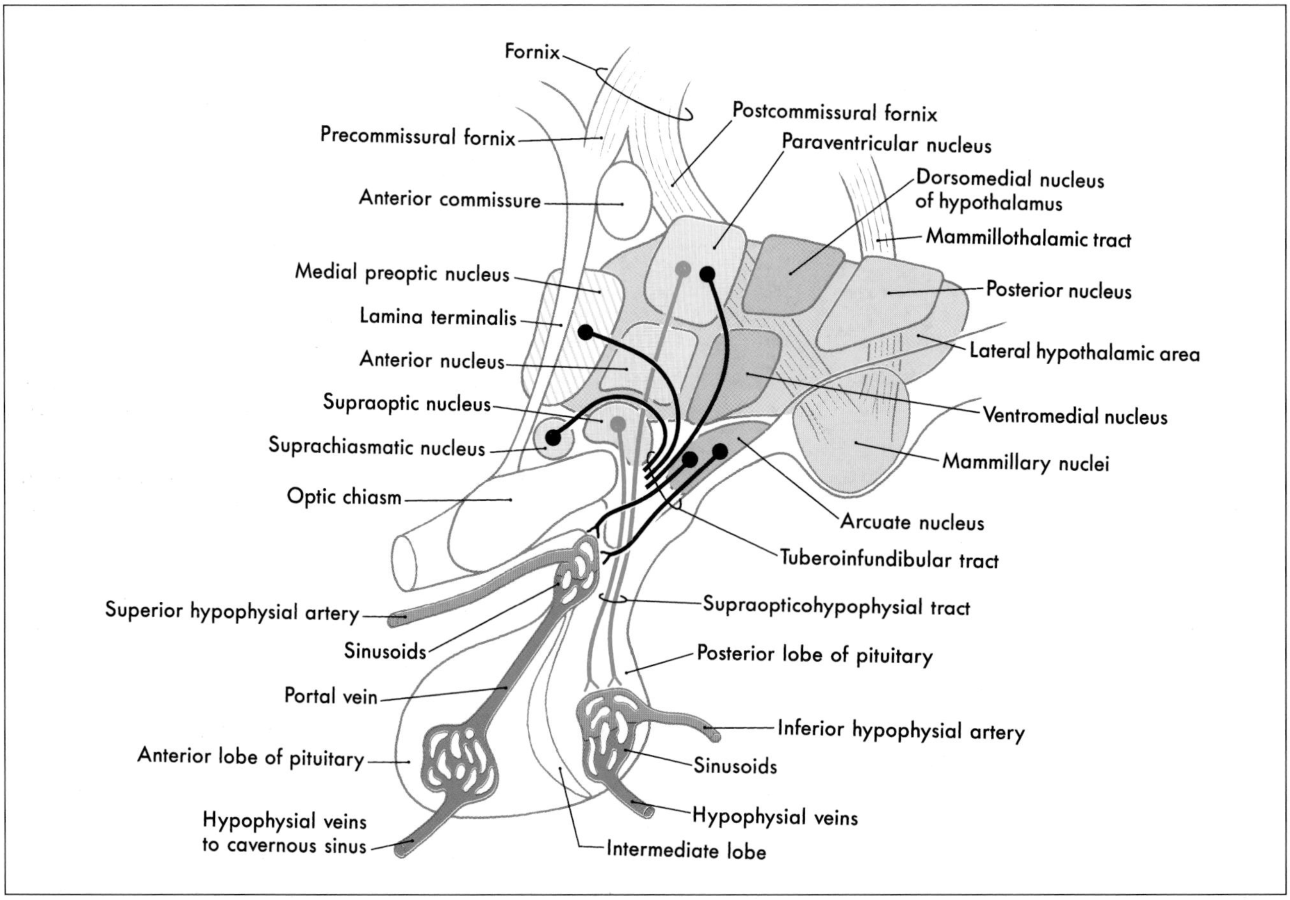

Fornix
Postcommissural fornix
Precommissural fornix
Paraventricular nucleus
Dorsomedial nucleus
of hypothalamus
Anterior commissure
Mammillothalamic tract
Medial preoptic nucleus
Posterior nucleus
Lamina terminalis
Lateral hypothalamic area
Anterior nucleus
Supraoptic nucleus
Ventromedial nucleus
Suprachiasmatic nucleus
Mammillary nuclei
Optic chiasm
Arcuate nucleus
Tuberoinfundibular tract
Superior hypophysial artery
Supraopticohypophysial tract
Sinusoids
Posterior lobe of pituitary
Portal vein
Inferior hypophysial artery
Anterior lobe of pituitary
Sinusoids
Hypophysial veins
Hypophysial veins
to cavernous sinus
Intermediate lobe

FIGURE 29-4 *Midsagittal view of the hypothalamus emphasizing the nuclei, which contribute to the tuberoinfundibular and supraopticohypophysial tracts, the hypophysial portal system, and the general relations of the fornix and mammillothalamic tract.*

The *lateral hypothalamic zone* comprises a large, diffuse population of neurons commonly called the *lateral hypothalamic area* as well as smaller condensations of cells located in its ventral portions. The latter cell groups are the *lateral hypothalamic nucleus* and the *tuberal nuclei*. The *lateral hypothalamic nucleus* is a loose aggregation of relatively large cells that extends throughout the anteroposterior extent of the lateral hypothalamic zone. This nucleus represents a "feeding center." Stimulation of this nucleus in laboratory animals promotes feeding behavior; destruction of it causes feeding behavior to attenuate and the animal loses weight. The *tuberal nuclei* consist of small clusters of neurons, each containing small, pale, multipolar cells. Some tuberal neurons project into the tuberoinfundibular tract and, therefore, may convey releasing hormones to the hypophysial portal system. Others send a histaminergic input to the cerebellum that may be involved in the regulation of motor activity.

Medial Zone The *medial zone* is a cell-rich region composed of many individual nuclei (Figs. 29-3, 29-4). It is divided into three regions: the *supraoptic* (*chiasmatic*) *region*, the *tuberal region*, and the *mammillary region* (Fig. 29-3). The *supraoptic region* is located dorsal to the optic chiasm. The *tuberal region* is the widest part of the hypothalamus and corresponds, in general, to the position of the *tuber cinereum*. The *mammillary region*, the most posterior of the three, corresponds to the location of the mammillary bodies.

The *supraoptic region* contains four nuclei: the *supraoptic*, *paraventricular*, *suprachiasmatic*, and *anterior nuclei* (Figs. 29-2A,B; 29-4). Neurons of the *supraoptic* and *paraventricular nuclei* contain oxytocin and antidiuretic hormone (vasopressin) and transmit these substances to the posterior pituitary by way of the *supraopticohypophysial tract* for release into the circulatory system (Fig. 29-4). The functions of these hormones are discussed later in this chapter. The *suprachiasmatic nucleus* receives direct input from the retina and can influence other hypothalamic structures such as the medial preoptic nucleus. It is believed that the suprachiasmic nucleus may mediate *circadian rhythms*, these being the hormonal fluctuations that are secondary to light-darkness cycles. The *anterior nucleus* is located immediately caudal to the preoptic area. Although this nucleus participates in a wide range of visceral and somatic functions, many of its neurons are involved in the maintenance of body temperature.

The *tuberal region* contains three nuclei: the *ventromedial*, *dorsomedial*, and *arcuate nuclei* (Figs. 29-2A,C; 29-4). The *ventromedial nucleus*, one of the largest and best defined of the hypothalamic nuclei, is considered to be a "satiety center." If this nucleus is stimulated in the laboratory, the experimental animal will not engage in feeding behavior. Conversely, a lesion to this nucleus causes the animal to eat excessively and gain weight. The *dorsomedial nucleus*, located immediately dorsal to the ventromedial nucleus, subserves a function relating to emotion or, at least, to emotional behavior. In laboratory animals, stimulation of the dorsomedial nucleus results in unusually aggressive behavior, which lasts only as long as the stimulation is present. This phenomenon, known as *sham rage*, can also be elicited by the stimulation of other hypothalamic and extra-hypothalamic sites. The *arcuate nucleus* is the primary location of neurons that contain releasing hormones or factors. These substances are transmitted to the anterior pituitary by way of the tuberoinfundibular tract and hypophysial portal system, whereupon they influence the release of various pituitary hormones (Fig. 29-4).

The *mammillary region* contains four nuclei: the *medial*, *intermediate*, and *lateral mammillary* and the *posterior hypothalamic nuclei* (Figs. 29-2A,D; 29-4). The *medial mammillary nucleus* is large and especially well developed in the human. It represents the primary termination point for the axons of the postcommissural fornix, which originate primarily from the subiculum of the hippocampal complex. The medial mammillary nucleus is also the source of axons that are directed to the anterior nucleus of the dorsal thalamus as the *mammillothalamic tract* (Fig. 29-4). The latter pathway represents an important part of the limbic system. The much smaller *intermediate* and *lateral mammillary nuclei* are located lateral to the medial mammillary nucleus. The lateral mammillary nucleus receives input from the medial aspects of the midbrain reticular formation by way of the *mammillary peduncle*.

Insight into the function of the mammillary nuclei comes from experimental and clinical observations. For example, lesions of the mammillary bodies tend to result in a loss of *short-term memory*. That is, the patient has difficulty in remembering things that have just occurred, although the memory of more distant events remains intact. Defects in short-term memory are characteristic of *Korsakoff's syndrome*, a condition that is caused by thiamine deficiency and is typically associated with chronic alcoholism. The memory deficits in this syndrome are caused by progressive degeneration in the mammillary bodies and in functionally related brain structures, such as the hippocampal complex and the dorsomedial thalamic nucleus. Patients with Korsakoff's syndrome may have difficulty in understanding written material and in conducting meaningful conversations because they tend to forget what was just read or said. An interesting feature of this syndrome is the patient's tendency to *confabulate*, that is, to string together fragmentary memories from various events into a synthesized memory of an "event" that never occurred.

The *posterior hypothalamic nucleus* merges imperceptibly with the midbrain periaqueductal gray. Accordingly, this nucleus is associated with the same myriad of emotional, cardiovascular, and analgesic functions that have been attributed to the periaqueductal gray.

Periventricular Zone This periventricular zone (see Fig. 29-3), not to be confused with the paraventricular nucleus, is a very thin region composed of small cell bodies lying medial to the medial zone and immediately subjacent to the ependymal cells of the third ventricle. Many neurons of the periventricular zone synthesize releasing hormones or factors. These neurons project by way of the *tuberoinfundibular tract* to the hypophysial portal system and thus influence the release of various hormones by the anterior pituitary. Consequently, many of the neurons in the periventricular zone serve a function similar to that of neurons located in the arcuate nucleus.

BLOOD SUPPLY OF THE HYPOTHALAMUS

The hypothalamus and some immediately adjacent structures are served by small perforating arteries that arise from the *circle of Willis* (Fig. 29-5). Those branches from the anterior communicating artery and the A_1 segment of the anterior cerebral artery constitute the *anteromedial group* of perforating arteries (see also Fig. 8-9). In general, these vessels serve the nuclei of the preoptic area and supraoptic region, the septal nuclei, and rostral portions of the lateral hypothalamic area. A few perforating arteries may also arise from the bifurcation of the internal carotid artery.

The small perforating arteries that originate from the posterior communicating artery and the P_1 segment of the posterior cerebral artery constitute the *posteromedial group* (Figs. 8-9; 29-5). These vessels serve primarily the nuclei of the tuberal and mammillary regions. Branches arising from the rostral portion of the posterior communicating artery distribute to the former region, whereas the latter region is served by branches of the caudal parts of the posterior communicating artery and of P_1. In addition, the posteromedial group also sends branches into the middle and caudal parts of the lateral hypothalamic area. The large *thalamoperforating arteries* usually arise from P_1. Although these vessels distribute mainly to rostral areas of the dorsal thalamus, they do give rise to some small branches that enter the posterior hypothalamus.

The *hypophysial arteries* arise from the internal carotid artery. The *inferior branches* (or rami) originate from the cavernous part of this large vessel, whereas the *superior branches* (Fig. 29-4) are from the cerebral (supraclinoid) part of the internal carotid. Although small, these are important vessels.

HYPOTHALAMIC AFFERENTS

The hypothalamus is connected to diverse sites, including the hippocampus, amygdala, brainstem tegmentum, various thalamic nuclei, septal nuclei, and even neocortical areas such as the infralimbic and cingulate cortex. With very few exceptions, these connections are reciprocal. The following are the most important input/output relationships of the hypothalamus.

Fornix The *fornix* arises from neurons in the *subiculum* and the *hippocampus* (two components of the *hippocampal complex*) and is the largest single input to the hypothalamus (Figs. 29-4, 29-6). As the fornix approaches the anterior commissure, it divides into a small *precommissural bundle*, derived largely from the hippocampus, and a large *postcommissural bundle*, arising mainly from the subiculum. The former passes to septal and preoptic nuclei and to the anterior hypothalamic region, whereas the latter projects primarily to the medial mammillary nucleus, with lesser inputs to the anterior thalamic nucleus and lateral hypothalamus (Fig. 29-7).

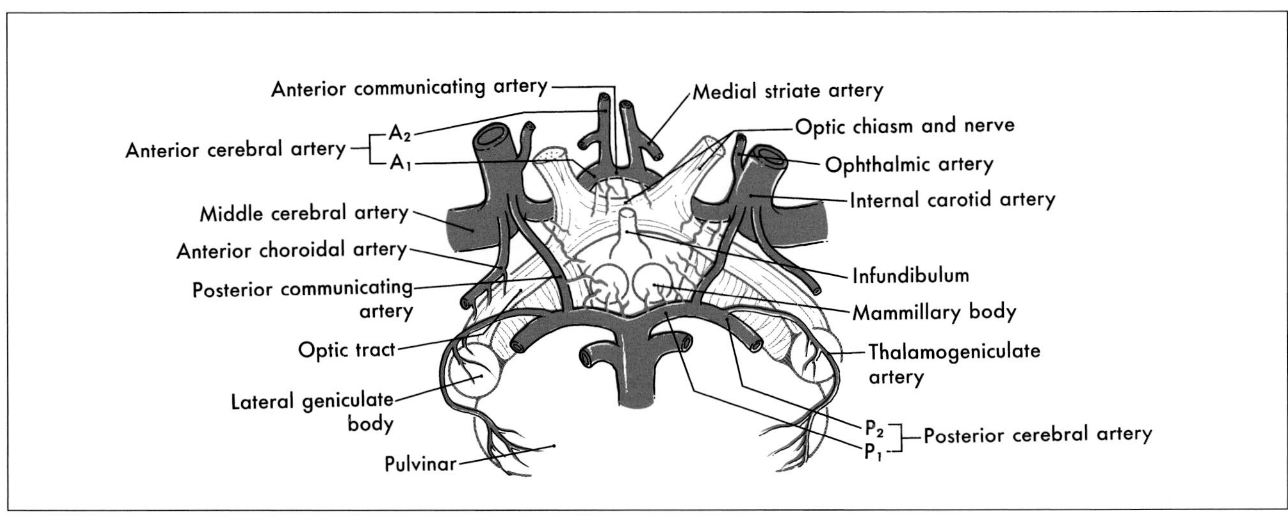

FIGURE 29-5 *The blood supply to the hypothalamus.*

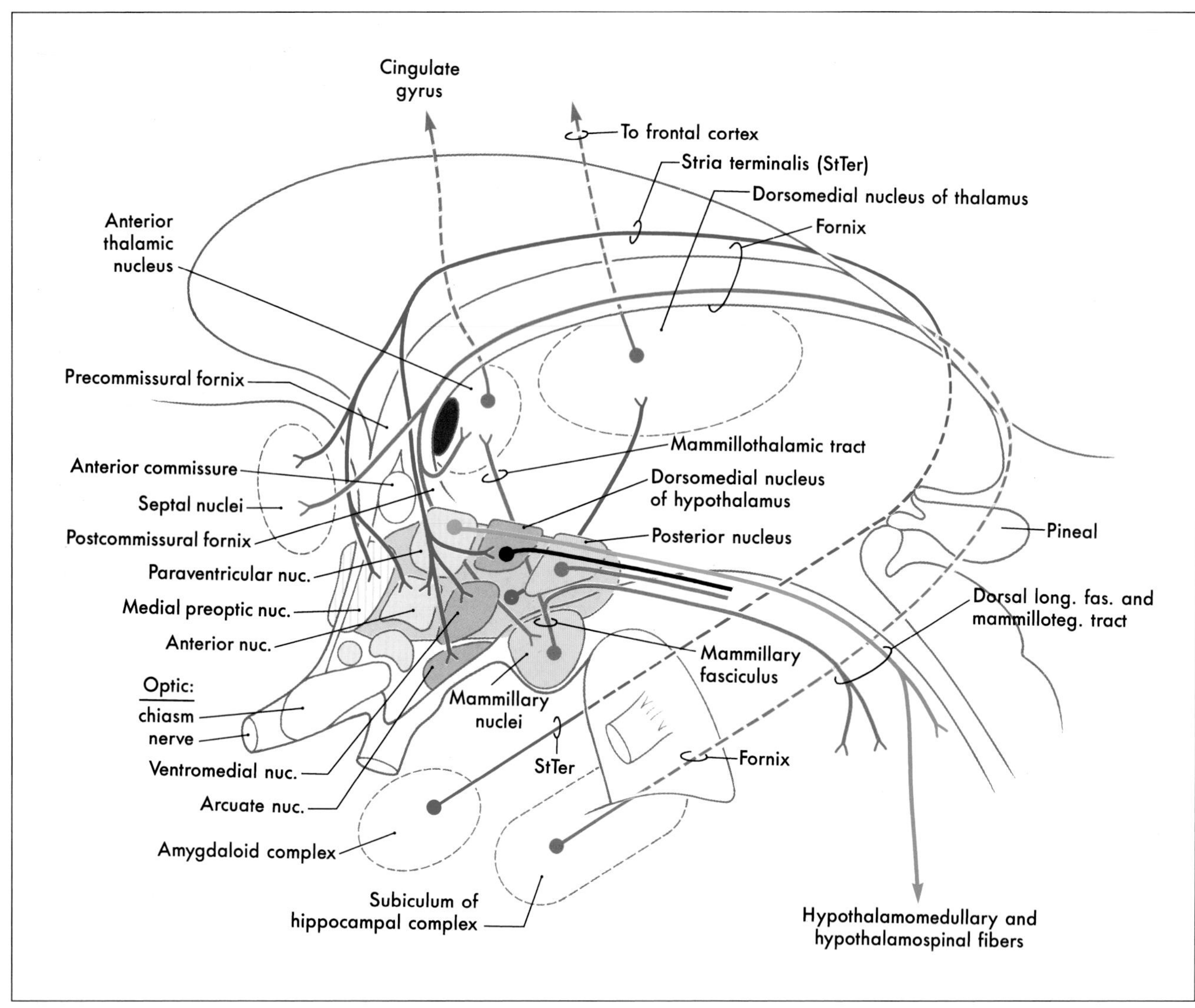

FIGURE 29-6 *Midsagittal view of the hypothalamus emphasizing afferent inputs from the amygdaloid complex and hippocampus. Also shown are the origin of descending fibers to the brainstem and spinal cord.*

Medial Forebrain Bundle The *medial forebrain bundle* is a diffuse, composite structure containing mainly fibers that course rostrocaudally through the lateral hypothalamic zone. It contains ascending and descending fibers that interconnect the septal nuclei (including the nucleus accumbens septi), hypothalamus, and midbrain tegmentum.

Amygdalohypothalamic Fibers Two major afferent fiber systems project to the hypothalamus from the amygdaloid complex: the stria terminalis, which is phylogenetically older, and a newer composite bundle called the ventral amygdalofugal pathway (Figs. 29-6, 29-7). The *stria terminalis* originates from the corticomedial portion of the amygdala and terminates in the septal nuclei, preoptic area, and medial hypothalamic zone. This bundle, which accompanies the terminal vein at the juncture of the caudate nucleus and the thalamus, follows the arched configuration of the caudate nucleus and thus takes a path quite similar to that of the fornix. The *ventral amygdalofugal pathway* originates from the basolateral portion of the amygdaloid complex and passes rostromedially beneath the lentiform nucleus and through the area of the substantia innominata to enter the hypothalamus (Fig. 29-7). The axons in this pathway terminate primarily in the lateral hypothalamic zone and the septal and preoptic nuclei. This bundle also contains fibers that pass from the hypothalamus to the amygdala.

Other Afferent Fibers The *mammillary peduncle* is a diffuse bundle of fibers that originates from medial portions of the midbrain reticular formation and terminates primarily in the lateral mammillary nucleus. Some of these axons enter the medial forebrain bundle and project to the septal nuclei. The dorsomedial nucleus of the thalamus gives rise to *thalamohypothalamic fibers* that

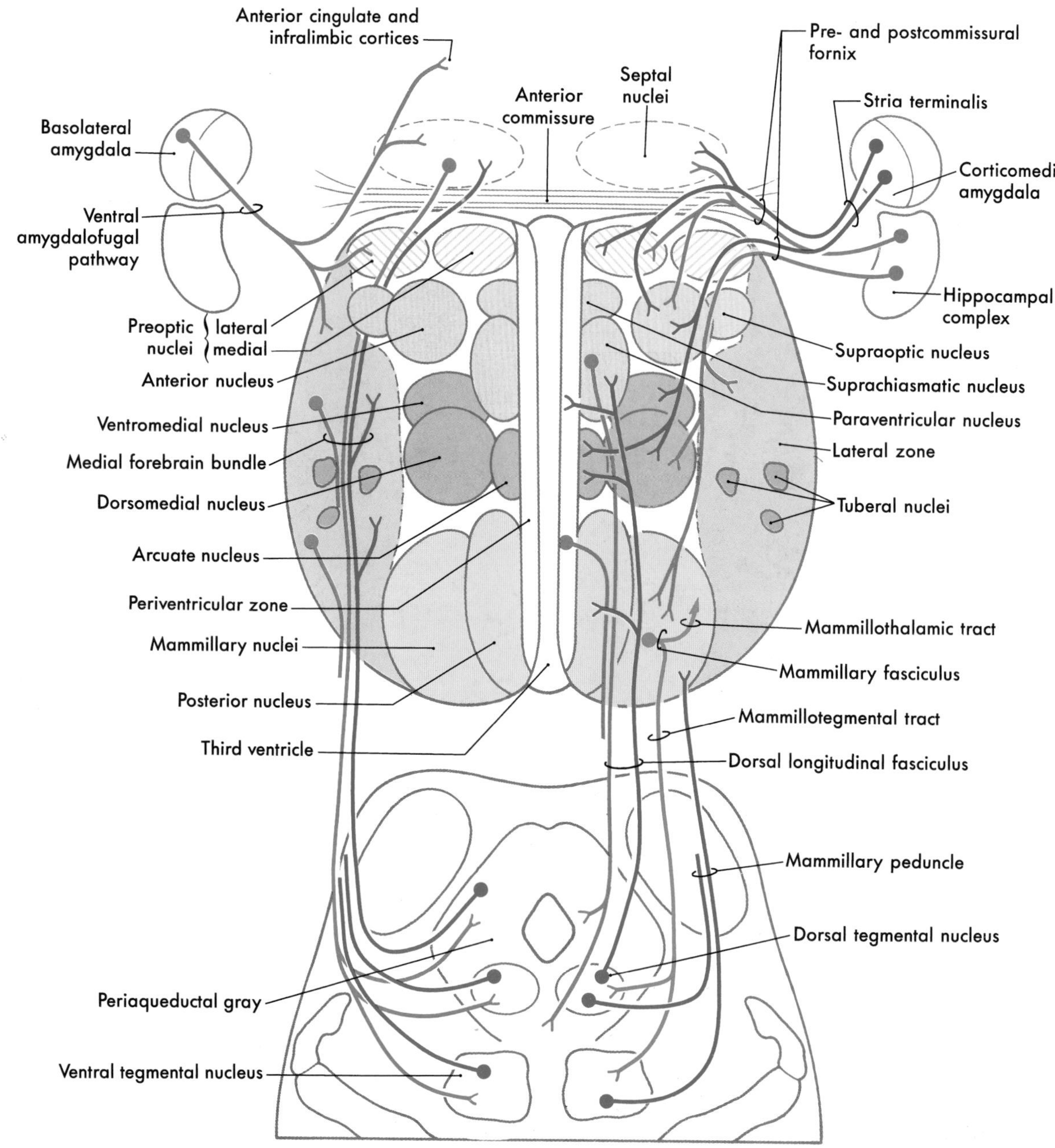

FIGURE 29-7 *Diagrammatic representation of hypothalamic connections as seen in the axial (horizontal) plane. Note the distribution of fibers of the ventral amygdalofugal pathway, the pre- and postcommissural fornix, and the stria terminalis.*

course ventrally to enter the lateral hypothalamus. The only direct neocortical projection to the hypothalamus originates from the prefrontal cortex. This *corticohypothalamic fiber* projection is sparse and terminates primarily in the lateral hypothalamic area. As mentioned earlier, the suprachiasmatic nucleus receives a projection from the retina that is involved in behavioral rhythms and light-dark cycles. These *retinohypothalamic fibers* arise as direct axons from the optic chiasm or as collaterals of retinogeniculate fibers.

HYPOTHALAMIC EFFERENTS

The various subdivisions of the hypothalamus have diffuse projections to numerous sites throughout the neuraxis. Only the major projections are summarized here. A useful generalization is that most structures projecting to the hypothalamus receive a reciprocal input from the hypothalamus. For example, the hypothalamus receives an amygdalohypothalamic projection and gives rise to hypothalamoamygdaloid fibers. For ease of discussion, the efferents are divided into those to forebrain structures (ascending) and those to brainstem and spinal targets (descending).

Ascending The *mammillary fasciculus* originates as a well-defined bundle from the medial mammillary nucleus (Figs. 29-6, 29-7). It passes dorsally for a short distance then bifurcates into the *mammillothalamic tract* and the *mammillotegmental tract* (Fig. 29-6). The former projects to the anterior nucleus of the thalamus and is an important part of the circuit of Papez (see Chapter 30). The latter (discussed below) turns caudally and distributes to the tegmental nuclei of the midbrain reticular formation, thus reciprocating the mammillary peduncle. *Hypothalamothalamic fibers* arise mainly from the lateral preoptic area and project to the dorsomedial nucleus of the thalamus. The hypothalamus also projects to the amygdaloid nucleus (*hypothalamoamygdaloid fibers*) via the stria terminalis and the ventral amygdalofugal pathway. These fibers originate from various hypothalamic nuclei and project primarily to corticomedial nuclei of the amygdaloid complex.

Descending There are four main descending projections from the hypothalamus. These are *hypothalamospinal* and *hypothalamomedullary* fibers, the *dorsal longitudinal fasciculus*, and *mammillotegmental tract* (Figs. 29-6, 29-7).

Hypothalamospinal and *hypothalamomedullary fibers* arise mainly from the paraventricular nucleus, although some originate from the dorsomedial and lateral hypothalamic nuclei (Fig. 29-6). These fibers descend through the periaqueductal gray and adjacent reticular formation of the midbrain and rostral pons, and then shift to a ventrolateral position in the medulla. *Hypothalamomedullary fibers* terminate in the solitary nucleus, dorsal vagal motor nucleus, nucleus ambiguus, and other nuclei of the ventrolateral medulla. *Hypothalamospinal fibers* traverse the ventrolateral medulla and lateral funiculus of the spinal cord to terminate on neurons of the intermediolateral cell column (general visceral efferent preganglionic cells). Hypothalamomedullary and hypothalamospinal fibers form an essential, *direct* link between the hypothalamus and autonomic nuclei of the medulla and spinal cord. Lesions in the ventrolateral medulla may disrupt these fibers. Although the functional effect of disrupting hypothalamomedullary fibers is not well understood, injury to hypothalamospinal fibers results in a loss of sympathetic outflow to the ipsilateral face and head (causing *Horner's syndrome*) and to the ipsilateral side of the body.

The *dorsal longitudinal fasciculus* originates from nuclei of the medial hypothalamic zone, whereas fibers of the *mammillotegmental tract* arise from the medial mammillary nucleus (Figs. 29-6, 29-7). Fibers of these pathways descend through, and largely terminate in, the periaqueductal gray. The mammillotegmental tract, being more ventrally located, also terminates on neurons of the *dorsal* and *ventral tegmental nuclei* situated in the periaqueductal gray of the caudal midbrain (Fig. 29-7). Some neurons of the periaqueductal gray function, at least in part, in relaying information to visceral areas of the brainstem, such as the solitary and dorsal motor vagal nuclei. Consequently, the dorsal longitudinal fasciculus and mammillotegmental tract are generally viewed as relatively short tracts that *indirectly* influence the autonomic nuclei of the brainstem.

INTRINSIC HYPOTHALAMIC CONNECTIONS

The pathways that interconnect the many nuclei of the hypothalamus are numerous and complex. Only two especially important ones are considered here: the *tuberoinfundibular tract* and the *supraopticohypophysial tract*. Both of these tracts link the hypothalamus to the pituitary (Fig. 29-4).

Tuberoinfundibular Tract Most of the input to the pituitary through the tuberoinfundibular tract comes from small (parvicellular) neurons located in the arcuate nucleus and the *periventricular zone*. Neurons of the paraventricular, suprachiasmatic, tuberal, and medial preoptic nuclei may also contribute to this tract (Fig. 29-4). These axons convey various *releasing hormones and/or factors* to the median eminence (the most inferior aspect of the tuberal area) and to the infundibulum of the pituitary gland. The substances are then released into a *primary plexus* of fenestrated capillaries (sinusoids), from which they are carried by *portal veins* to a *secondary plexus* of fenestrated capillaries in the pituitary (Fig. 29-4).

In the anterior lobe, the hypothalamic hormones and factors regulate the functioning of hormone-producing adenohypophysial cells. The hormones of the adenohypophysis include *growth hormone* (affecting all tissues of the body), *gonadotropins* (affecting the ovary and testis), *corticotropin* (affecting the cortex of the adrenal gland), and *thyrotropin* (affecting the thyroid gland). Hormones leave the anterior pituitary via hypophysial veins and are distributed in the systemic circulation.

Supraopticohypophysial Tract Two hormones are released by the posterior pituitary: *oxytocin* and *antidiuretic hormone* (ADH, vasopressin). These hormones are synthesized in large (magnocellular) neurons of the *supraoptic* and *paraventricular nuclei*. They are transported to the posterior pituitary in the axons of these neurons, which form the *supraopticohypophysial tract*. In the posterior pituitary, they are stored in specialized axon terminals, sometimes called *Herring bodies*, which release them in response to the arrival of action potentials from the nerve cell body. The activity of these hypothalamic neurons—and, thus, the release of the hormones—is regulated in response to appropriate stimuli. Once released, the hormones enter a capillary plexus in the posterior pituitary and are conveyed to the general circulation by hypophysial veins.

Neurons containing oxytocin release this hormone during coitus, nipple suckling, and periods in which there is an increased level of estrogen. The release of oxytocin induces the contraction of smooth muscle in the uterus and of the myoepithelial cells in the mammary gland. The effect of oxytocin on the uterus is critical during and after childbirth. A synthetic form of oxytocin called *pitocin* is often administered to hasten labor and delivery. After the baby is born, oxytocin continues to be important. During nursing, for example, the baby's suckling causes oxytocin to be released from the posterior pituitary, and oxytocin in turn causes the myoepithelial cells of the milk glands to contract, expelling milk. The oxytocin released during nursing also has beneficial effects on the postpartum uterus. Specifically, the contractions of uterine muscles caused by oxytocin help this organ to gradually regain its original form and size.

Neurons containing ADH are primarily influenced by fluctuations in the osmolarity of the blood. The relative concentration of sodium chloride in blood plasma is normally about 300 milliosmoles. This osmolarity is largely a function of how much water is retained within the body. It is important to note that, in the process of maintaining fluid balance homeostasis, small deviations from normal blood osmolarity occur throughout each day. These deviations serve as stimuli that influence the release of ADH from the posterior pituitary. Because these stimuli occur frequently, the neurons containing ADH (unlike those containing oxytocin) are tonically active. Thus, small amounts of ADH are released numerous times each day.

When the blood osmolarity is high, the release of ADH from the posterior pituitary is facilitated. Upon entering the systemic circulation, ADH has a primary effect on the kidneys. Specifically, ADH causes the collecting tubules to increase their resorption of water from the developing urine, thereby returning water to the circulatory system. This dilutes the blood and, in so doing, causes the blood osmolarity to be decreased. This process, however, results in the urine being more concentrated. As a consequence, urine output will be diminished, and the urine that is produced will have a darker color.

When the blood osmolarity is low, the release of ADH from the posterior pituitary is inhibited. Consequently, the amount of ADH in systemic circulation will be diminished. This causes the collecting tubules of the kidneys to decrease their resorption of water from the developing urine. Consequently, water remains in the urine and is not returned to the circulatory system. This has the effect of concentrating the blood and in so doing causes the blood osmolarity to be increased. As a result, there is also an increased output of pale colored (dilute) urine.

Lesions of the supraoptic nucleus or supraopticohypophysial tract produce a syndrome known as *diabetes insipidus*, which is characterized by *polyuria* (increased urination) and *polydipsia* (increased consumption of water). This condition is due to a deficit of circulating ADH. It is of interest that ethanol causes a decrease in the release of ADH from the posterior pituitary. This is the reason why consumption of alcoholic beverages tends to cause copious urination and consequent dehydration and thirst.

The main function of ADH (vasopressin) is to assist in the maintenance of normal blood osmolarity and blood pressure. Normally, ADH increases blood pressure by increasing blood volume. However, ADH at high levels will cause contraction of vascular smooth muscle and may also result in increased blood pressure. In this regard, the release of ADH from the posterior pituitary often occurs in those situations in which an increase of blood pressure would be beneficial. For example, the hypotension that occurs in conjunction with *hypovolemia* (decreased blood volume) represents a stimulus that promotes the release of ADH. As a result, arterial constriction takes place and blood pressure is elevated. Accordingly, the hypotensive state is partially alleviated.

REGIONAL FUNCTIONS

Because of the interrelationships existing among the autonomic, endocrine, and limbic systems (see Fig. 29-1), and because of the small size of the hypothalamus, it is difficult to assign a specific function to each of its individual nuclei. Some hypothalamic nuclei participate in functions that are shared with other nuclei within the same vicinity. Consequently, these latter nuclei may participate in functions that are not uniquely their own. Instead, they may act in concert to affect functions that are attributable to hypothalamic *regions*, rather than specific nuclei.

Although the functions of some hypothalamic nuclei are largely known, the functions of others are poorly understood. Because of this and the fact that many hypothalamic nuclei participate in regional functions, some believe that the hypothalamus should not be described as containing functional "centers," e.g., a "feeding center" or a "satiety center." Furthermore, because of the functional interrelationships that exist among certain nuclei within given hypothalamic regions, there is merit in thinking of the hypothalamus in terms of regional functions. Clinical observations are consistent with this view.

Based on clinical observations and experimental data, it is appropriate to divide the hypothalamus into two areas that share similar but opposing functions. These regions are the posterolateral and anteromedial area of the hypothalamus (Fig. 29-3). In general, the *posterolateral area* consists of the lateral hypothalamic zone and the mammillary region, and the *anteromedial area* consists of the supraoptic region and much of the tuberal regions (Fig. 29-7).

Posterolateral Hypothalamus *Activation* (stimulation) of the posterolateral hypothalamus produces behavioral manifestations that are generally associated with anxiety. These include (1) increased activity of the *sympathetic* division of the visceromotor system, (2) increased *aggressive* behavior (see Chapter 30), (3) increased *hunger*, and (4) *increased body temperature* (resulting from cutaneous vasoconstriction and shivering).

A *lesion* in the posterolateral hypothalamus typically results in manifestations opposite to those noted above. For example, damage in this area results in the inhibition of sympathetic activities and the reduction of body temperature.

Anteromedial Hypothalamus *Activation* (stimulation) of the anteromedial hypothalamus produces behavioral manifestations that are generally associated with contentment. These include (1) increased activity of the *parasympathetic* division of the visceromotor system, (2) increased *passive* behavior, (3) increased *satiety*, and (4) *decreased body temperature* (owing to cutaneous vasodilation and sweating).

The phenomenon of sweating is something of an oddity. Even though sweat glands are innervated by sympathetic nerve fibers, sweating is compatible with parasympathetic function in that it helps keep us cool. That the sympathetic terminals innervating most sweat glands are cholinergic, like the parasympathetic terminals innervating viscera, and that sweating can be elicited from the anteromedial hypothalamus support this observation.

Lesions in the anteromedial hypothalamus usually elicit behaviors opposite to those described above. For example, a lesion in this area results in the inhibition of parasympathetic activities and an increase in body temperature.

HYPOTHALAMIC REFLEXES

All the vital functions of the hypothalamus, including the maintenance of blood pressure, body temperature, and water balance, are controlled through reflexes and are *typically* not subject to conscious control. Through meditation or other means, some individuals can learn to alter certain hypothalamic responses. For example, *biofeedback* training enables some individuals to alter things such as blood pressure and body temperature that are generally under hypothalamic control. The neural mechanisms underlying these feats are largely unknown.

The internal environment is controlled partly through hypothalamic reflexes, which are typically mediated by the autonomic or endocrine systems (Fig. 29-1). Three examples are described in the following sections.

Baroreceptor Reflex The baroreceptor reflex (which is discussed in more detail in Chapter 18 and is illustrated in Fig. 18-8) regulates blood pressure in response to input from baroreceptors in the aortic arch and carotid sinus. These receptors are called extrinsic because they are outside the central nervous system. They sense variations in blood pressure and transmit the information to neurons in the solitary nucleus of the medulla. These neurons project to and activate cells in the dorsal vagal nucleus, which in turn project to the terminal ganglia of the heart and influence heart rate. A blood pressure level above normal activates the solitary nucleus and leads to a decrease in heart rate and force of cardiac contraction, and consequently a decrease in blood pressure; a blood pressure level below normal has the opposite effect.

This reflex involves the hypothalamus by a slightly more complex path. The solitary nucleus also contains cells that transmit baroreceptor information to the paraventricular, dorsomedial, and lateral hypothalamic nuclei. In turn, neurons in these hypothalamic areas project to the dorsal vagal nucleus of the medulla. By this route, the hypothalamus can powerfully modulate the responsiveness of the baroreceptor reflex.

Temperature-Regulation Reflex The reflex that maintains a constant body temperature depends on input from specialized temperature-sensing neurons in the hypothalamus (called *intrinsic receptors* because they are inside the central nervous system). When temperature of the blood reaching the hypothalamus rises above normal, these neurons stimulate regions in the *anterior hypothalamus* that are responsible for activating physiologic mechanisms for *heat dissipation*—sweating and cutaneous vasodilation. These effects are mediated by autonomic pathways. Conversely, when the blood temperature is below normal, the temperature sensors stimulate regions in the *posterior hypothalamus* that activate mechanisms for *heat conservation* (cutaneous vasoconstriction, mediated by autonomic pathways) and for *heat production* (shivering, mediated by reticulospinal pathways).

Water-Balance Reflex We have already mentioned the reflex by which the volume and osmolarity of the blood are kept constant. Unlike the previous two reflexes, which are entirely neural, this one is *neurohumoral*; its efferent limb consists of a hormonal signal carried by ADH. In brief, the reflex works as follows. The osmolarity of the blood is monitored by specialized osmolarity-sensitive neurons located in the anterior hypothalamus near the preoptic and paraventricular nuclei. The output from these receptors influences the release of

ADH by ADH-producing neurons in the supraoptic and paraventricular nuclei. When blood osmolarity is too high, more ADH is released and water resorption from the collecting tubules of the kidney is increased. When blood osmolarity is too low, the release of ADH is inhibited, and renal water resorption is reduced. Accordingly, water is not resorbed into the blood, but stays in the urine.

SOURCES AND ADDITIONAL READINGS

Haymaker W, Anderson E, Nauta WJH: *The Hypothalamus*. Charles C Thomas, Springfield, IL, 1969.

Koizumi K, Kollai M, Oomura Y, Yamashita H, Wayner MJ (eds): The hypothalamus: selected topics. Brain Res Bull 20:651–902, 1988.

Renaud LP: A neurophysiological approach to the identification, connections and pharmacology of the hypothalamic tuberoinfundibular system. Neuroendocrinology 33:186–191, 1981.

Saper CB, Lowey AD, Swanson LW, Cowan WM: Direct hypothalamo-autonomic connections. Brain Res 117:305–312, 1976.

Swanson LW, Sawchenko PE: Hypothalamic integration: organization of the paraventricular and supraoptic nuclei. Annu Rev Neurosci 6:269–324, 1983.

Ter Horst GJ, de Boer P, Luiten PGM, van Willigen JD: Ascending projections from the solitary tract nucleus to the hypothalamus. A phaseolus vulgaris lectin tracing study in the rat. Neuroscience 31:785–797, 1989.

van der Kooy D, Koda LY, McGinty JF, Gerfin CR, Bloom FE: The organization of projections from the cortex, amygdala, and hypothalamus to the nucleus of the solitary tract in rat. J Comp Neurol 224:1-24, 1984.

CHAPTER THIRTY

The Limbic System

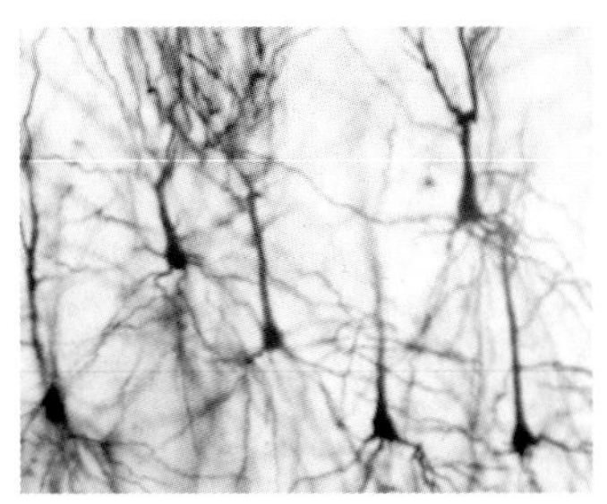

R. B. CHRONISTER • S. G. P. HARDY

Many complex brain systems are organized in a way that allows their function to be readily deduced. For example, even though the connections of the somatosensory pathways with the brainstem, thalamus, and cortex are complex, each component plays a fairly clear-cut role. The processing of somatosensory information is generally well understood. In contrast, some systems are interconnected in such a way that a given function may be carried out by several components acting in cooperation, and a given component may participate in several functions. The *limbic system* is a case in point. This system comprises structures that receive inputs from diverse areas of the neuraxis and participate in complicated and interrelated behaviors such as memory, learning, and social interactions. Thus, lesions involving the limbic system generally result in a wide range of deficits.

OVERVIEW

The concept of a "*limbic system*" actually encompasses two levels of structural/functional organization. The *first level* consists of the cortical structures on the most medial edge (the limbus) of the hemisphere; these collectively form the *limbic lobe* (Fig. 30-1A). Beginning just anterior to the lamina terminalis and proceeding caudally, these are the *subcallosal area*, containing the parolfactory and paraterminal gyri; the *cingulate gyrus*; the *isthmus of the cingulate gyrus*; the *parahippocampal gyrus*; and the *uncus* (Fig. 30-2). The limbic lobe also includes the *hippocampal formation*.

In 1878, Broca noted that the limbic lobe, which is present in all mammals, represents a relatively large part of the cerebral cortex in phylogenetically lower forms, and he postulated that it might be related to olfaction. Because of this latter point, the term rhinencephalon ("smell-brain") was later coined and used, interchangeably, with limbic lobe. However, it is now known that the limbic system has little olfactory function in humans. Thus, the term *rhinencephalon* is antiquated and has largely disappeared from use.

The *second level* includes structures of the limbic lobe plus a variety of subcortical nuclei and tracts that collectively form the *limbic system* (Fig. 30-1B). The subcortical nuclei of the limbic system include, among others, the *septal nuclei* and *nucleus accumbens* (*nucleus accumbens septi*), various nuclei of the hypothalamus, especially those associated with the *mammillary body*; the nuclei of the *amygdaloid complex* and adjacent *substantia innominata*; and parts of the dorsal thalamus, particularly the *anterior* and *dorsomedial nuclei*. Additional structures connected with the limbic system include the *habenular nuclei*, *ventral tegmental area*, and *periaqueductal gray*. Furthermore, the *prefrontal cortex* is considered by some to be an important component of the limbic system, primarily because of its potential influence on various other cortical and subcortical parts of the limbic system. Cortical targets of the prefrontal cortex include the cingulate gyrus, whereas the hypothalamus, dorsal thalamus, amygdaloid complex, and nuclei of the midbrain represent subcortical targets.

The main fiber bundles of the limbic system are the *fornix* (primarily efferents of the hippocampus and subiculum), the *stria terminalis* and *ventral amygdalofugal pathway* (both are mainly efferents of the amygdaloid complex), and the *mammillothalamic tract* (efferents of the medial mammillary nucleus (Fig. 30-1B). A few additional nuclei and smaller tracts will be introduced as connections and functions of the limbic system are described.

CYTOARCHITECTURAL DEFINITIONS OF THE LIMBIC CORTEX

The human cerebral cortex can be divided into several areas on the basis of the number of cell layers present. Most of the cerebral cortex (more than 90%) has six cell layers and is called the *neocortex* or *neopallium* (*isocortex*). Examples of neocortex include the primary sensory, motor, and association cortices. The cortical regions that have less than six layers are structurally and functionally associated with the limbic system and/or with olfaction and are classified as *allocortex*. Those structures that comprise three to five cellular layers are called the *paleocortex* (or *paleopallium*) and are represented by the cortex of the parahippocampal gyrus (the *entorhinal cortex*), the uncus (the *piriform cortex*), and the cortex overlying the termination of the lateral olfactory stria (lateral olfactory gyrus) (Fig. 30-2). The latter is directly rostromedial to the piriform cortex. Structures having only three cellular layers are classified as *archicortex* (or *archipallium*) and are represented by the dentate gyrus and hippocampus.

The separation between neocortex and allocortex is never sharp but instead consists of transitional areas where one cortical region blends into the next. Such areas are represented by caudal parts of the orbitofrontal cortex, the temporal pole, parts of the insula, and portions of the parahippocampal and cingulate gyri. They are especially important because they funnel input from association areas of the neocortex into the allocortex.

EARLY FUNCTIONAL CONCEPTS

In the late 1930s, two pivotal observations were made that formed the basis for the concept of a limbic system. First, based largely on the morphology of the brain, a circuit for the elaboration of emotion was proposed. This is now called the *Papez circuit* in recognition of

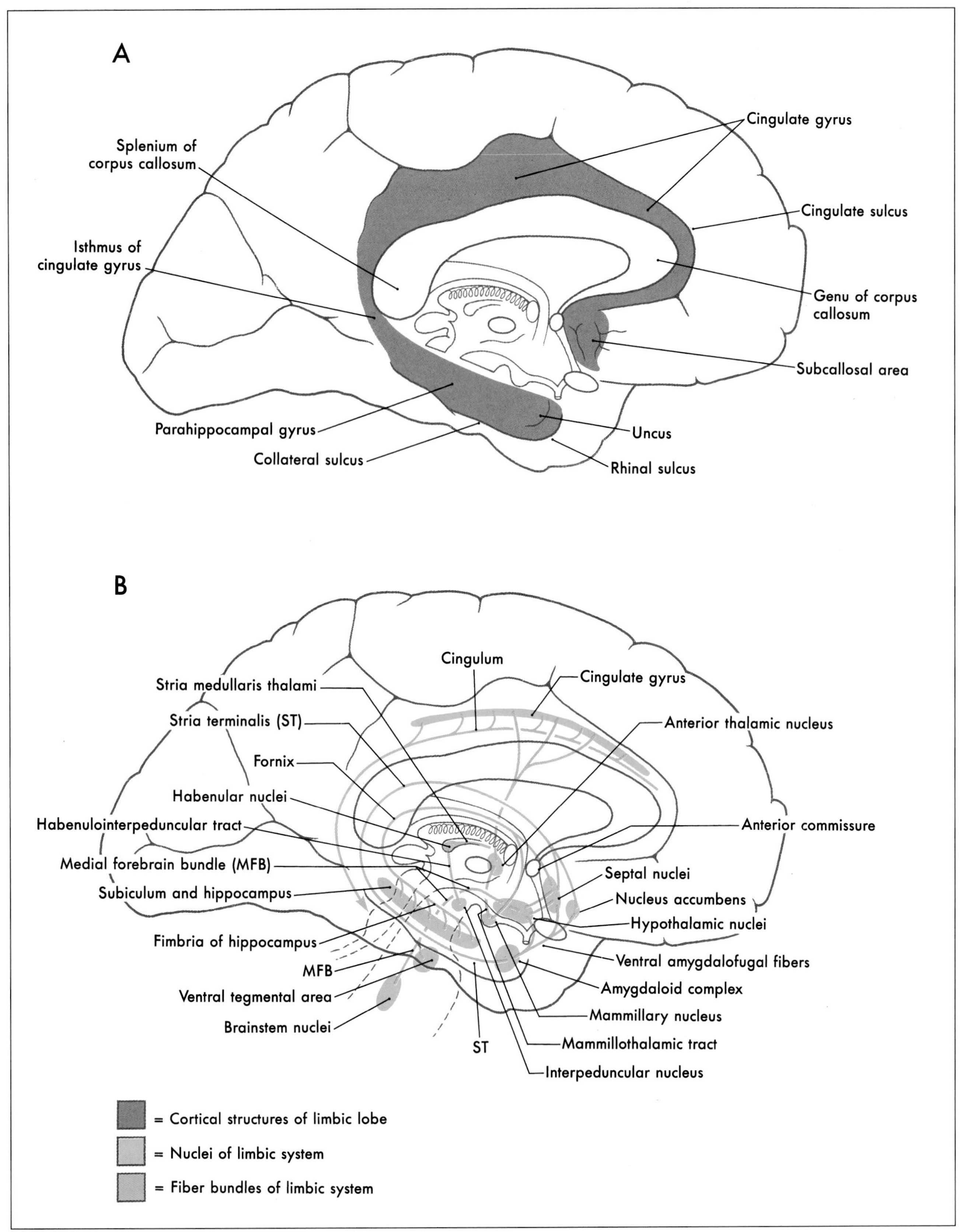

FIGURE 30-1 *Cortical structures forming the limbic lobe (A) and the subcortical structures (B, nuclei in red, fiber bundles in blue) which, with the lobe, represent the main components of the limbic system.*

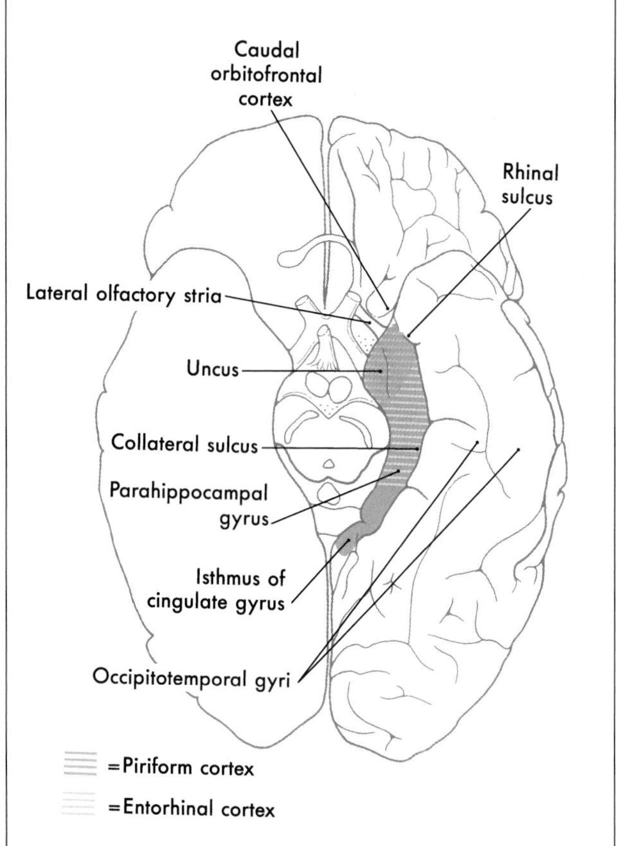

FIGURE 30-2 *Cortical structures of the limbic lobe as seen from a ventral view.*

James Papez, who initially described its components. This model, although surprisingly simple, has proved to be quite important in understanding limbic function. The circuit suggested that emotion, mediated through the hypothalamus, is controlled and modulated by fibers from the fornix. Specifically, the cortical control of emotional activity is presumed to originate from cingulate and hippocampal regions. These cortical influences are ultimately conveyed to the mammillary body of the hypothalamus via the fornix. In turn, the medial mammillary nucleus, via the mammillothalamic tract, projects to the anterior nucleus of the thalamus, and this cell group sends axons to the cingulate gyrus. This was the first time a specific anatomic substrate was proposed for a phenomenon as complex as emotion.

The second pivotal observation was that bilateral removal of large parts of the temporal lobe in monkeys resulted in a constellation of dysfunctions that came to be known as the *Klüver-Bucy syndrome*. This syndrome can also be caused in humans by temporal lobe injuries that involve primarily the amygdaloid complex. Its nature and significance are discussed later.

BLOOD SUPPLY TO THE LIMBIC SYSTEM

The blood supply to the limbic system originates from several sources. The main vessels that serve much of the limbic system are the *anterior* and *posterior cerebral arteries*, the *anterior choroidal artery*, and branches arising from the *circle of Willis*.

The subcallosal area and rostral parts of the cingulate gyrus are supplied by branches of the anterior cerebral artery as it loops around the genu of the corpus callosum (see Fig. 8-3). Most of the cingulate gyrus and its isthmus receive their blood supply via the *pericallosal artery*, a branch of the anterior cerebral artery. *Temporal branches* of the posterior cerebral artery supply the parahippocampal gyrus. Although the uncus may receive some small branches from the posterior cerebral artery, it is served primarily by *uncal arteries*, which are branches of the M_1 of the middle cerebral artery (see Fig. 8-2).

The anterior choroidal artery usually originates from the internal carotid artery and follows the general trajectory of the optic tract. En route it sends branches into the choroidal fissure of the temporal horn of the lateral ventricle. This vessel serves the choroid plexus of the temporal horn, the hippocampal formation, parts of the amygdaloid complex, and adjacent structures such as the tail of the caudate nucleus and the stria terminalis.

Vessels serving hypothalamic nuclei that are functionally associated with the limbic system originate from the circle of Willis. In general, rostral areas of the hypothalamus are served by branches from the anterior communicating artery and anterior cerebral artery and posterior areas by branches from the posterior communicating artery and proximal posterior cerebral artery (see Fig. 14-13). The anterior nucleus of the thalamus, an important synaptic station in the limbic system, is supplied by *thalamoperforating arteries* that arise from the P_1 segment of the posterior cerebral artery.

THE HIPPOCAMPAL FORMATION

The *hippocampal formation* is composed of the *subiculum*, *hippocampus* (also called the hippocampus proper or Ammon's horn), and the *dentate gyrus* (see Fig. 30-4). The subiculum is laterally continuous with the cortex of the parahippocampal gyrus. Medially the edge of the hippocampal formation is formed by the dentate gyrus and the fimbria of the hippocampus.

Developmentally, the hippocampal formation originates dorsally and migrates into its ventral and medial position in the temporal lobe. During this migration, small remnants of the hippocampal formation remain behind to form the *medial* and *lateral longitudinal stria* and their associated gray matter, the *indusium griseum* (Fig. 30-3). These structures are quite small in the human brain and extend rostrally along the dorsal aspect of the corpus callosum into the subcallosal area.

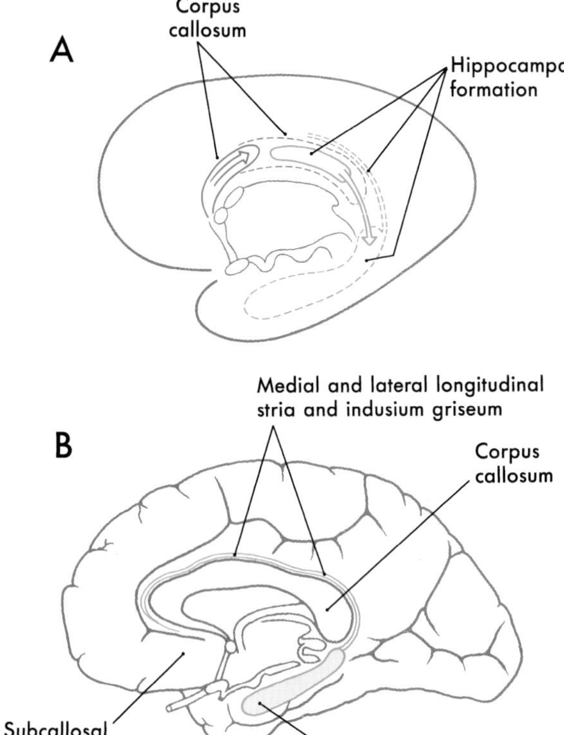

FIGURE 30-3 *Developmental relationships between the corpus callosum (in red) and the hippocampal formation (in green). As the corpus callosum expands caudally from the general area of the anterior commissure (A), the hippocampal primordium migrates into the temporal lobe. In the adult, remnants of the hippocampal formation left behind during development are located dorsal to the corpus callosum (B).*

Structure The *subiculum* of the hippocampal formation is the transitional area between the three-layered hippocampus (archicortex) and the five-layered entorhinal cortex (paleocortex) of the parahippocampal gyrus (Fig. 30-4). This transitional zone, although small, can be divided into a prosubiculum, subiculum proper, presubiculum, and parasubiculum. These areas are essential for the flow of information into the hippocampal formation.

The dentate gyrus and the hippocampus are each composed of three layers, which are characteristic of archicortex (Fig. 30-4). The external layer is called the *molecular layer* and contains afferent axons and dendrites of cells intrinsic to each structure. The middle layer, called the *granule cell layer* in the dentate gyrus and the *pyramidal layer* in the hippocampus, contains the efferent neurons of each structure (Fig. 30-4). These layers are named according to the shape of the cell body of the principal type of neuron found therein. The dendrites of granule and pyramidal cells radiate into the molecular layer. The inner layer, called the *polymorphic layer*, contains the axons of pyramidal and granule cells, a few intrinsic neurons, and many glial elements. In addition, the polymorphic layer of the hippocampus contains the elaborate basal dendrites of some larger pyramidal somata that are located in the pyramidal layer. These are called *double pyramid cells* because they have dendrites extending into both molecular and polymorphic layers (Fig. 30-4). The innermost part of the hippocampus borders on the wall of the lateral ventricle and is a layer of myelinated axons arising from cell bodies located in the subiculum and hippocampus. This layer, called the *alveus*, is continuous with the *fimbria of the hippocampus*, which, in turn, becomes the *fornix* (Fig. 30-4).

The hippocampus can be divided into four regions on the basis of a variety of cytoarchitectural criteria (Fig. 30-4). These areas are designated CA1 to CA4, where *CA* stands for cornu ammonis (Ammon's horn). Area CA1 is located at the subiculum-hippocampal interface, CA2 and CA3 within the hippocampus, and CA4 at the junction of the hippocampus with the dentate gyrus.

Afferents The major input to the hippocampal formation is from cells of the entorhinal cortex via a diffuse projection called the *perforant pathway* (Figs. 30-4, 30-5). Most fibers of the perforant pathway terminate in the molecular layer of the dentate gyrus, although a few terminate in the subiculum and hippocampus. Granule cells in the dentate gyrus project into the molecular layer of the CA3 region of the hippocampus. CA3 neurons project into CA1 of the hippocampus, which, in turn, provides an input to the subiculum. In addition, the subiculum also receives a modest projection from the amygdaloid complex. Although the fornix is mainly an efferent path from the hippocampus, it also conveys cholinergic septohippocampal projections to the hippocampal formation and entorhinal cortex.

Efferents The outflow of the hippocampal formation originates primarily from cells of the subiculum and, to a lesser degree, from pyramidal cells of the hippocampus (Figs. 30-4, 30-5). In both cases, axons of these neurons enter the alveus, coalesce to form the fimbria of the hippocampus, and then continue as the fornix. These glutaminergic fibers traverse the entire extent of the fornix, although some cross the midline in the *hippocampal decussation* just ventral to the splenium of the corpus callosum.

At the level of the anterior commissure, the column of the fornix divides into post- and precommissural parts (Fig. 30-5). The fibers originating in the subiculum mainly form the *postcommissural fornix*. Most of these fibers terminate in the *medial mammillary nucleus*, although some enter the ventromedial nucleus of the hypothalamus and the anterior nucleus of the dorsal thalamus (Fig. 30-5). The *precommissural fornix* is composed of fibers arising primarily in the hippocampus. These fibers are somewhat diffusely organized and distribute to the septal nuclei, the medial areas of the frontal cortex, the preoptic and anterior nuclei of the hypothalamus, and the nucleus accumbens (Fig. 30-5).

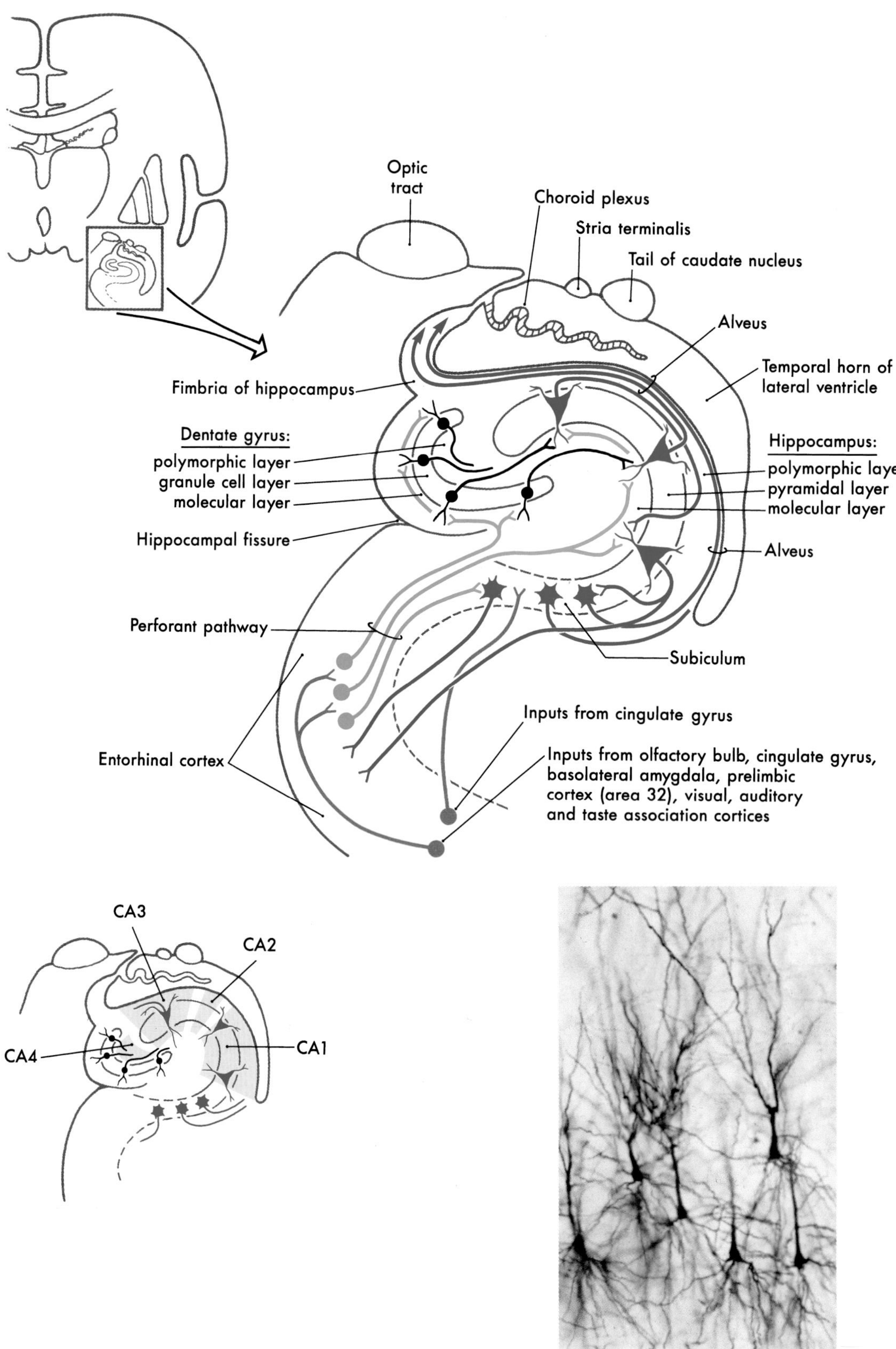

FIGURE 30-4 *The basic structure of the hippocampal formation and its relation to adjacent structures. The cell types of the dentate gyrus, hippocampus, and subiculum are shown diagrammatically. The general locations of fields C1 to C4 are shown in the lower left and a Golgi stain of double pyramid cells in the lower right. (Golgi stain courtesy of Dr. José Rafols, Wayne State University.)*

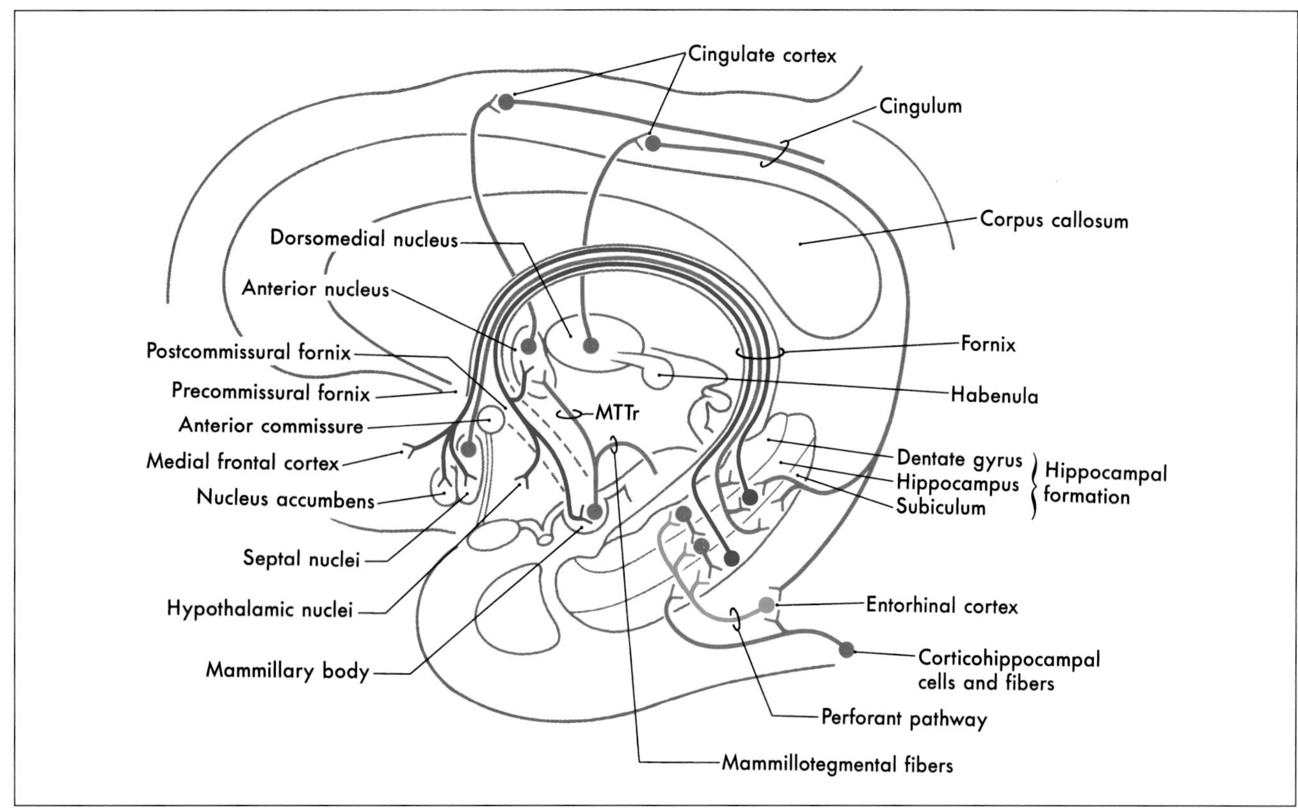

FIGURE 30-5 *Semidiagrammatic representation of afferent and efferent connections of the hippocampal formation. MTTr, mammillothalamic tract.*

The Complete Circuit of Papez As noted above, the initial segment of the *Papez circuit* is a projection primarily from the subiculum, to the medial mammillary nucleus via the postcommissural fornix. The circuit is completed by the following connections (Fig. 30-5): (1) a *mammillothalamic tract* that connects the *medial mammillary nucleus* to the *anterior nucleus* of the thalamus; (2) *thalamocortical fibers* from the anterior nucleus to broad expanses of the cortex of the *cingulate gyrus*; and (3) a projection from the cingulate cortex, via the cingulum, to the entorhinal cortex and also directly to the subiculum and hippocampus. The subiculum returns information to the mammillary body.

Other areas of the cerebral cortex are recruited into the various functions associated with the Papez circuit largely through connections of the cingulate gyrus. For example, the cingulate receives input from premotor and prefrontal areas and from visual, auditory, and somatosensory association cortices. In turn, the cingulate cortex is not only a major source of afferent fibers to the hippocampal formation but also projects to most cortical areas from which it receives input. The cingulate gyrus thus is not only an integral part of the Papez circuit, but also an important conduit through which a wide range of information can access the limbic system.

Dysfunctions and Korsakoff's Syndrome The basic function of the hippocampal formation appears to be the consolidation of long-term memories from immediate and short-term memories. *Immediate memory* and *short-term memory* refer to types of memory that persist for seconds and minutes, respectively. Normally, these memories can be incorporated into long-term memory, which can be recalled days, months, or years later. However, in individuals with hippocampal lesions, this conversion is not accomplished. Although patients may be able to perform a task for seconds or minutes, if distracted from the task they are unable to return to it. In other words, they do not "remember" what they were doing; the short-term experience is not incorporated into long-term memory. The redundancy and feedback in the hippocampus are ideal for this imprinting of memory.

One condition in which loss of memory and cognitive function is particularly obvious is *Alzheimer's disease*. This disease is characterized, in part, by the presence of neurofibrillary tangles, neuritic plaques, and neuronal loss in specific brain regions. The subiculum and entorhinal cortices are among the first sites in which these abnormalities appear. As a result, the relay of information through the hippocampal formation is markedly impeded. It is believed that this damage is at least partially responsible for the memory deficits characteristic of *Alzheimer's disease*.

As mentioned above, *Korsakoff's syndrome* (Korsakoff's psychosis) is a condition that is caused by prolonged thiamine deficiency and is typically seen in chronic alcoholics. The thiamine deficiency causes a characteristic pattern of degeneration in the brain. Typically, the mammillary bodies are involved, with some incursion into the dorsomedial nucleus of the thalamus and the columns of the fornix. There is also a loss of neurons in the hippocampal formation. These patients show a defect in short-term memory and consequently also in long-term memory for events occurring since the onset of disease. They may appear demented, and they are prone to *confabulation*; that is, they tend to string together fragments of memory from several different events to form a synthetic "memory" of an event that never occurred. In some chronic alcoholics, the memory loss and general confusion are accompanied by gaze palsies and ataxia, which occur secondary to cerebellar damage. When these deficits accompany profound memory losses and learning difficulties, the condition is called *Wernicke-Korsakoff syndrome*.

Bilateral damage to the hippocampal formation sometimes occurs in victims of heart attack or near drownings as a result of transient cerebral ischemia. The part of the hippocampal formation most vulnerable to anoxia during an ischemic episode is the CA1 area. These patients retain their long-term memories from the time before the ischemic event, but they have profound deficits in short-term memory. Consequently, they also have difficulty learning new skills because the new information is not retained (remembered) long enough to become a long-term memory.

Bilateral lesions of the anterior part of the cingulate gyrus greatly diminish the emotional responses of the patient and may result in *akinetic mutism*. This is a state in which the patient is immobile, mute, and unresponsive but not in a coma. Other patients with cingulate damage may be alert but have no idea of who they are. Patients may also be unable to recall the order in which past events occurred.

LONG-TERM POTENTIATION AND MEMORY

The process of *long-term potentiation* at individual synapses is the probable mechanism that underlies the consolidation of short-term into long-term memory. When several synapses are present on a single cell, the input from these synapses is *integrated*; that is, the small potential changes (epsp and ipsp) are added together. In long-term potentiation, one synapse fires in a particular temporal pattern (such as bursts or trains of action potentials). This changes the likelihood of that synapse and other synapses activating the target cell. This may be due to an increased probability that transmitter will be released from the presynaptic cell, or an increased response in the postsynaptic cell to the same amount of neurotransmitter, or both. Long-term potentiation has been demonstrated at terminals of the perforant pathway in the dentate gyrus and at the synapses of CA3 pyramidal cells on CA1 cells. These connections use the neurotransmitter glutamate.

According to one current model, the release of glutamate causes a change in the biochemistry of the N-methyl-D-aspartate (NMDA)-type glutamate receptors of the hippocampal cells, allowing an increased number of calcium ions to enter the cell. The calcium influx causes a second postsynaptic biochemical change: The gaseous neuromodulator, nitric oxide, is released and diffuses back to the presynaptic terminal. It acts on the presynaptic terminal to permanently increase the release of glutamate. At this type of synapse, *a brief, sustained increase in current synaptic activity increases the probability that future synaptic activity will take place.* That is, the more the circuit is activated, the easier it is to activate. This mechanism causes stimuli and responses to be paired in the process we call "memory." The increased probability of activation lasts in isolated preparations for hours; it cannot be measured in human brains but may be permanent.

THE AMYGDALOID COMPLEX

Structure The amygdaloid complex is an almond-shaped group of cells in the rostromedial part of the temporal lobe internal to the uncus (Fig. 30-1A,B). It is immediately rostral to the hippocampal formation and the anterior end of the temporal horn of the lateral ventricle. The amygdaloid complex is composed of a number of nuclei. For our purposes, these nuclei are grouped into a larger *basolateral group* and a smaller *corticomedial group* (including the *central nucleus*). The latter group is more closely related to olfaction, whereas the former has extensive interconnections with cortical structures.

Afferents The basolateral cell groups of the amygdala receive inputs from the dorsal thalamus, the prefrontal cortex, the cingulate and parahippocampal gyri, the temporal lobe and insular cortex, and the subiculum (Fig. 30-6A). These fibers supply a wide range of somatosensory, visual, and visceral information to the amygdaloid complex.

The corticomedial cell group receives olfactory input, fibers from the hypothalamus (ventromedial nucleus, lateral hypothalamic area), and fibers from the dorsomedial and medial nuclei of the dorsal thalamus (Fig. 30-6A). In addition, this cell group receives ascending input from nuclei in the brainstem known to be involved in visceral functions. Among others, these include the parabrachial nuclei, the solitary nucleus, and portions of the periaqueductal gray.

Efferents The two major efferent pathways of the amygdaloid complex are the stria terminalis and the ventral amygdalofugal pathway (Figs. 30-1B, 30-6B). The *stria terminalis* is a small fiber bundle that arises primarily

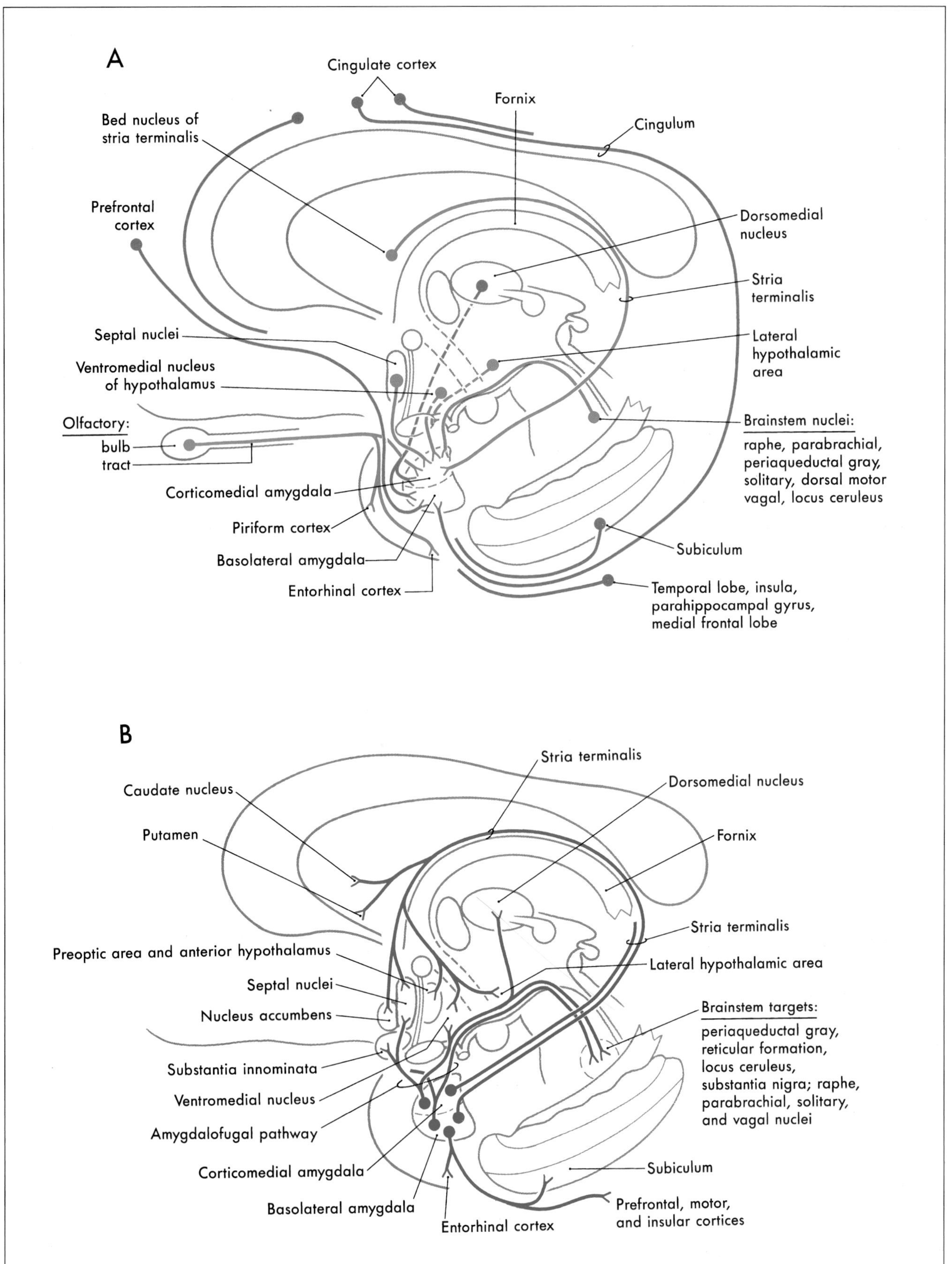

FIGURE 30-6 *Semidiagrammatic representation of the afferent (A) and efferent (B) connections of the amygdaloid complex.*

from cells of the corticomedial group. Through most of its course, this bundle lies in the groove between the caudate nucleus and the dorsal thalamus, where it is accompanied by the terminal vein. It is associated along its length with discontinuous aggregations of cells, which collectively are called the *bed nucleus of the stria terminalis*. This tract distributes to various nuclei of the *hypothalamus* (the preoptic nuclei, ventromedial nucleus, anterior nucleus, and lateral hypothalamic area), to the *nucleus accumbens* and the *septal nuclei*, and to the rostral areas of the caudate nucleus and putamen (Fig. 30-6B).

The *ventral amygdalofugal pathway* is the major efferent fiber bundle of the amygdaloid complex. These fibers arise from the basolateral cell group and the central nucleus of the corticomedial cell group and follow two general trajectories (Fig. 30-6B). Axons primarily from the basolateral cells pass medially through the substantia innominata (in which some of these fibers terminate) to eventually synapse in the hypothalamus and septal nuclei. The substantia innominata gives rise to a diffuse cholinergic projection to the cerebral cortex. It is probable that these fibers play a role in the activation of the cerebral cortex in response to behaviorally significant stimuli. In addition, cells of the basolateral group also project diffusely to frontal and prefrontal, cingulate, insular, and inferior temporal cortices (Fig. 30-6B). Other fibers, mainly from the central nucleus, turn caudally and descend diffusely in the brainstem to terminate in visceral (dorsal motor vagal) nuclei, raphe nuclei (magnus, obscurus, pallidus), and other areas such as the locus ceruleus, parabrachial nuclei, and periaqueductal gray (Fig. 30-6B). As noted above, most of those brainstem areas that receive input from the amygdala project back to this structure.

Another route by which hippocampal and amygdaloid efferents influence the brainstem is through the *stria medullaris thalami*. This bundle conveys fibers from the septal nuclei (targets of amygdaloid and hippocampal inputs) to the *habenular nuclei* (Fig. 30-7A). The latter cell groups, in turn, give rise to the *habenulointerpeduncular tract*, which projects to the interpeduncular nucleus and other midbrain sites, including the ventral tegmental area and periaqueductal gray.

Klüver-Bucy Syndrome As mentioned earlier, bilateral temporal lobe lesions that largely abolish the amygdaloid complex cause a set of behavioral changes called the *Klüver-Bucy syndrome*. These deficits were initially described in a series of animal experiments, but they have also been seen in patients as a result of either trauma to the temporal lobe or temporal lobe surgery for epilepsy. Damage to the amygdaloid complex frequently involves portions of adjacent structures and of the surrounding white matter, and these incursions into other structures may contribute to the clinical picture.

The Klüver-Bucy syndrome is characterized by the following. First, the patient is no longer able to recognize objects by sight (*visual agnosia*) and may also exhibit *tactile* and *auditory agnosia*. Second, there is the tendency to examine objects excessively by mouth (*hyperorality*) or to smell them. Even a harmful object such as a lit match may be examined by being brought to the lips or touched with the tongue. Third, the patient may have a compulsion to intensively explore the immediate environment (*hypermetamorphosis*) and to overreact to visual stimuli. Fourth, *placidity* is characteristically seen. The animal or the patient may no longer show fear or anger, even when such a reaction is appropriate. Fifth, the subject may eat in excessive amounts (*hyperphagia*), even when not hungry, or may eat objects that are not food, or food that is inappropriate to the species. For example, a monkey may eat raw meat, or a patient may eat leaves. Sixth, there is a striking augmentation in sexual behavior (*hypersexuality*). In humans, this takes the form of suggestive behavior, talk, and vague, ill-conceived attempts at sexual contact. In addition to these predictable deficits, these patients may also experience *amnesia*, *dementia*, and/or *aphasia*, depending on the extent of the lesion of the temporal lobe.

THE SEPTAL REGION

The septal region (septal nuclei), excluding the nucleus accumbens, is a small area just rostral to the anterior commissure and in the medial wall of the hemisphere (Figs. 30-1B, 30-7A). These nuclei extend into the base of the septum pellucidum. Despite their relatively small size, the septal nuclei have been implicated in a myriad of functions in animal models on the basis of the patterns of their inputs and outputs. In contrast, there is little clinical information regarding their function in humans. Rage behavior has been seen in a small group of patients with midline infarcts in this area.

The principal afferent pathways to the septal nuclei include fibers from the hippocampus (via the fornix), the amygdaloid complex (via the stria terminalis and ventral amygdalofugal pathways), and the ventral tegmental area of the midbrain (Fig. 30-7A). Fibers also originate from the preoptic, anterior, and paraventricular hypothalamic nuclei and from the lateral hypothalamic area. Many of the fibers in the stria terminalis and fornix also send branches into the nucleus accumbens.

The main efferent projections from the septal nuclei (Fig. 30-7A) are septohippocampal fibers (in the fornix), projections to the habenular nuclei, the medial thalamic nuclei (via the stria medullaris thalami), and the ventral tegmental area (via the medial forebrain bundle). The preoptic, anterior, and ventromedial nuclei and the lateral hypothalamic areas also receive input from the septal nuclei.

The *medial forebrain bundle* is a diffuse group of fibers that courses rostrocaudally through the lateral hypothalamic area (Fig. 30-7A). This bundle is complex, in that it

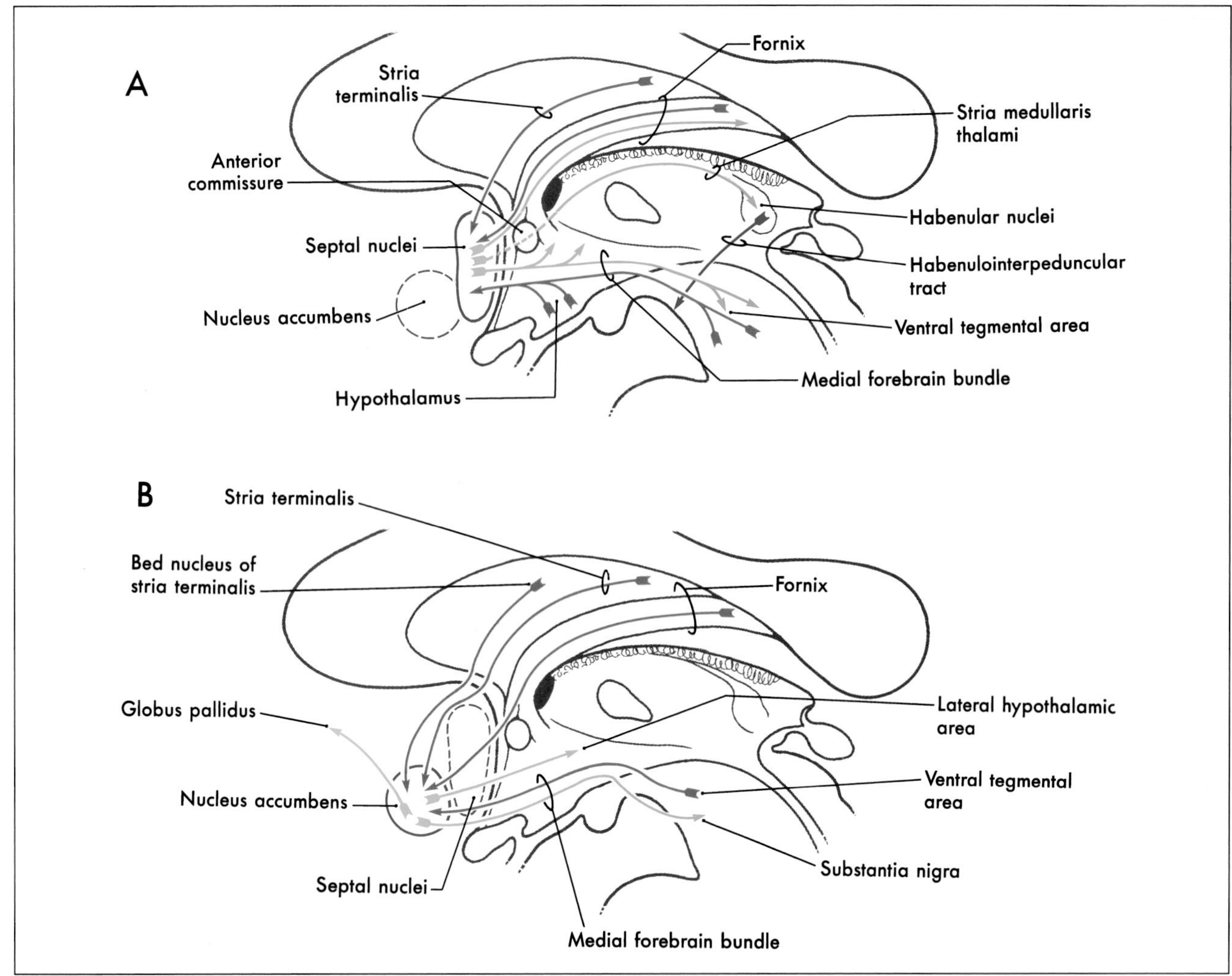

FIGURE 30-7 *Summary of the main afferent (red arrows) and efferent (green arrows) connections of the septal nuclei (A) and of the nucleus accumbens (B).*

conveys ascending inputs into the hypothalamus and through this area into the septal region. It also is a major conduit through which the septal nuclei and portions of the hypothalamus communicate with the brainstem (see Fig. 29-7). The dopamine-containing fibers in this area are thought to be related to perceptions of pleasure or drive reduction.

THE NUCLEUS ACCUMBENS

The nucleus accumbens (nucleus accumbens septi) is located in the rostral and ventral forebrain where the head of the caudate nucleus and the putamen are continuous (see Fig 25-3). These cells receive input from the amygdaloid complex (primarily via the ventral amygdalofugal pathway), from the hippocampal formation (through the precommissural fornix), and from cells of the bed nucleus of the stria terminalis (Fig. 30-7B). The ventral tegmental area also gives rise to ascending fibers that enter the nucleus accumbens via the medial forebrain bundle. In addition, amygdalofugal fibers traversing the stria terminalis also enter the nucleus accumbens.

Cells within the nucleus accumbens have receptors for a variety of neurotransmitters, including endogenous opiates. Studies in animal models indicate that the nucleus accumbens may play an important role in behaviors related to addiction.

Efferent projections of the nucleus accumbens include fibers to the hypothalamus, nuclei of the brainstem, and the globus pallidus (Fig. 30-7B). Nucleus accumbens fibers to the latter target represent an important route through which the limbic system may access the motor system.

THE LIMBIC SYSTEM AND EMOTIONS

In recent years, the term *limbic system* has been used mainly in reference to emotion-related areas of the brain and the pathways that interconnect them. These areas are generally composed of sites that function to alter one's emotions. These sites, which are often interspersed in a given region of the brain, are frequently called either *aversion centers* or *gratification centers*. If an aversion center is stimulated, the individual will experience fear or sorrow. On the other hand, stimulation of a gratification center will result in pleasure. Functional interconnections between aversion and gratifications centers probably contribute to emotional stability.

Although most limbic structures contain both gratification and aversion centers, in some structures one or the other type of center seems to predominate. For example, the hippocampus and amygdala have an abundance of aversion centers, whereas the nucleus accumbens contains an abundance of gratification centers. Consequently, stimulation of the amygdala may elicit fear, whereas stimulation of the nucleus accumbens results in feelings of joy or pleasure.

The emotion-related deficits resulting from small lesions in the limbic system are difficult to predict. However, the effects of relatively large lesions are more stereotypic. They typically result in the flattening of emotions, as reflected by the fact that emotional extremes (joy and anxiety) are reduced. This phenomenon, presumably resulting from the loss of both aversion and gratification centers, commonly results from large lesions in the amygdala, hippocampus, fornix, or in cingulate or prefrontal cortices.

THE LIMBIC SYSTEM AND COGNITIVE FUNCTION

There is a trend to look at the limbic system as a set of structures that influence not only emotion per se but also cognitive functions. The area on which there is the most agreement in this regard is memory. Other influences, however, are mediated through the nucleus basalis and are undoubtedly related to the control of cortical excitability. The full contribution of this upstream control is related to the transfer of information from limbic structures to the cerebral cortex.

SOURCES AND ADDITIONAL READINGS

Hodges JR, Patterson K: Is semantic memory consistently impaired early in the course of Alzheimer's disease? Neuroanatomical and diagnostic implications. Neuropsychologia 33:441–459, 1995.

Isaacson RL: *The Limbic System*. Plenum Press, New York, 1974.

Kalivas PW, Barnes CD (eds): *Limbic Motor Circuits and Neuropsychiatry*. CRC Press, Boca Raton, Florida, 1993.

Kötter R, Meyer N: The limbic system: a review of its empirical foundation. Behav Brain Res 52:105–127, 1992.

Masliah E, Mallory M, Hansen L, DeTeresa R, Alford M, Terry R: Synaptic and neuritic alterations during the progression of Alzheimer's disease. Neurosci Lett 174:67–72, 1994.

Nauta WJH: Hippocampal projections and related neural pathways to the midbrain in the cat. Brain 81:319–340, 1958.

Reep R: Relationship between prefrontal and limbic cortex: a comparative anatomical review. Brain Behav Evol 25:5–80, 1984.

Sandner G, Oberling P, Silveira MC, Di Scala G, Rocha B, Bagri A, Depoortere R: What brain structures are active during emotions? Effects of brain stimulation elicited aversion on c-fos immunoreactivity and behavior. Behav Brain Res 58:9–18, 1993.

Squire LR: *Memory and Brain*. Oxford University Press, New York, 1987.

Zola-Morgan S, Squire LR, Amaral DG: Human amnesia and the medial temporal region: enduring memory impairment following a bilateral lesion limited to field CA1 of the hippocampus. J Neurosci 6:2950–2967, 1986.

CHAPTER THIRTY-ONE

The Cerebral Cortex

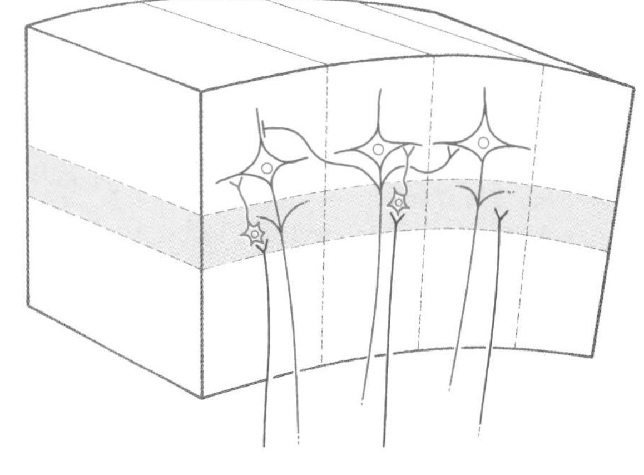

J. C. LYNCH

The cerebral cortex is the organ of thought. More than any other part of the nervous system, the cerebral cortex is the site of the intellectual functions that make us human and that make each of us a unique individual. These intellectual functions include the ability to use language and logic and to exercise imagination and judgment.

OVERVIEW

The cerebral cortex is a dense aggregation of neuron cell bodies that ranges from 2 to 4 mm in thickness and forms the surface of each cerebral hemisphere. The total area of the cerebral cortex is about 2,500 cm^2, a little larger than a single page of a newspaper. Neurons in the cortex receive input from many subcortical structures by way of the thalamus and also from other regions of the cortex via association fibers. Cortical neurons, in turn, project to a wide range of neural structures, including other areas of the cerebral cortex, the thalamus, the basal ganglia, the cerebellum via the pontine nuclei, many of the brainstem nuclei, and the spinal cord.

The cerebral cortex is divided into distinct functional areas, some of which are devoted to the processing of incoming sensory information, others to the organization of motor activity, and still others primarily to what are considered "higher intellectual functions." These functions include memory, judgment, the planning of complex activities, the processing of language, mathematical calculations, and the construction of an internal image of an individual's surroundings. This chapter considers (1) the basic internal organization of the cerebral cortex at the cellular level, (2) the parcellation of the cortex into distinct subregions on the basis of cellular organization and neural connections, and (3) the functional properties of some higher-order association cortical regions.

HISTOLOGY OF THE CEREBRAL CORTEX

The gray matter of the cerebral cortex is composed of neuron cell bodies of varying sizes and shapes, intermixed with myelinated and unmyelinated fibers (Figs. 31-1, 31-2A). These cell bodies may be visualized with

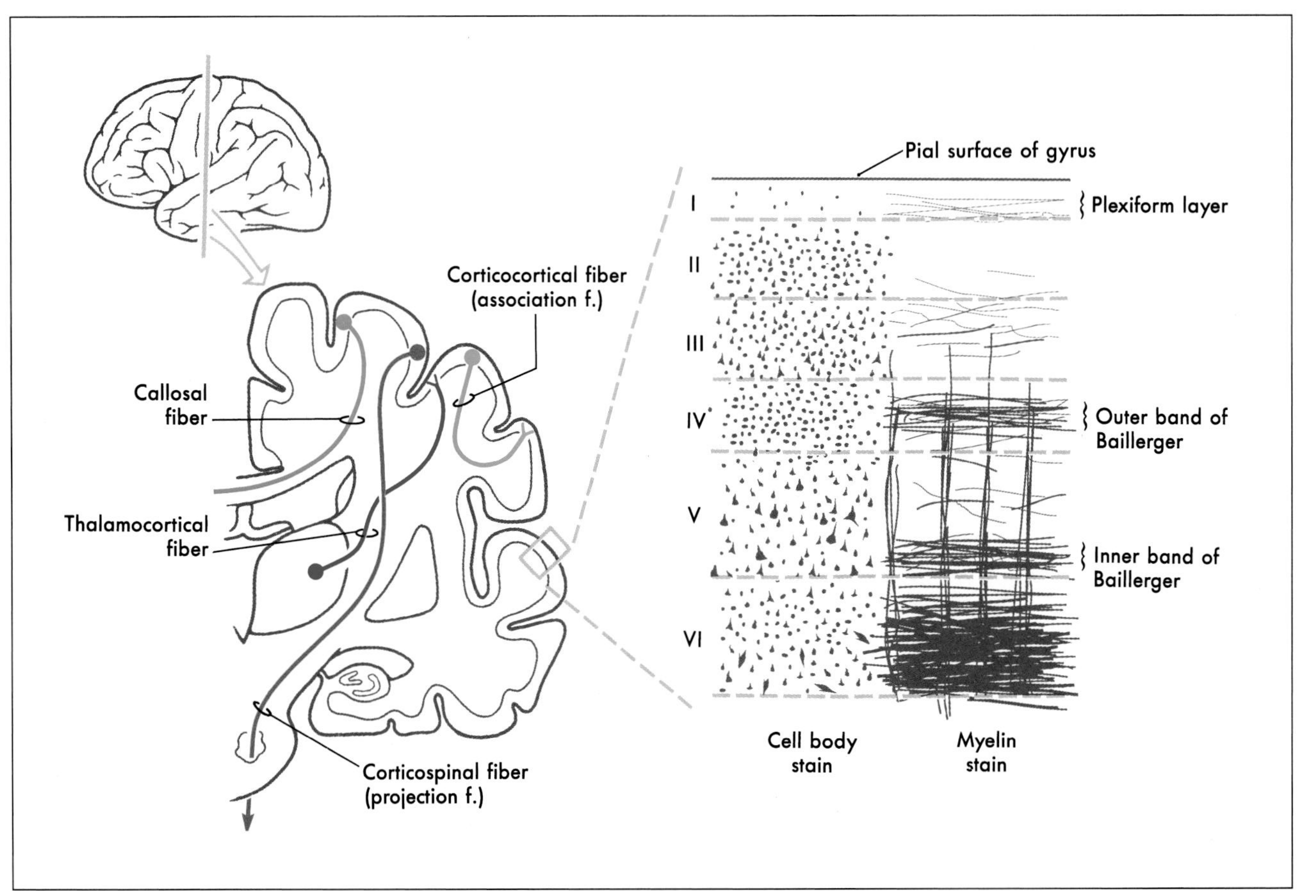

FIGURE 31-1 *A coronal section through the hemisphere (left) showing the major types of fibers projecting to and from the cerebral cortex. The representation on the right shows layers (I to VI) of the cerebral cortex as they appear after staining for cell bodies or the myelin sheath.*

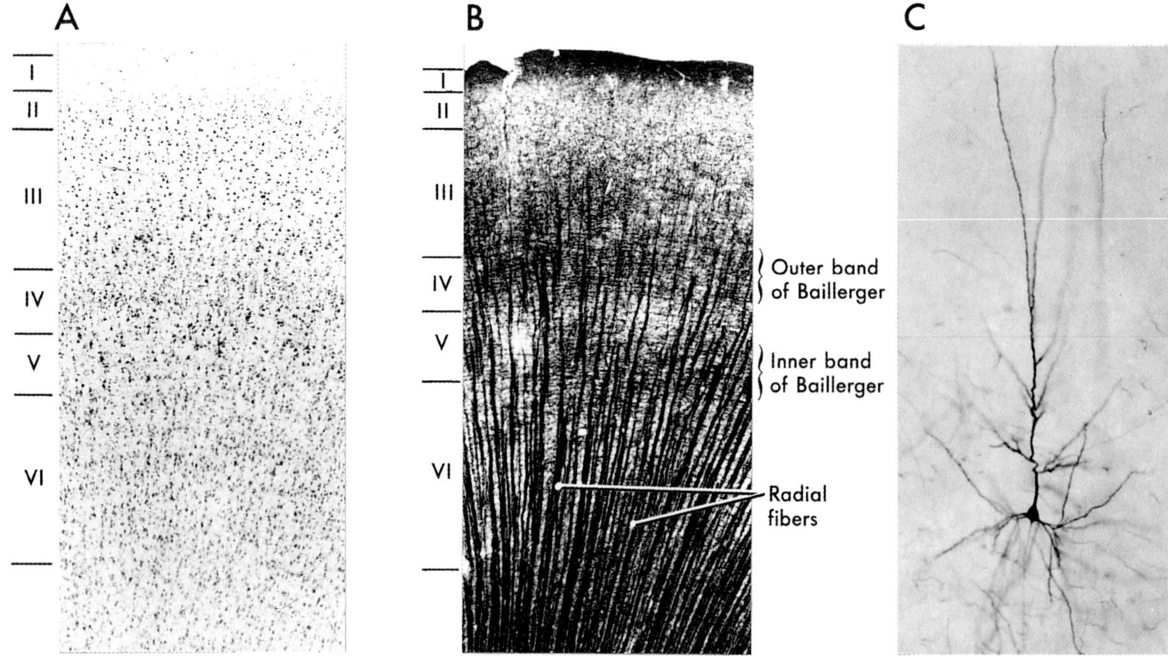

FIGURE 31-2 *Nissl (A) and myelin (B) stains of adjacent sections of the human cerebral cortex, and a Golgi impregnation (C) of a pyramidal neuron in the primate neocortex.(A and B courtesy of Drs. Grayzna Rajkowska and Patricia Goldman-Rakic, University of Mississippi Medical Center and Yale University; C courtesy of Dr. José Rafols, Wayne State University.)*

stains that bind to the rough endoplasmic reticulum (Nissl substance). Such stains leave the axons and dendrites almost invisible. Substances that bind to the lipoprotein of the myelin sheath surrounding some axons will make the myelinated portion of the fibers visible (Figs. 31-1, 31-2B). Yet another way of looking at cortical cells is to immerse small blocks of tissue in dilute silver salts, which precipitate on the membranes of the entire neuron. This causes the cell body, its dendrites, and portions of the axon to become visible (Fig. 31-2C); this technique is called the *Golgi method*. The basic connections of a given region of cortex include *projection fibers* to subcortical structures, *callosal fibers* to cortex in the opposite hemisphere, *association fibers* to cortex in the same hemisphere, and *thalamocortical fibers*, which provide virtually all of the neural input to the cortex that originates in noncortical structures (Fig. 31-1).

The pattern of distribution of neuron cell bodies is, in general, called *cytoarchitecture*. Specifically, the cytoarchitecture of the cerebral cortex is characterized by layers. Most of the cerebral cortex has six distinct layers of neurons and is classified as *neocortex*. Two regions of the cerebral cortex have fewer than six layers. The first contains only three layers, is classified as *archicortex*, and includes the hippocampal formation. The second contains from three to five layers, is classified as *paleocortex*, and includes the olfactory sensory area and the nearby entorhinal and periamygdaloid cortices. The following discussion concentrates primarily on the neocortex.

The neuronal layers in the neocortex are designated by Roman numerals, beginning at the pial surface (Fig. 31-1). Layer I, the *molecular layer*, contains very few neuron cell bodies and consists primarily of axons running parallel (horizontal) to the surface of the cortex. Layer II, the *external granular layer*, is composed of a mixture of small neurons called *granule cells* and slightly larger neurons, which are called *pyramidal cells* based on the shape of their cell body. Layer III, the *external pyramidal layer*, contains primarily small to medium-sized pyramidal cells, along with some neurons of other types. Layer IV, the *internal granule layer*, consists almost exclusively of *smooth (aspiny) stellate* (starlike) *neurons* and *spiny stellate neurons*, both of which have sometimes been categorized as "granule cells." Layer V, the *internal pyramidal layer*, consists predominantly of medium to large pyramidal cells. Layer VI, the *multiform layer*, contains an assortment of neuron types including some with *pyramidal* and *fusiform-shaped* cell bodies.

Two features of the myelinated fibers in the neocortex are noteworthy. First, there are prominent plexuses of horizontally running myelinated fibers in layers IV and V. These are called, the *outer* and *inner bands of Baillerger*, respectively (Fig. 31-1). In the primary visual cortex, bordering on the calcarine sulcus, the outer band of Baillerger is greatly expanded. This band can be seen with the naked eye in fresh and stained sections and is called the *stria (line) of Gennari* (see Fig. 19-17). Second, in most regions of the neocortex, there are many radially oriented bundles of axons passing between the subcortical white matter and various parts of the cortex or between inner and outer cortical layers (Fig. 31-2B).

NEURON TYPES IN THE CEREBRAL CORTEX

Pyramidal Cells The most common type of neuron in the cerebral cortex is the pyramidal cell (Figs. 31-2C, 31-3A). Pyramidal cells are found in all layers of the cortex with the exception of the molecular layer (layer I), and they are the predominant cell type in layers II, III, and V (Fig. 31-4). Pyramidal cells are characterized by (1) a roughly triangular cell body; (2) a single large *apical dendrite* that arises from the apex of the cell body and usually extends toward the molecular layer, giving off branches along the way; (3) an array of *basal dendrites* that run in a predominantly horizontal direction; and (4) an axon that originates from the base of the soma, leaves the cortex, and passes through the white matter.

The cell bodies of most pyramidal neurons range in size from 10 to 50 μm in height. The largest, called *giant pyramidal cells of Betz* or *Betz cells*, are found almost exclusively in the primary motor cortex of the precentral gyrus. Their somata may reach 100 μm in height. Betz cells are most common in the region of motor cortex that projects to the anterior horn of the lumbar spinal cord and hence are concerned with the control of leg movement. These cells are so large that they can be distinguished with the naked eye in Nissl-stained sections of the human brain.

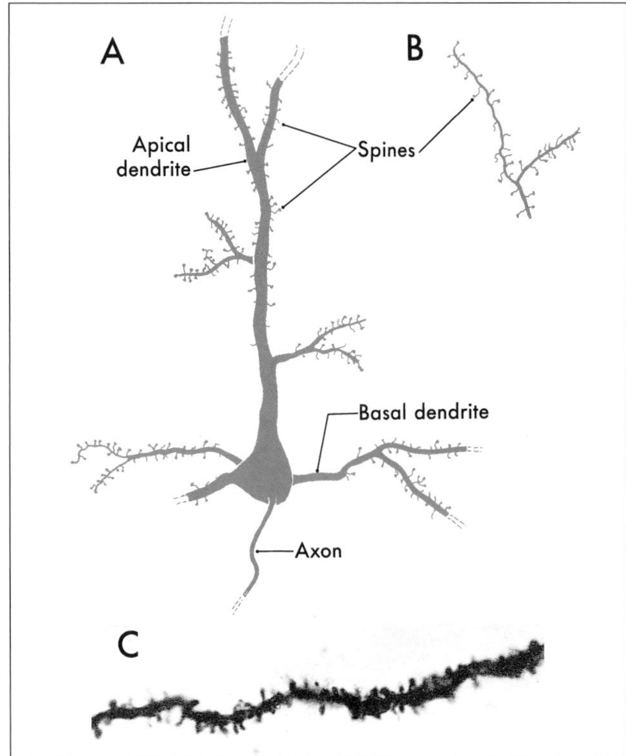

FIGURE 31-3 *Examples of spines on basal and apical dendrites (A, C) and on the terminal ramifications of apical dendrites (B).*

Both apical and basal dendrites of pyramidal cells are characterized by membrane specializations called *dendritic spines*. These spines are small outgrowths from the dendrite that give the impression of thorns on a rose bush (Fig. 31-3A-C). The vast majority of synaptic contacts received by a pyramidal cell are located on dendritic spines rather than directly on the dendrite shaft or on the cell body.

Pyramidal neurons represent virtually the only output pathway for the cerebral cortex. Almost all other cell types in the cortex are local circuit neurons that exert their influence within their own immediate vicinity. Axons of pyramidal cells may terminate in another region of the cortex in the same hemisphere (*association fibers*), decussate in the corpus callosum to terminate in the cerebral cortex of the opposite hemisphere (*callosal fibers*), or course through the white matter to any of the numerous subcortical targets in the forebrain, brainstem, or spinal cord (*projection fibers*).

Pyramidal cells display a laminar organization, with the cell bodies in a given layer projecting to specific neural targets (Fig. 31-4). In general, pyramidal neurons in layers II and III give rise to association and callosal fibers. Pyramidal cells in layer V project to many subcortical structures, including the spinal cord, as projection fibers. The neurons in layer VI send their axons to a variety of locations, including thalamic nuclei and other regions of cortex. Within the cortex, axons of pyramidal cells send off an extensive and relatively dense array of *axon collaterals*. These collaterals terminate in all cortical layers and extend through a horizontal area covering several millimeters around the cell body (Fig. 31-5).

Local Circuit Neurons As mentioned earlier, all the various nonpyramidal neurons of the cerebral cortex function as cortical *interneurons*; that is, their axons do not leave the immediate region of the cell body. These cells are often referred to as *local circuit neurons* or *intrinsic cortical neurons*.

Santiago Ramón y Cajal, working at the turn of the century, described a rich variety of intrinsic cortical neurons. However, by the 1950s it had become customary to refer to virtually all intrinsic cortical neurons as *stellate cells*, even though many were not actually star-shaped. Now the pendulum is swinging in the other direction, and a number of distinct morphologic types are recognized. Some of the more important of these are illustrated in Figure 31-4: spiny and aspiny stellate cells, basket cells, and chandelier cells.

Three types of intrinsic neurons receive thalamocortical axon terminals in layer IV: the *small spiny cells*, the *aspiny stellate cells*, and dendrites of the *large basket cells*. Of these, the spiny cells are believed to be excitatory, whereas basket cells and aspiny stellate cells use the neurotransmitter GABA (γ-aminobutyric acid) and are thus considered to be inhibitory interneurons.

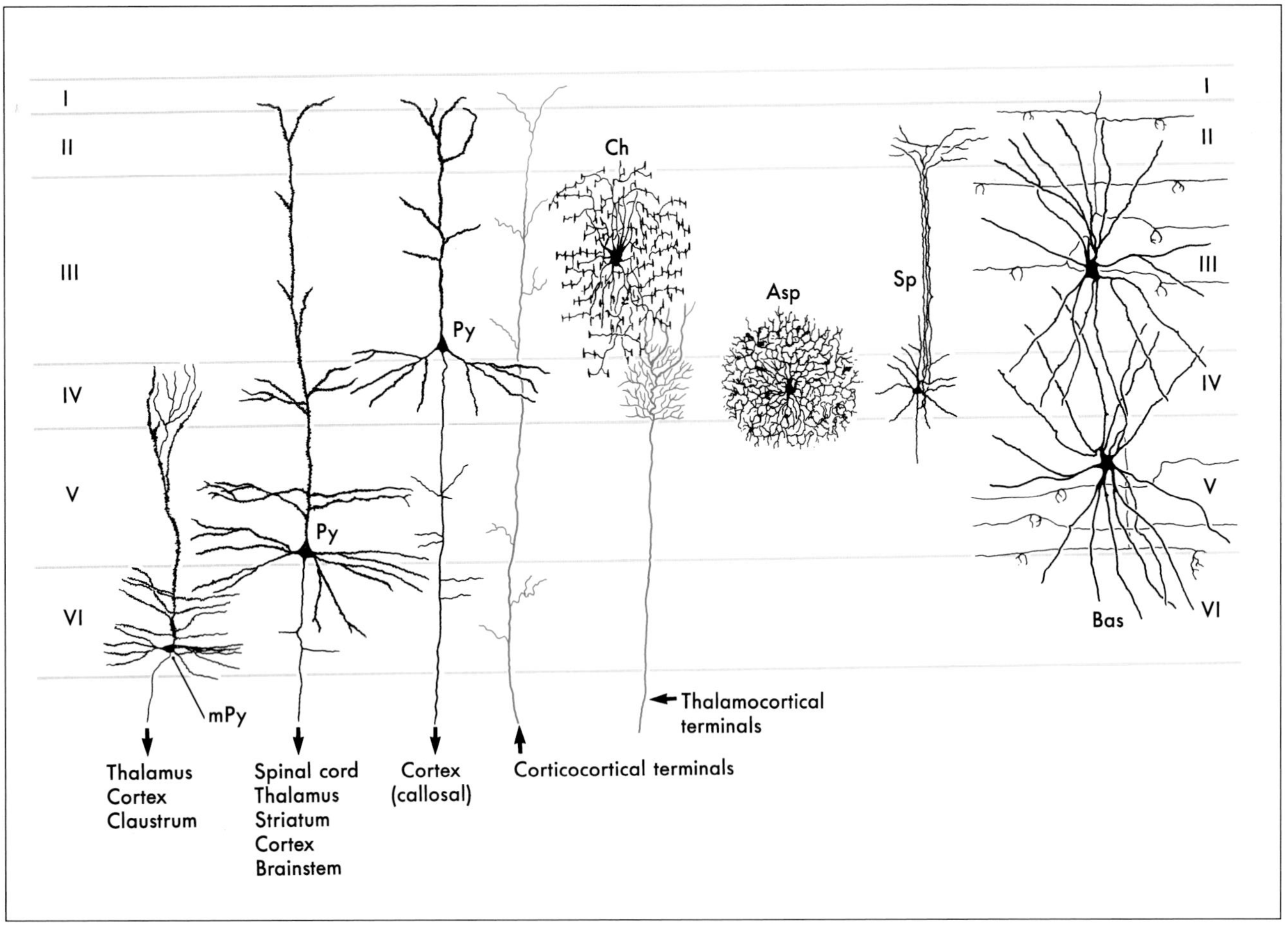

I
II
III
IV
V
VI
Ch
Py
Asp
Sp
Py
mPy
Bas
Thalamocortical terminals
Thalamus
Cortex
Claustrum
Spinal cord
Thalamus
Striatum
Cortex
Brainstem
Cortex
(callosal)
Corticocortical terminals

FIGURE 31-4 *Representative cell types in the cerebral cortex and the layers in which their cell bodies and dendrites are found. Dendrites of pyramidal cells (Py) of layers II, III, and V extend into layer I, whereas those of modified pyramidal cells (mPy) in layer VI extend only to about layer IV. Chandelier cells (Ch) are restricted almost entirely to layer III. The somata of aspiny and spiny stellate neurons (ASp, Sp) are in layer IV, although their processes extend into other layers. Basket cells (Bas) have processes that collectively extend into all cortical layers from cell bodies located mainly in layers III and V. (Adapted from Hendry and Jones, 1981, and Jones, 1984, with permission.)*

3.0 mm
0.5 mm
I
II
III
IV
V
VI
Corticospinal and corticobulbar axons
Corticocortical axon

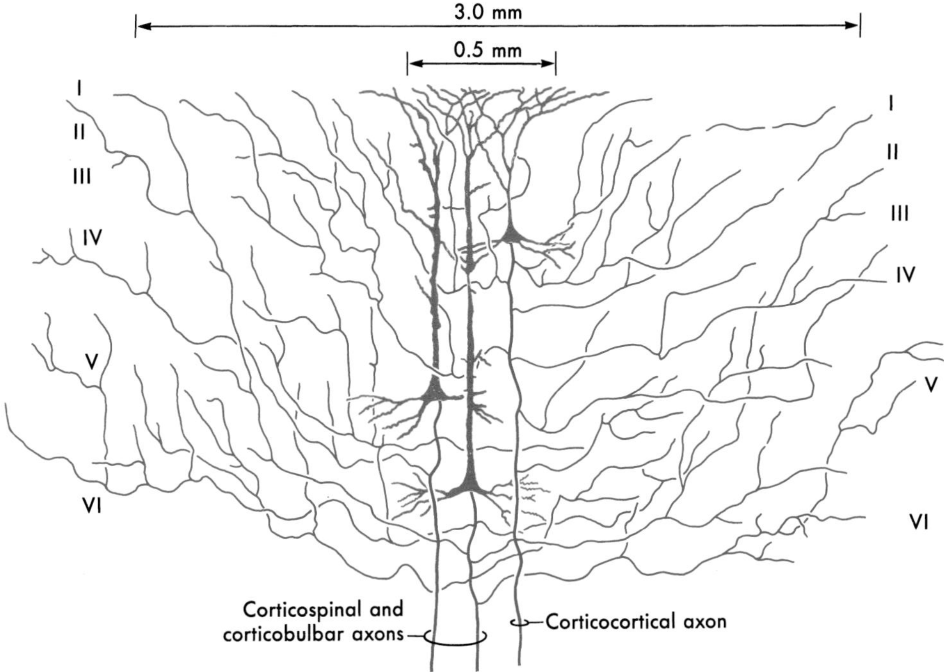

FIGURE 31-5 *The cell bodies and dendrites (in red) of three pyramidal cells in the cerebral cortex compared with the intracortical distribution of axons (in blue) arising from these cells. Axon collaterals distribute over a much wider area than do the dendrites. (Adapted from Scheibel and Scheibel, 1970, with permission.)*

Most other intrinsic neurons are presumed to be inhibitory. On the other hand, pyramidal neurons are uniformly associated with excitatory neurotransmitters, glutamate in particular.

LAMINAR ORGANIZATION

Intrinsic Circuitry of the Cerebral Cortex The ***basic framework*** of the internal circuit diagram of small regions of the cerebral cortex is well understood. In contrast, the *details* of this circuitry are only partially known and are, in fact, so complex as to defy the construction of a detailed circuit diagram like those used to represent a computer's electronic hardware. For example, a single axon may branch repeatedly and contact hundreds of other neurons. A single neuron may also receive synaptic contacts from thousands of other neurons. Within a small volume of cortex, there may be *millions* of neurons.

The basic framework of cortical circuitry consists of afferent fibers, local circuits for the processing of this afferent information, and efferent fibers that convey the processed information to another site (Fig. 31-6). Thalamocortical axons terminate primarily in layer IV and to a lesser extent in layers III and VI. In layer IV, they terminate on excitatory and inhibitory interneurons as well as on dendrites from neurons in other layers (Fig. 31-4). The axons of interneurons, in turn, may end on dendrites of pyramidal cells or of other interneurons. The local processing of information culminates in connections to pyramidal cells, which carry the information to other cortical or subcortical regions. A copy of the information also goes to neurons in the immediate vicinity via *axon collaterals* (Fig. 31-4).

The general pattern of termination of *corticocortical* axons is quite different from that of thalamocortical axons. Corticocortical axons branch repeatedly and make synaptic contacts on neurons in *all* layers of the cortex (Fig. 31-4).

The cerebral cortex receives a third set of inputs, called *diffuse inputs*, which consists of fibers that branch extensively and end diffusely over a wide area of cortex without respect for cytoarchitectural boundaries (Fig. 31-6). These inputs arise from a variety of sources, including certain *nonspecific nuclei of the thalamus* (for example, the ventral anterior, central lateral, and midline nuclei), the *locus ceruleus*, and the *basal nucleus (of Meynert)*. These structures are generally

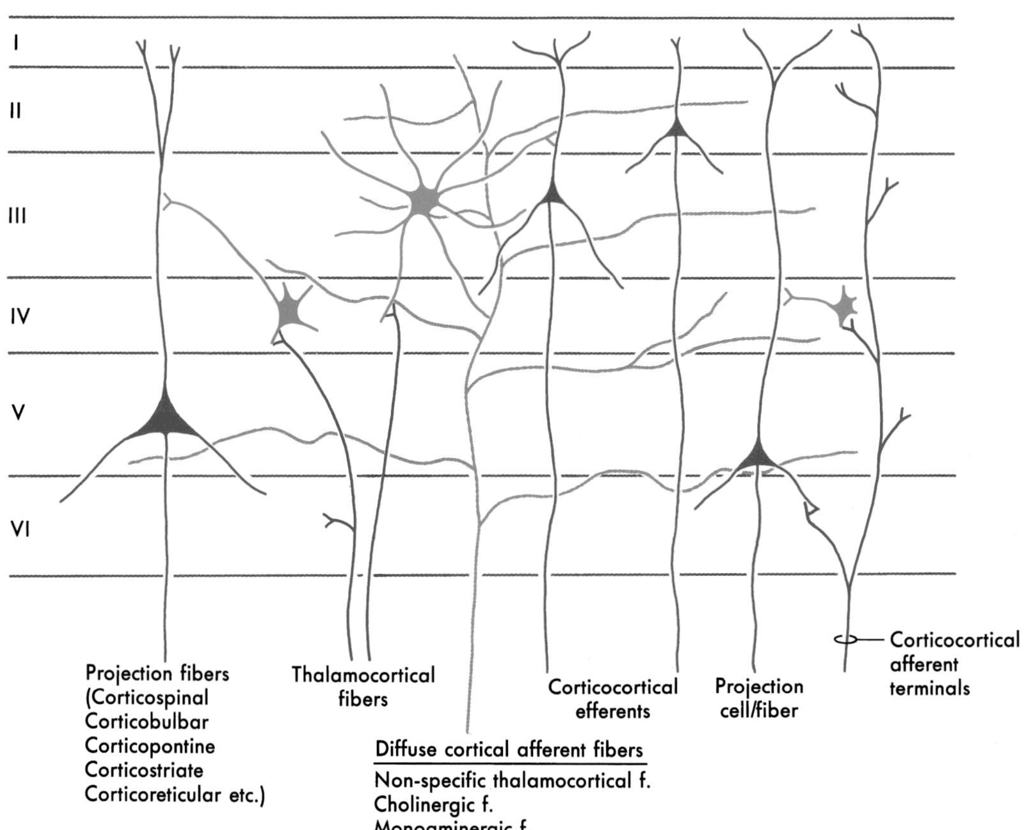

FIGURE 31-6 *Basic circuits in the cerebral cortex. Afferent fibers are shown in blue and gray, interneurons in green, and efferent fibers in red. Thalamocortical fibers terminate primarily in layer IV, whereas corticocortical fibers and diffuse cortical afferents synapse in all layers. Pyramidal cells in the outer layers give rise to corticocortical projections, and those in layer V project to a wide range of downstream targets.*

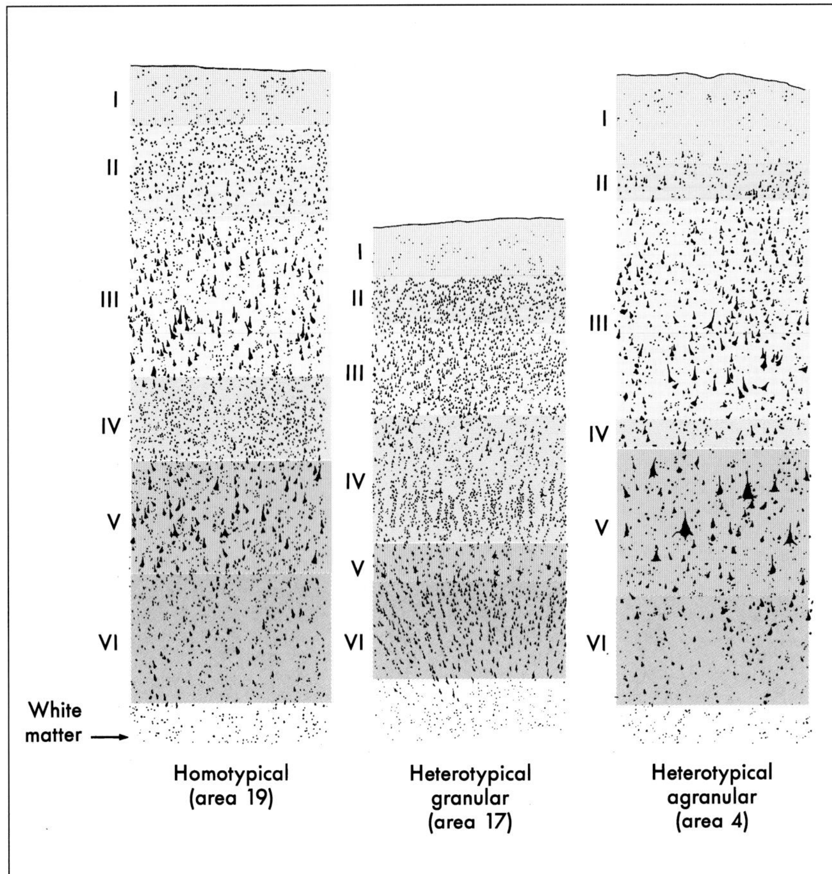

FIGURE 31-7 *Typical cytoarchitectural patterns for homotypical, heterotypical granular, and heterotypical agranular regions. (Adapted from Campbell AW:* Histological Studies on the Localization of Cerebral Function. *Cambridge University Press, Cambridge, England, 1905.)*

concerned with regulating overall levels of cortical excitability and the associated phenomena of arousal, sleep, and wakefulness.

Cytoarchitecture The cytoarchitecture of cortex differs from one area to another in ways that are related to function (Fig. 31-7). In primary sensory cortex, layer IV, the major *input* layer of cortex, is especially thick, whereas layer V, the major *projection layer*, is narrow and indistinct. Cortex with this pattern is called *heterotypical granular cortex*. In primary motor cortex, the pattern is reversed: layer IV is almost invisible, and layer V is very thick, seeming to merge directly with layer III. Thus, the *projection* layer is prominent, and the *input* layer is small. Cortex of this type is called *heterotypical agranular cortex*. In most other areas of the neocortex, including the association cortices, the six layers are all clearly represented and are of roughly equal thickness. This type of cortex is called *homotypical*.

The cerebral cortex has been subdivided on the basis on cytoarchitectural differences by many different investigators. The most famous of these, Korbinian Brodmann, worked in the early part of the twentieth century. He identified 47 distinct areas (Fig. 31-8), and his numbering scheme is still in common use today in both research and clinical settings. For example, the primary visual cortex is Brodmann's area 17, and the primary motor cortex is area 4. In most instances, Brodmann's cytoarchitectural areas are coextensive with cortical regions that have specific functional characteristics. This numbering scheme is used throughout the chapter.

COLUMNAR ORGANIZATION

A second, vertical pattern of organization is superimposed on the horizontal layered pattern described above. Unlike the cortical layers, this vertical pattern is not immediately obvious in histologic sections stained for neuron cell bodies (Nissl stains). However, when Golgi-stained material is studied, it is clear that neurons are often grouped together so that their cell bodies, axons, and apical dendrites form clusters that are oriented at right angles to the surface of the cortex.

Mountcastle was the first to demonstrate physiologically the existence of a vertical ("columnar") organization in the cerebral cortex by recording the activity of hundreds of individual neurons in the primary somatosensory cortex of cats and monkeys. Within an area of cortex a few millimeters in diameter, all neurons had overlapping or adjacent receptive fields. For example, in one cortical region, all neurons might have receptive fields on a finger, whereas in a nearby region, the neurons might have receptive fields on the wrist. Within a cortical region in which all neurons had about the same receptive field, the neurons responded to different *sensory submodalities*. Some neurons were activated by light touch on the skin, others by joint rotation, and yet others by strong pressure on deep tissue. However, when a microelectrode was inserted at right angles to the surface of the cortex, all the neurons encountered were activated by only one of these submodalities (Fig. 31-9B). In contrast, when a microelectrode was moved parallel or obliquely relative to the surface of the cortex, it encountered neurons of different submodalities as it moved from one functionally related group of neurons to another (Fig. 31-9A).

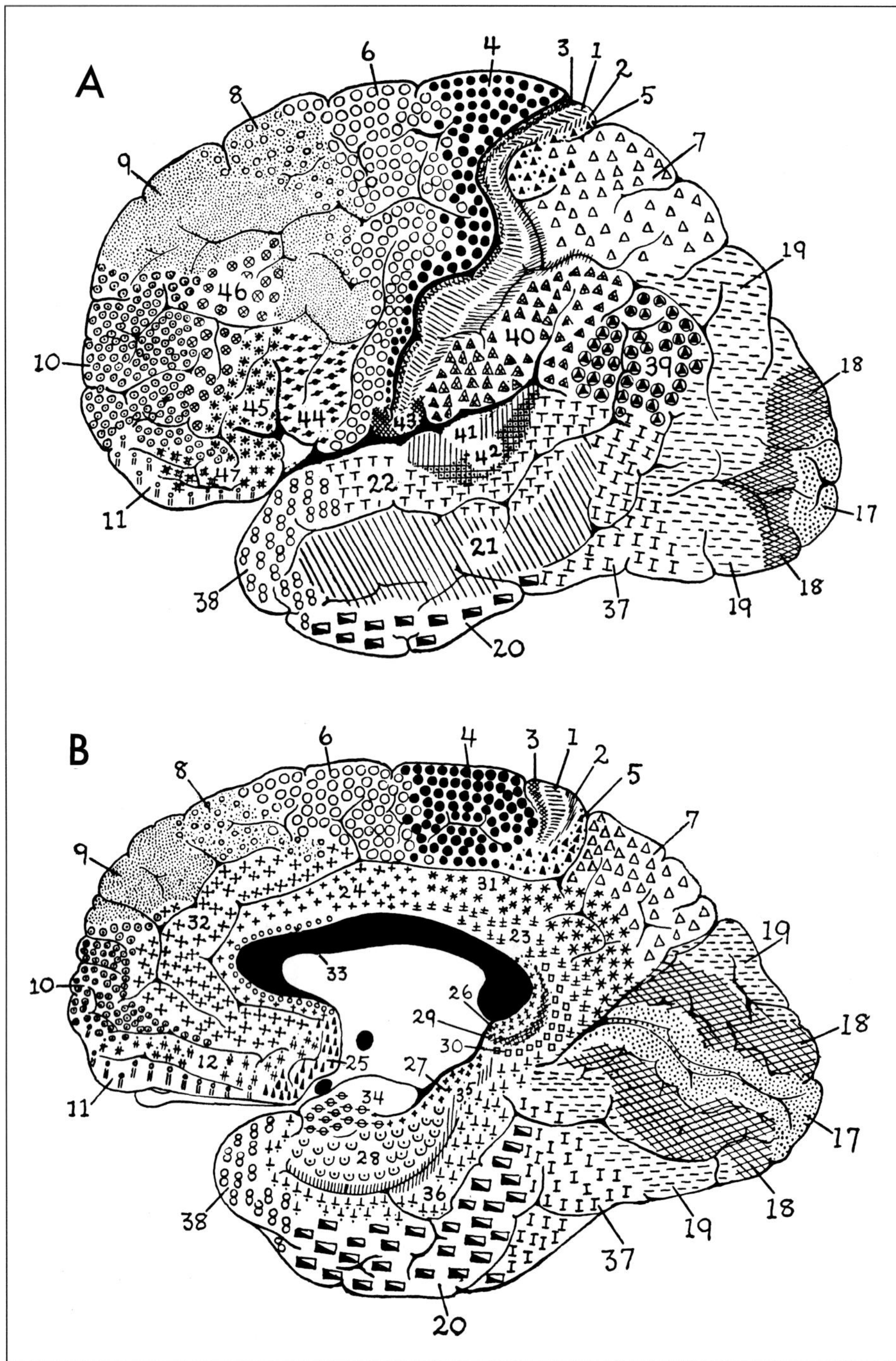

FIGURE 31-8 *Cytoarchitectural map showing Brodmann's areas on the lateral (A) and medial (B) surfaces of the hemisphere. (Modified after Brodmann K, from Carpenter MB, Sutin J:* Human Neuroanatomy. *Williams & Wilkins, Baltimore, 1983, with permission.)*

The basis of columnar organization in primary sensory cortices is selective input from relay nuclei of the thalamus. Obviously, if all of the cells in one column respond to maintained pressure on the skin while the cells in an adjacent column respond to joint position, the signals from the respective sensory receptors must have been continuously segregated all the way from the periphery through the dorsal column nuclei and the ventrobasal complex of the thalamus to terminate in the cortex.

The anatomic basis of the *columnar organization* of the cortex is understood in the greatest detail in the visual cortex. In this region, at least three types of regularly repeating features are superimposed on the laminar patterns of neurons: the stimulus orientation columns, the ocular dominance columns, and the cytochrome-oxidase-rich blobs. These features are discussed in detail in Chapter 19.

The ocular dominance columns in the visual cortex provide a clear example of the role of thalamic input in columnar organization. Neurons in the layers of the lateral geniculate nucleus that receive input from the right eye send their axons to layer IV of the right-eye-dominant columns (Fig. 31-10). Here, the axons terminate predominantly on spiny and aspiny stellate cells, which, in turn, project to pyramidal cells. Collaterals of pyramidal cell axons provide one pathway by which neural signals can spread from one column to influence activity in adjacent columns (Fig. 31-5). This influence may be either excitatory via direct connections or inhibitory via interneurons. The right eye, therefore, has a direct and strong influence on neurons in right-eye-dominant columns (R_L in Fig. 31-10) and an indirect and weaker influence on neurons in the adjacent left-eye-dominant columns (L_R in Fig. 31-10).

Connections between one region of the cortex and another, either through association fibers or callosal fibers, may also be arranged in a columnar pattern. For example, axons that originate in the inferior parietal lobule terminate in multiple columns in the ipsilateral and contralateral cingulate cortex. Columns of corticocortical axon terminals that originate in different functional regions may either overlap each other or interdigitate with each other.

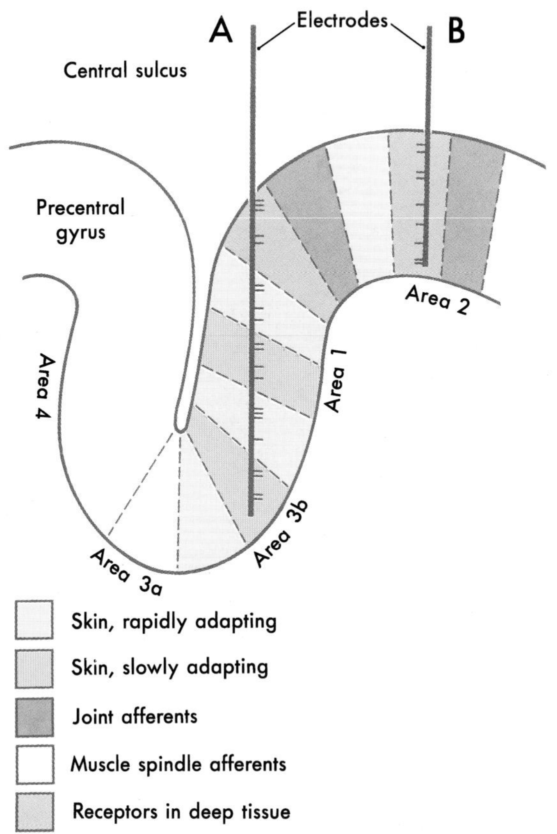

FIGURE 31-9 *Diagrammatic section through the pre- and postcentral gyri showing the organization of columns in the somatosensory cortex. The columns are shown as colored compartments oriented, in general, perpendicular to the surface of the cortex. An electrode (at A) passing parallel to the surface of the cortex will pass through several columns with resultant recordings of the several modalities represented by the types of afferent information arriving at each column. An electrode passing through one column (at B) passing perpendicular to the surface of the cortex penetrates only a single column. Therefore, it records activity related to the single submodality received by that column.*

SYNOPSIS OF THALAMOCORTICAL RELATIONS

The details of thalamocortical projections are described in the chapters devoted to specific systems. At this juncture, however, it is appropriate to briefly review what areas of the cortex are functionally related to which of the thalamic nuclei (Fig. 31-11).

The cortex of the frontal lobe encompasses Brodmann's areas 4, 6, 8 to 12, 32, and 44 to 47 (Fig. 31-8). The primary somatomotor cortex (area 4) and the premotor and supplementary motor cortices (area 6) receive input mainly from the ventral lateral nucleus of the thalamus and subserve important motor functions. Lateral, medial, and orbital aspects of the frontal lobe receive thalamocortical fibers mainly from the dorsomedial and anterior nuclei of the thalamus (Fig. 31-11). These latter cortical areas, through a variety of direct and indirect connections, relate primarily to functions of the limbic system. Of particular note are the partes orbitalis and triangularis of the inferior frontal gyrus, damage to which results in Broca's aphasia (discussed below).

Areas 3, 1, 2, 5, 7, 39, 40, and 43 are located in the parietal lobe (Fig. 31-8). The primary somatosensory cortex (areas 3, 1, and 2) receive inputs from the ventral posterolateral and ventral posteromedial nuclei. These thalamic nuclei receive a full range of somatosensory input through synaptic relays in the spinal cord and brainstem and transmit this information to the cerebral cortex. The inferior parietal lobule comprises, in general, areas 39 and 40. Along with area 22, these are the cortical regions associated with Wernicke's aphasia (discussed below).

The occipital and temporal lobes encompass areas 17 to 22, 36 to 38, and 41 and 42 (Fig. 31-8). These areas of the cortex, plus portions of the parietal lobe, have extensive connections with the pulvinar nucleus of the thalamus and are involved in the processing of visual and auditory information at several different functional levels (Fig. 31-11). Located in this geographic area are the primary sensory

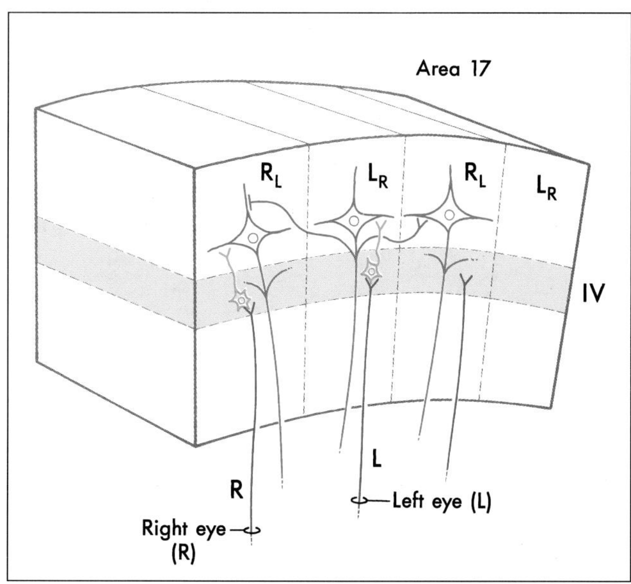

FIGURE 31-10 *Functional columnar organization of sensory cortices, using the visual cortex (ocular dominance columns) as an example. Axons from the layers of the lateral geniculate nucleus related to the right (R) and left (L) eyes terminate in alternating columns. In right-eye-dominant columns (R_L), cortical neurons are influenced predominantly by visual stimulation of the right retina, although they are also influenced to a lesser degree by stimulation of the left retina. The reverse is true for left-eye-dominant columns (L_R). The neurons in each column influence the adjacent columns (dominant for the other eye) via axon collaterals of pyramidal cells or through the action of cortical interneurons. In general, other sensory cortices are organized similarly in regard to their sensory inputs.*

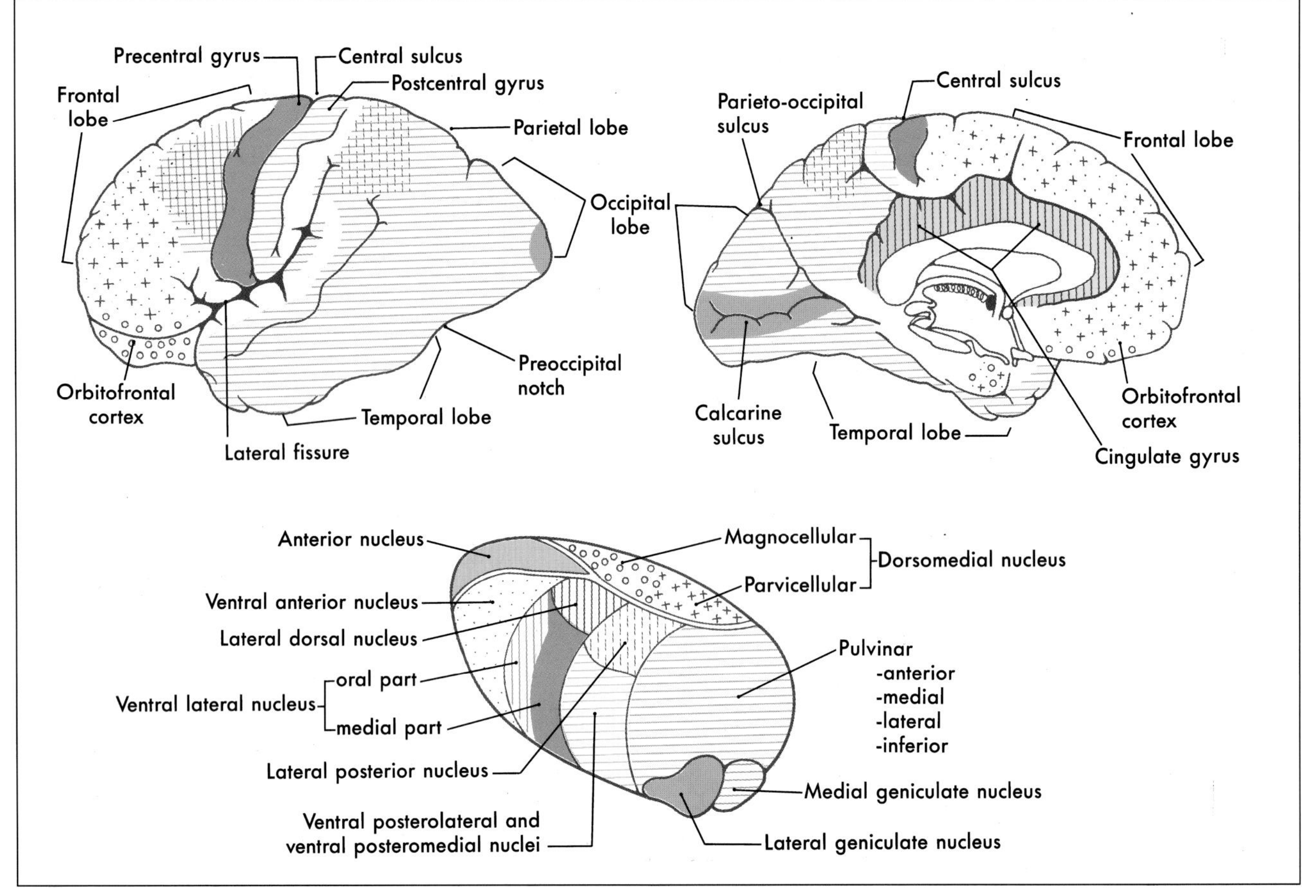

FIGURE 31-11 *Relationships of the thalamic nuclei to the cerebral cortex as revealed by the patterns of thalamocortical connections. Each thalamic nucleus is pattern-coded or color-coded to match its target area in the cerebral cortex.*

cortices for vision and hearing. Area 17, on the banks of the calcarine sulcus, is the primary visual cortex; areas 41 and 42, in the depth of the lateral fissure in the transverse temporal gyri, are the primary auditory cortex (Fig. 31-8). These cortical areas receive input from the lateral and medial geniculate nuclei of the thalamus, respectively.

The limbic lobe, which forms the most medial edge of the hemisphere, contains areas 23 to 31 and 33 to 35. The cingulate cortex receives fibers primarily from the anterior nucleus of the thalamus, but also from the lateral dorsal nucleus (Fig. 31-11). Other regions of the limbic lobe have some connections with the dorsomedial nucleus. However, many of the subcortical targets of the parahippocampal and uncal cortices are structures such as the hippocampal formation. This area, in turn, projects to a variety of thalamic and basal forebrain targets.

BLOOD SUPPLY TO THE CEREBRAL CORTEX

The blood supply to the cerebral cortex and to subcortical structures of the telencephalon, including the internal capsule, is discussed in Chapters 8 and 15. The goal here is to summarize the general nature of these patterns.

The cerebral cortex is served, in toto, by the *anterior*, *middle*, and *posterior cerebral arteries*. The anterior and middle cerebral arteries are the terminal branches of the internal carotid artery and the posterior cerebral artery is formed by the bifurcation of the basilar artery (see Figs. 8-3, 8-7).

The anterior cerebral artery is joined to its counterpart just anterior to the optic chiasm by the anterior communicating artery. Proximal to the anterior communicating artery, the anterior cerebral artery gives rise to small branches that serve rostral portions of the hypothalamus. Cortical branches of the anterior cerebral artery distribute to the medial surface of the hemisphere caudally to about the position of the parieto-occipital fissure. The distal portions of these branches arch over the edge of the hemisphere (from its medial to its lateral surface) for a short distance (see Figs. 8-3, 8-8, 15-8). Located in the domain of the branches of this major vessel are the foot, lower extremity, and hip areas of the primary somatomotor and primary somatosensory cortices.

The middle cerebral artery passes laterally from its origin and branches, in general, into superior and inferior trunks (these are M_2 branches) over the insular cortex. Terminal branches of the superior and inferior trunks serve the cortex on the lateral surface of the hemisphere above and below the lateral fissure, respectively (see Figs. 8-5, 15-8). In addition to lateral portions of the frontal cortex, parietal and temporal association cortices are served by branches of the middle cerebral artery. Also located in the distribution area of this vessel are the trunk, upper extremity, and head regions of the primary somatomotor and primary somatosensory cortices (via branches of the superior trunk) and the primary auditory cortex (inferior trunk).

The cortex forming the inferior surface of the temporal lobe and the medial aspect of the occipital lobe is served by branches of the posterior cerebral artery (Figs. 8-3, 8-7, 15-8). As with the anterior cerebral artery, the terminal rami of the posterior cerebral artery loop over the edge of the inferior and medial surface of the hemisphere to serve small portions of the lateral aspect of the hemisphere. The posterior cerebral artery serves large expanses of visual association cortex and some of the cortical structures associated with the limbic system. In addition, the calcarine artery supplies the primary visual cortex located on and in the banks of the calcarine sulcus.

The wedge-shaped areas of overlap between the distal branches of the anterior and middle cerebral arteries and the posterior and middle cerebral arteries form what are called *border zones* (Fig. 8-8). These areas are particularly susceptible to hypoperfusion during episodes of systemic hypotension. Such events may result in *watershed infarcts*.

HIGHER CORTICAL FUNCTIONS

The cerebral cortex is generally considered to be the seat of *higher intellectual functions*, those faculties of thought that have reached their most complex levels in humans. Although other brain structures, including the thalamus, striatum, claustrum, and cerebellum, contribute to these functions, the *multimodal association cortex* is closely linked to the most complex intellectual functions, such as logical analysis, judgment, language, and imagination.

The cerebral cortex can be divided into four general functional categories. These are *sensory*, *motor*, *unimodal association cortex*, and *multimodal association cortex* (Fig. 31-12). The primary sensory areas, except that for olfaction, receive thalamocortical fibers from diencephalic relay nuclei that are functionally related to each modality. For example, the ventral posterior complex of the thalamus projects to primary somatosensory cortex (Brodmann's areas 3, 1, and 2) in the postcentral gyrus. Similarly, the lateral geniculate nucleus projects to primary visual cor-

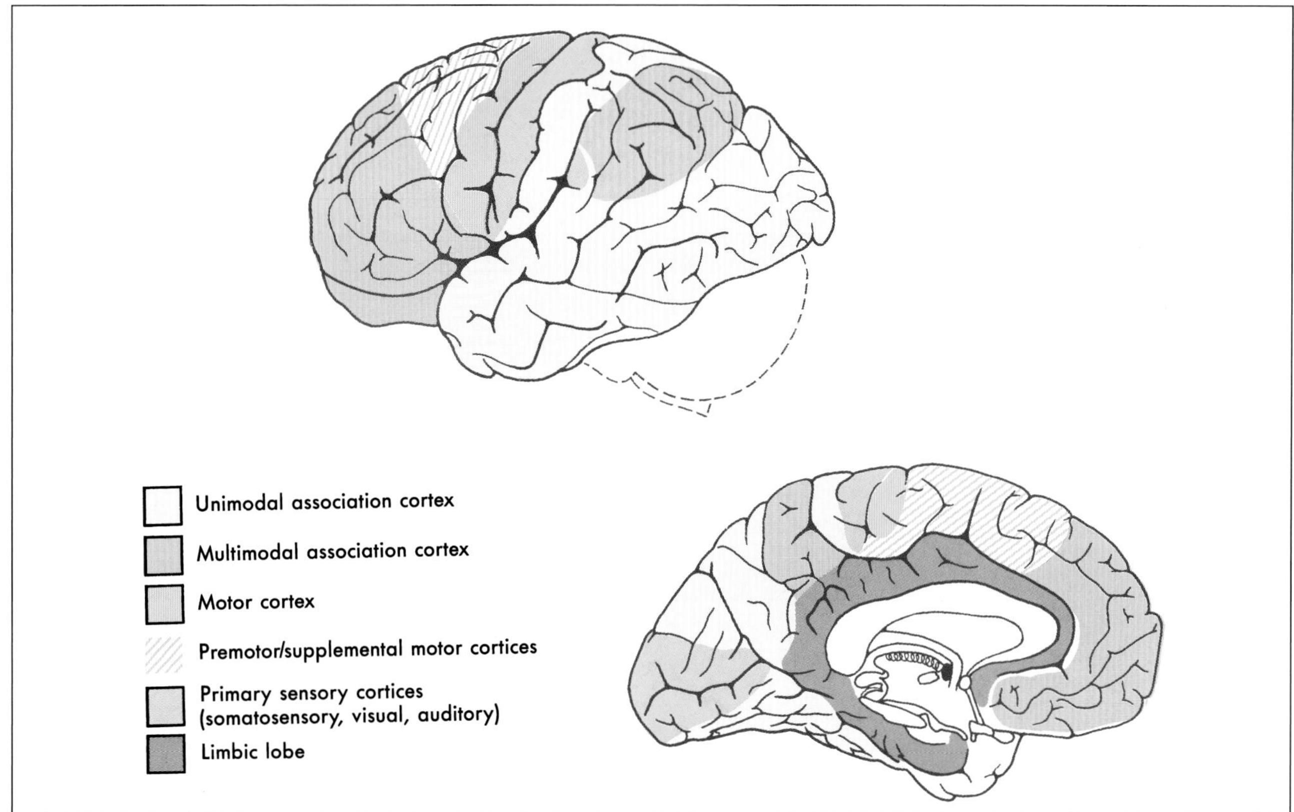

FIGURE 31-12 *Primary motor and sensory (blue), unimodal association (green), and multimodal association (red) areas of the cerebral cortex.*

tex (area 17) in the banks of the calcarine sulcus, and the medial geniculate nucleus projects to primary auditory cortex in the transverse temporal gyri (areas 41 and 42).

Adjacent to each primary sensory area is a region of cortex that is devoted to a higher level of information processing relevant to that specific sensory modality. These areas are called *unimodal association cortices* (Fig. 31-12). For example, *visual unimodal association cortex* (areas 18 and 19) occupies all of the occipital lobe outside area 17 (the primary visual sensory area), as well as much of the inferior gyrus of the temporal lobe. Within these visual association areas, the basic elements of visual sensation are molded into an overall perception of the visual world. Similarly, *somatosensory association cortex* lies just posterior to the postcentral gyrus in area 5, and the *auditory association cortex* is in the superior temporal gyrus (area 22) next to primary auditory cortex. In all of these examples, the primary sensory cortex (i.e., areas 3, 1, 2; 17; and 41 and 42) receive input from their respective thalamic relay nuclei. In turn, the primary sensory areas project, via corticocortical fibers, to their corresponding association cortices (i.e., areas 3, 1, 2 to area 5; area 17 to areas 18 and 19; and areas 41 and 42 to area 22).

The remaining portions of the cerebral cortex that are not motor in function are classified as *multimodal association cortex* (Fig. 31-12). These areas receive information from several different sensory modalities and create for us a complete experience of our surroundings. This allows us to communicate using language, to use reason to extrapolate future events on the basis of present experience, to make complex and long-range plans, and to imagine and create things that have never existed. An example of long-range planning is going to college so you can go to medical school so you can do a residency and become a physician. In this chapter, we will concentrate on the cortical areas responsible for three of these higher functions: language, the appreciation of space, and the planning of behavior.

The Dominant Hemisphere and Language The cerebral hemisphere that controls language is called the *dominant hemisphere*. In the vast majority of individuals, language functions are processed in the *left hemisphere*. Evidence of this is seen by the fact that brain lesions that adversely affect language are found in the left hemisphere in about 95% of cases. Almost all right-handed individuals and about half of left-handed individuals are *left-cerebral dominant*. It follows that the right cerebral hemisphere, in most of the general population, is the *nondominant hemisphere*.

Language is the faculty of communication using symbols organized by a system of grammar to describe things and events and to express ideas. In humans, the senses of vision and audition are closely linked to language, but language itself transcends any particular sensory system. Helen Keller was blind and deaf but used language eloquently to communicate very complex and subtle ideas. Language ability can be impaired selectively, with little or no change in the senses of vision or hearing, by brain damage in either the parietal/temporal junction or the frontal lobe. This impairment, termed *aphasia*, is a *disturbance of the comprehension and formulation of language, not a disorder of hearing, vision, or motor control.*

The two classic types of aphasia are *Broca's aphasia* and *Wernicke's aphasia*. Broca's aphasia, also termed *expressive aphasia*, consists of a loss of the ability to speak fluently. Lesions that produce this deficit are located in the inferior frontal gyrus, primarily in Brodmann's areas 44 and 45 (Fig. 31-13). Wernicke's aphasia is primarily a defect of the *comprehension* rather than the *expression* of language. This deficit is seen following injury to the supramarginal and angular gyri (areas 37, 39, and 40) and the posterior part of the superior temporal gyrus (area 22) (Fig. 31-13).

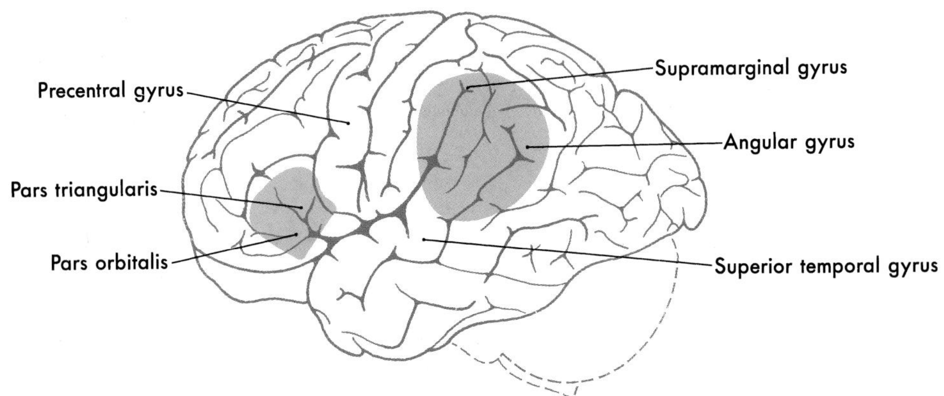

FIGURE 31-13 *Cortical areas that mediate the processing of language. Lesions in the partes orbitalis and triangularis of the inferior frontal lobe will result in Broca's aphasia, whereas damage in the supramarginal and angular gyri and adjacent superior temporal gyrus will result in Wernicke's aphasia.*

Patients with the most severe form of *Broca's aphasia* are unable to speak, although they are able to swallow, breathe, and make guttural sounds. Their problem is not one of paralysis of the vocal apparatus. Rather, it is a difficulty in turning a concept or thought into a sequence of meaningful sounds. In less severe cases, or in patients recovering from a stroke, limited speech is possible. However, it is slow and quite labored, enunciation is poor, and words like "the" or "and" are commonly omitted. Although the patient is able to communicate verbally, the extremely laborious nature of the process of communicating causes considerable frustration. Under particular emotional stress, patients may even use inappropriate or vulgar words or phrases to express their distress. The most common causes of Broca's aphasia are tumors and occlusions of frontal M_2 branches of the middle cerebral artery. Aphasia without other deficits indicates that the damage affects only cortical areas. On the other hand, these patients may also have *contralateral motor symptoms*, such as weakness (*paresis*) of the lower part of the face, lateral deviation of the tongue when protruded, and weakness of the arm. Aphasia plus these motor problems suggest an occlusion of branches from proximal parts of the middle cerebral artery (M_1), including the lenticulostriate arteries, which serve the internal capsule (see Fig. 15-9).

The second major type of aphasia is *Wernicke's aphasia*. Patients with this deficit are able to produce clear, fluent, melodic speech at a normal or even faster than normal rate. The content of the speech, however, may be unintelligible because of frequent errors of word choice, inappropriate use of words, or use of made-up nonsense words. An example of this type of speech pattern is "We went to drive in the bridge for red pymarids were crooking the lawn browsers." Such speech is sometimes termed "word salad." In addition, *paraphasias* frequently occur. For example, in trying to say "The cat has claws," a patient may use an incorrect but similar word ("The cat has clads," a *literal paraphasia*) or a word that seems appropriate to the patient but is incorrect ("The cat has tires," a *verbal paraphasia*). Patients with Wernicke's aphasia also have difficulty comprehending what others say to them. Surprisingly, these individuals seem much less aware of the extent of their disability than patients with Broca's aphasia, and they are usually less frustrated and depressed by it. Patients with Broca's aphasia are completely conscious of their communication problems.

Wernicke's aphasia may result from occlusion of temporal and parietal M_2 branches of the middle cerebral artery. In addition, hemorrhage into the thalamus (or tumors in the thalamus) may produce Wernicke's aphasia by extending laterally and caudally to invade the subcortical white matter. If this damage impinges on Meyer's loop and interrupts the optic radiations (see Chapter 19), a contralateral homonymous hemianopia may accompany the patient's other disabilities.

The severity of the aphasia depends on the severity of the associated brain damage. In mild cases, only one or two symptoms may be discernible, and those may resolve quickly. A major stroke or severe traumatic injury, however, could produce a full-blown set of symptoms that would never completely disappear.

Other, less common, types of aphasia have also been described. These include *conduction aphasia*, which results from interruption of the connections linking Broca's and Wernicke's areas. In this disorder, comprehension is normal and expression is fluent, but the patient has difficulty translating what someone has said to him into an appropriate reply. A more profound disorder is *global aphasia*, in which brain damage encroaches on both Broca's and Wernicke's areas, and the loss of language is virtually complete.

Several additional points should be mentioned. First, damage to the basal ganglia, particularly to the head of the caudate on the *left side*, has been associated with language disorders similar to Wernicke's aphasia. Second, although we have referred so far to spoken and written language (i.e., *verbal* language), aphasia can also affect *non-verbal* language. A deaf person who uses American Sign Language can lose the ability to use or understand sign language after focal brain damage in the left hemisphere. Third, although most aspects of language are processed in the left hemisphere, some features are influenced by lesions in the nondominant parietal lobe. In particular, a patient with a right parietal lesion may have difficulty appreciating the *prosody* of speech. This term refers to the variations in vocal inflections, emotional content, and melody that may alter the meaning of a spoken sentence, as in: "George is here." versus "*George* is here!" versus "George is *here*?" versus "George *is* here!"

Parietal Association Cortex—Space and Attention A completely different set of intellectual functions is mediated in the parietal association cortex of the nondominant hemisphere. Although the segregation of functions between the two parietal lobes is not complete, the parietal association cortex is nevertheless the most highly *lateralized* in the brain, with language functions concentrated in the left hemisphere and spatial relationships and related selective attention concentrated in the right hemisphere.

Much of our knowledge of the functional properties of different regions of the cerebral cortex has been gained from neurologic case studies of patients with cortical damage produced by stroke or head trauma. In this respect, the two great wars of the first half of the twentieth century led, inadvertently, to great progress in our understanding of the effects of brain injuries. One of the most striking symptoms of damage to *right parietal association cortex* (nondominant) is a defect of *attention*, in which the patient seems to be completely unaware of objects and events in the left half of his or her surrounding space. This symptom is termed *contralateral neglect* (Fig. 31-14A,B).

In its milder forms, contralateral neglect may simply be a tendency to ignore things on the left side of the patient's surroundings. For example, the patient might be asked to read a short passage and check off each word in the process. As the patient reads, words on the left side of the passage are progressively ignored, and only those on the right part are perceived (Fig. 31-14A). Another way to demonstrate contralateral neglect is to draw a circle and ask the patient to draw in the numbers of a clock face. Typically, the patient with right parietal damage will put all of the numbers (1 to 12) on the right side of the circle (the side ipsilateral to the lesion), completely ignoring the left (contralateral) side of the circle (Fig. 31-14B). A patient with contralateral neglect may not be aware of people standing to the left, may bump into large stationary objects on the left, and may not respond to sounds or words coming from the left. In extreme cases, the patient may not even recognize the left side of his or her own body. For example, the patient may ignore the left side when dressing or grooming, or, if in a hospital, may even demand that the staff get this "other person" (the left side of his or her own body) out of the bed.

A second characteristic group of symptoms of right parietal lobe lesions concerns a patient's ability to function successfully within his or her *spatial surroundings*. For example, the individual may be unable to describe his route between home and work, draw a floor plan of his house (Fig. 31-14C), or find a soft drink machine that is just down the hall. In extreme cases, the patient may not be able to navigate successfully from his or her bed to a chair that is just across the room and in full view.

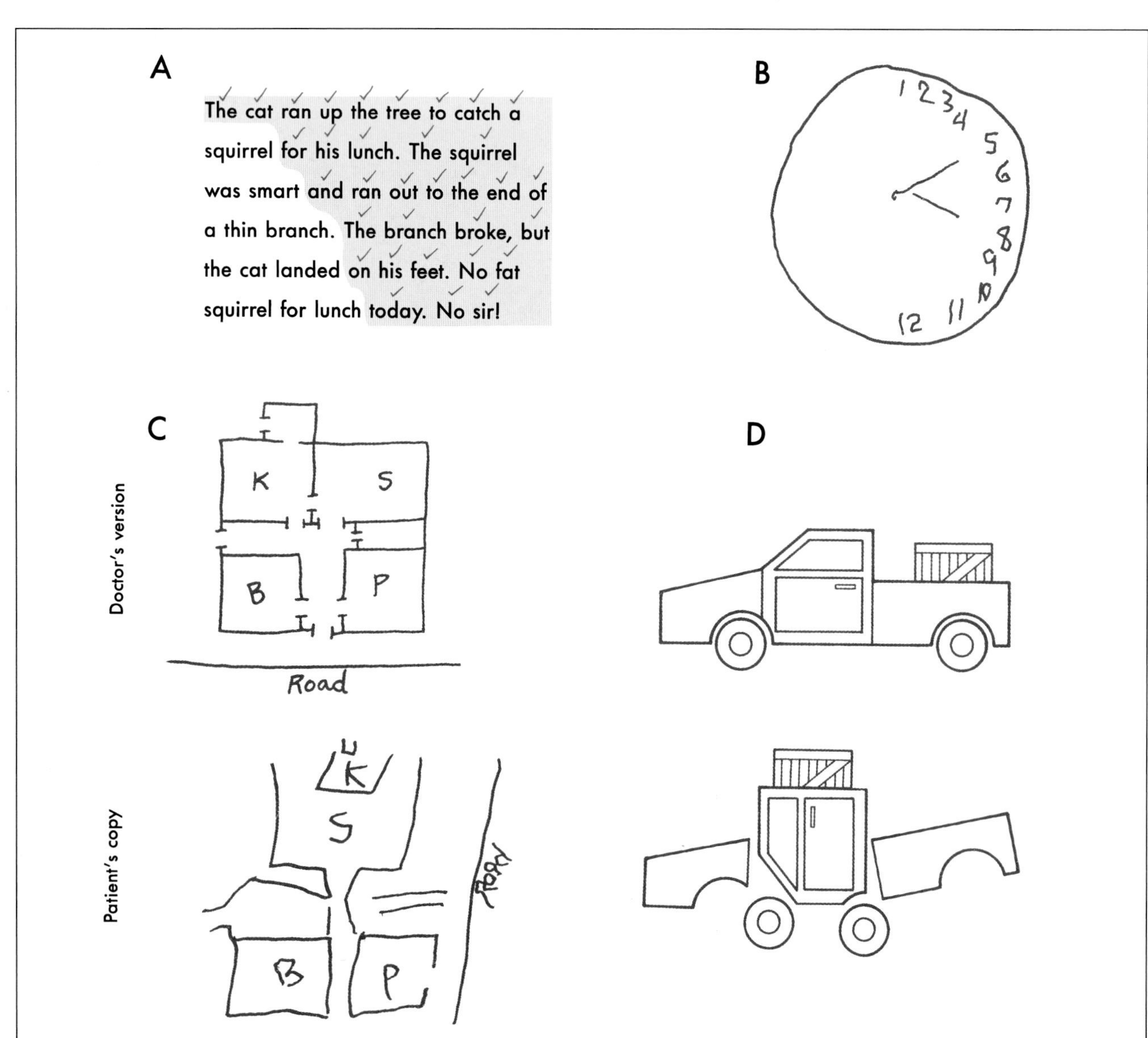

FIGURE 31-14 *Signs (A–D) of damage to the non-dominant parietal association cortex. See text for details.*

A third difficulty is an inability to successfully manipulate objects in space. A patient might be unable to duplicate a simple block construction while looking at a model (Fig. 31-14D). This difficulty is termed *constructional apraxia*; it is not related to visual acuity or to fine motor control but rather to an inability to internalize and duplicate the spatial relationships of the individual parts of the model. In addition, disorders of *affect* are common, including a reduced ability to understand and appreciate humor, a loss of the ability to appreciate the *prosody* of speech, and often an inappropriate cheerfulness and lack of concern for, or even awareness of, the implications of the illness. This may be the case even when symptoms are as serious as total left hemiplegia.

Prefrontal Cortex and "Plans for Future Operation" The other major region of multimodal association cortex is the large expanse anterior to the primary motor and premotor cortices, the *prefrontal association cortex*. This region has historically been connected with some of the most distinctly human intellectual traits, such as *judgment*, *foresight*, a sense of *purpose*, a sense of *responsibility*, and a sense of *social propriety*.

One of the earliest accounts of the effect of brain injury on higher intellectual functions described a series of events that began on September 13, 1848. A crew of railroad construction workers was blasting a right-of-way through the rugged granite mountains of Vermont. The well-liked young foreman of the crew, Phineas Gage, was in charge of placing a black powder charge in a deep hole drilled in the rock, adding a fuse, covering the powder with sand, and finally tamping the sand and powder down firmly with an iron rod before lighting the fuse and running for cover. On this day, something apparently distracted Gage, and he began to tamp down a charge before the sand had been added. The iron rod struck the granite wall of the hole and a spark ignited the powder. The 3.5-ft-long, 13-lb rod was propelled out of the hole like a giant bullet.

The rod struck Gage just beneath the left eye and exited through the top of his head, destroying most of his prefrontal cortex. Amazingly, Gage was not killed instantly, and even more incredibly, he survived the inevitable serious wound infection that followed. Eventually he recovered his health, or at least the physical portion of it. Mentally, however, he was changed forever. Although he did not suffer paralysis, language disorders, or memory loss, his personality was radically altered. John Harlow, one of the doctors who attended Gage, perceived the importance of this case with respect to the localization of intellectual functions in the brain. In an article describing the injury and Gage's persisting intellectual symptoms, Harlow said:

> His physical health is good, and I am inclined to say that he has recovered. . .The equilibrium or balance, so to speak, between his intellectual faculties and animal propensities seems to have been destroyed. He is fitful, irreverent, indulging at times in the grossest profanity (which was not previously his custom), manifesting but little deference for his fellows, impatient of restraint or advice when it conflicts with his desires, at times pertinaciously obstinate, yet capricious and vacillating, devising many plans for future operation, which are no sooner arranged than they are abandoned. . .In this regard his mind was radically changed, so decidedly that his friends and acquaintances said that he was 'no longer Gage.'

This passage, written more than 100 years ago, provides an accurate and insightful description of the major symptoms associated with destruction of prefrontal cortex. Patients with significant bilateral damage to the prefrontal cortex have a constellation of deficits that can be summarized as follows. First, they are *highly distractible*, turning from one activity to another according to the novelty of a new stimulus rather than according to a plan. This is sometimes described as a *lack of consistency of purpose*. Second, these individuals have a *lack of foresight*. They are not able to anticipate or predict future events on the basis of past events or present conditions. Third, they may be *unusually stubborn* in the face of advice with which they do not agree, and they may also *perseverate* in the performance of a task. Fourth, the patient with prefrontal damage displays a profound *lack of ambition*, a loss of the *sense of responsibility*, and a loss of a *sense of social propriety*. The first and third symptoms (*distractibility* versus *perseveration*) are obviously in conflict. It is impossible to predict which will dominate at a given moment, but *both* exemplify the individual's loss of the ability to govern his own actions and life according to a *plan*. He is instead imprisoned in a chaotic world, his actions governed by randomly changing whims.

It was this set of symptoms that prompted the Portuguese neurosurgeon Egas Moniz to develop the prefrontal lobotomy procedure in the late 1930s to treat a range of severe, intractable mental problems. At that time, mental hospitals ("insane asylums") all over the world contained many patients who were so immobilized by anxiety that they could not even take care of their own bodily needs. They were warehoused under reprehensible conditions. The discovery that a neurosurgical procedure could alleviate the anxiety to the extent that the patients could lead a somewhat more normal existence (albeit still within the confines of a mental institution) was hailed as a great breakthrough. In these desperate patients, the symptoms described above seemed a justifiable price to pay for freedom from the crushing anxiety that had immobilized them. Unfortunately, by the late 1940s and early 1950s, the procedure had acquired a popularity out of all proportion to its actual benefits, and it was widely misapplied (as in the movie "One Flew Over the Cuckoo's Nest"). The discovery of tranquilizers in the late 1950s provided a more effective method of treatment, having fewer undesirable side effects, and prefrontal lobotomy was rapidly abandoned as a method of treatment.

SOURCES AND ADDITIONAL READINGS

Blakemore C: *Mechanics of the Mind*. Cambridge University Press, Cambridge, 1977.

Damasio AR: Aphasia. N Engl J Med 326:531–538, 1992.

Damasio H, Grabowski T, Frank R, Galaburda AM, Damasio AR: The return of Phineas Gage: clues about the brain from the skull of a famous patient. Science 264:1102–1105, 1994.

Harlow JM: Recovery from the passage of an iron bar through the head. Pub Mass Med Soc 2:327–347, 1868.

Hendry SHC, Jones EG: Sizes and distributions of intrinsic neurons incorporating tritiated GABA in monkey sensory-motor cortex. J Neurosci 1:390–408, 1981.

Hubel DH, Wiesel TN: Functional architecture of macaque monkey visual cortex. Proc R Soc Lond B Biol Sci 198:1–59, 1977.

Jones EG: Varieties and distribution of non-pyramidal cells in the somatic sensory cortex of the squirrel monkey. J Comp Neurol 160:205–268, 1975.

Jones EG: Laminar distribution of cortical efferent cells. *Cerebral Cortex*, Vol. 1. Plenum Press, New York, 1984, pp 521–553.

Lynch JC: Parietal association cortex. *Encyclopedia of Neuroscience*, Vol. 2. Birkhauser, Boston, 1987, pp 925–926.

Lynch JC: Columnar organization of the cerebral cortex (cortical columns). *Neuroscience Year* (Supplement to *Encyclopedia of Neuroscience*.) Birkhauser, Boston, 1989, pp 37–40.

Mountcastle VB: Modality and topographic properties of single neurons of cat's somatic sensory cortex. J Neurophysiol 20:408–434, 1957.

Peters A, Jones EG: *Cerebral Cortex*, Vols 1–10. Plenum Press, New York, 1984.

Scheibel ME, Scheibel AB: Elementary processes in selected thalamic and cortical subsystems—the structural substrates. *The Neurosciences, Second Study Programs*, Vol 2. Rockefeller University Press, New York, 1970, pp 443–457.

Valenstein ES: *Great and Desperate Cures*. Basic Books, New York, 1986.

Index

Note: Page numbers followed by *f* indicate illustrations; those followed by *t* indicate tables. All anatomic structures are listed only under the noun; for example, the posterior cerebral artery is listed under *Artery(ies)*, the optic nerve is listed under *Nerve(s)*, and the corticospinal tract is listed under *Tract(s)*.